RSMeans Square Foot Costs

2015 36th annual edition

Marilyn Phelan, AIA, Senior Editor

Engineering Director
Bob Mewis, CCC (*1, 14, 41*)

Contributing Editors
Christopher Babbitt
Adrian C. Charest, PE
(*26, 27, 28, 48*)
Gary W. Christensen
Cheryl Elsmore
David Fiske
Robert Fortier, PE
(*2, 31, 32, 33, 34, 35, 44, 46*)
Robert J. Kuchta (*8*)

*Numbers in italics are the divisional responsibilities
for each editor. Please contact the designated editor
directly with any questions.

Thomas Lane (*6, 7*)
Robert C. McNichols
Genevieve Medeiros
Melville J. Mossman, PE (*21, 22, 23*)
Jeannene D. Murphy
Marilyn Phelan, AIA (*9, 10, 11, 12*)
Stephen C. Plotner (*3, 5*)
Siobhan Soucie (*4, 13*)

Vice President Data & Engineering
Chris Anderson

Cover Design
Sara Rutan

Product Manager
Andrea Sillah

Production Manager
Debbie Panarelli

Production
Jill Goodman
Jonathan Forgit
Sara Rutan
Mary Lou Geary

Technical Support
Gary L. Hoitt
Kathryn S. Rodriguez

RSMeans
Construction Publishers & Consultants
700 Longwater Drive
Norwell, MA 02061
United States of America
1-877-694-8329
www.rsmeans.com

Copyright 2014 by RSMeans
All rights reserved.
Cover photo ©
nikonaft/iStock/Thinkstock

Printed in the United States of America
ISSN 1540-6326
ISBN 978-1-940238-67-8

0055 $218.95 per copy (in United States)
Price is subject to change without prior notice

Related RSMeans Products and Services

The residential section of this book is aimed at new residential construction projects valued at up to $1,000,000. The labor rate used in the calculations is open shop (non-union). The commercial/industrial/institutional section is aimed primarily at new construction projects valued at $1,500,00 and up. The labor rate in the calculations is union.

The engineers at RSMeans suggest the following products and services as companion information resources to *RSMeans Square Foot Costs*:

Construction Cost Data Books

Assemblies Cost Data 2015
Building Construction Cost Data 2015
Residential Cost Data 2015
Light Commercial Cost Data 2015

Reference Books

Estimating Building Costs
RSMeans Estimating Handbook
Green Building: Project Planning & Estimating
How to Estimate with RSMeans Data
Plan Reading & Material Takeoff
Project Scheduling & Management for Construction
Universal Design for Style, Comfort & Safety

Seminars and In-House Training

Conceptual Estimating Using RSMeans CostWorks® CD
RSMeans Online® Training
Plan Reading & Material Take-Off

RSMeans Online Bookstore

Visit RSMeans at **www.rsmeans.com** for a one-stop portal for the most reliable and current resources available on the market. Here you can learn about our more than 20 annual cost data books and eBooks, also available on CD and online versions, as well as our libary of reference books and RSMeans seminars.

RSMeans Electronic Data

Get the information found in RSMeans cost books electronically at **RSMeansOnline.com**.

RSMeans Business Solutions

Cost engineers and analysts conduct facility life cycle and benchmark research studies and predictive cost modeling, as well as offer consultation services for conceptual estimating and real property management. Research studies are designed with the application of proprietary cost and project data from RSMeans' extensive North American databases. Analysts offer building product manufacturers qualitative and quantitative market research, as well as time/motion studies for new products, and Web-based dashboards for market opportunity analysis. Clients are from both the public and private sectors, including federal, state, and municipal agencies; corporations; institutions; construction management firms; hospitals; and associations.

Construction Costs for Software Applications

More than 25 unit price and assemblies cost databases are available through a number of leading estimating and facilities management software partners. For more information see the "Other RSMeans Products and Services" pages at the back of this publication.

RSMeans data is also available to federal, state, and local government agencies as multi-year, multi-seat licenses.

For information on our current partners, visit: **www.rsmeans.com/partners** or call 877.694.8329.

Table of Contents

Foreword

Our Mission

Since 1942, RSMeans has been actively engaged in construction cost publishing and consulting throughout North America.

Today, more than 70 years after RSMeans began, our primary objective remains the same: to provide you, the construction and facilities professional, with the most current and comprehensive construction cost data possible.

Whether you are a contractor, owner, architect, engineer, facilities manager, or anyone else who needs a reliable construction cost estimate, you'll find this publication to be a highly useful and necessary tool.

With the constant flow of new construction methods and materials today, it's difficult to find the time to look at and evaluate all the different construction cost possibilities. In addition, because labor and material costs keep changing, last year's cost information is not a reliable basis for today's estimate or budget.

That's why so many construction professionals turn to RSMeans. We keep track of the costs for you, along with a wide range of other key information, from city cost indexes . . . to productivity rates . . . to crew composition . . . to contractor's overhead and profit rates.

RSMeans performs these functions by collecting data from all facets of the industry and organizing it in a format that is instantly accessible to you. From the preliminary budget to the detailed unit price estimate, you'll find the data in this book useful for all phases of construction cost determination.

The Staff, the Organization, and Our Services

When you purchase one of RSMeans' publications, you are, in effect, hiring the services of a full-time staff of construction and engineering professionals.

Our thoroughly experienced and highly qualified staff works daily at collecting, analyzing, and disseminating comprehensive cost information for your needs. These staff members have years of practical construction experience and engineering training prior to joining the firm. As a result, you can count on them not only for accurate cost figures, but also for additional background reference information that will help you create a realistic estimate.

The RSMeans organization is always prepared to help you solve construction problems through its variety of data solutions, including online, CD, and print book formats, as well as cost estimating expertise available via our business solutions, training, and seminars.

Besides a full array of construction cost estimating books, RSMeans also publishes a number of other reference works for the construction industry. Subjects include construction estimating and project and business management, green building, and a library of facility management references.

In addition, you can access all of our construction cost data electronically in convenient CD format or on the web. Visit **RSMeansOnline.com** for more information on our 24/7 online cost data.

What's more, you can increase your knowledge and improve your construction estimating and management performance with an RSMeans construction seminar or in-house training program. These two-day seminar programs offer unparalleled opportunities for everyone in your organization to become updated on a wide variety of construction-related issues.

RSMeans is also a worldwide provider of construction cost management and analysis services for commercial and government owners.

In short, RSMeans can provide you with the tools and expertise for constructing accurate and dependable construction estimates and budgets in a variety of ways.

Robert Snow Means Established a Tradition of Quality That Continues Today

Robert Snow Means spent years building RSMeans, making certain he always delivered a quality product.

Today, at RSMeans, we do more than talk about the quality of our data and the usefulness of our books. We stand behind all of our data, from historical cost indexes to construction materials and techniques to current costs.

If you have any questions about our products or services, please call us toll-free at 1-877-694-8329. Our customer service representatives will be happy to assist you. You can also visit our website at **www.rsmeans.com**

How the Book Is Built: An Overview

A Powerful Construction Tool

You have in your hands one of the most powerful construction tools available today. A successful project is built on the foundation of an accurate and dependable estimate. This book will enable you to construct just such an estimate.

For the casual user the book is designed to be:

- quickly and easily understood so you can get right to your estimate.
- filled with valuable information so you can understand the necessary factors that go into the cost estimate.

For the regular user, the book is designed to be:

- a handy desk reference that can be quickly referred to for key costs.
- a comprehensive, fully reliable source of current construction costs for typical building structures when only minimum dimensions, specifications, and technical data are available. It is specifically useful in the conceptual or planning stage for preparing preliminary estimates which can be supported and developed into complete engineering estimates for construction.
- a source book for rapid estimating of replacement costs and preparing capital expenditure budgets.

To meet all of these requirements, we have organized the book into the following clearly defined sections.

How to Use the Book:
The Details

This section contains an in-depth explanation of how the book is arranged and how you can use it to determine a reliable construction cost estimate. It includes information about how we develop our cost figures and how to completely prepare your estimate.

Residential Models

Model buildings for four classes of construction—economy, average, custom, and luxury—are developed and shown with complete costs per square foot.

Commercial/Industrial/Institutional Models

This section contains complete costs for 77 typical model buildings expressed as costs per square foot.

Green Commercial/Industrial/Institutional Models

This section contains complete costs for 25 green model buildings expressed as costs per square foot.

Assemblies

The cost data in this section has been organized in an "Assemblies" format. These assemblies are the functional elements of a building and are arranged according to the 7 divisions of the UNIFORMAT II classification system. The costs in this section include standard markups for the overhead and profit of the installing contractor. The assemblies costs are those used to calculate the total building costs in the Commercial/Industrial/Institutional section.

References

This section includes information on General Conditions, Location Factors, Historical Cost Indexes, a Glossary, Square Foot Costs, and a listing of Abbreviations used in this book.

Location Factors: You can adjust the total project costs to over 900 locations throughout the U.S. and Canada by using the data in this section.

Historical Cost Indexes: These indexes provide you with data to adjust construction costs over time.

Glossary: A listing of common terms related to construction and their definitions is included.

Square Foot Costs: This section contains costs for 59 different building types that allows you to make a rough estimate for the overall cost of a project or its major components.

Abbreviations: A listing of the abbreviations used throughout this book, along with the terms they represent, is included.

The Scope of This Book

This book is designed to be as comprehensive and as easy to use as possible. To that end we have made certain assumptions and limited its scope in two key ways:

1. We have established material prices based on a national average.
2. We have computed labor costs based on a 30-city national average of union wage rates.

All costs and adjustments in this manual were developed from *RSMeans Building Construction Cost Data*, *Assemblies Cost Data*, *Residential Cost Data*, and *Light Commercial Cost Data*. Costs include labor, material, overhead and profit, and general conditions and architects' fees where specified. No allowance is made for sales tax, financing, premiums for material and labor, or unusual business and weather conditions.

What's Behind the Numbers? The Development of Cost Data

The staff at RSMeans continuously monitors developments in the construction industry in order to ensure reliable, thorough, and up-to-date cost information.

While overall construction costs may vary relative to general economic conditions, price fluctuations within the industry are dependent upon many factors. Individual price variations may, in fact, be opposite to overall economic trends. Therefore, costs are continually monitored and complete updates are published yearly. Also, new items are frequently added in response to changes in materials and methods.

Costs—$ (U.S.)

All costs represent U.S. national averages and are given in U.S. dollars. The RSMeans City Cost Indexes can be used to adjust costs to a particular location. The City Cost Indexes for Canada can be used to adjust U.S. national averages to local costs in Canadian dollars. No exchange rate conversion is necessary.

Factors Affecting Costs

Due to the almost limitless variation of building designs and combinations of construction methods and materials, costs in this book must be used with discretion. The editors have made every effort to develop costs for "typical" building types. By definition, a square foot estimate, due to the short time and limited amount of detail required, involves a relatively lower degree of accuracy than a detailed unit cost estimate, for example. The accuracy of an estimate

will increase as the project is defined in more detailed components. The user should examine closely and compare the specific requirements of the project being estimated with the components included in the models in this book. Any differences should be accounted for in the estimate.

In many building projects, there may be factors that increase or decrease the cost beyond the range shown in this manual. Some of these factors are:

- Substitution of building materials or systems for those used in the model
- Custom-designed finishes, fixtures, and/or equipment
- Special structural qualities (allowance for earthquake, future expansion, high winds, long spans, and unusual shape)
- Special foundations and substructures to compensate for unusual soil conditions
- Inclusion of typical "owner-supplied" equipment in the building cost (special testing and monitoring equipment for hospitals, kitchen equipment for hotels, and offices that have cafeteria accommodations for employees, etc.)
- An abnormal shortage of labor or materials
- Isolated building site or rough terrain that would affect the transportation of personnel, material, or equipment
- Unusual climatic conditions during the construction process

We strongly urge that, for maximum accuracy, these factors be considered each time a structure is being evaluated.

If users require greater accuracy than this manual can provide, the editors recommend

that the *RSMeans Building Construction Cost Data* or the *RSMeans Assemblies Cost Data* be consulted.

Final Checklist

Estimating can be a straightforward process provided you remember the basics. Here's a checklist of some of the steps you should remember to complete before finalizing your estimate.

Did you remember to . . .

- include the Location Factor for your city?
- mark up the entire estimate sufficiently for your purposes?
- include all components of your project in the final estimate?
- double check your figures for accuracy?
- call RSMeans if you have any questions about your estimate or the data you've found in our publications?

Remember, RSMeans stands behind its publications. If you have any questions about your estimate, about the costs you've used from our books, or even about the technical aspects of the job that may affect your estimate, feel free to call the RSMeans editors at 1-877-694-8329.

Free Quarterly Updates

Stay up-to-date throughout 2015 with RSMeans' free cost data updates four times a year. Sign up online to make sure you have access to the newest data. Every quarter we provide the city cost adjustment factors for hundreds of cities and key materials. Register at:
http://info.thegordiangroup.com/RSMeans.html.

Residential Section

Table of Contents

Introduction to the Residential Section

The Residential Section of this manual contains costs per square foot for four classes of construction in seven building types. Costs are listed for various exterior wall materials which are typical of the class and building type. There are cost tables for Wings and Ells with modification tables to adjust the base cost of each class of building. Non-standard items can easily be added to the standard structures.

Cost estimating for a residence is a three-step process:
1. Identification
2. Listing dimensions
3. Calculations

Guidelines and a sample cost estimating procedure are shown on the following pages.

Identification

To properly identify a residential building, the class of construction, type, and exterior wall material must be determined. Page 6 has drawings and guidelines for determining the class of construction. There are also detailed specifications and additional drawings at the beginning of each set of tables to further aid in proper building class and type identification.

Sketches for eight types of residential buildings and their configurations are shown on pages 8 and 9. Definitions of living area are next to each sketch. Sketches and definitions of garage types are on page 10.

Living Area

Base cost tables are prepared as costs per square foot of living area. The living area of a residence is that area which is suitable and normally designed for full time living. It does not include basement recreation rooms or finished attics, although these areas are often considered full time living areas by the owners.

Living area is calculated from the exterior dimensions without the need to adjust for exterior wall thickness. When calculating the living area of a 1-1/2 story, two story, three story or tri-level residence, overhangs and other differences in size and shape between floors must be considered.

Only the floor area with a ceiling height of seven feet or more in a 1-1/2 story residence is considered living area. In bi-levels and tri-levels, the areas that are below grade are considered living area, even when these areas may not be completely finished.

Base Tables and Modifications

Base cost tables show the base cost per square foot without a basement, with one full bath and one full kitchen for economy and average homes, and an additional half bath for custom and luxury models. Adjustments for finished and unfinished basements are part of the base cost tables. Adjustments for multi-family residences, additional bathrooms, townhouses, alternative roofs, and air conditioning and heating systems are listed in Modifications, Adjustments and Alternatives tables below the base cost tables.

Costs for other modifications, adjustments and alternatives, including garages, breezeways and site improvements, are on pages 58 to 61.

Listing of Dimensions

To use this section of the manual, only the dimensions used to calculate the horizontal area of the building and additions, modifications, adjustments, and alternatives are needed. The dimensions, normally the length and width, can come from drawings or field measurements. For ease in calculation, consider measuring in tenths of feet, i.e., 9 ft. 6 in. = 9.5 ft., 9 ft. 4 in. = 9.3 ft.

In all cases, make a sketch of the building. Any protrusions or other variations in shape should be noted on the sketch with dimensions.

Calculations

The calculations portion of the estimate is a two-step activity:
1. The selection of appropriate costs from the tables
2. Computations

Selection of Appropriate Costs

To select the appropriate cost from the base tables, the following information is needed:
1. Class of construction
   - ☐ Economy
   - ☐ Average
   - ☐ Custom
   - ☐ Luxury
2. Type of residence
   - ☐ 1 story
   - ☐ 1-1/2 story
   - ☐ 2 story
   - ☐ 3 story
   - ☐ Bi-level
   - ☐ Tri-level
3. Occupancy
   - ☐ One family
   - ☐ Two family
   - ☐ Three family
4. Building configuration
   - ☐ Detached
   - ☐ Town/Row house
   - ☐ Semi-detached
5. Exterior wall construction
   - ☐ Wood frame
   - ☐ Brick veneer
   - ☐ Solid masonry
6. Living areas

Modifications are classified by class, type, and size.

Computations

The computation process should take the following sequence:
1. Multiply the base cost by the area.
2. Add or subtract the modifications, adjustments, and alternatives.
3. Apply the location modifier.

When selecting costs, interpolate or use the cost that most nearly matches the structure under study. This applies to size, exterior wall construction, and class.

How to Use the Residential Square Foot Cost Pages

The following is a detailed explanation of a sample entry in the Residential Square Foot Cost Section. Each bold number below corresponds to the item being described on the facing page with the appropriate component of the sample entry following in parentheses.

Prices listed are costs that include overhead and profit of the installing contractor. Total model costs include an additional markup for General Contractors' overhead and profit, and fees specific to class of construction.

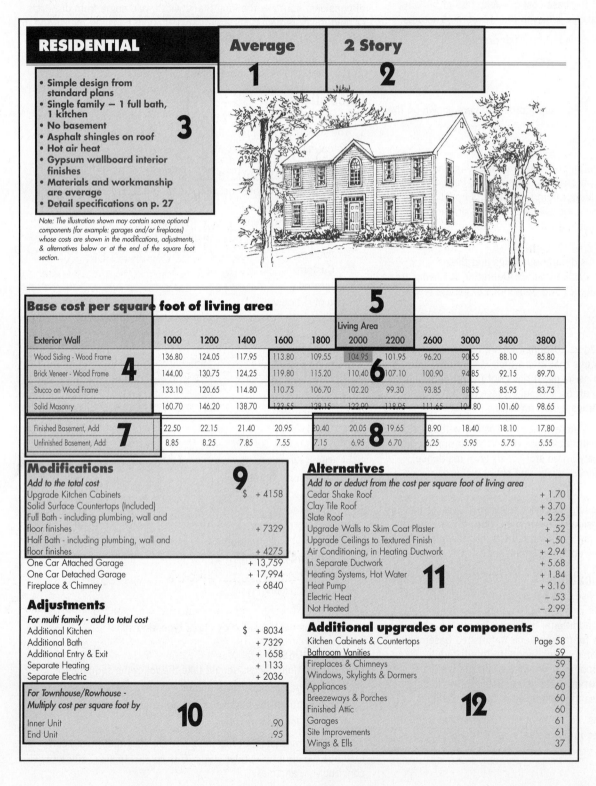

RESIDENTIAL

| Average | 2 Story |
| 1 | 2 |

3
- Simple design from standard plans
- Single family — 1 full bath, 1 kitchen
- No basement
- Asphalt shingles on roof
- Hot air heat
- Gypsum wallboard interior finishes
- Materials and workmanship are average
- Detail specifications on p. 27

Note: The illustration shown may contain some optional components (for example: garages and/or fireplaces) whose costs are shown in the modifications, adjustments, & alternatives below or at the end of the square foot section.

Base cost per square foot of living area

5

Exterior Wall	1000	1200	1400	1600	1800	Living Area 2000	2200	2600	3000	3400	3800
Wood Siding - Wood Frame	136.80	124.05	117.95	113.80	109.55	104.95	101.95	96.20	90.55	88.10	85.80
Brick Veneer - Wood Frame	144.00	130.75	124.25	119.80	115.20	110.40	107.10	100.90	94.85	92.15	89.70
Stucco on Wood Frame	133.10	120.65	114.80	110.75	106.70	102.20	99.30	93.85	88.35	85.95	83.75
Solid Masonry	160.70	146.20	138.70	133.55	128.15	122.90	118.95	111.65	104.80	101.60	98.65

4 **6**

	1000	1200	1400	1600	1800	2000	2200	2600	3000	3400	3800
Finished Basement, Add	22.50	22.15	21.40	20.95	20.40	20.05	19.65	18.90	18.40	18.10	17.80
Unfinished Basement, Add	8.85	8.25	7.85	7.55	7.15	6.95	6.70	6.25	5.95	5.75	5.55

7 **8**

Modifications

9

Add to the total cost

Upgrade Kitchen Cabinets	$ + 4158
Solid Surface Countertops (Included)	
Full Bath - including plumbing, wall and floor finishes	+ 7329
Half Bath - including plumbing, wall and floor finishes	+ 4275
One Car Attached Garage	+ 13,759
One Car Detached Garage	+ 17,994
Fireplace & Chimney	+ 6840

Adjustments

For multi family - add to total cost

Additional Kitchen	$ + 8034
Additional Bath	+ 7329
Additional Entry & Exit	+ 1658
Separate Heating	+ 1133
Separate Electric	+ 2036

For Townhouse/Rowhouse - Multiply cost per square foot by

10

Inner Unit	.90
End Unit	.95

Alternatives

Add to or deduct from the cost per square foot of living area

Cedar Shake Roof	+ 1.70
Clay Tile Roof	+ 3.70
Slate Roof	+ 3.25
Upgrade Walls to Skim Coat Plaster	+ .52
Upgrade Ceilings to Textured Finish	+ .50
Air Conditioning, in Heating Ductwork	+ 2.94
In Separate Ductwork	+ 5.68
Heating Systems, Hot Water	+ 1.84
Heat Pump	+ 3.16
Electric Heat	- .53
Not Heated	- 2.99

11

Additional upgrades or components

Kitchen Cabinets & Countertops	Page 58
Bathroom Vanities	59
Fireplaces & Chimneys	59
Windows, Skylights & Dormers	59
Appliances	60
Breezeways & Porches	60
Finished Attic	60
Garages	61
Site Improvements	61
Wings & Ells	37

12

1 Class of Construction (Average)
The class of construction depends upon the design and specifications of the plan. The four classes are economy, average, custom, and luxury.

2 Type of Residence (2 Story)
The building type describes the number of stories or levels in the model. The seven building types are 1 story, 1-1/2 story, 2 story, 2-1/2 story, 3 story, bi-level, and tri-level.

3 Specification Highlights (Hot Air Heat)
These specifications include information concerning the components of the model, including the number of baths, roofing types, HVAC systems, and materials and workmanship. If the components listed are not appropriate, modifications can be made by consulting the information shown lower on the page or the Assemblies Section.

4 Exterior Wall System (Wood Siding -Wood Frame)
This section includes the types of exterior wall systems and the structural frame used. The exterior wall systems shown are typical of the class of construction and the building type shown.

5 Living Areas (2000 SF)
The living area is that area of the residence which is suitable and normally designed for full time living. It does not include basement recreation rooms or finished attics. Living area is calculated from the exterior dimensions without the need to adjust for exterior wall thickness. When calculating the living area of a 1-1/2 story, 2 story, 3 story or tri-level residence, overhangs and other differences in size and shape between floors must be considered. Only the floor area with a ceiling height of seven feet or more in a 1-1/2 story residence is considered living area. In bi-levels and tri-levels, the areas that are below grade are considered living area, even when these areas may not be completely finished. A range of various living areas for the residential model are shown to aid in selection of values from the matrix.

6 Base Costs per Square Foot of Living Area ($104.95)
Base cost tables show the cost per square foot of living area without a basement, with one full bath and one full kitchen for economy and average homes, and an additional half bath for custom and luxury models. When selecting costs, interpolate or use the cost that most nearly matches the residence under consideration for size, exterior wall system, and class of construction. Prices listed are costs that include overhead and profit of the installing contractor, a general contractor markup, and an allowance for plans that vary by class of construction. For additional information on contractor overhead and architectural fees, see the Reference Section.

7 Basement Types (Finished)
The two types of basements are finished and unfinished. The specifications and components for both are shown on the Building Classes page in the Introduction to this section.

8 Additional Costs for Basements ($20.05 or $6.95)
These values indicate the additional cost per square foot of living area for either a finished or an unfinished basement.

9 Modifications and Adjustments (Upgrade Kitchen Cabinets $4158)
Modifications and Adjustments are costs added to or subtracted from the total cost of the residence. The total cost of the residence is equal to the cost per square foot of living area times the living area. Typical modifications and adjustments include kitchens, baths, garages, and fireplaces.

10 Multiplier for Townhouse/ Rowhouse (Inner Unit .90)
The multipliers shown adjust the base costs per square foot of living area for the common wall condition encountered in townhouses or rowhouses.

11 Alternatives (Skim Coat Plaster $.52)
Alternatives are costs added to or subtracted from the base cost per square foot of living area. Typical alternatives include variations in kitchens, baths, roofing, and air conditioning and heating systems.

12 Additional Upgrades or Components (Wings & Ells page 37)
Costs for additional upgrades or components, including wings or ells, breezeways, porches, finished attics, and site improvements, are shown in other locations in the Residential Square Foot Cost Section.

Building Classes

Economy Class

An economy class residence is usually built from stock plans. The materials and workmanship are sufficient to satisfy building codes. Low construction cost is more important than distinctive features. The overall shape of the foundation and structure is seldom other than square or rectangular.

An unfinished basement includes 7' high 8" thick foundation wall composed of either concrete block or cast-in-place concrete.

Included in the finished basement cost are inexpensive paneling or drywall as the interior finish on the foundation walls, a low cost sponge backed carpeting adhered to the concrete floor, a drywall ceiling, and overhead lighting.

Custom Class

A custom class residence is usually built from plans and specifications with enough features to give the building a distinction of design. Materials and workmanship are generally above average with obvious attention given to construction details. Construction normally exceeds building code requirements.

An unfinished basement includes a 7'-6" high 10" thick cast-in-place concrete foundation wall or a 7'-6" high 12" thick concrete block foundation wall.

A finished basement includes painted drywall on insulated 2" x4" wood furring as the interior finish to the concrete walls, a suspended ceiling, carpeting adhered to the concrete floor, overhead lighting and heating.

Average Class

An average class residence is a simple design and built from standard plans. Materials and workmanship are average, but often exceed minimum building codes. There are frequently special features that give the residence some distinctive characteristics.

An unfinished basement includes 7'-6" high 8" thick foundation wall composed of either cast-in-place concrete or concrete block.

Included in the finished basement are plywood paneling or drywall on furring that is fastened to the foundation walls, sponge backed carpeting adhered to the concrete floor, a suspended ceiling, overhead lighting and heating.

Luxury Class

A luxury class residence is built from an architect's plan for a specific owner. It is unique both in design and workmanship. There are many special features, and construction usually exceeds all building codes. It is obvious that primary attention is placed on the owner's comfort and pleasure. Construction is supervised by an architect.

An unfinished basement includes 8' high 12" thick foundation wall that is composed of cast-in-place concrete or concrete block.

A finished basement includes painted drywall on 2" x4" wood furring as the interior finish, suspended ceiling, tackless carpet on wood subfloor with sleepers, overhead lighting and heating.

Configurations

Detached House

This category of residence is a free-standing separate building with or without an attached garage. It has four complete walls.

Semi-Detached House

This category of residence has two living units side-by-side. The common wall is a fireproof wall. Semi-detached residences can be treated as a row house with two end units. Semi-detached residences can be any of the building types.

Town/Row House

This category of residence has a number of attached units made up of inner units and end units. The units are joined by common walls. The inner units have only two exterior walls. The common walls are fireproof. The end units have three walls and a common wall. Town houses/row houses can be any of the building types.

Building Types

One Story

This is an example of a one-story dwelling. The living area of this type of residence is confined to the ground floor. The headroom in the attic is usually too low for use as a living area.

One-and-a-half Story

The living area in the upper level of this type of residence is 50% to 90% of the ground floor. This is made possible by a combination of this design's high-peaked roof and/or dormers. Only the upper level area with a ceiling height of 7 feet or more is considered living area. The living area of this residence is the sum of the ground floor area plus the area on the second level with a ceiling height of 7 feet or more.

One Story with Finished Attic

The main living area in this type of residence is the ground floor. The upper level or attic area has sufficient headroom for use as a living area. This is made possible by a high peaked roof. The living area in the attic is less than 50% of the ground floor. The living area of this type of residence is the ground floor area only. The finished attic is considered an adjustment.

Two Story

This type of residence has a second floor or upper level area which is equal or nearly equal to the ground floor area. The upper level of this type of residence can range from 90% to 110% of the ground floor area, depending on setbacks or overhangs. The living area is the sum of the ground floor area and the upper level floor area.

Two-and-one-half Story

This type of residence has two levels of equal or nearly equal area and a third level which has a living area that is 50% to 90% of the ground floor. This is made possible by a high peaked roof, extended wall heights and/or dormers. Only the upper level area with a ceiling height of 7 feet or more is considered living area. The living area of this residence is the sum of the ground floor area, the second floor area and the area on the third level with a ceiling height of 7 feet or more.

Bi-level

This type of residence has two living areas, one above the other. One area is about 4 feet below grade and the second is about 4 feet above grade. Both areas are equal in size. The lower level in this type of residence is originally designed and built to serve as a living area and not as a basement. Both levels have full ceiling heights. The living area is the sum of the lower level area and the upper level area.

Three Story

This type of residence has three levels which are equal or nearly equal. As in the 2 story residence, the second and third floor areas may vary slightly depending on setbacks or overhangs. The living area is the sum of the ground floor area and the two upper level floor areas.

Tri-level

This type of residence has three levels of living area: one at grade level, one about 4 feet below grade, and one about 4 feet above grade. All levels are originally designed to serve as living areas. All levels have full ceiling heights. The living area is a sum of the areas of each of the three levels.

Garage Types

Attached Garage

Shares a common wall with the dwelling. Access is typically through a door between dwelling and garage.

Basement Garage

Constructed under the roof of the dwelling but below the living area.

Built-In Garage

Constructed under the second floor living space and above basement level of dwelling. Reduces gross square feet of living area.

Detached Garage

Constructed apart from the main dwelling. Shares no common area or wall with the dwelling.

Building Components

1. Excavation
2. Sill Plate
3. Basement Window
4. Floor Joist
5. Shoe Plate
6. Studs
7. Drywall
8. Plate
9. CeiHng Joists
10. Rafters
11. Collar Ties
12. Ridge Rafter
13. Roof Sheathing
14. Roof Felt
15. Roof Shingles
16. Flashing
17. Flue Lining
18. Chimney
19. Roof Shingles
20. Gutter
21. Fascia
22. Downspout
23. Shutter
24. Window
25. Wall Shingles
26. Weather Barrier
27. Wall Sheathing
28. Fire Stop
29. Dampproofing
30. Foundation Wall
31. Backfill
32. Drainage Stone
33. Drainage Tile
34. Wall Footing
35. Gravel
36. Concrete Slab
37. Column Footing
38. Pipe Column
39. Expansion Joint
40. Girder
41. Sub-floor
42. Finish Floor
43. Attic Insulation
44. Soffit
45. Ceiling Strapping
46. Wall Insulation
47. Cross Bridging
48. Bulkhead Stairs

Exterior Wall Construction

Typical Frame Construction

Typical wood frame construction consists of wood studs with insulation between them. A typical exterior surface is made up of sheathing, building paper and exterior siding consisting of wood, vinyl, aluminum or stucco over the wood sheathing.

Brick Veneer

Typical brick veneer construction consists of wood studs with insulation between them. A typical exterior surface is sheathing, building paper and an exterior of brick tied to the sheathing, with metal strips.

Stone

Typical solid masonry construction consists of a stone or block wall covered on the exterior with brick, stone or other masonry.

Residential Cost Estimate Worksheet

Worksheet Instructions

The residential cost estimate worksheet can be used as an outline for developing a residential construction or replacement cost. It is also useful for insurance appraisals. The design of the worksheet helps eliminate errors and omissions. To use the worksheet, follow the example below.

1. Fill out the owner's name, residence address, the estimator or appraiser's name, some type of project identifying number or code, and the date.

2. Determine from the plans, specifications, owner's description, photographs or any other means possible the class of construction. The models in this book use economy, average, custom, and luxury as classes. Fill in the appropriate box.

3. Fill in the appropriate box for the residence type, configuration, occupancy, and exterior wall. If you require clarification, the pages preceding this worksheet describe each of these.

4. Next, the living area of the residence must be established. The heated or air conditioned space of the residence, not including the basement, should be measured. It is easiest to break the structure up into separate components as shown in the example. The main house (A), a one-and-one-half story wing (B), and a one story wing (C). The breezeway (D), garage (E), and open covered porch (F) will be treated differently. Data entry blocks for the living area are included on the worksheet for your use. Keep each level of each component separate, and fill out the blocks as shown.

5. By using the information on the worksheet, find the model, wing or ell in the following square foot cost pages that best matches the class, type, exterior finish and size of the residence being estimated. Use the Modifications, Adjustments, and Alternatives to determine the adjusted cost per square foot of living area for each component.

6. For each component, multiply the cost per square foot by the living area square footage. If the residence is a townhouse/rowhouse, a multiplier should be applied based upon the configuration.

7. The second page of the residential cost estimate worksheet has space for the additional components of a house. The cost for additional bathrooms, finished attic space, breezeways, porches, fireplaces, appliance or cabinet upgrades, and garages should be added on this page. The information for each of these components is found with the model being used, or in the Modifications, Adjustments, and Alternatives pages.

8. Add the total from page one of the estimate worksheet and the items listed on page two. The sum is the adjusted total building cost.

9. Depending on the use of the final estimated cost, one of the remaining two boxes should be filled out. Any additional items or exclusions should be added or subtracted at this time. The data contained in this book is a national average. Construction costs are different throughout the country. To allow for this difference, a location factor based upon the first three digits of the residence's zip code must be applied. The location factor is a multiplier that increases or decreases the adjusted total building cost. Find the appropriate location factor and calculate the local cost. If depreciation is a concern, a dollar figure should be subtracted at this point.

10. No residence will match a model exactly. Many differences will be found. At this level of estimating, a variation of plus or minus 10% should be expected.

Adjustments Instructions

No residence matches a model exactly in shape, material, or specifications. The common differences are:

1. Two or more exterior wall systems:
   - Partial basement
   - Partly finished basement
2. Specifications or features that are between two classes
3. Crawl space instead of a basement

Examples

Below are quick examples. See pages 15–17 for complete examples of cost adjustments for these differences:

1. Residence "A" is an average one-story structure with 1,600 S.F. of living area and no basement. Three walls are wood siding on wood frame, and the fourth wall is brick veneer on wood frame. The brick veneer wall is 35% of the exterior wall area.

   Use page 28 to calculate the Base Cost per S.F. of Living Area. Wood Siding for 1,600 S.F. = $107.95 per S.F. Brick Veneer for 1,600 S.F. = $124.65 per S.F.

   .65 ($107.95) + .35 ($124.65) = $113.80 per S.F. of Living Area.

2a. Residence "B" is the same as Residence "A"; However, it has an unfinished basement under 50% of the building. To adjust the $113.80 per S.F. of living area for this partial basement, use page 28.

   $113.80 + .5 ($11.25) = $119.43 per S.F. of Living Area.

b. Residence "C" is the same as Residence "A"; However, it has a full basement under the entire building. 640 S.F. or 40% of the basement area is finished.

   Using Page 28:

   $113.80 + .40 ($32.35) + .60 ($11.25) = $133.49 per S.F. of Living Area.

3. When specifications or features of a building are between classes, estimate the percent deviation, and use two tables to calculate the cost per S.F.

   A two-story residence with wood siding and 1,800 S.F. of living area has features 30% better than Average, but 70% less than Custom.

   From pages 30 and 64:

Custom 1,800 S.F. Base Cost	**=**	**$134.25 per S.F.**
Average 1,800 S.F. Base Cost	**=**	**$109.55 per S.F.**
DIFFERENCE	**=**	**$24.70 per S.F.**
Cost is $109.55 + .30 ($24.70)	**=**	**$116.96 per S.F. of Living Area.**

4. To add the cost of a crawl space, use the cost of an unfinished basement as a maximum. For specific costs of components to be added or deducted, such as vapor barrier, underdrain, and floor, see the "Assemblies" section, pages 285 to 516.

Model Residence Example

First Floor Plan

E — 24' × 24'
D — 12' × 12', 8'
C — 12' × 16' (12')
A — 46' × 28', 16', 12', 20'
F — 18' × 12'
B — 18' × 20'

Second Floor Plan

A — 46' × 28'
B — 13' × 20', 5', 5', 5'

A = Main House
B = 1-1/2 Story Wing
C = 1 Story Wing
D = Breezeway
E = Garage
F = Open Covered Porch

RESIDENTIAL
COST ESTIMATE

OWNERS NAME:	**Albert Westenberg**	APPRAISER:	**Nicole Wojtowicz**
RESIDENCE ADDRESS:	**300 Sygiel Road**	PROJECT:	**# 55**
CITY, STATE, ZIP CODE:	**Three Rivers, MA 01080**	DATE:	**Jan. 1, 2015**

CLASS OF CONSTRUCTION	RESIDENCE TYPE	CONFIGURATION	EXTERIOR WALL SYSTEM
☐ ECONOMY	☐ 1 STORY	☑ DETACHED	☑ WOOD SIDING - WOOD FRAME
☑ AVERAGE	☐ 1 1/2 STORY	☐ TOWN/ROW HOUSE	☐ BRICK VENEER - WOOD FRAME
☐ CUSTOM	☑ 2 STORY	☐ SEMI-DETACHED	☐ STUCCO ON WOOD FRAME
☐ LUXURY	☐ 2 1/2 STORY		☐ PAINTED CONCRETE BLOCK
	☐ 3 STORY	OCCUPANCY	☐ SOLID MASONRY (AVERAGE & CUSTOM)
	☐ BI-LEVEL	☑ ONE FAMILY	☐ STONE VENEER - WOOD FRAME
	☐ TRI-LEVEL	☐ TWO FAMILY	☐ SOLID BRICK (LUXURY)
		☐ THREE FAMILY	☐ SOLID STONE (LUXURY)
		☐ OTHER	

* LIVING AREA (Main Building)			* LIVING AREA (Wing or Ell) (**B**)			* LIVING AREA (Wing or Ell) (**C**)		
First Level	**1288**	S.F.	First Level	**360**	S.F.	First Level	**192**	S.F.
Second level	**1288**	S.F.	Second level	**310**	S.F.	Second level		S.F.
Third Level		S.F.	Third Level		S.F.	Third Level		S.F.
Total	**2576**	S.F.	Total	**670**	S.F.	Total	**192**	S.F.

* Basement Area is not part of living area.

MAIN BUILDING			COSTS PER S.F. LIVING AREA	
Cost per Square Foot of Living Area, from Page	**30**		$	104.95
Basement Addition: _____ % Finished, **100** % Unfinished			+	6.95
Roof Cover Adjustment: **Cedar Shake** Type, Page **30** (Add or Deduct)			( )	1.70
Central Air Conditioning: ☐ Separate Ducts ☑ Heating Ducts, Page **30**			+	2.94
Heating System Adjustment: _____ Type, Page _____ (Add or Deduct)			( )	___
Main Building: Adjusted Cost per S.F. of Living Area			$ #	116.54

MAIN BUILDING TOTAL COST	$	**116.54**	/S.F.	x	**2,576**	S.F.	x	_____	=	$ **300,207**
		Cost per S.F. Living Area			Living Area			Town/Row House Multiplier (Use 1 for Detached)		TOTAL COST

WING OR ELL **(B)** **1 - 1/2** STORY			COSTS PER S.F. LIVING AREA	
Cost per Square Foot of Living Area, from Page	**37**		$	93.30
Basement Addition: **100** % Finished, _____ % Unfinished			+	29.25
Roof Cover Adjustment: _____ Type, Page _____ (Add or Deduct)			( )	___
Central Air Conditioning: ☐ Separate Ducts ☑ Heating Ducts, Page **30**			+	2.94
Heating System Adjustment: _____ Type, Page _____ (Add or Deduct)			( )	___
Wing or Ell (**B**): Adjusted Cost per S.F. of Living Area			$ #	125.49

WING OR ELL **(B)** TOTAL COST	$	**125.49**	/S.F.	x	**670**	S.F.		=	$ **84,078**
		Cost per S.F. Living Area			Living Area				TOTAL COST

WING OR ELL **(C)** **1** STORY			COSTS PER S.F. LIVING AREA	
Cost per Square Foot of Living Area, from Page	**# 37**	**(WOOD SIDING)**	$ #	157.35
Basement Addition: _____ % Finished, _____ % Unfinished			+	___
Roof Cover Adjustment: _____ Type, Page _____ (Add or Deduct)			( )	___
Central Air Conditioning: ☐ Separate Ducts ☐ Heating Ducts, Page _____			+	___
Heating System Adjustment: _____ Type, Page _____ (Add or Deduct)			( )	___
Wing or Ell **(C)** Adjusted Cost per S.F. of Living Area			$	157.35

WING OR ELL **(C)** TOTAL COST	$	**157.35**	/S.F.	x	**192**	S.F.		=	$ **30,211**
		Cost per S.F. Living Area			Living Area				TOTAL COST

TOTAL THIS PAGE	**414,496**

Page 1 of 2

RESIDENTIAL
COST ESTIMATE

					QUANTITY	UNIT COST		
Total Page 1							$	414,496
Additional Bathrooms:	**2**	Full,	**1**	Half 2 @ 7329 1 @ 4275			+	18,933
Finished Attic:	**N/A**	Ft. x		Ft.		S.F.		
Breezeway: ☑ Open ☐ closed	**12**	Ft. x	**12**	Ft.	144	S.F. 38.49	+	5,543
Covered Porch: ☑ Open ☐ Enclosed	**18**	Ft. x	**12**	Ft.	216	S.F. 36.47	+	7,878
Fireplace: ☑ Interior Chimney ☐ Exterior Chimney ☑ No. of Flues (**2**) ☑ Additional Fireplaces		**1 - 2nd Story**					+	11,467
Appliances:							+	—
Kitchen Cabinets Adjustments:				(±)				—
☑ Garage ☐ Carport: **2** Car(s) Description **Wood, Attached** (±)								23,447
Miscellaneous:							+	

ADJUSTED TOTAL BUILDING COST	$	481,764

REPLACEMENT COST			INSURANCE COST		
ADJUSTED TOTAL BUILDING COST	$	481,764	ADJUSTED TOTAL BUILDING COST	$	
Site Improvements			Insurance Exclusions		
(A) Paving & Sidewalks	$		(A) Footings, sitework, Underground Piping	-$	
(B) Landscaping	$		(B) Architects Fees	-$	
(C) Fences	$		Total Building Cost Less Exclusion	$	
(D) Swimming Pools	$		Location Factor	x	
(E) Miscellaneous	$				
TOTAL	$	481,764	LOCAL INSURABLE REPLACEMENT COST	$	
Location Factor	x	1.080			
Location Replacement Cost	$	520,305			
Depreciation	-$	52,030			
LOCAL DEPRECIATED COST	$	468,275			

SKETCH AND ADDITIONAL CALCULATIONS

1 Story

© Home Planners, Inc.

1-1/2 Story

2 Story

Bi-Level

Tri-Level

©Design Basics, Inc.

	Components
1 Site Work	Site preparation for slab or excavation for lower level; 4' deep trench excavation for foundation wall.
2 Foundations	Continuous reinforced concrete footing, 8" deep x 18" wide; dampproofed and insulated 8" thick reinforced concrete block foundation wall, 4' deep; 4" concrete slab on 4" crushed stone base and polyethylene vapor barrier, trowel finish.
3 Framing	Exterior walls—2 x 4 wood studs, 16" O.C.; 1/2" insulation board sheathing; wood truss roof 24" O.C. or 2 x 6 rafters 16" O.C. with 1/2" plywood sheathing, 4 in 12 or 8 in 12 roof pitch.
4 Exterior Walls	Beveled wood siding and #15 felt building paper on insulated wood frame walls; sliding sash or double hung wood windows; 2 flush solid core wood exterior doors. **Alternates:** • Brick veneer on wood frame, has 4" veneer of common brick. • Stucco on wood frame has 1" thick colored stucco finish. • Painted concrete block has 8" concrete block sealed and painted on exterior and furring on the interior for the drywall.
5 Roofing	20 year asphalt shingles; #15 felt building paper; aluminum gutters, downspouts, drip edge and flashings.
6 Interiors	Walls and ceilings—1/2" taped and finished drywall, primed and painted with 2 coats; painted baseboard and trim; rubber backed carpeting 80%, asphalt tile 20%; hollow core wood interior doors.
7 Specialties	Economy grade kitchen cabinets—6 L.F. wall and base with plastic laminate counter top and kitchen sink; 30 gallon electric water heater.
8 Mechanical	1 lavatory, white, wall hung; 1 water closet, white; 1 bathtub, enameled steel, white; gas fired warm air heat.
9 Electrical	100 Amp. service; romex wiring; incandescent lighting fixtures, switches, receptacles.
10 Overhead and Profit	General Contractor overhead and profit.

Adjustments

Unfinished Basement:
7' high 8" concrete block or cast-in-place concrete walls.

Finished Basement:
Includes inexpensive paneling or drywall on foundation walls. Inexpensive sponge backed carpeting on concrete floor, drywall ceiling, and lighting.

- **Mass produced from stock plans**
- **Single family — 1 full bath, 1 kitchen**
- **No basement**
- **Asphalt shingles on roof**
- **Hot air heat**
- **Gypsum wallboard interior finishes**
- **Materials and workmanship are sufficient to meet codes**
- **Detail specifications on page 19**

Note: The illustration shown may contain some optional components (for example: garages and/or fireplaces) whose costs are shown in the modifications, adjustments, & alternatives below or at the end of the square foot section.

©Home Planners, Inc.

Base cost per square foot of living area

Exterior Wall	Living Area										
	600	800	1000	1200	1400	1600	1800	2000	2400	2800	3200
Wood Siding - Wood Frame	127.95	115.85	106.60	99.20	92.65	88.50	86.50	83.75	78.15	73.95	71.20
Brick Veneer - Wood Frame	134.45	121.65	111.85	103.95	97.00	92.55	90.40	87.45	81.45	77.05	74.10
Stucco on Wood Frame	121.25	109.90	101.20	94.30	88.25	84.35	82.45	79.95	74.70	70.75	68.25
Painted Concrete Block	129.05	116.80	107.55	100.05	93.45	89.25	87.15	84.35	78.65	74.50	71.70
Finished Basement, Add	32.35	30.50	29.15	27.95	26.95	26.30	25.95	25.40	24.70	24.15	23.65
Unfinished Basement, Add	14.30	12.80	11.70	10.75	10.00	9.50	9.20	8.75	8.20	7.75	7.40

Modifications

Add to the total cost

Upgrade Kitchen Cabinets	$ + 946
Solid Surface Countertops	+ 678
Full Bath - including plumbing, wall and floor finishes	+ 5863
Half Bath - including plumbing, wall and floor finishes	+ 3420
One Car Attached Garage	+ 12,758
One Car Detached Garage	+ 16,393
Fireplace & Chimney	+ 5874

Adjustments

For multi family - add to total cost

Additional Kitchen	$ + 4417
Additional Bath	+ 5863
Additional Entry & Exit	+ 1658
Separate Heating	+ 1133
Separate Electric	+ 1138

For Townhouse/Rowhouse -
Multiply cost per square foot by

Inner Unit	.95
End Unit	.97

Alternatives

Add to or deduct from the cost per square foot of living area

Composition Roll Roofing	– 1.10
Cedar Shake Roof	+ 3.95
Upgrade Walls and Ceilings to Skim Coat Plaster	+ .70
Upgrade Ceilings to Textured Finish	+ .50
Air Conditioning, in Heating Ductwork	+ 4.70
In Separate Ductwork	+ 6.89
Heating Systems, Hot Water	+ 1.93
Heat Pump	+ 2.61
Electric Heat	– 1.23
Not Heated	– 3.84

Additional upgrades or components

Kitchen Cabinets & Countertops	Page 58
Bathroom Vanities	59
Fireplaces & Chimneys	59
Windows, Skylights & Dormers	59
Appliances	60
Breezeways & Porches	60
Finished Attic	60
Garages	61
Site Improvements	61
Wings & Ells	25

- **Mass produced from stock plans**
- **Single family — 1 full bath, 1 kitchen**
- **No basement**
- **Asphalt shingles on roof**
- **Hot air heat**
- **Gypsum wallboard interior finishes**
- **Materials and workmanship are sufficient to meet codes**
- **Detail specifications on page 19**

Note: The illustration shown may contain some optional components (for example: garages and/or fireplaces) whose costs are shown in the modifications, adjustments, & alternatives below or at the end of the square foot section.

Base cost per square foot of living area

Exterior Wall	Living Area										
	600	800	1000	1200	1400	1600	1800	2000	2400	2800	3200
Wood Siding - Wood Frame	149.60	124.05	110.65	104.35	99.80	92.95	89.60	86.20	79.00	76.25	73.25
Brick Veneer - Wood Frame	155.70	128.35	115.00	108.55	104.00	96.80	93.30	89.75	82.15	79.30	76.15
Stucco on Wood Frame	137.25	115.00	102.60	96.95	92.95	86.75	83.75	80.70	74.20	71.75	69.20
Painted Concrete Block	148.15	122.90	109.90	103.85	99.50	92.65	89.45	86.05	78.90	76.20	73.30
Finished Basement, Add	24.85	21.15	20.15	19.45	18.95	18.25	17.80	17.45	16.70	16.35	15.90
Unfinished Basement, Add	12.60	9.65	8.85	8.35	7.90	7.35	7.00	6.70	6.05	5.85	5.50

Modifications

Add to the total cost

Upgrade Kitchen Cabinets	$ + 946
Solid Surface Countertops	+ 678
Full Bath - including plumbing, wall and floor finishes	+ 5863
Half Bath - including plumbing, wall and floor finishes	+ 3420
One Car Attached Garage	+ 12,758
One Car Detached Garage	+ 16,393
Fireplace & Chimney	+ 5874

Adjustments

For multi family - add to total cost

Additional Kitchen	$ + 4417
Additional Bath	+ 5863
Additional Entry & Exit	+ 1658
Separate Heating	+ 1133
Separate Electric	+ 1138

For Townhouse/Rowhouse -
Multiply cost per square foot by

Inner Unit	.95
End Unit	.97

Alternatives

Add to or deduct from the cost per square foot of living area

Composition Roll Roofing	- .75
Cedar Shake Roof	+ 2.85
Upgrade Walls and Ceilings to Skim Coat Plaster	+ .71
Upgrade Ceilings to Textured Finish	+ .50
Air Conditioning, in Heating Ductwork	+ 3.50
In Separate Ductwork	+ 6.05
Heating Systems, Hot Water	+ 1.84
Heat Pump	+ 2.87
Electric Heat	- .98
Not Heated	- 3.53

Additional upgrades or components

Kitchen Cabinets & Countertops	Page 58
Bathroom Vanities	59
Fireplaces & Chimneys	59
Windows, Skylights & Dormers	59
Appliances	60
Breezeways & Porches	60
Finished Attic	60
Garages	61
Site Improvements	61
Wings & Ells	25

Important: See the Reference Section for Location Factors (to adjust for your city) and Estimating Forms.

- **Mass produced from stock plans**
- **Single family — 1 full bath, 1 kitchen**
- **No basement**
- **Asphalt shingles on roof**
- **Hot air heat**
- **Gypsum wallboard interior finishes**
- **Materials and workmanship are sufficient to meet codes**
- **Detail specifications on page 19**

Note: The illustration shown may contain some optional components (for example: garages and/or fireplaces) whose costs are shown in the modifications, adjustments, & alternatives below or at the end of the square foot section.

Base cost per square foot of living area

Exterior Wall	Living Area										
	1000	1200	1400	1600	1800	2000	2200	2600	3000	3400	3800
Wood Siding - Wood Frame	115.70	104.65	99.55	95.95	92.40	88.35	85.70	80.65	75.70	73.50	71.50
Brick Veneer - Wood Frame	122.75	111.15	105.65	101.80	97.90	93.55	90.75	85.20	79.85	77.50	75.30
Stucco on Wood Frame	108.50	97.95	93.30	90.00	86.75	82.90	80.60	76.00	71.35	69.40	67.60
Painted Concrete Block	116.95	105.75	100.60	96.95	93.35	89.25	86.55	81.45	76.35	74.10	72.15
Finished Basement, Add	16.95	16.25	15.65	15.30	14.90	14.65	14.35	13.75	13.35	13.10	12.90
Unfinished Basement, Add	7.75	7.15	6.70	6.40	6.05	5.85	5.65	5.20	4.90	4.70	4.50

Modifications

Add to the total cost

Upgrade Kitchen Cabinets	$ + 946
Solid Surface Countertops	+ 678
Full Bath - including plumbing, wall and floor finishes	+ 5863
Half Bath - including plumbing, wall and floor finishes	+ 3420
One Car Attached Garage	+ 12,758
One Car Detached Garage	+ 16,393
Fireplace & Chimney	+ 6491

Adjustments

For multi family - add to total cost

Additional Kitchen	$ + 4417
Additional Bath	+ 5863
Additional Entry & Exit	+ 1658
Separate Heating	+ 1133
Separate Electric	+ 1138

For Townhouse/Rowhouse -
Multiply cost per square foot by

Inner Unit	.93
End Unit	.96

Alternatives

Add to or deduct from the cost per square foot of living area

Composition Roll Roofing	− .55
Cedar Shake Roof	+ 2
Upgrade Walls and Ceilings to Skim Coat Plaster	+ .72
Upgrade Ceilings to Textured Finish	+ .50
Air Conditioning, in Heating Ductwork	+ 2.85
In Separate Ductwork	+ 5.55
Heating Systems, Hot Water	+ 1.79
Heat Pump	+ 3.05
Electric Heat	− .85
Not Heated	− 3.34

Additional upgrades or components

Kitchen Cabinets & Countertops	Page 58
Bathroom Vanities	59
Fireplaces & Chimneys	59
Windows, Skylights & Dormers	59
Appliances	60
Breezeways & Porches	60
Finished Attic	60
Garages	61
Site Improvements	61
Wings & Ells	25

- **Mass produced from stock plans**
- **Single family — 1 full bath, 1 kitchen**
- **No basement**
- **Asphalt shingles on roof**
- **Hot air heat**
- **Gypsum wallboard interior finishes**
- **Materials and workmanship are sufficient to meet codes**
- **Detail specifications on page 19**

Note: The illustration shown may contain some optional components (for example: garages and/or fireplaces) whose costs are shown in the modifications, adjustments, & alternatives below or at the end of the square foot section.

Base cost per square foot of living area

Exterior Wall	Living Area										
	1000	1200	1400	1600	1800	2000	2200	2600	3000	3400	3800
Wood Siding - Wood Frame	107.25	96.80	92.20	88.95	85.80	82.00	79.70	75.20	70.65	68.70	66.95
Brick Veneer - Wood Frame	112.50	101.75	96.80	93.35	89.95	85.95	83.45	78.60	73.80	71.65	69.75
Stucco on Wood Frame	101.80	91.80	87.55	84.50	81.60	77.95	75.85	71.70	67.40	65.60	64.00
Painted Concrete Block	108.20	97.65	93.00	89.70	86.55	82.60	80.30	75.80	71.15	69.15	67.45
Finished Basement, Add	16.95	16.25	15.65	15.30	14.90	14.65	14.35	13.75	13.35	13.10	12.90
Unfinished Basement, Add	7.75	7.15	6.70	6.40	6.05	5.85	5.65	5.20	4.90	4.70	4.50

Modifications

Add to the total cost

Upgrade Kitchen Cabinets	$ + 946
Solid Surface Countertops	+ 678
Full Bath - including plumbing, wall and floor finishes	+ 5863
Half Bath - including plumbing, wall and floor finishes	+ 3420
One Car Attached Garage	+ 12,758
One Car Detached Garage	+ 16,393
Fireplace & Chimney	+ 5874

Adjustments

For multi family - add to total cost

Additional Kitchen	$ + 4417
Additional Bath	+ 5863
Additional Entry & Exit	+ 1658
Separate Heating	+ 1133
Separate Electric	+ 1138

For Townhouse/Rowhouse -
Multiply cost per square foot by

Inner Unit	.94
End Unit	.97

Alternatives

Add to or deduct from the cost per square foot of living area

Composition Roll Roofing	– .55
Cedar Shake Roof	+ 2
Upgrade Walls and Ceilings to Skim Coat Plaster	+ .68
Upgrade Ceilings to Textured Finish	+ .50
Air Conditioning, in Heating Ductwork	+ 2.85
In Separate Ductwork	+ 5.55
Heating Systems, Hot Water	+ 1.79
Heat Pump	+ 3.05
Electric Heat	– .85
Not Heated	– 3.34

Additional upgrades or components

Kitchen Cabinets & Countertops	Page 58
Bathroom Vanities	59
Fireplaces & Chimneys	59
Windows, Skylights & Dormers	59
Appliances	60
Breezeways & Porches	60
Finished Attic	60
Garages	61
Site Improvements	61
Wings & Ells	25

Important: See the Reference Section for Location Factors (to adjust for your city) and Estimating Forms.

- **Mass produced from stock plans**
- **Single family — 1 full bath, 1 kitchen**
- **No basement**
- **Asphalt shingles on roof**
- **Hot air heat**
- **Gypsum wallboard interior finishes**
- **Materials and workmanship are sufficient to meet codes**
- **Detail specifications on page 19**

Note: The illustration shown may contain some optional components (for example: garages and/or fireplaces) whose costs are shown in the modifications, adjustments, & alternatives below or at the end of the square foot section.

©Design Basics, Inc.

Base cost per square foot of living area

Exterior Wall	Living Area										
	1200	1500	1800	2000	2200	2400	2800	3200	3600	4000	4400
Wood Siding - Wood Frame	99.10	90.90	84.90	82.50	78.85	76.05	73.70	70.70	67.20	65.90	63.15
Brick Veneer - Wood Frame	103.90	95.30	88.85	86.25	82.45	79.40	76.95	73.75	70.00	68.65	65.70
Stucco on Wood Frame	94.10	86.40	80.80	78.60	75.20	72.55	70.40	67.60	64.35	63.10	60.50
Solid Masonry	99.90	91.65	85.55	83.15	79.50	76.65	74.30	71.25	67.65	66.35	63.60
Finished Basement, Add*	20.30	19.45	18.65	18.30	17.95	17.55	17.30	16.85	16.50	16.30	16.05
Unfinished Basement, Add*	8.50	7.80	7.15	6.90	6.60	6.30	6.05	5.75	5.45	5.30	5.10

*Basement under middle level only.

Modifications

Add to the total cost

Upgrade Kitchen Cabinets	$ + 946
Solid Surface Countertops	+ 678
Full Bath - including plumbing, wall and floor finishes	+ 5863
Half Bath - including plumbing, wall and floor finishes	+ 3420
One Car Attached Garage	+ 12,758
One Car Detached Garage	+ 16,393
Fireplace & Chimney	+ 5874

Adjustments

For multi family - add to total cost

Additional Kitchen	$ + 4417
Additional Bath	+ 5863
Additional Entry & Exit	+ 1658
Separate Heating	+ 1133
Separate Electric	+ 1138

For Townhouse/Rowhouse - Multiply cost per square foot by

Inner Unit	.93
End Unit	.96

Alternatives

Add to or deduct from the cost per square foot of living area

Composition Roll Roofing	– .75
Cedar Shake Roof	+ 2.85
Upgrade Walls and Ceilings to Skim Coat Plaster	+ .62
Upgrade Ceilings to Textured Finish	+ .50
Air Conditioning, in Heating Ductwork	+ 2.44
In Separate Ductwork	+ 5.16
Heating Systems, Hot Water	+ 1.72
Heat Pump	+ 3.16
Electric Heat	– .74
Not Heated	– 3.24

Additional upgrades or components

Kitchen Cabinets & Countertops	Page 58
Bathroom Vanities	59
Fireplaces & Chimneys	59
Windows, Skylights & Dormers	59
Appliances	60
Breezeways & Porches	60
Finished Attic	60
Garages	61
Site Improvements	61
Wings & Ells	25

Important: See the Reference Section for Location Factors (to adjust for your city) and Estimating Forms.

1 Story — Base cost per square foot of living area

Exterior Wall	Living Area							
	50	100	200	300	400	500	600	700
Wood Siding - Wood Frame	178.75	136.40	118.05	97.95	91.95	88.30	85.90	86.45
Brick Veneer - Wood Frame	195.15	148.15	127.80	104.45	97.85	93.75	91.10	91.50
Stucco on Wood Frame	161.85	124.40	108.00	91.25	85.90	82.70	80.60	81.30
Painted Concrete Block	181.55	138.45	119.70	99.00	92.95	89.25	86.80	87.30
Finished Basement, Add	49.50	40.40	36.65	30.35	29.05	28.30	27.80	27.45
Unfinished Basement, Add	27.40	20.35	17.40	12.50	11.50	10.95	10.55	10.30

1-1/2 Story — Base cost per square foot of living area

Exterior Wall	Living Area							
	100	200	300	400	500	600	700	800
Wood Siding - Wood Frame	140.85	113.50	96.65	86.20	81.15	78.65	75.55	74.70
Brick Veneer - Wood Frame	155.50	125.20	106.40	93.85	88.20	85.30	81.85	80.90
Stucco on Wood Frame	125.80	101.40	86.60	78.40	73.95	71.85	69.10	68.30
Painted Concrete Block	143.35	115.50	98.35	87.50	82.35	79.85	76.65	75.80
Finished Basement, Add	33.05	29.30	26.75	24.00	23.25	22.75	22.30	22.20
Unfinished Basement, Add	16.90	13.95	12.00	9.80	9.20	8.85	8.50	8.40

2 Story — Base cost per square foot of living area

Exterior Wall	Living Area							
	100	200	400	600	800	1000	1200	1400
Wood Siding - Wood Frame	143.80	107.05	91.05	74.80	69.60	66.40	64.35	65.15
Brick Veneer - Wood Frame	160.20	118.85	100.80	81.35	75.45	71.90	69.55	70.15
Stucco on Wood Frame	126.90	95.05	81.00	68.10	63.60	60.80	59.05	59.95
Painted Concrete Block	146.60	109.05	92.70	75.95	70.65	67.35	65.25	66.00
Finished Basement, Add	24.75	20.25	18.35	15.15	14.60	14.20	13.95	13.75
Unfinished Basement, Add	13.70	10.15	8.70	6.30	5.80	5.50	5.30	5.15

Base costs do not include bathroom or kitchen facilities. Use Modifications/Adjustments/Alternatives on pages 58-61 where appropriate.

Important: See the Reference Section for Location Factors (to adjust for your city) and Estimating Forms.

1 Story

1-1/2 Story

2 Story

2-1/2 Story

Bi-Level

Tri-Level

	Components
1 Site Work	Site preparation for slab or excavation for lower level; 4' deep trench excavation for foundation wall.
2 Foundations	Continuous reinforced concrete footing, 8" deep x 18" wide; dampproofed and insulated 8" thick reinforced concrete foundation wall, 4' deep; 4" concrete slab on 4" crushed stone base and polyethylene vapor barrier, trowel finish.
3 Framing	Exterior walls—2 x 4 wood studs, 16" O.C.; 1/2" plywood sheathing; 2 x 6 rafters 16" O.C. with 1/2" plywood sheathing, 4 in 12 or 8 in 12 roof pitch, 2 x 6 ceiling joists; 1/2" plywood subfloor on 1 x 2 wood sleepers 16" O.C.; 2 x 8 floor joists with 5/8" plywood subfloor on models with more than one level.
4 Exterior Walls	Beveled wood siding and #15 felt building paper on insulated wood frame walls; double hung windows; 3 flush solid core wood exterior doors with storms. **Alternates:** • Brick veneer on wood frame, has 4" veneer of common brick. • Stucco on wood frame has 1" thick colored stucco finish. • Solid masonry has a 6" concrete block load bearing wall with insulation and a brick or stone exterior. Also solid brick or stone walls.
5 Roofing	25 year asphalt shingles; #15 felt building paper; aluminum gutters, downspouts, drip edge and flashings.
6 Interiors	Walls and ceilings—1/2" taped and finished drywall, primed and painted with 2 coats; painted baseboard and trim, finished hardwood floor 40%, carpet with 1/2" underlayment 40%, vinyl tile with 1/2" underlayment 15%, ceramic tile with 1/2" underlayment 5%; hollow core and louvered interior doors.
7 Specialties	Average grade kitchen cabinets—14 L.F. wall and base with plastic laminate counter top and kitchen sink; 40 gallon electric water heater.
8 Mechanical	1 lavatory, white, wall hung; 1 water closet, white; 1 bathtub, enameled steel, white; gas fired warm air heat.
9 Electrical	100 Amp. service; romex wiring; incandescent lighting fixtures, switches, receptacles.
10 Overhead and Profit	General Contractor overhead and profit.

Adjustments

Unfinished Basement:
7'-6" high cast-in-place concrete walls 8" thick or 8" concrete block.

Finished Basement:
Includes plywood paneling or drywall on furring on foundation walls, sponge backed carpeting on concrete floor, suspended ceiling, lighting and heating.

- **Simple design from standard plans**
- **Single family — 1 full bath, 1 kitchen**
- **No basement**
- **Asphalt shingles on roof**
- **Hot air heat**
- **Gypsum wallboard interior finishes**
- **Materials and workmanship are average**
- **Detail specifications on p. 27**

Note: The illustration shown may contain some optional components (for example: garages and/or fireplaces) whose costs are shown in the modifications, adjustments, & alternatives below or at the end of the square foot section.

©Home Planners, Inc.

Base cost per square foot of living area

Exterior Wall	Living Area										
	600	800	1000	1200	1400	1600	1800	2000	2400	2800	3200
Wood Siding - Wood Frame	155.00	140.20	129.15	120.35	112.85	107.95	105.40	102.25	95.80	91.10	87.90
Brick Veneer - Wood Frame	174.20	158.70	147.05	137.75	129.85	124.65	121.90	118.50	111.75	106.80	103.40
Stucco on Wood Frame	164.10	149.65	138.85	130.35	123.05	118.30	115.85	112.80	106.50	101.90	98.90
Solid Masonry	189.65	172.45	159.55	149.05	140.05	134.25	131.15	127.20	119.70	114.15	110.20
Finished Basement, Add	39.25	37.90	36.15	34.50	33.20	32.35	31.85	31.15	30.20	29.45	28.80
Unfinished Basement, Add	16.25	14.75	13.60	12.60	11.80	11.25	10.95	10.55	9.95	9.50	9.10

Modifications

Add to the total cost

Upgrade Kitchen Cabinets	$ + 4158
Solid Surface Countertops (Included)	
Full Bath - including plumbing, wall and floor finishes	+ 7329
Half Bath - including plumbing, wall and floor finishes	+ 4275
One Car Attached Garage	+ 13,759
One Car Detached Garage	+ 17,994
Fireplace & Chimney	+ 6193

Adjustments

For multi family - add to total cost

Additional Kitchen	$ + 8034
Additional Bath	+ 7329
Additional Entry & Exit	+ 1658
Separate Heating	+ 1133
Separate Electric	+ 2036

For Townhouse/Rowhouse -
Multiply cost per square foot by

Inner Unit	.92
End Unit	.96

Alternatives

Add to or deduct from the cost per square foot of living area

Cedar Shake Roof	+ 3.35
Clay Tile Roof	+ 7.45
Slate Roof	+ 6.45
Upgrade Walls to Skim Coat Plaster	+ .44
Upgrade Ceilings to Textured Finish	+ .50
Air Conditioning, in Heating Ductwork	+ 4.86
In Separate Ductwork	+ 7.15
Heating Systems, Hot Water	+ 1.97
Heat Pump	+ 2.71
Electric Heat	– .68
Not Heated	– 3.21

Additional upgrades or components

Kitchen Cabinets & Countertops	Page 58
Bathroom Vanities	59
Fireplaces & Chimneys	59
Windows, Skylights & Dormers	59
Appliances	60
Breezeways & Porches	60
Finished Attic	60
Garages	61
Site Improvements	61
Wings & Ells	37

- **Simple design from standard plans**
- **Single family — 1 full bath, 1 kitchen**
- **No basement**
- **Asphalt shingles on roof**
- **Hot air heat**
- **Gypsum wallboard interior finishes**
- **Materials and workmanship are average**
- **Detail specifications on p. 27**

Note: The illustration shown may contain some optional components (for example: garages and/or fireplaces) whose costs are shown in the modifications, adjustments, & alternatives below or at the end of the square foot section.

©By Designer

Base cost per square foot of living area

Exterior Wall	Living Area										
	600	800	1000	1200	1400	1600	1800	2000	2400	2800	3200
Wood Siding - Wood Frame	173.40	144.90	129.85	122.65	117.55	109.90	106.15	102.25	94.30	91.20	87.90
Brick Veneer - Wood Frame	182.75	151.75	136.10	128.55	123.15	115.05	110.95	106.85	98.35	95.05	91.45
Stucco on Wood Frame	168.60	141.50	126.65	119.70	114.70	107.30	103.65	99.90	92.25	89.30	86.10
Solid Masonry	204.30	167.40	150.60	142.10	136.05	126.75	122.10	117.40	107.65	103.85	99.60
Finished Basement, Add	31.95	27.80	26.60	25.75	25.10	24.10	23.50	23.05	22.00	21.55	21.00
Unfinished Basement, Add	14.00	11.00	10.25	9.65	9.25	8.60	8.25	8.00	7.35	7.05	6.75

Modifications

Add to the total cost

Upgrade Kitchen Cabinets	$ + 4158
Solid Surface Countertops (Included)	
Full Bath - including plumbing, wall and floor finishes	+ 7329
Half Bath - including plumbing, wall and floor finishes	+ 4275
One Car Attached Garage	+ 13,759
One Car Detached Garage	+ 17,994
Fireplace & Chimney	+ 6193

Adjustments

For multi family - add to total cost

Additional Kitchen	$ + 8034
Additional Bath	+ 7329
Additional Entry & Exit	+ 1658
Separate Heating	+ 1133
Separate Electric	+ 2036

For Townhouse/Rowhouse -
Multiply cost per square foot by

Inner Unit	.92
End Unit	.96

Alternatives

Add to or deduct from the cost per square foot of living area

Cedar Shake Roof	+ 2.45
Clay Tile Roof	+ 5.40
Slate Roof	+ 4.65
Upgrade Walls to Skim Coat Plaster	+ .50
Upgrade Ceilings to Textured Finish	+ .50
Air Conditioning, in Heating Ductwork	+ 3.69
In Separate Ductwork	+ 6.26
Heating Systems, Hot Water	+ 1.88
Heat Pump	+ 3
Electric Heat	– .62
Not Heated	– 3.08

Additional upgrades or components

Kitchen Cabinets & Countertops	Page 58
Bathroom Vanities	59
Fireplaces & Chimneys	59
Windows, Skylights & Dormers	59
Appliances	60
Breezeways & Porches	60
Finished Attic	60
Garages	61
Site Improvements	61
Wings & Ells	37

- Simple design from standard plans
- Single family — 1 full bath, 1 kitchen
- No basement
- Asphalt shingles on roof
- Hot air heat
- Gypsum wallboard interior finishes
- Materials and workmanship are average
- Detail specifications on p. 27

Note: The illustration shown may contain some optional components (for example: garages and/or fireplaces) whose costs are shown in the modifications, adjustments, & alternatives below or at the end of the square foot section.

Base cost per square foot of living area

Exterior Wall	Living Area										
	1000	1200	1400	1600	1800	2000	2200	2600	3000	3400	3800
Wood Siding - Wood Frame	136.80	124.05	117.95	113.80	109.55	104.95	101.95	96.20	90.55	88.10	85.80
Brick Veneer - Wood Frame	144.00	130.75	124.25	119.80	115.20	110.40	107.10	100.90	94.85	92.15	89.70
Stucco on Wood Frame	133.10	120.65	114.80	110.75	106.70	102.20	99.30	93.85	88.35	85.95	83.75
Solid Masonry	160.70	146.20	138.70	133.55	128.15	122.90	118.95	111.65	104.80	101.60	98.65
Finished Basement, Add	22.50	22.15	21.40	20.95	20.40	20.05	19.65	18.90	18.40	18.10	17.80
Unfinished Basement, Add	8.85	8.25	7.85	7.55	7.15	6.95	6.70	6.25	5.95	5.75	5.55

Modifications

Add to the total cost

Upgrade Kitchen Cabinets	$ + 4158
Solid Surface Countertops (Included)	
Full Bath - including plumbing, wall and floor finishes	+ 7329
Half Bath - including plumbing, wall and floor finishes	+ 4275
One Car Attached Garage	+ 13,759
One Car Detached Garage	+ 17,994
Fireplace & Chimney	+ 6840

Adjustments

For multi family - add to total cost

Additional Kitchen	$ + 8034
Additional Bath	+ 7329
Additional Entry & Exit	+ 1658
Separate Heating	+ 1133
Separate Electric	+ 2036

For Townhouse/Rowhouse -
Multiply cost per square foot by

Inner Unit	.90
End Unit	.95

Alternatives

Add to or deduct from the cost per square foot of living area

Cedar Shake Roof	+ 1.70
Clay Tile Roof	+ 3.70
Slate Roof	+ 3.25
Upgrade Walls to Skim Coat Plaster	+ .52
Upgrade Ceilings to Textured Finish	+ .50
Air Conditioning, in Heating Ductwork	+ 2.94
In Separate Ductwork	+ 5.68
Heating Systems, Hot Water	+ 1.84
Heat Pump	+ 3.16
Electric Heat	– .53
Not Heated	– 2.99

Additional upgrades or components

Kitchen Cabinets & Countertops	Page 58
Bathroom Vanities	59
Fireplaces & Chimneys	59
Windows, Skylights & Dormers	59
Appliances	60
Breezeways & Porches	60
Finished Attic	60
Garages	61
Site Improvements	61
Wings & Ells	37

- **Simple design from standard plans**
- **Single family — 1 full bath, 1 kitchen**
- **No basement**
- **Asphalt shingles on roof**
- **Hot air heat**
- **Gypsum wallboard interior finishes**
- **Materials and workmanship are average**
- **Detail specifications on p. 27**

Note: The illustration shown may contain some optional components (for example: garages and/or fireplaces) whose costs are shown in the modifications, adjustments, & alternatives below or at the end of the square foot section.

Base cost per square foot of living area

Exterior Wall	Living Area										
	1200	1400	1600	1800	2000	2400	2800	3200	3600	4000	4400
Wood Siding - Wood Frame	135.80	127.55	116.45	114.35	110.25	103.75	98.60	93.40	90.75	85.95	84.45
Brick Veneer - Wood Frame	143.50	134.50	122.90	120.80	116.25	109.25	103.75	98.15	95.25	90.10	88.50
Stucco on Wood Frame	131.90	123.95	113.25	111.10	107.15	101.00	95.95	91.00	88.45	83.80	82.40
Solid Masonry	161.25	150.60	137.65	135.60	130.15	121.80	115.75	109.00	105.50	99.65	97.75
Finished Basement, Add	18.95	18.60	17.90	17.75	17.35	16.55	16.25	15.75	15.40	15.05	14.90
Unfinished Basement, Add	7.45	6.85	6.40	6.35	6.05	5.60	5.40	5.05	4.85	4.65	4.55

Modifications

Add to the total cost

Upgrade Kitchen Cabinets	$ + 4158
Solid Surface Countertops (Included)	
Full Bath - including plumbing, wall and floor finishes	+ 7329
Half Bath - including plumbing, wall and floor finishes	+ 4275
One Car Attached Garage	+ 13,759
One Car Detached Garage	+ 17,994
Fireplace & Chimney	+ 7504

Adjustments

For multi family - add to total cost

Additional Kitchen	$ + 8034
Additional Bath	+ 7329
Additional Entry & Exit	+ 1658
Separate Heating	+ 1133
Separate Electric	+ 2036

For Townhouse/Rowhouse -
Multiply cost per square foot by

Inner Unit	.90
End Unit	.95

Alternatives

Add to or deduct from the cost per square foot of living area

Cedar Shake Roof	+ 1.45
Clay Tile Roof	+ 3.25
Slate Roof	+ 2.80
Upgrade Walls to Skim Coat Plaster	+ .50
Upgrade Ceilings to Textured Finish	+ .50
Air Conditioning, in Heating Ductwork	+ 2.67
In Separate Ductwork	+ 5.49
Heating Systems, Hot Water	+ 1.66
Heat Pump	+ 3.25
Electric Heat	- .95
Not Heated	- 3.49

Additional upgrades or components

Kitchen Cabinets & Countertops	Page 58
Bathroom Vanities	59
Fireplaces & Chimneys	59
Windows, Skylights & Dormers	59
Appliances	60
Breezeways & Porches	60
Finished Attic	60
Garages	61
Site Improvements	61
Wings & Ells	37

Important: See the Reference Section for Location Factors (to adjust for your city) and Estimating Forms.

- **Simple design from standard plans**
- **Single family — 1 full bath, 1 kitchen**
- **No basement**
- **Asphalt shingles on roof**
- **Hot air heat**
- **Gypsum wallboard interior finishes**
- **Materials and workmanship are average**
- **Detail specifications on p. 27**

Note: The illustration shown may contain some optional components (for example: garages and/or fireplaces) whose costs are shown in the modifications, adjustments, & alternatives below or at the end of the square foot section.

Base cost per square foot of living area

Exterior Wall	Living Area										
	1500	1800	2100	2500	3000	3500	4000	4500	5000	5500	6000
Wood Siding - Wood Frame	125.85	114.30	109.20	105.20	97.60	94.35	89.70	84.70	83.15	81.35	79.45
Brick Veneer - Wood Frame	133.05	121.05	115.45	111.15	103.00	99.50	94.40	89.05	87.35	85.35	83.25
Stucco on Wood Frame	122.15	110.90	106.00	102.15	94.80	91.75	87.35	82.50	81.00	79.30	77.55
Solid Masonry	149.75	136.45	129.90	124.95	115.50	111.30	105.15	98.95	96.95	94.60	91.95
Finished Basement, Add	16.25	16.05	15.60	15.30	14.65	14.40	13.95	13.60	13.40	13.25	13.05
Unfinished Basement, Add	6.00	5.65	5.30	5.15	4.75	4.55	4.30	4.10	3.95	3.85	3.75

Modifications

Add to the total cost

Upgrade Kitchen Cabinets	$ + 4158
Solid Surface Countertops (Included)	
Full Bath - including plumbing, wall and floor finishes	+ 7329
Half Bath - including plumbing, wall and floor finishes	+ 4275
One Car Attached Garage	+ 13,759
One Car Detached Garage	+ 17,994
Fireplace & Chimney	+ 7504

Adjustments

For multi family - add to total cost

Additional Kitchen	$ + 8034
Additional Bath	+ 7329
Additional Entry & Exit	+ 1658
Separate Heating	+ 1133
Separate Electric	+ 2036

For Townhouse/Rowhouse -
Multiply cost per square foot by

Inner Unit	.88
End Unit	.94

Alternatives

Add to or deduct from the cost per square foot of living area

Cedar Shake Roof	+ 1.10
Clay Tile Roof	+ 2.50
Slate Roof	+ 2.15
Upgrade Walls to Skim Coat Plaster	+ .52
Upgrade Ceilings to Textured Finish	+ .50
Air Conditioning, in Heating Ductwork	+ 2.67
In Separate Ductwork	+ 5.49
Heating Systems, Hot Water	+ 1.66
Heat Pump	+ 3.25
Electric Heat	- .74
Not Heated	- 3.25

Additional upgrades or components

Kitchen Cabinets & Countertops	Page 58
Bathroom Vanities	59
Fireplaces & Chimneys	59
Windows, Skylights & Dormers	59
Appliances	60
Breezeways & Porches	60
Finished Attic	60
Garages	61
Site Improvements	61
Wings & Ells	37

Important: See the Reference Section for Location Factors (to adjust for your city) and Estimating Forms.

- **Simple design from standard plans**
- **Single family — 1 full bath, 1 kitchen**
- **No basement**
- **Asphalt shingles on roof**
- **Hot air heat**
- **Gypsum wallboard interior finishes**
- **Materials and workmanship are average**
- **Detail specifications on p. 27**

Note: The illustration shown may contain some optional components (for example: garages and/or fireplaces) whose costs are shown in the modifications, adjustments, & alternatives below or at the end of the square foot section.

Base cost per square foot of living area

Exterior Wall	Living Area										
	1000	1200	1400	1600	1800	2000	2200	2600	3000	3400	3800
Wood Siding - Wood Frame	127.70	115.55	110.10	106.30	102.45	98.15	95.50	90.30	85.15	82.90	80.85
Brick Veneer - Wood Frame	133.10	120.65	114.80	110.80	106.65	102.20	99.35	93.85	88.35	86.00	83.75
Stucco on Wood Frame	124.95	113.05	107.75	104.00	100.30	96.05	93.50	88.55	83.50	81.40	79.40
Solid Masonry	145.65	132.20	125.65	121.05	116.45	111.60	108.25	101.90	95.85	93.10	90.55
Finished Basement, Add	22.50	22.15	21.40	20.95	20.40	20.05	19.65	18.90	18.40	18.10	17.80
Unfinished Basement, Add	8.85	8.25	7.85	7.55	7.15	6.95	6.70	6.25	5.95	5.75	5.55

Modifications

Add to the total cost

Upgrade Kitchen Cabinets	$ + 4158
Solid Surface Countertops (Included)	
Full Bath - including plumbing, wall and floor finishes	+ 7329
Half Bath - including plumbing, wall and floor finishes	+ 4275
One Car Attached Garage	+ 13,759
One Car Detached Garage	+ 17,994
Fireplace & Chimney	+ 6193

Adjustments

For multi family - add to total cost

Additional Kitchen	$ + 8034
Additional Bath	+ 7329
Additional Entry & Exit	+ 1658
Separate Heating	+ 1133
Separate Electric	+ 2036

For Townhouse/Rowhouse -
Multiply cost per square foot by

Inner Unit	.91
End Unit	.96

Alternatives

Add to or deduct from the cost per square foot of living area

Cedar Shake Roof	+ 1.70
Clay Tile Roof	+ 3.70
Slate Roof	+ 3.25
Upgrade Walls to Skim Coat Plaster	+ .48
Upgrade Ceilings to Textured Finish	+ .50
Air Conditioning, in Heating Ductwork	+ 2.94
In Separate Ductwork	+ 5.68
Heating Systems, Hot Water	+ 1.84
Heat Pump	+ 3.16
Electric Heat	– .53
Not Heated	– 2.99

Additional upgrades or components

Kitchen Cabinets & Countertops	Page 58
Bathroom Vanities	59
Fireplaces & Chimneys	59
Windows, Skylights & Dormers	59
Appliances	60
Breezeways & Porches	60
Finished Attic	60
Garages	61
Site Improvements	61
Wings & Ells	37

- **Simple design from standard plans**
- **Single family — 1 full bath, 1 kitchen**
- **No basement**
- **Asphalt shingles on roof**
- **Hot air heat**
- **Gypsum wallboard interior finishes**
- **Materials and workmanship are average**
- **Detail specifications on p. 27**

Note: The illustration shown may contain some optional components (for example: garages and/or fireplaces) whose costs are shown in the modifications, adjustments, & alternatives below or at the end of the square foot section.

©Design Basics, Inc.

Base cost per square foot of living area

Exterior Wall	Living Area										
	1200	1500	1800	2100	2400	2700	3000	3400	3800	4200	4600
Wood Siding - Wood Frame	121.90	112.45	105.40	99.25	95.30	93.10	90.65	88.30	84.35	81.25	79.65
Brick Veneer - Wood Frame	126.90	116.95	109.45	102.95	98.80	96.40	93.75	91.35	87.20	83.90	82.25
Stucco on Wood Frame	119.35	110.10	103.25	97.30	93.55	91.35	89.00	86.75	82.90	79.90	78.40
Solid Masonry	138.40	127.40	118.80	111.45	106.80	104.15	101.00	98.35	93.75	90.05	88.20
Finished Basement, Add*	25.90	25.45	24.35	23.45	22.85	22.55	22.15	21.85	21.40	20.95	20.70
Unfinished Basement, Add*	9.90	9.15	8.45	7.95	7.60	7.40	7.15	6.95	6.70	6.45	6.30

*Basement under middle level only.

Modifications

Add to the total cost

Upgrade Kitchen Cabinets	$ + 4158
Solid Surface Countertops (Included)	
Full Bath - including plumbing, wall and floor finishes	+ 7329
Half Bath - including plumbing, wall and floor finishes	+ 4275
One Car Attached Garage	+ 13,759
One Car Detached Garage	+ 17,994
Fireplace & Chimney	+ 6193

Adjustments

For multi family - add to total cost

Additional Kitchen	$ + 8034
Additional Bath	+ 7329
Additional Entry & Exit	+ 1658
Separate Heating	+ 1133
Separate Electric	+ 2036

For Townhouse/Rowhouse -
Multiply cost per square foot by

Inner Unit	.90
End Unit	.95

Alternatives

Add to or deduct from the cost per square foot of living area

Cedar Shake Roof	+ 2.45
Clay Tile Roof	+ 5.40
Slate Roof	+ 4.65
Upgrade Walls to Skim Coat Plaster	+ .42
Upgrade Ceilings to Textured Finish	+ .50
Air Conditioning, in Heating Ductwork	+ 2.48
In Separate Ductwork	+ 5.35
Heating Systems, Hot Water	+ 1.78
Heat Pump	+ 3.30
Electric Heat	- .45
Not Heated	- 2.88

Additional upgrades or components

Kitchen Cabinets & Countertops	Page 58
Bathroom Vanities	59
Fireplaces & Chimneys	59
Windows, Skylights & Dormers	59
Appliances	60
Breezeways & Porches	60
Finished Attic	60
Garages	61
Site Improvements	61
Wings & Ells	37

- Post and beam frame
- Log exterior walls
- Simple design from standard plans
- Single family — 1 full bath, 1 kitchen
- No basement
- Asphalt shingles on roof
- Hot air heat
- Gypsum wallborad interior finishes
- Materials and workmanship are average
- Detail specification on page 27

Note: The illustration shown may contain some optional components (for example: garages and/or fireplaces) whose costs are shown in the modifications, adjustments, & alternatives below or at the end of the square foot section.

Base cost per square foot of living area

Exterior Wall	Living Area										
	600	800	1000	1200	1400	1600	1800	2000	2400	2800	3200
6" Log - Solid Wall	173.55	157.60	145.70	136.15	127.95	122.65	119.85	116.35	109.35	104.30	100,75
8" Log - Solid Wall	163.05	148.25	137.20	128.45	120.95	116.10	113.55	110.40	103.95	99.25	96.15
Finished Basement, Add	39.25	37.90	36.15	34.50	33.20	32.35	31.85	31.15	30.20	29.45	28.80
Unfinished Basement, Add	16.25	14.75	13.60	12.60	11.80	11.25	10.95	10.55	9.95	9.50	9.10

Modifications

Add to the total cost

Upgrade Kitchen Cabinets	$	+ 4158
Solid Surface Countertops (Included)		
Full Bath - including plumbing, wall and floor finishes		+ 7329
Half Bath - including plumbing, wall and floor finishes		+ 4275
One Car Attached Garage		+ 13,759
One Car Detached Garage		+ 17,994
Fireplace & Chimney		+ 6193

Adjustments

For multi family - add to total cost

Additional Kitchen	$	+ 8034
Additional Bath		+ 7329
Additional Entry & Exit		+ 1658
Separate Heating		+ 1133
Separate Electric		+ 2036

For Townhouse/Rowhouse - Multiply cost per square foot by

Inner Unit	.92
End Unit	.96

Alternatives

Add to or deduct from the cost per square foot of living area

Cedar Shake Roof	+ 3.35
Air Conditioning, in Heating Ductwork	+ 4.86
In Separate Ductwork	+ 7.14
Heating Systems, Hot Water	+ 1.97
Heat Pump	+ 2.70
Electric Heat	– .71
Not Heated	– 3.21

Additional upgrades or components

Kitchen Cabinets & Countertops	Page 58
Bathroom Vanities	59
Fireplaces & Chimneys	59
Windows, Skylights & Dormers	59
Appliances	60
Breezeways & Porches	60
Finished Attic	60
Garages	61
Site Improvements	61
Wings & Ells	37

- **Post and beam frame**
- **Log exterior walls**
- **Simple design from standard plans**
- **Single family — 1 full bath, 1 kitchen**
- **No basement**
- **Asphalt shingles on roof**
- **Hot air heat**
- **Gypsum wallboard interior finishes**
- **Materials and workmanship are average**
- **Detail specification on page 27**

Note: The illustration shown may contain some optional components (for example: garages and/or fireplaces) whose costs are shown in the modifications, adjustments, & alternatives below or at the end of the square foot section.

Base cost per square foot of living area

Exterior Wall	Living Area										
	1000	1200	1400	1600	1800	2000	2200	2600	3000	3400	3800
6" Log - Solid Wall	154.10	140.45	133.75	129.20	124.40	119.50	116.05	109.55	103.35	100.60	97.95
8" Log - Solid Wall	142.70	129.95	123.95	119.80	115.55	110.95	108.00	102.25	96.65	94.15	91.85
Finished Basement, Add	22.50	22.15	21.40	20.95	20.40	20.05	19.65	18.90	18.40	18.10	17.80
Unfinished Basement, Add	8.85	8.25	7.85	7.55	7.15	6.95	6.70	6.25	5.95	5.75	5.55

Modifications

Add to the total cost

Upgrade Kitchen Cabinets	$ + 4158
Solid Surface Countertops (Included)	
Full Bath - including plumbing, wall and floor finishes	+ 7329
Half Bath - including plumbing, wall and floor finishes	+ 4275
One Car Attached Garage	+ 13,759
One Car Detached Garage	+ 17,994
Fireplace & Chimney	+ 6840

Adjustments

For multi family - add to total cost

Additional Kitchen	$ + 8034
Additional Bath	+ 7329
Additional Entry & Exit	+ 1658
Separate Heating	+ 1133
Separate Electric	+ 2036

For Townhouse/Rowhouse - Multiply cost per square foot by

Inner Unit	.92
End Unit	.96

Alternatives

Add to or deduct from the cost per square foot of living area

Cedar Shake Roof	+ 1.70
Air Conditioning, in Heating Ductwork	+ 2.94
In Separate Ductwork	+ 5.68
Heating Systems, Hot Water	+ 1.84
Heat Pump	+ 3.16
Electric Heat	− .53
Not Heated	− 2.99

Additional upgrades or components

Kitchen Cabinets & Countertops	Page 58
Bathroom Vanities	59
Fireplaces & Chimneys	59
Windows, Skylights & Dormers	59
Appliances	60
Breezeways & Porches	60
Finished Attic	60
Garages	61
Site Improvements	61
Wings & Ells	37

1 Story — Base cost per square foot of living area

Exterior Wall	Living Area							
	50	100	200	300	400	500	600	700
Wood Siding - Wood Frame	203.65	157.35	137.20	115.90	109.30	105.35	102.75	103.55
Brick Veneer - Wood Frame	208.00	156.90	134.75	110.10	102.85	98.50	95.60	96.20
Stucco on Wood Frame	194.75	150.90	131.80	112.25	106.00	102.20	99.75	100.70
Solid Masonry	259.40	197.15	170.45	138.05	129.25	123.95	122.05	122.05
Finished Basement, Add	62.90	52.10	46.85	38.15	36.40	35.35	34.65	34.15
Unfinished Basement, Add	29.75	22.45	19.40	14.40	13.40	12.80	12.40	12.10

1-1/2 Story — Base cost per square foot of living area

Exterior Wall	Living Area							
	100	200	300	400	500	600	700	800
Wood Siding - Wood Frame	163.00	131.70	112.90	101.75	95.95	93.30	89.75	88.80
Brick Veneer - Wood Frame	222.45	165.95	137.70	120.65	112.10	107.50	102.55	100.75
Stucco on Wood Frame	199.60	147.70	122.50	108.75	101.15	97.15	92.80	91.10
Solid Masonry	257.10	193.70	160.85	138.70	128.75	123.20	117.50	115.50
Finished Basement, Add	42.40	38.30	34.80	30.95	29.95	29.25	28.60	28.50
Unfinished Basement, Add	18.65	15.65	13.60	11.40	10.80	10.40	10.00	9.95

2 Story — Base cost per square foot of living area

Exterior Wall	Living Area							
	100	200	400	600	800	1000	1200	1400
Wood Siding - Wood Frame	162.80	122.45	104.75	87.45	81.70	78.25	75.90	76.95
Brick Veneer - Wood Frame	226.05	157.65	126.35	101.95	93.50	88.50	85.15	85.45
Stucco on Wood Frame	200.50	139.40	111.10	91.75	84.45	80.00	77.05	77.65
Solid Masonry	264.90	185.40	149.50	117.35	107.45	101.45	97.50	97.35
Finished Basement, Add	33.40	28.05	25.45	21.10	20.20	19.70	19.35	19.10
Unfinished Basement, Add	15.00	11.40	9.90	7.35	6.85	6.55	6.35	6.20

Base costs do not include bathroom or kitchen facilities. Use Modifications/Adjustments/Alternatives on pages 58-61 where appropriate.

Important: See the Reference Section for Location Factors (to adjust for your city) and Estimating Forms.

1 Story

1-1/2 Story

2 Story

2-1/2 Story

Bi-Level

Tri-Level

	Components
1 Site Work	Site preparation for slab or excavation for lower level; 4' deep trench excavation for foundation wall.
2 Foundations	Continuous reinforced concrete footing, 8" deep x 18" wide; dampproofed and insulated 8" thick reinforced concrete foundation wall, 4' deep; 4" concrete slab on 4" crushed stone base and polyethylene vapor barrier, trowel finish.
3 Framing	Exterior walls—2 x 6 wood studs, 16" O.C.; 1/2" plywood sheathing; 2 x 8 rafters 16" O.C. with 1/2" plywood sheathing, 4 in 12, 6 in 12 or 8 in 12 roof pitch, 2 x 6 or 2 x 8 ceiling joists; 1/2" plywood subfloor on 1 x 3 wood sleepers 16" O.C.; 2 x 10 floor joists with 5/8" plywood subfloor on models with more than one level.
4 Exterior Walls	Horizontal beveled wood siding and #15 felt building paper on insulated wood frame walls; double hung windows; 3 flush solid core wood exterior doors with storms. **Alternates:** • Brick veneer on wood frame, has 4" veneer high quality face brick or select common brick. • Stone veneer on wood frame has exterior veneer of field stone or 2" thick limestone. • Solid masonry has an 8" concrete block wall with insulation and a brick or stone facing. It may be a solid brick or stone structure.
5 Roofing	30 year asphalt shingles; #15 felt building paper; aluminum gutters, downspouts and drip edge; copper flashings.
6 Interiors	Walls and ceilings—5/8" drywall, skim coat plaster, primed and painted with 2 coats; hardwood baseboard and trim; sanded and finished, hardwood floor 70%, ceramic tile with 1/2" underlayment 20%, vinyl tile with 1/2" underlayment 10%; wood panel interior doors, primed and painted with 2 coats.
7 Specialties	Custom grade kitchen cabinets—20 L.F. wall and base with plastic laminate counter top and kitchen sink; 4 L.F. bathroom vanity; 75 gallon electric water heater; medicine cabinet.
8 Mechanical	Gas fired warm air heat/air conditioning; one full bath including bathtub, corner shower, built in lavatory and water closet; one 1/2 bath including built in lavatory and water closet.
9 Electrical	100 Amp. service; romex wiring; incandescent lighting fixtures, switches, receptacles.
10 Overhead and Profit	General Contractor overhead and profit.

Adjustments

Unfinished Basement:
7'-6" high cast-in-place concrete walls 10" thick or 12" concrete block.

Finished Basement:
Includes painted drywall on 2 x 4 wood furring with insulation, suspended ceiling, carpeting on concrete floor, heating and lighting.

- A distinct residence from designer's plans
- Single family — 1 full bath, 1 half bath, 1 kitchen
- No basement
- Asphalt shingles on roof
- Forced hot air heat/air conditioning
- Gypsum wallboard interior finishes
- Materials and workmanship are above average
- Detail specifications on page 39

Note: The illustration shown may contain some optional components (for example: garages and/or fireplaces) whose costs are shown in the modifications, adjustments, & alternatives below or at the end of the square foot section.

©Design Basics, Inc.

Base cost per square foot of living area

Exterior Wall	Living Area										
	800	1000	1200	1400	1600	1800	2000	2400	2800	3200	3600
Wood Siding - Wood Frame	186.65	168.40	154.40	143.00	135.25	130.95	125.95	116.45	109.70	104.95	100.05
Brick Veneer - Wood Frame	208.10	189.05	174.30	162.25	154.20	149.60	144.25	134.35	127.25	122.15	116.90
Stone Veneer - Wood Frame	217.00	197.15	181.60	168.90	160.35	155.60	149.90	139.50	132.00	126.50	121.00
Solid Masonry	220.45	200.35	184.45	171.45	162.85	157.95	152.05	141.50	133.80	128.25	122.65
Finished Basement, Add	60.95	60.95	58.45	56.50	55.15	54.40	53.30	51.90	50.75	49.75	48.85
Unfinished Basement, Add	25.00	23.60	22.35	21.30	20.60	20.25	19.65	18.95	18.30	17.80	17.35

Modifications

Add to the total cost

Upgrade Kitchen Cabinets	$ + 1314
Solid Surface Countertops (Included)	
Full Bath - including plumbing, wall and floor finishes	+ 8795
Half Bath - including plumbing, wall and floor finishes	+ 5130
Two Car Attached Garage	+ 27,346
Two Car Detached Garage	+ 31,142
Fireplace & Chimney	+ 6494

Adjustments

For multi family - add to total cost

Additional Kitchen	$ + 16,920
Additional Full Bath & Half Bath	+ 13,925
Additional Entry & Exit	+ 1658
Separate Heating & Air Conditioning	+ 7328
Separate Electric	+ 2036

For Townhouse/Rowhouse -
Multiply cost per square foot by

Inner Unit	.90
End Unit	.95

Alternatives

Add to or deduct from the cost per square foot of living area

Cedar Shake Roof	+ 2.45
Clay Tile Roof	+ 6.55
Slate Roof	+ 5.60
Upgrade Ceilings to Textured Finish	+ .50
Air Conditioning, in Heating Ductwork	Base System
Heating Systems, Hot Water	+ 2.02
Heat Pump	+ 2.69
Electric Heat	− 1.99
Not Heated	− 4.02

Additional upgrades or components

Kitchen Cabinets & Countertops	Page 58
Bathroom Vanities	59
Fireplaces & Chimneys	59
Windows, Skylights & Dormers	59
Appliances	60
Breezeways & Porches	60
Finished Attic	60
Garages	61
Site Improvements	61
Wings & Ells	47

- **A distinct residence from designer's plans**
- **Single family — 1 full bath, 1 half bath, 1 kitchen**
- **No basement**
- **Asphalt shingles on roof**
- **Forced hot air heat/air conditioning**
- **Gypsum wallboard interior finishes**
- **Materials and workmanship are above average**
- **Detail specifications on page 39**

Note: The illustration shown may contain some optional components (for example: garages and/or fireplaces) whose costs are shown in the modifications, adjustments, & alternatives below or at the end of the square foot section.

●Donald A. Gardner Architects, Inc.

Base cost per square foot of living area

Exterior Wall	Living Area										
	1000	1200	1400	1600	1800	2000	2400	2800	3200	3600	4000
Wood Siding - Wood Frame	167.80	155.80	147.30	137.10	131.20	125.60	114.75	109.80	105.20	101.75	96.85
Brick Veneer - Wood Frame	176.65	164.10	155.20	144.25	138.00	132.05	120.45	115.20	110.20	106.60	101.35
Stone Veneer - Wood Frame	186.00	172.85	163.55	151.85	145.20	138.85	126.45	120.90	115.50	111.65	106.05
Solid Masonry	189.65	176.25	166.80	154.85	148.05	141.55	128.80	123.15	117.55	113.70	107.90
Finished Basement, Add	40.30	40.60	39.65	38.05	37.20	36.40	34.80	34.20	33.30	32.90	32.20
Unfinished Basement, Add	16.75	16.05	15.55	14.75	14.30	13.90	13.05	12.75	12.30	12.10	11.75

Modifications

Add to the total cost

Upgrade Kitchen Cabinets	$ + 1314
Solid Surface Countertops (Included)	
Full Bath - including plumbing, wall and floor finishes	+ 8795
Half Bath - including plumbing, wall and floor finishes	+ 5130
Two Car Attached Garage	+ 27,346
Two Car Detached Garage	+ 31,142
Fireplace & Chimney	+ 6494

Adjustments

For multi family - add to total cost

Additional Kitchen	$ + 16,920
Additional Full Bath & Half Bath	+ 13,925
Additional Entry & Exit	+ 1658
Separate Heating & Air Conditioning	+ 7328
Separate Electric	+ 2036

For Townhouse/Rowhouse -
Multiply cost per square foot by

Inner Unit	.90
End Unit	.95

Alternatives

Add to or deduct from the cost per square foot of living area

Cedar Shake Roof	+ 1.80
Clay Tile Roof	+ 4.75
Slate Roof	+ 4
Upgrade Ceilings to Textured Finish	+ .50
Air Conditioning, in Heating Ductwork	Base System
Heating Systems, Hot Water	+ 1.93
Heat Pump	+ 2.81
Electric Heat	− 1.76
Not Heated	− 3.71

Additional upgrades or components

Kitchen Cabinets & Countertops	Page 58
Bathroom Vanities	59
Fireplaces & Chimneys	59
Windows, Skylights & Dormers	59
Appliances	60
Breezeways & Porches	60
Finished Attic	60
Garages	61
Site Improvements	61
Wings & Ells	47

- **A distinct residence from designer's plans**
- **Single family — 1 full bath, 1 half bath, 1 kitchen**
- **No basement**
- **Asphalt shingles on roof**
- **Forced hot air heat/air conditioning**
- **Gypsum wallboard interior finishes**
- **Materials and workmanship are above average**
- **Detail specifications on page 39**

Note: The illustration shown may contain some optional components (for example: garages and/or fireplaces) whose costs are shown in the modifications, adjustments, & alternatives below or at the end of the square foot section.

Base cost per square foot of living area

Exterior Wall	Living Area										
	1200	1400	1600	1800	2000	2400	2800	3200	3600	4000	4400
Wood Siding - Wood Frame	156.70	147.20	140.35	134.25	127.90	118.65	110.80	105.65	102.70	99.45	96.65
Brick Veneer - Wood Frame	166.15	156.05	148.80	142.20	135.55	125.55	117.10	111.50	108.40	104.75	101.85
Stone Veneer - Wood Frame	176.15	165.40	157.75	150.60	143.65	132.85	123.70	117.75	114.35	110.40	107.30
Solid Masonry	180.00	169.00	161.20	153.90	146.80	135.70	126.30	120.25	116.75	112.55	109.45
Finished Basement, Add	32.40	32.75	31.90	31.05	30.50	29.25	28.30	27.60	27.25	26.70	26.40
Unfinished Basement, Add	13.45	12.90	12.50	12.05	11.80	11.20	10.65	10.30	10.10	9.85	9.70

Modifications

Add to the total cost

Upgrade Kitchen Cabinets	$ + 1314
Solid Surface Countertops (Included)	
Full Bath - including plumbing, wall and floor finishes	+ 8795
Half Bath - including plumbing, wall and floor finishes	+ 5130
Two Car Attached Garage	+ 27,346
Two Car Detached Garage	+ 31,142
Fireplace & Chimney	+ 7329

Adjustments

For multi family - add to total cost

Additional Kitchen	$ + 16,920
Additional Full Bath & Half Bath	+ 13,925
Additional Entry & Exit	+ 1658
Separate Heating & Air Conditioning	+ 7328
Separate Electric	+ 2036

For Townhouse/Rowhouse -
Multiply cost per square foot by

Inner Unit	.87
End Unit	.93

Alternatives

Add to or deduct from the cost per square foot of living area

Cedar Shake Roof	+ 1.25
Clay Tile Roof	+ 3.30
Slate Roof	+ 2.80
Upgrade Ceilings to Textured Finish	+ .50
Air Conditioning, in Heating Ductwork	Base System
Heating Systems, Hot Water	+ 1.88
Heat Pump	+ 3.14
Electric Heat	– 1.76
Not Heated	– 3.50

Additional upgrades or components

Kitchen Cabinets & Countertops	Page 58
Bathroom Vanities	59
Fireplaces & Chimneys	59
Windows, Skylights & Dormers	59
Appliances	60
Breezeways & Porches	60
Finished Attic	60
Garages	61
Site Improvements	61
Wings & Ells	47

- **A distinct residence from designer's plans**
- **Single family — 1 full bath, 1 half bath, 1 kitchen**
- **No basement**
- **Asphalt shingles on roof**
- **Forced hot air heat/air conditioning**
- **Gypsum wallboard interior finishes**
- **Materials and workmanship are above average**
- **Detail specifications on page 39**

Note: The illustration shown may contain some optional components (for example: garages and/or fireplaces) whose costs are shown in the modifications, adjustments, & alternatives below or at the end of the square foot section.

Base cost per square foot of living area

Exterior Wall	Living Area										
	1500	1800	2100	2400	2800	3200	3600	4000	4500	5000	5500
Wood Siding - Wood Frame	153.75	138.35	129.10	123.50	116.35	109.55	105.90	100.05	96.90	94.10	91.30
Brick Veneer - Wood Frame	163.60	147.40	137.25	131.25	123.65	116.25	112.25	105.85	102.45	99.40	96.35
Stone Veneer - Wood Frame	173.95	156.95	145.80	139.35	131.40	123.25	118.85	112.05	108.25	105.00	101.70
Solid Masonry	178.05	160.65	149.15	142.50	134.45	126.00	121.45	114.50	110.60	107.10	103.75
Finished Basement, Add	25.70	25.70	24.45	23.85	23.30	22.40	21.95	21.45	21.00	20.65	20.35
Unfinished Basement, Add	10.80	10.25	9.55	9.25	9.00	8.60	8.35	8.05	7.80	7.65	7.50

Modifications

Add to the total cost

Upgrade Kitchen Cabinets	$ + 1314
Solid Surface Countertops (Included)	
Full Bath - including plumbing, wall and floor finishes	+ 8795
Half Bath - including plumbing, wall and floor finishes	+ 5130
Two Car Attached Garage	+ 27,346
Two Car Detached Garage	+ 31,142
Fireplace & Chimney	+ 8277

Adjustments

For multi family - add to total cost

Additional Kitchen	$ + 16,920
Additional Full Bath & Half Bath	+ 13,925
Additional Entry & Exit	+ 1658
Separate Heating & Air Conditioning	+ 7328
Separate Electric	+ 2036

For Townhouse/Rowhouse -
Multiply cost per square foot by

Inner Unit	.87
End Unit	.94

Alternatives

Add to or deduct from the cost per square foot of living area

Cedar Shake Roof	+ 1.10
Clay Tile Roof	+ 2.85
Slate Roof	+ 2.40
Upgrade Ceilings to Textured Finish	+ .50
Air Conditioning, in Heating Ductwork	Base System
Heating Systems, Hot Water	+ 1.69
Heat Pump	+ 3.25
Electric Heat	– 3.09
Not Heated	– 3.50

Additional upgrades or components

Kitchen Cabinets & Countertops	Page 58
Bathroom Vanities	59
Fireplaces & Chimneys	59
Windows, Skylights & Dormers	59
Appliances	60
Breezeways & Porches	60
Finished Attic	60
Garages	61
Site Improvements	61
Wings & Ells	47

Important: See the Reference Section for Location Factors (to adjust for your city) and Estimating Forms.

- A distinct residence from designer's plans
- Single family — 1 full bath, 1 half bath, 1 kitchen
- No basement
- Asphalt shingles on roof
- Forced hot air heat/air conditioning
- Gypsum wallboard interior finishes
- Materials and workmanship are above average
- Detail specifications on page 39

Note: The illustration shown may contain some optional components (for example: garages and/or fireplaces) whose costs are shown in the modifications, adjustments, & alternatives below or at the end of the square foot section.

Base cost per square foot of living area

Exterior Wall	Living Area										
	1500	1800	2100	2500	3000	3500	4000	4500	5000	5500	6000
Wood Siding - Wood Frame	153.05	137.75	130.30	124.05	114.30	109.55	103.70	97.60	95.35	93.05	90.65
Brick Veneer - Wood Frame	163.25	147.25	139.05	132.40	121.95	116.80	110.35	103.70	101.30	98.70	96.00
Stone Veneer - Wood Frame	174.05	157.20	148.40	141.40	130.05	124.40	117.30	110.10	107.45	104.70	101.60
Solid Masonry	178.30	161.10	152.05	144.80	133.20	127.45	120.00	112.60	109.95	107.00	103.80
Finished Basement, Add	22.50	22.55	21.75	21.20	20.30	19.80	19.10	18.60	18.35	18.10	17.75
Unfinished Basement, Add	9.45	8.95	8.55	8.30	7.80	7.60	7.25	7.00	6.85	6.75	6.55

Modifications

Add to the total cost

Upgrade Kitchen Cabinets	$ + 1314
Solid Surface Countertops (Included)	
Full Bath - including plumbing, wall and floor finishes	+ 8795
Half Bath - including plumbing, wall and floor finishes	+ 5130
Two Car Attached Garage	+ 27,346
Two Car Detached Garage	+ 31,142
Fireplace & Chimney	+ 8277

Adjustments

For multi family - add to total cost

Additional Kitchen	$ + 16,920
Additional Full Bath & Half Bath	+ 13,925
Additional Entry & Exit	+ 1658
Separate Heating & Air Conditioning	+ 7328
Separate Electric	+ 2036

For Townhouse/Rowhouse -
Multiply cost per square foot by

Inner Unit	.85
End Unit	.93

Alternatives

Add to or deduct from the cost per square foot of living area

Cedar Shake Roof	+ .80
Clay Tile Roof	+ 2.20
Slate Roof	+ 1.85
Upgrade Ceilings to Textured Finish	+ .50
Air Conditioning, in Heating Ductwork	Base System
Heating Systems, Hot Water	+ 1.69
Heat Pump	+ 3.25
Electric Heat	- 3.09
Not Heated	- 3.39

Additional upgrades or components

Kitchen Cabinets & Countertops	Page 58
Bathroom Vanities	59
Fireplaces & Chimneys	59
Windows, Skylights & Dormers	59
Appliances	60
Breezeways & Porches	60
Finished Attic	60
Garages	61
Site Improvements	61
Wings & Ells	47

- A distinct residence from designer's plans
- Single family — 1 full bath, 1 half bath, 1 kitchen
- No basement
- Asphalt shingles on roof
- Forced hot air heat/air conditioning
- Gypsum wallboard interior finishes
- Materials and workmanship are above average
- Detail specifications on page 39

Note: The illustration shown may contain some optional components (for example: garages and/or fireplaces) whose costs are shown in the modifications, adjustments, & alternatives below or at the end of the square foot section.

Base cost per square foot of living area

Exterior Wall	Living Area										
	1200	1400	1600	1800	2000	2400	2800	3200	3600	4000	4400
Wood Siding - Wood Frame	148.15	139.20	132.70	127.15	121.00	112.40	105.15	100.35	97.60	94.60	92.05
Brick Veneer - Wood Frame	155.25	145.85	139.05	133.10	126.75	117.65	109.85	104.75	101.85	98.60	95.85
Stone Veneer - Wood Frame	162.70	152.80	145.75	139.40	132.85	123.10	114.80	109.40	106.35	102.80	100.00
Solid Masonry	165.65	155.60	148.35	141.85	135.15	125.25	116.75	111.20	108.05	104.45	101.60
Finished Basement, Add	32.40	32.75	31.90	31.05	30.50	29.25	28.30	27.60	27.25	26.70	26.40
Unfinished Basement, Add	13.45	12.90	12.50	12.05	11.80	11.20	10.65	10.30	10.10	9.85	9.70

Modifications

Add to the total cost

Upgrade Kitchen Cabinets	$ + 1314
Solid Surface Countertops (Included)	
Full Bath - including plumbing, wall and floor finishes	+ 8795
Half Bath - including plumbing, wall and floor finishes	+ 5130
Two Car Attached Garage	+ 27,346
Two Car Detached Garage	+ 31,142
Fireplace & Chimney	+ 6494

Adjustments

For multi family - add to total cost

Additional Kitchen	$ + 16,920
Additional Full Bath & Half Bath	+ 13,925
Additional Entry & Exit	+ 1658
Separate Heating & Air Conditioning	+ 7328
Separate Electric	+ 2036

For Townhouse/Rowhouse -
Multiply cost per square foot by

Inner Unit	.89
End Unit	.95

Alternatives

Add to or deduct from the cost per square foot of living area

Cedar Shake Roof	+ 1.80
Clay Tile Roof	+ 3.30
Slate Roof	+ 2.80
Upgrade Ceilings to Textured Finish	+ .50
Air Conditioning, in Heating Ductwork	Base System
Heating Systems, Hot Water	+ 1.88
Heat Pump	+ 3.14
Electric Heat	– 1.76
Not Heated	– 3.39

Additional upgrades or components

Kitchen Cabinets & Countertops	Page 58
Bathroom Vanities	59
Fireplaces & Chimneys	59
Windows, Skylights & Dormers	59
Appliances	60
Breezeways & Porches	60
Finished Attic	60
Garages	61
Site Improvements	61
Wings & Ells	47

Important: See the Reference Section for Location Factors (to adjust for your city) and Estimating Forms.

- **A distinct residence from designer's plans**
- **Single family — 1 full bath, 1 half bath, 1 kitchen**
- **No basement**
- **Asphalt shingles on roof**
- **Forced hot air heat/air conditioning**
- **Gypsum wallboard interior finishes**
- **Materials and workmanship are above average**
- **Detail specifications on page 39**

Note: The illustration shown may contain some optional components (for example: garages and/or fireplaces) whose costs are shown in the modifications, adjustments, & alternatives below or at the end of the square foot section.

eDesign Basics, Inc.

Base cost per square foot of living area

	Living Area										
Exterior Wall	1200	1500	1800	2100	2400	2800	3200	3600	4000	4500	5000
Wood Siding - Wood Frame	152.60	138.30	127.70	118.90	112.95	108.60	103.55	98.30	95.95	90.85	88.15
Brick Veneer - Wood Frame	159.65	144.70	133.50	124.20	117.90	113.30	107.95	102.40	99.90	94.55	91.60
Stone Veneer - Wood Frame	167.05	151.45	139.50	129.70	123.00	118.30	112.50	106.65	104.05	98.35	95.20
Solid Masonry	170.00	154.05	141.90	131.80	125.00	120.20	114.30	108.30	105.65	99.80	96.60
Finished Basement, Add*	40.65	40.65	38.95	37.65	36.75	36.15	35.30	34.55	34.25	33.45	32.95
Unfinished Basement, Add*	16.65	15.70	14.90	14.20	13.75	13.45	13.00	12.65	12.45	12.05	11.80

*Basement under middle level only.

Modifications

Add to the total cost

Upgrade Kitchen Cabinets	$ + 1314
Solid Surface Countertops (Included)	
Full Bath - including plumbing, wall and floor finishes	+ 8795
Half Bath - including plumbing, wall and floor finishes	+ 5130
Two Car Attached Garage	+ 27,346
Two Car Detached Garage	+ 31,142
Fireplace & Chimney	+ 6494

Adjustments

For multi family - add to total cost

Additional Kitchen	$ + 16,920
Additional Full Bath & Half Bath	+ 13,925
Additional Entry & Exit	+ 1658
Separate Heating & Air Conditioning	+ 7328
Separate Electric	+ 2036

For Townhouse/Rowhouse -
Multiply cost per square foot by

Inner Unit	.87
End Unit	.94

Alternatives

Add to or deduct from the cost per square foot of living area

Cedar Shake Roof	+ 1.25
Clay Tile Roof	+ 4.75
Slate Roof	+ 4
Upgrade Ceilings to Textured Finish	+ .50
Air Conditioning, in Heating Ductwork	Base System
Heating Systems, Hot Water	+ 1.81
Heat Pump	+ 3.28
Electric Heat	− 1.56
Not Heated	− 3.39

Additional upgrades or components

Kitchen Cabinets & Countertops	Page 58
Bathroom Vanities	59
Fireplaces & Chimneys	59
Windows, Skylights & Dormers	59
Appliances	60
Breezeways & Porches	60
Finished Attic	60
Garages	61
Site Improvements	61
Wings & Ells	47

1 Story — Base cost per square foot of living area

Exterior Wall	Living Area							
	50	100	200	300	400	500	600	700
Wood Siding - Wood Frame	237.35	184.95	162.20	137.75	130.30	125.75	122.80	123.70
Brick Veneer - Wood Frame	261.20	202.00	176.45	147.20	138.80	133.70	130.35	130.95
Stone Veneer - Wood Frame	286.35	219.95	191.35	157.20	147.80	142.10	138.35	138.65
Solid Masonry	296.20	226.95	197.25	161.05	151.25	145.35	141.45	141.70
Finished Basement, Add	95.80	82.35	74.95	62.60	60.15	58.65	57.65	56.95
Unfinished Basement, Add	68.05	47.05	37.45	28.80	26.45	25.10	24.20	23.50

1-1/2 Story — Base cost per square foot of living area

Exterior Wall	Living Area							
	100	200	300	400	500	600	700	800
Wood Siding - Wood Frame	188.65	155.05	134.60	122.30	116.15	113.15	109.40	108.35
Brick Veneer - Wood Frame	209.95	172.05	148.80	133.35	126.35	122.80	118.50	117.40
Stone Veneer - Wood Frame	232.40	190.05	163.75	145.00	137.15	132.95	128.15	126.95
Solid Masonry	241.15	197.10	169.65	149.60	141.40	136.95	131.90	130.65
Finished Basement, Add	64.45	60.00	55.05	49.60	48.10	47.10	46.25	46.10
Unfinished Basement, Add	40.70	31.10	26.50	22.55	21.20	20.30	19.50	19.25

2 Story — Base cost per square foot of living area

Exterior Wall	Living Area							
	100	200	400	600	800	1000	1200	1400
Wood Siding - Wood Frame	187.70	143.65	124.30	105.55	99.25	95.40	92.95	94.15
Brick Veneer - Wood Frame	211.50	160.65	138.50	115.05	107.80	103.35	100.45	101.45
Stone Veneer - Wood Frame	236.65	178.65	153.45	125.00	116.75	111.70	108.45	109.10
Solid Masonry	246.55	185.65	159.30	128.85	120.30	115.00	111.55	112.15
Finished Basement, Add	48.00	41.25	37.50	31.35	30.05	29.35	28.85	28.50
Unfinished Basement, Add	34.05	23.55	18.70	14.40	13.25	12.55	12.10	11.80

Base costs do not include bathroom or kitchen facilities. Use Modifications/Adjustments/Alternatives on pages 58-61 where appropriate.

Important: See the Reference Section for Location Factors (to adjust for your city) and Estimating Forms.

1 Story

1-1/2 Story

2 Story

2-1/2 Story

Bi-Level

Tri-Level

	Components
1 Site Work	Site preparation for slab or excavation for lower level; 4' deep trench excavation for foundation wall.
2 Foundations	Continuous reinforced concrete footing, 8" deep x 18" wide; dampproofed and insulated 12" thick reinforced concrete foundation wall, 4' deep; 4" concrete slab on 4" crushed stone base and polyethylene vapor barrier, trowel finish.
3 Framing	Exterior walls—2 x 6 wood studs, 16" O.C.; 1/2" plywood sheathing; 2 x 8 rafters 16" O.C. with 1/2" plywood sheathing, 4 in 12, 6 in 12 or 8 in 12 roof pitch, 2 x 6 or 2 x 8 ceiling joists; 1/2" plywood subfloor on 1 x 3 wood sleepers 16" O.C.; 2 x 10 floor joists with 5/8" plywood subfloor on models with more than one level.
4 Exterior Walls	Face brick veneer and #15 felt building paper on insulated wood frame walls; double hung windows; 3 flush solid core wood exterior doors with storms. **Alternates:** • Wood siding on wood frame, has top quality cedar or redwood siding or hand split cedar shingles or shakes. • Solid brick may have solid brick exterior wall or brick on concrete block. • Solid stone concrete block with selected fieldstone or limestone exterior.
5 Roofing	Red cedar shingles; #15 felt building paper; aluminum gutters, downspouts and drip edge; copper flashings.
6 Interiors	Walls and ceilings—5/8" drywall, skim coat plaster, primed and painted with 2 coats; hardwood baseboard and trim; sanded and finished, hardwood floor 70%, ceramic tile with 1/2" underlayment 20%, vinyl tile with 1/2" underlayment 10%; wood panel interior doors, primed and painted with 2 coats.
7 Specialties	Luxury grade kitchen cabinets—25 L.F. wall and base with plastic laminate counter top and kitchen sink; 6 L.F. bathroom vanity; 75 gallon electric water heater; medicine cabinet.
8 Mechanical	Gas fired warm air heat/air conditioning; one full bath including bathtub, corner shower, built in lavatory and water closet; one 1/2 bath including built in lavatory and water closet.
9 Electrical	100 Amp. service; romex wiring; incandescent lighting fixtures, switches, receptacles.
10 Overhead and Profit	General Contractor overhead and profit.

Adjustments

Unfinished Basement:
8" high cast-in-place concrete walls 12" thick or 12" concrete block.

Finished Basement:
Includes painted drywall on 2 x 4 wood furring with insulation, suspended ceiling, carpeting on subfloor with sleepers, heating and lighting.

- **Unique residence built from an architect's plan**
- **Single family — 1 full bath, 1 half bath, 1 kitchen**
- **No basement**
- **Cedar shakes on roof**
- **Forced hot air heat/air conditioning**
- **Gypsum wallboard interior finishes**
- **Many special features**
- **Extraordinary materials and workmanship**
- **Detail specifications on page 49**

Note: The illustration shown may contain some optional components (for example: garages and/or fireplaces) whose costs are shown in the modifications, adjustments, & alternatives below or at the end of the square foot section.

eHome Planners, Inc.

Base cost per square foot of living area

	Living Area										
Exterior Wall	1000	1200	1400	1600	1800	2000	2400	2800	3200	3600	4000
Wood Siding - Wood Frame	215.90	199.20	185.70	176.60	171.30	165.40	154.40	146.50	140.90	135.20	130.50
Brick Veneer - Wood Frame	224.70	207.15	192.90	183.35	177.85	171.50	160.00	151.65	145.70	139.70	134.65
Solid Brick	239.65	220.70	205.20	194.85	188.90	181.95	169.55	160.55	153.85	147.25	141.80
Solid Stone	246.60	226.95	210.90	200.25	194.10	186.75	174.00	164.55	157.65	150.80	145.10
Finished Basement, Add	58.80	63.40	60.95	59.30	58.35	57.05	55.20	53.80	52.50	51.40	50.55
Unfinished Basement, Add	25.60	24.05	22.85	22.05	21.55	20.90	19.95	19.20	18.60	18.00	17.55

Modifications

Add to the total cost

Upgrade Kitchen Cabinets	$ + 1735
Solid Surface Countertops (Included)	
Full Bath - including plumbing, wall and floor finishes	+ 10,554
Half Bath - including plumbing, wall and floor finishes	+ 6157
Two Car Attached Garage	+ 31,466
Two Car Detached Garage	+ 35,593
Fireplace & Chimney	+ 9357

Adjustments

For multi family - add to total cost

Additional Kitchen	$ + 21,730
Additional Full Bath & Half Bath	+ 16,711
Additional Entry & Exit	+ 2306
Separate Heating & Air Conditioning	+ 7328
Separate Electric	+ 2036

For Townhouse/Rowhouse -
Multiply cost per square foot by

Inner Unit	.90
End Unit	.95

Alternatives

Add to or deduct from the cost per square foot of living area

Heavyweight Asphalt Shingles	– 2.45
Clay Tile Roof	+ 4.10
Slate Roof	+ 3.10
Upgrade Ceilings to Textured Finish	+ .50
Air Conditioning, in Heating Ductwork	Base System
Heating Systems, Hot Water	+ 2.18
Heat Pump	+ 2.90
Electric Heat	– 1.76
Not Heated	– 4.38

Additional upgrades or components

Kitchen Cabinets & Countertops	Page 58
Bathroom Vanities	59
Fireplaces & Chimneys	59
Windows, Skylights & Dormers	59
Appliances	60
Breezeways & Porches	60
Finished Attic	60
Garages	61
Site Improvements	61
Wings & Ells	57

Important: See the Reference Section for Location Factors (to adjust for your city) and Estimating Forms.

- **Unique residence built from an architect's plan**
- **Single family — 1 full bath, 1 half bath, 1 kitchen**
- **No basement**
- **Cedar shakes on roof**
- **Forced hot air heat/air conditioning**
- **Gypsum wallboard interior finishes**
- **Many special features**
- **Extraordinary materials and workmanship**
- **Detail specifications on page 49**

Note: The illustration shown may contain some optional components (for example: garages and/or fireplaces) whose costs are shown in the modifications, adjustments, & alternatives below or at the end of the square foot section.

©Larry E. Belk Designs

Base cost per square foot of living area

Exterior Wall	Living Area										
	1000	1200	1400	1600	1800	2000	2400	2800	3200	3600	4000
Wood Siding - Wood Frame	199.70	185.25	174.95	162.95	155.80	149.25	136.50	130.70	125.25	121.20	115.50
Brick Veneer - Wood Frame	209.90	194.75	184.00	171.20	163.65	156.65	143.10	136.90	131.00	126.75	120.65
Solid Brick	227.20	211.00	199.50	185.20	176.95	169.35	154.25	147.45	140.80	136.20	129.50
Solid Stone	235.30	218.55	206.60	191.70	183.15	175.15	159.40	152.40	145.35	140.55	133.55
Finished Basement, Add	41.15	44.70	43.45	41.55	40.55	39.50	37.55	36.75	35.60	35.10	34.30
Unfinished Basement, Add	18.40	17.55	16.95	16.00	15.50	15.00	14.00	13.55	13.05	12.80	12.40

Modifications

Add to the total cost

Upgrade Kitchen Cabinets	$ + 1735
Solid Surface Countertops (Included)	
Full Bath - including plumbing, wall and floor finishes	+ 10,554
Half Bath - including plumbing, wall and floor finishes	+ 6157
Two Car Attached Garage	+ 31,466
Two Car Detached Garage	+ 35,593
Fireplace & Chimney	+ 9357

Adjustments

For multi family - add to total cost

Additional Kitchen	$ + 21,730
Additional Full Bath & Half Bath	+ 16,711
Additional Entry & Exit	+ 2306
Separate Heating & Air Conditioning	+ 7328
Separate Electric	+ 2036

For Townhouse/Rowhouse -
Multiply cost per square foot by

Inner Unit	.90
End Unit	.95

Alternatives

Add to or deduct from the cost per square foot of living area

Heavyweight Asphalt Shingles	– 1.80
Clay Tile Roof	+ 2.95
Slate Roof	+ 2.25
Upgrade Ceilings to Textured Finish	+ .50
Air Conditioning, in Heating Ductwork	Base System
Heating Systems, Hot Water	+ 2.09
Heat Pump	+ 3.23
Electric Heat	– 1.76
Not Heated	– 4.04

Additional upgrades or components

Kitchen Cabinets & Countertops	Page 58
Bathroom Vanities	59
Fireplaces & Chimneys	59
Windows, Skylights & Dormers	59
Appliances	60
Breezeways & Porches	60
Finished Attic	60
Garages	61
Site Improvements	61
Wings & Ells	57

Important: See the Reference Section for Location Factors (to adjust for your city) and Estimating Forms.

51

- **Unique residence built from an architect's plan**
- **Single family — 1 full bath, 1 half bath, 1 kitchen**
- **No basement**
- **Cedar shakes on roof**
- **Forced hot air heat/air conditioning**
- **Gypsum wallboard interior finishes**
- **Many special features**
- **Extraordinary materials and workmanship**
- **Detail specifications on page 49**

Note: The illustration shown may contain some optional components (for example: garages and/or fireplaces) whose costs are shown in the modifications, adjustments, & alternatives below or at the end of the square foot section.

Base cost per square foot of living area

Exterior Wall	Living Area										
	1200	1400	1600	1800	2000	2400	2800	3200	3600	4000	4400
Wood Siding - Wood Frame	184.50	173.10	165.00	157.80	150.30	139.45	130.35	124.35	120.90	116.95	113.95
Brick Veneer - Wood Frame	195.40	183.25	174.70	166.95	159.10	147.40	137.60	131.15	127.40	123.15	119.80
Solid Brick	213.85	200.55	191.20	182.45	174.15	160.90	149.85	142.65	138.45	133.55	129.90
Solid Stone	222.50	208.50	198.85	189.65	181.05	167.15	155.60	148.05	143.60	138.40	134.60
Finished Basement, Add	33.10	36.00	35.10	34.00	33.30	31.80	30.55	29.75	29.30	28.60	28.25
Unfinished Basement, Add	14.75	14.10	13.65	13.15	12.80	12.05	11.40	11.00	10.80	10.45	10.25

Modifications

Add to the total cost

Upgrade Kitchen Cabinets	$ + 1735
Solid Surface Countertops (Included)	
Full Bath - including plumbing, wall and floor finishes	+ 10,554
Half Bath - including plumbing, wall and floor finishes	+ 6157
Two Car Attached Garage	+ 31,466
Two Car Detached Garage	+ 35,593
Fireplace & Chimney	+ 10,256

Adjustments

For multi family - add to total cost

Additional Kitchen	$ + 21,730
Additional Full Bath & Half Bath	+ 16,711
Additional Entry & Exit	+ 2306
Separate Heating & Air Conditioning	+ 7328
Separate Electric	+ 2036

For Townhouse/Rowhouse -
Multiply cost per square foot by

Inner Unit	.86
End Unit	.93

Alternatives

Add to or deduct from the cost per square foot of living area

Heavyweight Asphalt Shingles	– 1.25
Clay Tile Roof	+ 2.05
Slate Roof	+ 1.55
Upgrade Ceilings to Textured Finish	+ .50
Air Conditioning, in Heating Ductwork	Base System
Heating Systems, Hot Water	+ 2.02
Heat Pump	+ 3.40
Electric Heat	– 1.59
Not Heated	– 3.82

Additional upgrades or components

Kitchen Cabinets & Countertops	Page 58
Bathroom Vanities	59
Fireplaces & Chimneys	59
Windows, Skylights & Dormers	59
Appliances	60
Breezeways & Porches	60
Finished Attic	60
Garages	61
Site Improvements	61
Wings & Ells	57

- Unique residence built from an architect's plan
- Single family — 1 full bath, 1 half bath, 1 kitchen
- No basement
- Cedar shakes on roof
- Forced hot air heat/air conditioning
- Gypsum wallboard interior finishes
- Many special features
- Extraordinary materials and workmanship
- Detail specifications on page 49

Note: The illustration shown may contain some optional components (for example: garages and/or fireplaces) whose costs are shown in the modifications, adjustments, & alternatives below or at the end of the square foot section.

©Larry W. Garnett & Associates, Inc

Base cost per square foot of living area

Exterior Wall	Living Area										
	1500	1800	2100	2500	3000	3500	4000	4500	5000	5500	6000
Wood Siding - Wood Frame	179.05	161.05	150.35	142.40	131.65	123.95	116.55	112.85	109.65	106.45	102.95
Brick Veneer - Wood Frame	190.35	171.50	159.65	151.20	139.60	131.20	123.25	119.20	115.70	112.25	108.45
Solid Brick	209.55	189.20	175.50	166.25	153.20	143.60	134.70	130.05	126.00	122.05	117.90
Solid Stone	218.50	197.40	182.90	173.15	159.55	149.30	140.00	135.05	130.85	126.65	122.25
Finished Basement, Add	26.35	28.40	26.85	26.05	24.85	23.80	23.05	22.60	22.10	21.75	21.40
Unfinished Basement, Add	11.90	11.25	10.45	10.05	9.45	9.00	8.60	8.35	8.15	7.95	7.75

Modifications

Add to the total cost

Upgrade Kitchen Cabinets	$ + 1735
Solid Surface Countertops (Included)	
Full Bath - including plumbing, wall and floor finishes	+ 10,554
Half Bath - including plumbing, wall and floor finishes	+ 6157
Two Car Attached Garage	+ 31,466
Two Car Detached Garage	+ 35,593
Fireplace & Chimney	+ 11,225

Adjustments

For multi family - add to total cost

Additional Kitchen	$ + 21,730
Additional Full Bath & Half Bath	+ 16,711
Additional Entry & Exit	+ 2306
Separate Heating & Air Conditioning	+ 7328
Separate Electric	+ 2036

For Townhouse/Rowhouse -
Multiply cost per square foot by

Inner Unit	.86
End Unit	.93

Alternatives

Add to or deduct from the cost per square foot of living area

Heavyweight Asphalt Shingles	– 1.10
Clay Tile Roof	+ 1.80
Slate Roof	+ 1.35
Upgrade Ceilings to Textured Finish	+ .50
Air Conditioning, in Heating Ductwork	Base System
Heating Systems, Hot Water	+ 1.83
Heat Pump	+ 3.50
Electric Heat	– 3.12
Not Heated	– 3.82

Additional upgrades or components

Kitchen Cabinets & Countertops	Page 58
Bathroom Vanities	59
Fireplaces & Chimneys	59
Windows, Skylights & Dormers	59
Appliances	60
Breezeways & Porches	60
Finished Attic	60
Garages	61
Site Improvements	61
Wings & Ells	57

Important: See the Reference Section for Location Factors (to adjust for your city) and Estimating Forms.

- **Unique residence built from an architect's plan**
- **Single family — 1 full bath, 1 half bath, 1 kitchen**
- **No basement**
- **Cedar shakes on roof**
- **Forced hot air heat/air conditioning**
- **Gypsum wallboard interior finishes**
- **Many special features**
- **Extraordinary materials and workmanship**
- **Detail specifications on page 49**

Note: The illustration shown may contain some optional components (for example: garages and/or fireplaces) whose costs are shown in the modifications, adjustments, & alternatives below or at the end of the square foot section.

Base cost per square foot of living area

Exterior Wall	Living Area										
	1500	1800	2100	2500	3000	3500	4000	4500	5000	5500	6000
Wood Siding - Wood Frame	177.70	160.00	151.10	143.75	132.50	126.90	120.15	113.15	110.60	107.85	105.00
Brick Veneer - Wood Frame	189.45	170.90	161.25	153.45	141.30	135.20	127.70	120.20	117.35	114.35	111.15
Solid Brick	209.45	189.40	178.50	169.95	156.30	149.35	140.65	132.05	128.85	125.40	121.60
Solid Stone	218.75	198.00	186.55	177.60	163.25	155.90	146.65	137.60	134.20	130.55	126.40
Finished Basement, Add	23.15	24.95	24.00	23.30	22.15	21.55	20.75	20.10	19.75	19.45	19.00
Unfinished Basement, Add	10.45	9.85	9.40	9.05	8.55	8.25	7.80	7.45	7.35	7.15	6.95

Modifications

Add to the total cost

Upgrade Kitchen Cabinets	$ + 1735
Solid Surface Countertops (Included)	
Full Bath - including plumbing, wall and floor finishes	+ 10,554
Half Bath - including plumbing, wall and floor finishes	+ 6157
Two Car Attached Garage	+ 31,466
Two Car Detached Garage	+ 35,593
Fireplace & Chimney	+ 11,225

Adjustments

For multi family - add to total cost

Additional Kitchen	$ + 21,730
Additional Full Bath & Half Bath	+ 16,711
Additional Entry & Exit	+ 2306
Separate Heating & Air Conditioning	+ 7328
Separate Electric	+ 2036

For Townhouse/Rowhouse -
Multiply cost per square foot by

Inner Unit	.84
End Unit	.92

Alternatives

Add to or deduct from the cost per square foot of living area

Heavyweight Asphalt Shingles	– .80
Clay Tile Roof	+ 1.35
Slate Roof	+ 1.05
Upgrade Ceilings to Textured Finish	+ .50
Air Conditioning, in Heating Ductwork	Base System
Heating Systems, Hot Water	+ 1.83
Heat Pump	+ 3.50
Electric Heat	– 3.12
Not Heated	– 3.71

Additional upgrades or components

Kitchen Cabinets & Countertops	Page 58
Bathroom Vanities	59
Fireplaces & Chimneys	59
Windows, Skylights & Dormers	59
Appliances	60
Breezeways & Porches	60
Finished Attic	60
Garages	61
Site Improvements	61
Wings & Ells	57

- **Unique residence built from an architect's plan**
- **Single family — 1 full bath, 1 half bath, 1 kitchen**
- **No basement**
- **Cedar shakes on roof**
- **Forced hot air heat/air conditioning**
- **Gypsum wallboard interior finishes**
- **Many special features**
- **Extraordinary materials and workmanship**
- **Detail specifications on page 49**

Note: The illustration shown may contain some optional components (for example: garages and/or fireplaces) whose costs are shown in the modifications, adjustments, & alternatives below or at the end of the square foot section.

Base cost per square foot of living area

Exterior Wall	Living Area										
	1200	1400	1600	1800	2000	2400	2800	3200	3600	4000	4400
Wood Siding - Wood Frame	174.95	164.20	156.45	149.75	142.65	132.50	124.05	118.45	115.10	111.60	108.70
Brick Veneer - Wood Frame	183.10	171.80	163.75	156.65	149.20	138.45	129.45	123.55	120.00	116.25	113.15
Solid Brick	197.00	184.85	176.15	168.35	160.45	148.60	138.70	132.20	128.30	124.05	120.75
Solid Stone	203.45	190.80	181.90	173.75	165.60	153.25	143.00	136.20	132.20	127.70	124.30
Finished Basement, Add	33.10	36.00	35.10	34.00	33.30	31.80	30.55	29.75	29.30	28.60	28.25
Unfinished Basement, Add	14.75	14.10	13.65	13.15	12.80	12.05	11.40	11.00	10.80	10.45	10.25

Modifications

Add to the total cost

Upgrade Kitchen Cabinets	$ + 1735
Solid Surface Countertops (Included)	
Full Bath - including plumbing, wall and floor finishes	+ 10,554
Half Bath - including plumbing, wall and floor finishes	+ 6157
Two Car Attached Garage	+ 31,466
Two Car Detached Garage	+ 35,593
Fireplace & Chimney	+ 9357

Adjustments

For multi family - add to total cost

Additional Kitchen	$ + 21,730
Additional Full Bath & Half Bath	+ 16,711
Additional Entry & Exit	+ 2306
Separate Heating & Air Conditioning	+ 7328
Separate Electric	+ 2036

For Townhouse/Rowhouse -
Multiply cost per square foot by

Inner Unit	.89
End Unit	.94

Alternatives

Add to or deduct from the cost per square foot of living area

Heavyweight Asphalt Shingles	– 1.25
Clay Tile Roof	+ 2.05
Slate Roof	+ 1.55
Upgrade Ceilings to Textured Finish	+ .50
Air Conditioning, in Heating Ductwork	Base System
Heating Systems, Hot Water	+ 2.02
Heat Pump	+ 3.40
Electric Heat	– 1.59
Not Heated	– 3.82

Additional upgrades or components

Kitchen Cabinets & Countertops	Page 58
Bathroom Vanities	59
Fireplaces & Chimneys	59
Windows, Skylights & Dormers	59
Appliances	60
Breezeways & Porches	60
Finished Attic	60
Garages	61
Site Improvements	61
Wings & Ells	57

Important: See the Reference Section for Location Factors (to adjust for your city) and Estimating Forms.

- **Unique residence built from an architect's plan**
- **Single family — 1 full bath, 1 half bath, 1 kitchen**
- **No basement**
- **Cedar shakes on roof**
- **Forced hot air heat/air conditioning**
- **Gypsum wallboard interior finishes**
- **Many special features**
- **Extraordinary materials and workmanship**
- **Detail specifications on page 49**

Note: The illustration shown may contain some optional components (for example: garages and/or fireplaces) whose costs are shown in the modifications, adjustments, & alternatives below or at the end of the square foot section.

©Home Planners, Inc.

Base cost per square foot of living area

Exterior Wall	Living Area										
	1500	1800	2100	2400	2800	3200	3600	4000	4500	5000	5500
Wood Siding - Wood Frame	164.45	151.95	141.65	134.65	129.50	123.60	117.50	114.75	108.95	105.80	102.25
Brick Veneer - Wood Frame	171.85	158.60	147.70	140.35	134.90	128.65	122.20	119.25	113.15	109.70	105.95
Solid Brick	184.35	169.85	157.90	149.95	144.15	137.20	130.10	126.95	120.25	116.45	112.30
Solid Stone	190.10	175.00	162.65	154.40	148.45	141.10	133.80	130.50	123.55	119.55	115.20
Finished Basement, Add*	39.20	42.25	40.60	39.50	38.80	37.75	36.80	36.40	35.45	34.85	34.25
Unfinished Basement, Add*	17.10	16.05	15.20	14.65	14.30	13.80	13.35	13.10	12.65	12.30	12.05

*Basement under middle level only.

Modifications

Add to the total cost
Upgrade Kitchen Cabinets	$ + 1735
Solid Surface Countertops (Included)	
Full Bath - including plumbing, wall and floor finishes	+ 10,554
Half Bath - including plumbing, wall and floor finishes	+ 6157
Two Car Attached Garage	+ 31,466
Two Car Detached Garage	+ 35,593
Fireplace & Chimney	+ 9357

Adjustments

For multi family - add to total cost
Additional Kitchen	$ + 21,730
Additional Full Bath & Half Bath	+ 16,711
Additional Entry & Exit	+ 2306
Separate Heating & Air Conditioning	+ 7328
Separate Electric	+ 2036

For Townhouse/Rowhouse -
Multiply cost per square foot by
Inner Unit	.86
End Unit	.93

Alternatives

Add to or deduct from the cost per square foot of living area
Heavyweight Asphalt Shingles	– 1.80
Clay Tile Roof	+ 2.95
Slate Roof	+ 2.25
Upgrade Ceilings to Textured Finish	+ .50
Air Conditioning, in Heating Ductwork	Base System
Heating Systems, Hot Water	+ 1.96
Heat Pump	+ 3.53
Electric Heat	– 1.40
Not Heated	– 3.71

Additional upgrades or components
Kitchen Cabinets & Countertops	Page 58
Bathroom Vanities	59
Fireplaces & Chimneys	59
Windows, Skylights & Dormers	59
Appliances	60
Breezeways & Porches	60
Finished Attic	60
Garages	61
Site Improvements	61
Wings & Ells	57

Important: See the Reference Section for Location Factors (to adjust for your city) and Estimating Forms.

1 Story — Base cost per square foot of living area

Exterior Wall	Living Area							
	50	100	200	300	400	500	600	700
Wood Siding - Wood Frame	261.15	202.65	177.15	149.80	141.50	136.45	133.20	134.10
Brick Veneer - Wood Frame	288.55	222.20	193.50	160.75	151.25	145.60	141.90	142.45
Solid Brick	335.15	255.50	221.20	179.25	167.90	161.15	156.65	156.75
Solid Stone	356.80	270.95	234.10	187.85	175.70	168.35	163.50	163.35
Finished Basement, Add	103.45	94.30	84.80	69.05	65.95	64.00	62.80	61.85
Unfinished Basement, Add	53.05	41.00	35.95	27.60	25.90	24.90	24.25	23.75

1-1/2 Story — Base cost per square foot of living area

Exterior Wall	Living Area							
	100	200	300	400	500	600	700	800
Wood Siding - Wood Frame	206.75	169.30	146.40	132.70	125.80	122.50	118.30	117.10
Brick Veneer - Wood Frame	231.20	188.85	162.75	145.45	137.50	133.60	128.75	127.50
Solid Brick	272.80	222.15	190.45	167.05	157.50	152.45	146.65	145.10
Solid Stone	292.20	237.60	203.35	177.10	166.75	161.20	154.90	153.35
Finished Basement, Add	68.40	67.80	61.55	54.65	52.75	51.50	50.35	50.20
Unfinished Basement, Add	33.80	28.75	25.40	21.75	20.75	20.05	19.45	19.35

2 Story — Base cost per square foot of living area

Exterior Wall	Living Area							
	100	200	400	600	800	1000	1200	1400
Wood Siding - Wood Frame	202.90	153.70	132.05	111.10	104.05	99.80	97.05	98.35
Brick Veneer - Wood Frame	230.30	173.25	148.35	122.00	113.85	108.95	105.75	106.70
Solid Brick	276.95	206.50	176.10	140.50	130.45	124.50	120.50	121.00
Solid Stone	298.55	221.95	189.00	149.10	138.25	131.75	127.35	127.60
Finished Basement, Add	51.75	47.25	42.50	34.60	33.05	32.10	31.45	31.05
Unfinished Basement, Add	26.55	20.50	18.00	13.75	12.95	12.45	12.10	11.85

Base costs do not include bathroom or kitchen facilities. Use Modifications/Adjustments/Alternatives on pages 58-61 where appropriate.

Important: See the Reference Section for Location Factors (to adjust for your city) and Estimating Forms.

Kitchen cabinets - Base units, hardwood (Cost per Unit)

	Economy	Average	Custom	Luxury
24" deep, 35" high,				
One top drawer,				
One door below				
12" wide	$ 248	$ 330	$ 439	$ 578
15" wide	255	340	452	595
18" wide	278	370	492	648
21" wide	285	380	505	665
24" wide	330	440	585	770
Four drawers				
12" wide	255	340	452	595
15" wide	259	345	459	604
18" wide	289	385	512	674
24" wide	315	420	559	735
Two top drawers,				
Two doors below				
27" wide	353	470	625	823
30" wide	383	510	678	893
33" wide	394	525	698	919
36" wide	409	545	725	954
42" wide	431	575	765	1006
48" wide	461	615	818	1076
Range or sink base				
(Cost per unit)				
Two doors below				
30" wide	319	425	565	744
33" wide	341	455	605	796
36" wide	360	480	638	840
42" wide	375	500	665	875
48" wide	394	525	698	919
Corner Base Cabinet				
(Cost per unit)				
36" wide	555	740	984	1295
Lazy Susan (Cost per unit)				
With revolving door	735	980	1303	1715

Kitchen cabinets - Wall cabinets, hardwood (Cost per Unit)

	Economy	Average	Custom	Luxury
12" deep, 2 doors				
12" high				
30" wide	$ 223	$ 297	$ 395	$ 520
36" wide	259	345	459	604
15" high				
30" wide	225	300	399	525
33" wide	278	370	492	648
36" wide	270	360	479	630
24" high				
30" wide	296	395	525	691
36" wide	323	430	572	753
42" wide	368	490	652	858
30" high, 1 door				
12" wide	209	278	370	487
15" wide	229	305	406	534
18" wide	251	335	446	586
24" wide	289	385	512	674
30" high, 2 doors				
27" wide	315	420	559	735
30" wide	330	440	585	770
36" wide	371	495	658	866
42" wide	405	540	718	945
48" wide	450	600	798	1050
Corner wall, 30" high				
24" wide	330	440	585	770
30" wide	349	465	618	814
36" wide	394	525	698	919
Broom closet				
84" high, 24" deep				
18" wide	604	805	1071	1409
Oven Cabinet				
84" high, 24" deep				
27" wide	900	1200	1596	2100

Kitchen countertops (Cost per L.F.)

	Economy	Average	Custom	Luxury
Solid Surface				
24" wide, no backsplash	$ 110	$ 147	$ 196	$ 257
with backsplash	130	173	230	303
Stock plastic laminate, 24" wide				
with backsplash	25	33	45	59
Custom plastic laminate, no splash				
7/8" thick, alum. molding	36	47	63	83
1-1/4" thick, no splash	41	55	73	96
Marble				
1/2" - 3/4" thick w/splash	54	71	95	125
Maple, laminated				
1-1/2" thick w/splash	86	114	152	200
Stainless steel				
(per S.F.)	140	186	247	326
Cutting blocks, recessed				
16" x 20" x 1" (each)	117	156	207	273

Vanity bases *(Cost per Unit)*

2 door, 30" high, 21" deep	Economy	Average	Custom	Luxury
24" wide	$ 289	$ 385	$ 512	$ 674
30" wide	345	460	612	805
36" wide	345	460	612	805
48" wide	446	595	791	1041

Solid surface vanity tops *(Cost Each)*

Center bowl	Economy	Average	Custom	Luxury
22" x 25"	$ 425	$ 459	$ 496	$ 536
22" x 31"	490	529	571	617
22" x 37"	565	610	659	712
22" x 49"	700	756	816	881

Fireplaces & Chimneys *(Cost per Unit)*

	1-1/2 Story	2 Story	3 Story
Economy (prefab metal)			
Exterior chimney & 1 fireplace	$ 5873	$ 6489	$ 7118
Interior chimney & 1 fireplace	5628	6257	6546
Average (masonry)			
Exterior chimney & 1 fireplace	6191	6841	7504
Interior chimney & 1 fireplace	5933	6596	6901
For more than 1 flue, add	421	717	1202
For more than 1 fireplace, add	4154	4154	4154
Custom (masonry)			
Exterior chimney & 1 fireplace	6493	7329	8276
Interior chimney & 1 fireplace	6088	6891	7427
For more than 1 flue, add	509	881	1475
For more than 1 fireplace, add	4664	4664	4664
Luxury (masonry)			
Exterior chimney & 1 fireplace	9357	10257	11224
Interior chimney & 1 fireplace	8927	9760	10331
For more than 1 flue, add	772	1289	1800
For more than 1 fireplace, add	7375	7375	7375

Windows and Skylights *(Cost Each)*

	Economy	Average	Custom	Luxury
Fixed Picture Windows				
3'-6" x 4'-0"	$ 358	$ 477	$ 635	$ 836
4'-0" x 6'-0"	620	827	1100	1447
5'-0" x 6'-0"	705	939	1250	1645
6'-0" x 6'-0"	705	939	1250	1645
Bay/Bow Windows				
8'-0" x 5'-0"	874	1165	1550	2039
10'-0" x 5'-0"	930	1240	1650	2171
10'-0" x 6'-0"	1466	1954	2600	3421
12'-0" x 6'-0"	1903	2537	3375	4441
Palladian Windows				
3'-2" x 6'-4"		1484	1975	2599
4'-0" x 6'-0"		1729	2300	3026
5'-5" x 6'-10"		2255	3000	3947
8'-0" x 6'-0"		2669	3550	4671
Skylights				
46" x 21-1/2"	595	915	1120	1232
46" x 28"	618	950	1160	1276
57" x 44"	679	1045	1175	1293

Dormers *(Cost/S.F. of plan area)*

Framing and Roofing Only	Economy	Average	Custom	Luxury
Gable dormer, 2" x 6" roof frame	$ 30	$ 34	$ 36	$ 60
2" x 8" roof frame	31	35	37	63
Shed dormer, 2" x 6" roof frame	19	22	25	38
2" x 8" roof frame	21	23	26	39
2" x 10" roof frame	22	25	27	40

Appliances *(Cost per Unit)*

	Economy	Average	Custom	Luxury
Range				
30" free standing, 1 oven	$ 550	$ 1475	$ 1938	$ 2400
2 oven	1400	2138	2507	2875
30" built-in, 1 oven	830	1390	1670	1950
2 oven	1675	1927	2053	2178
21" free standing				
1 oven	565	613	637	660
Counter Top Ranges				
4 burner standard	405	1003	1302	1600
As above with griddle	1450	2838	3532	4225
Microwave Oven	252	501	626	750
Compactor				
4 to 1 compaction	800	1038	1157	1275
Deep Freeze				
15 to 23 C.F.	735	880	953	1025
30 C.F.	855	1040	1133	1225
Dehumidifier, portable, auto.				
15 pint	287	330	352	373
30 pint	360	414	441	468
Washing Machine, automatic	700	1438	1807	2175
Water Heater				
Electric, glass lined				
30 gal.	595	730	798	865
80 gal.	1675	2075	2275	2475
Water Heater, Gas, glass lined				
30 gal.	1100	1338	1457	1575
50 gal.	1325	1638	1794	1950
Dishwasher, built-in				
2 cycles	420	608	702	795
4 or more cycles	575	710	1118	1525
Dryer, automatic	680	1578	2027	2475
Garage Door Opener	525	628	679	730
Garbage Disposal	159	232	269	305
Heater, Electric, built-in				
1250 watt ceiling type	238	294	322	350
1250 watt wall type	310	338	352	365
Wall type w/blower				
1500 watt	310	357	380	403
3000 watt	545	627	668	709
Hood For Range, 2 speed				
30" wide	190	620	835	1050
42" wide	297	1349	1875	2400
Humidifier, portable				
7 gal. per day	146	168	179	190
15 gal. per day	232	267	285	302
Ice Maker, automatic				
13 lb. per day	1050	1208	1287	1365
51 lb. per day	2000	2300	2450	2600
Refrigerator, no frost				
10-12 C.F.	530	568	587	605
14-16 C.F.	630	723	769	815
18-20 C.F.	770	1260	1505	1750
21-29 C.F.	1175	2550	3238	3925
Sump Pump, 1/3 H.P.	320	448	512	575

Breezeway *(Cost per S.F.)*

Class	Type	Area (S.F.)			
		50	100	150	200
Economy	Open	$ 45.40	$ 37.91	$ 33.41	$ 31.17
	Enclosed	165.23	117.54	96.91	83.44
Average	Open	53.61	43.76	38.49	35.86
	Enclosed	169.10	122.41	101.69	91.10
Custom	Open	64.56	53.10	46.86	43.74
	Enclosed	227.04	164.06	135.81	121.44
Luxury	Open	72.93	60.18	53.50	50.16
	Enclosed	268.54	191.24	156.86	139.42

Porches *(Cost per S.F.)*

Class	Type	Area (S.F.)				
		25	50	100	200	300
Economy	Open	$ 81.31	$ 61.19	$ 46.11	$ 35.43	$ 31.98
	Enclosed	173.76	138.22	90.89	68.90	60.25
Average	Open	85.36	65.72	48.30	36.47	32.63
	Enclosed	187.75	149.79	95.97	70.72	60.99
Custom	Open	137.43	100.35	77.02	57.89	56.56
	Enclosed	230.93	186.65	122.76	92.87	99.67
Luxury	Open	156.39	114.20	86.50	63.27	57.39
	Enclosed	272.13	246.00	158.04	115.21	95.51

Finished attic *(Cost per S.F.)*

Class	Area (S.F.)				
	400	500	600	800	1000
Economy	$ 20.57	$ 19.88	$ 19.05	$ 18.73	$ 18.04
Average	31.89	31.20	30.43	30.05	29.21
Custom	40.89	39.97	39.06	38.46	37.68
Luxury	52.48	51.20	50.01	48.82	48.01

Alarm system *(Cost per System)*

	Burglar Alarm	Smoke Detector
Economy	$ 425	$ 71
Average	485	84
Custom	835	199
Luxury	1200	258

Sauna, prefabricated
(Cost per unit, including heater and controls—7' high)

Size	Cost
6' x 4'	$ 5750
6' x 5'	6450
6' x 6'	6900
6' x 9'	8700
8' x 10'	11,200
8' x 12'	12,700
10' x 12'	14,700

For customer support on your Square Foot Cost Data, call 877.694.8329.

Garages *

(Costs include exterior wall systems comparable with the quality of the residence. Included in the cost is an allowance for one personnel door, manual overhead door(s) and electrical fixture.)

Class	Type									
	Detached			Attached			Built-in		Basement	
	One Car	Two Car	Three Car	One Car	Two Car	Three Car	One Car	Two Car	One Car	Two Car
Economy										
Wood	$16,393	$25,160	$33,926	$12,758	$22,043	$30,810	$-1720	$-3439	$1798	$2508
Masonry	22,375	32,646	42,916	16,501	27,290	37,561	-2454	-4908		
Average										
Wood	17,994	27,163	36,332	13,759	23,447	32,616	-1916	-3832	2034	2980
Masonry	22,416	32,697	42,978	16,527	27,326	37,607	-2459	-3896		
Custom										
Wood	20,221	31,142	42,063	15,749	27,346	38,266	-3109	-2889	2930	4773
Masonry	24,500	36,496	48,493	18,426	31,099	43,095	-3635	-3940		
Luxury										
Wood	22,777	35,593	48,409	17,974	31,466	44,282	-3218	-3106	3909	6280
Masonry	29,636	44,176	58,717	22,266	37,483	52,023	-4060	-4791		

*See the Introduction to this section for definitions of garage types.

Swimming pools *(Cost per S.F.)*

Residential		
In-ground		$ 34.00 - 83.50
Deck equipment		1.30
Paint pool, preparation & 3 coats (epoxy)		5.31
Rubber base paint		4.56
Pool Cover		1.79
Swimming Pool Heaters		
(not including wiring, external piping, base or pad)		
Gas		
155 MBH		$ 3100.00
190 MBH		3700.00
500 MBH		13,200.
Electric		
15 KW	7200 gallon pool	2825.00
24 KW	9600 gallon pool	3275.00
54 KW	24,000 gallon pool	5100.00

Wood and coal stoves

Wood Only		
Free Standing (minimum)		$ 1975
Fireplace Insert (minimum)		2087
Coal Only		
Free Standing		$ 2359
Fireplace Insert		2582
Wood and Coal		
Free Standing		$ 4851
Fireplace Insert		4972

Sidewalks *(Cost per S.F.)*

Concrete, 3000 psi with wire mesh	4" thick	$ 3.78
	5" thick	4.58
	6" thick	5.13
Precast concrete patio blocks (natural)	2" thick	6.55
Precast concrete patio blocks (colors)	2" thick	6.95
Flagstone, bluestone	1" thick	16.75
Flagstone, bluestone	1-1/2" thick	22.10
Slate (natural, irregular)	3/4" thick	18.65
Slate (random rectangular)	1/2" thick	29.50
Seeding		
Fine grading & seeding includes lime, fertilizer & seed	per S.Y.	2.99
Lawn Sprinkler System	per S.F.	1.10

Fencing *(Cost per L.F.)*

Chain Link, 4' high, galvanized	$ 16.10
Gate, 4' high (each)	209.00
Cedar Picket, 3' high, 2 rail	14.70
Gate (each)	230.00
3 Rail, 4' high	18.00
Gate (each)	249.00
Cedar Stockade, 3 Rail, 6' high	17.95
Gate (each)	251.00
Board & Battens, 2 sides 6' high, pine	27.50
6' high, cedar	35.50
No. 1 Cedar, basketweave, 6' high	36.50
Gate, 6' high (each)	255.00

Carport *(Cost per S.F.)*

Economy	$ 9.41
Average	14.09
Custom	20.70
Luxury	24.20

For customer support on your Square Foot Cost Data, call 877.694.8329.

61

Insurance Exclusions

Insurance exclusions are a matter of policy coverage and are not standard or universal. When making an appraisal for insurance purposes, it is recommended that some time be taken in studying the insurance policy to determine the specific items to be excluded. Most homeowners insurance policies have, as part of their provisions, statements that read "the policy permits the insured to exclude from value, in determining whether the amount of the insurance equals or exceeds 80% of its replacement cost, such items as building excavation, basement footings, underground piping and wiring, piers and other supports which are below the undersurface of the lowest floor or where there is no basement below the ground." Loss to any of these items, however, is covered.

Costs for Excavation, Spread and Strip Footings and Underground Piping

This chart shows excluded items expressed as a percentage of total building cost.

Class	Number of Stories			
	1	1-1/2	2	3
Luxury	3.4%	2.7%	2.4%	1.9%
Custom	3.5%	2.7%	2.4%	1.9%
Average	5.0%	4.3%	3.8%	3.3%
Economy	5.0%	4.4%	4.1%	3.5%

Architect/Designer Fees

Architect/designer fees as presented in the following chart are typical ranges for 4 classes of residences. Factors affecting these ranges include economic conditions, size and scope of project, site selection, and standardization. Where superior quality and detail is required or where closer supervision is required, fees may run higher than listed. Lower quality or simplicity may dictate lower fees.

The editors have included architect/designer fees from the "mean" column in calculating costs.

Listed below are average costs expressed as a range and mean of total building cost for 4 classes of residences.

Class		Range	Mean
Luxury	— Architecturally designed and supervised	5%–15%	10%
Custom	— Architecturally modified designer plans	$2000–$3500	$2500
Average	— Designer plans	$800–$1500	$1100
Economy	— Stock plans	$100–$500	$300

Depreciation as generally defined is "the loss of value due to any cause." Specifically, depreciation can be broken down into three categories: **Physical Depreciation, Functional Obsolescence,** and **Economic Obsolescence.**

Physical Depreciation is the loss of value from the wearing out of the building's components. Such causes may be decay, dry rot, cracks, or structural defects.

Functional Obsolescence is the loss of value from features which render the residence less useful or desirable. Such features may be higher than normal ceilings, design and/or style changes, or outdated mechanical systems.

Economic Obsolescence is the loss of value from external features which render the residence less useful or desirable. These features may be changes in neighborhood socio-economic grouping, zoning changes, legislation, etc.

Depreciation as it applies to the residential portion of this manual deals with the observed physical condition of the residence being appraised. It is in essence *"the cost to cure."*

The ultimate method to arrive at *"the cost to cure"* is to analyze each component of the residence.

For example:

> Component, Roof Covering – Cost = $2400
> Average Life = 15 years
> Actual Life = 5 years
> Depreciation = 5 ÷ 15 = .33 or 33%
>
> $2400 × 33% = $792 = Amount of Depreciation

The following table, however, can be used as a guide in estimating the % depreciation using the actual age and general condition of the residence.

Depreciation Table — Residential

Age in Years	Good	Average	Poor
2	2%	3%	10%
5	4	6	20
10	7	10	25
15	10	15	30
20	15	20	35
25	18	25	40
30	24	30	45
35	28	35	50
40	32	40	55
45	36	45	60
50	40	50	65

Commercial/Industrial/ Institutional Models

Table of Contents

Introduction to the Commercial/ Industrial/Institutional Section

General

The Commercial/Industrial/Institutional section of this manual contains base building costs per square foot of floor area for 75 model buildings. Each model has a table of square foot costs for combinations of exterior wall and framing systems. This table is supplemented by a list of common additives and their unit costs. A breakdown of the component costs used to develop the base cost for the model is on the opposite page. The total cost derived from the base cost and additives must be modified using the appropriate location factor from the Reference Section.

This section may be used directly to estimate the construction cost of most types of buildings knowing only the floor area, exterior wall construction and framing systems. To adjust the base cost for components which are different than the model, use the tables from the Assemblies Section.

Building Identification & Model Selection

The building models in this section represent structures by use. Occupancy, however, does not necessarily identify the building, i.e., a restaurant could be a converted warehouse. In all instances, the building should be described and identified by its own physical characteristics. The model selection should also be guided by comparing specifications with the model. In the case of converted use, data from one model may be used to supplement data from another.

Green Models

Consistent with expanding green trends in the design and construction industry, RSMeans introduced 25 new green building models. Although similar to our standard models in building type and structural system, the new green models meet or exceed Energy Star requirements and address many of the items necessary to obtain LEED certification. Although our models do not include site-specific information, our design assumption is that they are located in climate zone 5. DOE's eQuest software was used to perform an energy analysis of the model buildings. By reducing energy use, we were able to reduce the size of the service entrances, switchgear, power feeds, and generators. Each of the following building systems was researched and analyzed: building envelope, HVAC, plumbing fixtures, lighting, and electrical service. These systems were targeted because of their impact on energy usage and green building.

Wall and roof insulation was increased to reduce heat loss and a recycled vapor barrier was added to the foundation. White roofs were specified for models with flat roofs. Interior finishes include materials with recycled content and low VOC paint. Stainless steel toilet partitions were selected because it is a recycled material and is easy to maintain. Plumbing fixtures were selected to conserve water. Water closets are low-flow and equipped with auto-sensor valves. Waterless urinals are specified throughout. Faucets use auto-sensor flush valves and are powered by a hydroelectric power unit inside the faucet. Energy efficient, low flow water coolers were specified to reduce energy consumption and water usage throughout each building model.

Lighting efficiency was achieved by specifying LED fixtures in place of standard fluorescents. Daylight on/off lighting control systems were incorporated into the models. These controls are equipped with sensors that automatically turn off the lights when sufficient daylight is available. Energy monitoring systems were also included in all green models.

Adjustments

The base cost tables represent the base cost per square foot of floor area for buildings without a basement and without unusual special features. Basement costs and other common additives are listed below the base cost table. Cost adjustments can also be made to the model by using the tables from the Assemblies Section. This table is for example only.

Dimensions

All base cost tables are developed so that measurement can be readily made during the inspection process. Areas are calculated from exterior dimensions and story heights are measured from the top surface of one floor to the top surface of the floor above. Roof areas are measured by horizontal area covered and costs related to inclines are converted with appropriate factors. The precision of measurement is a matter of the user's choice and discretion. For ease in calculation, consideration should be given to measuring in tenths of a foot, i.e., 9 ft. 6 in. = 9.5 ft., 9 ft. 4 in. = 9.3 ft.

Floor Area

The term "Floor Area" as used in this section includes the sum of floor plate at grade level and above. This dimension is measured from the outside face of the foundation wall. Basement costs are calculated separately. The user must exercise his/her own judgment, where the lowest level floor is slightly below grade, whether to consider it at grade level or make the basement adjustment.

How to Use the Commercial/ Industrial/Institutional Section

The following is a detailed explanation of a sample entry in the Commercial/Industrial/Institutional Square Foot Cost Section. Each bold number below corresponds to the described item on the following page with the appropriate component or cost of the sample entry following in parentheses.

Prices listed are costs that include overhead and profit of the installing contractor and additional markups for General Conditions and Architects' Fees.

COMMERCIAL/INDUSTRIAL/ INSTITUTIONAL — M.010 **1** — Apartment, 1-3 Story **2**

Costs per square foot of floor area

Exterior Wall		S.F. Area	8000	12000	15000	19000	22500	25000	29000	32000	36000
		L.F. Perimeter	213	280	330	350	400	433	442	480	520
Face Brick with Concrete Block Back-up	Wood Joists		191.05	182.00	178.35	170.85	169.10	167.95	163.85	163.20	161.90
	Steel Joists		189.40	181.65	178.50	172.40	170.90	169.	166.65	166.05	165.00
Stucco on Concrete Block **3**	Wood Joists		170.60	163.75	161.00	155.85	154.50	153.	150.90	150.40	149.50
	Steel Joists		182.25	175.20	172.35	167.05	165.65	164.80	161.95	161.40	160.45
Wood Siding	Wood Frame		169.60	162.95	160.25	155.35	154.05	153.25	150.65	150.15	149.25
Brick Veneer	Wood Frame		179.00	171.10	167.90	161.70	160.15	159.20	155.75	155.20	154.10
Perimeter Adj., Add or Deduct **6**	Per 100 L.F.		22.15	14.75	11.85	9.35	7.90	7.15	6.10	5.50	4.90
Story Hgt. Adj., Add or Deduct	Per 1 Ft.		4.20	3.65	3.45	2.95	2.80	2.80	2.40	2.35	2.25

4 **5** **8** **7**

For Basement, add $36.80 per square foot of basement area

The above costs were calculated using the basic specifications shown on the facing page. These costs should be adjusted where necessary for design alternatives and owner's requirements. Reported completed project costs, for this type of structure, range from $62.60 to $232.75 per S.F. **9**

Common additives

Description	Unit	$ Cost
Appliances		
Cooking range, 30" free standing		
1 oven	Ea.	580 - 2450
2 oven	Ea.	1425 - 2900
30" built-in		
1 oven	Ea.	860 - 2075
2 oven	Ea.	1750 - 2550
Counter top cook tops, 4 burner **10**	Ea.	435 - 1675
Microwave oven	Ea.	296 - 840
Combination range, refrig. & sink, 30" wide	Ea.	1900 - 3000
72" wide	Ea.	2900
Combination range, refrigerator, sink,		
microwave oven & icemaker	Ea.	6700
Compactor, residential, 4-1 compaction	Ea.	825 - 1325
Dishwasher, built-in, 2 cycles	Ea.	600 - 1150
4 cycles	Ea.	755 - 1875
Garbage disposer, sink type	Ea.	232 - 380
Hood for range, 2 speed, vented, 30" wide	Ea.	330 - 1275
42" wide	Ea.	435 - 2650
Refrigerator, no frost 10-12 C.F.	Ea.	560 - 650
18-20 C.F.	Ea.	805 - 1800

Description	Unit	$ Cost
Closed Circuit Surveillance, One station		
Camera and monitor	Ea.	1975
For additional camera stations, add	Ea.	1075
Elevators, Hydraulic passenger, 2 stops		
2000# capacity	Ea.	66,600
2500# capacity	Ea.	68,800
3500# capacity	Ea.	73,300
Additional stop, add	Ea.	7825
Emergency Lighting, 25 watt, battery operated		
Lead battery	Ea.	330
Nickel cadmium	Ea.	780
Laundry Equipment		
Dryer, gas, 16 lb. capacity	Ea.	975
30 lb. capacity	Ea.	3850
Washer, 4 cycle	Ea.	1200
Commercial	Ea.	1600
Smoke Detectors		
Ceiling type	Ea.	227
Duct type	Ea.	565

1 ## Model Number (M.010)

"M" distinguishes this section of the book and stands for model. The number designation is a sequential number. Our newest models are green and are designated with the letter "G" followed by the building type number.

2 ## Type of Building (Apartment, 1–3 Story)

There are 50 different types of commercial/industrial/institutional buildings highlighted in this section.

3 ## Exterior Wall Construction and Building Framing Options (Face Brick with Concrete Block Back-up and Open Web Steel Bar Joists)

Three or more commonly used exterior walls and, in most cases, two typical building framing systems are presented for each type of building. The model selected should be based on the actual characteristics of the building being estimated.

4 ## Total Square Foot of Floor Area and Base Perimeter Used to Compute Base Costs (22,500 Square Feet and 400 Linear Feet)

Square foot of floor area is the total gross area of all floors at grade, and above, and does not include a basement. The perimeter in linear feet used for the base cost is generally for a rectangular, economical building shape.

5 ## Cost per Square Foot of Floor Area ($170.90)

The highlighted cost is for a building of the selected exterior wall and framing system and floor area. Costs for buildings with floor areas other than those calculated may be interpolated between the costs shown.

6 ## Building Perimeter and Story Height Adjustments

Square-foot costs for a building with a perimeter or floor-to-floor story height significantly different from the model used to calculate the base cost may be adjusted (add or deduct) to reflect the actual building geometry.

7 ## Cost per Square Foot of Floor Area for the Perimeter and/or Height Adjustment ($7.90 for Perimeter Difference and $2.80 for Story Height Difference)

Add (or deduct) $7.90 to the base square-foot cost for each 100 feet of perimeter difference between the model and the actual building. Add (or deduct) $2.80 to the base square-foot cost for each 1 foot of story height difference between the model and the actual building.

8 ## Optional Cost per Square Foot of Basement Floor Area ($36.80)

The cost of an unfinished basement for the building being estimated is $36.80 times the gross floor area of the basement.

9 ## Range of Cost per Square Foot of Floor Area for Similar Buildings ($62.60 to $232.75)

Many different buildings of the same type have been built using similar materials and systems. RSMeans historical cost data of actual construction projects indicates a range of $62.60 to $232.75 for this type of building.

10 ## Common Additives

Common components and/or systems used in this type of building are listed. These costs should be added to the total building cost. Additional selections may be found in the Assemblies Section.

The following is a detailed explanation of the specification and costs for a model building in the Commercial/Industrial/Institutional Square Foot Cost Section. Each bold number below corresponds to the described item on the following page with the appropriate component of the sample entry following in parentheses.

Prices listed are costs that include overhead and profit of the installing contractor.

Model costs calculated for a 3 story building with 10' story height and 22,500 square feet of floor area — **1**

2 — **Apartment, 1-3 Story**

				Unit	Unit Cost	Cost Per S.F.	% Of Sub-Total
A. SUBSTRUCTURE							
1010	Standard Foundations	Poured concrete; strip and spread footings		S.F. Ground	6.51	2.17	
1020	Special Foundations	N/A		–	–	–	
1030	Slab on Grade	4" reinforced concrete with vapor barrier and granular base		S.F. Slab	5.44	1.81	4.5%
2010	Basement Excavation	Site preparation for slab and trench for foundation wall and footing		S.F. Ground	.32	.11	
2020	Basement Walls	4' foundation wall		L.F. Wall	79	1.62	
B. SHELL							
B10 Superstructure							
1010	Floor Construction	Open web steel joists, slab form, concrete, interior steel columns		S.F. Floor	20.24	13.49	13.3%
1020	Roof Construction	Open web steel joists with rib metal deck, interior steel columns		S.F. Roof	10.35	3.45	
B20 Exterior Enclosure							
2010	Exterior Walls	Face brick with concrete block backup	88% of wall	S.F. Wall	23.84	11.19	
2020	Exterior Windows	Aluminum horizontal sliding	12% of wall	Each	528	2.26	10.8%
2030	Exterior Doors	Aluminum and glass		Each	2020	.36	
B30 Roofing							
3010	Roof Coverings	Built-up tar and gravel with flashing; perlite/EPS composite insulation		S.F. Roof	6.78	2.26	1.8%
3020	Roof Openings	N/A		–	–	–	
C. INTERIORS							
1010	Partitions	Gypsum board and sound deadening board on metal studs	10 S.F. of Floor/L.F. Partition	S.F. Partition	7.03	6.25	
1020	Interior Doors	15% solid core wood, 85% hollow core wood	80 S.F. Floor/Door	Each	549	6.87	
1030	Fittings	Kitchen Cabinets		S.F. Floor	3.27	3.27	
2010	Stair Construction	Concrete filled metal pan		Flight	14,500	3.22	24.1%
3010	Wall Finishes	70% paint, 25% vinyl wall covering, 5% ceramic tile		S.F. Surface	1.46	2.60	
3020	Floor Finishes	60% carpet, 30% vinyl composition tile, 10% ceramic tile		S.F. Floor	4.69	4.69	
3030	Ceiling Finishes	Painted gypsum board on resilient channels		S.F. Ceiling	3.90	3.90	
D. SERVICES							
D10 Conveying							
1010	Elevators & Lifts	One hydraulic passenger elevator		Each	108,450	4.82	3.8%
1020	Escalators & Moving Walks	N/A		–	–	–	
D20 Plumbing							
2010	Plumbing Fixtures	Kitchen, bathroom and service fixtures, supply and drainage	1 Fixture/200 S.F. Floor	Each	2928	14.64	
2020	Domestic Water Distribution	Gas fired water heater		S.F. Floor	4.78	4.78	15.6%
2040	Rain Water Drainage	Roof drains		S.F. Roof	1.50	.50	
D30 HVAC							
3010	Energy Supply	Oil fired hot water, baseboard radiation		S.F. Floor	8.90	8.90	
3020	Heat Generating Systems	N/A		–	–	–	
3030	Cooling Generating Systems	Chilled water, air cooled condenser system		S.F. Floor	9.62	9.62	14.5%
3050	Terminal & Package Units	N/A		–	–	–	
3090	Other HVAC Sys. & Equipment	N/A		–	–	–	
D40 Fire Protection							
4010	Sprinklers	Wet pipe sprinkler system		S.F. Floor	3.60	3.60	2.8%
4020	Standpipes	N/A		–	–	–	
D50 Electrical							
5010	Electrical Service/Distribution	600 ampere service, panel board and feeders		S.F. Floor	2.14	2.14	
5020	Lighting & Branch Wiring	Incandescent fixtures, receptacles, switches, A.C. and misc. power		S.F. Floor	7.27	7.27	8.9%
5030	Communications & Security	Addressable alarm systems, internet wiring, and emergency lighting		S.F. Floor	1.78	1.78	
5090	Other Electrical Systems	Emergency generator, 11.5 kW		S.F. Floor	.19	.19	
E. EQUIPMENT & FURNISHINGS							
1010	Commercial Equipment	N/A		–	–	–	
1020	Institutional Equipment	N/A		–	–	–	0.0%
1030	Vehicular Equipment	N/A		–	–	–	
1090	Other Equipment	N/A		–	–	–	
F. SPECIAL CONSTRUCTION							
1020	Integrated Construction	N/A		–	–	–	0.0%
1040	Special Facilities	N/A		–	–	–	
G. BUILDING SITEWORK	**N/A**						
				Sub-Total		127.76	100%
CONTRACTOR FEES (General Requirements: 10%, Overhead: 5%, Profit: 10%)					25%	31.96	
ARCHITECT FEES					7%	11.18	
				Total Building Cost		170.90	

1 Building Description (Model costs are calculated for a 3-story apartment building with 10' story height and 22,500 square feet of floor area)

The model highlighted is described in terms of building type, number of stories, typical story height, and square footage.

2 Type of Building (Apartment, 1–3 Story)

3 Division C Interiors (C1020 Interior Doors)

System costs are presented in divisions according to the 7-division UNIFORMAT II classifications. Each of the component systems is listed.

4 Specification Highlights (15% solid core wood; 85% hollow core wood)

All systems in each subdivision are described with the material and proportions used.

5 Quantity Criteria (80 S.F. Floor/Door)

The criteria used in determining quantities for the calculations are shown.

6 Unit (Each)

The unit of measure shown in this column is the unit of measure of the particular system shown that corresponds to the unit cost.

7 Unit Cost ($549)

The cost per unit of measure of each system subdivision.

8 Cost per Square Foot ($6.87)

The cost per square foot for each system is the unit cost of the system times the total number of units, divided by the total square feet of building area.

9 % of Sub-Total (24.1%)

The percent of sub-total is the total cost per square foot of all systems in the division divided by the sub-total cost per square foot of the building.

10 Sub-Total ($127.76)

The sub-total is the total of all the system costs per square foot.

11 Project Fees (Contractor Fees) (25%) (Architects' Fees) (7%)

Contractor Fees to cover the general requirements, overhead, and profit of the General Contractor are added as a percentage of the sub-total. An Architect's Fee, also as a percentage of the sub-total, is also added. These values vary with the building type.

12 Total Building Cost ($170.90)

The total building cost per square foot of building area is the sum of the square foot costs of all the systems, the General Contractor's general requirements, overhead and profit, and the Architect's fee. The total building cost is the amount which appears shaded in the Cost per Square Foot of Floor Area table shown previously.

Examples

Example 1
This example illustrates the use of the base cost tables. The base cost is adjusted for different exterior wall systems, different story height and a partial basement.

CII APPRAISAL FIELD DATA FORM

COMMERCIAL
INDUSTRIAL
INSTITUTIONAL

Use sheets A & D for
App. Form 2
Use sheets A,B & C for
App. Form 3

CII APPRAISAL

		5. DATE: **Jan. 1, 2015**
1. SUBJECT PROPERTY: **Westernberg Holdings, Inc.**		6. APPRAISER: **SJS**
2. BUILDING: **# 623**		
3. ADDRESS: **55 Pine wood Rd., Bondsville, N.J. 07410**		
4. BUILDING USE: **Apartments**		7. YEAR BUILT: **1991**

8. EXTERIOR WALL CONSTRUCTION: **North wall - Face brick w/concrete block back-up**
East, West & South walls - Decorative concrete block

9. FRAME: **Steel**

10. GROUND FLOOR AREA: **60' x 191'   11,460** S.F.

11. GROSS FLOOR AREA (EXCL. BASEMENT): **80, 220 (7 Floors)** S.F.

12. NUMBER OF STORIES: **7**

13. STORY HEIGHT: **11 ' - 4"**

14. PERIMETER: **502** L.F.

15. BASEMENT AREA: **4200 S.F.** S.F.

16. GENERAL COMMENTS:

Example 1 (continued)

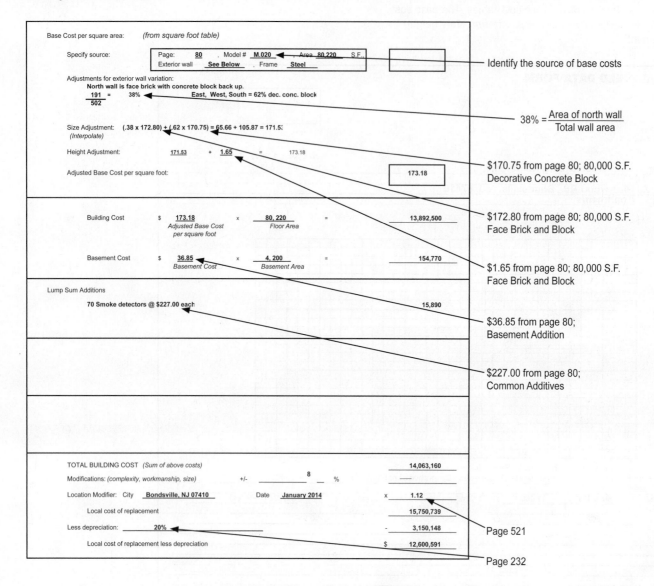

Base Cost per square area: *(from square foot table)*

Specify source:
Page: **80** , Model # **M.020** , Area **80,220** S.F.
Exterior wall **See Below** , Frame **Steel**

— Identify the source of base costs

Adjustments for exterior wall variation:
North wall is face brick with concrete block back up.
$\frac{191}{502}$ = 38%   **East, West, South = 62% dec. conc. block**

$38\% = \dfrac{\text{Area of north wall}}{\text{Total wall area}}$

Size Adjustment: (.38 x 172.80) + (.62 x 170.75) = 65.66 + 105.87 = 171.5:
(Interpolate)

Height Adjustment: **171.53** + **1.65** = 173.18

Adjusted Base Cost per square foot: **173.18**

$170.75 from page 80; 80,000 S.F.
Decorative Concrete Block

Building Cost $ **173.18** x **80, 220** = 13,892,500
Adjusted Base Cost per square foot   *Floor Area*

$172.80 from page 80; 80,000 S.F.
Face Brick and Block

Basement Cost $ **36.85** x **4, 200** = 154,770
Basement Cost   *Basement Area*

$1.65 from page 80; 80,000 S.F.
Face Brick and Block

Lump Sum Additions
70 Smoke detectors @ $227.00 each   15,890

$36.85 from page 80;
Basement Addition

$227.00 from page 80;
Common Additives

TOTAL BUILDING COST *(Sum of above costs)*   14,063,160
Modifications: *(complexity, workmanship, size)* +/- **8** %   —
Location Modifier: City **Bondsville, NJ 07410**   Date **January 2014**   x **1.12**
Local cost of replacement   15,750,739
Less depreciation: **20%**   - 3,150,148
Local cost of replacement less depreciation $ 12,600,591

Page 521

Page 232

74

Example 2

This example shows how to modify a model building.

Model #M.020, 4–7 Story Apartment, page 80, matches the example quite closely. The model specifies a 6-story building with a 10' story height.

The following adjustments must be made:

- add stone ashlar wall
- adjust model to a 5-story building
- change partitions
- change floor finish

CII APPRAISAL

1. SUBJECT PROPERTY: **Westernberg Holdings, Inc.**	5. DATE: **Jan. 2, 2015**
2. BUILDING: **# 623**	6. APPRAISER: **SJS**
3. ADDRESS: **55 Pinewood Rd., Bondsville, N.J. 07410**	
4. BUILDING USE: **Apartments**	7. YEAR BUILT: **1992**

N
170'

W
64.7'

E

S

North wall - 4" stone Ashlar w/8" block back up (split Face)

8. EXTERIOR WALL CONSTRUCTION:	**East, West & South walls - 12" Decorative concrete block**		
9. FRAME:	**Steel**		
10. GROUND FLOOR AREA:	**11,000**	11. GROSS FLOOR AREA (EXCL. BASEMENT):	**55,000** S.F.
12. NUMBER OF STORIES:	**5**	13. STORY HEIGHT:	**11 Ft.**
14. PERIMETER:	**469.4**	15. BASEMENT AREA:	None S.F.
16. GENERAL COMMENTS:			

Two exterior wall types

Example 2 (continued)

Square foot area (excluding basement) from item 11 ___55,000___ S.F.
Perimeter from item 14 ___469.4___ Lin. Ft.
Item 17-Model square foot costs (from model sub-total) ___$134.83___

NO.	DESCRIPTION			UNIT	UNIT COST	NEW SF COST	MODEL SF COST	+/- CHANGE
A SUBSTRUCTURE								
A 1010	Standard Foundations		Bay size: *Same as model*	S.F. Gnd.	7.44	1.49	1.24	+ 0.25
A 1030	Slab on Grade	Material:	Thickness: *Same as model*	S.F. Slab	5.44	1.09	0.91	+ 0.18
A 2010	Basement Excavation	Depth:	Area: *11,000 S.F.*	S.F. Gnd.	0.32	0.06	0.05	+ 0.01
A 2020	Basement Walls			L.F. Walls	84	0.72	0.70	+ 0.02
B SHELL								
B10 Superstructure								
B 1010	Floor Construction	Elevated floors: *Same as model*		S.F. Floor	24.59	19.67	21.81	- 2.14
B 1020	Roof Construction			S.F. Roof	9.66	1.93	1.59	+ 0.34
B20 Exterior Closure								
B 2010	Exterior Walls	Material: *North wall- 4" stone ashlar on 8" CMU*	Thickness: *36% of wall*	S.F. Walls	30.11	12.16	10.59	+ 1.57
		Material: *East, West, & South walls- 12" split rib block*	Thickness: *64% of wall*	S.F. Walls				
B 2020	Exterior Windows	Type:	% of wall	S.F. Wind.				
		Type:	14% of wall Each	S.F. Wind.	528	2.31	2.55	- 0.24
B 2030	Exterior Doors	Type: *Same as model*	Number: 4	Each	4888	0.36	0.33	+ 0.03
		Type:	Number:	Each				
B30 Roofing								
B 3010	Roof Coverings	Material: *Same as model*		S.F. Roof	7.26	1.45	1.21	+ 0.24
B 3020	Roof Openings	N/A		S.F. Opng.				
C INTERIORS								
C 1010	Partitions: *High*	Material: *Gyp board on metal studs* Density: *8 S.F. Floor/L.F. Part*		S.F. Part.	7.42	8.35	7.42	+ 0.93
		Material:	Density:					
C 1020	Interior Doors	Type:	Number: 80 SF	Each	682	8.53	8.53	---
C 1030	Fittings			S.F. Floor	3.92	3.92	3.92	---
C 2010	Stair Construction	14 Flights		Flight	11600	2.95	3.09	- 0.14
C 3010	Wall Finishes	Material: *Same as model* % of Wall 9'-0" *High*		S.F. Walls	1.63	3.67	3.25	+ 0.42
		Material:						
C 3020	Floor Finishes	Material: *Hardwood - 60% of floor*		S.F. Floor	5.42	8.75	5.42	+ 3.33
		Material: *Carpet - 30% of floor*						
		Material: *Tile -10% of floor*						
C 3030	Ceiling Finishes	Material: *Same as model*		S.F. Ceil.	4.23	4.23	4.23	---

From page 81, Model #M.020 Sub-total

$.32/S.F. ground area
6 stories

$.32/S.F. ground area
5 stories

Area of Elevated Floors
$24.59/S.F. x 4 x 11,000 S.F.
55,000
Building Floor Area
9.66/S.F. roof
5 stories

Windows are 14% of wall area,
15 S.F./window

528 x .14 x 469.4 x 11 x 5 ÷ 55,000 = 2.31
15

From Table C1010 124 page 399, C1010 126
on page 400 and C1010 128 page 403

$7.42/S.F. partition x 9 ft. high
8 S.F. floor/L.F. partition

From Table C3020 410 page 415 & 416

Oak Strip, (max price) @$10.37/S.F.
Carpet (26 oz.) @$4.18/S.F.
Ceramic Tile, (max price) @$12.75/S.F.

.60 x 10.37 = 6.22
.30 x 4.18 = 1.25
.10 x 12.75 = 1.28
8.75

North wall is 36% of wall area, e.g.
From Table B2010 128 2350 page 360
$\frac{170 \times 11}{469.4 \times 11}$
Stone ashiar @ $45.05/S.F. wall
East, west and south walls are 64% of wall area
From Table B2010 113 1530 page 354
Split ribbed block @$21.70/S.F. wall

New wall cost
(.36 x $45.05/S.F.) + (.64 x $21.70/S.F.) = $30.11/S.F. wall

Convert Cost per S.F. wall to cost per S.F. of floor
Windows are 14% of wall area

$12.16 S.F. floor = $\frac{\$30.11/S.F. wall \times .86 \times 469.4 L.F. \times 11 ft. high \times 5 stories}{55,000 S.F. of floor}$

Example 2 (continued)

NO.	SYSTEM/COMPONENT	DESCRIPTION	UNIT	UNIT COST	NEW S.F. COST	MODEL S.F. COST	+/- CHANGE
D SERVICES							
D10 Conveying							
D 1010	Elevators & Lifts 2	Type: *Same as model* Capacity: Stops:	Each	229,500	8.35	7.65	+ 0.70
D 1020	Escalators & Moving Walks	Type:	Each				
D20 Plumbing							
D 2010	Plumbing *Kitchen, bath, + service fixtures*	1 Fixture/215 S.F. Floor	Each	2872	13.36	13.36	---
D 2020	Domestic Water Distribution		S.F. Floor	4.60	4.60	4.60	---
D 2040	Rain Water Drainage		S.F. Roof	1.44	0.29	0.24	+ 0.05
D30 HVAC							
D 3010	Energy Supply		S.F. Floor	7.83	7.83	7.83	---
D 3020	Heat Generating Systems	Type: *Same as model*	S.F. Floor				
D 3030	Cooling Generating Systems	Type: *Same as model*	S.F. Floor	9.54	9.54	9.54	---
D 3090	Other HVAC Sys. & Equipment		Each				
D40 Fire Protection							
D 4010	Sprinklers	*Same as model*	S.F. Floor	2.85	2.85	2.85	---
D 4020	Standpipes	*Same as model*	S.F. Floor	0.89	0.89	0.89	---
D50 Electrical							
D 5010	Electrical Service/Distribution	*Same as model*	S.F. Floor	1.94	1.94	1.94	---
D 5020	Lighting & Branch Wiring		S.F. Floor	7.22	7.22	7.22	---
D 5030	Communications & Security		S.F. Floor	1.69	1.69	1.69	---
D 5090	Other Electrical Systems		S.F. Floor	0.18	0.18	0.18	---
E EQUIPMENT & FURNISHINGS							
E 1010	Commercial Equipment		Each				
E 1020	Institutional Equipment		Each				
E 1030	Vehicular Equipment		Each				
E 1090	Other Equipment		Each				
F SPECIAL CONSTRUCTION							
F 1020	Integrated Construction	N/A	S.F.				
F 1040	Special Facilities	N/A	S.F.				
G BUILDING SITEWORK							

ITEM			Total Change	+ 5.55
17	Model 020 total $ **134.83**			
18	Adjusted S.F. cost $ **140.38** item 17 +/- changes			
19	Building area - from item 11 **55,000** S.F. x adjusted S.F. cost	$	7,720,900
20	Basement area - from item 15 S.F. x S.F. cost $	$	
21	Base building sub-total - item 20 + item 19	$	7,720,900
22	Miscellaneous addition (quality, etc.)	$	
23	Sub-total - item 22 + 21	$	7,720,900
24	General conditions -25 % of item 23	$	1,930,225
25	Sub-total - item 24 + item 23	$	9,651,125
26	Architects fees **5.5** % of item 25	$	530,812
27	Sub-total - item 26 + item 25	$	10,181,937
28	Location modifier .... **zip code 07410**	x	1.12
29	Local replacement cost - item 28 x item 27	$	11,403,769
30	Depreciation **20** % of item 29	$	2,280,754
31	Depreciated local replacement cost - item 29 less item 30	$	9,123,015
32	Exclusions	$	
33	Net depreciated replacement cost - item 31 less item 32	$	9,123,015

From page 521

Use table on page 232

Costs per square foot of floor area

Exterior Wall	S.F. Area	8000	12000	15000	19000	22500	25000	29000	32000	36000
	L.F. Perimeter	213	280	330	350	400	433	442	480	520
Face Brick with Concrete Block Back-up	Wood Joists	191.05	182.00	178.35	170.85	169.10	167.95	163.85	163.20	161.90
	Steel Joists	189.40	181.65	178.50	172.40	170.90	169.95	166.65	166.05	165.00
Stucco on Concrete Block	Wood Joists	170.60	163.75	161.00	155.85	154.50	153.70	150.90	150.40	149.50
	Steel Joists	182.25	175.20	172.35	167.05	165.65	164.80	161.95	161.40	160.45
Wood Siding	Wood Frame	169.60	162.95	160.25	155.35	154.05	153.25	150.65	150.15	149.25
Brick Veneer	Wood Frame	179.00	171.10	167.90	161.70	160.15	159.20	155.75	155.20	154.10
Perimeter Adj., Add or Deduct	Per 100 L.F.	22.15	14.75	11.85	9.35	7.90	7.15	6.10	5.50	4.90
Story Hgt. Adj., Add or Deduct	Per 1 Ft.	4.20	3.65	3.45	2.95	2.80	2.80	2.40	2.35	2.25

For Basement, add $36.80 per square foot of basement area

The above costs were calculated using the basic specifications shown on the facing page. These costs should be adjusted where necessary for design alternatives and owner's requirements. Reported completed project costs, for this type of structure, range from $62.60 to $232.75 per S.F.

Common additives

Description	Unit	$ Cost
Appliances		
Cooking range, 30" free standing		
1 oven	Ea.	580 - 2450
2 oven	Ea.	1425 - 2900
30" built-in		
1 oven	Ea.	860 - 2075
2 oven	Ea.	1750 - 2550
Counter top cook tops, 4 burner	Ea.	435 - 1675
Microwave oven	Ea.	296 - 840
Combination range, refrig. & sink, 30" wide	Ea.	1900 - 3000
72" wide	Ea.	2900
Combination range, refrigerator, sink,		
microwave oven & icemaker	Ea.	6700
Compactor, residential, 4-1 compaction	Ea.	825 - 1325
Dishwasher, built-in, 2 cycles	Ea.	600 - 1150
4 cycles	Ea.	755 - 1875
Garbage disposer, sink type	Ea.	232 - 380
Hood for range, 2 speed, vented, 30" wide	Ea.	330 - 1275
42" wide	Ea.	435 - 2650
Refrigerator, no frost 10-12 C.F.	Ea.	560 - 650
18-20 C.F.	Ea.	805 - 1800

Description	Unit	$ Cost
Closed Circuit Surveillance, One station		
Camera and monitor	Ea.	1975
For additional camera stations, add	Ea.	1075
Elevators, Hydraulic passenger, 2 stops		
2000# capacity	Ea.	66,600
2500# capacity	Ea.	68,800
3500# capacity	Ea.	73,300
Additional stop, add	Ea.	7825
Emergency Lighting, 25 watt, battery operated		
Lead battery	Ea.	330
Nickel cadmium	Ea.	780
Laundry Equipment		
Dryer, gas, 16 lb. capacity	Ea.	975
30 lb. capacity	Ea.	3850
Washer, 4 cycle	Ea.	1200
Commercial	Ea.	1600
Smoke Detectors		
Ceiling type	Ea.	227
Duct type	Ea.	565

Important: See the Reference Section for Location Factors.

Model costs calculated for a 3 story building with 10' story height and 22,500 square feet of floor area

Apartment, 1-3 Story

				Unit	Unit Cost	Cost Per S.F.	% Of Sub-Total
A. SUBSTRUCTURE							
1010	Standard Foundations	Poured concrete; strip and spread footings		S.F. Ground	6.51	2.17	
1020	Special Foundations	N/A		—	—	—	
1030	Slab on Grade	4" reinforced concrete with vapor barrier and granular base		S.F. Slab	5.44	1.81	4.5%
2010	Basement Excavation	Site preparation for slab and trench for foundation wall and footing		S.F. Ground	.32	.11	
2020	Basement Walls	4' foundation wall		L.F. Wall	79	1.62	
B. SHELL							
B10 Superstructure							
1010	Floor Construction	Open web steel joists, slab form, concrete, interior steel columns		S.F. Floor	20.24	13.49	13.3%
1020	Roof Construction	Open web steel joists with rib metal deck, interior steel columns		S.F. Roof	10.35	3.45	
B20 Exterior Enclosure							
2010	Exterior Walls	Face brick with concrete block backup	88% of wall	S.F. Wall	23.84	11.19	
2020	Exterior Windows	Aluminum horizontal sliding	12% of wall	Each	528	2.26	10.8%
2030	Exterior Doors	Aluminum and glass		Each	2020	.36	
B30 Roofing							
3010	Roof Coverings	Built-up tar and gravel with flashing; perlite/EPS composite insulation		S.F. Roof	6.78	2.26	1.8%
3020	Roof Openings	N/A		—	—	—	
C. INTERIORS							
1010	Partitions	Gypsum board and sound deadening board on metal studs	10 S.F. of Floor/L.F. Partition	S.F. Partition	7.03	6.25	
1020	Interior Doors	15% solid core wood, 85% hollow core wood	80 S.F. Floor/Door	Each	549	6.87	
1030	Fittings	Kitchen Cabinets		S.F. Floor	3.27	3.27	
2010	Stair Construction	Concrete filled metal pan		Flight	14,500	3.22	24.1%
3010	Wall Finishes	70% paint, 25% vinyl wall covering, 5% ceramic tile		S.F. Surface	1.46	2.60	
3020	Floor Finishes	60% carpet, 30% vinyl composition tile, 10% ceramic tile		S.F. Floor	4.69	4.69	
3030	Ceiling Finishes	Painted gypsum board on resilient channels		S.F. Ceiling	3.90	3.90	
D. SERVICES							
D10 Conveying							
1010	Elevators & Lifts	One hydraulic passenger elevator		Each	108,450	4.82	3.8%
1020	Escalators & Moving Walks	N/A		—	—	—	
D20 Plumbing							
2010	Plumbing Fixtures	Kitchen, bathroom and service fixtures, supply and drainage	1 Fixture/200 S.F. Floor	Each	2928	14.64	
2020	Domestic Water Distribution	Gas fired water heater		S.F. Floor	4.78	4.78	15.6%
2040	Rain Water Drainage	Roof drains		S.F. Roof	1.50	.50	
D30 HVAC							
3010	Energy Supply	Oil fired hot water, baseboard radiation		S.F. Floor	8.90	8.90	
3020	Heat Generating Systems	N/A		—	—	—	
3030	Cooling Generating Systems	Chilled water, air cooled condenser system		S.F. Floor	9.62	9.62	14.5%
3050	Terminal & Package Units	N/A		—	—	—	
3090	Other HVAC Sys. & Equipment	N/A		—	—	—	
D40 Fire Protection							
4010	Sprinklers	Wet pipe sprinkler system		S.F. Floor	3.60	3.60	2.8%
4020	Standpipes	N/A		—	—	—	
D50 Electrical							
5010	Electrical Service/Distribution	600 ampere service, panel board and feeders		S.F. Floor	2.14	2.14	
5020	Lighting & Branch Wiring	Incandescent fixtures, receptacles, switches, A.C. and misc. power		S.F. Floor	7.27	7.27	8.9%
5030	Communications & Security	Addressable alarm systems, internet wiring, and emergency lighting		S.F. Floor	1.78	1.78	
5090	Other Electrical Systems	Emergency generator, 11.5 kW		S.F. Floor	.19	.19	
E. EQUIPMENT & FURNISHINGS							
1010	Commercial Equipment	N/A		—	—	—	
1020	Institutional Equipment	N/A		—	—	—	
1030	Vehicular Equipment	N/A		—	—	—	0.0%
1090	Other Equipment	N/A		—	—	—	
F. SPECIAL CONSTRUCTION							
1020	Integrated Construction	N/A		—	—	—	0.0%
1040	Special Facilities	N/A		—	—	—	
G. BUILDING SITEWORK	**N/A**						

			Sub-Total	127.76	100%
CONTRACTOR FEES (General Requirements: 10%, Overhead: 5%, Profit: 10%)		25%	31.96		
ARCHITECT FEES		7%	11.18		

Total Building Cost	**170.90**

For customer support on your Square Foot Cost Data, call 877.694.8329.

79

Costs per square foot of floor area

Exterior Wall	S.F. Area	40000	45000	50000	55000	60000	70000	80000	90000	100000
	L.F. Perimeter	366	400	433	466	500	510	530	550	594
Face Brick with Concrete Block Back-up	Steel Frame	183.00	181.55	180.35	179.40	178.60	175.05	172.80	171.05	170.20
	R/Conc. Frame	191.70	190.05	188.70	187.55	186.70	182.50	179.80	177.70	176.80
Decorative Concrete Block	Steel Frame	179.65	178.30	177.15	176.30	175.55	172.65	170.75	169.25	168.55
	R/Conc. Frame	176.20	174.85	173.75	172.85	172.10	169.15	167.20	165.65	164.90
Precast Concrete Panels	Steel Frame	194.55	192.80	191.30	190.15	189.15	184.55	181.55	179.20	178.20
	R/Conc. Frame	190.80	189.05	187.60	186.35	185.45	180.70	177.75	175.40	174.40
Perimeter Adj., Add or Deduct	Per 100 L.F.	7.05	6.25	5.70	5.10	4.70	4.00	3.50	3.10	2.80
Story Hgt. Adj., Add or Deduct	Per 1 Ft.	2.35	2.25	2.20	2.10	2.05	1.85	1.65	1.50	1.50

For Basement, add $36.85 per square foot of basement area

The above costs were calculated using the basic specifications shown on the facing page. These costs should be adjusted where necessary for design alternatives and owner's requirements. Reported completed project costs, for this type of structure, range from $74.95 to $213.15 per S.F.

Common additives

Description	Unit	$ Cost	Description	Unit	$ Cost
Appliances			Closed Circuit Surveillance, One station		
Cooking range, 30" free standing			Camera and monitor	Ea.	1975
1 oven	Ea.	580 - 2450	For additional camera stations, add	Ea.	1075
2 oven	Ea.	1425 - 2900	Elevators, Electric passenger, 5 stops		
30" built-in			2000# capacity	Ea.	167,000
1 oven	Ea.	860 - 2075	3500# capacity	Ea.	173,000
2 oven	Ea.	1750 - 2550	5000# capacity	Ea.	179,500
Counter top cook tops, 4 burner	Ea.	435 - 1675	Additional stop, add	Ea.	10,200
Microwave oven	Ea.	296 - 840	Emergency Lighting, 25 watt, battery operated		
Combination range, refrig. & sink, 30" wide	Ea.	1900 - 3000	Lead battery	Ea.	330
72" wide	Ea.	2900	Nickel cadmium	Ea.	780
Combination range, refrigerator, sink,			Laundry Equipment		
microwave oven & icemaker	Ea.	6700	Dryer, gas, 16 lb. capacity	Ea.	975
Compactor, residential, 4-1 compaction	Ea.	825 - 1325	30 lb. capacity	Ea.	3850
Dishwasher, built-in, 2 cycles	Ea.	600 - 1150	Washer, 4 cycle	Ea.	1200
4 cycles	Ea.	755 - 1875	Commercial	Ea.	1600
Garbage disposer, sink type	Ea.	232 - 380	Smoke Detectors		
Hood for range, 2 speed, vented, 30" wide	Ea.	330 - 1275	Ceiling type	Ea.	227
42" wide	Ea.	435 - 2650	Duct type	Ea.	565
Refrigerator, no frost 10-12 C.F.	Ea.	560 - 650			
18-20 C.F.	Ea.	805 - 1800			

Important: See the Reference Section for Location Factors.

Model costs calculated for a 6 story building with 10'-4" story height and 60,000 square feet of floor area

Apartment, 4-7 Story

				Unit	Unit Cost	Cost Per S.F.	% Of Sub-Total
A. SUBSTRUCTURE							
1010	Standard Foundations	Poured concrete; strip and spread footings and 4' foundation wall		S.F. Ground	7.44	1.24	
1020	Special Foundations	N/A		—	—	—	
1030	Slab on Grade	4" reinforced concrete with vapor barrier and granular base		S.F. Slab	5.44	.91	2.2%
2010	Basement Excavation	Site preparation for slab and trench for foundation wall and footing		S.F. Ground	.32	.05	
2020	Basement Walls	4' foundation wall		L.F. Wall	84	.70	
B. SHELL							
	B10 Superstructure						
1010	Floor Construction	Open web steel joists, slab form, concrete, steel columns		S.F. Floor	24.59	20.49	16.6%
1020	Roof Construction	Open web steel joists with rib metal deck, steel columns		S.F. Roof	9.66	1.61	
	B20 Exterior Enclosure						
2010	Exterior Walls	Face brick with concrete block backup	86% of wall	S.F. Wall	23.83	10.59	
2020	Exterior Windows	Aluminum horizontal sliding	14% of wall	Each	528	2.55	10.1%
2030	Exterior Doors	Aluminum and glass		Each	4888	.33	
	B30 Roofing						
3010	Roof Coverings	Built-up tar and gravel with flashing; perlite/EPS composite insulation		S.F. Roof	7.26	1.21	0.9%
3020	Roof Openings	N/A		—	—	—	
C. INTERIORS							
1010	Partitions	Gypsum board and sound deadening board on metal studs	8 S.F. Floor/L.F. Partitions	S.F. Partition	7.42	7.42	
1020	Interior Doors	15% solid core wood, 85% hollow core wood	80 S.F Floor/Door	Each	682	8.53	
1030	Fittings	Kitchen cabinets		S.F. Floor	3.92	3.92	
2010	Stair Construction	Concrete filled metal pan		Flight	11,600	3.09	26.9%
3010	Wall Finishes	70% paint, 25% vinyl wall covering, 5% ceramic tile		S.F. Surface	1.63	3.25	
3020	Floor Finishes	60% carpet, 30% vinyl composition tile, 10% ceramic tile		S.F. Floor	5.42	5.42	
3030	Ceiling Finishes	Painted gypsum board on resilient channels		S.F. Ceiling	4.23	4.23	
D. SERVICES							
	D10 Conveying						
1010	Elevators & Lifts	Two geared passenger elevators		Each	229,500	7.65	5.7%
1020	Escalators & Moving Walks	N/A		—	—	—	
	D20 Plumbing						
2010	Plumbing Fixtures	Kitchen, bathroom and service fixtures, supply and drainage	1 Fixture/215 S.F. Floor	Each	2872	13.36	
2020	Domestic Water Distribution	Gas fired water heater		S.F. Floor	4.60	4.60	13.6%
2040	Rain Water Drainage	Roof drains		S.F. Roof	1.44	.24	
	D30 HVAC						
3010	Energy Supply	Oil fired hot water, baseboard radiation		S.F. Floor	7.83	7.83	
3020	Heat Generating Systems	N/A		—	—	—	
3030	Cooling Generating Systems	Chilled water, air cooled condenser system		S.F. Floor	9.54	9.54	13.0%
3050	Terminal & Package Units	N/A		—	—	—	
3090	Other HVAC Sys. & Equipment	N/A		—	—	—	
	D40 Fire Protection						
4010	Sprinklers	Wet pipe sprinkler system		S.F. Floor	2.85	2.85	2.8%
4020	Standpipes	Standpipe, fire pump		S.F. Floor	.89	.89	
	D50 Electrical						
5010	Electrical Service/Distribution	1600 ampere service, panel board and feeders		S.F. Floor	1.94	1.94	
5020	Lighting & Branch Wiring	Incandescent fixtures, receptacles, switches, A.C. and misc. power		S.F. Floor	7.22	7.22	8.3%
5030	Communications & Security	Addressable alarm system, internet wiring, emergency lighting, and intercom		S.F. Floor	1.69	1.69	
5090	Other Electrical Systems	Emergency generator, 30 kW		S.F. Floor	.18	.18	
E. EQUIPMENT & FURNISHINGS							
1010	Commercial Equipment	N/A		—	—	—	
1020	Institutional Equipment	N/A		—	—	—	
1030	Vehicular Equipment	N/A		—	—	—	0.0 %
1090	Other Equipment	N/A		—	—	—	
F. SPECIAL CONSTRUCTION							
1020	Integrated Construction	N/A		—	—	—	
1040	Special Facilities	N/A		—	—	—	0.0 %
G. BUILDING SITEWORK	**N/A**						

			Sub-Total	133.53	100%
CONTRACTOR FEES (General Requirements: 10%, Overhead: 5%, Profit: 10%)			25%	33.39	
ARCHITECT FEES			7%	11.68	

Total Building Cost	**178.60**

Costs per square foot of floor area

Exterior Wall	S.F. Area	95000	112000	129000	145000	170000	200000	275000	400000	600000
	L.F. Perimeter	345	386	406	442	480	510	530	660	800
Ribbed Precast Concrete Panel	Steel Frame	239.45	236.15	231.90	230.15	226.95	223.25	215.25	211.30	207.10
	R/Conc. Frame	230.85	227.70	223.75	222.10	219.05	215.60	208.20	204.50	200.65
Face Brick with Concrete Block Back-up	Steel Frame	219.90	217.20	213.80	212.40	209.85	207.00	200.95	197.80	194.60
	R/Conc. Frame	223.00	220.30	217.00	215.60	213.05	210.15	204.10	201.00	197.75
Stucco on Concrete Block	Steel Frame	208.95	206.80	204.35	203.25	201.40	199.30	195.15	192.85	190.65
	R/Conc. Frame	212.10	209.95	207.50	206.35	204.50	202.45	198.25	196.00	193.75
Perimeter Adj., Add or Deduct	Per 100 L.F.	11.90	10.20	8.80	7.90	6.65	5.70	4.10	2.85	1.90
Story Hgt. Adj., Add or Deduct	Per 1 Ft.	3.75	3.55	3.30	3.20	2.95	2.65	2.00	1.70	1.35
For Basement, add $37.75 per square foot of basement area										

The above costs were calculated using the basic specifications shown on the facing page. These costs should be adjusted where necessary for design alternatives and owner's requirements. Reported completed project costs, for this type of structure, range from $95.95 to $225.20 per S.F.

Common additives

Description	Unit	$ Cost
Appliances		
Cooking range, 30" free standing		
1 oven	Ea.	580 - 2450
2 oven	Ea.	1425 - 2900
30" built-in		
1 oven	Ea.	860 - 2075
2 oven	Ea.	1750 - 2550
Counter top cook tops, 4 burner	Ea.	435 - 1675
Microwave oven	Ea.	296 - 840
Combination range, refrig. & sink, 30" wide	Ea.	1900 - 3000
72" wide	Ea.	2900
Combination range, refrigerator, sink,		
microwave oven & icemaker	Ea.	6700
Compactor, residential, 4-1 compaction	Ea.	825 - 1325
Dishwasher, built-in, 2 cycles	Ea.	600 - 1150
4 cycles	Ea.	755 - 1875
Garbage disposer, sink type	Ea.	232 - 380
Hood for range, 2 speed, vented, 30" wide	Ea.	330 - 1275
42" wide	Ea.	435 - 2650
Refrigerator, no frost 10-12 C.F.	Ea.	560 - 650
18-20 C.F.	Ea.	805 - 1800

Description	Unit	$ Cost
Closed Circuit Surveillance, One station		
Camera and monitor	Ea.	1975
For additional camera stations, add	Ea.	1075
Elevators, Electric passenger, 10 stops		
3000# capacity	Ea.	403,500
4000# capacity	Ea.	406,000
5000# capacity	Ea.	411,000
Additional stop, add	Ea.	10,200
Emergency Lighting, 25 watt, battery operated		
Lead battery	Ea.	330
Nickel cadmium	Ea.	780
Laundry Equipment		
Dryer, gas, 16 lb. capacity	Ea.	975
30 lb. capacity	Ea.	3850
Washer, 4 cycle	Ea.	1200
Commercial	Ea.	1600
Smoke Detectors		
Ceiling type	Ea.	227
Duct type	Ea.	565

Important: See the Reference Section for Location Factors.

Model costs calculated for a 15 story building with 10'-6" story height and 145,000 square feet of floor area

				Unit	Unit Cost	Cost Per S.F.	% Of Sub-Total
A.	**SUBSTRUCTURE**						
1010	Standard Foundations	CIP concrete pile caps		S.F. Ground	9.60	.64	
1020	Special Foundations	Steel H-piles, concrete grade beams		S.F. Ground	205	13.67	
1030	Slab on Grade	4" reinforced concrete with vapor barrier and granular base		S.F. Slab	5.44	.36	8.6%
2010	Basement Excavation	Site preparation for slab, piles and grade beam		S.F. Ground	.32	.02	
2020	Basement Walls	4' Foundation wall		L.F. Wall	79	.28	
B.	**SHELL**						
	B10 Superstructure						
1010	Floor Construction	Open web steel joists, slab form, concrete, interior steel columns		S.F. Floor	22.55	21.05	12.4%
1020	Roof Construction	Open web steel joists with rib metal deck, interior steel columns		S.F. Roof	7.20	.48	
	B20 Exterior Enclosure						
2010	Exterior Walls	Ribbed precast concrete panel	87% of wall	S.F. Wall	46.28	19.33	
2020	Exterior Windows	Aluminum horizontal sliding	13% of wall	Each	528	2.20	14.0%
2030	Exterior Doors	Aluminum and glass		Each	3240	2.77	
	B30 Roofing						
3010	Roof Coverings	Built-up tar and gravel with flashing; perlite/EPS composite insulation		S.F. Roof	6.90	.46	0.3%
3020	Roof Openings	N/A		—	—	—	
C.	**INTERIORS**						
1010	Partitions	Gypsum board on concrete block and metal studs	10 S.F. of Floor/L.F. Partition	S.F. Partition	14.86	14.86	
1020	Interior Doors	15% solid core wood, 85% hollow core wood	80 S.F. Floor/Door	Each	698	8.72	
1030	Fittings	Kitchen cabinets		S.F. Floor	4.06	4.06	
2010	Stair Construction	Concrete filled metal pan		Flight	11,600	3.44	25.3%
3010	Wall Finishes	70% paint, 25% vinyl wall covering, 5% ceramic tile		S.F. Surface	1.61	3.22	
3020	Floor Finishes	60% carpet, 30% vinyl composition tile, 10% ceramic tile		S.F. Floor	5.42	5.42	
3030	Ceiling Finishes	Painted gypsum board on resilient channels		S.F. Ceiling	4.23	4.23	
D.	**SERVICES**						
	D10 Conveying						
1010	Elevators & Lifts	Four geared passenger elevators		Each	464,725	12.82	7.4%
1020	Escalators & Moving Walks	N/A		—	—	—	
	D20 Plumbing						
2010	Plumbing Fixtures	Kitchen, bathroom and service fixtures, supply and drainage	1 Fixture/210 S.F. Floor	Each	3219	15.33	
2020	Domestic Water Distribution	Gas fired water heater		S.F. Floor	5.30	5.30	12.0%
2040	Rain Water Drainage	Roof drains		S.F. Roof	2.85	.19	
	D30 HVAC						
3010	Energy Supply	Oil fired hot water, baseboard radiation		S.F. Floor	7.83	7.83	
3020	Heat Generating Systems	N/A		—	—	—	
3030	Cooling Generating Systems	Chilled water, air cooled condenser system		S.F. Floor	9.54	9.54	10.0%
3050	Terminal & Package Units	N/A		—	—	—	
3090	Other HVAC Sys. & Equipment	N/A		—	—	—	
	D40 Fire Protection						
4010	Sprinklers	Wet pipe sprinkler system		S.F. Floor	2.88	2.88	2.7%
4020	Standpipes	Standpipe, fire pumps		S.F. Floor	1.75	1.75	
	D50 Electrical						
5010	Electrical Service/Distribution	4000 ampere service, panel board and feeders		S.F. Floor	1.69	1.69	
5020	Lighting & Branch Wiring	Incandescent fixtures, receptacles, switches, A.C. and misc. power		S.F. Floor	7.82	7.82	
5030	Communications & Security	Addressable alarm systems, internet wiring, em. lighting, intercom & security TV		S.F. Floor	3.17	3.17	7.4%
5090	Other Electrical Systems	Emergency generator, 30 kW		S.F. Floor	.19	.19	
E.	**EQUIPMENT & FURNISHINGS**						
1010	Commercial Equipment	N/A		—	—	—	
1020	Institutional Equipment	N/A		—	—	—	
1030	Vehicular Equipment	N/A		—	—	—	0.0%
1090	Other Equipment	N/A		—	—	—	
F.	**SPECIAL CONSTRUCTION**						
1020	Integrated Construction	N/A		—	—	—	
1040	Special Facilities	N/A		—	—	—	0.0%
G.	**BUILDING SITEWORK**	**N/A**					

	Sub-Total	173.72	100%
CONTRACTOR FEES (General Requirements: 10%, Overhead: 5%, Profit: 10%)	25%	43.40	
ARCHITECT FEES	6%	13.03	
Total Building Cost		**230.15**	

Costs per square foot of floor area

Exterior Wall	S.F. Area	2000	3000	4000	5000	7500	10000	15000	25000	50000
	L.F. Perimeter	188	224	260	300	356	400	504	700	1200
Brick Veneer	Wood Frame	242.00	215.30	201.95	194.45	180.75	173.10	166.20	160.40	156.10
	Steel Frame	268.90	242.65	229.45	222.05	208.60	201.10	194.35	188.65	184.40
Brick Veneer on Block	Wood Truss	245.45	217.40	203.40	195.55	180.90	172.80	165.45	159.25	154.70
	Bearing Wall	280.40	252.35	238.30	230.45	215.80	207.65	200.35	194.10	189.55
Wood Siding	Wood Frame	227.00	203.05	191.05	184.25	172.20	165.60	159.55	154.50	150.70
Vinyl Siding	Wood Frame	226.90	203.35	191.55	184.80	173.10	166.65	160.80	155.90	152.25
Perimeter Adj., Add or Deduct	Per 100 L.F.	31.25	20.80	15.60	12.45	8.35	6.30	4.15	2.50	1.25
Story Hgt. Adj., Add or Deduct	Per 1 Ft.	3.85	3.05	2.70	2.45	1.90	1.60	1.35	1.10	1.00
For Basement, add $33.20 per square foot of basement area										

The above costs were calculated using the basic specifications shown on the facing page. These costs should be adjusted where necessary for design alternatives and owner's requirements. Reported completed project costs, for this type of structure, range from $94.25 to $236.40 per S.F.

Common additives

Description	Unit	$ Cost
Appliances		
Cooking range, 30" free standing		
1 oven	Ea.	580 - 2450
2 oven	Ea.	1425 - 2900
30" built-in		
1 oven	Ea.	860 - 2075
2 oven	Ea.	1750 - 2550
Counter top cook tops, 4 burner	Ea.	435 - 1675
Microwave oven	Ea.	296 - 840
Combination range, refrig. & sink, 30" wide	Ea.	1900 - 3000
60" wide	Ea.	2675
72" wide	Ea.	2900
Combination range, refrigerator, sink, microwave oven & icemaker	Ea.	6700
Compactor, residential, 4-1 compaction	Ea.	825 - 1325
Dishwasher, built-in, 2 cycles	Ea.	600 - 1150
4 cycles	Ea.	755 - 1875
Garbage disposer, sink type	Ea.	232 - 380
Hood for range, 2 speed, vented, 30" wide	Ea.	330 - 1275
42" wide	Ea.	435 - 2650

Description	Unit	$ Cost
Appliances, cont.		
Refrigerator, no frost 10-12 C.F.	Ea.	560 - 650
14-16 C.F.	Ea.	660 - 870
18-20 C.F.	Ea.	805 - 1800
Laundry Equipment		
Dryer, gas, 16 lb. capacity	Ea.	975
30 lb. capacity	Ea.	3850
Washer, 4 cycle	Ea.	1200
Commercial	Ea.	1600
Sound System		
Amplifier, 250 watts	Ea.	2050
Speaker, ceiling or wall	Ea.	213
Trumpet	Ea.	410

Important: See the Reference Section for Location Factors.

Model costs calculated for a 1 story building with 10' story height and 10,000 square feet of floor area

Assisted - Senior Living

				Unit	Unit Cost	Cost Per S.F.	% Of Sub-Total
A.	**SUBSTRUCTURE**						
1010	Standard Foundations	Poured concrete strip footings		S.F. Ground	2.62	2.62	
1020	Special Foundations	N/A		—	—	—	
1030	Slab on Grade	4" reinforced concrete with vapor barrier and granular base		S.F. Slab	5.44	5.44	9.6%
2010	Basement Excavation	Site preparation for slab and trench for foundation wall and footing		S.F. Ground	.57	.57	
2020	Basement Walls	4' foundation wall		L.F. Wall	84	3.34	
B.	**SHELL**						
	B10 Superstructure						
1010	Floor Construction	N/A		—	—	—	6.5%
1020	Roof Construction	Plywood on wood trusses (pitched); fiberglass batt insulation		S.F. Roof	7.46	7.46	
	B20 Exterior Enclosure						
2010	Exterior Walls	Brick veneer on wood studs, insulated	80% of wall	S.F. Wall	24.59	7.87	
2020	Exterior Windows	Double hung wood	20% of wall	Each	638	3.19	10.1%
2030	Exterior Doors	Aluminum and glass; 18 ga. steel flush		Each	2467	1.48	
	B30 Roofing						
3010	Roof Coverings	Asphalt shingles with flashing (pitched); al. gutters & downspouts		S.F. Roof	2.93	2.93	2.3%
3020	Roof Openings	N/A		—	—	—	
C.	**INTERIORS**						
1010	Partitions	Gypsum board on wood studs	6 S.F. Floor/L.F. Partition	S.F. Partition	5.39	7.19	
1020	Interior Doors	Single leaf wood	80 S.F. Floor/Door	Each	616	7.25	
1030	Fittings	N/A		—	—	—	
2010	Stair Construction	N/A		—	—	—	26.5%
3010	Wall Finishes	Paint on drywall		S.F. Surface	.85	2.27	
3020	Floor Finishes	80% carpet, 20% ceramic tile		S.F. Floor	7.30	7.30	
3030	Ceiling Finishes	Gypsum board on wood furring, painted		S.F. Ceiling	9	9	
D.	**SERVICES**						
	D10 Conveying						
1010	Elevators & Lifts	N/A		—	—	—	0.0%
1020	Escalators & Moving Walks	N/A		—	—	—	
	D20 Plumbing						
2010	Plumbing Fixtures	Kitchen, toilet and service fixtures, supply and drainage	1 Fixture/200 S.F. Floor	Each	2964	14.82	
2020	Domestic Water Distribution	Gas water heater		S.F. Floor	2.28	2.28	13.7%
2040	Rain Water Drainage	N/A		—	—	—	
	D30 HVAC						
3010	Energy Supply	N/A		—	—	—	
3020	Heat Generating Systems	N/A		—	—	—	
3030	Cooling Generating Systems	N/A		—	—	—	3.5%
3050	Terminal & Package Units	Gas fired hot water boiler; terminal package heat/A/C units		S.F. Floor	4.35	4.35	
3090	Other HVAC Sys. & Equipment	N/A		—	—	—	
	D40 Fire Protection						
4010	Sprinklers	Wet pipe sprinkler system		S.F. Floor	3.67	3.67	3.6%
4020	Standpipes	Standpipes		S.F. Floor	.86	.86	
	D50 Electrical						
5010	Electrical Service/Distribution	600 ampere service, panel board, feeders and branch wiring		S.F. Floor	8.55	8.55	
5020	Lighting & Branch Wiring	Fuorescent & incandescent fixtures, receptacles, switches, A.C. and misc. power		S.F. Floor	11.55	11.55	23.0%
5030	Communications & Security	Addressable alarm, fire detection and command center, communication system		S.F. Floor	8.38	8.38	
5090	Other Electrical Systems	Emergency generator, 7.5 kW		S.F. Floor	.19	.19	
E.	**EQUIPMENT & FURNISHINGS**						
1010	Commercial Equipment	Residential washers and dryers		S.F. Floor	.44	.44	
1020	Institutional Equipment	N/A		—	—	—	1.3%
1030	Vehicular Equipment	N/A		—	—	—	
1090	Other Equipment	Residential ranges & refrigerators		S.F. Floor	1.15	1.15	
F.	**SPECIAL CONSTRUCTION**						
1020	Integrated Construction	N/A		—	—	—	0.0%
1040	Special Facilities	N/A		—	—	—	
G.	**BUILDING SITEWORK**	**N/A**					

	Sub-Total	124.75	100%
CONTRACTOR FEES (General Requirements: 10%, Overhead: 5%, Profit: 10%)	25%	31.20	
ARCHITECT FEES	11%	17.15	
Total Building Cost		**173.10**	

Costs per square foot of floor area

Exterior Wall	S.F. Area	12000	15000	18000	21000	24000	27000	30000	33000	36000
	L.F. Perimeter	440	500	540	590	640	665	700	732	770
Face Brick with Concrete Block Back-up	Steel Frame	203.45	197.70	192.00	188.65	186.20	182.80	180.60	178.60	177.30
	Bearing Wall	196.35	191.20	186.15	183.20	180.95	177.95	175.95	174.20	173.00
Precast Concrete	Steel Frame	206.85	200.75	194.70	191.20	188.55	184.95	182.60	180.50	179.15
Decorative Concrete Block	Bearing Wall	183.25	179.35	175.45	173.25	171.55	169.25	167.80	166.40	165.60
Concrete Block	Steel Frame	186.65	183.85	181.05	179.50	178.25	176.70	175.60	174.65	174.05
	Bearing Wall	179.05	175.55	172.05	170.05	168.50	166.45	165.10	163.85	163.15
Perimeter Adj., Add or Deduct	Per 100 L.F.	13.80	11.00	9.15	7.90	6.95	6.15	5.50	5.05	4.55
Story Hgt. Adj., Add or Deduct	Per 1 Ft.	2.15	1.95	1.75	1.75	1.65	1.45	1.35	1.30	1.25
For Basement, add $30.15 per square foot of basement area										

The above costs were calculated using the basic specifications shown on the facing page. These costs should be adjusted where necessary for design alternatives and owner's requirements. Reported completed project costs, for this type of structure, range from $95.00 to $240.45 per S.F.

Common additives

Description	Unit	$ Cost
Closed Circuit Surveillance, One station		
Camera and monitor	Ea.	1975
For additional camera stations, add	Ea.	1075
Emergency Lighting, 25 watt, battery operated		
Lead battery	Ea.	330
Nickel cadmium	Ea.	780
Seating		
Auditorium chair, all veneer	Ea.	310
Veneer back, padded seat	Ea.	320
Upholstered, spring seat	Ea.	320
Classroom, movable chair & desk	Set	81 - 171
Lecture hall, pedestal type	Ea.	263 - 620
Smoke Detectors		
Ceiling type	Ea.	227
Duct type	Ea.	565
Sound System		
Amplifier, 250 watts	Ea.	2050
Speaker, ceiling or wall	Ea.	213
Trumpet	Ea.	410

Model costs calculated for a 1 story building with 24' story height and 24,000 square feet of floor area

				Unit	Unit Cost	Cost Per S.F.	% Of Sub-Total
A.	**SUBSTRUCTURE**						
1010	Standard Foundations	Poured concrete; strip and spread footings		S.F. Ground	1.49	1.49	
1020	Special Foundations	N/A		—	—	—	
1030	Slab on Grade	6" reinforced concrete with vapor barrier and granular base		S.F. Slab	6.47	6.47	7.5%
2010	Basement Excavation	Site preparation for slab and trench for foundation wall and footing		S.F. Ground	.18	.18	
2020	Basement Walls	4' foundation wall		L.F. Wall	93	2.48	
B.	**SHELL**						
	B10 Superstructure						
1010	Floor Construction	Open web steel joists, slab form, concrete	(balcony)	S.F. Floor	21.60	2.70	8.8%
1020	Roof Construction	Metal deck on steel truss		S.F. Roof	9.76	9.76	
	B20 Exterior Enclosure						
2010	Exterior Walls	Precast concrete panel	80% of wall (adjusted for end walls)	S.F. Wall	44.57	22.82	
2020	Exterior Windows	Glass curtain wall	20% of wall	Each	43.90	5.62	21.7%
2030	Exterior Doors	Double aluminum and glass and hollow metal		Each	4203	2.10	
	B30 Roofing						
3010	Roof Coverings	Built-up tar and gravel with flashing; perlite/EPS composite insulation		S.F. Roof	6.24	6.24	4.6%
3020	Roof Openings	Roof hatches		S.F. Roof	.23	.23	
C.	**INTERIORS**						
1010	Partitions	Concrete Block and toilet partitions	40 S.F. Floor/L.F. Partition	S.F. Partition	9.65	3.86	
1020	Interior Doors	Single leaf hollow metal	400 S.F. Floor/Door	Each	1157	2.90	
1030	Fittings	N/A		—	—	—	
2010	Stair Construction	Concrete filled metal pan		Flight	17,300	2.17	18.8%
3010	Wall Finishes	70% paint, 30% epoxy coating		S.F. Surface	5.74	4.59	
3020	Floor Finishes	70% vinyl tile, 30% carpet		S.F. Floor	7.30	7.30	
3030	Ceiling Finishes	Fiberglass board, suspended		S.F. Ceiling	5.70	5.70	
D.	**SERVICES**						
	D10 Conveying						
1010	Elevators & Lifts	One hydraulic passenger elevator		Each	92,640	3.86	2.7%
1020	Escalators & Moving Walks	N/A		—	—	—	
	D20 Plumbing						
2010	Plumbing Fixtures	Toilet and service fixtures, supply and drainage	1 Fixture/800 S.F. Floor	Each	6880	8.60	
2020	Domestic Water Distribution	Gas fired water heater		S.F. Floor	3.75	3.75	9.5%
2040	Rain Water Drainage	Roof drains		S.F. Roof	1.10	1.10	
	D30 HVAC						
3010	Energy Supply	N/A		—	—	—	
3020	Heat Generating Systems	Included in D3050		—	—	—	
3030	Cooling Generating Systems	N/A		—	—	—	9.6 %
3050	Terminal & Package Units	Single zone rooftop unit, gas heating, electric cooling		S.F. Floor	13.55	13.55	
3090	Other HVAC Sys. & Equipment	N/A		—	—	—	
	D40 Fire Protection						
4010	Sprinklers	Wet pipe sprinkler system		S.F. Floor	3.67	3.67	2.9%
4020	Standpipes	Standpipe		S.F. Floor	.36	.36	
	D50 Electrical						
5010	Electrical Service/Distribution	800 ampere service, panel board and feeders		S.F. Floor	1.90	1.90	
5020	Lighting & Branch Wiring	Fluorescent fixtures, receptacles, switches, A.C. and misc. power		S.F. Floor	12.25	12.25	13.8%
5030	Communications & Security	Addressable alarm systems, emergency lighting and public address system		S.F. Floor	4.13	4.13	
5090	Other Electrical Systems	Emergency generator, 100 kW		S.F. Floor	1.21	1.21	
E.	**EQUIPMENT & FURNISHINGS**						
1010	Commercial Equipment	N/A		—	—	—	
1020	Institutional Equipment	N/A		—	—	—	0.0 %
1030	Vehicular Equipment	N/A		—	—	—	
1090	Other Equipment	N/A		—	—	—	
F.	**SPECIAL CONSTRUCTION**						
1020	Integrated Construction	N/A		—	—	—	0.0 %
1040	Special Facilities	N/A		—	—	—	
G.	**BUILDING SITEWORK**	**N/A**					

						Sub-Total	140.99	100%

CONTRACTOR FEES (General Requirements: 10%, Overhead: 5%, Profit: 10%)		25%	35.22
ARCHITECT FEES		7%	12.34

Total Building Cost | **188.55**

Costs per square foot of floor area

Exterior Wall	S.F. Area	2000	2700	3400	4100	4800	5500	6200	6900	7600
	L.F. Perimeter	180	208	236	256	280	303	317	337	357
Face Brick with Concrete Block Back-up	Steel Frame	265.55	249.80	240.55	232.50	227.65	223.85	219.40	216.85	214.65
	R/Conc. Frame	281.40	265.65	256.40	248.35	243.50	239.70	235.30	232.70	230.55
Precast Concrete Panel	Steel Frame	269.50	253.05	243.45	235.05	229.95	226.00	221.45	218.70	216.40
	R/Conc. Frame	290.00	271.50	260.70	251.45	245.85	241.45	236.55	233.50	231.05
Limestone with Concrete Block Back-up	Steel Frame	295.25	275.25	263.50	253.15	246.85	242.00	236.35	233.00	230.15
	R/Conc. Frame	311.15	291.10	279.35	269.00	262.80	257.90	252.25	248.85	246.00
Perimeter Adj., Add or Deduct	Per 100 L.F.	49.20	36.45	28.90	24.00	20.50	17.90	15.90	14.25	12.95
Story Hgt. Adj., Add or Deduct	Per 1 Ft.	4.65	4.00	3.60	3.20	3.00	2.85	2.70	2.50	2.45
For Basement, add $34.85 per square foot of basement area										

The above costs were calculated using the basic specifications shown on the facing page. These costs should be adjusted where necessary for design alternatives and owner's requirements. Reported completed project costs, for this type of structure, range from $144.85 to $356.50 per S.F.

Common additives

Description	Unit	$ Cost	Description	Unit	$ Cost
Bulletproof Teller Window, 44" x 60"	Ea.	5175	Service Windows, Pass thru, steel		
60" x 48"	Ea.	7075	24" x 36"	Ea.	3850
Closed Circuit Surveillance, One station			48" x 48"	Ea.	4200
Camera and monitor	Ea.	1975	72" x 40"	Ea.	6500
For additional camera stations, add	Ea.	1075	Smoke Detectors		
Counters, Complete	Station	6275	Ceiling type	Ea.	227
Door & Frame, 3' x 6'-8", bullet resistant steel			Duct type	Ea.	565
with vision panel	Ea.	7450 - 9625	Twenty-four Hour Teller		
Drive-up Window, Drawer & micr., not incl. glass	Ea.	9075 - 13,000	Automatic deposit cash & memo	Ea.	54,500
Emergency Lighting, 25 watt, battery operated			Vault Front, Door & frame		
Lead battery	Ea.	330	1 hour test, 32"x 78"	Opng.	8275
Nickel cadmium	Ea.	780	2 hour test, 32" door	Opng.	10,100
Night Depository	Ea.	9775 - 14,700	40" door	Opng.	11,700
Package Receiver, painted	Ea.	1975	4 hour test, 32" door	Opng.	10,800
stainless steel	Ea.	3025	40" door	Opng.	12,600
Partitions, Bullet resistant to 8' high	L.F.	335 - 535	Time lock, two movement, add	Ea.	2300
Pneumatic Tube Systems, 2 station	Ea.	33,800			
With TV viewer	Ea.	62,000			

Important: See the Reference Section for Location Factors.

Model costs calculated for a 1 story building with 14' story height and 4,100 square feet of floor area

Bank

				Unit	Unit Cost	Cost Per S.F.	% Of Sub-Total
A. SUBSTRUCTURE							
1010	Standard Foundations	Poured concrete; strip and spread footings		S.F. Ground	4.33	4.33	
1020	Special Foundations	N/A		—	—	—	
1030	Slab on Grade	4" reinforced concrete with vapor barrier and granular base		S.F. Slab	5.44	5.44	8.9%
2010	Basement Excavation	Site preparation for slab and trench for foundation wall and footing		S.F. Ground	.32	.32	
2020	Basement Walls	4' Foundation wall		L.F. Wall	84	6.25	
B. SHELL							
	B10 Superstructure						
1010	Floor Construction	Cast-in-place columns		L.F. Column	124	6.05	12.0%
1020	Roof Construction	Cast-in-place concrete flat plate		S.F. Roof	16.10	16.10	
	B20 Exterior Enclosure						
2010	Exterior Walls	Face brick with concrete block backup	80% of wall	S.F. Wall	33.45	23.39	
2020	Exterior Windows	Horizontal aluminum sliding	20% of wall	Each	588	6.85	17.3%
2030	Exterior Doors	Double aluminum and glass and hollow metal		Each	3338	1.63	
	B30 Roofing						
3010	Roof Coverings	Built-up tar and gravel with flashing; perlite/EPS composite insulation		S.F. Roof	7.80	7.80	4.2%
3020	Roof Openings	N/A		—	—	—	
C. INTERIORS							
1010	Partitions	Gypsum board on metal studs	20 S.F. of Floor/L.F. Partition	S.F. Partition	10.80	5.40	
1020	Interior Doors	Single leaf hollow metal	200 S.F. Floor/Door	Each	1157	5.79	
1030	Fittings	N/A		—	—	—	
2010	Stair Construction	N/A		—	—	—	13.6%
3010	Wall Finishes	50% vinyl wall covering, 50% paint		S.F. Surface	1.45	1.45	
3020	Floor Finishes	50% carpet tile, 40% vinyl composition tile, 10% quarry tile		S.F. Floor	5.34	5.34	
3030	Ceiling Finishes	Mineral fiber tile on concealed zee bars		S.F. Ceiling	7.03	7.03	
D. SERVICES							
	D10 Conveying						
1010	Elevators & Lifts	N/A		—	—	—	0.0 %
1020	Escalators & Moving Walks	N/A		—	—	—	
	D20 Plumbing						
2010	Plumbing Fixtures	Toilet and service fixtures, supply and drainage	1 Fixture/580 S.F. Floor	Each	5574	9.61	
2020	Domestic Water Distribution	Gas fired water heater		S.F. Floor	1.33	1.33	6.8%
2040	Rain Water Drainage	Roof drains		S.F. Roof	1.59	1.59	
	D30 HVAC						
3010	Energy Supply	N/A		—	—	—	
3020	Heat Generating Systems	Included in D3050		—	—	—	
3030	Cooling Generating Systems	N/A		—	—	—	6.7 %
3050	Terminal & Package Units	Single zone rooftop unit, gas heating, electric cooling		S.F. Floor	12.40	12.40	
3090	Other HVAC Sys. & Equipment	N/A		—	—	—	
	D40 Fire Protection						
4010	Sprinklers	Wet pipe sprinkler system		S.F. Floor	4.98	4.98	4.3%
4020	Standpipes	Standpipe		S.F. Floor	2.94	2.94	
	D50 Electrical						
5010	Electrical Service/Distribution	200 ampere service, panel board and feeders		S.F. Floor	2.70	2.70	
5020	Lighting & Branch Wiring	Fluorescent fixtures, receptacles, switches, A.C. and misc. power		S.F. Floor	9.59	9.59	13.0%
5030	Communications & Security	Alarm systems, internet and phone wiring, emergency lighting, and security television		S.F. Floor	10.91	10.91	
5090	Other Electrical Systems	Emergency generator, 15 kW, Uninterruptible power supply		S.F. Floor	.63	.63	
E. EQUIPMENT & FURNISHINGS							
1010	Commercial Equipment	Automatic teller, drive up window, night depository		S.F. Floor	8.02	8.02	
1020	Institutional Equipment	Closed circuit TV monitoring system		S.F. Floor	2.93	2.93	6.0%
1030	Vehicular Equipment	N/A		—	—	—	
1090	Other Equipment	N/A		—	—	—	
F. SPECIAL CONSTRUCTION							
1020	Integrated Construction	N/A		—	—	—	7.2 %
1040	Special Facilities	Security vault door		S.F. Floor	13.17	13.17	
G. BUILDING SITEWORK	**N/A**						

		Sub-Total	183.97	100%
CONTRACTOR FEES (General Requirements: 10%, Overhead: 5%, Profit: 10%)		25%	45.98	
ARCHITECT FEES		8%	18.40	

	Total Building Cost	**248.35**

For customer support on your Square Foot Cost Data, call 877.694.8329.

89

Costs per square foot of floor area

Exterior Wall	S.F. Area	12000	14000	16000	18000	20000	22000	24000	26000	28000
	L.F. Perimeter	460	491	520	540	566	593	620	645	670
Concrete Block	Steel Roof Deck	173.90	170.75	168.25	166.05	164.45	163.20	162.15	161.15	160.35
Decorative Concrete Block	Steel Roof Deck	178.45	174.95	172.10	169.65	167.90	166.40	165.25	164.20	163.25
Face Brick on Concrete Block	Steel Roof Deck	186.80	182.55	179.25	176.15	174.00	172.25	170.85	169.55	168.45
Jumbo Brick on Concrete Block	Steel Roof Deck	183.25	179.30	176.20	173.40	171.40	169.80	168.45	167.30	166.25
Stucco on Concrete Block	Steel Roof Deck	177.40	173.95	171.30	168.85	167.10	165.70	164.55	163.50	162.65
Precast Concrete Panel	Steel Roof Deck	182.30	178.45	175.40	172.70	170.70	169.15	167.85	166.65	165.65
Perimeter Adj., Add or Deduct	Per 100 L.F.	5.30	4.55	4.05	3.55	3.20	2.90	2.70	2.55	2.35
Story Hgt. Adj., Add or Deduct	Per 1 Ft.	1.20	1.05	1.00	0.90	0.90	0.80	0.80	0.80	0.80
For Basement, add $34.35 per square foot of basement area										

The above costs were calculated using the basic specifications shown on the facing page. These costs should be adjusted where necessary for design alternatives and owner's requirements. Reported completed project costs, for this type of structure, range from $64.05 to $168.70 per S.F.

Common additives

Description	Unit	$ Cost
Bowling Alleys, Incl. alley, pinsetter,		
scorer, counter & misc. supplies, average	Lane	63,000
For automatic scorer, add (maximum)	Lane	11,000
Emergency Lighting, 25 watt, battery operated		
Lead battery	Ea.	330
Nickel cadmium	Ea.	780
Lockers, Steel, Single tier, 60" or 72"	Opng.	221 - 370
2 tier, 60" or 72" total	Opng.	120 - 165
5 tier, box lockers	Opng.	77 - 86.50
Locker bench, lam., maple top only	L.F.	34
Pedestals, steel pipe	Ea.	76
Seating		
Auditorium chair, all veneer	Ea.	310
Veneer back, padded seat	Ea.	320
Upholstered, spring seat	Ea.	320
Sound System		
Amplifier, 250 watts	Ea.	2050
Speaker, ceiling or wall	Ea.	213
Trumpet	Ea.	410

Important: See the Reference Section for Location Factors.

Model costs calculated for a 1 story building with 14' story height and 20,000 square feet of floor area

Bowling Alley

				Unit	Unit Cost	Cost Per S.F.	% Of Sub-Total
A. SUBSTRUCTURE							
1010	Standard Foundations	Poured concrete; strip and spread footings		S.F. Ground	1.27	1.27	
1020	Special Foundations	N/A		—	—	—	
1030	Slab on Grade	4" reinforced concrete with vapor barrier and granular base		S.F. Slab	5.44	5.44	7.8%
2010	Basement Excavation	Site preparation for slab and trench for foundation wall and footing		S.F. Ground	.57	.57	
2020	Basement Walls	4' Foundation wall		L.F. Wall	84	2.36	
B. SHELL							
	B10 Superstructure						
1010	Floor Construction	N/A		—	—	—	7.9 %
1020	Roof Construction	Metal deck on open web steel joists with columns and beams		S.F. Roof	9.84	9.84	
	B20 Exterior Enclosure						
2010	Exterior Walls	Concrete block	90% of wall	S.F. Wall	13.24	4.72	
2020	Exterior Windows	Horizontal pivoted	10% of wall	Each	2020	3.33	7.0%
2030	Exterior Doors	Double aluminum and glass and hollow metal		Each	2460	.62	
	B30 Roofing						
3010	Roof Coverings	Built-up tar and gravel with flashing; perlite/EPS composite insulation		S.F. Roof	6	6	5.0%
3020	Roof Openings	Roof hatches		S.F. Roof	.17	.17	
C. INTERIORS							
1010	Partitions	Concrete block	50 S.F. Floor/L.F. Partition	S.F. Partition	10.75	2.15	
1020	Interior Doors	Hollow metal single leaf	1000 S.F. Floor/Door	Each	1157	1.16	
1030	Fittings	N/A		—	—	—	
2010	Stair Construction	N/A		—	—	—	6.4%
3010	Wall Finishes	Paint and block filler		S.F. Surface	6.33	2.53	
3020	Floor Finishes	Vinyl tile	25% of floor	S.F. Floor	2.52	.63	
3030	Ceiling Finishes	Suspended fiberglass board	25% of area	S.F. Ceiling	5.70	1.42	
D. SERVICES							
	D10 Conveying						
1010	Elevators & Lifts	N/A		—	—	—	0.0 %
1020	Escalators & Moving Walks	N/A		—	—	—	
	D20 Plumbing						
2010	Plumbing Fixtures	Toilet and service fixtures, supply and drainage	1 Fixture/2200 S.F. Floor	Each	3564	1.62	
2020	Domestic Water Distribution	Gas fired water heater		S.F. Floor	.24	.24	2.1%
2040	Rain Water Drainage	Roof drains		S.F. Roof	.70	.70	
	D30 HVAC						
3010	Energy Supply	N/A		—	—	—	
3020	Heat Generating Systems	Included in D3050		—	—	—	
3030	Cooling Generating Systems	N/A		—	—	—	12.4 %
3050	Terminal & Package Units	Single zone rooftop unit, gas heating, electric cooling		S.F. Floor	15.35	15.35	
3090	Other HVAC Sys. & Equipment	N/A		—	—	—	
	D40 Fire Protection						
4010	Sprinklers	Sprinklers, light hazard		S.F. Floor	3.67	3.67	3.4%
4020	Standpipes	Standpipe, wet, Class III		S.F. Floor	.61	.61	
	D50 Electrical						
5010	Electrical Service/Distribution	800 ampere service, panel board and feeders		S.F. Floor	2.28	2.28	
5020	Lighting & Branch Wiring	High efficiency fluorescent fixtures, receptacles, switches, A.C. and misc. power		S.F. Floor	11.12	11.12	11.9%
5030	Communications & Security	Addressable alarm systems and emergency lighting		S.F. Floor	1.35	1.35	
5090	Other Electrical Systems	Emergency generator, 7.5 kW		S.F. Floor	.08	.08	
E. EQUIPMENT & FURNISHINGS							
1010	Commercial Equipment	N/A		—	—	—	
1020	Institutional Equipment	N/A		—	—	—	36.2 %
1030	Vehicular Equipment	N/A		—	—	—	
1090	Other Equipment	Bowling Alley, ballrack, automatic scorers		S.F. Floor	44.90	44.90	
F. SPECIAL CONSTRUCTION							
1020	Integrated Construction	N/A		—	—	—	0.0 %
1040	Special Facilities	N/A		—	—	—	
G. BUILDING SITEWORK	**N/A**						

		Sub-Total	124.13	100%
CONTRACTOR FEES (General Requirements: 10%, Overhead: 5%, Profit: 10%)		25%	31.01	
ARCHITECT FEES		6%	9.31	

Total Building Cost	**164.45**

For customer support on your Square Foot Cost Data, call 877.694.8329.

Costs per square foot of floor area

Exterior Wall	S.F. Area	6000	8000	10000	12000	14000	16000	18000	20000	22000
	L.F. Perimeter	320	386	453	475	505	530	570	600	630
Face Brick with Concrete Block Back-up	Bearing Walls	180.35	174.05	170.40	164.70	161.15	158.25	156.60	154.95	153.55
	Steel Frame	183.40	177.25	173.60	168.05	164.50	161.60	160.05	158.45	157.05
Decorative Concrete Block	Bearing Walls	171.30	165.90	162.70	158.00	155.05	152.60	151.30	149.90	148.70
	Steel Frame	174.35	169.10	165.95	161.30	158.40	156.00	154.70	153.35	152.20
Precast Concrete Panels	Bearing Walls	179.10	173.00	169.30	163.80	160.30	157.45	155.90	154.35	152.90
	Steel Frame	182.15	176.15	172.50	167.10	163.65	160.85	159.30	157.75	156.35
Perimeter Adj., Add or Deduct	Per 100 L.F.	14.30	10.80	8.55	7.15	6.15	5.30	4.80	4.35	3.95
Story Hgt. Adj., Add or Deduct	Per 1 Ft.	2.35	2.20	2.00	1.75	1.60	1.45	1.45	1.40	1.30
Basement—Not Applicable										

The above costs were calculated using the basic specifications shown on the facing page. These costs should be adjusted where necessary for design alternatives and owner's requirements. Reported completed project costs, for this type of structure, range from $77.40 to $186.60 per S.F.

Common additives

Description	Unit	$ Cost
Directory Boards, Plastic, glass covered		
30" x 20"	Ea.	600
36" x 48"	Ea.	1500
Aluminum, 24" x 18"	Ea.	635
36" x 24"	Ea.	755
48" x 32"	Ea.	1050
48" x 60"	Ea.	2225
Emergency Lighting, 25 watt, battery operated		
Lead battery	Ea.	330
Nickel cadmium	Ea.	780
Benches, Hardwood	L.F.	133 - 245
Ticket Printer	Ea.	7475
Turnstiles, One way		
4 arm, 46" dia., manual	Ea.	2100
Electric	Ea.	3250
High security, 3 arm		
65" dia., manual	Ea.	7625
Electric	Ea.	11,000
Gate with horizontal bars		
65" dia., 7' high transit type	Ea.	7350

Important: See the Reference Section for Location Factors.

Model costs calculated for a 1 story building with 14′ story height and 12,000 square feet of floor area

					Unit	Unit Cost	Cost Per S.F.	% Of Sub-Total
A. SUBSTRUCTURE								
1010	Standard Foundations	Poured concrete; strip and spread footings			S.F. Ground	2.45	2.45	
1020	Special Foundations	N/A			—	—	—	
1030	Slab on Grade	6″ reinforced concrete with vapor barrier and granular base			S.F. Slab	8.05	8.05	11.4%
2010	Basement Excavation	Site preparation for slab and trench for foundation wall and footing			S.F. Ground	.32	.32	
2020	Basement Walls	4′ Foundation wall			L.F. Wall	86	3.41	
B. SHELL								
B10 Superstructure								
1010	Floor Construction	N/A			—	—	—	5.4 %
1020	Roof Construction	Metal deck on open web steel joists with columns and beams			S.F. Roof	6.77	6.77	
B20 Exterior Enclosure								
2010	Exterior Walls	Face brick with concrete block backup	70% of wall		S.F. Wall	33.44	12.97	
2020	Exterior Windows	Store front	30% of wall		Each	31.90	5.31	16.3%
2030	Exterior Doors	Double aluminum and glass			Each	5825	1.94	
B30 Roofing								
3010	Roof Coverings	Built-up tar and gravel with flashing; perlite/EPS composite insulation			S.F. Roof	6.81	6.81	5.5%
3020	Roof Openings	N/A			—	—	—	
C. INTERIORS								
1010	Partitions	Lightweight concrete block	15 S.F. Floor/L.F. Partition		S.F. Partition	10.74	7.16	
1020	Interior Doors	Hollow metal	150 S.F. Floor/Door		Each	1157	7.71	
1030	Fittings	N/A			—	—	—	
2010	Stair Construction	N/A			—	—	—	27.3%
3010	Wall Finishes	Glazed coating			S.F. Surface	1.98	2.64	
3020	Floor Finishes	Quarry tile and vinyl composition tile			S.F. Floor	9.38	9.38	
3030	Ceiling Finishes	Mineral fiber tile on concealed zee bars			S.F. Ceiling	7.03	7.03	
D. SERVICES								
D10 Conveying								
1010	Elevators & Lifts	N/A			—	—	—	0.0 %
1020	Escalators & Moving Walks	N/A			—	—	—	
D20 Plumbing								
2010	Plumbing Fixtures	Toilet and service fixtures, supply and drainage	1 Fixture/850 S.F. Floor		Each	7336	8.63	
2020	Domestic Water Distribution	Electric water heater			S.F. Floor	1.57	1.57	9.2%
2040	Rain Water Drainage	Roof drains			S.F. Roof	1.23	1.23	
D30 HVAC								
3010	Energy Supply	N/A			—	—	—	
3020	Heat Generating Systems	Included in D3050			—	—	—	
3030	Cooling Generating Systems	N/A			—	—	—	9.3 %
3050	Terminal & Package Units	Single zone rooftop unit, gas heating, electric cooling			S.F. Floor	11.50	11.50	
3090	Other HVAC Sys. & Equipment	N/A			—	—	—	
D40 Fire Protection								
4010	Sprinklers	Wet pipe sprinkler system			S.F. Floor	4.73	4.73	4.6%
4020	Standpipes	Standpipe			S.F. Floor	1.01	1.01	
D50 Electrical								
5010	Electrical Service/Distribution	400 ampere service, panel board and feeders			S.F. Floor	1.84	1.84	
5020	Lighting & Branch Wiring	Fluorescent fixtures, receptacles, switches, A.C. and misc. power			S.F. Floor	9.41	9.41	11.0%
5030	Communications & Security	Alarm systems and emergency lighting			S.F. Floor	2.06	2.06	
5090	Other Electrical Systems	Emergency generator, 7.5 kW			S.F. Floor	.39	.39	
E. EQUIPMENT & FURNISHINGS								
1010	Commercial Equipment	N/A			—	—	—	
1020	Institutional Equipment	N/A			—	—	—	
1030	Vehicular Equipment	N/A			—	—	—	0.0 %
1090	Other Equipment	N/A			—	—	—	
F. SPECIAL CONSTRUCTION								
1020	Integrated Construction	N/A			—	—	—	0.0 %
1040	Special Facilities	N/A			—	—	—	
G. BUILDING SITEWORK	**N/A**							

Sub-Total		124.32	**100%**
CONTRACTOR FEES (General Requirements: 10%, Overhead: 5%, Profit: 10%)		25%	31.06
ARCHITECT FEES		6%	9.32
Total Building Cost		**164.70**	

For customer support on your Square Foot Cost Data, call 877.694.8329.

93

Costs per square foot of floor area

Exterior Wall	S.F. Area	600	800	1000	1200	1600	2000	2400	3000	4000
	L.F. Perimeter	100	114	128	139	164	189	210	235	280
Brick with Concrete Block Back-up	Steel Frame	348.70	333.95	325.00	317.55	309.35	304.50	300.15	294.60	289.45
	Bearing Walls	341.65	326.85	317.95	310.50	302.30	297.40	293.10	287.50	282.40
Concrete Block	Steel Frame	304.90	296.30	291.15	286.75	282.05	279.25	276.70	273.50	270.50
	Bearing Walls	298.70	290.15	285.00	280.60	275.90	273.10	270.55	267.35	264.35
Galvanized Steel Siding	Steel Frame	308.65	299.55	294.05	289.40	284.40	281.35	278.75	275.25	272.10
Metal Sandwich Panel	Steel Frame	303.05	293.65	288.10	283.35	278.25	275.15	272.40	268.90	265.65
Perimeter Adj., Add or Deduct	Per 100 L.F.	102.20	76.60	61.30	51.05	38.30	30.65	25.60	20.40	15.35
Story Hgt. Adj., Add or Deduct	Per 1 Ft.	6.20	5.30	4.80	4.30	3.85	3.50	3.25	2.95	2.60
Basement—Not Applicable										

The above costs were calculated using the basic specifications shown on the facing page. These costs should be adjusted where necessary for design alternatives and owner's requirements. Reported completed project costs, for this type of structure, range from $ 138.90 to $ 394.55 per S.F.

Common additives

Description	Unit	$ Cost
Air Compressors, Electric		
1-1/2 H.P., standard controls	Ea.	1600
Dual controls	Ea.	2000
5 H.P., 115/230 volt, standard control	Ea.	4475
Dual controls	Ea.	5375
Emergency Lighting, 25 watt, battery operated		
Lead battery	Ea.	330
Nickel cadmium	Ea.	780
Fence, Chain link, 6' high		
9 ga. wire, galvanized	L.F.	28
6 ga. wire	L.F.	30
Gate	Ea.	390
Product Dispenser with vapor recovery		
for 6 nozzles	Ea.	27,200
Laundry Equipment		
Dryer, gas, 16 lb. capacity	Ea.	975
30 lb. capacity	Ea.	3850
Washer, 4 cycle	Ea.	1200
Commercial	Ea.	1600

Description	Unit	$ Cost
Lockers, Steel, single tier, 60" or 72"	Opng.	221 - 370
2 tier, 60" or 72" total	Opng.	120 - 165
5 tier, box lockers	Opng.	77 - 86.50
Locker bench, lam. maple top only	L.F.	34
Pedestals, steel pipe	Ea.	76
Paving, Bituminous		
Wearing course plus base course	S.Y.	6.55
Safe, Office type, 1 hour rating		
30" x 18" x 18"	Ea.	2500
60" x 36" x 18", double door	Ea.	8125
Sidewalks, Concrete 4" thick	S.F.	4.48
Yard Lighting,		
20' aluminum pole		
with 400 watt		
high pressure sodium		
fixture	Ea.	3175

Important: See the Reference Section for Location Factors.

Model costs calculated for a 1 story building with 12' story height and 800 square feet of floor area

Car Wash

				Unit	Unit Cost	Cost Per S.F.	% Of Sub-Total
A.	**SUBSTRUCTURE**						
1010	Standard Foundations	Poured concrete; strip and spread footings		S.F. Ground	3.96	3.96	
1020	Special Foundations	N/A		—	—	—	
1030	Slab on Grade	5" reinforced concrete with vapor barrier and granular base		S.F. Slab	7.13	7.13	9.2%
2010	Basement Excavation	Site preparation for slab and trench for foundation wall and footing		S.F. Ground	1.07	1.07	
2020	Basement Walls	4' Foundation wall		L.F. Wall	75	10.69	
B.	**SHELL**						
	B10 Superstructure						
1010	Floor Construction	N/A		—	—	—	
1020	Roof Construction	Metal deck, open web steel joists, beams, columns		S.F. Roof	9.11	9.11	3.6 %
	B20 Exterior Enclosure						
2010	Exterior Walls	Face brick with concrete block backup	70% of wall	S.F. Wall	33.45	40.04	
2020	Exterior Windows	Horizontal pivoted steel	5% of wall	Each	790	7.51	26.2%
2030	Exterior Doors	Steel overhead hollow metal		Each	3548	17.74	
	B30 Roofing						
3010	Roof Coverings	Built-up tar and gravel with flashing; perlite/EPS composite insulation		S.F. Roof	8.91	8.91	3.6%
3020	Roof Openings	N/A		—	—	—	
C.	**INTERIORS**						
1010	Partitions	Concrete block	20 S.F. Floor/S.F. Partition	S.F. Partition	10.54	5.27	
1020	Interior Doors	Hollow metal	600 S.F. Floor/Door	Each	1157	1.93	
1030	Fittings	N/A		—	—	—	
2010	Stair Construction	N/A		—	—	—	3.4%
3010	Wall Finishes	Paint		S.F. Surface	1.32	1.32	
3020	Floor Finishes	N/A		—	—	—	
3030	Ceiling Finishes	N/A		—	—	—	
D.	**SERVICES**						
	D10 Conveying						
1010	Elevators & Lifts	N/A		—	—	—	0.0 %
1020	Escalators & Moving Walks	N/A		—	—	—	
	D20 Plumbing						
2010	Plumbing Fixtures	Toilet and service fixtures, supply and drainage	1 Fixture/160 S.F. Floor	Each	2837	17.73	
2020	Domestic Water Distribution	Gas fired water heater		S.F. Floor	35.17	35.17	23.0%
2040	Rain Water Drainage	Roof drains		S.F. Roof	4.42	4.42	
	D30 HVAC						
3010	Energy Supply	N/A		—	—	—	
3020	Heat Generating Systems	N/A		—	—	—	
3030	Cooling Generating Systems	N/A		—	—	—	4.2 %
3050	Terminal & Package Units	Single zone rooftop unit, gas heating, electric cooling		S.F. Floor	10.37	10.37	
3090	Other HVAC Sys. & Equipment	N/A		—	—	—	
	D40 Fire Protection						
4010	Sprinklers	N/A		—	—	—	0.0 %
4020	Standpipes	N/A		—	—	—	
	D50 Electrical						
5010	Electrical Service/Distribution	400 ampere service, panel board and feeders		S.F. Floor	24.70	24.70	
5020	Lighting & Branch Wiring	Fluorescent fixtures, receptacles, switches and misc. power		S.F. Floor	40.89	40.89	27.0%
5030	Communications & Security	N/A		—	—	—	
5090	Other Electrical Systems	Emergency generator, 7.5 kW		S.F. Floor	1.70	1.70	
E.	**EQUIPMENT & FURNISHINGS**						
1010	Commercial Equipment	N/A		—	—	—	
1020	Institutional Equipment	N/A		—	—	—	
1030	Vehicular Equipment	N/A		—	—	—	0.0 %
1090	Other Equipment	N/A		—	—	—	
F.	**SPECIAL CONSTRUCTION**						
1020	Integrated Construction	N/A		—	—	—	
1040	Special Facilities	N/A		—	—	—	0.0 %
G.	**BUILDING SITEWORK**	**N/A**					

		Sub-Total	249.66	100%
CONTRACTOR FEES (General Requirements: 10%, Overhead: 5%, Profit: 10%)			25%	62.44
ARCHITECT FEES			7%	21.85
		Total Building Cost	**333.95**	

For customer support on your Square Foot Cost Data, call 877.694.8329.

95

Costs per square foot of floor area

Exterior Wall	S.F. Area	2000	7000	12000	17000	22000	27000	32000	37000	42000
	L.F. Perimeter	180	340	470	540	640	740	762	793	860
Decorative Concrete Block	Wood Arch	265.05	196.60	181.80	170.95	166.85	164.25	159.20	155.85	154.45
	Steel Truss	265.75	197.30	182.50	171.60	167.50	164.90	159.85	156.55	155.10
Stone with Concrete Block Back-up	Wood Arch	329.15	231.15	209.70	193.55	187.50	183.75	176.15	171.15	169.00
	Steel Truss	322.75	224.80	203.35	187.20	181.15	177.40	169.80	164.75	162.65
Face Brick with Concrete Block Back-up	Wood Arch	296.40	213.50	195.45	182.00	177.00	173.75	167.50	163.30	161.55
	Steel Truss	290.00	207.15	189.10	175.65	170.60	167.40	161.10	156.95	155.20
Perimeter Adj., Add or Deduct	Per 100 L.F.	67.75	19.35	11.30	7.95	6.10	5.00	4.20	3.65	3.20
Story Hgt. Adj., Add or Deduct	Per 1 Ft.	3.95	2.10	1.80	1.35	1.25	1.20	1.00	0.90	0.85

For Basement, add $35.90 per square foot of basement area

The above costs were calculated using the basic specifications shown on the facing page. These costs should be adjusted where necessary for design alternatives and owner's requirements. Reported completed project costs, for this type of structure, range from $88.65 to $362.55 per S.F.

Common additives

Description	Unit	$ Cost
Altar, Wood, custom design, plain	Ea.	3250
Deluxe	Ea.	16,500
Granite or marble, average	Ea.	16,800
Deluxe	Ea.	47,100
Ark, Prefabricated, plain	Ea.	12,200
Deluxe	Ea.	158,500
Baptistry, Fiberglass, incl. plumbing	Ea.	7800 - 12,500
Bells & Carillons, 48 bells	Ea.	1,095,500
24 bells	Ea.	515,500
Confessional, Prefabricated wood		
Single, plain	Ea.	4475
Deluxe	Ea.	11,100
Double, plain	Ea.	8450
Deluxe	Ea.	23,800
Emergency Lighting, 25 watt, battery operated		
Lead battery	Ea.	330
Nickel cadmium	Ea.	780
Lecterns, Wood, plain	Ea.	1050
Deluxe	Ea.	7050

Description	Unit	$ Cost
Pews/Benches, Hardwood	L.F.	133 - 245
Pulpits, Prefabricated, hardwood	Ea.	1925 - 11,500
Railing, Hardwood	L.F.	252
Steeples, translucent fiberglass		
30" square, 15' high	Ea.	11,000
25' high	Ea.	12,700
Painted fiberglass, 24" square, 14' high	Ea.	9475
28' high	Ea.	9025
Aluminum		
20' high, 3'- 6" base	Ea.	10,500
35' high, 8'- 0" base	Ea.	40,400
60' high, 14'- 0" base	Ea.	85,000

Important: See the Reference Section for Location Factors.

Model costs calculated for a 1 story building with 24' story height and 17,000 square feet of floor area

Church

			Unit	Unit Cost	Cost Per S.F.	% Of Sub-Total
A. SUBSTRUCTURE						
1010	Standard Foundations	Poured concrete; strip and spread footings	S.F. Ground	4.09	4.09	
1020	Special Foundations	N/A	—	—	—	
1030	Slab on Grade	4" reinforced concrete with vapor barrier and granular base	S.F. Slab	5.44	5.44	10.3%
2010	Basement Excavation	Site preparation for slab and trench for foundation wall and footing	S.F. Ground	.18	.18	
2020	Basement Walls	4' Foundation wall	L.F. Wall	84	3.98	
B. SHELL						
	B10 Superstructure					
1010	Floor Construction	N/A	—	—	—	15.7 %
1020	Roof Construction	Wood deck on laminated wood arches	S.F. Roof	19.50	20.81	
	B20 Exterior Enclosure					
2010	Exterior Walls	Face brick with concrete block backup 80% of wall (adjusted for end walls)	S.F. Wall	39.73	24.23	
2020	Exterior Windows	Aluminum, top hinged, in-swinging and curtain wall panels 20% of wall	S.F. Window	32.47	4.95	22.4%
2030	Exterior Doors	Double hollow metal swinging, single hollow metal	Each	1465	.52	
	B30 Roofing					
3010	Roof Coverings	Asphalt shingles with flashing; polystyrene insulation	S.F. Roof	6.36	6.36	4.8%
3020	Roof Openings	N/A	—	—	—	
C. INTERIORS						
1010	Partitions	Plaster on metal studs 40 S.F. Floor/L.F. Partitions	S.F. Partition	15.12	9.07	
1020	Interior Doors	Hollow metal 400 S.F. Floor/Door	Each	1157	2.90	
1030	Fittings	N/A	—	—	—	
2010	Stair Construction	N/A	—	—	—	14.2%
3010	Wall Finishes	Paint	S.F. Surface	1.20	1.44	
3020	Floor Finishes	Carpet	S.F. Floor	5.36	5.36	
3030	Ceiling Finishes	N/A	—	—	—	
D. SERVICES						
	D10 Conveying					
1010	Elevators & Lifts	N/A	—	—	—	0.0 %
1020	Escalators & Moving Walks	N/A	—	—	—	
	D20 Plumbing					
2010	Plumbing Fixtures	Kitchen, toilet and service fixtures, supply and drainage 1 Fixture/2430 S.F. Floor	Each	8335	3.43	
2020	Domestic Water Distribution	Gas fired hot water heater	S.F. Floor	.32	.32	2.8%
2040	Rain Water Drainage	N/A	—	—	—	
	D30 HVAC					
3010	Energy Supply	Oil fired hot water, wall fin radiation	S.F. Floor	10.74	10.74	
3020	Heat Generating Systems	N/A	—	—	—	
3030	Cooling Generating Systems	N/A	—	—	—	16.7%
3050	Terminal & Package Units	Split systems with air cooled condensing units	S.F. Floor	11.31	11.31	
3090	Other HVAC Sys. & Equipment	N/A	—	—	—	
	D40 Fire Protection					
4010	Sprinklers	Wet pipe sprinkler system	S.F. Floor	3.67	3.67	3.7%
4020	Standpipes	Standpipe	S.F. Floor	1.21	1.21	
	D50 Electrical					
5010	Electrical Service/Distribution	400 ampere service, panel board and feeders	S.F. Floor	.81	.81	
5020	Lighting & Branch Wiring	Fluorescent fixtures, receptacles, switches, A.C. and misc. power	S.F. Floor	8.41	8.41	9.3%
5030	Communications & Security	Alarm systems, sound system and emergency lighting	S.F. Floor	3.03	3.03	
5090	Other Electrical Systems	Emergency generator, 7.5 kW	S.F. Floor	.11	.11	
E. EQUIPMENT & FURNISHINGS						
1010	Commercial Equipment	N/A	—	—	—	
1020	Institutional Equipment	N/A	—	—	—	
1030	Vehicular Equipment	N/A	—	—	—	0.0 %
1090	Other Equipment	N/A	—	—	—	
F. SPECIAL CONSTRUCTION						
1020	Integrated Construction	N/A	—	—	—	0.0 %
1040	Special Facilities	N/A	—	—	—	
G. BUILDING SITEWORK	**N/A**					

	Sub-Total	132.37	100%
CONTRACTOR FEES (General Requirements: 10%, Overhead: 5%, Profit: 10%)	25%	33.08	
ARCHITECT FEES	10%	16.55	
Total Building Cost		**182**	

Costs per square foot of floor area

Exterior Wall	S.F. Area	2000	4000	6000	8000	12000	15000	18000	20000	22000
	L.F. Perimeter	180	280	340	386	460	535	560	600	640
Stone Ashlar Veneer On Concrete Block	Wood Truss	258.65	230.75	215.15	205.70	194.95	191.85	187.10	185.95	184.85
	Steel Joists	264.25	235.70	219.70	210.00	198.95	195.75	190.85	189.65	188.55
Stucco on Concrete Block	Wood Truss	230.55	208.90	197.50	190.70	183.00	180.70	177.45	176.60	175.80
	Steel Joists	236.65	214.30	202.45	195.40	187.45	185.05	181.60	180.75	179.90
Brick Veneer	Wood Frame	260.90	220.95	202.75	192.40	180.95	177.30	172.85	171.50	170.35
Wood Shingles	Wood Frame	249.20	211.80	195.40	186.10	175.95	172.70	168.80	167.65	166.55
Perimeter Adj., Add or Deduct	Per 100 L.F.	46.75	23.35	15.55	11.75	7.80	6.20	5.25	4.65	4.25
Story Hgt. Adj., Add or Deduct	Per 1 Ft.	5.40	4.20	3.40	2.90	2.30	2.15	1.90	1.80	1.75

For Basement, add $31.50 per square foot of basement area

The above costs were calculated using the basic specifications shown on the facing page. These costs should be adjusted where necessary for design alternatives and owner's requirements. Reported completed project costs, for this type of structure, range from $87.70 to $291.80 per S.F.

Common additives

Description	Unit	$ Cost
Bar, Front bar	L.F.	425
Back bar	L.F.	340
Booth, Upholstered, custom, straight	L.F.	250 - 460
"L" or "U" shaped	L.F.	259 - 440
Fireplaces, Brick not incl. chimney		
or foundation, 30" x 24" opening	Ea.	3175
Chimney, standard brick		
Single flue 16" x 20"	V.L.F.	106
20" x 20"	V.L.F.	125
2 flue, 20" x 32"	V.L.F.	191
Lockers, Steel, single tier, 60" or 72"	Opng.	221 - 370
2 tier, 60" or 72" total	Opng.	120 - 165
5 tier, box lockers	Opng.	77 - 86.50
Locker bench, lam. maple top only	L.F.	34
Pedestals, steel pipe	Ea.	76
Refrigerators, Prefabricated, walk-in		
7'-6" High, 6' x 6'	S.F.	202
10' x 10'	S.F.	159
12' x 14'	S.F.	142
12' x 20'	S.F.	125

Description	Unit	$ Cost
Sauna, Prefabricated, complete, 6' x 4'	Ea.	6575
6' x 9'	Ea.	10,000
8' x 8'	Ea.	11,500
10' x 12'	Ea.	15,500
Smoke Detectors		
Ceiling type	Ea.	227
Duct type	Ea.	565
Sound System		
Amplifier, 250 watts	Ea.	2050
Speaker, ceiling or wall	Ea.	213
Trumpet	Ea.	410
Steam Bath, Complete, to 140 C.F.	Ea.	3000
To 300 C.F.	Ea.	3325
To 800 C.F.	Ea.	7100
To 2500 C.F.	Ea.	9300
Swimming Pool Complete, gunite	S.F.	97 - 120
Tennis Court, Complete with fence		
Bituminous	Ea.	47,000 - 82,000
Clay	Ea.	50,000 - 83,500

Important: See the Reference Section for Location Factors.

Model costs calculated for a 1 story building with 12' story height and 6,000 square feet of floor area

				Unit	Unit Cost	Cost Per S.F.	% Of Sub-Total
A.	**SUBSTRUCTURE**						
1010	Standard Foundations	Poured concrete; strip and spread footings		S.F. Ground	2.53	2.53	
1020	Special Foundations	N/A		—	—	—	
1030	Slab on Grade	4" reinforced concrete with vapor barrier and granular base		S.F. Slab	5.44	5.44	9.1%
2010	Basement Excavation	Site preparation for slab and trench for foundation wall and footing		S.F. Ground	.32	.32	
2020	Basement Walls	4' Foundation wall		L.F. Wall	84	6.15	
B.	**SHELL**						
	B10 Superstructure						
1010	Floor Construction	N/A		—	—	—	4.2 %
1020	Roof Construction	Wood truss with plywood sheathing		S.F. Roof	6.65	6.65	
	B20 Exterior Enclosure						
2010	Exterior Walls	Stone ashlar veneer on concrete block	65% of wall	S.F. Wall	45.07	19.92	
2020	Exterior Windows	Aluminum horizontal sliding	35% of wall	Each	528	8.38	19.5%
2030	Exterior Doors	Double aluminum and glass, hollow metal		Each	2850	2.85	
	B30 Roofing						
3010	Roof Coverings	Asphalt shingles		S.F. Roof	18.17	3.14	2.0%
3020	Roof Openings	N/A		—	—	—	
C.	**INTERIORS**						
1010	Partitions	Gypsum board on metal studs, load bearing	14 S.F. Floor/L.F. Partition	S.F. Partition	8.20	5.86	
1020	Interior Doors	Single leaf wood	140 S.F. Floor/Door	Each	616	4.40	
1030	Fittings	N/A		—	—	—	
2010	Stair Construction	N/A		—	—	—	16.7%
3010	Wall Finishes	40% vinyl wall covering, 40% paint, 20% ceramic tile		S.F. Surface	2.60	3.72	
3020	Floor Finishes	50% carpet, 30% hardwood tile, 20% ceramic tile		S.F. Floor	8.06	8.06	
3030	Ceiling Finishes	Gypsum plaster on wood furring		S.F. Ceiling	4.51	4.51	
D.	**SERVICES**						
	D10 Conveying						
1010	Elevators & Lifts	N/A		—	—	—	0.0 %
1020	Escalators & Moving Walks	N/A		—	—	—	
	D20 Plumbing						
2010	Plumbing Fixtures	Kitchen, toilet and service fixtures, supply and drainage	1 Fixture/125 S.F. Floor	Each	2198	17.58	
2020	Domestic Water Distribution	Gas fired hot water		S.F. Floor	7.98	7.98	16.0%
2040	Rain Water Drainage	N/A		—	—	—	
	D30 HVAC						
3010	Energy Supply	N/A		—	—	—	
3020	Heat Generating Systems	Included in D3050		—	—	—	
3030	Cooling Generating Systems	N/A		—	—	—	18.8 %
3050	Terminal & Package Units	Multizone rooftop unit, gas heating, electric cooling		S.F. Floor	29.89	29.89	
3090	Other HVAC Sys. & Equipment	N/A		—	—	—	
	D40 Fire Protection						
4010	Sprinklers	Wet pipe sprinkler system		S.F. Floor	5.92	5.92	4.8%
4020	Standpipes	Standpipe		S.F. Floor	1.72	1.72	
	D50 Electrical						
5010	Electrical Service/Distribution	400 ampere service, panel board and feeders		S.F. Floor	3.54	3.54	
5020	Lighting & Branch Wiring	High efficiency fluorescent fixtures, receptacles, switches, A.C. and misc. power		S.F. Floor	8.56	8.56	9.0%
5030	Communications & Security	Addressable alarm systems and emergency lighting		S.F. Floor	1.98	1.98	
5090	Other Electrical Systems	Emergency generator, 11.5 kW		S.F. Floor	.22	.22	
E.	**EQUIPMENT & FURNISHINGS**						
1010	Commercial Equipment	N/A		—	—	—	
1020	Institutional Equipment	N/A		—	—	—	0.0 %
1030	Vehicular Equipment	N/A		—	—	—	
1090	Other Equipment	N/A		—	—	—	
F.	**SPECIAL CONSTRUCTION**						
1020	Integrated Construction	N/A		—	—	—	0.0 %
1040	Special Facilities	N/A		—	—	—	
G.	**BUILDING SITEWORK**	**N/A**					

		Sub-Total	159.38	**100%**
CONTRACTOR FEES (General Requirements: 10%, Overhead: 5%, Profit: 10%)			25%	39.83
ARCHITECT FEES			8%	15.94
		Total Building Cost	**215.15**	

For customer support on your Square Foot Cost Data, call 877.694.8329.

99

Costs per square foot of floor area

Exterior Wall	S.F. Area	4000	8000	12000	17000	22000	27000	32000	37000	42000
	L.F. Perimeter	280	386	520	585	640	740	840	940	940
Stone Ashlar on Concrete Block	Steel Joists	218.65	185.85	177.30	166.10	159.50	157.05	155.40	154.25	150.80
	Wood Joists	206.50	178.55	171.55	161.40	155.50	153.35	152.00	151.00	147.80
Face Brick on Concrete Block	Steel Joists	203.65	177.45	170.85	161.45	156.00	154.00	152.70	151.75	148.80
	Wood Joists	198.35	172.90	166.45	157.40	152.10	150.20	148.90	148.00	145.15
Decorative Concrete Block	Steel Joists	194.15	170.90	164.95	156.80	152.00	150.30	149.10	148.30	145.75
	Wood Joists	186.95	164.60	158.85	151.05	146.50	144.80	143.70	142.90	140.45
Perimeter Adj., Add or Deduct	Per 100 L.F.	25.60	12.85	8.55	6.00	4.65	3.80	3.20	2.75	2.40
Story Hgt. Adj., Add or Deduct	Per 1 Ft.	4.65	3.20	2.90	2.25	1.95	1.80	1.70	1.70	1.50

For Basement, add $32.10 per square foot of basement area

The above costs were calculated using the basic specifications shown on the facing page. These costs should be adjusted where necessary for design alternatives and owner's requirements. Reported completed project costs, for this type of structure, range from $89.70 to $257.70 per S.F.

Common additives

Description	Unit	$ Cost
Bar, Front bar	L.F.	425
Back bar	L.F.	340
Booth, Upholstered, custom, straight	L.F.	250 - 460
"L" or "U" shaped	L.F.	259 - 440
Emergency Lighting, 25 watt, battery operated		
Lead battery	Ea.	330
Nickel cadmium	Ea.	780
Flagpoles, Complete		
Aluminum, 20' high	Ea.	1850
40' High	Ea.	4200
70' High	Ea.	11,000
Fiberglass, 23' High	Ea.	1325
39'-5" High	Ea.	3450
59' High	Ea.	6850
Kitchen Equipment		
Broiler	Ea.	4050
Coffee urn, twin 6 gallon	Ea.	3000
Cooler, 6 ft. long, reach-in	Ea.	4175
Dishwasher, 10-12 racks per hr.	Ea.	4275
Food warmer, counter 1.2 kw	Ea.	735

Description	Unit	$ Cost
Kitchen Equipment, cont.		
Freezer, 44 C.F., reach-in	Ea.	5175
Ice cube maker, 50 lb. per day	Ea.	2000
Lockers, Steel, single tier, 60" or 72"	Opng.	221 - 370
2 tier, 60" or 72" total	Opng.	120 - 165
5 tier, box lockers	Opng.	77 - 86.50
Locker bench, lam. maple top only	L.F.	34
Pedestals, steel pipe	Ea.	76
Refrigerators, Prefabricated, walk-in		
7'-6" High, 6' x 6'	S.F.	202
10' x 10'	S.F.	159
12' x 14'	S.F.	142
12' x 20'	S.F.	125
Smoke Detectors		
Ceiling type	Ea.	227
Duct type	Ea.	565
Sound System		
Amplifier, 250 watts	Ea.	2050
Speaker, ceiling or wall	Ea.	213
Trumpet	Ea.	410

Important: See the Reference Section for Location Factors.

Model costs calculated for a 1 story building with 12' story height and 22,000 square feet of floor area

Club, Social

				Unit	Unit Cost	Cost Per S.F.	% Of Sub-Total
A. SUBSTRUCTURE							
1010	Standard Foundations	Poured concrete; strip and spread footings		S.F. Ground	1.42	1.42	
1020	Special Foundations	N/A		—	—	—	
1030	Slab on Grade	4" reinforced concrete with vapor barrier and granular base		S.F. Slab	5.44	5.44	7.9%
2010	Basement Excavation	Site preparation for slab and trench for foundation wall and footing		S.F. Ground	.18	.18	
2020	Basement Walls	4' Foundation wall		L.F. Wall	84	2.43	
B. SHELL							
B10 Superstructure							
1010	Floor Construction	N/A		—	—	—	6.4%
1020	Roof Construction	Metal deck on open web steel joists		S.F. Roof	6.97	6.97	
B20 Exterior Enclosure							
2010	Exterior Walls	Stone, ashlar veneer on concrete block	65% of wall	S.F. Wall	45.04	10.22	
2020	Exterior Windows	Window wall	35% of wall	Each	49.40	6.04	14.3%
2030	Exterior Doors	Double aluminum and glass doors		Each	2850	.78	
B30 Roofing							
3010	Roof Coverings	Built-up tar and gravel with flashing; perlite/EPS composite insulation		S.F. Roof	6.36	6.36	5.3%
3020	Roof Openings	N/A		—	—	—	
C. INTERIORS							
1010	Partitions	Lightweight concrete block	14 S.F. Floor/L.F. Partition	S.F. Partition	12.18	8.70	
1020	Interior Doors	Single leaf wood	140 S.F. Floor/Door	Each	616	4.40	
1030	Fittings	N/A		—	—	—	
2010	Stair Construction	N/A		—	—	—	25.2%
3010	Wall Finishes	65% paint, 25% vinyl wall covering, 10% ceramic tile		S.F. Surface	1.78	2.54	
3020	Floor Finishes	60% carpet, 35% hardwood, 15% ceramic tile		S.F. Floor	7.43	7.43	
3030	Ceiling Finishes	Mineral fiber tile on concealed zee bars		S.F. Ceiling	7.03	7.03	
D. SERVICES							
D10 Conveying							
1010	Elevators & Lifts	N/A		—	—	—	0.0%
1020	Escalators & Moving Walks	N/A		—	—	—	
D20 Plumbing							
2010	Plumbing Fixtures	Kitchen, toilet and service fixtures, supply and drainage	1 Fixture/1050 S.F. Floor	Each	8904	8.48	
2020	Domestic Water Distribution	Gas fired hot water heater		S.F. Floor	.28	.28	8.0%
2040	Rain Water Drainage	Roof drains		S.F. Roof	.81	.81	
D30 HVAC							
3010	Energy Supply	N/A		—	—	—	
3020	Heat Generating Systems	Included in D3050		—	—	—	
3030	Cooling Generating Systems	N/A		—	—	—	18.7%
3050	Terminal & Package Units	Multizone rooftop unit, gas heating, electric cooling		S.F. Floor	22.35	22.35	
3090	Other HVAC Sys. & Equipment	N/A		—	—	—	
D40 Fire Protection							
4010	Sprinklers	Wet pipe sprinkler system		S.F. Floor	5.08	5.08	4.7%
4020	Standpipes	Standpipe		S.F. Floor	.47	.47	
D50 Electrical							
5010	Electrical Service/Distribution	400 ampere service, panel board and feeders		S.F. Floor	1.22	1.22	
5020	Lighting & Branch Wiring	High efficiency fluorescent fixtures, receptacles, switches, A.C. and misc. power		S.F. Floor	8.37	8.37	9.4%
5030	Communications & Security	Addressable alarm systems and emergency lighting		S.F. Floor	1.49	1.49	
5090	Other Electrical Systems	Emergency generator, 11.5 kW		S.F. Floor	.12	.12	
E. EQUIPMENT & FURNISHINGS							
1010	Commercial Equipment	N/A		—	—	—	
1020	Institutional Equipment	N/A		—	—	—	0.0%
1030	Vehicular Equipment	N/A		—	—	—	
1090	Other Equipment	N/A		—	—	—	
F. SPECIAL CONSTRUCTION							
1020	Integrated Construction	N/A		—	—	—	0.0%
1040	Special Facilities	N/A		—	—	—	
G. BUILDING SITEWORK	**N/A**						

			Sub-Total	119.26	100%
CONTRACTOR FEES (General Requirements: 10%, Overhead: 5%, Profit: 10%)		25%	29.80		
ARCHITECT FEES		7%	10.44		
	Total Building Cost		159.50		

For customer support on your Square Foot Cost Data, call 877.694.8329.

101

Costs per square foot of floor area

Exterior Wall	S.F. Area	15000	20000	28000	38000	50000	65000	85000	100000	150000
	L.F. Perimeter	350	400	480	550	630	660	750	825	1035
Face Brick with Concrete Block Back-up	Steel Frame	230.85	221.55	213.50	207.15	202.70	197.85	194.95	193.65	190.80
	Bearing Walls	229.60	218.95	209.75	202.30	197.10	191.25	187.90	186.30	182.95
Decorative Concrete Block	Steel Frame	222.65	214.55	207.50	202.05	198.30	194.25	191.85	190.75	188.35
	Bearing Walls	221.40	211.90	203.70	197.20	192.70	187.70	184.75	183.35	180.45
Stucco on Concrete Block	Steel Frame	222.10	214.05	207.10	201.70	197.95	194.00	191.65	190.55	188.20
	Bearing Walls	220.85	211.45	203.30	196.85	192.40	187.40	184.55	183.20	180.30
Perimeter Adj., Add or Deduct	Per 100 L.F.	9.75	7.30	5.25	3.80	2.90	2.25	1.75	1.40	0.95
Story Hgt. Adj., Add or Deduct	Per 1 Ft.	2.40	2.10	1.75	1.45	1.35	1.00	0.90	0.85	0.65

For Basement, add $37.40 per square foot of basement area

The above costs were calculated using the basic specifications shown on the facing page. These costs should be adjusted where necessary for design alternatives and owner's requirements. Reported completed project costs, for this type of structure, range from $129.75 to $304.90 per S.F.

Common additives

Description	Unit	$ Cost
Carrels Hardwood	Ea.	975 - 2350
Clock System		
20 Room	Ea.	19,900
50 Room	Ea.	46,800
Elevators, Hydraulic passenger, 2 stops		
1500# capacity	Ea.	65,700
2500# capacity	Ea.	68,800
3500# capacity	Ea.	73,300
Additional stop, add	Ea.	7825
Emergency Lighting, 25 watt, battery operated		
Lead battery	Ea.	330
Nickel cadmium	Ea.	780
Flagpoles, Complete		
Aluminum, 20' high	Ea.	1850
40' High	Ea.	4200
70' High	Ea.	11,000
Fiberglass, 23' High	Ea.	1325
39'-5" High	Ea.	3450
59' High	Ea.	6850

Description	Unit	$ Cost
Lockers, Steel, single tier, 60" or 72"	Opng.	221 - 370
2 tier, 60" or 72" total	Opng.	120 - 165
5 tier, box lockers	Opng.	77 - 86.50
Locker bench, lam. maple top only	L.F.	34
Pedestals, steel pipe	Ea.	76
Seating		
Auditorium chair, all veneer	Ea.	310
Veneer back, padded seat	Ea.	320
Upholstered, spring seat	Ea.	320
Classroom, movable chair & desk	Set	81 - 171
Lecture hall, pedestal type	Ea.	263 - 620
Smoke Detectors		
Ceiling type	Ea.	227
Duct type	Ea.	565
Sound System		
Amplifier, 250 watts	Ea.	2050
Speaker, ceiling or wall	Ea.	213
Trumpet	Ea.	410
TV Antenna, Master system, 12 outlet	Outlet	255
30 outlet	Outlet	221
100 outlet	Outlet	209

Important: See the Reference Section for Location Factors.

Model costs calculated for a 2 story building with 12' story height and 50,000 square feet of floor area

				Unit	Unit Cost	Cost Per S.F.	% Of Sub-Total
A. SUBSTRUCTURE							
1010	Standard Foundations	Poured concrete; strip and spread footings		S.F. Ground	1.38	.69	
1020	Special Foundations	N/A		—	—	—	
1030	Slab on Grade	4" reinforced concrete with vapor barrier and granular base		S.F. Slab	5.44	2.72	3.6%
2010	Basement Excavation	Site preparation for slab and trench for foundation wall and footing		S.F. Ground	.32	.16	
2020	Basement Walls	4' Foundation wall		L.F. Wall	79	1.61	
B. SHELL							
B10 Superstructure							
1010	Floor Construction	Open web steel joists, slab form, concrete		S.F. Floor	17.36	8.68	9.5%
1020	Roof Construction	Metal deck on open web steel joists, columns		S.F. Roof	9.96	4.98	
B20 Exterior Enclosure							
2010	Exterior Walls	Decorative concrete block	65% of wall	S.F. Wall	16.59	3.26	
2020	Exterior Windows	Window wall	35% of wall	Each	41.80	4.43	5.8%
2030	Exterior Doors	Double glass and aluminum with transom		Each	5825	.70	
B30 Roofing							
3010	Roof Coverings	Built-up tar and gravel with flashing; perlite/EPS composite insulation		S.F. Roof	6.22	3.11	2.2%
3020	Roof Openings	N/A		—	—	—	
C. INTERIORS							
1010	Partitions	Concrete block	20 S.F. Floor/L.F. Partition	S.F. Partition	15.90	7.95	
1020	Interior Doors	Single leaf hollow metal	200 S.F. Floor/Door	Each	1157	5.79	
1030	Fittings	Chalkboards, counters, cabinets		S.F. Floor	6.03	6.03	
2010	Stair Construction	Concrete filled metal pan		Flight	17,300	3.46	26.7%
3010	Wall Finishes	95% paint, 5% ceramic tile		S.F. Surface	4.21	4.21	
3020	Floor Finishes	70% vinyl composition tile, 25% carpet, 5% ceramic tile		S.F. Floor	4.02	4.02	
3030	Ceiling Finishes	Mineral fiber tile on concealed zee bars		S.F. Ceiling	7.03	7.03	
D. SERVICES							
D10 Conveying							
1010	Elevators & Lifts	Two hydraulic passenger elevators		Each	83,500	3.34	2.3%
1020	Escalators & Moving Walks	N/A		—	—	—	
D20 Plumbing							
2010	Plumbing Fixtures	Toilet and service fixtures, supply and drainage	1 Fixture/455 S.F. Floor	Each	8772	19.28	
2020	Domestic Water Distribution	Oil fired hot water heater		S.F. Floor	2.76	2.76	15.8%
2040	Rain Water Drainage	Roof drains		S.F. Roof	1.40	.70	
D30 HVAC							
3010	Energy Supply	N/A		—	—	—	
3020	Heat Generating Systems	Included in D3050		—	—	—	
3030	Cooling Generating Systems	N/A		—	—	—	14.4%
3050	Terminal & Package Units	Multizone unit, gas heating, electric cooling		S.F. Floor	20.70	20.70	
3090	Other HVAC Sys. & Equipment	N/A		—	—	—	
D40 Fire Protection							
4010	Sprinklers	Sprinklers, light hazard		S.F. Floor	3.19	3.19	2.5%
4020	Standpipes	Standpipe, dry, Class III		S.F. Floor	.36	.36	
D50 Electrical							
5010	Electrical Service/Distribution	2000 ampere service, panel board and feeders		S.F. Floor	3.83	3.83	
5020	Lighting & Branch Wiring	High efficiency fluorescent fixtures, receptacles, switches, A.C. and misc. power		S.F. Floor	13.21	13.21	17.3%
5030	Communications & Security	Addressable alarm systems, internet wiring, communications systems and emergency lighting		S.F. Floor	7.22	7.22	
5090	Other Electrical Systems	Emergency generator, 100 kW		S.F. Floor	.65	.65	
E. EQUIPMENT & FURNISHINGS							
1010	Commercial Equipment	N/A		—	—	—	
1020	Institutional Equipment	N/A		—	—	—	0.0%
1030	Vehicular Equipment	N/A		—	—	—	
1090	Other Equipment	N/A		—	—	—	
F. SPECIAL CONSTRUCTION							
1020	Integrated Construction	N/A		—	—	—	0.0%
1040	Special Facilities	N/A		—	—	—	
G. BUILDING SITEWORK	**N/A**						

	Sub-Total	144.07	**100%**
CONTRACTOR FEES (General Requirements: 10%, Overhead: 5%, Profit: 10%)	25%	36.02	
ARCHITECT FEES	7%	12.61	
	Total Building Cost	**192.70**	

For customer support on your Square Foot Cost Data, call 877.694.8329.

103

Costs per square foot of floor area

Exterior Wall	S.F. Area	10000	15000	25000	40000	55000	70000	80000	90000	100000
	L.F. Perimeter	260	320	400	476	575	628	684	721	772
Face Brick with Concrete Block Back-up	R/Conc. Frame	236.10	219.05	202.35	190.65	186.25	182.35	181.15	179.75	178.90
	Steel Frame	248.75	231.70	214.95	203.30	198.90	195.00	193.80	192.40	191.55
Decorative Concrete Block	R/Conc. Frame	219.20	205.15	191.90	182.90	179.45	176.50	175.55	174.55	173.90
	Steel Frame	231.85	217.80	204.55	195.55	192.15	189.15	188.20	187.20	186.55
Precast Concrete Panels	R/Conc. Frame	241.60	223.60	205.65	193.20	188.45	184.25	182.95	181.45	180.55
	Steel Frame	254.25	236.15	218.30	205.80	201.10	196.90	195.55	194.10	193.20
Perimeter Adj., Add or Deduct	Per 100 L.F.	19.60	13.05	7.80	4.95	3.55	2.80	2.45	2.15	1.95
Story Hgt. Adj., Add or Deduct	Per 1 Ft.	3.75	3.10	2.30	1.75	1.55	1.30	1.25	1.20	1.15

For Basement, add $36.40 per square foot of basement area

The above costs were calculated using the basic specifications shown on the facing page. These costs should be adjusted where necessary for design alternatives and owner's requirements. Reported completed project costs, for this type of structure, range from $83.80 to $246.95 per S.F.

Common additives

Description	Unit	$ Cost
Carrels Hardwood	Ea.	975 - 2350
Closed Circuit Surveillance, One station		
Camera and monitor	Ea.	1975
For additional camera stations, add	Ea.	1075
Elevators, Hydraulic passenger, 2 stops		
2000# capacity	Ea.	66,600
2500# capacity	Ea.	68,800
3500# capacity	Ea.	73,300
Additional stop, add	Ea.	7825
Emergency Lighting, 25 watt, battery operated		
Lead battery	Ea.	330
Nickel cadmium	Ea.	780
Furniture	Student	2800 - 5350
Intercom System, 25 station capacity		
Master station	Ea.	3250
Intercom outlets	Ea.	189
Handset	Ea.	520

Description	Unit	$ Cost
Kitchen Equipment		
Broiler	Ea.	4050
Coffee urn, twin 6 gallon	Ea.	3000
Cooler, 6 ft. long	Ea.	4175
Dishwasher, 10-12 racks per hr.	Ea.	4275
Food warmer	Ea.	735
Freezer, 44 C.F., reach-in	Ea.	5175
Ice cube maker, 50 lb. per day	Ea.	2000
Range with 1 oven	Ea.	2925
Laundry Equipment		
Dryer, gas, 16 lb. capacity	Ea.	975
30 lb. capacity	Ea.	3850
Washer, 4 cycle	Ea.	1200
Commercial	Ea.	1600
Smoke Detectors		
Ceiling type	Ea.	227
Duct type	Ea.	565
TV Antenna, Master system, 12 outlet	Outlet	255
30 outlet	Outlet	221
100 outlet	Outlet	209

Important: See the Reference Section for Location Factors.

Model costs calculated for a 3 story building with 12' story height and 25,000 square feet of floor area

College, Dormitory, 2-3 Story

				Unit	Unit Cost	Cost Per S.F.	% Of Sub-Total
A. SUBSTRUCTURE							
1010	Standard Foundations	Poured concrete; strip and spread footings		S.F. Ground	6.39	2.13	
1020	Special Foundations	N/A		—	—	—	
1030	Slab on Grade	4" reinforced concrete with vapor barrier and granular base		S.F. Slab	5.44	1.81	3.5%
2010	Basement Excavation	Site preparation for slab and trench for foundation wall and footing		S.F. Ground	.18	.06	
2020	Basement Walls	4' Foundation wall		L.F. Wall	84	1.33	
B. SHELL							
	B10 Superstructure						
1010	Floor Construction	Concrete flat plate		S.F. Floor	25.59	17.06	14.4%
1020	Roof Construction	Concrete flat plate		S.F. Roof	13.98	4.66	
	B20 Exterior Enclosure						
2010	Exterior Walls	Face brick with concrete block backup	80% of wall	S.F. Wall	33.46	15.42	
2020	Exterior Windows	Aluminum horizontal sliding	20% of wall	Each	750	3.75	13.6%
2030	Exterior Doors	Double glass & aluminum doors		Each	5975	1.43	
	B30 Roofing						
3010	Roof Coverings	Single ply membrane, loose laid and ballasted; perlite/EPS composite insulation		S.F. Roof	5.79	1.93	1.3%
3020	Roof Openings	N/A		—	—	—	
C. INTERIORS							
1010	Partitions	Gypsum board on metal studs, concrete block	9 S.F. Floor/L.F. Partition	S.F. Partition	6.44	7.16	
1020	Interior Doors	Single leaf wood	90 S.F. Floor/Door	Each	616	6.84	
1030	Fittings	Closet shelving, mirrors, bathroom accessories		S.F. Floor	1.96	1.96	
2010	Stair Construction	Cast in place concrete		Flight	5375	2.79	20.3%
3010	Wall Finishes	95% paint, 5% ceramic tile		S.F. Surface	1.92	4.26	
3020	Floor Finishes	80% carpet, 10% vinyl composition tile, 10% ceramic tile		S.F. Floor	6.56	6.56	
3030	Ceiling Finishes	90% paint, 10% suspended fiberglass board		S.F. Ceiling	1.15	1.15	
D. SERVICES							
	D10 Conveying						
1010	Elevators & Lifts	One hydraulic passenger elevator		Each	114,000	4.56	3.0%
1020	Escalators & Moving Walks	N/A		—	—	—	
	D20 Plumbing						
2010	Plumbing Fixtures	Toilet and service fixtures, supply and drainage	1 Fixture/455 S.F. Floor	Each	10,228	22.48	
2020	Domestic Water Distribution	Electric water heater		S.F. Floor	3.62	3.62	17.6%
2040	Rain Water Drainage	Roof drains		S.F. Roof	1.47	.49	
	D30 HVAC						
3010	Energy Supply	N/A		—	—	—	
3020	Heat Generating Systems	Included in D3050		—	—	—	
3030	Cooling Generating Systems	N/A		—	—	—	8.6 %
3050	Terminal & Package Units	Rooftop multizone unit system		S.F. Floor	13	13	
3090	Other HVAC Sys. & Equipment	N/A		—	—	—	
	D40 Fire Protection						
4010	Sprinklers	Wet pipe sprinkler system		S.F. Floor	3.02	3.02	2.6%
4020	Standpipes	Standpipe, dry, Class III		S.F. Floor	.88	.88	
	D50 Electrical						
5010	Electrical Service/Distribution	800 ampere service, panel board and feeders		S.F. Floor	2.05	2.05	
5020	Lighting & Branch Wiring	High efficiency fluorescent fixtures, receptacles, switches, A.C. and misc. power		S.F. Floor	9.96	9.96	12.9%
5030	Communications & Security	Alarm systems, internet and phone wiring, communications systems and emergency lighting		S.F. Floor	7.44	7.44	
5090	Other Electrical Systems	Emergency generator, 7.5 kW		S.F. Floor	.08	.08	
E. EQUIPMENT & FURNISHINGS							
1010	Commercial Equipment	N/A		—	—	—	
1020	Institutional Equipment	N/A		—	—	—	
1030	Vehicular Equipment	N/A		—	—	—	2.2 %
2020	Moveable Furnishings	Dormitory furniture		S.F. Floor	3.38	3.38	
F. SPECIAL CONSTRUCTION							
1020	Integrated Construction	N/A		—	—	—	0.0 %
1040	Special Facilities	N/A		—	—	—	
G. BUILDING SITEWORK	**N/A**						

			Sub-Total	151.26	100%
CONTRACTOR FEES (General Requirements: 10%, Overhead: 5%, Profit: 10%)		25%	37.85	
ARCHITECT FEES		7%	13.24	
	Total Building Cost		**202.35**	

For customer support on your Square Foot Cost Data, call 877.694.8329.

105

Costs per square foot of floor area

Exterior Wall	S.F. Area	20000	35000	45000	65000	85000	110000	135000	160000	200000
	L.F. Perimeter	260	340	400	440	500	540	560	590	640
Face Brick with Concrete Block Back-up	R/Conc. Frame	231.85	215.80	211.65	202.50	198.45	194.35	191.30	189.20	187.20
	Steel Frame	246.50	230.45	226.25	217.15	213.10	209.00	205.90	203.85	201.85
Decorative Concrete Block	R/Conc. Frame	216.65	204.45	201.20	194.60	191.60	188.65	186.40	184.90	183.50
	Steel Frame	231.25	219.05	215.85	209.25	206.25	203.25	201.05	199.55	198.15
Precast Concrete Panels With Exposed Aggregate	R/Conc. Frame	236.10	218.95	214.50	204.70	200.40	195.95	192.60	190.45	188.25
	Steel Frame	250.60	233.50	229.05	219.25	214.95	210.55	207.20	205.05	202.85
Perimeter Adj., Add or Deduct	Per 100 L.F.	19.00	10.90	8.45	5.80	4.50	3.50	2.75	2.40	1.95
Story Hgt. Adj., Add or Deduct	Per 1 Ft.	3.80	2.90	2.60	2.00	1.80	1.45	1.15	1.15	0.95

For Basement, add $36.40 per square foot of basement area

The above costs were calculated using the basic specifications shown on the facing page. These costs should be adjusted where necessary for design alternatives and owner's requirements. Reported completed project costs, for this type of structure, range from $119.10 to $287.25 per S.F.

Common additives

Description	Unit	$ Cost
Carrels Hardwood	Ea.	975 - 2350
Closed Circuit Surveillance, One station		
Camera and monitor	Ea.	1975
For additional camera stations, add	Ea.	1075
Elevators, Electric passenger, 5 stops		
2000# capacity	Ea.	167,000
2500# capacity	Ea.	171,000
3500# capacity	Ea.	171,500
Additional stop, add	Ea.	10,200
Emergency Lighting, 25 watt, battery operated		
Lead battery	Ea.	330
Nickel cadmium	Ea.	780
Furniture	Student	2800 - 5350
Intercom System, 25 station capacity		
Master station	Ea.	3250
Intercom outlets	Ea.	189
Handset	Ea.	520

Description	Unit	$ Cost
Kitchen Equipment		
Broiler	Ea.	4050
Coffee urn, twin, 6 gallon	Ea.	3000
Cooler, 6 ft. long	Ea.	4175
Dishwasher, 10-12 racks per hr.	Ea.	4275
Food warmer	Ea.	735
Freezer, 44 C.F., reach-in	Ea.	5175
Ice cube maker, 50 lb. per day	Ea.	2000
Range with 1 oven	Ea.	2925
Laundry Equipment		
Dryer, gas, 16 lb. capacity	Ea.	975
30 lb. capacity	Ea.	3850
Washer, 4 cycle	Ea.	1200
Commercial	Ea.	1600
Smoke Detectors		
Ceiling type	Ea.	227
Duct type	Ea.	565
TV Antenna, Master system, 12 outlet	Outlet	255
30 outlet	Outlet	221
100 outlet	Outlet	209

Important: See the Reference Section for Location Factors.

Model costs calculated for a 6 story building with 12' story height and 85,000 square feet of floor area

College, Dormitory, 4-8 Story

				Unit	Unit Cost	Cost Per S.F.	% Of Sub-Total
A. SUBSTRUCTURE							
1010	Standard Foundations	Poured concrete; strip and spread footings		S.F. Ground	9.36	1.56	
1020	Special Foundations	N/A		—	—	—	
1030	Slab on Grade	4" reinforced concrete with vapor barrier and granular base		S.F. Slab	5.44	.91	2.1%
2010	Basement Excavation	Site preparation for slab and trench for foundation wall and footing		S.F. Ground	.32	.05	
2020	Basement Walls	4' foundation wall		L.F. Wall	79	.68	
B. SHELL							
	B10 Superstructure						
1010	Floor Construction	Concrete slab with metal deck and beams		S.F. Floor	30.26	25.22	18.8%
1020	Roof Construction	Concrete slab with metal deck and beams		S.F. Roof	22.74	3.79	
	B20 Exterior Enclosure						
2010	Exterior Walls	Decorative concrete block	80% of wall	S.F. Wall	18.24	6.18	
2020	Exterior Windows	Aluminum horizontal sliding	20% of wall	Each	528	2.98	6.2%
2030	Exterior Doors	Double glass & aluminum doors		Each	3694	.34	
	B30 Roofing						
3010	Roof Coverings	Single ply membrane, loose laid and ballasted; perlite/EPS composite insulation		S.F. Roof	5.22	.87	0.6%
3020	Roof Openings	N/A		—	—	—	
C. INTERIORS							
1010	Partitions	Concrete block	9 S.F. Floor/L.F. Partition	S.F. Partition	10.74	11.93	
1020	Interior Doors	Single leaf wood	90 S.F. Floor/Door	Each	616	6.84	
1030	Fittings	Closet shelving, mirrors, bathroom accessories		S.F. Floor	1.60	1.60	
2010	Stair Construction	Concrete filled metal pan		Flight	17,300	3.66	24.4%
3010	Wall Finishes	95% paint, 5% ceramic tile		S.F. Surface	2.64	5.87	
3020	Floor Finishes	80% carpet, 10% vinyl composition tile, 10% ceramic tile		S.F. Floor	6.56	6.56	
3030	Ceiling Finishes	Mineral fiber tile on concealed zee bars, paint		S.F. Ceiling	1.15	1.15	
D. SERVICES							
	D10 Conveying						
1010	Elevators & Lifts	Four geared passenger elevators		Each	243,100	11.44	7.4%
1020	Escalators & Moving Walks	N/A		—	—	—	
	D20 Plumbing						
2010	Plumbing Fixtures	Toilet and service fixtures, supply and drainage	1 Fixture/390 S.F. Floor	Each	8732	22.39	
2020	Domestic Water Distribution	Oil fired water heater		S.F. Floor	1.95	1.95	16.0%
2040	Rain Water Drainage	Roof drains		S.F. Roof	2.16	.36	
	D30 HVAC						
3010	Energy Supply	Oil fired hot water, wall fin radiation		S.F. Floor	4.59	4.59	
3020	Heat Generating Systems	N/A		—	—	—	
3030	Cooling Generating Systems	Chilled water, air cooled condenser system		S.F. Floor	9.54	9.54	9.2%
3050	Terminal & Package Units	N/A		—	—	—	
3090	Other HVAC Sys. & Equipment	N/A		—	—	—	
	D40 Fire Protection						
4010	Sprinklers	Sprinklers, light hazard		S.F. Floor	2.93	2.93	2.3%
4020	Standpipes	Standpipe, dry, Class III		S.F. Floor	.65	.65	
	D50 Electrical						
5010	Electrical Service/Distribution	1200 ampere service, panel board and feeders		S.F. Floor	.99	.99	
5020	Lighting & Branch Wiring	High efficiency fluorescent fixtures, receptacles, switches, A.C. and misc. power		S.F. Floor	9.78	9.78	11.3%
5030	Communications & Security	Alarm systems, internet and phone wiring, communications systems and emergency lighting		S.F. Floor	6.41	6.41	
5090	Other Electrical Systems	Emergency generator, 30 kW		S.F. Floor	.20	.20	
E. EQUIPMENT & FURNISHINGS							
1010	Commercial Equipment	N/A		—	—	—	
1020	Institutional Equipment	N/A		—	—	—	
1030	Vehicular Equipment	N/A		—	—	—	1.8%
2020	Moveable Furnishings	Built-in dormitory furnishings		S.F. Floor	2.77	2.77	
F. SPECIAL CONSTRUCTION							
1020	Integrated Construction	N/A		—	—	—	0.0%
1040	Special Facilities	N/A		—	—	—	
G. BUILDING SITEWORK	**N/A**						

			Sub-Total	154.19	100%
CONTRACTOR FEES (General Requirements: 10%, Overhead: 5%, Profit: 10%)			25%	38.57	
ARCHITECT FEES			7%	13.49	
		Total Building Cost		**206.25**	

For customer support on your Square Foot Cost Data, call 877.694.8329.

107

Costs per square foot of floor area

Exterior Wall	S.F. Area	12000	20000	28000	37000	45000	57000	68000	80000	92000
	L.F. Perimeter	470	600	698	793	900	1000	1075	1180	1275
Face Brick with Concrete Brick Back-up	Steel Frame	319.00	256.15	228.25	210.90	201.75	191.85	185.70	181.20	177.65
	Bearing Walls	320.45	257.00	228.75	211.10	201.90	191.85	185.55	180.95	177.40
Decorative Concrete Block	Steel Frame	311.50	250.40	223.40	206.70	197.85	188.40	182.50	178.25	174.90
	Bearing Walls	313.15	251.40	224.10	207.10	198.15	188.60	182.65	178.25	174.80
Stucco on Concrete Block	Steel Frame	310.15	249.40	222.60	206.05	197.20	187.90	182.05	177.80	174.50
	Bearing Walls	311.75	250.40	223.25	206.40	197.50	188.00	182.10	177.75	174.35
Perimeter Adj., Add or Deduct	Per 100 L.F.	7.25	4.45	3.15	2.30	1.85	1.55	1.35	1.05	0.95
Story Hgt. Adj., Add or Deduct	Per 1 Ft.	1.95	1.60	1.30	1.10	1.05	0.95	0.85	0.75	0.75

For Basement, add $23.55 per square foot of basement area

The above costs were calculated using the basic specifications shown on the facing page. These costs should be adjusted where necessary for design alternatives and owner's requirements. Reported completed project costs, for this type of structure, range from $171.55 to $320.35 per S.F.

Common additives

Description	Unit	$ Cost
Cabinets, Base, door units, metal	L.F.	320
Drawer units	L.F.	630
Tall storage cabinets, open	L.F.	605
With doors	L.F.	875
Wall, metal 12-1/2" deep, open	L.F.	241
With doors	L.F.	440
Carrels Hardwood	Ea.	975 - 2350
Countertops, not incl. base cabinets, acid proof	S.F.	58.50 - 71.50
Stainless steel	S.F.	158
Fume Hood, Not incl. ductwork	L.F.	765 - 1600
Ductwork	Hood	5850 - 9625
Glassware Washer, Distilled water rinse	Ea.	7700 - 16,000
Seating		
Auditorium chair, all veneer	Ea.	310
Veneer back, padded seat	Ea.	320
Upholstered, spring seat	Ea.	320
Classroom, movable chair & desk	Set	81 - 171
Lecture hall, pedestal type	Ea.	263 - 620

Description	Unit	$ Cost
Safety Equipment, Eye wash, hand held	Ea.	450
Deluge shower	Ea.	850
Sink, One piece plastic		
Flask wash, freestanding	Ea.	2600
Tables, acid resist. top, drawers	L.F.	219
Titration Unit, Four 2000 ml reservoirs	Ea.	6125

Important: See the Reference Section for Location Factors.

Model costs calculated for a 1 story building with 12' story height and 45,000 square feet of floor area

				Unit	Unit Cost	Cost Per S.F.	% Of Sub-Total
A. SUBSTRUCTURE							
1010	Standard Foundations	Poured concrete; strip and spread footings		S.F. Ground	1.81	1.81	
1020	Special Foundations	N/A		—	—	—	
1030	Slab on Grade	4" reinforced concrete with vapor barrier and granular base		S.F. Slab	5.44	5.44	6.4%
2010	Basement Excavation	Site preparation for slab and trench for foundation wall and footing		S.F. Ground	.18	.18	
2020	Basement Walls	4' foundation wall		L.F. Wall	79	2.04	
B. SHELL							
	B10 Superstructure						
1010	Floor Construction	Metal deck on open web steel joists	(5680 S.F.)	S.F. Floor	6.54	.85	5.9%
1020	Roof Construction	Metal deck on open web steel joists		S.F. Roof	7.84	7.84	
	B20 Exterior Enclosure						
2010	Exterior Walls	Face brick with concrete block backup	75% of wall	S.F. Wall	33.44	6.02	
2020	Exterior Windows	Window wall	25% of wall	Each	41.80	2.51	7.3%
2030	Exterior Doors	Glass and metal doors and entrances with transom		Each	5003	2.23	
	B30 Roofing						
3010	Roof Coverings	Built-up tar and gravel with flashing; perlite/EPS composite insulation		S.F. Roof	5.97	5.97	4.3%
3020	Roof Openings	Skylight		S.F. Roof	.35	.35	
C. INTERIORS							
1010	Partitions	Concrete block	10 S.F. Floor/L.F. Partition	S.F. Partition	12.47	12.47	
1020	Interior Doors	Single leaf-kalamein fire doors	820 S.F. Floor/Door	Each	1275	1.55	
1030	Fittings	Lockers		S.F. Floor	.05	.05	
2010	Stair Construction	N/A		—	—	—	22.5%
3010	Wall Finishes	60% paint, 40% epoxy coating		S.F. Surface	3.10	6.19	
3020	Floor Finishes	60% epoxy, 20% carpet, 20% vinyl composition tile		S.F. Floor	5.99	5.99	
3030	Ceiling Finishes	Mineral fiber tile on concealed zee runners		S.F. Ceiling	7.03	7.03	
D. SERVICES							
	D10 Conveying						
1010	Elevators & Lifts	N/A		—	—	—	0.0 %
1020	Escalators & Moving Walks	N/A		—	—	—	
	D20 Plumbing						
2010	Plumbing Fixtures	Toilet and service fixtures, supply and drainage	1 Fixture/260 S.F. Floor	Each	8809	33.88	
2020	Domestic Water Distribution	Gas fired water heater		S.F. Floor	3.06	3.06	25.4%
2040	Rain Water Drainage	Roof drains		S.F. Roof	.75	.75	
	D30 HVAC						
3010	Energy Supply	N/A		—	—	—	
3020	Heat Generating Systems	Included in D3050		—	—	—	
3030	Cooling Generating Systems	N/A		—	—	—	14.0 %
3050	Terminal & Package Units	Multizone unit, gas heating, electric cooling		S.F. Floor	20.70	20.70	
3090	Other HVAC Sys. & Equipment	N/A		—	—	—	
	D40 Fire Protection						
4010	Sprinklers	Sprinklers, light hazard		S.F. Floor	2.94	2.94	2.2%
4020	Standpipes	Standpipe, wet, Class III		S.F. Floor	.33	.33	
	D50 Electrical						
5010	Electrical Service/Distribution	1000 ampere service, panel board and feeders		S.F. Floor	1.62	1.62	
5020	Lighting & Branch Wiring	High efficiency fluorescent fixtures, receptacles, switches, A.C. and misc. power		S.F. Floor	11.18	11.18	11.2%
5030	Communications & Security	Addressable alarm systems, internet wiring, and emergency lighting		S.F. Floor	3.56	3.56	
5090	Other Electrical Systems	Emergency generator, 11.5 kW, Uninterruptible power supply		S.F. Floor	.22	.22	
E. EQUIPMENT & FURNISHINGS							
1010	Commercial Equipment	N/A		—	—	—	
1020	Institutional Equipment	Cabinets, fume hoods, lockers, glassware washer		S.F. Floor	1.41	1.41	1.0 %
1030	Vehicular Equipment	N/A		—	—	—	
1090	Other Equipment	N/A		—	—	—	
F. SPECIAL CONSTRUCTION							
1020	Integrated Construction	N/A		—	—	—	0.0 %
1040	Special Facilities	N/A		—	—	—	
G. BUILDING SITEWORK	**N/A**						

			Sub-Total	148.17	100%
CONTRACTOR FEES (General Requirements: 10%, Overhead: 5%, Profit: 10%)			25%	37.06	
ARCHITECT FEES			9%	16.67	

Total Building Cost	**201.90**

For customer support on your Square Foot Cost Data, call 877.694.8329.

109

Costs per square foot of floor area

Exterior Wall	S.F. Area	15000	20000	25000	30000	35000	40000	45000	50000	55000
	L.F. Perimeter	354	425	457	513	568	583	629	644	683
Brick Face with Concrete Block Back-up	Steel Frame	191.75	186.05	180.25	177.65	175.70	172.75	171.55	169.60	168.70
	R/Conc. Frame	180.85	175.15	169.40	166.75	164.80	161.85	160.65	158.75	157.80
Precast Concrete Panel	Steel Frame	197.25	191.15	184.70	181.85	179.70	176.40	175.00	172.80	171.80
	R/Conc. Frame	187.05	180.75	174.20	171.25	169.10	165.75	164.40	162.15	161.05
Limestone Face Concrete Block Back-up	Steel Frame	204.15	197.20	189.90	186.65	184.20	180.45	178.90	176.40	175.20
	R/Conc. Frame	193.25	186.30	179.00	175.75	173.35	169.55	168.00	165.50	164.30
Perimeter Adj., Add or Deduct	Per 100 L.F.	9.80	7.30	5.85	4.85	4.25	3.70	3.30	2.95	2.65
Story Hgt. Adj., Add or Deduct	Per 1 Ft.	2.40	2.15	1.90	1.75	1.70	1.55	1.45	1.35	1.30
For Basement, add $39.05 per square foot of basement area										

The above costs were calculated using the basic specifications shown on the facing page. These costs should be adjusted where necessary for design alternatives and owner's requirements. Reported completed project costs, for this type of structure, range from $140.55 to $286.75 per S.F.

Common additives

Description	Unit	$ Cost
Carrels Hardwood	Ea.	975 - 2350
Elevators, Hydraulic passenger, 2 stops		
2000# capacity	Ea.	66,600
2500# capacity	Ea.	68,800
3500# capacity	Ea.	73,300
Emergency Lighting, 25 watt, battery operated		
Lead battery	Ea.	330
Nickel cadmium	Ea.	780
Escalators, Metal		
32" wide, 10' story height	Ea.	153,500
20' story height	Ea.	187,500
48" wide, 10' Story height	Ea.	162,000
20' story height	Ea.	197,000
Glass		
32" wide, 10' story height	Ea.	145,500
20' story height	Ea.	178,500
48" wide, 10' story height	Ea.	153,000
20' story height	Ea.	187,500

Description	Unit	$ Cost
Lockers, Steel, Single tier, 60" or 72"	Opng.	221 - 370
2 tier, 60" or 72" total	Opng.	120 - 165
5 tier, box lockers	Opng.	77 - 86.50
Locker bench, lam. maple top only	L.F.	34
Pedestals, steel pipe	Ea.	76
Sound System		
Amplifier, 250 watts	Ea.	2050
Speaker, ceiling or wall	Ea.	213
Trumpet	Ea.	410

Important: See the Reference Section for Location Factors.

Model costs calculated for a 2 story building with 12' story height and 25,000 square feet of floor area

				Unit	Unit Cost	Cost Per S.F.	% Of Sub-Total
A.	**SUBSTRUCTURE**						
1010	Standard Foundations	Poured concrete; strip and spread footings		S.F. Ground	4.72	2.36	
1020	Special Foundations	N/A		—	—	—	
1030	Slab on Grade	4" reinforced concrete with vapor barrier and granular base		S.F. Slab	5.44	2.72	5.3%
2010	Basement Excavation	Site preparation for slab and trench for foundation wall and footing		S.F. Ground	.18	.09	
2020	Basement Walls	4' foundation wall		L.F. Wall	84	1.52	
B.	**SHELL**						
	B10 Superstructure						
1010	Floor Construction	Concrete flat plate		S.F. Floor	27.28	13.64	16.9%
1020	Roof Construction	Concrete flat plate		S.F. Roof	15.50	7.75	
	B20 Exterior Enclosure						
2010	Exterior Walls	Face brick with concrete block backup	75% of wall	S.F. Wall	33.46	11.01	
2020	Exterior Windows	Window wall	25% of wall	Each	41.80	4.58	12.7%
2030	Exterior Doors	Double aluminum and glass		Each	2700	.43	
	B30 Roofing						
3010	Roof Coverings	Built-up tar and gravel with flashing; perlite/EPS composite insulation		S.F. Roof	6.70	3.35	2.7%
3020	Roof Openings	Roof hatches		S.F. Roof	.12	.06	
C.	**INTERIORS**						
1010	Partitions	Gypsum board on metal studs	14 S.F. Floor/L.F. Partition	S.F. Partition	6.31	4.51	
1020	Interior Doors	Single leaf hollow metal	140 S.F. Floor/Door	Each	1157	8.26	
1030	Fittings	N/A		—	—	—	
2010	Stair Construction	Cast in place concrete		Flight	8225	1.31	22.1%
3010	Wall Finishes	50% paint, 50% vinyl wall covering		S.F. Surface	2.47	3.53	
3020	Floor Finishes	50% carpet, 50% vinyl composition tile		S.F. Floor	4.71	4.71	
3030	Ceiling Finishes	Suspended fiberglass board		S.F. Ceiling	5.70	5.70	
D.	**SERVICES**						
	D10 Conveying						
1010	Elevators & Lifts	One hydraulic passenger elevator		Each	83,500	3.34	2.6%
1020	Escalators & Moving Walks	N/A		—	—	—	
	D20 Plumbing						
2010	Plumbing Fixtures	Toilet and service fixtures, supply and drainage	1 Fixture/1040 S.F. Floor	Each	3182	3.06	
2020	Domestic Water Distribution	Gas fired water heater		S.F. Floor	.89	.89	3.6%
2040	Rain Water Drainage	Roof drains		S.F. Roof	1.20	.60	
	D30 HVAC						
3010	Energy Supply	N/A		—	—	—	
3020	Heat Generating Systems	Included in D3050		—	—	—	
3030	Cooling Generating Systems	N/A		—	—	—	16.3 %
3050	Terminal & Package Units	Multizone unit, gas heating, electric cooling		S.F. Floor	20.70	20.70	
3090	Other HVAC Sys. & Equipment	N/A		—	—	—	
	D40 Fire Protection						
4010	Sprinklers	Wet pipe sprinkler system		S.F. Floor	3.19	3.19	3.2%
4020	Standpipes	Standpipe, wet, Class III		S.F. Floor	.89	.89	
	D50 Electrical						
5010	Electrical Service/Distribution	600 ampere service, panel board and feeders		S.F. Floor	1.81	1.81	
5020	Lighting & Branch Wiring	High efficiency fluorescent fixtures, receptacles, switches, A.C. and misc. power		S.F. Floor	12.91	12.91	14.6%
5030	Communications & Security	Addressable alarm systems, internet wiring, communications systems and emergency lighting		S.F. Floor	3.56	3.56	
5090	Other Electrical Systems	Emergency generator, 11.5 kW		S.F. Floor	.15	.15	
E.	**EQUIPMENT & FURNISHINGS**						
1010	Commercial Equipment	N/A		—	—	—	
1020	Institutional Equipment	N/A		—	—	—	
1030	Vehicular Equipment	N/A		—	—	—	0.0 %
1090	Other Equipment	N/A		—	—	—	
F.	**SPECIAL CONSTRUCTION**						
1020	Integrated Construction	N/A		—	—	—	0.0 %
1040	Special Facilities	N/A		—	—	—	
G.	**BUILDING SITEWORK**	**N/A**					
				Sub-Total		126.63	100%
	CONTRACTOR FEES (General Requirements: 10%, Overhead: 5%, Profit: 10%)				25%	31.69	
	ARCHITECT FEES				7%	11.08	
				Total Building Cost		**169.40**	

Costs per square foot of floor area

Exterior Wall	S.F. Area	4000	6000	8000	10000	12000	14000	16000	18000	20000
	L.F. Perimeter	260	340	420	453	460	510	560	610	600
Face Brick with Concrete Block Back-up	Bearing Walls	168.65	160.70	156.80	150.60	144.85	143.20	141.90	140.85	137.75
	Steel Frame	173.25	165.05	161.05	154.70	148.70	147.00	145.65	144.60	141.40
Decorative Concrete Block	Bearing Walls	147.05	141.10	138.10	133.60	129.40	128.15	127.20	126.45	124.15
	Steel Frame	159.05	152.70	149.55	144.80	140.35	139.00	137.95	137.15	134.80
Tilt Up Concrete Wall Panels	Bearing Walls	149.95	144.35	141.60	137.50	133.80	132.65	131.75	131.05	129.10
	Steel Frame	154.50	148.70	145.85	141.60	137.65	136.45	135.50	134.80	132.75
Perimeter Adj., Add or Deduct	Per 100 L.F.	19.85	13.25	9.90	7.95	6.65	5.60	4.90	4.45	3.95
Story Hgt. Adj., Add or Deduct	Per 1 Ft.	3.00	2.60	2.30	2.10	1.75	1.60	1.50	1.60	1.35

For Basement, add $34.15 per square foot of basement area

The above costs were calculated using the basic specifications shown on the facing page. These costs should be adjusted where necessary for design alternatives and owner's requirements. Reported completed project costs, for this type of structure, range from $85.75 to $270.95 per S.F.

Common additives

Description	Unit	$ Cost
Bar, Front bar	L.F.	425
Back bar	L.F.	340
Booth, Upholstered, custom straight	L.F.	250 - 460
"L" or "U" shaped	L.F.	259 - 440
Bowling Alleys, incl. alley, pinsetter		
Scorer, counter & misc. supplies, average	Lane	63,000
For automatic scorer, add	Lane	11,000
Emergency Lighting, 25 watt, battery operated		
Lead battery	Ea.	330
Nickel cadmium	Ea.	780
Kitchen Equipment		
Broiler	Ea.	4050
Coffee urn, twin 6 gallon	Ea.	3000
Cooler, 6 ft. long	Ea.	4175
Dishwasher, 10-12 racks per hr.	Ea.	4275
Food warmer	Ea.	735
Freezer, 44 C.F., reach-in	Ea.	5175
Ice cube maker, 50 lb. per day	Ea.	2000
Range with 1 oven	Ea.	2925

Description	Unit	$ Cost
Movie Equipment		
Projector, 35mm	Ea.	12,300 - 16,900
Screen, wall or ceiling hung	S.F.	8.80 - 13.40
Partitions, Folding leaf, wood		
Acoustic type	S.F.	82 - 135
Seating		
Auditorium chair, all veneer	Ea.	310
Veneer back, padded seat	Ea.	320
Upholstered, spring seat	Ea.	320
Classroom, movable chair & desk	Set	81 - 171
Lecture hall, pedestal type	Ea.	263 - 620
Sound System		
Amplifier, 250 watts	Ea.	2050
Speaker, ceiling or wall	Ea.	213
Trumpet	Ea.	410
Stage Curtains, Medium weight	S.F.	10.75 - 40
Curtain Track, Light duty	L.F.	88.50
Swimming Pools, Complete, gunite	S.F.	97 - 120

Model costs calculated for a 1 story building with 12' story height and 10,000 square feet of floor area

					Unit	Unit Cost	Cost Per S.F.	% Of Sub-Total

A. SUBSTRUCTURE

					Unit	Unit Cost	Cost Per S.F.	% Of Sub-Total
1010	Standard Foundations	Poured concrete; strip and spread footings			S.F. Ground	1.93	1.93	
1020	Special Foundations	N/A			—	—	—	
1030	Slab on Grade	4" reinforced concrete with vapor barrier and granular base			S.F. Slab	5.44	5.44	11.0%
2010	Basement Excavation	Site preparation for slab and trench for foundation wall and footing			S.F. Ground	.32	.32	
2020	Basement Walls	4' foundation wall			L.F. Wall	79	4.63	

B. SHELL

B10 Superstructure

					Unit	Unit Cost	Cost Per S.F.	% Of Sub-Total
1010	Floor Construction	N/A			—	—	—	7.0%
1020	Roof Construction	Metal deck on open web steel joists			S.F. Roof	7.84	7.84	

B20 Exterior Enclosure

					Unit	Unit Cost	Cost Per S.F.	% Of Sub-Total
2010	Exterior Walls	Face brick with concrete block backup		80% of wall	S.F. Wall	33.43	14.54	
2020	Exterior Windows	Aluminum sliding		20% of wall	Each	629	2.14	15.9%
2030	Exterior Doors	Double aluminum and glass and hollow metal			Each	2606	1.05	

B30 Roofing

					Unit	Unit Cost	Cost Per S.F.	% Of Sub-Total
3010	Roof Coverings	Built-up tar and gravel with flashing; perlite/EPS composite insulation			S.F. Roof	7.05	7.05	6.4%
3020	Roof Openings	Roof hatches			S.F. Roof	.09	.09	

C. INTERIORS

					Unit	Unit Cost	Cost Per S.F.	% Of Sub-Total
1010	Partitions	Gypsum board on metal studs	14 S.F. Floor/L.F. Partition		S.F. Partition	8.68	6.20	
1020	Interior Doors	Single leaf hollow metal	140 S.F. Floor/Door		Each	1157	8.26	
1030	Fittings	Toilet partitions, directory board, mailboxes			S.F. Floor	1.69	1.69	
2010	Stair Construction	N/A			—	—	—	27.3%
3010	Wall Finishes	Paint			S.F. Surface	1.86	2.66	
3020	Floor Finishes	50% carpet, 50% vinyl tile			S.F. Floor	4.59	4.59	
3030	Ceiling Finishes	Mineral fiber tile on concealed zee bars			S.F. Ceiling	7.03	7.03	

D. SERVICES

D10 Conveying

					Unit	Unit Cost	Cost Per S.F.	% Of Sub-Total
1010	Elevators & Lifts	N/A			—	—	—	0.0%
1020	Escalators & Moving Walks	N/A			—	—	—	

D20 Plumbing

					Unit	Unit Cost	Cost Per S.F.	% Of Sub-Total
2010	Plumbing Fixtures	Kitchen, toilet and service fixtures, supply and drainage	1 Fixture/910 S.F. Floor		Each	3203	3.52	
2020	Domestic Water Distribution	Electric water heater			S.F. Floor	6.72	6.72	9.7%
2040	Rain Water Drainage	Roof drains			S.F. Roof	.57	.57	

D30 HVAC

					Unit	Unit Cost	Cost Per S.F.	% Of Sub-Total
3010	Energy Supply	N/A			—	—	—	
3020	Heat Generating Systems	Included in D3050			—	—	—	
3030	Cooling Generating Systems	N/A			—	—	—	9.5%
3050	Terminal & Package Units	Single zone rooftop unit, gas heating, electric cooling			S.F. Floor	10.63	10.63	
3090	Other HVAC Sys. & Equipment	N/A			—	—	—	

D40 Fire Protection

					Unit	Unit Cost	Cost Per S.F.	% Of Sub-Total
4010	Sprinklers	Wet pipe sprinkler system			S.F. Floor	3.67	3.67	3.3%
4020	Standpipes	N/A			—	—	—	

D50 Electrical

					Unit	Unit Cost	Cost Per S.F.	% Of Sub-Total
5010	Electrical Service/Distribution	200 ampere service, panel board and feeders			S.F. Floor	1.12	1.12	
5020	Lighting & Branch Wiring	High efficiency fluorescent fixtures, receptacles, switches, A.C. and misc. power			S.F. Floor	5.90	5.90	7.8%
5030	Communications & Security	Addressable alarm systems and emergency lighting			S.F. Floor	1.52	1.52	
5090	Other Electrical Systems	Emergency generator, 15 kW			S.F. Floor	.17	.17	

E. EQUIPMENT & FURNISHINGS

					Unit	Unit Cost	Cost Per S.F.	% Of Sub-Total
1010	Commercial Equipment	Freezer, chest type			S.F. Floor	.95	.95	
1020	Institutional Equipment	N/A			—	—	—	2.1%
1030	Vehicular Equipment	N/A			—	—	—	
1090	Other Equipment	Kitchen equipment, directory board, mailboxes, built-in coat racks			S.F. Floor	1.34	1.34	

F. SPECIAL CONSTRUCTION

					Unit	Unit Cost	Cost Per S.F.	% Of Sub-Total
1020	Integrated Construction	N/A			—	—	—	0.0%
1040	Special Facilities	N/A			—	—	—	

G. BUILDING SITEWORK   N/A

	Sub-Total		111.57	**100%**
CONTRACTOR FEES (General Requirements: 10%, Overhead: 5%, Profit: 10%)		25%	27.87	
ARCHITECT FEES		8%	11.16	
	Total Building Cost		**150.60**	

For customer support on your Square Foot Cost Data, call 877.694.8329.

113

Costs per square foot of floor area

Exterior Wall	S.F. Area	10000	12500	15000	17500	20000	22500	25000	30000	40000
	L.F. Perimeter	400	450	500	550	600	625	650	700	800
Brick Veneer	Steel Frame	324.55	321.10	318.75	317.15	315.90	314.00	312.45	310.20	307.35
	Bearing Wall	338.80	333.95	330.65	328.35	326.65	323.95	321.75	318.60	314.60
Tilt Up Concrete	Steel Frame	324.55	321.10	318.75	317.15	315.90	314.00	312.45	310.20	307.35
Concrete Panels	Bearing Wall	346.40	340.75	337.05	334.35	332.35	329.30	326.80	323.10	318.45
E.I.F.S. System on Metal Studs	Steel Frame	318.45	315.60	313.70	312.35	311.30	309.75	308.50	306.60	304.30
	Bearing Wall	323.40	320.10	317.90	316.25	315.15	313.30	311.85	309.70	307.00
Perimeter Adj., Add or Deduct	Per 100 L.F.	8.50	6.80	5.65	4.80	4.25	3.75	3.35	2.75	2.10
Story Hgt. Adj., Add or Deduct	Per 1 Ft.	1.50	1.30	1.25	1.15	1.10	1.05	0.95	0.85	0.70
For Basement, add $37.55 per square foot of basement area										

The above costs were calculated using the basic specifications shown on the facing page. These costs should be adjusted where necessary for design alternatives and owner's requirements. Reported completed project costs, for this type of structure, range from $177.00 to $359.15 per S.F.

Common additives

Description	Unit	$ Cost	Description	Unit	$ Cost
Clock System			Smoke Detectors		
20 room	Ea.	19,900	Ceiling type	Ea.	227
50 room	Ea.	46,800	Duct type	Ea.	565
Closed Circuit Surveillance, one station			Sound System		
Camera and monitor	Total	1975	Amplifier, 250 watts	Ea.	2050
For additional camera stations, add	Ea.	1075	Speakers, ceiling or wall	Ea.	213
For zoom lens - remote control, add	Ea.	3100 - 10,400	Trumpets	Ea.	410
For automatic iris for low light, add	Ea.	2750			
Directory Boards, plastic, glass covered					
30" x 20"	Ea.	600			
36" x 48"	Ea.	1500			
Aluminum, 24" x 18"	Ea.	635			
36" x 24"	Ea.	755			
48" x 32"	Ea.	1050			
48" x 60"	Ea.	2225			
Emergency Lighting, 25 watt, battery operated					
Lead battery	Ea.	330			
Nickel cadmium	Ea.	780			

Important: See the Reference Section for Location Factors.

Model costs calculated for a 1 story building with 16'-6" story height and 22,500 square feet of floor area

Computer Data Center

				Unit	Unit Cost	Cost Per S.F.	% Of Sub-Total
A. SUBSTRUCTURE							
1010	Standard Foundations	Poured concrete; strip and spread footings		S.F. Ground	1.69	1.69	
1020	Special Foundations	N/A		—	—	—	
1030	Slab on Grade	4" concrete with vapor barrier and granular base		S.F. Slab	5.44	5.44	4.1%
2010	Basement Excavation	Site preparation for slab and trench for foundation wall and footing		S.F. Ground	.18	.20	
2020	Basement Walls	4' foundation wall		L.F. Wall	84	2.32	
B. SHELL							
B10 Superstructure							
1010	Floor Construction	N/A		—	—	—	
1020	Roof Construction	Steel joist, girders & deck on columns		S.F. Roof	9.87	9.87	4.2 %
B20 Exterior Enclosure							
2010	Exterior Walls	Brick veneer on metal studs		S.F. Wall	24.51	10.67	
2020	Exterior Windows	Aluminum sliding	5% of Wall	Each	588	.90	5.4%
2030	Exterior Doors	Aluminum and glass		Each	3908	1.04	
B30 Roofing							
3010	Roof Coverings	Single ply membrane, loose laid and ballasted; polyisocyanurate insulation		S.F. Roof	6.23	6.23	2.7%
3020	Roof Openings	Roof hatch		S.F. Roof	.11	.11	
C. INTERIORS							
1010	Partitions	Gypsum board on metal studs	13 S.F. Floor/L.F. Partition	S.F. Partition	4.97	4.59	
1020	Interior Doors	Solid core wood, metal frame	370 S.F. Floor/Door	Each	771	2.08	
1030	Fittings	Toilet partitions		S.F. Floor	.48	.48	
2010	Stair Construction	N/A		—	—	—	10.3%
3010	Wall Finishes	90% paint, 10% ceramic tile		S.F. Surface	2.09	3.85	
3020	Floor Finishes	65% carpet tile, 10% ceramic tile, 10% quarry tile		S.F. Floor	6.08	6.08	
3030	Ceiling Finishes	Mineral fiber tile on concealed zee bars		S.F. Ceiling	7.03	7.03	
D. SERVICES							
D10 Conveying							
1010	Elevators & Lifts	N/A		—	—	—	0.0 %
1020	Escalators & Moving Walks	N/A		—	—	—	
D20 Plumbing							
2010	Plumbing Fixtures	Toilet and service fixtures, supply and drainage	1 fixture/1180 S.F. Floor	Each	5227	4.43	
2020	Domestic Water Distribution	Gas fired water heater		S.F. Floor	.70	.70	2.7%
2040	Rain Water Drainage	Roof drains		S.F. Floor	1.16	1.16	
D30 HVAC							
3010	Energy Supply	Hot water reheat		S.F. Floor	6.54	6.54	
3020	Heat Generating Systems	Boiler, oil fired hot water		Each	51,875	10.06	
3030	Cooling Generating Systems	Cooling tower		S.F. Floor	9.50	9.50	40.0%
3050	Terminal & Package Units	N/A		—	—	—	
3090	Other HVAC Sys. & Equipment	Heat exchanger, ductwork, air handling units, VAV terminals		S.F. Floor	68	67.78	
D40 Fire Protection							
4010	Sprinklers	85% wet pipe sprinkler system, 15% preaction sprinkler system		S.F. Floor	4.36	4.36	2.1%
4020	Standpipes	Standpipes and hose systems		S.F. Floor	.65	.65	
D50 Electrical							
5010	Electrical Service/Distribution	1200 ampere service, panel board and feeders		S.F. Floor	3.75	3.75	
5020	Lighting & Branch Wiring	High efficiency fluorescent fixtures, receptacles, A.C. and misc. power		S.F. Floor	16.76	16.76	27.6%
5030	Communications & Security	Addressable alarm & telephone systems, internet wiring and emergency lighting		S.F. Floor	31.53	31.53	
5090	Other Electrical Systems	Emergency generator, 350 kW; UPS system		S.F. Floor	12.64	12.64	
E. EQUIPMENT & FURNISHINGS							
1010	Commercial Equipment	N/A		—	—	—	
1020	Institutional Equipment	N/A		—	—	—	0.0 %
1030	Vehicular Equipment	N/A		—	—	—	
1090	Other Equipment	N/A		—	—	—	
F. SPECIAL CONSTRUCTION & DEMOLITION							
1020	Integrated Construction	Pedestal access floor	15% of Floor	S.F. Floor	2.30	2.30	1.0%
1040	Special Facilities	N/A		—	—	—	
G. BUILDING SITEWORK	**N/A**						

		Sub-Total	234.74	100%
CONTRACTOR FEES (General Requirements: 10%, Overhead: 5%, Profit: 10%)		25%	58.72	
ARCHITECT FEES		7%	20.54	
	Total Building Cost		314	

For customer support on your Square Foot Cost Data, call 877.694.8329.

115

Costs per square foot of floor area

Exterior Wall	S.F. Area	16000	23000	30000	37000	44000	51000	58000	65000	72000
	L.F. Perimeter	597	763	821	968	954	1066	1090	1132	1220
Limestone with Concrete Block Back-up	R/Conc. Frame	224.85	216.10	206.95	204.25	197.90	196.30	193.30	191.20	190.35
	Steel Frame	226.80	218.05	208.90	206.25	199.85	198.25	195.25	193.15	192.25
Face Brick with Concrete Block Back-up	R/Conc. Frame	212.60	205.25	198.00	195.70	190.80	189.40	187.15	185.50	184.80
	Steel Frame	214.50	207.15	199.95	197.65	192.75	191.40	189.05	187.45	186.70
Stone with Concrete Block Back-up	R/Conc. Frame	218.65	210.60	202.45	199.95	194.35	192.85	190.20	188.35	187.50
	Steel Frame	220.60	212.55	204.35	201.90	196.25	194.80	192.10	190.25	189.50
Perimeter Adj., Add or Deduct	Per 100 L.F.	7.70	5.45	4.20	3.40	2.85	2.40	2.10	1.95	1.70
Story Hgt. Adj., Add or Deduct	Per 1 Ft.	2.70	2.40	2.05	1.95	1.60	1.55	1.30	1.25	1.20

For Basement, add $ 32.25 per square foot of basement area

The above costs were calculated using the basic specifications shown on the facing page. These costs should be adjusted where necessary for design alternatives and owner's requirements. Reported completed project costs, for this type of structure, range from $ 151.00 to $ 281.30 per S.F.

Common additives

Description	Unit	$ Cost	Description	Unit	$ Cost
Benches, Hardwood	L.F.	133 - 245	Flagpoles, Complete		
Clock System			Aluminum, 20' high	Ea.	1850
20 room	Ea.	19,900	40' high	Ea.	4200
50 room	Ea.	46,800	70' high	Ea.	11,000
Closed Circuit Surveillance, One station			Fiberglass, 23' high	Ea.	1325
Camera and monitor	Ea.	1975	39'-5" high	Ea.	3450
For additional camera stations, add	Ea.	1075	59' high	Ea.	6850
Directory Boards, Plastic, glass covered			Intercom System, 25 station capacity		
30" x 20"	Ea.	600	Master station	Ea.	3250
36" x 48"	Ea.	1500	Intercom outlets	Ea.	189
Aluminum, 24" x 18"	Ea.	635	Handset	Ea.	520
36" x 24"	Ea.	755	Safe, Office type, 1 hour rating		
48" x 32"	Ea.	1050	30" x 18" x 18"	Ea.	2500
48" x 60"	Ea.	2225	60" x 36" x 18", double door	Ea.	8125
Emergency Lighting, 25 watt, battery operated			Smoke Detectors		
Lead battery	Ea.	330	Ceiling type	Ea.	227
Nickel cadmium	Ea.	780	Duct type	Ea.	565

Important: See the Reference Section for Location Factors.

Model costs calculated for a 1 story building with 14' story height and 30,000 square feet of floor area

Courthouse, 1 Story

				Unit	Unit Cost	Cost Per S.F.	% Of Sub-Total
A.	**SUBSTRUCTURE**						
1010	Standard Foundations	Poured concrete; strip and spread footings		S.F. Ground	1.87	1.87	
1020	Special Foundations	N/A		—	—	—	
1030	Slab on Grade	4" reinforced concrete with vapor barrier and granular base		S.F. Slab	5.44	5.44	6.4%
2010	Basement Excavation	Site preparation for slab and trench for foundation wall and footing		S.F. Ground	.18	.18	
2020	Basement Walls	4' foundation wall		L.F. Wall	84	2.44	
B.	**SHELL**						
	B10 Superstructure						
1010	Floor Construction	Cast-in-place columns		L.F. Column	74	1.65	14.1%
1020	Roof Construction	Cast-in-place concrete waffle slab		S.F. Roof	20.20	20.20	
	B20 Exterior Enclosure						
2010	Exterior Walls	Limestone panels with concrete block backup	75% of wall	S.F. Wall	57	16.32	
2020	Exterior Windows	Aluminum with insulated glass	25% of wall	Each	750	3.13	12.9%
2030	Exterior Doors	Double wood		Each	1911	.44	
	B30 Roofing						
3010	Roof Coverings	Built-up tar and gravel with flashing; perlite/EPS composite insulation		S.F. Roof	5.97	5.97	3.9%
3020	Roof Openings	Roof hatches		S.F. Roof	.09	.09	
C.	**INTERIORS**						
1010	Partitions	Plaster on metal studs	10 S.F. Floor/L.F. Partition	S.F. Partition	13.76	16.51	
1020	Interior Doors	Single leaf wood	100 S.F. Floor/Door	Each	616	6.16	
1030	Fittings	Toilet partitions		S.F. Floor	.49	.49	
2010	Stair Construction	N/A		—	—	—	33.7%
3010	Wall Finishes	70% paint, 20% wood paneling, 10% vinyl wall covering		S.F. Surface	2.16	5.18	
3020	Floor Finishes	60% hardwood, 20% carpet, 20% terrazzo		S.F. Floor	13.49	13.49	
3030	Ceiling Finishes	Gypsum plaster on metal lath, suspended		S.F. Ceiling	10.31	10.31	
D.	**SERVICES**						
	D10 Conveying						
1010	Elevators & Lifts	N/A		—	—	—	0.0 %
1020	Escalators & Moving Walks	N/A		—	—	—	
	D20 Plumbing						
2010	Plumbing Fixtures	Toilet and service fixtures, supply and drainage	1 Fixture/1110 S.F. Floor	Each	4829	4.35	
2020	Domestic Water Distribution	Electric hot water heater		S.F. Floor	2.41	2.41	5.1%
2040	Rain Water Drainage	Roof drains		S.F. Roof	1.17	1.17	
	D30 HVAC						
3010	Energy Supply	N/A		—	—	—	
3020	Heat Generating Systems	Included in D3050		—	—	—	
3030	Cooling Generating Systems	N/A		—	—	—	13.4 %
3050	Terminal & Package Units	Multizone unit, gas heating, electric cooling		S.F. Floor	20.70	20.70	
3090	Other HVAC Sys. & Equipment	N/A		—	—	—	
	D40 Fire Protection						
4010	Sprinklers	Wet pipe sprinkler system		S.F. Floor	2.94	2.94	2.2%
4020	Standpipes	Standpipe, wet, Class III		S.F. Floor	.49	.49	
	D50 Electrical						
5010	Electrical Service/Distribution	400 ampere service, panel board and feeders		S.F. Floor	1.03	1.03	
5020	Lighting & Branch Wiring	High efficiency fluorescent fixtures, receptacles, switches, A.C. and misc. power		S.F. Floor	9.96	9.96	8.3%
5030	Communications & Security	Addressable alarm systems, internet wiring, and emergency lighting		S.F. Floor	1.70	1.70	
5090	Other Electrical Systems	Emergency generator, 11.5 kW		S.F. Floor	.12	.12	
E.	**EQUIPMENT & FURNISHINGS**						
1010	Commercial Equipment	N/A		—	—	—	
1020	Institutional Equipment	N/A		—	—	—	0.0 %
1030	Vehicular Equipment	N/A		—	—	—	
1090	Other Equipment	N/A		—	—	—	
F.	**SPECIAL CONSTRUCTION**						
1020	Integrated Construction	N/A		—	—	—	0.0 %
1040	Special Facilities	N/A		—	—	—	
G.	**BUILDING SITEWORK**	**N/A**					

Sub-Total		154.74	**100%**
CONTRACTOR FEES (General Requirements: 10%, Overhead: 5%, Profit: 10%)	25%	38.67	
ARCHITECT FEES	7%	13.54	
Total Building Cost		206.95	

For customer support on your Square Foot Cost Data, call 877.694.8329.

117

Costs per square foot of floor area

Exterior Wall	S.F. Area	30000	40000	45000	50000	60000	70000	80000	90000	100000
	L.F. Perimeter	410	493	535	533	600	666	733	800	795
Limestone with Concrete Block Back-up	R/Conc. Frame	239.30	230.45	227.55	222.20	218.25	215.35	213.20	211.55	207.75
	Steel Frame	245.20	236.35	233.40	228.05	224.10	221.25	219.10	217.40	213.65
Face Brick with Concrete Block Back-up	R/Conc. Frame	227.85	220.10	217.50	213.20	209.80	207.30	205.50	204.05	201.10
	Steel Frame	233.80	226.05	223.45	219.15	215.75	213.30	211.50	210.00	207.00
Stone with Concrete Block Back-up	R/Conc. Frame	233.55	225.30	222.50	217.70	214.05	211.30	209.35	207.80	204.45
	Steel Frame	239.45	231.15	228.35	223.60	219.95	217.20	215.25	213.70	210.30
Perimeter Adj., Add or Deduct	Per 100 L.F.	11.45	8.60	7.60	6.90	5.70	4.95	4.30	3.85	3.45
Story Hgt. Adj., Add or Deduct	Per 1 Ft.	3.70	3.35	3.15	2.85	2.70	2.60	2.50	2.40	2.15

For Basement, add $32.95 per square foot of basement area

The above costs were calculated using the basic specifications shown on the facing page. These costs should be adjusted where necessary for design alternatives and owner's requirements. Reported completed project costs, for this type of structure, range from $151.70 to $282.95 per S.F.

Common additives

Description	Unit	$ Cost	Description	Unit	$ Cost
Benches, Hardwood	L.F.	133 - 245	Emergency Lighting, 25 watt, battery operated		
Clock System			Lead battery	Ea.	330
20 room	Ea.	19,900	Nickel cadmium	Ea.	780
50 room	Ea.	46,800	Flagpoles, Complete		
Closed Circuit Surveillance, One station			Aluminum, 20' high	Ea.	1850
Camera and monitor	Ea.	1975	40' high	Ea.	4200
For additional camera stations, add	Ea.	1075	70' high	Ea.	11,000
Directory Boards, Plastic, glass covered			Fiberglass, 23' high	Ea.	1325
30" x 20"	Ea.	600	39'-5" high	Ea.	3450
36" x 48"	Ea.	1500	59' high	Ea.	6850
Aluminum, 24" x 18"	Ea.	635	Intercom System, 25 station capacity		
36" x 24"	Ea.	755	Master station	Ea.	3250
48" x 32"	Ea.	1050	Intercom outlets	Ea.	189
48" x 60"	Ea.	2225	Handset	Ea.	520
Elevators, Hydraulic passenger, 2 stops			Safe, Office type, 1 hour rating		
1500# capacity	Ea.	65,700	30" x 18" x 18"	Ea.	2500
2500# capacity	Ea.	68,800	60" x 36" x 18", double door	Ea.	8125
3500# capacity	Ea.	73,300	Smoke Detectors		
Additional stop, add	Ea.	7825	Ceiling type	Ea.	227
			Duct type	Ea.	565

Important: See the Reference Section for Location Factors.

Model costs calculated for a 3 story building with 12' story height and 60,000 square feet of floor area

				Unit	Unit Cost	Cost Per S.F.	% Of Sub-Total
A.	**SUBSTRUCTURE**						
1010	Standard Foundations	Poured concrete; strip and spread footings		S.F. Ground	2.01	.67	
1020	Special Foundations	N/A		—	—	—	
1030	Slab on Grade	4" reinforced concrete with vapor barrier and granular base		S.F. Slab	5.44	1.81	2.1%
2010	Basement Excavation	Site preparation for slab and trench for foundation wall and footing		S.F. Ground	.18	.06	
2020	Basement Walls	4' foundation wall		L.F. Wall	79	.96	
B.	**SHELL**						
	B10 Superstructure						
1010	Floor Construction	Concrete slab with metal deck and beams		S.F. Floor	36.12	24.08	19.0%
1020	Roof Construction	Concrete slab with metal deck and beams		S.F. Roof	20.76	6.92	
	B20 Exterior Enclosure						
2010	Exterior Walls	Face brick with concrete block backup	75% of wall	S.F. Wall	36.07	9.74	
2020	Exterior Windows	Horizontal pivoted steel	25% of wall	Each	758	7.58	10.9%
2030	Exterior Doors	Double aluminum and glass and hollow metal		Each	3583	.36	
	B30 Roofing						
3010	Roof Coverings	Built-up tar and gravel with flashing; perlite/EPS composite insulation		S.F. Roof	6.42	2.14	1.3%
3020	Roof Openings	N/A		—	—	—	
C.	**INTERIORS**						
1010	Partitions	Plaster on metal studs	10 S.F. Floor/L.F. Partition	S.F. Partition	13.92	13.92	
1020	Interior Doors	Single leaf wood	100 S.F. Floor/Door	Each	616	6.16	
1030	Fittings	Toilet partitions		S.F. Floor	.25	.25	
2010	Stair Construction	Concrete filled metal pan		Flight	20,200	1.69	30.8%
3010	Wall Finishes	70% paint, 20% wood paneling, 10% vinyl wall covering		S.F. Surface	2.16	4.32	
3020	Floor Finishes	60% hardwood, 20% terrazzo, 20% carpet		S.F. Floor	13.49	13.49	
3030	Ceiling Finishes	Gypsum plaster on metal lath, suspended		S.F. Ceiling	10.31	10.31	
D.	**SERVICES**						
	D10 Conveying						
1010	Elevators & Lifts	Five hydraulic passenger elevators		Each	154,560	12.88	7.9%
1020	Escalators & Moving Walks	N/A		—	—	—	
	D20 Plumbing						
2010	Plumbing Fixtures	Toilet and service fixtures, supply and drainage	1 Fixture/665 S.F. Floor	Each	3312	4.98	
2020	Domestic Water Distribution	Electric water heater		S.F. Floor	2.77	2.77	5.2%
2040	Rain Water Drainage	Roof drains		S.F. Roof	2.13	.71	
	D30 HVAC						
3010	Energy Supply	N/A		—	—	—	
3020	Heat Generating Systems	Included in D3050		—	—	—	
3030	Cooling Generating Systems	N/A		—	—	—	12.7 %
3050	Terminal & Package Units	Multizone unit, gas heating, electric cooling		S.F. Floor	20.70	20.70	
3090	Other HVAC Sys. & Equipment	N/A		Unit	—	—	
	D40 Fire Protection						
4010	Sprinklers	Wet pipe sprinkler system		S.F. Floor	3.10	3.10	2.2%
4020	Standpipes	Standpipe, wet, Class III		S.F. Floor	.45	.45	
	D50 Electrical						
5010	Electrical Service/Distribution	800 ampere service, panel board and feeders		S.F. Floor	.95	.95	
5020	Lighting & Branch Wiring	High efficiency fluorescent fixtures, receptacles, switches, A.C. and misc. power		S.F. Floor	10.10	10.10	7.9%
5030	Communications & Security	Addressable alarm systems, internet wiring, and emergency lighting		S.F. Floor	1.53	1.53	
5090	Other Electrical Systems	Emergency generator, 15 kW		S.F. Floor	.21	.21	
E.	**EQUIPMENT & FURNISHINGS**						
1010	Commercial Equipment	N/A		—	—	—	
1020	Institutional Equipment	N/A		—	—	—	
1030	Vehicular Equipment	N/A		—	—	—	0.0 %
1090	Other Equipment	N/A		—	—	—	
F.	**SPECIAL CONSTRUCTION**						
1020	Integrated Construction	N/A		—	—	—	
1040	Special Facilities	N/A		—	—	—	0.0 %
G.	**BUILDING SITEWORK**	**N/A**					

		Sub-Total	162.84	100%
CONTRACTOR FEES (General Requirements: 10%, Overhead: 5%, Profit: 10%)		25%	40.70	
ARCHITECT FEES		6%	12.21	
		Total Building Cost	**215.75**	

For customer support on your Square Foot Cost Data, call 877.694.8329.

119

Costs per square foot of floor area

Exterior Wall	S.F. Area	2000	5000	7000	10000	12000	15000	18000	21000	25000
	L.F. Perimeter	200	310	360	440	480	520	560	600	660
Tiltup Concrete Panel	Steel Joists	214.05	183.40	175.75	170.30	167.55	164.10	161.70	160.10	158.65
Decorative Concrete Block	Bearing Walls	219.40	186.10	177.65	171.70	168.70	164.80	162.20	160.35	158.75
Brick on Block	Steel Joists	253.15	207.25	195.50	187.15	182.95	177.45	173.80	171.20	168.95
Stucco on Concrete Block	Wood Truss	214.10	185.10	177.65	172.35	169.65	166.15	163.80	162.15	160.70
Brick Veneer	Wood Frame	223.55	191.15	182.80	176.80	173.75	169.80	167.15	165.25	163.70
Wood Siding	Wood Frame	205.35	180.05	173.70	169.15	166.75	163.85	161.85	160.45	159.30
Perimeter Adj., Add or Deduct	Per 100 L.F.	27.50	11.05	7.80	5.55	4.60	3.70	3.10	2.65	2.25
Story Hgt. Adj., Add or Deduct	Per 1 Ft.	2.70	1.75	1.35	1.20	1.10	0.95	0.85	0.75	0.75

For Basement, add $32.05 per square foot of basement area

The above costs were calculated using the basic specifications shown on the facing page. These costs should be adjusted where necessary for design alternatives and owner's requirements. Reported completed project costs, for this type of structure, range from $65.40 to $238.35 per S.F.

Common additives

Description	Unit	$ Cost
Emergency Lighting, 25 watt, battery operated		
Lead battery	Ea.	330
Nickel cadmium	Ea.	780
Flagpoles, Complete		
Aluminum, 20' high	Ea.	1850
40' high	Ea.	4200
70' high	Ea.	11,000
Fiberglass, 23' high	Ea.	1325
39'-5" high	Ea.	3450
59' high	Ea.	6850
Gym Floor, Incl. sleepers and finish, maple	S.F.	14.55
Intercom System, 25 Station capacity		
Master station	Ea.	3250
Intercom outlets	Ea.	189
Handset	Ea.	520

Description	Unit	$ Cost
Lockers, Steel, single tier, 60" to 72"	Opng.	221 - 370
2 tier, 60" to 72" total	Opng.	120 - 165
5 tier, box lockers	Opng.	77 - 86.50
Locker bench, lam. maple top only	L.F.	34
Pedestals, steel pipe	Ea.	76
Smoke Detectors		
Ceiling type	Ea.	227
Duct type	Ea.	565
Sound System		
Amplifier, 250 watts	Ea.	2050
Speaker, ceiling or wall	Ea.	213
Trumpet	Ea.	410

Important: See the Reference Section for Location Factors.

Model costs calculated for a 1 story building with 12' story height and 10,000 square feet of floor area

Day Care Center

				Unit	Unit Cost	Cost Per S.F.	% Of Sub-Total
A. SUBSTRUCTURE							
1010	Standard Foundations	Poured concrete; strip and spread footings		S.F. Ground	2.17	2.17	
1020	Special Foundations	N/A		—	—	—	
1030	Slab on Grade	4" concrete with vapor barrier and granular base		S.F. Slab	4.90	4.90	8.5%
2010	Basement Excavation	Site preparation for slab and trench for foundation wall and footing		S.F. Ground	.32	.32	
2020	Basement Walls	4' foundation wall		L.F. Wall	84	3.68	
B. SHELL							
B10 Superstructure							
1010	Floor Construction	Wood beams on columns		S.F. Floor	.22	.22	6.7%
1020	Roof Construction	Wood trusses		S.F. Roof	8.48	8.48	
B20 Exterior Enclosure							
2010	Exterior Walls	Brick veneer on wood studs	85% of wall	S.F. Wall	25	11.22	
2020	Exterior Windows	Window wall	15% of wall	Each	49.40	3.91	15.4%
2030	Exterior Doors	Aluminum and glass; steel		Each	2541	4.83	
B30 Roofing							
3010	Roof Coverings	Asphalt shingles, 9" fiberglass batt insulation, gutters and downspouts		S.F. Roof	4.64	4.64	3.6%
3020	Roof Openings	N/A		—	—	—	
C. INTERIORS							
1010	Partitions	Gypsum board on wood studs	8 S.F. Floor/S.F. Partition	S.F. Partition	7.98	5.32	
1020	Interior Doors	Single leaf hollow metal	700 S.F. Floor/Door	Each	1157	3.09	
1030	Fittings	Toilet partitions		S.F. Floor	.44	.44	
2010	Stair Construction	N/A					15.1%
3010	Wall Finishes	Paint		S.F. Surface	1.60	2.13	
3020	Floor Finishes	5% quarry tile, 95% vinyl composition tile		S.F. Floor	2.94	2.94	
3030	Ceiling Finishes	Fiberglass tile on tee grid		S.F. Ceiling	5.70	5.70	
D. SERVICES							
D10 Conveying							
1010	Elevators & Lifts	N/A		—	—	—	0.0 %
1020	Escalators & Moving Walks	N/A		—	—	—	
D20 Plumbing							
2010	Plumbing Fixtures	Toilet and service fixtures, supply and drainage	1 Fixture/455 S.F. Floor	Each	3012	16.92	
2020	Domestic Water Distribution	Electric water heater		S.F. Floor	2.03	2.03	14.6%
2040	Rain Water Drainage	N/A		—	—	—	
D30 HVAC							
3010	Energy Supply	Oil fired hot water, wall fin radiation		S.F. Floor	11.82	11.82	
3020	Heat Generating Systems	—		—	—	—	
3030	Cooling Generating Systems	N/A		—	—	—	20.5%
3050	Terminal & Package Units	Split systems with air cooled condensing units		S.F. Floor	14.75	14.75	
3090	Other HVAC Sys. & Equipment	N/A		—	—	—	
D40 Fire Protection							
4010	Sprinklers	Sprinkler, light hazard		S.F. Floor	3.67	3.67	3.5%
4020	Standpipes	Standpipe		S.F. Floor	.86	.86	
D50 Electrical							
5010	Electrical Service/Distribution	200 ampere service, panel board and feeders		S.F. Floor	1.12	1.12	
5020	Lighting & Branch Wiring	High efficiency fluorescent fixtures, receptacles, switches, A.C. and misc. power		S.F. Floor	8.27	8.27	8.5%
5030	Communications & Security	Addressable alarm systems and emergency lighting		S.F. Floor	1.55	1.55	
5090	Other Electrical Systems	Emergency generator, 15 kW		S.F. Floor	.08	.08	
E. EQUIPMENT & FURNISHINGS							
1010	Commercial Equipment	N/A		—	—	—	
1020	Institutional Equipment	Cabinets and countertop		S.F. Floor	4.72	4.72	3.6 %
1030	Vehicular Equipment	N/A		—	—	—	
1090	Other Equipment	N/A		—	—	—	
F. SPECIAL CONSTRUCTION							
1020	Integrated Construction	N/A		—	—	—	0.0 %
1040	Special Facilities	N/A		—	—	—	
G. BUILDING SITEWORK	**N/A**						

		Sub-Total	129.78	100%
CONTRACTOR FEES (General Requirements: 10%, Overhead: 5%, Profit: 10%)		25%	32.42	
ARCHITECT FEES		9%	14.60	
		Total Building Cost	**176.80**	

For customer support on your Square Foot Cost Data, call 877.694.8329.

121

Costs per square foot of floor area

Exterior Wall	S.F. Area	12000	18000	24000	30000	36000	42000	48000	54000	60000
	L.F. Perimeter	460	580	713	730	826	880	965	1006	1045
Concrete Block	Steel Frame	139.70	132.50	129.25	124.75	123.20	121.40	120.50	119.20	118.25
	Bearing Walls	137.25	130.05	126.85	122.30	120.75	118.95	118.05	116.75	115.80
Precast Concrete Panels	Steel Frame	147.20	138.80	135.10	129.55	127.70	125.45	124.45	122.85	121.60
Insulated Metal Panels	Steel Frame	142.00	134.45	131.10	126.25	124.60	122.65	121.70	120.35	119.25
Face Brick on Common Brick	Steel Frame	158.15	147.95	143.60	136.45	134.25	131.40	130.15	128.15	126.60
Tilt-up Concrete Panel	Steel Frame	141.45	133.95	130.65	125.85	124.25	122.35	121.40	120.10	119.00
Perimeter Adj., Add or Deduct	Per 100 L.F.	5.70	3.80	2.90	2.25	1.90	1.60	1.40	1.30	1.10
Story Hgt. Adj., Add or Deduct	Per 1 Ft.	0.85	0.70	0.65	0.50	0.50	0.40	0.40	0.40	0.35

For Basement, add $32.65 per square foot of basement area

The above costs were calculated using the basic specifications shown on the facing page. These costs should be adjusted where necessary for design alternatives and owner's requirements. Reported completed project costs, for this type of structure, range from $48.55 to $187.90 per S.F.

Common additives

Description	Unit	$ Cost
Clock System		
20 room	Ea.	19,900
50 room	Ea.	46,800
Dock Bumpers, Rubber blocks		
4-1/2" thick, 10" high, 14" long	Ea.	92
24" long	Ea.	133
36" long	Ea.	120
12" high, 14" long	Ea.	135
24" long	Ea.	152
36" long	Ea.	176
6" thick, 10" high, 14" long	Ea.	141
24" long	Ea.	167
36" long	Ea.	252
20" high, 11" long	Ea.	208
Dock Boards, Heavy		
60" x 60" Aluminum, 5,000# cap.	Ea.	1600
9000# cap.	Ea.	1775
15,000# cap.	Ea.	1925

Description	Unit	$ Cost
Dock Levelers, Hinged 10 ton cap.		
6' x 8'	Ea.	7075
7' x 8'	Ea.	8100
Partitions, Woven wire, 10 ga., 1-1/2" mesh		
4' wide x 7' high	Ea.	204
8' high	Ea.	230
10' High	Ea.	273
Platform Lifter, Portable, 6' x 6'		
3000# cap.	Ea.	10,200
4000# cap.	Ea.	12,500
Fixed, 6' x 8', 5000# cap.	Ea.	13,400

Important: See the Reference Section for Location Factors.

Model costs calculated for a 1 story building with 20' story height and 30,000 square feet of floor area

Factory, 1 Story

					Unit	Unit Cost	Cost Per S.F.	% Of Sub-Total
A.	**SUBSTRUCTURE**							
1010	Standard Foundations	Poured concrete; strip and spread footings			S.F. Ground	1.56	1.56	
1020	Special Foundations	N/A			—	—	—	
1030	Slab on Grade	4" reinforced concrete with vapor barrier and granular base			S.F. Slab	7.13	7.13	12.1%
2010	Basement Excavation	Site preparation for slab and trench for foundation wall and footing			S.F. Ground	.18	.18	
2020	Basement Walls	4' foundation wall			L.F. Wall	84	2.44	
B.	**SHELL**							
	B10 Superstructure							
1010	Floor Construction	N/A			—	—	—	
1020	Roof Construction	Metal deck, open web steel joists, beams and columns			S.F. Roof	9.83	9.83	10.5 %
	B20 Exterior Enclosure							
2010	Exterior Walls	Concrete block		75% of wall	S.F. Wall	8.85	3.23	
2020	Exterior Windows	Industrial horizontal pivoted steel		25% of wall	Each	864	3.29	8.3%
2030	Exterior Doors	Double aluminum and glass, hollow metal, steel overhead			Each	2493	1.24	
	B30 Roofing							
3010	Roof Coverings	Built-up tar and gravel with flashing; perlite/EPS composite insulation			S.F. Roof	6.14	6.14	7.0%
3020	Roof Openings	Roof hatches			S.F. Roof	.37	.37	
C.	**INTERIORS**							
1010	Partitions	Concrete block		60 S.F. Floor/L.F. Partition	S.F. Partition	9	1.80	
1020	Interior Doors	Single leaf hollow metal and fire doors		600 S.F. Floor/Door	Each	1157	1.93	
1030	Fittings	Toilet partitions			S.F. Floor	.97	.97	
2010	Stair Construction	N/A			—	—	—	8.2%
3010	Wall Finishes	Paint			S.F. Surface	4.93	1.97	
3020	Floor Finishes	Vinyl composition tile		10% of floor	S.F. Floor	2.70	.27	
3030	Ceiling Finishes	Fiberglass board on exposed grid system		10% of area	S.F. Ceiling	7.03	.71	
D.	**SERVICES**							
	D10 Conveying							
1010	Elevators & Lifts	N/A			—	—	—	0.0 %
1020	Escalators & Moving Walks	N/A			—	—	—	
	D20 Plumbing							
2010	Plumbing Fixtures	Toilet and service fixtures, supply and drainage		1 Fixture/1000 S.F. Floor	Each	5130	5.13	
2020	Domestic Water Distribution	Gas fired water heater			S.F. Floor	.60	.60	7.2%
2040	Rain Water Drainage	Roof drains			S.F. Roof	.98	.98	
	D30 HVAC							
3010	Energy Supply	Oil fired hot water, unit heaters			S.F. Floor	10.19	10.19	
3020	Heat Generating Systems	N/A			—	—	—	
3030	Cooling Generating Systems	Chilled water, air cooled condenser system			S.F. Floor	12.03	12.03	23.8%
3050	Terminal & Package Units	N/A			—	—	—	
3090	Other HVAC Sys. & Equipment	N/A			—	—	—	
	D40 Fire Protection							
4010	Sprinklers	Sprinklers, ordinary hazard			S.F. Floor	4.25	4.25	5.2%
4020	Standpipes	Standpipe, wet, Class III			S.F. Floor	.62	.62	
	D50 Electrical							
5010	Electrical Service/Distribution	600 ampere service, panel board and feeders			S.F. Floor	1.06	1.06	
5020	Lighting & Branch Wiring	High intensity discharge fixtures, receptacles, switches, A.C. and misc. power			S.F. Floor	13.35	13.35	17.0%
5030	Communications & Security	Addressable alarm systems and emergency lighting			S.F. Floor	1.49	1.49	
5090	Other Electrical Systems	N/A			—	—	—	
E.	**EQUIPMENT & FURNISHINGS**							
1010	Commercial Equipment	N/A			—	—	—	
1020	Institutional Equipment	N/A			—	—	—	
1030	Vehicular Equipment	Dock shelters			Each	.51	.51	0.5 %
1090	Other Equipment	N/A			—	—	—	
F.	**SPECIAL CONSTRUCTION**							
1020	Integrated Construction	N/A			—	—	—	0.0 %
1040	Special Facilities	N/A			—	—	—	
G.	**BUILDING SITEWORK**	**N/A**						

		Sub-Total	93.27	100%
CONTRACTOR FEES (General Requirements: 10%, Overhead: 5%, Profit: 10%)			25%	23.32
ARCHITECT FEES			7%	8.16

Total Building Cost	**124.75**

Costs per square foot of floor area

Exterior Wall	S.F. Area	20000	30000	40000	50000	60000	70000	80000	90000	100000
	L.F. Perimeter	625	750	850	950	1015	1095	1150	1275	1300
Face Brick Common Brick Back-up	Steel Frame	209.20	190.15	179.00	172.30	166.30	162.60	159.05	158.25	155.05
	Concrete Frame	199.20	180.15	168.95	162.30	156.30	152.55	149.00	**148.25**	145.05
Face Brick Concrete Block Back-up	Steel Frame	208.00	189.20	178.15	171.55	165.65	161.95	158.45	157.70	154.55
	Concrete Frame	198.00	179.15	168.15	161.55	155.60	151.95	148.45	147.70	144.50
Stucco on Concrete Block	Steel Frame	189.20	174.15	165.40	160.10	155.45	152.55	149.85	149.20	146.70
	Concrete Frame	179.20	164.10	155.40	150.10	145.45	142.55	139.80	139.20	136.70
Perimeter Adj., Add or Deduct	Per 100 L.F.	12.85	8.50	6.35	5.10	4.25	3.65	3.15	2.85	2.55
Story Hgt. Adj., Add or Deduct	Per 1 Ft.	6.15	4.90	4.15	3.70	3.30	3.05	2.80	2.80	2.55

For Basement, add $35.60 per square foot of basement area

The above costs were calculated using the basic specifications shown on the facing page. These costs should be adjusted where necessary for design alternatives and owner's requirements. Reported completed project costs, for this type of structure, range from $48.55 to $187.95 per S.F.

Common additives

Description	Unit	$ Cost
Clock System		
20 room	Ea.	19,900
50 room	Ea.	46,800
Dock Bumpers, Rubber blocks		
4-1/2" thick, 10" high, 14" long	Ea.	92
24" long	Ea.	133
36" long	Ea.	120
12" high, 14" long	Ea.	135
24" long	Ea.	152
36" long	Ea.	176
6" thick, 10" high, 14" long	Ea.	141
24" long	Ea.	167
36" long	Ea.	252
20" high, 11" long	Ea.	208
Dock Boards, Heavy		
60" x 60" Aluminum, 5,000# cap.	Ea.	1600
9000# cap.	Ea.	1775
15,000# cap.	Ea.	1925

Description	Unit	$ Cost
Dock Levelers, Hinged 10 ton cap.		
6' x 8'	Ea.	7075
7' x 8'	Ea.	8100
Elevator, Hydraulic freight, 2 stops		
3500# capacity	Ea.	142,400
4000# capacity	Ea.	146,900
Additional stop, add	Ea.	9125
Partitions, Woven wire, 10 ga., 1-1/2" mesh		
4' Wide x 7' high	Ea.	204
8' High	Ea.	230
10' High	Ea.	273
Platform Lifter, Portable, 6' x 6'		
3000# cap.	Ea.	10,200
4000# cap.	Ea.	12,500
Fixed, 6' x 8', 5000# cap.	Ea.	13,400

Important: See the Reference Section for Location Factors.

Model costs calculated for a 3 story building with 12' story height and 90,000 square feet of floor area

Factory, 3 Story

					Unit	Unit Cost	Cost Per S.F.	% Of Sub-Total
A. SUBSTRUCTURE								
1010	Standard Foundations	Poured concrete; strip and spread footings			S.F. Ground	2.28	.76	
1020	Special Foundations	N/A			—	—	—	
1030	Slab on Grade	4" reinforced concrete with vapor barrier and granular base			S.F. Slab	11.05	3.68	5.1%
2010	Basement Excavation	Site preparation for slab and trench for foundation wall and footing			S.F. Ground	.18	.06	
2020	Basement Walls	4' foundation wall			L.F. Wall	84	1.18	
B. SHELL								
B10 Superstructure								
1010	Floor Construction	Concrete flat slab			S.F. Floor	22.23	14.82	18.0%
1020	Roof Construction	Concrete flat slab			S.F. Roof	15.99	5.33	
B20 Exterior Enclosure								
2010	Exterior Walls	Face brick with common brick backup	70% of wall		S.F. Wall	34.62	12.36	
2020	Exterior Windows	Industrial, horizontal pivoted steel	30% of wall		Each	1338	12.79	22.9%
2030	Exterior Doors	Double aluminum & glass, hollow metal, overhead doors			Each	2360	.44	
B30 Roofing								
3010	Roof Coverings	Built-up tar and gravel with flashing; perlite/EPS composite insulation			S.F. Roof	6.75	2.25	2.1%
3020	Roof Openings	Roof hatches			S.F. Roof	.39	.13	
C. INTERIORS								
1010	Partitions	Gypsum board on metal studs	50 S.F. Floor/L.F. Partition		S.F. Partition	4.25	.85	
1020	Interior Doors	Single leaf fire doors	500 S.F. Floor/Door		Each	1275	2.55	
1030	Fittings	Toilet partitions			S.F. Floor	.49	.49	
2010	Stair Construction	Concrete			Flight	8225	.36	8.0%
3010	Wall Finishes	Paint			S.F. Surface	.85	.34	
3020	Floor Finishes	90% metallic hardener, 10% vinyl composition tile			S.F. Floor	3.80	3.80	
3030	Ceiling Finishes	Fiberglass board on exposed grid systems	10% of area		S.F. Ceiling	5.70	.57	
D. SERVICES								
D10 Conveying								
1010	Elevators & Lifts	Two hydraulic freight elevators			Each	206,550	4.59	4.1%
1020	Escalators & Moving Walks	N/A			—	—	—	
D20 Plumbing								
2010	Plumbing Fixtures	Toilet and service fixtures, supply and drainage	1 Fixture/1345 S.F. Floor		Each	13,934	10.36	
2020	Domestic Water Distribution	Gas fired water heater			S.F. Floor	.36	.36	10.0%
2040	Rain Water Drainage	Roof drains			S.F. Roof	1.29	.43	
D30 HVAC								
3010	Energy Supply	Oil fired hot water, unit heaters			S.F. Floor	4.84	4.84	
3020	Heat Generating Systems	N/A			—	—	—	
3030	Cooling Generating Systems	Chilled water, air cooled condenser system			S.F. Floor	12.03	12.03	15.1%
3050	Terminal & Package Units	N/A			—	—	—	
3090	Other HVAC Sys. & Equipment	N/A			—	—	—	
D40 Fire Protection								
4010	Sprinklers	Wet pipe sprinkler system			S.F. Floor	3.74	3.74	3.9%
4020	Standpipes	Standpipe, wet, Class III			S.F. Floor	.59	.59	
D50 Electrical								
5010	Electrical Service/Distribution	800 ampere service, panel board and feeders			S.F. Floor	.65	.65	
5020	Lighting & Branch Wiring	High intensity discharge fixtures, switches, A.C. and misc. power			S.F. Floor	9.96	9.96	
5030	Communications & Security	Addressable alarm systems and emergency lighting			S.F. Floor	1.22	1.22	10.7%
5090	Other Electrical Systems	Emergency generator, 30 kW			S.F. Floor	.18	.18	
E. EQUIPMENT & FURNISHINGS								
1010	Commercial Equipment	N/A			—	—	—	
1020	Institutional Equipment	N/A			—	—	—	
1030	Vehicular Equipment	Dock Shelters			Each	.17	.17	0.2%
1090	Other Equipment	N/A			—	—	—	
F. SPECIAL CONSTRUCTION								
1020	Integrated Construction	N/A			—	—	—	
1040	Special Facilities	N/A			—	—	—	0.0%
G. BUILDING SITEWORK	**N/A**							

		Sub-Total	111.88	100%
CONTRACTOR FEES (General Requirements: 10%, Overhead: 5%, Profit: 10%)		25%	27.98	
ARCHITECT FEES		6%	8.39	

Total Building Cost	**148.25**

For customer support on your Square Foot Cost Data, call 877.694.8329.

125

Costs per square foot of floor area

Exterior Wall	S.F. Area	4000	4500	5000	5500	6000	6500	7000	7500	8000
	L.F. Perimeter	260	280	300	310	320	336	353	370	386
Face Brick Concrete Block Back-up	Steel Joists	190.70	187.05	184.10	180.20	176.95	174.90	173.35	171.95	170.65
	Bearing Walls	176.20	172.55	169.65	165.70	162.40	160.40	158.85	157.45	156.15
Decorative Concrete Block	Steel Joists	175.40	172.40	170.00	166.95	164.40	162.80	161.55	160.40	159.40
	Bearing Walls	160.90	157.90	155.50	152.45	149.90	148.30	147.05	145.90	144.90
Limestone with Concrete Block Back-up	Steel Joists	211.15	206.55	202.95	197.95	193.75	191.20	189.25	187.55	185.90
	Bearing Walls	196.60	192.05	188.45	183.40	179.25	176.65	174.75	173.00	171.35
Perimeter Adj., Add or Deduct	Per 100 L.F.	20.90	18.50	16.65	15.20	13.90	12.90	11.95	11.15	10.45
Story Hgt. Adj., Add or Deduct	Per 1 Ft.	2.80	2.70	2.55	2.40	2.25	2.25	2.20	2.15	2.10

For Basement, add $39.90 per square foot of basement area

The above costs were calculated using the basic specifications shown on the facing page. These costs should be adjusted where necessary for design alternatives and owner's requirements. Reported completed project costs, for this type of structure, range from $78.60 to $232.50 per S.F.

Common additives

Description	Unit	$ Cost
Appliances		
Cooking range, 30" free standing		
1 oven	Ea.	580 - 2450
2 oven	Ea.	1425 - 2900
30" built-in		
1 oven	Ea.	860 - 2075
2 oven	Ea.	1750 - 2550
Counter top cook tops, 4 burner	Ea.	435 - 1675
Microwave oven	Ea.	296 - 840
Combination range, refrig. & sink, 30" wide	Ea.	1900 - 3000
60" wide	Ea.	2675
72" wide	Ea.	2900
Combination range refrigerator, sink		
microwave oven & icemaker	Ea.	6700
Compactor, residential, 4-1 compaction	Ea.	825 - 1325
Dishwasher, built-in, 2 cycles	Ea.	600 - 1150
4 cycles	Ea.	755 - 1875
Garbage disposer, sink type	Ea.	232 - 380
Hood for range, 2 speed, vented, 30" wide	Ea.	330 - 1275
42" wide	Ea.	435 - 2650

Description	Unit	$ Cost
Appliances, cont.		
Refrigerator, no frost 10-12 C.F.	Ea.	560 - 650
14-16 C.F.	Ea.	660 - 870
18-20 C.F.	Ea.	805 - 1800
Lockers, Steel, single tier, 60" or 72"	Opng.	221 - 370
2 tier, 60" or 72" total	Opng.	120 - 165
5 tier, box lockers	Opng.	77 - 86.50
Locker bench, lam. maple top only	L.F.	34
Pedestals, steel pipe	Ea.	76
Sound System		
Amplifier, 250 watts	Ea.	2050
Speaker, ceiling or wall	Ea.	213
Trumpet	Ea.	410

Important: See the Reference Section for Location Factors.

Fire Station, 1 Story

Model costs calculated for a 1 story building with 14' story height and 6,000 square feet of floor area

				Unit	Unit Cost	Cost Per S.F.	% Of Sub-Total
A.	**SUBSTRUCTURE**						
1010	Standard Foundations	Poured concrete; strip and spread footings		S.F. Ground	2.73	2.73	
1020	Special Foundations	N/A		—	—	—	
1030	Slab on Grade	6" reinforced concrete with vapor barrier and granular base		S.F. Slab	7.13	7.13	11.4%
2010	Basement Excavation	Site preparation for slab and trench for foundation wall and footing		S.F. Ground	.57	.57	
2020	Basement Walls	4' foundation wall		L.F. Wall	84	4.45	
B.	**SHELL**						
	B10 Superstructure						
1010	Floor Construction	N/A		—	—	—	
1020	Roof Construction	Metal deck, open web steel joists, beams on columns		S.F. Roof	10.82	10.82	8.3 %
	B20 Exterior Enclosure						
2010	Exterior Walls	Face brick with concrete block backup	75% of wall	S.F. Wall	33.45	18.73	
2020	Exterior Windows	Aluminum insulated glass	10% of wall	Each	864	2.02	19.3%
2030	Exterior Doors	Single aluminum and glass, overhead, hollow metal	15% of wall	S.F. Door	40.36	4.52	
	B30 Roofing						
3010	Roof Coverings	Built-up tar and gravel with flashing; perlite/EPS composite insulation		S.F. Roof	7.41	7.41	5.8%
3020	Roof Openings	Skylights, roof hatches		S.F. Roof	.16	.16	
C.	**INTERIORS**						
1010	Partitions	Concrete block	17 S.F. Floor/L.F. Partition	S.F. Partition	9.66	5.68	
1020	Interior Doors	Single leaf hollow metal	500 S.F. Floor/Door	Each	1157	2.31	
1030	Fittings	Toilet partitions		S.F. Floor	.49	.49	
2010	Stair Construction	N/A		—	—	—	12.9%
3010	Wall Finishes	Paint		S.F. Surface	2.17	2.55	
3020	Floor Finishes	50% vinyl tile, 50% paint		S.F. Floor	2.29	2.29	
3030	Ceiling Finishes	Fiberglass board on exposed grid, suspended	50% of area	S.F. Ceiling	7.03	3.52	
D.	**SERVICES**						
	D10 Conveying						
1010	Elevators & Lifts	N/A		—	—	—	0.0 %
1020	Escalators & Moving Walks	N/A		—	—	—	
	D20 Plumbing						
2010	Plumbing Fixtures	Kitchen, toilet and service fixtures, supply and drainage	1 Fixture/375 S.F. Floor	Each	3885	10.36	
2020	Domestic Water Distribution	Gast fired water heater		S.F. Floor	3	3	11.2%
2040	Rain Water Drainage	Roof drains		S.F. Roof	1.35	1.35	
	D30 HVAC						
3010	Energy Supply	N/A		—	—	—	
3020	Heat Generating Systems	Included in D3050		—	—	—	
3030	Cooling Generating Systems	N/A		—	—	—	18.4 %
3050	Terminal & Package Units	Rooftop multizone unit system		S.F. Floor	24.10	24.10	
3090	Other HVAC Sys. & Equipment	N/A		—	—	—	
	D40 Fire Protection						
4010	Sprinklers	Wet pipe sprinkler system		S.F. Floor	4.98	4.98	4.9%
4020	Standpipes	Standpipe, wet, Class III		S.F. Floor	1.43	1.43	
	D50 Electrical						
5010	Electrical Service/Distribution	200 ampere service, panel board and feeders		S.F. Floor	1.62	1.62	
5020	Lighting & Branch Wiring	High efficiency fluorescent fixtures, receptacles, switches, A.C. and misc. power		S.F. Floor	6.85	6.85	8.0%
5030	Communications & Security	Addressable alarm systems		S.F. Floor	1.98	1.98	
5090	Other Electrical Systems	N/A		—	—	—	
E.	**EQUIPMENT & FURNISHINGS**						
1010	Commercial Equipment	N/A		—	—	—	
1020	Institutional Equipment	N/A		—	—	—	
1030	Vehicular Equipment	N/A		—	—	—	0.0 %
1090	Other Equipment	N/A		—	—	—	
F.	**SPECIAL CONSTRUCTION**						
1020	Integrated Construction	N/A		—	—	—	
1040	Special Facilities	N/A		—	—	—	0.0 %
G.	**BUILDING SITEWORK**	**N/A**					

			Sub-Total	131.05	100%
CONTRACTOR FEES (General Requirements: 10%, Overhead: 5%, Profit: 10%)		25%	32.80		
ARCHITECT FEES		8%	13.10		
	Total Building Cost		**176.95**		

For customer support on your Square Foot Cost Data, call 877.694.8329.

127

Costs per square foot of floor area

Exterior Wall	S.F. Area	6000	7000	8000	9000	10000	11000	12000	13000	14000
	L.F. Perimeter	220	240	260	280	286	303	320	336	353
Face Brick with Concrete Block Back-up	Steel Joists	201.35	194.90	190.15	186.35	181.40	178.70	176.50	174.45	172.90
	Precast Conc.	207.80	201.45	196.65	192.95	187.90	185.25	183.05	181.05	179.45
Decorative Concrete Block	Steel Joists	185.30	179.80	175.60	172.40	168.30	166.05	164.10	162.40	161.10
	Precast Conc.	194.25	188.75	184.60	181.40	177.30	175.05	173.20	171.45	170.10
Limestone with Concrete Block Back-up	Steel Joists	221.60	213.65	207.80	203.20	196.55	193.25	190.50	188.00	186.05
	Precast Conc.	230.35	222.50	216.55	211.95	205.40	202.05	199.35	196.80	194.85
Perimeter Adj., Add or Deduct	Per 100 L.F.	24.55	21.10	18.45	16.45	14.65	13.40	12.25	11.35	10.55
Story Hgt. Adj., Add or Deduct	Per 1 Ft.	3.20	3.00	2.80	2.80	2.45	2.40	2.30	2.35	2.25
For Basement, add $38.75 per square foot of basement area										

The above costs were calculated using the basic specifications shown on the facing page. These costs should be adjusted where necessary for design alternatives and owner's requirements. Reported completed project costs, for this type of structure, range from $78.35 to $232.50 per S.F.

Common additives

Description	Unit	$ Cost
Appliances		
Cooking range, 30" free standing		
1 oven	Ea.	580 - 2450
2 oven	Ea.	1425 - 2900
30" built-in		
1 oven	Ea.	860 - 2075
2 oven	Ea.	1750 - 2550
Counter top cook tops, 4 burner	Ea.	435 - 1675
Microwave oven	Ea.	296 - 840
Combination range, refrig. & sink, 30" wide	Ea.	1900 - 3000
60" wide	Ea.	2675
72" wide	Ea.	2900
Combination range, refrigerator, sink,		
microwave oven & icemaker	Ea.	6700
Compactor, residential, 4-1 compaction	Ea.	825 - 1325
Dishwasher, built-in, 2 cycles	Ea.	600 - 1150
4 cycles	Ea.	755 - 1875
Garbage diposer, sink type	Ea.	232 - 380
Hood for range, 2 speed, vented, 30" wide	Ea.	330 - 1275
42" wide	Ea.	435 - 2650

Description	Unit	$ Cost
Appliances, cont.		
Refrigerator, no frost 10-12 C.F.	Ea.	560 - 650
14-16 C.F.	Ea.	660 - 870
18-20 C.F.	Ea.	805 - 1800
Elevators, Hydraulic passenger, 2 stops		
1500# capacity	Ea.	65,700
2500# capacity	Ea.	68,800
3500# capacity	Ea.	73,300
Lockers, Steel, single tier, 60" or 72"	Opng.	221 - 370
2 tier, 60" or 72" total	Opng.	120 - 165
5 tier, box lockers	Opng.	77 - 86.50
Locker bench, lam. maple top only	L.F.	34
Pedestals, steel pipe	Ea.	76
Sound System		
Amplifier, 250 watts	Ea.	2050
Speaker, ceiling or wall	Ea.	213
Trumpet	Ea.	410

Important: See the Reference Section for Location Factors.

Model costs calculated for a 2 story building with 14′ story height and 10,000 square feet of floor area

					Unit	Unit Cost	Cost Per S.F.	% Of Sub-Total
A. SUBSTRUCTURE								
1010	Standard Foundations	Poured concrete; strip and spread footings			S.F. Ground	2.66	1.33	
1020	Special Foundations	N/A			—	—	—	
1030	Slab on Grade	6″ reinforced concrete with vapor barrier and granular base			S.F. Slab	7.13	3.57	6.3%
2010	Basement Excavation	Site preparation for slab and trench for foundation wall and footing			S.F. Ground	.57	.29	
2020	Basement Walls	4′ foundation wall			L.F. Wall	93	2.66	
B. SHELL								
B10 Superstructure								
1010	Floor Construction	Open web steel joists, slab form, concrete			S.F. Floor	15.82	7.91	8.4%
1020	Roof Construction	Metal deck on open web steel joists			S.F. Roof	5.14	2.57	
B20 Exterior Enclosure								
2010	Exterior Walls	Decorative concrete block		75% of wall	S.F. Wall	20.40	12.25	
2020	Exterior Windows	Aluminum insulated glass		10% of wall	Each	750	2.62	14.9%
2030	Exterior Doors	Single aluminum and glass, steel overhead, hollow metal		15% of wall	S.F. Door	31.14	3.74	
B30 Roofing								
3010	Roof Coverings	Built-up tar and gravel with flashing; perlite/EPS composite insulation			S.F. Roof	7.58	3.79	3.0%
3020	Roof Openings	N/A			—	—	—	
C. INTERIORS								
1010	Partitions	Concrete block		10 S.F. Floor/L.F. Partition	S.F. Partition	9.66	5.68	
1020	Interior Doors	Single leaf hollow metal		500 S.F. Floor/Door	Each	1275	2.55	
1030	Fittings	Toilet partitions			S.F. Floor	.44	.44	
2010	Stair Construction	Concrete filled metal pan			Flight	20,200	4.04	17.8%
3010	Wall Finishes	Paint			S.F. Surface	3.67	4.32	
3020	Floor Finishes	50% vinyl tile, 50% paint			S.F. Floor	2.29	2.29	
3030	Ceiling Finishes	Fiberglass board on exposed grid, suspended		50% of area	S.F. Ceiling	5.70	2.86	
D. SERVICES								
D10 Conveying								
1010	Elevators & Lifts	One hydraulic passenger elevator			Each	83,600	8.36	6.7%
1020	Escalators & Moving Walks	N/A			—	—	—	
D20 Plumbing								
2010	Plumbing Fixtures	Kitchen toilet and service fixtures, supply and drainage		1 Fixture/400 S.F. Floor	Each	4024	10.06	
2020	Domestic Water Distribution	Gas fired water heater			S.F. Floor	2.43	2.43	10.7%
2040	Rain Water Drainage	Roof drains			S.F. Roof	1.76	.88	
D30 HVAC								
3010	Energy Supply	N/A			—	—	—	
3020	Heat Generating Systems	Included in D3050			—	—	—	
3030	Cooling Generating Systems	N/A			—	—	—	19.3%
3050	Terminal & Package Units	Rooftop multizone unit system			S.F. Floor	24.10	24.10	
3090	Other HVAC Sys. & Equipment	N/A			—	—	—	
D40 Fire Protection								
4010	Sprinklers	Wet pipe sprinkler system			S.F. Floor	3.96	3.96	4.3%
4020	Standpipes	Standpipe, wet, Class III			S.F. Floor	1.46	1.46	
D50 Electrical								
5010	Electrical Service/Distribution	400 ampere service, panel board and feeders			S.F. Floor	1.65	1.65	
5020	Lighting & Branch Wiring	High efficiency fluorescent fixtures, receptacles, switches, A.C. and misc. power			S.F. Floor	7.07	7.07	8.4%
5030	Communications & Security	Addressable alarm systems and emergency lighting			S.F. Floor	1.55	1.55	
5090	Other Electrical Systems	Emergency generator, 15 kW			S.F. Floor	.23	.23	
E. EQUIPMENT & FURNISHINGS								
1010	Commercial Equipment	N/A			—	—	—	
1020	Institutional Equipment	N/A			—	—	—	
1030	Vehicular Equipment	N/A			—	—	—	0.0%
1090	Other Equipment	N/A			—	—	—	
F. SPECIAL CONSTRUCTION								
1020	Integrated Construction	N/A			—	—	—	
1040	Special Facilities	N/A			—	—	—	0.0%
G. BUILDING SITEWORK	**N/A**							

				Sub-Total	124.66	100%
CONTRACTOR FEES (General Requirements: 10%, Overhead: 5%, Profit: 10%)				25%	31.17	
ARCHITECT FEES				8%	12.47	

Total Building Cost	**168.30**

For customer support on your Square Foot Cost Data, call 877.694.8329.

129

Costs per square foot of floor area

Exterior Wall	S.F. Area	4000	5000	6000	8000	10000	12000	14000	16000	18000
	L.F. Perimeter	180	205	230	260	300	340	353	386	420
Cedar Beveled Siding	Wood Frame	217.65	209.15	203.45	194.70	190.15	187.15	183.65	181.85	180.60
Aluminum Siding	Wood Frame	214.75	206.50	201.00	192.60	188.25	185.35	182.00	180.30	179.10
Board and Batten	Wood Frame	212.85	205.75	201.05	193.50	189.70	187.20	184.05	182.50	181.45
Face Brick on Block	Bearing Wall	220.55	210.35	203.55	192.30	186.65	183.00	178.15	175.85	174.25
Stucco on Block	Bearing Wall	203.80	195.00	189.15	180.10	175.40	172.25	168.60	166.80	165.45
Decorative Block	Wood Joists	222.40	213.45	207.50	198.15	193.35	190.15	186.35	184.40	183.05
Perimeter Adj., Add or Deduct	Per 100 L.F.	16.70	13.35	11.15	8.35	6.65	5.55	4.75	4.20	3.65
Story Hgt. Adj., Add or Deduct	Per 1 Ft.	2.15	1.95	1.85	1.60	1.50	1.40	1.25	1.20	1.10

For Basement, add $24.65 per square foot of basement area

The above costs were calculated using the basic specifications shown on the facing page. These costs should be adjusted where necessary for design alternatives and owner's requirements. Reported completed project costs, for this type of structure, range from $113.95 to $249.25 per S.F.

Common additives

Description	Unit	$ Cost
Appliances		
Cooking range, 30" free standing		
1 oven	Ea.	580 - 2450
2 oven	Ea.	1425 - 2900
30" built-in		
1 oven	Ea.	860 - 2075
2 oven	Ea.	1750 - 2550
Counter top cook tops, 4 burner	Ea.	435 - 1675
Microwave oven	Ea.	296 - 840
Combination range, refrig. & sink, 30" wide	Ea.	1900 - 3000
60" wide	Ea.	2675
72" wide	Ea.	2900
Combination range, refrigerator, sink,		
microwave oven & icemaker	Ea.	6700
Compactor, residential, 4-1 compaction	Ea.	825 - 1325
Dishwasher, built-in, 2 cycles	Ea.	600 - 1150
4 cycles	Ea.	755 - 1875
Garbage disposer, sink type	Ea.	232 - 380
Hood for range, 2 speed, vented, 30" wide	Ea.	330 - 1275
42" wide	Ea.	435 - 2650

Description	Unit	$ Cost
Appliances, cont.		
Refrigerator, no frost 10-12 C.F.	Ea.	560 - 650
14-16 C.F.	Ea.	660 - 870
18-20 C.F.	Ea.	805 - 1800
Elevators, Hydraulic passenger, 2 stops		
1500# capacity	Ea.	65,700
2500# capacity	Ea.	68,800
3500# capacity	Ea.	73,300
Laundry Equipment		
Dryer, gas, 16 lb. capacity	Ea.	975
30 lb. capacity	Ea.	3850
Washer, 4 cycle	Ea.	1200
Commercial	Ea.	1600
Sound System		
Amplifier, 250 watts	Ea.	2050
Speaker, ceiling or wall	Ea.	213
Trumpet	Ea.	410

Important: See the Reference Section for Location Factors.

Model costs calculated for a 2 story building with 10' story height and 10,000 square feet of floor area

Fraternity/Sorority House

					Unit	Unit Cost	Cost Per S.F.	% Of Sub-Total
A. SUBSTRUCTURE								
1010	Standard Foundations	Poured concrete; strip and spread footings			S.F. Ground	2.98	1.49	
1020	Special Foundations	N/A			—	—	—	
1030	Slab on Grade	4" reinforced concrete with vapor barrier and granular base			S.F. Slab	5.44	2.72	4.8%
2010	Basement Excavation	Site preparation for slab and trench for foundation wall and footing			S.F. Ground	.57	.29	
2020	Basement Walls	4' foundation wall			L.F. Wall	75	2.26	
B. SHELL								
	B10 Superstructure							
1010	Floor Construction	Plywood on wood joists			S.F. Floor	41.62	20.81	16.4%
1020	Roof Construction	Plywood on wood rafters (pitched)			S.F. Roof	3.68	2.06	
	B20 Exterior Enclosure							
2010	Exterior Walls	Cedar bevel siding on wood studs, insulated	80% of wall		S.F. Wall	13.40	6.43	
2020	Exterior Windows	Double hung wood	20% of wall		Each	638	3.06	8.0%
2030	Exterior Doors	Solid core wood			Each	2360	1.66	
	B30 Roofing							
3010	Roof Coverings	Asphalt shingles with flashing (pitched); rigid fiberglass insulation			S.F. Roof	5.26	2.63	1.9%
3020	Roof Openings	N/A			—	—	—	
C. INTERIORS								
1010	Partitions	Gypsum board on wood studs	25 S.F. Floor/L.F. Partition		S.F. Partition	8.44	2.70	
1020	Interior Doors	Single leaf wood	200 S.F. Floor/Door		Each	616	3.08	
1030	Fittings	N/A			—	—	—	
2010	Stair Construction	Wood			Flight	2730	.82	13.9%
3010	Wall Finishes	Paint			S.F. Surface	.84	.54	
3020	Floor Finishes	10% hardwood, 70% carpet, 20% ceramic tile			S.F. Floor	7.76	7.76	
3030	Ceiling Finishes	Gypsum board on wood furring			S.F. Ceiling	4.51	4.51	
D. SERVICES								
	D10 Conveying							
1010	Elevators & Lifts	One hydraulic passenger elevator			Each	83,600	8.36	6.0%
1020	Escalators & Moving Walks	N/A			—	—	—	
	D20 Plumbing							
2010	Plumbing Fixtures	Kitchen toilet and service fixtures, supply and drainage	1 Fixture/150 S.F. Floor		Each	2222	14.81	
2020	Domestic Water Distribution	Gas fired water heater			S.F. Floor	3.27	3.27	13.0%
2040	Rain Water Drainage	N/A			—	—	—	
	D30 HVAC							
3010	Energy Supply	Oil fired hot water, baseboard radiation			S.F. Floor	7.98	7.98	
3020	Heat Generating Systems	N/A			—	—	—	
3030	Cooling Generating Systems	N/A			—	—	—	11.7%
3050	Terminal & Package Units	Split system with air cooled condensing unit			S.F. Floor	8.39	8.39	
3090	Other HVAC Sys. & Equipment	N/A			—	—	—	
	D40 Fire Protection							
4010	Sprinklers	Wet pipe sprinkler system			S.F. Floor	3.96	3.96	3.6%
4020	Standpipes	Standpipe			S.F. Floor	1.05	1.05	
	D50 Electrical							
5010	Electrical Service/Distribution	600 ampere service, panel board and feeders			S.F. Floor	4	4	
5020	Lighting & Branch Wiring	High efficiency fluorescent fixtures, receptacles, switches, A.C. and misc. power			S.F. Floor	10.24	10.24	20.7%
5030	Communications & Security	Addressable alarm, communication system, internet and phone wiring, and generator set			S.F. Floor	14.51	14.51	
5090	Other Electrical Systems	Emergency generator, 7.5 kW			S.F. Floor	.19	.19	
E. EQUIPMENT & FURNISHINGS								
1010	Commercial Equipment	N/A			—	—	—	
1020	Institutional Equipment	N/A			—	—	—	
1030	Vehicular Equipment	N/A			—	—	—	0.0 %
1090	Other Equipment	N/A			—	—	—	
F. SPECIAL CONSTRUCTION								
1020	Integrated Construction	N/A			—	—	—	
1040	Special Facilities	N/A			—	—	—	0.0 %
G. BUILDING SITEWORK	**N/A**							

Sub-Total		139.58	**100%**
CONTRACTOR FEES (General Requirements: 10%, Overhead: 5%, Profit: 10%)	25%	34.87	
ARCHITECT FEES	9%	15.70	
Total Building Cost		**190.15**	

For customer support on your Square Foot Cost Data, call 877.694.8329.

131

Costs per square foot of floor area

Exterior Wall	S.F. Area	4000	6000	8000	10000	12000	14000	16000	18000	20000
	L.F. Perimeter	260	320	384	424	460	484	510	540	576
Vertical Redwood Siding	Wood Frame	160.35	149.10	143.70	139.30	136.25	133.70	131.85	130.50	129.60
Brick Veneer	Wood Frame	194.05	173.40	163.35	155.80	150.60	146.30	143.15	140.85	139.25
Aluminum Siding	Wood Frame	179.50	161.50	152.65	146.35	141.95	138.55	136.00	134.15	132.80
Brick on Block	Wood Truss	202.35	180.30	169.70	161.45	155.80	151.00	147.55	145.05	143.30
Limestone on Block	Wood Truss	222.55	196.80	184.45	174.45	167.45	161.60	157.30	154.15	152.00
Stucco on Block	Wood Truss	185.60	166.55	157.30	150.50	145.85	142.10	139.35	137.35	135.85
Perimeter Adj., Add or Deduct	Per 100 L.F.	11.35	7.50	5.60	4.50	3.80	3.25	2.85	2.55	2.25
Story Hgt. Adj., Add or Deduct	Per 1 Ft.	1.55	1.25	1.05	1.00	0.95	0.85	0.75	0.75	0.70

For Basement, add $32.85 per square foot of basement area

The above costs were calculated using the basic specifications shown on the facing page. These costs should be adjusted where necessary for design alternatives and owner's requirements. Reported completed project costs, for this type of structure, range from $111.55 to $343.05 per S.F.

Common additives

Description	Unit	$ Cost	Description	Unit	$ Cost
Autopsy Table, Standard	Ea.	11,400	Planters, Precast concrete		
Deluxe	Ea.	17,800	48" diam., 24" high	Ea.	755
Directory Boards, Plastic, glass covered			7" diam., 36" high	Ea.	1825
30" x 20"	Ea.	600	Fiberglass, 36" diam., 24" high	Ea.	855
36" x 48"	Ea.	1500	60" diam., 24" high	Ea.	1325
Aluminum, 24" x 18"	Ea.	635	Smoke Detectors		
36" x 24"	Ea.	755	Ceiling type	Ea.	227
48" x 32"	Ea.	1050	Duct type	Ea.	565
48" x 60"	Ea.	2225			
Emergency Lighting, 25 watt, battery operated					
Lead battery	Ea.	330			
Nickel cadmium	Ea.	780			
Mortuary Refrigerator, End operated					
Two capacity	Ea.	13,500			
Six capacity	Ea.	24,400			

Important: See the Reference Section for Location Factors.

Model costs calculated for a 1 story building with 12′ story height and 10,000 square feet of floor area

Funeral Home

				Unit	Unit Cost	Cost Per S.F.	% Of Sub-Total
A. SUBSTRUCTURE							
1010	Standard Foundations	Poured concrete; strip and spread footings		S.F. Ground	2.63	2.63	
1020	Special Foundations	N/A		—	—	—	
1030	Slab on Grade	4″ reinforced concrete with vapor barrier and granular base		S.F. Slab	5.44	5.44	11.3%
2010	Basement Excavation	Site preparation for slab and trench for foundation wall and footing		S.F. Ground	.32	.32	
2020	Basement Walls	4′ foundation wall		L.F. Wall	75	3.18	
B. SHELL							
B10 Superstructure							
1010	Floor Construction	N/A		—	—	—	
1020	Roof Construction	Plywood on wood truss		S.F. Roof	5.22	5.22	5.9 %
B20 Exterior Enclosure							
2010	Exterior Walls	1″ x 4″ vertical T & G redwood siding on wood studs	90% of wall	S.F. Wall	12.49	5.72	
2020	Exterior Windows	Double hung wood	10% of wall	Each	511	2.16	8.9%
2030	Exterior Doors	Wood swinging double doors, single leaf hollow metal		Each	1985	1.19	
B30 Roofing							
3010	Roof Coverings	Single ply membrane, fully adhered; polyisocyanurate sheets		S.F. Roof	7.42	7.42	7.3%
3020	Roof Openings	N/A		—	—	—	
C. INTERIORS							
1010	Partitions	Gypsum board on wood studs with sound deadening board	15 S.F. Floor/L.F. Partition	S.F. Partition	9.58	5.11	
1020	Interior Doors	Single leaf wood	150 S.F. Floor/Door	Each	616	4.11	
1030	Fittings	N/A		—	—	—	
2010	Stair Construction	N/A		—	—	—	26.6%
3010	Wall Finishes	50% wallpaper, 25% wood paneling, 25% paint		S.F. Surface	3.47	3.70	
3020	Floor Finishes	70% carpet, 30% ceramic tile		S.F. Floor	8.54	8.54	
3030	Ceiling Finishes	Fiberglass board on exposed grid, suspended		S.F. Ceiling	5.70	5.70	
D. SERVICES							
D10 Conveying							
1010	Elevators & Lifts	N/A		—	—	—	
1020	Escalators & Moving Walks	N/A		—	—	—	0.0 %
D20 Plumbing							
2010	Plumbing Fixtures	Toilet and service fixtures, supply and drainage	1 Fixture/770 S.F. Floor	Each	2895	3.76	
2020	Domestic Water Distribution	Electric water heater		S.F. Floor	3.41	3.41	7.0%
2040	Rain Water Drainage	Roof drain		S.F. Roof			
D30 HVAC							
3010	Energy Supply	N/A		—	—	—	
3020	Heat Generating Systems	Included in D3030		—	—	—	
3030	Cooling Generating Systems	Multizone rooftop unit, gas heating, electric cooling		S.F. Floor	18.70	18.70	18.3 %
3050	Terminal & Package Units	N/A		—	—	—	
3090	Other HVAC Sys. & Equipment	N/A		—	—	—	
D40 Fire Protection							
4010	Sprinklers	Wet pipe sprinkler system		S.F. Floor	3.67	3.67	
4020	Standpipes	Standpipe, wet, Class III		S.F. Floor	.86	.86	4.4%
D50 Electrical							
5010	Electrical Service/Distribution	400 ampere service, panel board and feeders		S.F. Floor	1.93	1.93	
5020	Lighting & Branch Wiring	High efficiency fluorescent fixtures, receptacles, switches, A.C. and misc. power		S.F. Floor	7	7	
5030	Communications & Security	Addressable alarm systems and emergency lighting		S.F. Floor	1.55	1.55	10.3%
5090	Other Electrical Systems	Emergency generator, 15 kW		S.F. Floor	.08	.08	
E. EQUIPMENT & FURNISHINGS							
1010	Commercial Equipment	N/A		—	—	—	
1020	Institutional Equipment	N/A		—	—	—	
1030	Vehicular Equipment	N/A		—	—	—	0.0 %
1090	Other Equipment	N/A		—	—	—	
F. SPECIAL CONSTRUCTION							
1020	Integrated Construction	N/A		—	—	—	
1040	Special Facilities	N/A		—	—	—	0.0 %
G. BUILDING SITEWORK	**N/A**						

	Sub-Total	102.25	100%
CONTRACTOR FEES (General Requirements: 10%, Overhead: 5%, Profit: 10%)	25%	25.55	
ARCHITECT FEES	9%	11.50	
Total Building Cost		**139.30**	

For customer support on your Square Foot Cost Data, call 877.694.8329.

133

Costs per square foot of floor area

Exterior Wall	S.F. Area	12000	14000	16000	19000	21000	23000	26000	28000	30000
	L.F. Perimeter	440	474	510	556	583	607	648	670	695
E.I.F.S. on Concrete Block	Steel Frame	132.25	129.45	127.40	124.85	123.40	122.10	120.65	119.80	119.05
Tilt-up Concrete Wall	Steel Frame	129.80	127.20	125.30	122.90	121.60	120.35	119.05	118.15	117.55
Face Brick with Concrete Block Back-up	Bearing Walls	132.15	128.80	126.35	123.35	121.55	119.95	118.25	117.15	116.30
	Steel Frame	139.15	135.80	133.35	130.35	128.55	127.00	125.25	124.15	123.30
Stucco on Concrete Block	Bearing Walls	123.70	121.00	119.05	116.60	115.15	113.90	112.60	111.65	110.95
	Steel Frame	131.05	128.30	126.40	123.95	122.50	121.20	119.90	119.00	118.30
Perimeter Adj., Add or Deduct	Per 100 L.F.	6.95	5.95	5.25	4.40	4.00	3.60	3.25	2.95	2.80
Story Hgt. Adj., Add or Deduct	Per 1 Ft.	1.45	1.35	1.25	1.20	1.10	1.05	1.05	0.90	0.95

For Basement, add $37.35 per square foot of basement area

The above costs were calculated using the basic specifications shown on the facing page. These costs should be adjusted where necessary for design alternatives and owner's requirements. Reported completed project costs, for this type of structure, range from $59.35 to $172.70 per S.F.

Common additives

Description	Unit	$ Cost
Emergency Lighting, 25 watt, battery operated		
Lead battery	Ea.	330
Nickel cadmium	Ea.	780
Smoke Detectors		
Ceiling type	Ea.	227
Duct type	Ea.	565
Sound System		
Amplifier, 250 watts	Ea.	2050
Speaker, ceiling or wall	Ea.	213
Trumpet	Ea.	410

Important: See the Reference Section for Location Factors.

Model costs calculated for a 1 story building with 14' story height and 21,000 square feet of floor area

				Unit	Unit Cost	Cost Per S.F.	% Of Sub-Total
A. SUBSTRUCTURE							
1010	Standard Foundations	Poured concrete; strip and spread footings		S.F. Ground	1.49	1.49	
1020	Special Foundations	N/A		—	—	—	
1030	Slab on Grade	4" reinforced concrete with vapor barrier and granular base		S.F. Slab	8.05	8.05	13.2%
2010	Basement Excavation	Site preparation for slab and trench for foundation wall and footing		S.F. Ground	.32	.32	
2020	Basement Walls	4' foundation wall		L.F. Wall	84	2.32	
B. SHELL							
B10 Superstructure							
1010	Floor Construction	N/A		—	—	—	
1020	Roof Construction	Metal deck, open web steel joists, beams, columns		S.F. Roof	12.03	12.03	13.0 %
B20 Exterior Enclosure							
2010	Exterior Walls	E.I.F.S.	70% of wall	S.F. Wall	17.86	4.86	
2020	Exterior Windows	Window wall	30% of wall	Each	56	6.59	16.4%
2030	Exterior Doors	Double aluminum and glass, hollow metal, steel overhead		Each	4241	3.64	
B30 Roofing							
3010	Roof Coverings	Built-up tar and gravel with flashing; perlite/EPS composite insulation		S.F. Roof	7	7	7.6%
3020	Roof Openings	N/A		—	—	—	
C. INTERIORS							
1010	Partitions	Gypsum board on metal studs	28 S.F. Floor/L.F. Partition	S.F. Partition	5.04	2.16	
1020	Interior Doors	Hollow metal	280 S.F. Floor/Door	Each	1157	4.13	
1030	Fittings	N/A		—	—	—	
2010	Stair Construction	N/A		—	—	—	13.4%
3010	Wall Finishes	Paint		S.F. Surface	1.12	.96	
3020	Floor Finishes	50% vinyl tile, 50% paint		S.F. Floor	2.29	2.29	
3030	Ceiling Finishes	Fiberglass board on exposed grid, suspended	50% of area	S.F. Ceiling	5.70	2.86	
D. SERVICES							
D10 Conveying							
1010	Elevators & Lifts	N/A		—	—	—	
1020	Escalators & Moving Walks	N/A		—	—	—	0.0 %
D20 Plumbing							
2010	Plumbing Fixtures	Toilet and service fixtures, supply and drainage	1 Fixture/1500 S.F. Floor	Each	3000	2	
2020	Domestic Water Distribution	Gas fired water heater		S.F. Floor	2	2	7.1%
2040	Rain Water Drainage	Roof drains		S.F. Roof	2.51	2.51	
D30 HVAC							
3010	Energy Supply	N/A		—	—	—	
3020	Heat Generating Systems	N/A		—	—	—	
3030	Cooling Generating Systems	N/A		—	—	—	10.6 %
3050	Terminal & Package Units	Single zone rooftop unit, gas heating, electric cooling		S.F. Floor	9.09	9.09	
3090	Other HVAC Sys. & Equipment	Underfloor garage exhaust system		S.F. Floor	.67	.67	
D40 Fire Protection							
4010	Sprinklers	Wet pipe sprinkler system		S.F. Floor	4.73	4.73	
4020	Standpipes	Standpipe		S.F. Floor	.45	.45	5.6%
D50 Electrical							
5010	Electrical Service/Distribution	200 ampere service, panel board and feeders		S.F. Floor	.48	.48	
5020	Lighting & Branch Wiring	T-8 fluorescent fixtures, receptacles, switches, A.C. and misc. power		S.F. Floor	6.75	6.75	
5030	Communications & Security	Addressable alarm systems, partial internet wiring and emergency lighting		S.F. Floor	3.38	3.38	11.6%
5090	Other Electrical Systems	Emergency generator, 7.5 kW		S.F. Floor	.06	.06	
E. EQUIPMENT & FURNISHINGS							
1010	Commercial Equipment	N/A		—	—	—	
1020	Institutional Equipment	N/A		—	—	—	
1030	Vehicular Equipment	Hoists, compressor, fuel pump		S.F. Floor	1.45	1.45	1.6 %
1090	Other Equipment	N/A		—	—	—	
F. SPECIAL CONSTRUCTION							
1020	Integrated Construction	N/A		—	—	—	
1040	Special Facilities	N/A		—	—	—	0.0 %
G. BUILDING SITEWORK	**N/A**						

		Sub-Total	92.27	100%
CONTRACTOR FEES (General Requirements: 10%, Overhead: 5%, Profit: 10%)		25%	23.06	
ARCHITECT FEES		7%	8.07	
	Total Building Cost		**123.40**	

For customer support on your Square Foot Cost Data, call 877.694.8329.

135

Costs per square foot of floor area

Exterior Wall	S.F. Area	85000	115000	145000	175000	205000	235000	265000	295000	325000
	L.F. Perimeter	529	638	723	823	875	910	975	1027	1075
Face Brick with Concrete Block Back-up	Steel Frame	81.70	80.05	78.80	78.15	77.30	76.55	76.15	75.75	75.45
	Precast Concrete	69.70	68.00	66.80	66.10	65.30	64.50	64.15	63.75	63.40
Precast Concrete	Steel Frame	89.70	87.15	85.20	84.15	82.70	81.50	80.85	80.25	79.70
	Precast Concrete	76.25	73.90	72.15	71.20	69.95	68.85	68.20	67.65	67.15
Reinforced Concrete	Steel Frame	79.05	77.65	76.65	76.10	75.45	74.90	74.55	74.30	74.05
	R/Conc. Frame	70.40	69.00	68.05	67.50	66.80	66.25	65.95	65.65	65.35
Perimeter Adj., Add or Deduct	Per 100 L.F.	1.80	1.35	1.10	0.90	0.70	0.65	0.50	0.55	0.45
Story Hgt. Adj., Add or Deduct	Per 1 Ft.	0.85	0.75	0.65	0.60	0.55	0.50	0.45	0.45	0.45
Basement—Not Applicable										

The above costs were calculated using the basic specifications shown on the facing page. These costs should be adjusted where necessary for design alternatives and owner's requirements. Reported completed project costs, for this type of structure, range from $36.45 to $140.95 per S.F.

Common additives

Description	Unit	$ Cost
Elevators, Electric passenger, 5 stops		
2000# capacity	Ea.	167,000
3500# capacity	Ea.	173,000
5000# capacity	Ea.	179,500
Barrier gate w/programmable controller	Ea.	4150
Booth for attendant, average	Ea.	13,800
Fee computer	Ea.	16,000
Ticket spitter with time/date stamp	Ea.	7475
Mag strip encoding	Ea.	21,400
Collection station, pay on foot	Ea.	128,500
Parking control software	Ea.	27,500 - 118,500
Painting, Parking stalls	Stall	8.60
Parking Barriers		
Timber with saddles, 4" x 4"	L.F.	7.70
Precast concrete, 6" x 10" x 6'	Ea.	63
Traffic Signs, directional, 12" x 18", high density	Ea.	97.50

Important: See the Reference Section for Location Factors.

Model costs calculated for a 5 story building with 10' story height and 145,000 square feet of floor area

Garage, Parking

				Unit	Unit Cost	Cost Per S.F.	% Of Sub-Total
A.	**SUBSTRUCTURE**						
1010	Standard Foundations	Poured concrete; strip and spread footings		S.F. Ground	2.05	.41	
1020	Special Foundations	N/A		—	—	—	
1030	Slab on Grade	6" reinforced concrete with vapor barrier and granular base		S.F. Slab	7	1.40	4.6%
2010	Basement Excavation	Site preparation for slab and trench for foundation wall and footing		S.F. Ground	.18	.04	
2020	Basement Walls	4' foundation wall		L.F. Wall	84	.46	
B.	**SHELL**						
	B10 Superstructure						
1010	Floor Construction	Double tee precast concrete slab, precast concrete columns		S.F. Floor	28.03	22.42	44.5%
1020	Roof Construction	N/A		—	—	—	
	B20 Exterior Enclosure						
2010	Exterior Walls	Face brick with concrete block backup	40% of story height	S.F. Wall	50	5	
2020	Exterior Windows	N/A		—	—	—	9.9%
2030	Exterior Doors	N/A		—	—	—	
	B30 Roofing						
3010	Roof Coverings	N/A		—	—	—	0.0%
3020	Roof Openings	N/A		—	—	—	
C.	**INTERIORS**						
1010	Partitions	Concrete block		S.F. Partition	33.80	1.30	
1020	Interior Doors	Hollow metal		Each	23,140	.16	
1030	Fittings	N/A		—	—	—	
2010	Stair Construction	Concrete		Flight	11,600	1.28	15.1%
3010	Wall Finishes	Paint		S.F. Surface	1.82	.14	
3020	Floor Finishes	Parking deck surface coating		S.F. Floor	4.73	4.73	
3030	Ceiling Finishes	N/A		—	—	—	
D.	**SERVICES**						
	D10 Conveying						
1010	Elevators & Lifts	Two hydraulic passenger elevators		Each	164,575	2.27	4.5%
1020	Escalators & Moving Walks	N/A		—	—	—	
	D20 Plumbing						
2010	Plumbing Fixtures	Toilet and service fixtures, supply and drainage	1 Fixture/18,125 S.F. Floor	Each	725	.04	
2020	Domestic Water Distribution	Electric water heater		S.F. Floor	.07	.07	3.2%
2040	Rain Water Drainage	Roof drains		S.F. Roof	7.50	1.50	
	D30 HVAC						
3010	Energy Supply	N/A		—	—	—	
3020	Heat Generating Systems	N/A		—	—	—	
3030	Cooling Generating Systems	N/A		—	—	—	0.0%
3050	Terminal & Package Units	N/A		—	—	—	
3090	Other HVAC Sys. & Equipment	N/A		—	—	—	
	D40 Fire Protection						
4010	Sprinklers	Dry pipe sprinkler system		S.F. Floor	4.28	4.28	8.7%
4020	Standpipes	Standpipes and hose systems		S.F. Floor	.09	.09	
	D50 Electrical						
5010	Electrical Service/Distribution	400 ampere service, panel board and feeders		S.F. Floor	.25	.25	
5020	Lighting & Branch Wiring	T-8 fluorescent fixtures, receptacles, switches and misc. power		S.F. Floor	3.07	3.07	6.9%
5030	Communications & Security	Addressable alarm systems and emergency lighting		S.F. Floor	.12	.12	
5090	Other Electrical Systems	Emergency generator, 7.5 kW		S.F. Floor	.05	.05	
E.	**EQUIPMENT & FURNISHINGS**						
1010	Commercial Equipment	N/A		—	—	—	
1020	Institutional Equipment	N/A		—	—	—	2.6%
1030	Vehicular Equipment	Ticket dispensers, booths, automatic gates		S.F. Floor	1.32	1.32	
1090	Other Equipment	N/A		—	—	—	
F.	**SPECIAL CONSTRUCTION**						
1020	Integrated Construction	N/A		—	—	—	0.0%
1040	Special Facilities	N/A		—	—	—	
G.	**BUILDING SITEWORK**	**N/A**					

		Sub-Total	50.40	100%
CONTRACTOR FEES (General Requirements: 10%, Overhead: 5%, Profit: 10%)		25%	12.62	
ARCHITECT FEES		6%	3.78	

Total Building Cost   66.80

For customer support on your Square Foot Cost Data, call 877.694.8329.

137

Costs per square foot of floor area

Exterior Wall	S.F. Area	20000	30000	40000	50000	75000	100000	125000	150000	175000
	L.F. Perimeter	400	500	600	650	775	900	1000	1100	1185
Reinforced Concrete	R/Conc. Frame	103.45	96.60	93.20	90.00	85.60	83.45	81.95	80.90	80.10
Perimeter Adj., Add or Deduct	Per 100 L.F.	5.95	4.00	3.00	2.35	1.65	1.15	0.90	0.85	0.65
Story Hgt. Adj., Add or Deduct	Per 1 Ft.	2.25	1.90	1.75	1.45	1.15	1.05	0.90	0.85	0.70
Basement—Not Applicable										

The above costs were calculated using the basic specifications shown on the facing page. These costs should be adjusted where necessary for design alternatives and owner's requirements. Reported completed project costs, for this type of structure, range from $52.95 to $126.15 per S.F.

Common additives

Description	Unit	$ Cost
Elevators, Hydraulic passenger, 2 stops		
1500# capacity	Ea.	65,700
2500# capacity	Ea.	68,800
3500# capacity	Ea.	73,300
Barrier gate w/programmable controller	Ea.	4150
Booth for attendant, average	Ea.	13,800
Fee computer	Ea.	16,000
Ticket spitter with time/date stamp	Ea.	7475
Mag strip encoding	Ea.	21,400
Collection station, pay on foot	Ea.	128,500
Parking control software	Ea.	27,500 - 118,500
Painting, Parking stalls	Stall	8.60
Parking Barriers		
Timber with saddles, 4" x 4"	L.F.	7.70
Precast concrete, 6" x 10" x 6'	Ea.	63
Traffic Signs, directional, 12" x 18"	Ea.	97.50

Important: See the Reference Section for Location Factors.

Model costs calculated for a 2 story building with 10' story height and 100,000 square feet of floor area

Garage, Underground Parking

			Unit	Unit Cost	Cost Per S.F.	% Of Sub-Total
A. SUBSTRUCTURE						
1010	Standard Foundations	Poured concrete; strip and spread footings and waterproofing	S.F. Ground	8.02	4.01	
1020	Special Foundations	N/A	—	—	—	
1030	Slab on Grade	5" reinforced concrete with vapor barrier and granular base	S.F. Slab	7.13	3.57	21.2%
2010	Basement Excavation	Excavation 24' deep	S.F. Ground	11.05	5.53	
2020	Basement Walls	N/A	—	—	—	
B. SHELL						
B10 Superstructure						
1010	Floor Construction	Cast-in-place concrete beam and slab, concrete columns	S.F. Floor	31.14	15.57	46.1%
1020	Roof Construction	Cast-in-place concrete beam and slab, concrete columns	S.F. Roof	25.90	12.95	
B20 Exterior Enclosure						
2010	Exterior Walls	Cast-in place concrete	S.F. Wall	22.78	4.10	
2020	Exterior Windows	N/A	—	—	—	6.9%
2030	Exterior Doors	Steel overhead, hollow metal	Each	4513	.19	
B30 Roofing						
3010	Roof Coverings	Neoprene membrane traffic deck	S.F. Roof	5.40	2.70	4.4%
3020	Roof Openings	N/A	—	—	—	
C. INTERIORS						
1010	Partitions	Concrete block	S.F. Partition	44.20	.85	
1020	Interior Doors	Hollow metal	Each	9256	.09	
1030	Fittings	N/A	—	—	—	
2010	Stair Construction	Concrete	Flight	6800	.34	2.2%
3010	Wall Finishes	Paint	S.F. Surface	2.60	.10	
3020	Floor Finishes	N/A	—	—	—	
3030	Ceiling Finishes	N/A	—	—	—	
D. SERVICES						
D10 Conveying						
1010	Elevators & Lifts	Two hydraulic passenger elevators	Each	83,500	1.67	2.7%
1020	Escalators & Moving Walks	N/A	—	—	—	
D20 Plumbing						
2010	Plumbing Fixtures	Drainage in parking areas, toilets, & service fixtures 1 Fixture/5000 S.F. Floor	Each	.05	.05	
2020	Domestic Water Distribution	Electric water heater	S.F. Floor	.11	.11	2.2%
2040	Rain Water Drainage	Roof drains	S.F. Roof	2.44	1.22	
D30 HVAC						
3010	Energy Supply	N/A	—	—	—	
3020	Heat Generating Systems	N/A	—	—	—	
3030	Cooling Generating Systems	N/A	—	—	—	0.3 %
3050	Terminal & Package Units	Exhaust fans	S.F. Floor	.17	.17	
3090	Other HVAC Sys. & Equipment	N/A	—	—	—	
D40 Fire Protection						
4010	Sprinklers	Dry pipe sprinkler system	S.F. Floor	4.35	4.35	7.3%
4020	Standpipes	Dry standpipe system, class 3	S.F. Floor	.16	.16	
D50 Electrical						
5010	Electrical Service/Distribution	200 ampere service, panel board and feeders	S.F. Floor	.12	.12	
5020	Lighting & Branch Wiring	T-8 fluorescent fixtures, receptacles, switches and misc. power	S.F. Floor	3.31	3.31	5.9%
5030	Communications & Security	Addressable alarm systems and emergency lighting	S.F. Floor	.18	.18	
5090	Other Electrical Systems	Emergency generator, 11.5 kW	S.F. Floor	.06	.06	
E. EQUIPMENT & FURNISHINGS						
1010	Commercial Equipment	N/A	—	—	—	
1020	Institutional Equipment	N/A	—	—	—	
1030	Vehicular Equipment	Ticket dispensers, booths, automatic gates	S.F. Floor	.42	.42	0.7 %
1090	Other Equipment	N/A	—	—	—	
F. SPECIAL CONSTRUCTION						
1020	Integrated Construction	N/A	—	—	—	
1040	Special Facilities	N/A	—	—	—	0.0 %
G. BUILDING SITEWORK	**N/A**					

		Sub-Total	61.82	100%
CONTRACTOR FEES (General Requirements: 10%, Overhead: 5%, Profit: 10%)		25%	15.45	
ARCHITECT FEES		8%	6.18	
	Total Building Cost		**83.45**	

For customer support on your Square Foot Cost Data, call 877.694.8329.

139

Costs per square foot of floor area

Exterior Wall	S.F. Area	2000	4000	6000	8000	10000	12000	14000	16000	18000
	L.F. Perimeter	180	260	340	420	450	500	550	575	600
Concrete Block	Wood Joists	161.00	141.55	135.10	131.80	127.30	125.05	123.55	121.50	120.00
	Steel Joists	169.15	147.10	139.75	136.00	130.95	128.50	126.75	124.50	122.85
Cast in Place Concrete	Wood Joists	172.40	150.05	142.65	138.95	133.40	130.70	128.85	126.35	124.50
	Steel Joists	182.00	156.35	147.80	143.50	137.40	134.45	132.35	129.60	127.60
Stucco	Wood Frame	166.90	144.20	136.65	132.85	127.15	124.45	122.50	120.00	118.10
Insulated Metal Panels	Steel Frame	185.05	158.25	149.30	144.90	138.55	135.55	133.30	130.55	128.40
Perimeter Adj., Add or Deduct	Per 100 L.F.	26.35	13.20	8.80	6.65	5.25	4.45	3.75	3.30	2.95
Story Hgt. Adj., Add or Deduct	Per 1 Ft.	2.10	1.50	1.35	1.30	1.05	1.05	0.90	0.85	0.80

For Basement, add $35.15 per square foot of basement area

The above costs were calculated using the basic specifications shown on the facing page. These costs should be adjusted where necessary for design alternatives and owner's requirements. Reported completed project costs, for this type of structure, range from $75.95 to $228.15 per S.F.

Common additives

Description	Unit	$ Cost
Air Compressors		
Electric 1-1/2 H.P., standard controls	Ea.	1600
Dual controls	Ea.	2000
5 H.P. 115/230 Volt, standard controls	Ea.	4475
Dual controls	Ea.	5375
Product Dispenser		
with vapor recovery for 6 nozzles	Ea.	27,200
Lifts, Single post		
8000# cap., swivel arm	Ea.	10,200
Two post, adjustable frames, 12,000# cap.	Ea.	5275
24,000# cap.	Ea.	11,400
30,000# cap.	Ea.	52,000
Four post, roll on ramp, 25,000# cap.	Ea.	20,800
Lockers, Steel, single tier, 60" or 72"	Opng.	221 - 370
2 tier, 60" or 72" total	Opng.	120 - 165
5 tier, box lockers	Opng.	77 - 86.50
Locker bench, lam. maple top only	L.F.	34
Pedestals, steel pipe	Ea.	76
Lube Equipment		
3 reel type, with pumps, no piping	Ea.	13,000
Spray Painting Booth, 26' long, complete	Ea.	22,100

Important: See the Reference Section for Location Factors.

Model costs calculated for a 1 story building with 14' story height and 10,000 square feet of floor area

					Unit	Unit Cost	Cost Per S.F.	% Of Sub-Total
A. SUBSTRUCTURE								
1010	Standard Foundations	Poured concrete; strip and spread footings			S.F. Ground	2.72	2.72	
1020	Special Foundations	N/A			—	—	—	
1030	Slab on Grade	6" reinforced concrete with vapor barrier and granular base			S.F. Slab	8.05	8.05	15.3%
2010	Basement Excavation	Site preparation for slab and trench for foundation wall and footing			S.F. Ground	.32	.32	
2020	Basement Walls	4' foundation wall			L.F. Wall	84	3.76	
B. SHELL								
B10 Superstructure								
1010	Floor Construction	N/A			—	—	—	
1020	Roof Construction	Metal deck on open web steel joists			S.F. Roof	5.70	5.70	5.9 %
B20 Exterior Enclosure								
2010	Exterior Walls	Concrete block	80% of wall		S.F. Wall	13.23	6.67	
2020	Exterior Windows	Hopper type commercial steel	5% of wall		Each	528	1.11	10.1%
2030	Exterior Doors	Steel overhead and hollow metal	15% of wall		S.F. Door	21.69	2.05	
B30 Roofing								
3010	Roof Coverings	Built-up tar and gravel; perlite/EPS composite insulation			S.F. Roof	6.82	6.82	7.1%
3020	Roof Openings	Skylight			S.F. Roof	.02	.02	
C. INTERIORS								
1010	Partitions	Concrete block	50 S.F. Floor/L.F. Partition		S.F. Partition	23.35	4.67	
1020	Interior Doors	Single leaf hollow metal	3000 S.F. Floor/Door		Each	1157	.39	
1030	Fittings	Toilet partitions			S.F. Floor	.15	.15	
2010	Stair Construction	N/A			—	—	—	10.3%
3010	Wall Finishes	Paint			S.F. Surface	7.58	3.03	
3020	Floor Finishes	90% metallic floor hardener, 10% vinyl composition tile			S.F. Floor	1.33	1.33	
3030	Ceiling Finishes	Gypsum board on wood joists in office and washrooms	10% of area		S.F. Ceiling	4.13	.41	
D. SERVICES								
D10 Conveying								
1010	Elevators & Lifts	N/A			—	—	—	0.0 %
1020	Escalators & Moving Walks	N/A			—	—	—	
D20 Plumbing								
2010	Plumbing Fixtures	Toilet and service fixtures, supply and drainage	1 Fixture/500 S.F. Floor		Each	1610	3.22	
2020	Domestic Water Distribution	Gas fired water heater			S.F. Floor	.54	.54	6.4%
2040	Rain Water Drainage	Roof drains			S.F. Roof	2.42	2.42	
D30 HVAC								
3010	Energy Supply	N/A			—	—	—	
3020	Heat Generating Systems	N/A			—	—	—	
3030	Cooling Generating Systems	N/A			—	—	—	10.6 %
3050	Terminal & Package Units	Single zone AC unit			S.F. Floor	9.32	9.32	
3090	Other HVAC Sys. & Equipment	Garage exhaust system			S.F. Floor	.96	.96	
D40 Fire Protection								
4010	Sprinklers	Sprinklers, ordinary hazard			S.F. Floor	4.73	4.73	5.8%
4020	Standpipes	Standpipe			S.F. Floor	.94	.94	
D50 Electrical								
5010	Electrical Service/Distribution	200 ampere service, panel board and feeders			S.F. Floor	.49	.49	
5020	Lighting & Branch Wiring	T-8 fluorescent fixtures, receptacles, switches, A.C. and misc. power			S.F. Floor	8.23	8.23	12.7%
5030	Communications & Security	Addressable alarm systems, partial internet wiring and emergency lighting			S.F. Floor	3.55	3.55	
5090	Other Electrical Systems	Emergency generator, 15 kW			S.F. Floor	.08	.08	
E. EQUIPMENT & FURNISHINGS								
1010	Commercial Equipment	N/A			—	—	—	
1020	Institutional Equipment	N/A			—	—	—	15.8 %
1030	Vehicular Equipment	Hoists			S.F. Floor	15.34	15.34	
1090	Other Equipment	N/A			—	—	—	
F. SPECIAL CONSTRUCTION								
1020	Integrated Construction	N/A			—	—	—	0.0 %
1040	Special Facilities	N/A			—	—	—	
G. BUILDING SITEWORK	**N/A**							

					Sub-Total	97.02	**100%**
CONTRACTOR FEES (General Requirements: 10%, Overhead: 5%, Profit: 10%)					25%	24.23	
ARCHITECT FEES					8%	9.70	

Total Building Cost 130.95

For customer support on your Square Foot Cost Data, call 877.694.8329.

141

Costs per square foot of floor area

Exterior Wall	S.F. Area	600	800	1000	1200	1400	1600	1800	2000	2200
	L.F. Perimeter	100	120	126	140	153	160	170	180	190
Face Brick with Concrete Block Back-up	Wood Truss	284.40	266.55	245.45	236.30	229.35	221.25	216.25	212.20	208.90
	Steel Joists	305.70	286.85	264.55	254.95	247.50	238.95	233.70	229.50	225.95
Enameled Sandwich Panel	Steel Frame	269.70	252.55	232.85	224.25	217.60	210.05	205.35	201.60	198.50
Tile on Concrete Block	Steel Joists	322.80	302.25	277.50	266.95	258.75	249.25	243.35	238.70	234.80
Aluminum Siding	Wood Frame	243.35	228.95	213.15	206.10	200.65	194.60	190.80	187.85	185.25
Wood Siding	Wood Frame	244.60	230.10	214.15	206.95	201.50	195.35	191.55	188.50	185.95
Perimeter Adj., Add or Deduct	Per 100 L.F.	123.45	92.60	74.05	61.80	52.90	46.30	41.10	37.05	33.65
Story Hgt. Adj., Add or Deduct	Per 1 Ft.	7.95	7.15	6.00	5.60	5.20	4.80	4.50	4.35	4.05
Basement—Not Applicable										

The above costs were calculated using the basic specifications shown on the facing page. These costs should be adjusted where necessary for design alternatives and owner's requirements. Reported completed project costs, for this type of structure, range from $ 54.70 to $ 276.35 per S.F.

Common additives

Description	Unit	$ Cost
Air Compressors		
Electric 1-1/2 H.P., standard controls	Ea.	1600
Dual controls	Ea.	2000
5 H.P. 115/230 volt, standard controls	Ea.	4475
Dual controls	Ea.	5375
Product Dispenser		
with vapor recovery for 6 nozzles	Ea.	27,200
Lifts, Single post		
8000# cap. swivel arm	Ea.	10,200
Two post, adjustable frames, 12,000# cap.	Ea.	5275
24,000# cap.	Ea.	11,400
30,000# cap.	Ea.	52,000
Four post, roll on ramp, 25,000# cap.	Ea.	20,800
Lockers, Steel, single tier, 60" or 72"	Opng.	221 - 370
2 tier, 60" or 72" total	Opng.	120 - 165
5 tier, box lockers	Ea.	77 - 86.50
Locker bench, lam. maple top only	L.F.	34
Pedestals, steel pipe	Ea.	76
Lube Equipment		
3 reel type, with pumps, no piping	Ea.	13,000

Important: See the Reference Section for Location Factors.

Model costs calculated for a 1 story building with 12' story height and 1,400 square feet of floor area

				Unit	Unit Cost	Cost Per S.F.	% Of Sub-Total
A. SUBSTRUCTURE							
1010	Standard Foundations	Poured concrete; strip and spread footings		S.F. Ground	4.40	4.40	
1020	Special Foundations	N/A		—	—	—	
1030	Slab on Grade	6" reinforced concrete with vapor barrier and granular base		S.F. Slab	8.05	8.05	12.8%
2010	Basement Excavation	Site preparation for slab and trench for foundation wall and footing		S.F. Ground	1.07	1.07	
2020	Basement Walls	4' foundation wall		L.F. Wall	75	8.20	
B. SHELL							
B10 Superstructure							
1010	Floor Construction	N/A		—	—	—	
1020	Roof Construction	Plywood on wood trusses		S.F. Roof	7.01	7.01	4.1 %
B20 Exterior Enclosure							
2010	Exterior Walls	Face brick with concrete block backup	60% of wall	S.F. Wall	33.45	26.32	
2020	Exterior Windows	Store front and metal top hinged outswinging	20% of wall	Each	41.80	10.97	28.9%
2030	Exterior Doors	Steel overhead, aluminum & glass and hollow metal	20% of wall	Each	44.76	11.74	
B30 Roofing							
3010	Roof Coverings	Asphalt shingles with flashing; perlite/EPS composite insulation		S.F. Roof	5.36	5.36	3.2%
3020	Roof Openings	N/A		—	—	—	
C. INTERIORS							
1010	Partitions	Concrete block	25 S.F. Floor/L.F. Partition	S.F. Partition	7.91	2.53	
1020	Interior Doors	Single leaf hollow metal	700 S.F. Floor/Door	Each	1157	1.65	
1030	Fittings	Toilet partitions		S.F. Floor	2.09	2.09	
2010	Stair Construction	N/A		—	—	—	7.4%
3010	Wall Finishes	Paint		S.F. Surface	5.97	3.82	
3020	Floor Finishes	Vinyl composition tile	35% of floor area	S.F. Floor	2.69	.94	
3030	Ceiling Finishes	Painted gypsum board on furring in sales area & washrooms	35% of floor area	S.F. Ceiling	4.51	1.58	
D. SERVICES							
D10 Conveying							
1010	Elevators & Lifts	N/A		—	—	—	0.0 %
1020	Escalators & Moving Walks	N/A		—	—	—	
D20 Plumbing							
2010	Plumbing Fixtures	Toilet and service fixtures, supply and drainage	1 Fixture/235 S.F. Floor	Each	2270	9.66	
2020	Domestic Water Distribution	Gas fired water heater		S.F. Floor	3.88	3.88	8.0%
2040	Rain Water Drainage	N/A		—	—	—	
D30 HVAC							
3010	Energy Supply	N/A		—	—	—	
3020	Heat Generating Systems	N/A		—	—	—	
3030	Cooling Generating Systems	N/A		—	—	—	15.8 %
3050	Terminal & Package Units	Single zone AC unit		S.F. Floor	21.14	21.14	
3090	Other HVAC Sys. & Equipment	Underfloor exhaust system		S.F. Floor	5.64	5.64	
D40 Fire Protection							
4010	Sprinklers	Wet pipe sprinkler system		S.F. Floor	9.39	9.39	8.0%
4020	Standpipes	Standpipe		S.F. Floor	4.14	4.14	
D50 Electrical							
5010	Electrical Service/Distribution	200 ampere service, panel board and feeders		S.F. Floor	4.91	4.91	
5020	Lighting & Branch Wiring	High efficiency fluorescent fixtures, receptacles, switches, A.C. and misc. power		S.F. Floor	6.20	6.20	11.9%
5030	Communications & Security	Addressable alarm systems and emergency lighting		S.F. Floor	8.59	8.59	
5090	Other Electrical Systems	Emergency generator		S.F. Floor	.58	.58	
E. EQUIPMENT & FURNISHINGS							
1010	Commercial Equipment	N/A		—	—	—	
1020	Institutional Equipment	N/A		—	—	—	0.0 %
1030	Vehicular Equipment	N/A		—	—	—	
1090	Other Equipment	N/A		—	—	—	
F. SPECIAL CONSTRUCTION							
1020	Integrated Construction	N/A		—	—	—	0.0 %
1040	Special Facilities	N/A		—	—	—	
G. BUILDING SITEWORK	**N/A**						

		Sub-Total	169.86	100%
CONTRACTOR FEES (General Requirements: 10%, Overhead: 5%, Profit: 10%)			25%	42.50	
ARCHITECT FEES			8%	16.99	

Total Building Cost	229.35

For customer support on your Square Foot Cost Data, call 877.694.8329.

143

Costs per square foot of floor area

Exterior Wall	S.F. Area	12000	16000	20000	25000	30000	35000	40000	45000	50000
	L.F. Perimeter	440	520	600	700	708	780	841	910	979
Reinforced Concrete Block	Lam. Wood Arches	175.75	171.10	168.35	166.10	162.00	160.60	159.35	158.50	157.85
	Rigid Steel Frame	184.35	179.50	176.65	174.25	170.00	168.50	167.20	166.35	165.65
Face Brick with Concrete Block Back-up	Lam. Wood Arches	199.15	191.80	187.50	183.95	177.05	174.80	172.75	171.45	170.35
	Rigid Steel Frame	206.35	199.00	194.65	191.15	184.25	181.95	179.95	178.55	177.55
Metal Sandwich Panels	Lam. Wood Arches	185.85	180.00	176.60	173.80	168.50	166.70	165.15	164.10	163.25
	Rigid Steel Frame	196.75	190.50	186.80	183.80	178.00	176.15	174.45	173.30	172.40
Perimeter Adj., Add or Deduct	Per 100 L.F.	7.10	5.25	4.20	3.40	2.80	2.45	2.10	1.90	1.70
Story Hgt. Adj., Add or Deduct	Per 1 Ft.	1.00	0.85	0.85	0.80	0.65	0.60	0.60	0.60	0.60
Basement—Not Applicable										

The above costs were calculated using the basic specifications shown on the facing page. These costs should be adjusted where necessary for design alternatives and owner's requirements. Reported completed project costs, for this type of structure, range from $83.00 to $247.65 per S.F.

Common additives

Description	Unit	$ Cost
Bleachers, Telescoping, manual		
To 15 tier	Seat	137 - 190
16-20 tier	Seat	281 - 345
21-30 tier	Seat	299 - 390
For power operation, add	Seat	55 - 86
Gym Divider Curtain, Mesh top		
Manual roll-up	S.F.	13.20
Gym Mats		
2" naugahyde covered	S.F.	4.08
2" nylon	S.F.	8.85
1-1/2" wall pads	S.F.	8.70
1" wrestling mats	S.F.	6.15
Scoreboard		
Basketball, one side	Ea.	3525 - 9800
Basketball Backstop		
Wall mtd., 6' extended, fixed	Ea.	2550 - 3125
Swing up, wall mtd.	Ea.	2600 - 4150

Description	Unit	$ Cost
Lockers, Steel, single tier, 60" or 72"	Opng.	221 - 370
2 tier, 60" or 72" total	Opng.	120 - 165
5 tier, box lockers	Opng.	77 - 86.50
Locker bench, lam. maple top only	L.F.	34
Pedestals, steel pipe	Ea.	76
Sound System		
Amplifier, 250 watts	Ea.	2050
Speaker, ceiling or wall	Ea.	213
Trumpet	Ea.	410
Emergency Lighting, 25 watt, battery operated		
Lead battery	Ea.	330
Nickel cadmium	Ea.	780

Important: See the Reference Section for Location Factors.

Model costs calculated for a 1 story building with 25' story height and 20,000 square feet of floor area

Gymnasium

					Unit	Unit Cost	Cost Per S.F.	% Of Sub-Total
A. SUBSTRUCTURE								
1010	Standard Foundations	Poured concrete; strip and spread footings			S.F. Ground	1.57	1.57	
1020	Special Foundations	N/A			—	—	—	
1030	Slab on Grade	4" reinforced concrete with vapor barrier and granular base			S.F. Slab	5.44	5.44	7.5%
2010	Basement Excavation	Site preparation for slab and trench for foundation wall and footing			S.F. Ground	.18	.18	
2020	Basement Walls	4' foundation wall			L.F. Wall	75	2.26	
B. SHELL								
B10 Superstructure								
1010	Floor Construction	N/A			—	—	—	
1020	Roof Construction	Wood deck on laminated wood arches			S.F. Roof	19.25	19.25	17.1 %
B20 Exterior Enclosure								
2010	Exterior Walls	Reinforced concrete block (end walls included)		90% of wall	S.F. Wall	12.25	8.27	
2020	Exterior Windows	Metal horizontal pivoted		10% of wall	Each	676	5.08	11.1%
2030	Exterior Doors	Aluminum and glass, hollow metal, steel overhead			Each	2131	.63	
B30 Roofing								
3010	Roof Coverings	EPDM, 60 mils, fully adhered; polyisocyanurate insulation			S.F. Roof	5.50	5.50	
3020	Roof Openings	N/A			—	—	—	4.4%
C. INTERIORS								
1010	Partitions	Concrete block	50 S.F. Floor/L.F. Partition		S.F. Partition	9.65	1.93	
1020	Interior Doors	Single leaf hollow metal	500 S.F. Floor/Door		Each	1157	2.31	
1030	Fittings	Toilet partitions			S.F. Floor	.30	.30	
2010	Stair Construction	N/A			—	—	—	19.8%
3010	Wall Finishes	50% paint, 50% ceramic tile			S.F. Surface	10.20	4.08	
3020	Floor Finishes	90% hardwood, 10% ceramic tile			S.F. Floor	15.29	15.29	
3030	Ceiling Finishes	Mineral fiber tile on concealed zee bars	15% of area		S.F. Ceiling	7.03	1.05	
D. SERVICES								
D10 Conveying								
1010	Elevators & Lifts	N/A			—	—	—	
1020	Escalators & Moving Walks	N/A			—	—	—	0.0 %
D20 Plumbing								
2010	Plumbing Fixtures	Toilet and service fixtures, supply and drainage	1 Fixture/515 S.F. Floor		Each	4692	9.11	
2020	Domestic Water Distribution	Electric water heater			S.F. Floor	4.79	4.79	11.0%
2040	Rain Water Drainage	N/A			—	—	—	
D30 HVAC								
3010	Energy Supply	N/A			—	—	—	
3020	Heat Generating Systems	Included in D3050			—	—	—	
3030	Cooling Generating Systems	N/A			—	—	—	9.1 %
3050	Terminal & Package Units	Single zone rooftop unit, gas heating, electric cooling			S.F. Floor	11.50	11.50	
3090	Other HVAC Sys. & Equipment	N/A			—	—	—	
D40 Fire Protection								
4010	Sprinklers	Wet pipe sprinkler system			S.F. Floor	3.67	3.67	
4020	Standpipes	Standpipe			S.F. Floor	1.03	1.03	3.7%
D50 Electrical								
5010	Electrical Service/Distribution	400 ampere service, panel board and feeders			S.F. Floor	1.12	1.12	
5020	Lighting & Branch Wiring	High efficiency fluorescent fixtures, receptacles, switches, A.C. and misc. power			S.F. Floor	9.36	9.36	
5030	Communications & Security	Addressable alarm systems, sound system and emergency lighting			S.F. Floor	2.92	2.92	10.8%
5090	Other Electrical Systems	Emergency generator, 7.5 kW			S.F. Floor	.19	.19	
E. EQUIPMENT & FURNISHINGS								
1010	Commercial Equipment	N/A			—	—	—	
1020	Institutional Equipment	N/A			—	—	—	
1030	Vehicular Equipment	N/A			—	—	—	5.3 %
1090	Other Equipment	Bleachers, sauna, weight room			S.F. Floor	6.72	6.72	
F. SPECIAL CONSTRUCTION								
1020	Integrated Construction	N/A			—	—	—	
1040	Special Facilities	N/A			—	—	—	0.0 %
G. BUILDING SITEWORK	**N/A**							

	Sub-Total	125.87	100%
CONTRACTOR FEES (General Requirements: 10%, Overhead: 5%, Profit: 10%)	25%	31.47	
ARCHITECT FEES	7%	11.01	

Total Building Cost	**168.35**

For customer support on your Square Foot Cost Data, call 877.694.8329.

145

Costs per square foot of floor area

Exterior Wall	S.F. Area	5000	10000	15000	20000	30000	40000	50000	75000	100000
	L.F. Perimeter	300	410	500	580	710	830	930	1150	1300
Concrete Block Reinforced	Steel Frame	181.55	147.55	134.85	127.95	120.10	115.85	112.95	108.55	105.75
	Bearing Walls	193.00	158.80	145.95	139.10	131.10	126.85	123.90	119.50	116.60
Precast Concrete	Steel Frame	217.20	171.90	154.60	145.15	134.10	128.15	123.95	117.65	113.35
	Bearing Walls	223.25	177.70	160.25	150.80	139.65	133.65	129.40	123.05	118.75
Galv. Steel Siding	Steel Frame	175.40	143.35	131.40	125.00	117.65	113.75	111.00	106.95	104.35
Metal Sandwich Panel	Steel Frame	186.60	150.35	136.80	129.50	121.20	116.75	113.65	109.15	106.15
Perimeter Adj., Add or Deduct	Per 100 L.F.	20.70	10.40	6.95	5.20	3.45	2.65	2.05	1.40	1.00
Story Hgt. Adj., Add or Deduct	Per 1 Ft.	2.10	1.40	1.15	1.00	0.80	0.75	0.60	0.55	0.40
Basement—Not Applicable										

The above costs were calculated using the basic specifications shown on the facing page. These costs should be adjusted where necessary for design alternatives and owner's requirements. Reported completed project costs, for this type of structure, range from $47.65 to $217.25 per S.F.

Common additives

Description	Unit	$ Cost
Closed Circuit Surveillance, One station		
Camera and monitor	Ea.	1975
For additional camera stations, add	Ea.	1075
Emergency Lighting, 25 watt, battery operated		
Lead battery	Ea.	330
Nickel cadmium	Ea.	780
Lockers, Steel, single tier, 60" or 72"	Opng.	221 - 370
2 tier, 60" or 72" total	Opng.	120 - 165
5 tier, box lockers	Opng.	77 - 86.50
Locker bench, lam. maple top only	L.F.	34
Pedestals, steel pipe	Ea.	76
Safe, Office type, 1 hour rating		
30" x 18" x 18"	Ea.	2500
60" x 36" x 18", double door	Ea.	8125
Sound System		
Amplifier, 250 watts	Ea.	2050
Speaker, ceiling or wall	Ea.	213
Trumpet	Ea.	410

Model costs calculated for a 1 story building with 24' story height and 20,000 square feet of floor area

Hangar, Aircraft

					Unit	Unit Cost	Cost Per S.F.	% Of Sub-Total
A. SUBSTRUCTURE								
1010	Standard Foundations	Poured concrete; strip and spread footings and			S.F. Ground	1.47	1.47	
1020	Special Foundations	N/A			—	—	—	
1030	Slab on Grade	6" reinforced concrete with vapor barrier and granular base			S.F. Slab	8.05	8.05	13.0%
2010	Basement Excavation	Site preparation for slab and trench for foundation wall and footing			S.F. Ground	.18	.18	
2020	Basement Walls	4' foundation wall			L.F. Wall	84	2.42	
B. SHELL								
	B10 Superstructure							
1010	Floor Construction	N/A			—	—	—	
1020	Roof Construction	Metal deck, open web steel joists, beams, columns			S.F. Roof	9.87	9.87	10.6 %
	B20 Exterior Enclosure							
2010	Exterior Walls	Galvanized steel siding			S.F. Wall	11.12	3.87	
2020	Exterior Windows	Industrial horizontal pivoted steel		20% of wall	Each	2020	11.72	26.1%
2030	Exterior Doors	Steel overhead and sliding		30% of wall	Each	42.19	8.81	
	B30 Roofing							
3010	Roof Coverings	Elastomeric membrane; fiberboard insulation			S.F. Roof	5.34	5.34	5.9%
3020	Roof Openings	Roof hatches			S.F. Roof	.13	.13	
C. INTERIORS								
1010	Partitions	Concrete block	200 S.F. Floor/L.F. Partition		S.F. Partition	10.40	.52	
1020	Interior Doors	Single leaf hollow metal	5000 S.F. Floor/Door		Each	1157	.24	
1030	Fittings	Toilet partitions			S.F. Floor	.26	.26	
2010	Stair Construction	N/A			—	—	—	1.2%
3010	Wall Finishes	Paint			S.F. Surface	1.30	.13	
3020	Floor Finishes	N/A			—	—	—	
3030	Ceiling Finishes	N/A			—	—	—	
D. SERVICES								
	D10 Conveying							
1010	Elevators & Lifts	N/A			—	—	—	0.0 %
1020	Escalators & Moving Walks	N/A			—	—	—	
	D20 Plumbing							
2010	Plumbing Fixtures	Toilet and service fixtures, supply and drainage	1 Fixture/1000 S.F. Floor		Each	2690	2.69	
2020	Domestic Water Distribution	Gas fired water heater			S.F. Floor	2.39	2.39	6.8%
2040	Rain Water Drainage	N/A			—	—	—	
	D30 HVAC							
3010	Energy Supply	N/A			—	—	—	
3020	Heat Generating Systems	Unit heaters			Each	7.72	7.72	
3030	Cooling Generating Systems	N/A			—	—	—	8.6 %
3050	Terminal & Package Units	Exhaust fan			S.F. Floor	.28	.28	
3090	Other HVAC Sys. & Equipment	N/A			—			
	D40 Fire Protection							
4010	Sprinklers	Sprinklers, extra hazard, deluge system			S.F. Floor	8.02	8.02	13.9%
4020	Standpipes	Standpipe and fire pump			S.F. Floor	4.98	4.98	
	D50 Electrical							
5010	Electrical Service/Distribution	200 ampere service, panel board and feeders			S.F. Floor	.99	.99	
5020	Lighting & Branch Wiring	High intensity discharge fixtures, receptacles, switches and misc. power			S.F. Floor	10.63	10.63	14.0%
5030	Communications & Security	Addressable alarm systems and emergency lighting			S.F. Floor	1.35	1.35	
5090	Other Electrical Systems	Emergency generator, 7.5 kW			S.F. Floor	.12	.12	
E. EQUIPMENT & FURNISHINGS								
1010	Commercial Equipment	N/A			—	—	—	
1020	Institutional Equipment	N/A			—	—	—	
1030	Vehicular Equipment	N/A			—	—	—	0.0 %
1090	Other Equipment	N/A			—	—	—	
F. SPECIAL CONSTRUCTION								
1020	Integrated Construction	N/A			—	—	—	0.0 %
1040	Special Facilities	N/A			—	—	—	
G. BUILDING SITEWORK	**N/A**							

		Sub-Total	93.45	100%
CONTRACTOR FEES (General Requirements: 10%, Overhead: 5%, Profit: 10%)		25%	23.37	
ARCHITECT FEES		7%	8.18	

Total Building Cost   125

For customer support on your Square Foot Cost Data, call 877.694.8329.

147

Costs per square foot of floor area

Exterior Wall	S.F. Area	25000	40000	55000	70000	85000	100000	115000	130000	145000
	L.F. Perimeter	388	520	566	666	766	866	878	962	1045
Face Brick with Structural Facing Tile	Steel Frame	352.10	343.55	335.85	333.35	331.70	330.55	327.85	327.10	326.55
	R/Conc. Frame	357.85	349.30	341.55	339.10	337.40	336.25	333.60	332.80	332.25
Face Brick with Concrete Block Back-up	Steel Frame	343.45	336.35	330.10	328.10	326.70	325.70	323.60	323.00	322.50
	R/Conc. Frame	341.80	334.65	328.40	326.40	325.00	324.05	321.90	321.30	320.85
Precast Concrete Panels	Steel Frame	357.70	348.20	339.55	336.80	334.95	333.65	330.60	329.75	329.10
	R/Conc. Frame	363.40	353.90	345.25	342.50	340.65	339.35	336.30	335.50	334.85
Perimeter Adj., Add or Deduct	Per 100 L.F.	9.75	6.10	4.45	3.50	2.85	2.40	2.15	1.85	1.65
Story Hgt. Adj., Add or Deduct	Per 1 Ft.	2.85	2.45	1.85	1.75	1.65	1.55	1.45	1.35	1.30

For Basement, add $36.70 per square foot of basement area

The above costs were calculated using the basic specifications shown on the facing page. These costs should be adjusted where necessary for design alternatives and owner's requirements. Reported completed project costs, for this type of structure, range from $183.15 to $462.60 per S.F.

Common additives

Description	Unit	$ Cost
Cabinet, Base, door units, metal	L.F.	320
Drawer units	L.F.	630
Tall storage cabinets, 7' high, open	L.F.	605
With doors	L.F.	875
Wall, metal 12-1/2" deep, open	L.F.	241
With doors	L.F.	440
Closed Circuit TV (Patient monitoring)		
One station camera & monitor	Ea.	1975
For additional camera, add	Ea.	1075
For automatic iris for low light, add	Ea.	2750
Hubbard Tank, with accessories		
Stainless steel, 125 GPM 45 psi	Ea.	29,900
For electric hoist, add	Ea.	3250
Mortuary Refrigerator, End operated		
2 capacity	Ea.	13,500
6 capacity	Ea.	24,400

Description	Unit	$ Cost
Nurses Call Station		
Single bedside call station	Ea.	335
Ceiling speaker station	Ea.	156
Emergency call station	Ea.	162
Pillow speaker	Ea.	277
Double bedside call station	Ea.	320
Duty station	Ea.	325
Standard call button	Ea.	178
Master control station for 20 stations	Ea.	6400
Sound System		
Amplifier, 250 watts	Ea.	2050
Speaker, ceiling or wall	Ea.	213
Trumpet	Ea.	410
Station, Dietary with ice	Ea.	18,400
Sterilizers		
Single door, steam	Ea.	134,500
Double door, steam	Ea.	227,000
Portable, countertop, steam	Ea.	3900 - 4750
Gas	Ea.	43,700
Automatic washer/sterilizer	Ea.	60,500

Important: See the Reference Section for Location Factors.

Model costs calculated for a 3 story building with 12' story height and 55,000 square feet of floor area

Hospital, 2-3 Story

					Unit	Unit Cost	Cost Per S.F.	% Of Sub-Total
A.	**SUBSTRUCTURE**							
1010	Standard Foundations	Poured concrete; strip and spread footings			S.F. Ground	5.52	1.84	
1020	Special Foundations	N/A			—	—	—	
1030	Slab on Grade	4" reinforced concrete with vapor barrier and granular base			S.F. Slab	5.44	1.81	1.8%
2010	Basement Excavation	Site preparation for slab and trench for foundation wall and footing			S.F. Ground	.18	.06	
2020	Basement Walls	4' foundation wall			L.F. Wall	84	.86	
B.	**SHELL**							
	B10 Superstructure							
1010	Floor Construction	Cast-in-place concrete beam and slab			S.F. Floor	24.57	16.38	9.1%
1020	Roof Construction	Cast-in-place concrete beam and slab			S.F. Roof	19.65	6.55	
	B20 Exterior Enclosure							
2010	Exterior Walls	Face brick and structural facing tile	85% of wall		S.F. Wall	46.74	14.72	
2020	Exterior Windows	Aluminum sliding	15% of wall		Each	750	1.81	6.7%
2030	Exterior Doors	Double aluminum and glass and sliding doors			Each	3288	.36	
	B30 Roofing							
3010	Roof Coverings	Built-up tar and gravel with flashing; perlite/EPS composite insulation			S.F. Roof	6.42	2.14	0.9%
3020	Roof Openings	Roof hatches			S.F. Roof	.03	.01	
C.	**INTERIORS**							
1010	Partitions	Concrete block, gypsum board on metal studs	9 S.F. Floor/L.F. Partition		S.F. Partition	8.75	9.72	
1020	Interior Doors	Single leaf hollow metal	90 S.F. Floor/Door		Each	1203	13.37	
1030	Fittings	Hospital curtains			S.F. Floor	1.10	1.10	
2010	Stair Construction	Concrete filled metal pan			Flight	8225	.90	19.6%
3010	Wall Finishes	40% vinyl wall covering, 35% ceramic tile, 25% epoxy coating			S.F. Surface	3.62	8.05	
3020	Floor Finishes	60% vinyl tile, 20% ceramic, 20% terrazzo			S.F. Floor	8.99	8.99	
3030	Ceiling Finishes	Plaster on suspended metal lath			S.F. Ceiling	7.03	7.03	
D.	**SERVICES**							
	D10 Conveying							
1010	Elevators & Lifts	Two hydraulic hospital elevators			Each	125,400	4.56	1.8%
1020	Escalators & Moving Walks	N/A			—	—	—	
	D20 Plumbing							
2010	Plumbing Fixtures	Medical, patient & specialty fixtures, supply and drainage	1 Fixture/224 S.F. Floor		Each	6975	31.14	
2020	Domestic Water Distribution	Electric water heater			S.F. Floor	11.88	11.88	17.4%
2040	Rain Water Drainage	Roof drains			S.F. Roof	1.80	.60	
	D30 HVAC							
3010	Energy Supply	Conditioned air with hot water reheat system			S.F. Floor	4.87	4.87	
3020	Heat Generating Systems	Boiler			Each	127,875	3.65	
3030	Cooling Generating Systems	Conditioned air, chillers, cooling towers			S.F. Floor	6.79	6.79	22.8%
3050	Terminal & Package Units	N/A			—	—	—	
3090	Other HVAC Sys. & Equipment	Conditioned air, ductwork, surgery air curtain			S.F. Floor	41.86	41.86	
	D40 Fire Protection							
4010	Sprinklers	Wet pipe sprinkler system			S.F. Floor	3.10	3.10	1.6%
4020	Standpipes	Standpipe			S.F. Floor	.79	.79	
	D50 Electrical							
5010	Electrical Service/Distribution	1200 ampere service, panel board and feeders			S.F. Floor	1.54	1.54	
5020	Lighting & Branch Wiring	High efficiency hospital grade light fixtures, receptacles, switches, A.C. and misc. power			S.F. Floor	17.07	17.07	
5030	Communications & Security	Addressable alarm systems, internet wiring, communications systems, emergency lighting			S.F. Floor	5.45	5.45	11.4%
5090	Other Electrical Systems	Emergency generator, 200 kW with fuel tank, Uninterruptible power supply			S.F. Floor	4.61	4.61	
E.	**EQUIPMENT & FURNISHINGS**							
1010	Commercial Equipment	N/A			—	—	—	
1020	Institutional Equipment	Medical gas system, kitchen equip, lab equip., curtain partitions			S.F. Floor	11.27	11.27	
1030	Vehicular Equipment	N/A			—	—	—	6.8 %
2020	Moveable Furnishings	Patient wall systems			S.F. Floor	5.80	5.80	
F.	**SPECIAL CONSTRUCTION**							
1020	Integrated Construction	N/A			—	—	—	
1040	Special Facilities	N/A			—	—	—	0.0 %
G.	**BUILDING SITEWORK**	**N/A**						

			Sub-Total	250.68	100%
CONTRACTOR FEES (General Requirements: 10%, Overhead: 5%, Profit: 10%)			25%	62.67	
ARCHITECT FEES			9%	28.20	
		Total Building Cost		**341.55**	

For customer support on your Square Foot Cost Data, call 877.694.8329.

149

Costs per square foot of floor area

Exterior Wall	S.F. Area	100000	125000	150000	175000	200000	225000	250000	275000	300000
	L.F. Perimeter	594	705	816	783	866	950	1033	1116	1200
Face Brick with Structural Facing Tile	Steel Frame	304.85	301.50	299.25	293.60	292.20	291.20	290.30	289.55	289.00
	R/Conc. Frame	309.35	305.85	303.55	297.85	296.40	295.35	294.45	293.65	293.15
Face Brick with Concrete Block Back-up	Steel Frame	299.70	296.50	294.40	289.60	288.25	287.25	286.50	285.70	285.30
	R/Conc. Frame	303.95	300.75	298.60	293.80	292.50	291.45	290.65	289.95	289.45
Precast Concrete Panels With Exposed Aggregate	Steel Frame	303.10	299.75	297.50	292.10	290.70	289.65	288.80	288.05	287.50
	R/Conc. Frame	307.30	303.90	301.65	296.30	294.90	293.85	293.05	292.25	291.75
Perimeter Adj., Add or Deduct	Per 100 L.F.	4.90	3.85	3.20	2.80	2.40	2.10	1.95	1.85	1.65
Story Hgt. Adj., Add or Deduct	Per 1 Ft.	2.20	2.05	2.00	1.70	1.60	1.50	1.50	1.50	1.50

For Basement, add $37.85 per square foot of basement area

The above costs were calculated using the basic specifications shown on the facing page. These costs should be adjusted where necessary for design alternatives and owner's requirements. Reported completed project costs, for this type of structure, range from $187.35 to $456.80 per S.F.

Common additives

Description	Unit	$ Cost
Cabinets, Base, door units, metal	L.F.	320
Drawer units	L.F.	630
Tall storage cabinets, 7' high, open	L.F.	605
With doors	L.F.	875
Wall, metal 12-1/2" deep, open	L.F.	241
With doors	L.F.	440
Closed Circuit TV (Patient monitoring)		
One station camera & monitor	Ea.	1975
For additional camera add	Ea.	1075
For automatic iris for low light add	Ea.	2750
Hubbard Tank, with accessories		
Stainless steel, 125 GPM 45 psi	Ea.	29,900
For electric hoist, add	Ea.	3250
Mortuary Refrigerator, End operated		
2 capacity	Ea.	13,500
6 capacity	Ea.	24,400

Description	Unit	$ Cost
Nurses Call Station		
Single bedside call station	Ea.	335
Ceiling speaker station	Ea.	156
Emergency call station	Ea.	162
Pillow speaker	Ea.	277
Double bedside call station	Ea.	320
Duty station	Ea.	325
Standard call button	Ea.	178
Master control station for 20 stations	Ea.	6400
Sound System		
Amplifier, 250 watts	Ea.	2050
Speaker, ceiling or wall	Ea.	213
Trumpet	Ea.	410
Station, Dietary with ice	Ea.	18,400
Sterilizers		
Single door, steam	Ea.	134,500
Double door, steam	Ea.	227,000
Portable, counter top, steam	Ea.	3900 - 4750
Gas	Ea.	43,700
Automatic washer/sterilizer	Ea.	60,500

Important: See the Reference Section for Location Factors.

Model costs calculated for a 6 story building with 12' story height and 200,000 square feet of floor area

Hospital, 4-8 Story

						Unit	Unit Cost	Cost Per S.F.	% Of Sub-Total
A. SUBSTRUCTURE									
1010	Standard Foundations		Poured concrete; strip and spread footings			S.F. Ground	11.10	1.85	
1020	Special Foundations		N/A			—	—	—	
1030	Slab on Grade		4" reinforced concrete with vapor barrier and granular base			S.F. Slab	8.05	1.35	1.7%
2010	Basement Excavation		Site preparation for slab and trench for foundation wall and footing			S.F. Ground	.18	.03	
2020	Basement Walls		4' foundation wall			L.F. Wall	84	.36	
B. SHELL									
B10 Superstructure									
1010	Floor Construction		Concrete slab with metal deck and beams, steel columns			S.F. Floor	23.02	19.18	9.7%
1020	Roof Construction		Metal deck, open web steel joists, beams, interior columns			S.F. Roof	9.36	1.56	
B20 Exterior Enclosure									
2010	Exterior Walls		Face brick and structural facing tile	70% of wall		S.F. Wall	46.74	10.20	
2020	Exterior Windows		Aluminum sliding	30% of wall		Each	588	3.67	6.9%
2030	Exterior Doors		Double aluminum and glass and sliding doors			Each	5923	.84	
B30 Roofing									
3010	Roof Coverings		Built-up tar and gravel with flashing; perlite/EPS composite insulation			S.F. Roof	7.92	1.32	0.6%
3020	Roof Openings		Roof hatches			S.F. Roof	.18	.03	
C. INTERIORS									
1010	Partitions		Gypsum board on metal studs with sound deadening board	9 S.F. Floor/L.F. Partition		S.F. Partition	7.72	8.58	
1020	Interior Doors		Single leaf hollow metal	90 S.F. Floor/Door		Each	1203	13.37	
1030	Fittings		Hospital curtains			S.F. Floor	1.21	1.21	
2010	Stair Construction		Concrete filled metal pan			Flight	11,600	1.40	23.0%
3010	Wall Finishes		40% vinyl wall covering, 35% ceramic tile, 25% epoxy coating			S.F. Surface	3.62	8.04	
3020	Floor Finishes		60% vinyl tile, 20% ceramic, 20% terrazzo			S.F. Floor	8.53	8.53	
3030	Ceiling Finishes		Plaster on suspended metal lath			S.F. Ceiling	8.24	8.24	
D. SERVICES									
D10 Conveying									
1010	Elevators & Lifts		Six geared hospital elevators			Each	224,333	6.73	3.1%
1020	Escalators & Moving Walks		N/A			—	—	—	
D20 Plumbing									
2010	Plumbing Fixtures		Kitchen, toilet and service fixtures, supply and drainage	1 Fixture/416S.F. Floor		Each	5836	14.03	
2020	Domestic Water Distribution		Electric water heater			S.F. Floor	5.04	5.04	9.2%
2040	Rain Water Drainage		Roof drains			S.F. Floor	3.96	.66	
D30 HVAC									
3010	Energy Supply		Oil fired hot water, wall fin radiation			S.F. Floor	4.16	4.16	
3020	Heat Generating Systems		Hot water boilers, steam boiler for services			Each	43,325	.54	
3030	Cooling Generating Systems		Chilled water units			S.F. Floor	3.05	3.05	21.4%
3050	Terminal & Package Units		N/A			—	—	—	
3090	Other HVAC Sys. & Equipment		Conditioned air with reheat, operating room air curtains			S.F. Floor	38.11	38.11	
D40 Fire Protection									
4010	Sprinklers		Wet pipe sprinkler system			S.F. Floor	2.90	2.90	1.6%
4020	Standpipes		Standpipe			S.F. Floor	.50	.50	
D50 Electrical									
5010	Electrical Service/Distribution		4000 ampere service, panel board and feeders			S.F. Floor	3.29	3.29	
5020	Lighting & Branch Wiring		High efficiency hospital grade light fixtures, receptacles, switches, A.C. and misc. power			S.F. Floor	19.95	19.95	14.1%
5030	Communications & Security		Addressable alarm systems, internet wiring, communications system, emergency lighting			S.F. Floor	2.50	2.50	
5090	Other Electrical Systems		Emergency generator, 800 kW with fuel tank, uninterruptible power supply			S.F. Floor	4.54	4.54	
E. EQUIPMENT & FURNISHINGS									
1010	Commercial Equipment		N/A			—	—	—	
1020	Institutional Equipment		Medical gases, curtain partitions			S.F. Floor	14.88	14.88	
1030	Vehicular Equipment		N/A			—	—	—	8.7 %
2020	Moveable Furnishings		Patient wall systems			S.F. Floor	3.83	3.83	
F. SPECIAL CONSTRUCTION									
1020	Integrated Construction		N/A			—	—	—	
1040	Special Facilities		N/A			—	—	—	0.0 %
G. BUILDING SITEWORK		**N/A**							

		Sub-Total	214.47	100%
CONTRACTOR FEES (General Requirements: 10%, Overhead: 5%, Profit: 10%)		25%	53.60	
ARCHITECT FEES		9%	24.13	
	Total Building Cost		**292.20**	

For customer support on your Square Foot Cost Data, call 877.694.8329.

151

Costs per square foot of floor area

Exterior Wall	S.F. Area	35000	55000	75000	95000	115000	135000	155000	175000	195000
	L.F. Perimeter	314	401	497	555	639	722	754	783	850
Face Brick with Concrete Block Back-up	Steel Frame	203.30	194.50	190.90	187.35	185.85	184.75	182.80	181.25	180.70
	R/Conc. Frame	203.75	194.95	191.35	187.80	186.30	185.20	183.25	181.75	181.15
Glass and Metal Curtain Walls	Steel Frame	216.15	204.65	199.95	195.20	193.25	191.85	189.20	187.10	186.25
	R/Conc. Frame	216.85	205.35	200.65	195.90	193.90	192.55	189.90	187.75	186.95
Precast Concrete Panels	Steel Frame	213.80	202.75	198.25	193.65	191.80	190.45	187.90	185.95	185.15
	R/Conc. Frame	215.85	204.50	199.90	195.20	193.35	191.95	189.35	187.25	186.55
Perimeter Adj., Add or Deduct	Per 100 L.F.	10.00	6.35	4.65	3.70	3.00	2.60	2.25	1.95	1.70
Story Hgt. Adj., Add or Deduct	Per 1 Ft.	2.90	2.40	2.15	1.85	1.80	1.80	1.55	1.45	1.35

For Basement, add $36.05 per square foot of basement area

The above costs were calculated using the basic specifications shown on the facing page. These costs should be adjusted where necessary for design alternatives and owner's requirements. Reported completed project costs, for this type of structure, range from $123.55 to $237.20 per S.F.

Common additives

Description	Unit	$ Cost	Description	Unit	$ Cost
Bar, Front bar	L.F.	425	Laundry Equipment		
Back bar	L.F.	340	Folders, blankets & sheets, king size	Ea.	72,500
Booth, Upholstered, custom, straight	L.F.	250 - 460	Ironers, 110" single roll	Ea.	38,600
"L" or "U" shaped	L.F.	259 - 440	Combination washer extractor 50#	Ea.	13,500
Closed Circuit Surveillance, One station			125#	Ea.	36,900
Camera and monitor	Ea.	1975	Sauna, Prefabricated, complete		
For additional camera stations, add	Ea.	1075	6' x 4'	Ea.	6575
Directory Boards, Plastic, glass covered			6' x 6'	Ea.	7925
30" x 20"	Ea.	600	6' x 9'	Ea.	10,000
36" x 48"	Ea.	1500	8' x 8'	Ea.	11,500
Aluminum, 24" x 18"	Ea.	635	10' x 12'	Ea.	15,500
48" x 32"	Ea.	1050	Smoke Detectors		
48" x 60"	Ea.	2225	Ceiling type	Ea.	227
Elevators, Electric passenger, 5 stops			Duct type	Ea.	565
3500# capacity	Ea.	173,000	Sound System		
5000# capacity	Ea.	179,500	Amplifier, 250 watts	Ea.	2050
Additional stop, add	Ea.	10,200	Speaker, ceiling or wall	Ea.	213
Emergency Lighting, 25 watt, battery operated			Trumpet	Ea.	410
Lead battery	Ea.	330	TV Antenna, Master system, 12 outlet	Outlet	255
Nickel cadmium	Ea.	780	30 outlet	Outlet	221
			100 outlet	Outlet	209

Important: See the Reference Section for Location Factors.

Model costs calculated for a 6 story building with 10' story height and 135,000 square feet of floor area

				Unit	Unit Cost	Cost Per S.F.	% Of Sub-Total

A. SUBSTRUCTURE

				Unit	Unit Cost	Cost Per S.F.	% Of Sub-Total
1010	Standard Foundations	Poured concrete; strip and spread footings		S.F. Ground	2.10	.35	
1020	Special Foundations	N/A		—	—	—	
1030	Slab on Grade	4" reinforced concrete with vapor barrier and granular base		S.F. Slab	5.44	.91	1.3%
2010	Basement Excavation	Site preparation for slab and trench for foundation wall and footing		S.F. Ground	.18	.03	
2020	Basement Walls	4' foundation wall		L.F. Wall	84	.49	

B. SHELL

B10 Superstructure

				Unit	Unit Cost	Cost Per S.F.	% Of Sub-Total
1010	Floor Construction	Concrete slab with metal deck, beams, columns		S.F. Floor	21.64	18.03	14.2%
1020	Roof Construction	Metal deck, open web steel joists, beams, columns		S.F. Roof	10.86	1.81	

B20 Exterior Enclosure

				Unit	Unit Cost	Cost Per S.F.	% Of Sub-Total
2010	Exterior Walls	Face brick with concrete block backup	80% of wall	S.F. Wall	33.42	8.58	
2020	Exterior Windows	Window wall	20% of wall	Each	52	3.32	8.7%
2030	Exterior Doors	Glass amd metal doors and entrances		Each	3630	.27	

B30 Roofing

				Unit	Unit Cost	Cost Per S.F.	% Of Sub-Total
3010	Roof Coverings	Built-up tar and gravel with flashing; perlite/EPS composite insulation		S.F. Roof	6.60	1.10	0.8%
3020	Roof Openings	Roof hatches		S.F. Roof	.06	.01	

C. INTERIORS

				Unit	Unit Cost	Cost Per S.F.	% Of Sub-Total
1010	Partitions	Gypsum board and sound deadening board, steel studs	9 S.F. Floor/L.F. Partition	S.F. Partition	7.47	6.64	
1020	Interior Doors	Single leaf hollow metal	90 S.F. Floor/Door	Each	1157	12.85	
1030	Fittings	N/A		—	—	—	
2010	Stair Construction	Concrete filled metal pan		Flight	14,500	1.28	24.8%
3010	Wall Finishes	20% paint, 75% vinyl cover, 5% ceramic tile		S.F. Surface	2.06	3.66	
3020	Floor Finishes	80% carpet, 10% vinyl composition tile, 10% ceramic tile		S.F. Floor	5.39	5.39	
3030	Ceiling Finishes	Mineral fiber tile applied with adhesive		S.F. Ceiling	4.76	4.76	

D. SERVICES

D10 Conveying

				Unit	Unit Cost	Cost Per S.F.	% Of Sub-Total
1010	Elevators & Lifts	Four geared passenger elevators		Each	229,500	6.80	4.9%
1020	Escalators & Moving Walks	N/A		—	—	—	

D20 Plumbing

				Unit	Unit Cost	Cost Per S.F.	% Of Sub-Total
2010	Plumbing Fixtures	Kitchen, toilet and service fixtures, supply and drainage	1 Fixture/155 S.F. Floor	Each	3971	25.62	
2020	Domestic Water Distribution	Oil fired water heater		S.F. Floor	.62	.62	19.0%
2040	Rain Water Drainage	Roof drains		S.F. Roof	1.62	.27	

D30 HVAC

				Unit	Unit Cost	Cost Per S.F.	% Of Sub-Total
3010	Energy Supply	Gas fired hot water, wall fin radiation		S.F. Floor	5.05	5.05	
3020	Heat Generating Systems	N/A		—	—	—	
3030	Cooling Generating Systems	Chilled water, fan coil units		S.F. Floor	13.80	13.80	13.5%
3050	Terminal & Package Units	N/A		—	—	—	
3090	Other HVAC Sys. & Equipment	N/A		—	—	—	

D40 Fire Protection

				Unit	Unit Cost	Cost Per S.F.	% Of Sub-Total
4010	Sprinklers	Sprinkler system, light hazard		S.F. Floor	2.91	2.91	2.4%
4020	Standpipes	Standpipes and hose systems		S.F. Floor	.38	.38	

D50 Electrical

				Unit	Unit Cost	Cost Per S.F.	% Of Sub-Total
5010	Electrical Service/Distribution	2000 ampere service, panel board and feeders		S.F. Floor	1.08	1.08	
5020	Lighting & Branch Wiring	High efficiency fluorescent fixtures, receptacles, switches, A.C. and misc. power		S.F. Floor	8.76	8.76	10.4%
5030	Communications & Security	Addressable alarm systems, internet wiring, communications systems and emergency lighting		S.F. Floor	4.25	4.25	
5090	Other Electrical Systems	Emergency generator, 250 kW		S.F. Floor	.43	.43	

E. EQUIPMENT & FURNISHINGS

				Unit	Unit Cost	Cost Per S.F.	% Of Sub-Total
1010	Commercial Equipment	N/A		—	—	—	
1020	Institutional Equipment	N/A		—	—	—	
1030	Vehicular Equipment	N/A		—	—	—	0.0 %
1090	Other Equipment	N/A		—	—	—	

F. SPECIAL CONSTRUCTION

				Unit	Unit Cost	Cost Per S.F.	% Of Sub-Total
1020	Integrated Construction	N/A		—	—	—	0.0 %
1040	Special Facilities	N/A		—	—	—	

G. BUILDING SITEWORK   N/A

				Sub-Total	139.45	100%

	CONTRACTOR FEES (General Requirements: 10%, Overhead: 5%, Profit: 10%)		25%	34.84	
	ARCHITECT FEES		6%	10.46	

	Total Building Cost	**184.75**

Costs per square foot of floor area

Exterior Wall	S.F. Area	140000	243000	346000	450000	552000	655000	760000	860000	965000
	L.F. Perimeter	403	587	672	800	936	1073	1213	1195	1312
Face Brick with Concrete Block Back-up	Steel Frame	183.15	176.75	171.65	169.75	168.70	168.00	167.45	165.55	165.15
	R/Conc. Frame	187.10	180.70	175.65	173.70	172.65	171.90	171.40	169.50	169.10
Face Brick Veneer On Steel Studs	Steel Frame	179.50	173.70	169.20	167.50	166.55	165.90	165.40	163.75	163.50
	R/Conc. Frame	186.50	180.65	176.15	174.45	173.50	172.85	172.35	170.75	170.40
Glass and Metal Curtain Walls	Steel Frame	230.35	216.35	207.25	203.40	201.30	199.80	198.75	195.85	195.10
	R/Conc. Frame	225.80	212.00	203.00	199.25	197.15	195.65	194.60	191.70	191.00
Perimeter Adj., Add or Deduct	Per 100 L.F.	6.15	3.50	2.50	1.90	1.55	1.25	1.10	0.95	0.95
Story Hgt. Adj., Add or Deduct	Per 1 Ft.	2.35	1.90	1.60	1.40	1.40	1.30	1.30	1.10	1.10

For Basement, add $37.40 per square foot of basement area

The above costs were calculated using the basic specifications shown on the facing page. These costs should be adjusted where necessary for design alternatives and owner's requirements. Reported completed project costs, for this type of structure, range from $137.15 to $239.80 per S.F.

Common additives

Description	Unit	$ Cost
Bar, Front bar	L.F.	425
Back bar	L.F.	340
Booth, Upholstered, custom, straight	L.F.	250 - 460
"L" or "U" shaped	L.F.	259 - 440
Closed Circuit Surveillance, One station		
Camera and monitor	Ea.	1975
For additional camera stations, add	Ea.	1075
Directory Boards, Plastic, glass covered		
30" x 20"	Ea.	600
36" x 48"	Ea.	1500
Aluminum, 24" x 18"	Ea.	635
48" x 32"	Ea.	1050
48" x 60"	Ea.	2225
Elevators, Electric passenger, 10 stops		
3500# capacity	Ea.	403,500
5000# capacity	Ea.	411,000
Additional stop, add	Ea.	10,200
Emergency Lighting, 25 watt, battery operated		
Lead battery	Ea.	330
Nickel cadmium	Ea.	780

Description	Unit	$ Cost
Laundry Equipment		
Folders, blankets & sheets, king size	Ea.	72,500
Ironers, 110" single roll	Ea.	38,600
Combination washer & extractor 50#	Ea.	13,500
125#	Ea.	36,900
Sauna, Prefabricated, complete		
6' x 4'	Ea.	6575
6' x 6'	Ea.	7925
6' x 9'	Ea.	10,000
8' x 8'	Ea.	11,500
10' x 12'	Ea.	15,500
Smoke Detectors		
Ceiling type	Ea.	227
Duct type	Ea.	565
Sound System		
Amplifier, 250 watts	Ea.	2050
Speaker, ceiling or wall	Ea.	213
Trumpet	Ea.	410
TV Antenna, Master system, 12 outlet	Outlet	255
30 outlet	Outlet	221
100 outlet	Outlet	209

Important: See the Reference Section for Location Factors.

Model costs calculated for a 15 story building with 10' story height and 450,000 square feet of floor area

				Unit	Unit Cost	Cost Per S.F.	% Of Sub-Total
A.	**SUBSTRUCTURE**						
1010	Standard Foundations	CIP concrete pile caps		S.F. Ground	6.15	.41	
1020	Special Foundations	Steel H-piles, concrete grade beams		S.F. Ground	114	7.63	
1030	Slab on Grade	4" reinforced concrete with vapor barrier and granular base		S.F. Slab	5.44	.36	5.6%
2010	Basement Excavation	Site preparation for slab, piles and grade beams		S.F. Ground	.18	.01	
2020	Basement Walls	4' foundation wall		L.F. Wall	84	.16	
B.	**SHELL**						
	B10 Superstructure						
1010	Floor Construction	Open web steel joists, slab form, concrete, columns		S.F. Floor	21.59	20.15	13.5%
1020	Roof Construction	Metal deck, open web steel joists, beams, columns		S.F. Roof	9.45	.63	
	B20 Exterior Enclosure						
2010	Exterior Walls	N/A		—	—	—	
2020	Exterior Windows	Glass and metal curtain walls	100% of wall	Each	57	15.08	10.0 %
2030	Exterior Doors	Glass and metal doors and entrances		Each	3308	.23	
	B30 Roofing						
3010	Roof Coverings	Built-up tar and gravel with flashing; perlite/EPS composite insulation		S.F. Roof	6.30	.42	0.3%
3020	Roof Openings	N/A		—	—	—	
C.	**INTERIORS**						
1010	Partitions	Gypsum board and sound deadening board, steel studs	9 S.F. Floor/L.F. Partition	S.F. Partition	6.40	5.69	
1020	Interior Doors	Single leaf hollow metal	90 S.F. Floor/Door	Each	1157	12.85	
1030	Fittings	N/A		—	—	—	
2010	Stair Construction	Concrete filled metal pan		Flight	14,500	.96	21.3%
3010	Wall Finishes	20% paint, 75% vinyl cover, 5% ceramic tile		S.F. Surface	2.04	3.63	
3020	Floor Finishes	80% carpet tile, 10% vinyl composition tile, 10% ceramic tile		S.F. Floor	5.39	5.39	
3030	Ceiling Finishes	Gypsum board on resilient channel		S.F. Ceiling	4.23	4.23	
D.	**SERVICES**						
	D10 Conveying						
1010	Elevators & Lifts	One geared freight, six geared passenger elevators		Each	483,000	6.44	4.2%
1020	Escalators & Moving Walks	N/A		—	—	—	
	D20 Plumbing						
2010	Plumbing Fixtures	Kitchen, toilet and service fixtures, supply and drainage	1 Fixture/165 S.F. Floor	Each	4437	26.89	
2020	Domestic Water Distribution	Electric water heater		S.F. Floor	3.41	3.41	19.8%
2040	Rain Water Drainage	Roof drains		S.F. Roof	2.55	.17	
	D30 HVAC						
3010	Energy Supply	Gas fired hot water, wall fin radiation		S.F. Floor	2.69	2.69	
3020	Heat Generating Systems	N/A		—	—	—	
3030	Cooling Generating Systems	Chilled water, fan coil units		S.F. Floor	13.80	13.80	10.7%
3050	Terminal & Package Units	N/A		—	—	—	
3090	Other HVAC Sys. & Equipment	N/A		—	—	—	
	D40 Fire Protection						
4010	Sprinklers	Sprinkler system, light hazard		S.F. Floor	4.12	4.12	5.1%
4020	Standpipes	Standpipes, fire pump and hose systems		S.F. Floor	3.77	3.77	
	D50 Electrical						
5010	Electrical Service/Distribution	6000 ampere service, panel board and feeders		S.F. Floor	1.22	1.22	
5020	Lighting & Branch Wiring	High efficiency fluorescent fixtures, receptacles, switches, A.C. and misc. power		S.F. Floor	8.63	8.63	9.4%
5030	Communications & Security	Addressable alarm systems, internet wiring, communications systems and emergency lighting		S.F. Floor	4.24	4.24	
5090	Other Electrical Systems	Emergency generator, 500 kW		S.F. Floor	.32	.32	
E.	**EQUIPMENT & FURNISHINGS**						
1010	Commercial Equipment	N/A		—	—	—	
1020	Institutional Equipment	N/A		—	—	—	
1030	Vehicular Equipment	N/A		—	—	—	0.0 %
1090	Other Equipment	N/A		—	—	—	
F.	**SPECIAL CONSTRUCTION**						
1020	Integrated Construction	N/A		—	—	—	
1040	Special Facilities	N/A		—	—	—	0.0 %
G.	**BUILDING SITEWORK**	**N/A**					

			Sub-Total	153.53	**100%**
CONTRACTOR FEES (General Requirements: 10%, Overhead: 5%, Profit: 10%)		25%	38.36		
ARCHITECT FEES		6%	11.51		

Total Building Cost	**203.40**

For customer support on your Square Foot Cost Data, call 877.694.8329.

155

Costs per square foot of floor area

Exterior Wall	S.F. Area	5500	12000	20000	30000	40000	60000	80000	100000	145000
	L.F. Perimeter	310	330	400	490	530	590	680	760	920
Face Brick with Concrete Block Back-up	Steel Frame	493.65	373.95	340.00	323.20	311.30	298.30	292.90	289.40	284.60
	R/Conc. Frame	484.10	364.40	330.45	313.65	301.70	288.75	283.35	279.90	275.00
Stucco on Concrete Block	Steel Frame	462.35	359.10	329.45	314.70	304.55	293.55	288.95	285.95	281.80
	R/Conc. Frame	451.30	348.40	318.80	304.15	294.00	283.05	278.40	275.40	271.30
Reinforced Concrete	Steel Frame	468.75	361.80	331.20	316.00	305.45	294.00	289.20	286.10	281.80
	R/Conc. Frame	459.15	352.25	321.60	306.45	295.85	284.45	279.65	276.55	272.25
Perimeter Adj., Add or Deduct	Per 100 L.F.	52.60	24.10	14.50	9.65	7.20	4.85	3.65	2.90	2.00
Story Hgt. Adj., Add or Deduct	Per 1 Ft.	12.55	6.15	4.50	3.65	2.95	2.20	1.90	1.70	1.40

For Basement, add $51.75 per square foot of basement area

The above costs were calculated using the basic specifications shown on the facing page. These costs should be adjusted where necessary for design alternatives and owner's requirements. Reported completed project costs, for this type of structure, range from $199.55 to $530.00 per S.F.

Common additives

Description	Unit	$ Cost
Clock System		
20 room	Ea.	19,900
50 room	Ea.	46,800
Closed Circuit Surveillance, One station		
Camera and monitor	Ea.	1975
For additional camera stations, add	Ea.	1075
Elevators, Hydraulic passenger, 2 stops		
1500# capacity	Ea.	65,700
2500# capacity	Ea.	68,800
3500# capacity	Ea.	73,300
Additional stop, add	Ea.	7825
Emergency Generators, Complete system, gas		
15 KW	Ea.	16,300
70 KW	Ea.	30,900
85 KW	Ea.	34,300
115 KW	Ea.	65,500
170 KW	Ea.	85,000
Diesel, 50 KW	Ea.	26,000
100 KW	Ea.	38,600
150 KW	Ea.	44,100
350 KW	Ea.	70,000

Description	Unit	$ Cost
Emergency Lighting, 25 watt, battery operated		
Nickel cadmium	Ea.	780
Flagpoles, Complete		
Aluminum, 20' high	Ea.	1850
40' High	Ea.	4200
70' High	Ea.	11,000
Fiberglass, 23' High	Ea.	1325
39'-5" High	Ea.	3450
59' High	Ea.	6850
Laundry Equipment		
Folders, blankets, & sheets	Ea.	72,500
Ironers, 110" single roll	Ea.	38,600
Combination washer & extractor, 50#	Ea.	13,500
125#	Ea.	36,900
Safe, Office type, 1 hour rating		
30" x 18" x 18"	Ea.	2500
60" x 36" x 18", double door	Ea.	8125
Sound System		
Amplifier, 250 watts	Ea.	2050
Speaker, ceiling or wall	Ea.	213
Trumpet	Ea.	410

Important: See the Reference Section for Location Factors.

Model costs calculated for a 3 story building with 12' story height and 40,000 square feet of floor area

				Unit	Unit Cost	Cost Per S.F.	% Of Sub-Total
A.	**SUBSTRUCTURE**						
1010	Standard Foundations	Poured concrete; strip and spread footings		S.F. Ground	2.88	.96	
1020	Special Foundations	N/A		—	—	—	
1030	Slab on Grade	4" reinforced concrete with vapor barrier and granular base		S.F. Slab	5.44	1.81	1.7%
2010	Basement Excavation	Site preparation for slab and trench for foundation wall and footing		S.F. Ground	.32	.11	
2020	Basement Walls	4' foundation wall		L.F. Wall	84	1.11	
B.	**SHELL**						
	B10 Superstructure						
1010	Floor Construction	Concrete slab with metal deck and beams		S.F. Floor	30.26	20.17	12.1%
1020	Roof Construction	Concrete slab with metal deck and beams		S.F. Roof	23.76	7.92	
	B20 Exterior Enclosure						
2010	Exterior Walls	Face brick with reinforced concrete block backup	85% of wall	S.F. Wall	33.47	13.57	
2020	Exterior Windows	Bullet resisting and metal horizontal pivoted	15% of wall	Each	161	11.54	10.9%
2030	Exterior Doors	Metal		Each	5533	.27	
	B30 Roofing						
3010	Roof Coverings	Built-up tar and gravel with flashing; perlite/EPS composite insulation		S.F. Roof	6.60	2.20	1.0%
3020	Roof Openings	Roof hatches		S.F. Roof	.09	.03	
C.	**INTERIORS**						
1010	Partitions	Concrete block	45 S.F. of Floor/L.F. Partition	S.F. Partition	10.73	2.86	
1020	Interior Doors	Hollow metal fire doors	930 S.F. Floor/Door	Each	1157	1.24	
1030	Fittings	N/A		—	—	—	
2010	Stair Construction	Concrete filled metal pan		Flight	17,300	2.17	6.2%
3010	Wall Finishes	Paint		S.F. Surface	4.39	2.34	
3020	Floor Finishes	70% vinyl composition tile, 20% carpet, 10% ceramic tile		S.F. Floor	3.63	3.63	
3030	Ceiling Finishes	Mineral tile on zee runners	30% of area	S.F. Ceiling	7.03	2.11	
D.	**SERVICES**						
	D10 Conveying						
1010	Elevators & Lifts	One hydraulic passenger elevator		Each	165,200	4.13	1.8%
1020	Escalators & Moving Walks	N/A		—	—	—	
	D20 Plumbing						
2010	Plumbing Fixtures	Toilet and service fixtures, supply and drainage	1 Fixture/78 S.F. Floor	Each	4324	55.43	
2020	Domestic Water Distribution	Electric water heater		S.F. Floor	4.22	4.22	26.0%
2040	Rain Water Drainage	Roof drains		S.F. Roof	2.82	.94	
	D30 HVAC						
3010	Energy Supply	N/A		—	—	—	
3020	Heat Generating Systems	Included in D3050		—	—	—	
3030	Cooling Generating Systems	N/A		—	—	—	7.0%
3050	Terminal & Package Units	Rooftop multizone unit systems		S.F. Floor	16.26	16.26	
3090	Other HVAC Sys. & Equipment	N/A		—	—	—	
	D40 Fire Protection						
4010	Sprinklers	Sprinkler system, light hazard		S.F. Floor	3.03	3.03	1.6%
4020	Standpipes	Standpipes and hose systems		S.F. Floor	.62	.62	
	D50 Electrical						
5010	Electrical Service/Distribution	600 ampere service, panel board and feeders		S.F. Floor	1	1	
5020	Lighting & Branch Wiring	Fluorescent fixtures, receptacles, switches, A.C. and misc. power		S.F. Floor	9.92	9.92	6.9%
5030	Communications & Security	Addressable alarm systems, intercom, and emergency lighting		S.F. Floor	4.46	4.46	
5090	Other Electrical Systems	Emergency generator, 80 kW		S.F. Floor	.68	.68	
E.	**EQUIPMENT & FURNISHINGS**						
1010	Commercial Equipment	N/A		—	—	—	
1020	Institutional Equipment	Prefabricated cells, visitor cubicles		S.F. Floor	58	58	24.9%
1030	Vehicular Equipment	N/A		—	—	—	
1090	Other Equipment	N/A		—	—	—	
F.	**SPECIAL CONSTRUCTION**						
1020	Integrated Construction	N/A		—	—	—	0.0%
1040	Special Facilities	N/A		—	—	—	
G.	**BUILDING SITEWORK**	**N/A**					

		Sub-Total	232.73	100%
CONTRACTOR FEES (General Requirements: 10%, Overhead: 5%, Profit: 10%)		25%	58.21	
ARCHITECT FEES		7%	20.36	
Total Building Cost			**311.30**	

For customer support on your Square Foot Cost Data, call 877.694.8329.

157

Costs per square foot of floor area

Exterior Wall	S.F. Area	1000	2000	3000	4000	5000	10000	15000	20000	25000
	L.F. Perimeter	126	179	219	253	283	400	490	568	632
Decorative Concrete Block	Steel Frame	299.40	250.50	231.70	221.35	214.75	199.40	193.30	189.80	187.40
	Bearing Walls	296.25	246.95	227.85	217.50	210.70	195.20	188.95	185.40	183.00
Face Brick with Concrete Block Back-up	Steel Frame	330.00	272.25	249.30	236.70	228.35	208.95	201.05	196.55	193.35
	Bearing Walls	326.95	268.80	245.65	232.90	224.55	204.95	196.95	192.40	189.20
Metal Sandwich Panel	Steel Frame	299.15	250.45	231.70	221.45	214.80	199.50	193.40	190.00	187.60
Precast Concrete Panel	Bearing Walls	291.60	243.65	225.15	215.15	208.65	193.70	187.75	184.40	182.10
Perimeter Adj., Add or Deduct	Per 100 L.F.	59.55	29.80	19.85	14.90	11.85	5.95	3.90	2.95	2.40
Story Hgt. Adj., Add or Deduct	Per 1 Ft.	3.90	2.80	2.25	2.00	1.75	1.25	0.95	0.85	0.80

For Basement, add $34.25 per square foot of basement area

The above costs were calculated using the basic specifications shown on the facing page. These costs should be adjusted where necessary for design alternatives and owner's requirements. Reported completed project costs, for this type of structure, range from $111.10 to $271.60 per S.F.

Common additives

Description	Unit	$ Cost
Closed Circuit Surveillance, One station		
Camera and monitor	Ea.	1975
For additional camera stations, add	Ea.	1075
Emergency Lighting, 25 watt, battery operated		
Lead battery	Ea.	330
Nickel cadmium	Ea.	780
Laundry Equipment		
Dryers, coin operated 30 lb.	Ea.	3850
Double stacked	Ea.	8350
50 lb.	Ea.	4175
Dry cleaner 20 lb.	Ea.	43,800
30 lb.	Ea.	65,500
Washers, coin operated	Ea.	1600
Washer/extractor 20 lb.	Ea.	7225
30 lb.	Ea.	11,500
50 lb.	Ea.	13,500
75 lb.	Ea.	26,000
Smoke Detectors		
Ceiling type	Ea.	227
Duct type	Ea.	565

Important: See the Reference Section for Location Factors.

Model costs calculated for a 1 story building with 12′ story height and 3,000 square feet of floor area

Laundromat

				Unit	Unit Cost	Cost Per S.F.	% Of Sub-Total
A.	**SUBSTRUCTURE**						
1010	Standard Foundations	Poured concrete; strip and spread footings		S.F. Ground	3.55	3.55	
1020	Special Foundations	N/A		—	—	—	
1030	Slab on Grade	5″ reinforced concrete with vapor barrier and granular base		S.F. Slab	5.87	5.87	8.9%
2010	Basement Excavation	Site preparation for slab and trench for foundation wall and footing		S.F. Ground	.57	.57	
2020	Basement Walls	4′ Foundation wall		L.F. Wall	75	5.48	
B.	**SHELL**						
	B10 Superstructure						
1010	Floor Construction	N/A		—	—	—	6.5 %
1020	Roof Construction	Metal deck, open web steel joists, beams, columns		S.F. Roof	10.39	10.39	
	B20 Exterior Enclosure						
2010	Exterior Walls	Decorative concrete block	90% of wall	S.F. Wall	16.11	12.70	
2020	Exterior Windows	Store front	10% of wall	Each	44	3.85	10.9%
2030	Exterior Doors	Double aluminum and glass		Each	3575	2.38	
	B30 Roofing						
3010	Roof Coverings	Built-up tar and gravel with flashing; perlite/EPS composite insulation		S.F. Roof	8.89	8.89	5.1%
3020	Roof Openings	N/A		—	—	—	
C.	**INTERIORS**						
1010	Partitions	Gypsum board on metal studs	60 S.F. Floor/L.F. Partition	S.F. Partition	26.76	4.46	
1020	Interior Doors	Single leaf wood	750 S.F. Floor/Door	Each	616	.82	
1030	Fittings	N/A		—	—	—	
2010	Stair Construction	N/A		—	—	—	6.9%
3010	Wall Finishes	Paint		S.F. Surface	.84	.28	
3020	Floor Finishes	Vinyl composition tile		S.F. Floor	2.31	2.31	
3030	Ceiling Finishes	Fiberglass board on exposed grid system		S.F. Ceiling	4.13	4.13	
D.	**SERVICES**						
	D10 Conveying						
1010	Elevators & Lifts	N/A		—	—	—	0.0 %
1020	Escalators & Moving Walks	N/A		—	—	—	
	D20 Plumbing						
2010	Plumbing Fixtures	Toilet and service fixtures, supply and drainage	1 Fixture/600 S.F. Floor	Each	6888	11.48	
2020	Domestic Water Distribution	Gas fired hot water heater		S.F. Floor	47.86	47.86	35.5%
2040	Rain Water Drainage	Roof drains		S.F. Roof	2.08	2.08	
	D30 HVAC						
3010	Energy Supply	N/A		—	—	—	
3020	Heat Generating Systems	Included in D3050		—	—	—	
3030	Cooling Generating Systems	N/A		—	—	—	5.4 %
3050	Terminal & Package Units	Rooftop single zone unit systems		S.F. Floor	9.37	9.37	
3090	Other HVAC Sys. & Equipment	N/A		—	—	—	
	D40 Fire Protection						
4010	Sprinklers	Sprinkler, ordinary hazard		S.F. Floor	5.43	5.43	3.1%
4020	Standpipes	N/A		—	—	—	
	D50 Electrical						
5010	Electrical Service/Distribution	200 ampere service, panel board and feeders		S.F. Floor	4.06	4.06	
5020	Lighting & Branch Wiring	High efficiency fluorescent fixtures, receptacles, switches, A.C. and misc. power		S.F. Floor	22.74	22.74	17.5%
5030	Communications & Security	Addressable alarm systems and emergency lighting		S.F. Floor	3.12	3.12	
5090	Other Electrical Systems	Emergency generator, 7.5 kW		S.F. Floor	.46	.46	
E.	**EQUIPMENT & FURNISHINGS**						
1010	Commercial Equipment	N/A		—	—	—	
1020	Institutional Equipment	N/A		—	—	—	
1030	Vehicular Equipment	N/A		—	—	—	0.0 %
1090	Other Equipment	N/A		—	—	—	
F.	**SPECIAL CONSTRUCTION**						
1020	Integrated Construction	N/A		—	—	—	0.0 %
1040	Special Facilities	N/A		—	—	—	
G.	**BUILDING SITEWORK**	**N/A**					

		Sub-Total	173.22	100%
CONTRACTOR FEES (General Requirements: 10%, Overhead: 5%, Profit: 10%)		25%	43.32	
ARCHITECT FEES		7%	15.16	

Total Building Cost	**231.70**

For customer support on your Square Foot Cost Data, call 877.694.8329.

159

Costs per square foot of floor area

Exterior Wall	S.F. Area	7000	10000	13000	16000	19000	22000	25000	28000	31000
	L.F. Perimeter	240	300	336	386	411	435	472	510	524
Face Brick with Concrete Block Back-up	R/Conc. Frame	195.30	185.65	177.35	173.55	168.80	165.30	163.50	162.15	159.70
	Steel Frame	196.10	186.45	178.10	174.40	169.60	166.05	164.25	162.90	160.55
Limestone with Concrete Block	R/Conc. Frame	225.85	210.70	198.00	192.20	185.00	179.75	177.00	174.85	171.35
	Steel Frame	221.20	208.40	197.00	192.05	185.45	180.55	178.10	176.25	172.85
Precast Concrete Panels	R/Conc. Frame	207.75	194.85	184.35	179.45	173.60	169.30	167.00	165.25	162.40
	Steel Frame	208.55	195.60	185.15	180.25	174.40	170.10	167.80	166.05	163.20
Perimeter Adj., Add or Deduct	Per 100 L.F.	24.05	16.85	12.95	10.60	8.90	7.65	6.75	6.00	5.45
Story Hgt. Adj., Add or Deduct	Per 1 Ft.	3.60	3.20	2.65	2.60	2.30	2.10	2.00	1.85	1.80

For Basement, add $49.05 per square foot of basement area

The above costs were calculated using the basic specifications shown on the facing page. These costs should be adjusted where necessary for design alternatives and owner's requirements. Reported completed project costs, for this type of structure, range from $106.75 to $273.95 per S.F.

Common additives

Description	Unit	$ Cost
Carrels Hardwood	Ea.	975 - 2350
Closed Circuit Surveillance, One station		
Camera and monitor	Ea.	1975
For additional camera stations, add	Ea.	1075
Elevators, Hydraulic passenger, 2 stops		
1500# capacity	Ea.	65,700
2500# capacity	Ea.	68,800
3500# capacity	Ea.	73,300
Emergency Lighting, 25 watt, battery operated		
Lead battery	Ea.	330
Nickel cadmium	Ea.	780
Flagpoles, Complete		
Aluminum, 20' high	Ea.	1850
40' high	Ea.	4200
70' high	Ea.	11,000
Fiberglass, 23' high	Ea.	1325
39'-5" high	Ea.	3450
59' high	Ea.	6850

Description	Unit	$ Cost
Library Furnishings		
Bookshelf, 90" high, 10" shelf double face	L.F.	216
single face	L.F.	185
Charging desk, built-in with counter		
Plastic laminated top	L.F.	420
Reading table, laminated		
top 60" x 36"	Ea.	775

Important: See the Reference Section for Location Factors.

Model costs calculated for a 2 story building with 14′ story height and 22,000 square feet of floor area

Library

					Unit	Unit Cost	Cost Per S.F.	% Of Sub-Total
A. SUBSTRUCTURE								
1010	Standard Foundations	Poured concrete; strip and spread footings			S.F. Ground	3.94	1.97	
1020	Special Foundations	N/A			—	—	—	
1030	Slab on Grade	4" reinforced concrete with vapor barrier and granular base			S.F. Slab	5.44	2.72	5.3%
2010	Basement Excavation	Site preparation for slab and trench for foundation wall and footing			S.F. Ground	.32	.16	
2020	Basement Walls	4′ foundation wall			L.F. Wall	84	1.65	
B. SHELL								
B10 Superstructure								
1010	Floor Construction	Concrete waffle slab			S.F. Floor	25.82	12.91	
1020	Roof Construction	Concrete waffle slab			S.F. Roof	19.66	9.83	18.6%
B20 Exterior Enclosure								
2010	Exterior Walls	Face brick with concrete block backup		90% of wall	S.F. Wall	32.95	16.42	
2020	Exterior Windows	Window wall		10% of wall	Each	54	3	16.3%
2030	Exterior Doors	Double aluminum and glass, single leaf hollow metal			Each	5825	.53	
B30 Roofing								
3010	Roof Coverings	Single ply membrane, EPDM, fully adhered; perlite/EPS composite insulation			S.F. Roof	5.54	2.77	
3020	Roof Openings	Roof hatches			S.F. Roof	.08	.04	2.3%
C. INTERIORS								
1010	Partitions	Gypsum board on metal studs		30 S.F. Floor/L.F. Partition	S.F. Partition	12.18	4.87	
1020	Interior Doors	Single leaf wood		300 S.F. Floor/Door	Each	616	2.05	
1030	Fittings	N/A			—	—	—	
2010	Stair Construction	Concrete filled metal pan			Flight	9650	.87	15.8%
3010	Wall Finishes	Paint			S.F. Surface	.85	.68	
3020	Floor Finishes	50% carpet, 50% vinyl tile			S.F. Floor	3.84	3.84	
3030	Ceiling Finishes	Mineral fiber on concealed zee bars			S.F. Ceiling	7.03	7.03	
D. SERVICES								
D10 Conveying								
1010	Elevators & Lifts	One hydraulic passenger elevator			Each	87,120	3.96	
1020	Escalators & Moving Walks	N/A			—	—	—	3.2%
D20 Plumbing								
2010	Plumbing Fixtures	Toilet and service fixtures, supply and drainage		1 Fixture/1835 S.F. Floor	Each	7560	4.12	
2020	Domestic Water Distribution	Gas fired water heater			S.F. Floor	1.28	1.28	4.9%
2040	Rain Water Drainage	Roof drains			S.F. Roof	1.24	.62	
D30 HVAC								
3010	Energy Supply	N/A			—	—	—	
3020	Heat Generating Systems	Included in D3050			—	—	—	
3030	Cooling Generating Systems	N/A			—	—	—	17.9 %
3050	Terminal & Package Units	Multizone unit, gas heating, electric cooling			S.F. Floor	21.90	21.90	
3090	Other HVAC Sys. & Equipment	N/A			—	—	—	
D40 Fire Protection								
4010	Sprinklers	Wet pipe sprinkler system			S.F. Floor	3.19	3.19	
4020	Standpipes	Standpipe			S.F. Floor	.98	.98	3.4%
D50 Electrical								
5010	Electrical Service/Distribution	400 ampere service, panel board and feeders			S.F. Floor	1.19	1.19	
5020	Lighting & Branch Wiring	High efficiency fluorescent fixtures, receptacles, switches, A.C. and misc. power			S.F. Floor	11.48	11.48	12.3%
5030	Communications & Security	Addressable alarm systems, internet wiring, and emergency lighting			S.F. Floor	2.10	2.10	
5090	Other Electrical Systems	Emergency generator, 7.5 kW, Uninterruptible power supply			S.F. Floor	.26	.26	
E. EQUIPMENT & FURNISHINGS								
1010	Commercial Equipment	N/A			—	—	—	
1020	Institutional Equipment	N/A			—	—	—	
1030	Vehicular Equipment	N/A			—	—	—	0.0 %
1090	Other Equipment	N/A			—	—	—	
F. SPECIAL CONSTRUCTION								
1020	Integrated Construction	N/A			—	—	—	
1040	Special Facilities	N/A			—	—	—	0.0 %
G. BUILDING SITEWORK	**N/A**							

			Sub-Total	122.42	100%
CONTRACTOR FEES (General Requirements: 10%, Overhead: 5%, Profit: 10%)			25%	30.64	
ARCHITECT FEES			8%	12.24	
		Total Building Cost		**165.30**	

For customer support on your Square Foot Cost Data, call 877.694.8329.

Costs per square foot of floor area

Exterior Wall	S.F. Area	4000	5500	7000	8500	10000	11500	13000	14500	16000
	L.F. Perimeter	280	320	380	440	453	503	510	522	560
Face Brick with Concrete Block Back-up	Steel Joists	231.30	222.15	218.90	216.75	212.05	210.75	207.50	205.15	204.40
	Wood Truss	226.70	217.15	213.75	211.50	206.60	205.25	201.85	199.40	198.60
Stucco on Concrete Block	Steel Joists	219.65	212.45	209.85	208.05	204.50	203.45	200.95	199.15	198.55
	Wood Truss	215.00	207.40	204.65	202.85	199.05	197.95	195.25	193.40	192.75
Brick Veneer	Wood Truss	218.90	210.25	207.15	205.15	200.75	199.50	196.40	194.20	193.50
Wood Siding	Wood Frame	211.10	204.25	201.65	200.00	196.55	195.55	193.15	191.40	190.85
Perimeter Adj., Add or Deduct	Per 100 L.F.	17.25	12.55	9.80	8.05	6.90	6.00	5.30	4.75	4.25
Story Hgt. Adj., Add or Deduct	Per 1 Ft.	3.65	3.00	2.80	2.65	2.30	2.25	2.05	1.85	1.80

For Basement, add $31.95 per square foot of basement area

The above costs were calculated using the basic specifications shown on the facing page. These costs should be adjusted where necessary for design alternatives and owner's requirements. Reported completed project costs, for this type of structure, range from $89.95 to $231.40 per S.F.

Common additives

Description	Unit	$ Cost
Cabinets, Hospital, base		
Laminated plastic	L.F.	415
Stainless steel	L.F.	670
Counter top, laminated plastic	L.F.	81
Stainless steel	L.F.	198
For drop-in sink, add	Ea.	1100
Nurses station, door type		
Laminated plastic	L.F.	465
Enameled steel	L.F.	450
Stainless steel	L.F.	785
Wall cabinets, laminated plastic	L.F.	305
Enameled steel	L.F.	355
Stainless steel	L.F.	630

Description	Unit	$ Cost
Directory Boards, Plastic, glass covered		
30" x 20"	Ea.	600
36" x 48"	Ea.	1500
Aluminum, 24" x 18"	Ea.	635
36" x 24"	Ea.	755
48" x 32"	Ea.	1050
48" x 60"	Ea.	2225
Heat Therapy Unit		
Humidified, 26" x 78" x 28"	Ea.	3975
Smoke Detectors		
Ceiling type	Ea.	227
Duct type	Ea.	565
Tables, Examining, vinyl top with base cabinets	Ea.	1500 - 6750
Utensil Washer, Sanitizer	Ea.	12,800
X-Ray, Mobile	Ea.	18,500 - 85,500

Important: See the Reference Section for Location Factors.

Model costs calculated for a 1 story building with 10' story height and 7,000 square feet of floor area

			Unit	Unit Cost	Cost Per S.F.	% Of Sub-Total
A. SUBSTRUCTURE						
1010	Standard Foundations	Poured concrete; strip and spread footings	S.F. Ground	2.73	2.73	
1020	Special Foundations	N/A	—	—	—	
1030	Slab on Grade	4" reinforced concrete with vapor barrier and granular base	S.F. Slab	5.44	5.44	8.3%
2010	Basement Excavation	Site preparation for slab and trench for foundation wall and footing	S.F. Ground	.32	.32	
2020	Basement Walls	4' foundation wall	L.F. Wall	84	4.54	
B. SHELL						
B10 Superstructure						
1010	Floor Construction	N/A	—	—	—	4.2 %
1020	Roof Construction	Plywood on wood trusses	S.F. Roof	6.65	6.65	
B20 Exterior Enclosure						
2010	Exterior Walls	Face brick with concrete block backup 70% of wall	S.F. Wall	32.95	12.52	
2020	Exterior Windows	Wood double hung 30% of wall	Each	638	6.11	12.7%
2030	Exterior Doors	Aluminum and glass doors and entrance with transoms	Each	1465	1.26	
B30 Roofing						
3010	Roof Coverings	Asphalt shingles with flashing (Pitched); rigid fiber glass insulation, gutters	S.F. Roof	4.89	4.89	3.1%
3020	Roof Openings	N/A	—	—	—	
C. INTERIORS						
1010	Partitions	Gypsum bd. & sound deadening bd. on wood studs w/insul. 6 S.F. Floor/L.F. Partition	S.F. Partition	9.32	12.43	
1020	Interior Doors	Single leaf wood 60 S.F. Floor/Door	Each	616	10.27	
1030	Fittings	N/A	—	—	—	
2010	Stair Construction	N/A	—	—	—	24.0%
3010	Wall Finishes	50% paint, 50% vinyl wall covering	S.F. Surface	1.43	3.82	
3020	Floor Finishes	50% carpet, 50% vinyl composition tile	S.F. Floor	4.71	4.71	
3030	Ceiling Finishes	Mineral fiber tile on concealed zee bars	S.F. Ceiling	6.40	6.40	
D. SERVICES						
D10 Conveying						
1010	Elevators & Lifts	N/A	—	—	—	0.0 %
1020	Escalators & Moving Walks	N/A	—	—	—	
D20 Plumbing						
2010	Plumbing Fixtures	Toilet, exam room and service fixtures, supply and drainage 1 Fixture/195 S.F. Floor	Each	4276	21.93	
2020	Domestic Water Distribution	Gas fired water heater	S.F. Floor	3.36	3.36	16.1%
2040	Rain Water Drainage	Roof drains	S.F. Roof			
D30 HVAC						
3010	Energy Supply	N/A	—	—	—	
3020	Heat Generating Systems	Included in D3050	—	—	—	
3030	Cooling Generating Systems	N/A	—	—	—	9.9 %
3050	Terminal & Package Units	Multizone unit, gas heating, electric cooling	S.F. Floor	15.55	15.55	
3090	Other HVAC Sys. & Equipment	N/A	—	—	—	
D40 Fire Protection						
4010	Sprinklers	Wet pipe sprinkler system	S.F. Floor	4.98	4.98	4.0%
4020	Standpipes	Standpipe	S.F. Floor	1.23	1.23	
D50 Electrical						
5010	Electrical Service/Distribution	200 ampere service, panel board and feeders	S.F. Floor	1.61	1.61	
5020	Lighting & Branch Wiring	High efficiency fluorescent fixtures, receptacles, switches, A.C. and misc. power	S.F. Floor	7.70	7.70	11.6%
5030	Communications & Security	Alarm systems, internet and phone wiring, intercom system, and emergency lighting	S.F. Floor	8	8	
5090	Other Electrical Systems	Emergency generator, 7.5 kW	S.F. Floor	.82	.82	
E. EQUIPMENT & FURNISHINGS						
1010	Commercial Equipment	N/A	—	—	—	
1020	Institutional Equipment	Exam room casework and countertops	S.F. Floor	9.58	9.58	6.1 %
1030	Vehicular Equipment	N/A	—	—	—	
1090	Other Equipment	N/A	—	—	—	
F. SPECIAL CONSTRUCTION						
1020	Integrated Construction	N/A	—	—	—	0.0 %
1040	Special Facilities	N/A	—	—	—	
G. BUILDING SITEWORK	**N/A**					

		Sub-Total	156.85	100%
CONTRACTOR FEES (General Requirements: 10%, Overhead: 5%, Profit: 10%)		25%	39.25	
ARCHITECT FEES		9%	17.65	
	Total Building Cost		**213.75**	

For customer support on your Square Foot Cost Data, call 877.694.8329.

163

Costs per square foot of floor area

Exterior Wall	S.F. Area	4000	5500	7000	8500	10000	11500	13000	14500	16000
	L.F. Perimeter	180	210	240	270	286	311	336	361	386
Face Brick with Concrete Block Back-up	Steel Joists	268.10	255.80	248.85	244.25	239.35	236.70	234.65	233.05	231.65
	Wood Joists	267.65	255.65	248.80	244.35	239.55	236.90	234.95	233.40	232.05
Stucco on Concrete Block	Steel Joists	256.00	245.55	239.60	235.70	231.65	229.40	227.70	226.35	225.20
	Wood Joists	255.65	245.50	239.70	235.90	231.95	229.70	228.05	226.75	225.60
Brick Veneer	Wood Frame	260.45	249.50	243.25	239.20	234.95	232.55	230.80	229.40	228.20
Wood Siding	Wood Frame	249.05	239.85	234.65	231.15	227.75	225.75	224.30	223.05	222.10
Perimeter Adj., Add or Deduct	Per 100 L.F.	30.95	22.50	17.65	14.55	12.35	10.75	9.50	8.50	7.75
Story Hgt. Adj., Add or Deduct	Per 1 Ft.	4.60	3.90	3.45	3.20	2.90	2.70	2.60	2.55	2.50

For Basement, add $35.40 per square foot of basement area

The above costs were calculated using the basic specifications shown on the facing page. These costs should be adjusted where necessary for design alternatives and owner's requirements. Reported completed project costs, for this type of structure, range from $92.35 to $315.45 per S.F.

Common additives

Description	Unit	$ Cost
Cabinets, Hospital, base		
Laminated plastic	L.F.	415
Stainless steel	L.F.	670
Counter top, laminated plastic	L.F.	81
Stainless steel	L.F.	198
For drop-in sink, add	Ea.	1100
Nurses station, door type		
Laminated plastic	L.F.	465
Enameled steel	L.F.	450
Stainless steel	L.F.	785
Wall cabinets, laminated plastic	L.F.	305
Enameled steel	L.F.	355
Stainless steel	L.F.	630
Elevators, Hydraulic passenger, 2 stops		
1500# capacity	Ea.	65,700
2500# capacity	Ea.	68,800
3500# capacity	Ea.	73,300

Description	Unit	$ Cost
Directory Boards, Plastic, glass covered		
30" x 20"	Ea.	600
36" x 48"	Ea.	1500
Aluminum, 24" x 18"	Ea.	635
36" x 24"	Ea.	755
48" x 32"	Ea.	1050
48" x 60"	Ea.	2225
Emergency Lighting, 25 watt, battery operated		
Lead battery	Ea.	330
Nickel cadmium	Ea.	780
Heat Therapy Unit		
Humidified, 26" x 78" x 28"	Ea.	3975
Smoke Detectors		
Ceiling type	Ea.	227
Duct type	Ea.	565
Tables, Examining, vinyl top		
with base cabinets	Ea.	1500 - 6750
Utensil Washer, Sanitizer	Ea.	12,800
X-Ray, Mobile	Ea.	18,500 - 85,500

Model costs calculated for a 2 story building with 10' story height and 7,000 square feet of floor area

Medical Office, 2 Story

				Unit	Unit Cost	Cost Per S.F.	% Of Sub-Total
A. SUBSTRUCTURE							
1010	Standard Foundations	Poured concrete; strip and spread footings		S.F. Ground	3.50	1.75	
1020	Special Foundations	N/A		—	—	—	
1030	Slab on Grade	4" reinforced concrete with vapor barrier and granular base		S.F. Slab	5.44	2.72	4.3%
2010	Basement Excavation	Site preparation for slab and trench for foundation wall and footing		S.F. Ground	.57	.29	
2020	Basement Walls	4' foundation wall		L.F. Wall	84	2.87	
B. SHELL							
	B10 Superstructure						
1010	Floor Construction	Open web steel joists, slab form, concrete, columns		S.F. Floor	13.76	6.88	5.7%
1020	Roof Construction	Metal deck, open web steel joists, beams, columns		S.F. Roof	6.24	3.12	
	B20 Exterior Enclosure						
2010	Exterior Walls	Stucco on concrete block		S.F. Wall	19.35	9.29	
2020	Exterior Windows	Outward projecting metal	30% of wall	Each	528	7.25	10.4%
2030	Exterior Doors	Aluminum and glass doors with transoms		Each	5825	1.67	
	B30 Roofing						
3010	Roof Coverings	Built-up tar and gravel with flashing; perlite/EPS composite insulation		S.F. Roof	7.74	3.87	2.3%
3020	Roof Openings	Roof hatches		S.F. Roof	.26	.13	
C. INTERIORS							
1010	Partitions	Gypsum bd. & acous. insul. on metal studs	6 S.F. Floor/L.F. Partition	S.F. Partition	7.82	10.42	
1020	Interior Doors	Single leaf wood	60 S.F. Floor/Door	Each	616	10.27	
1030	Fittings	N/A		—	—	—	
2010	Stair Construction	Concrete filled metal pan		Flight	14,500	6.22	23.8%
3010	Wall Finishes	45% paint, 50% vinyl wall coating, 5% ceramic tile		S.F. Surface	1.43	3.82	
3020	Floor Finishes	50% carpet, 50% vinyl composition tile		S.F. Floor	4.71	4.71	
3030	Ceiling Finishes	Mineral fiber tile on concealed zee bars		S.F. Ceiling	6.40	6.40	
D. SERVICES							
	D10 Conveying						
1010	Elevators & Lifts	One hydraulic hospital elevator		Each	106,610	15.23	8.7%
1020	Escalators & Moving Walks	N/A		—	—	—	
	D20 Plumbing						
2010	Plumbing Fixtures	Toilet, exam room and service fixtures, supply and drainage	1 Fixture/160 S.F. Floor	Each	3926	24.54	
2020	Domestic Water Distribution	Gas fired water heater		S.F. Floor	3.36	3.36	16.5%
2040	Rain Water Drainage	Roof drains		S.F. Roof	2.28	1.14	
	D30 HVAC						
3010	Energy Supply	N/A		—	—	—	
3020	Heat Generating Systems	Included in D3050		—	—	—	
3030	Cooling Generating Systems	N/A		—	—	—	8.8 %
3050	Terminal & Package Units	Multizone unit, gas heating, electric cooling		S.F. Floor	15.55	15.55	
3090	Other HVAC Sys. & Equipment	N/A		—	—	—	
	D40 Fire Protection						
4010	Sprinklers	Wet pipe sprinkler system		S.F. Floor	3.96	3.96	3.3%
4020	Standpipes	Standpipe		S.F. Floor	1.86	1.86	
	D50 Electrical						
5010	Electrical Service/Distribution	400 ampere service, panel board and feeders		S.F. Floor	2.44	2.44	
5020	Lighting & Branch Wiring	High efficiency fluorescent fixtures, receptacles, switches, A.C. and misc. power		S.F. Floor	7.70	7.70	10.8%
5030	Communications & Security	Alarm system, internet and phone wiring, intercom system, and emergency lighting		S.F. Floor	8	8	
5090	Other Electrical Systems	Emergency generator, 7.5 kW		S.F. Floor	.82	.82	
E. EQUIPMENT & FURNISHINGS							
1010	Commercial Equipment	N/A		—	—	—	
1020	Institutional Equipment	Exam room casework and contertops		S.F. Floor	9.58	9.58	5.4 %
1030	Vehicular Equipment	N/A		—	—	—	
1090	Other Equipment	N/A		—	—	—	
F. SPECIAL CONSTRUCTION							
1020	Integrated Construction	N/A		—	—	—	0.0 %
1040	Special Facilities	N/A		—	—	—	
G. BUILDING SITEWORK	**N/A**						

			Sub-Total	175.86	100%
CONTRACTOR FEES (General Requirements: 10%, Overhead: 5%, Profit: 10%)			25%	43.96	
ARCHITECT FEES			9%	19.78	

	Total Building Cost	239.60

Costs per square foot of floor area

Exterior Wall	S.F. Area	2000	3000	4000	6000	8000	10000	12000	14000	16000
	L.F. Perimeter	240	260	280	380	480	560	580	660	740
Brick Veneer	Wood Frame	195.55	171.75	159.85	153.40	150.20	147.20	142.40	141.40	140.55
Aluminum Siding	Wood Frame	178.00	159.10	149.60	144.15	141.40	139.00	135.35	134.50	133.80
Wood Siding	Wood Frame	177.85	158.95	149.50	144.05	141.35	138.95	135.25	134.40	133.75
Wood Shingles	Wood Frame	181.15	161.35	151.35	145.75	142.90	140.35	136.55	135.65	134.90
Precast Concrete Block	Wood Truss	176.90	158.30	149.00	143.55	140.85	138.50	134.95	134.05	133.40
Brick on Concrete Block	Wood Truss	205.35	178.80	165.55	158.55	155.10	151.75	146.35	145.20	144.30
Perimeter Adj., Add or Deduct	Per 100 L.F.	27.40	18.20	13.65	9.10	6.75	5.40	4.55	3.85	3.40
Story Hgt. Adj., Add or Deduct	Per 1 Ft.	5.00	3.60	2.90	2.65	2.50	2.30	2.00	1.95	1.90

For Basement, add $24.40 per square foot of basement area

The above costs were calculated using the basic specifications shown on the facing page. These costs should be adjusted where necessary for design alternatives and owner's requirements. Reported completed project costs, for this type of structure, range from $74.50 to $352.85 per S.F.

Common additives

Description	Unit	$ Cost
Closed Circuit Surveillance, One station		
Camera and monitor	Ea.	1975
For additional camera stations, add	Ea.	1075
Emergency Lighting, 25 watt, battery operated		
Lead battery	Ea.	330
Nickel cadmium	Ea.	780
Laundry Equipment		
Dryer, gas, 16 lb. capacity	Ea.	975
30 lb. capacity	Ea.	3850
Washer, 4 cycle	Ea.	1200
Commercial	Ea.	1600
Sauna, Prefabricated, complete		
6' x 4'	Ea.	6575
6' x 6'	Ea.	7925
6' x 9'	Ea.	10,000
8' x 8'	Ea.	11,500
8' x 10'	Ea.	12,800
10' x 12'	Ea.	15,500
Smoke Detectors		
Ceiling type	Ea.	227
Duct type	Ea.	565

Description	Unit	$ Cost
Swimming Pools, Complete, gunite	S.F.	97 - 120
TV Antenna, Master system, 12 outlet	Outlet	255
30 outlet	Outlet	221
100 outlet	Outlet	209

Important: See the Reference Section for Location Factors.

Model costs calculated for a 1 story building with 9' story height and 8,000 square feet of floor area

			Unit	Unit Cost	Cost Per S.F.	% Of Sub-Total
A. SUBSTRUCTURE						
1010	Standard Foundations	Poured concrete; strip and spread footings	S.F. Ground	2.42	2.42	
1020	Special Foundations	N/A	—	—	—	
1030	Slab on Grade	4" reinforced concrete with vapor barrier and granular base	S.F. Slab	5.44	5.44	11.5%
2010	Basement Excavation	Site preparation for slab and trench for foundation wall and footing	S.F. Ground	.32	.32	
2020	Basement Walls	4' foundation wall	L.F. Wall	79	4.74	
B. SHELL						
	B10 Superstructure					
1010	Floor Construction	N/A	—	—	—	5.9 %
1020	Roof Construction	Plywood on wood trusses	S.F. Roof	6.65	6.65	
	B20 Exterior Enclosure					
2010	Exterior Walls	Face brick on wood studs with sheathing, insulation and paper 80% of wall	S.F. Wall	25	10.80	
2020	Exterior Windows	Wood double hung 20% of wall	Each	511	3.95	16.7%
2030	Exterior Doors	Wood solid core	Each	1465	4.03	
	B30 Roofing					
3010	Roof Coverings	Asphalt shingles with flashing (pitched); rigid fiberglass insulation	S.F. Roof	4.51	4.51	4.0%
3020	Roof Openings	N/A	—	—	—	
C. INTERIORS						
1010	Partitions	Gypsum bd. and sound deadening bd. on wood studs 9 S.F. Floor/L.F. Partition	S.F. Partition	10.28	9.14	
1020	Interior Doors	Single leaf hollow core wood 300 S.F. Floor/Door	Each	537	1.79	
1030	Fittings	N/A	—	—	—	
2010	Stair Construction	N/A	—	—	—	22.5%
3010	Wall Finishes	90% paint, 10% ceramic tile	S.F. Surface	1.49	2.64	
3020	Floor Finishes	85% carpet, 15% ceramic tile	S.F. Floor	7.21	7.21	
3030	Ceiling Finishes	Painted gypsum board on furring	S.F. Ceiling	4.51	4.51	
D. SERVICES						
	D10 Conveying					
1010	Elevators & Lifts	N/A	—	—	—	0.0 %
1020	Escalators & Moving Walks	N/A	—	—	—	
	D20 Plumbing					
2010	Plumbing Fixtures	Toilet and service fixtures, supply and drainage 1 Fixture/90 S.F. Floor	Each	1838	20.42	
2020	Domestic Water Distribution	Gas fired water heater	S.F. Floor	1.82	1.82	19.8%
2040	Rain Water Drainage	Roof Drains	S.F. Roof			
	D30 HVAC					
3010	Energy Supply	N/A	—	—	—	
3020	Heat Generating Systems	Included in D3050	—	—	—	
3030	Cooling Generating Systems	N/A	—	—	—	3.4 %
3050	Terminal & Package Units	Through the wall electric heating and cooling units	S.F. Floor	3.77	3.77	
3090	Other HVAC Sys. & Equipment	N/A	—	—	—	
	D40 Fire Protection					
4010	Sprinklers	Wet pipe sprinkler system	S.F. Floor	4.98	4.98	5.4%
4020	Standpipes	Standpipe, wet, Class III	S.F. Floor	1.08	1.08	
	D50 Electrical					
5010	Electrical Service/Distribution	200 ampere service, panel board and feeders	S.F. Floor	1.32	1.32	
5020	Lighting & Branch Wiring	Fluorescent fixtures, receptacles, switches and misc. power	S.F. Floor	7.05	7.05	9.9%
5030	Communications & Security	Addressable alarm systems	S.F. Floor	2.73	2.73	
5090	Other Electrical Systems	N/A	—	—	—	
E. EQUIPMENT & FURNISHINGS						
1010	Commercial Equipment	Laundry equipment	S.F. Floor	.97	.97	
1020	Institutional Equipment	N/A	—	—	—	
1030	Vehicular Equipment	N/A	—	—	—	0.9%
1090	Other Equipment	N/A	—	—	—	
F. SPECIAL CONSTRUCTION						
1020	Integrated Construction	N/A	—	—	—	0.0 %
1040	Special Facilities	N/A	—	—	—	
G. BUILDING SITEWORK	**N/A**					

			Sub-Total		112.29	100%
CONTRACTOR FEES (General Requirements: 10%, Overhead: 5%, Profit: 10%)				25%	28.08	
ARCHITECT FEES				7%	9.83	
			Total Building Cost		**150.20**	

For customer support on your Square Foot Cost Data, call 877.694.8329.

Costs per square foot of floor area

Exterior Wall	S.F. Area	25000	37000	49000	61000	73000	81000	88000	96000	104000
	L.F. Perimeter	433	593	606	720	835	911	978	1054	1074
Decorative Concrete Block	Wood Joists	161.95	159.10	154.35	153.25	152.60	152.20	152.00	151.70	150.90
	Precast Conc.	178.50	174.45	169.00	167.55	166.60	166.10	165.80	165.40	164.50
Stucco on Concrete Block	Wood Joists	164.15	160.25	154.75	153.40	152.45	152.00	151.70	151.30	150.45
	Precast Conc.	179.75	175.30	169.45	167.90	166.80	166.30	166.00	165.55	164.55
Wood Siding	Wood Frame	158.45	155.85	151.85	150.95	150.30	149.95	149.75	149.50	148.90
Brick Veneer	Wood Frame	167.05	162.90	156.90	155.35	154.35	153.90	153.55	153.15	152.15
Perimeter Adj., Add or Deduct	Per 100 L.F.	4.40	2.95	2.20	1.80	1.50	1.40	1.25	1.15	1.05
Story Hgt. Adj., Add or Deduct	Per 1 Ft.	1.75	1.60	1.20	1.20	1.15	1.15	1.10	1.15	1.05
For Basement, add $32.00 per square foot of basement area										

The above costs were calculated using the basic specifications shown on the facing page. These costs should be adjusted where necessary for design alternatives and owner's requirements. Reported completed project costs, for this type of structure, range from $68.65 to $352.85 per S.F.

Common additives

Description	Unit	$ Cost		Description	Unit	$ Cost
Closed Circuit Surveillance, One station				Sauna, Prefabricated, complete		
Camera and monitor	Ea.	1975		6' x 4'	Ea.	6575
For additional camera station, add	Ea.	1075		6' x 6'	Ea.	7925
Elevators, Hydraulic passenger, 2 stops				6' x 9'	Ea.	10,000
1500# capacity	Ea.	65,700		8' x 8'	Ea.	11,500
2500# capacity	Ea.	68,800		8' x 10'	Ea.	12,800
3500# capacity	Ea.	73,300		10' x 12'	Ea.	15,500
Additional stop, add	Ea.	7825		Smoke Detectors		
Emergency Lighting, 25 watt, battery operated				Ceiling type	Ea.	227
Lead battery	Ea.	330		Duct type	Ea.	565
Nickel cadmium	Ea.	780		Swimming Pools, Complete, gunite	S.F.	97 - 120
Laundry Equipment				TV Antenna, Master system, 12 outlet	Outlet	255
Dryer, gas, 16 lb. capacity	Ea.	975		30 outlet	Outlet	221
30 lb. capacity	Ea.	3850		100 outlet	Outlet	209
Washer, 4 cycle	Ea.	1200				
Commercial	Ea.	1600				

Important: See the Reference Section for Location Factors.

Model costs calculated for a 3 story building with 9' story height and 49,000 square feet of floor area

					Unit	Unit Cost	Cost Per S.F.	% Of Sub-Total
A. SUBSTRUCTURE								
1010	Standard Foundations	Poured concrete; strip and spread footings			S.F. Ground	2.61	.87	
1020	Special Foundations	N/A			—	—	—	
1030	Slab on Grade	4" reinforced concrete with vapor barrier and granular base			S.F. Slab	5.44	1.81	3.0%
2010	Basement Excavation	Site preparation for slab and trench for foundation wall and footing			S.F. Ground	.32	.11	
2020	Basement Walls	4' foundation wall			L.F. Wall	84	1.03	
B. SHELL								
B10 Superstructure								
1010	Floor Construction	Precast concrete plank			S.F. Floor	10.40	6.93	
1020	Roof Construction	Precast concrete plank			S.F. Roof	10.32	3.44	8.1%
B20 Exterior Enclosure								
2010	Exterior Walls	Decorative concrete block	85% of wall		S.F. Wall	18.25	5.18	
2020	Exterior Windows	Aluminum sliding	15% of wall		Each	588	1.96	10.7%
2030	Exterior Doors	Aluminum and glass doors and entrance with transom			Each	1973	6.53	
B30 Roofing								
3010	Roof Coverings	Built-up tar and gravel with flashing; perlite/EPS composite insulation			S.F. Roof	6.54	2.18	
3020	Roof Openings	Roof hatches			S.F. Roof	.12	.04	1.7%
C. INTERIORS								
1010	Partitions	Concrete block	7 S.F. Floor/L.F. Partition		S.F. Partition	20.10	22.97	
1020	Interior Doors	Wood hollow core	70 S.F. Floor/Door		Each	537	7.67	
1030	Fittings	N/A			—	—	—	
2010	Stair Construction	Concrete filled metal pan			Flight	14,500	3.55	38.7%
3010	Wall Finishes	90% paint, 10% ceramic tile			S.F. Surface	1.49	3.41	
3020	Floor Finishes	85% carpet, 5% vinyl composition tile, 10% ceramic tile			S.F. Floor	7.21	7.21	
3030	Ceiling Finishes	Textured finish			S.F. Ceiling	4.51	4.51	
D. SERVICES								
D10 Conveying								
1010	Elevators & Lifts	Two hydraulic passenger elevators			Each	108,045	4.41	
1020	Escalators & Moving Walks	N/A			—	—	—	3.5%
D20 Plumbing								
2010	Plumbing Fixtures	Toilet and service fixtures, supply and drainage	1 Fixture/180 S.F. Floor		Each	4261	23.67	
2020	Domestic Water Distribution	Gas fired water heater			S.F. Floor	1.23	1.23	19.9%
2040	Rain Water Drainage	Roof drains			S.F. Roof	1.41	.47	
D30 HVAC								
3010	Energy Supply	N/A			—	—	—	
3020	Heat Generating Systems	Included in D3050			—	—	—	
3030	Cooling Generating Systems	N/A			—	—	—	2.7 %
3050	Terminal & Package Units	Through the wall electric heating and cooling units			S.F. Floor	3.40	3.40	
3090	Other HVAC Sys. & Equipment	N/A			—	—	—	
D40 Fire Protection								
4010	Sprinklers	Sprinklers, wet, light hazard			S.F. Floor	3.03	3.03	
4020	Standpipes	Standpipe, wet, Class III			S.F. Floor	.23	.23	2.6%
D50 Electrical								
5010	Electrical Service/Distribution	800 ampere service, panel board and feeders			S.F. Floor	1.59	1.59	
5020	Lighting & Branch Wiring	Fluorescent fixtures, receptacles, switches and misc. power			S.F. Floor	7.77	7.77	
5030	Communications & Security	Addressable alarm systems and emergency lighting			S.F. Floor	1.89	1.89	8.9%
5090	Other Electrical Systems	Emergency generator, 7.5 kW			S.F. Floor	.11	.11	
E. EQUIPMENT & FURNISHINGS								
1010	Commercial Equipment	Commercial laundry equipment			S.F. Floor	.32	.32	
1020	Institutional Equipment	N/A			—	—	—	
1030	Vehicular Equipment	N/A			—	—	—	0.3%
1090	Other Equipment	N/A			—	—	—	
F. SPECIAL CONSTRUCTION								
1020	Integrated Construction	N/A			—	—	—	
1040	Special Facilities	N/A			—	—	—	0.0 %
G. BUILDING SITEWORK	**N/A**							

		Sub-Total	127.52	100%
CONTRACTOR FEES (General Requirements: 10%, Overhead: 5%, Profit: 10%)		25%	31.92	
ARCHITECT FEES		6%	9.56	
	Total Building Cost		169	

For customer support on your Square Foot Cost Data, call 877.694.8329.

169

Costs per square foot of floor area

Exterior Wall	S.F. Area	9000	10000	12000	13000	14000	15000	16000	18000	20000
	L.F. Perimeter	385	410	440	460	480	500	510	547	583
Decorative Concrete Block	Steel Joists	187.30	182.45	173.60	170.60	168.05	165.75	163.20	159.75	157.15
Painted Concrete Block	Steel Joists	180.75	176.20	167.95	165.20	162.80	160.65	158.30	155.10	152.65
Face Brick on Conc. Block	Steel Joists	199.25	193.90	183.85	180.50	177.60	175.05	172.10	168.20	165.30
Precast Concrete Panels	Steel Joists	206.80	201.20	190.30	186.75	183.70	180.95	177.70	173.60	170.45
Tilt-up Panels	Steel Joists	184.85	180.15	171.45	168.55	166.05	163.85	161.35	158.00	155.50
Metal Sandwich Panels	Steel Joists	183.70	179.05	170.50	167.65	165.15	162.95	160.50	157.20	154.70
Perimeter Adj., Add or Deduct	Per 100 L.F.	11.40	10.25	8.55	7.90	7.30	6.90	6.40	5.70	5.10
Story Hgt. Adj., Add or Deduct	Per 1 Ft.	1.70	1.70	1.50	1.45	1.40	1.35	1.30	1.25	1.15
Basement—Not Applicable										

The above costs were calculated using the basic specifications shown on the facing page. These costs should be adjusted where necessary for design alternatives and owner's requirements. Reported completed project costs, for this type of structure, range from $87.80 to $229.65 per S.F.

Common additives

Description	Unit	$ Cost
Emergency Lighting, 25 watt, battery operated		
Lead battery	Ea.	330
Nickel cadmium	Ea.	780
Seating		
Auditorium chair, all veneer	Ea.	310
Veneer back, padded seat	Ea.	320
Upholstered, spring seat	Ea.	320
Classroom, movable chair & desk	Set	81 - 171
Lecture hall, pedestal type	Ea.	263 - 620
Smoke Detectors		
Ceiling type	Ea.	227
Duct type	Ea.	565
Sound System		
Amplifier, 250 watts	Ea.	2050
Speaker, ceiling or wall	Ea.	213
Trumpet	Ea.	410

Important: See the Reference Section for Location Factors.

Model costs calculated for a 1 story building with 20' story height and 12,000 square feet of floor area

Movie Theater

				Unit	Unit Cost	Cost Per S.F.	% Of Sub-Total
A. SUBSTRUCTURE							
1010	Standard Foundations	Poured concrete; strip and spread footings		S.F. Ground	1.53	1.53	
1020	Special Foundations	N/A		—	—	—	
1030	Slab on Grade	4" reinforced concrete with vapor barrier and granular base		S.F. Slab	5.44	5.44	8.0%
2010	Basement Excavation	Site preparation for slab and trench for foundation wall and footing		S.F. Ground	.32	.32	
2020	Basement Walls	4' foundation wall		L.F. Wall	84	3.06	
B. SHELL							
B10 Superstructure							
1010	Floor Construction	Open web steel joists, slab form, concrete	mezzanine 2250 S.F.	S.F. Floor	11.76	2.20	10.0%
1020	Roof Construction	Metal deck on open web steel joists		S.F. Roof	10.76	10.76	
B20 Exterior Enclosure							
2010	Exterior Walls	Decorative concrete block	100% of wall	S.F. Wall	16.32	11.97	
2020	Exterior Windows	N/A		—	—	—	16.6%
2030	Exterior Doors	Sliding mallfront aluminum and glass and hollow metal		Each	2533	1.28	
B30 Roofing							
3010	Roof Coverings	Built-up tar and gravel with flashing; perlite/EPS composite insulation		S.F. Roof	6.49	6.49	5.0%
3020	Roof Openings	N/A		—	—	—	
C. INTERIORS							
1010	Partitions	Concrete block	40 S.F. Floor/L.F. Partition	S.F. Partition	9.65	3.86	
1020	Interior Doors	Single leaf hollow metal	705 S.F. Floor/Door	Each	1157	1.64	
1030	Fittings	Toilet partitions		S.F. Floor	.63	.63	
2010	Stair Construction	Concrete filled metal pan		Flight	17,300	2.88	20.2%
3010	Wall Finishes	Paint		S.F. Surface	4.31	3.45	
3020	Floor Finishes	Carpet 50%, ceramic tile 5%	50% of area	S.F. Floor	13.42	6.71	
3030	Ceiling Finishes	Mineral fiber tile on concealed zee runners suspended		S.F. Ceiling	7.03	7.03	
D. SERVICES							
D10 Conveying							
1010	Elevators & Lifts	N/A		—	—	—	0.0 %
1020	Escalators & Moving Walks	N/A		—	—	—	
D20 Plumbing							
2010	Plumbing Fixtures	Toilet and service fixtures, supply and drainage	1 Fixture/500 S.F. Floor	Each	5030	10.06	
2020	Domestic Water Distribution	Gas fired water heater		S.F. Floor	.45	.45	9.1%
2040	Rain Water Drainage	Roof drains		S.F. Roof	1.32	1.32	
D30 HVAC							
3010	Energy Supply	N/A		—	—	—	
3020	Heat Generating Systems	Included in D3050		—	—	—	
3030	Cooling Generating Systems	N/A		—	—	—	8.2 %
3050	Terminal & Package Units	Single zone rooftop unit, gas heating, electric cooling		S.F. Floor	10.63	10.63	
3090	Other HVAC Sys. & Equipment	N/A		—	—	—	
D40 Fire Protection							
4010	Sprinklers	Wet pipe sprinkler system		S.F. Floor	4.39	4.39	4.1%
4020	Standpipes	Standpipe		S.F. Floor	.88	.88	
D50 Electrical							
5010	Electrical Service/Distribution	400 ampere service, panel board and feeders		S.F. Floor	1.78	1.78	
5020	Lighting & Branch Wiring	High efficiency fluorescent fixtures, receptacles, switches, A.C. and misc. power		S.F. Floor	6.65	6.65	8.4%
5030	Communications & Security	Addressable alarm systems, sound system and emergency lighting		S.F. Floor	2.33	2.33	
5090	Other Electrical Systems	Emergency generator, 7.5 kW		S.F. Floor	.16	.16	
E. EQUIPMENT & FURNISHINGS							
1010	Commercial Equipment	N/A		—	—	—	
1020	Institutional Equipment	Projection equipment, screen		S.F. Floor	9.10	9.10	
1030	Vehicular Equipment	N/A		—	—	—	10.5 %
2010	Fixed Furnishings	Movie theater seating		S.F. Floor	4.51	4.51	
F. SPECIAL CONSTRUCTION							
1020	Integrated Construction	N/A		—	—	—	0.0 %
1040	Special Facilities	N/A		—	—	—	
G. BUILDING SITEWORK	**N/A**						

			Sub-Total	129.78	100%
CONTRACTOR FEES (General Requirements: 10%, Overhead: 5%, Profit: 10%)		25%	32.46		
ARCHITECT FEES		7%	11.36		

Total Building Cost 173.60

For customer support on your Square Foot Cost Data, call 877.694.8329.

Costs per square foot of floor area

Exterior Wall	S.F. Area	10000	15000	20000	25000	30000	35000	40000	45000	50000
	L.F. Perimeter	286	370	453	457	513	568	624	680	735
Precast Concrete Panels	Bearing Walls	229.15	220.35	215.85	208.15	205.70	203.95	202.70	201.70	200.85
	Steel Frame	235.70	226.95	222.50	214.85	212.40	210.70	209.45	208.40	207.65
Face Brick with Concrete Block Back-up	Bearing Walls	220.55	212.95	209.00	202.65	200.50	199.10	198.00	197.10	196.40
	Steel Joists	228.00	220.40	216.50	210.20	208.15	206.65	205.65	204.75	204.05
Stucco on Concrete Block	Bearing Walls	211.50	205.10	201.80	196.85	195.10	193.95	193.05	192.30	191.75
	Steel Joists	218.95	212.60	209.35	204.45	202.75	201.55	200.70	199.95	199.40
Perimeter Adj., Add or Deduct	Per 100 L.F.	15.85	10.55	7.85	6.30	5.30	4.55	3.95	3.50	3.15
Story Hgt. Adj., Add or Deduct	Per 1 Ft.	3.90	3.35	3.05	2.50	2.35	2.20	2.10	2.00	2.00

For Basement, add $33.20 per square foot of basement area

The above costs were calculated using the basic specifications shown on the facing page. These costs should be adjusted where necessary for design alternatives and owner's requirements. Reported completed project costs, for this type of structure, range from $94.25 to $236.35 per S.F.

Common additives

Description	Unit	$ Cost
Beds, Manual	Ea.	880 - 2800
Elevators, Hydraulic passenger, 2 stops		
1500# capacity	Ea.	65,700
2500# capacity	Ea.	68,800
3500# capacity	Ea.	73,300
Emergency Lighting, 25 watt, battery operated		
Lead battery	Ea.	330
Nickel cadmium	Ea.	780
Intercom System, 25 station capacity		
Master station	Ea.	3250
Intercom outlets	Ea.	189
Handset	Ea.	520
Kitchen Equipment		
Broiler	Ea.	4050
Coffee urn, twin 6 gallon	Ea.	3000
Cooler, 6 ft. long	Ea.	4175
Dishwasher, 10-12 racks per hr.	Ea.	4275
Food warmer	Ea.	735
Freezer, 44 C.F., reach-in	Ea.	5175

Description	Unit	$ Cost
Kitchen Equipment, cont.		
Ice cube maker, 50 lb. per day	Ea.	2000
Range with 1 oven	Ea.	2925
Laundry Equipment		
Dryer, gas, 16 lb. capacity	Ea.	975
30 lb. capacity	Ea.	3850
Washer, 4 cycle	Ea.	1200
Commercial	Ea.	1600
Nurses Call System		
Single bedside call station	Ea.	335
Pillow speaker	Ea.	277
Refrigerator, Prefabricated, walk-in		
7'-6" high, 6' x 6'	S.F.	202
10' x 10'	S.F.	159
12' x 14'	S.F.	142
12' x 20'	S.F.	125
TV Antenna, Master system, 12 outlet	Outlet	255
30 outlet	Outlet	221
100 outlet	Outlet	209
Whirlpool Bath, Mobile, 18" x 24" x 60"	Ea.	5400
X-Ray, Mobile	Ea.	18,500 - 85,500

Important: See the Reference Section for Location Factors.

Nursing Home

Model costs calculated for a 2 story building with 10′ story height and 25,000 square feet of floor area

				Unit	Unit Cost	Cost Per S.F.	% Of Sub-Total
A. SUBSTRUCTURE							
1010	Standard Foundations	Poured concrete; strip and spread footings		S.F. Ground	1.84	.92	
1020	Special Foundations	N/A		—	—	—	
1030	Slab on Grade	4″ reinforced concrete with vapor barrier and granular base		S.F. Slab	5.44	2.72	3.4%
2010	Basement Excavation	Site preparation for slab and trench for foundation wall and footing		S.F. Ground	.32	.16	
2020	Basement Walls	4′ foundation wall		L.F. Wall	75	1.37	
B. SHELL							
	B10 Superstructure						
1010	Floor Construction	Pre-cast double tees with concrete topping		S.F. Floor	12.86	6.43	7.9%
1020	Roof Construction	Pre-cast double tees		S.F. Roof	10.82	5.41	
	B20 Exterior Enclosure						
2010	Exterior Walls	Precast concrete panels		S.F. Wall	45.76	14.22	
2020	Exterior Windows	Wood double hung	15% of wall	Each	588	2.15	11.2%
2030	Exterior Doors	Double aluminum & glass doors, single leaf hollow metal		Each	2337	.47	
	B30 Roofing						
3010	Roof Coverings	Built-up tar and gravel with flashing; perlite/EPS composite insulation		S.F. Roof	6.52	3.26	2.2%
3020	Roof Openings	Roof hatches		S.F. Roof	.08	.04	
C. INTERIORS							
1010	Partitions	Gypsum board on metal studs	8 S.F. Floor/L.F. Partition	S.F. Partition	9.18	9.18	
1020	Interior Doors	Single leaf wood	80 S.F. Floor/Door	Each	616	7.70	
1030	Fittings	N/A		—	—	—	
2010	Stair Construction	Concrete stair		Flight	6800	1.09	19.2%
3010	Wall Finishes	50% vinyl wall coverings, 45% paint, 5% ceramic tile		S.F. Surface	1.76	3.52	
3020	Floor Finishes	95% vinyl tile, 5% ceramic tile		S.F. Floor	3.13	3.13	
3030	Ceiling Finishes	Painted gypsum board		S.F. Ceiling	4.23	4.23	
D. SERVICES							
	D10 Conveying						
1010	Elevators & Lifts	One hydraulic hospital elevator		Each	106,500	4.26	2.8%
1020	Escalators & Moving Walks	N/A		—	—	—	
	D20 Plumbing						
2010	Plumbing Fixtures	Kitchen, toilet and service fixtures, supply and drainage	1 Fixture/230 S.F. Floor	Each	9982	43.40	
2020	Domestic Water Distribution	Oil fired water heater		S.F. Floor	1.31	1.31	30.2%
2040	Rain Water Drainage	Roof drains		S.F. Roof	1.18	.59	
	D30 HVAC						
3010	Energy Supply	Oil fired hot water, wall fin radiation		S.F. Floor	8.49	8.49	
3020	Heat Generating Systems	N/A		—	—	—	
3030	Cooling Generating Systems	N/A		—	—	—	11.4%
3050	Terminal & Package Units	Split systems with air cooled condensing units		S.F. Floor	8.55	8.55	
3090	Other HVAC Sys. & Equipment	N/A		—	—	—	
	D40 Fire Protection						
4010	Sprinklers	Sprinkler, light hazard		S.F. Floor	3.19	3.19	2.4%
4020	Standpipes	Standpipe		S.F. Floor	.44	.44	
	D50 Electrical						
5010	Electrical Service/Distribution	800 ampere service, panel board and feeders		S.F. Floor	1.64	1.64	
5020	Lighting & Branch Wiring	High efficiency fluorescent fixtures, receptacles, switches, A.C. and misc. power		S.F. Floor	10.24	10.24	9.2%
5030	Communications & Security	Addressable alarm systems and emergency lighting		S.F. Floor	1.41	1.41	
5090	Other Electrical Systems	Emergency generator, 15 kW		S.F. Floor	.50	.50	
E. EQUIPMENT & FURNISHINGS							
1010	Commercial Equipment	N/A		—	—	—	
1020	Institutional Equipment	N/A		—	—	—	
1030	Vehicular Equipment	N/A		—	—	—	0.0%
1090	Other Equipment	N/A		—	—	—	
F. SPECIAL CONSTRUCTION							
1020	Integrated Construction	N/A		—	—	—	
1040	Special Facilities	N/A		—	—	—	0.0%
G. BUILDING SITEWORK	**N/A**						

		Sub-Total	150.02	100%
CONTRACTOR FEES (General Requirements: 10%, Overhead: 5%, Profit: 10%)		25%	37.50	
ARCHITECT FEES		11%	20.63	

Total Building Cost   208.15

For customer support on your Square Foot Cost Data, call 877.694.8329.

173

Costs per square foot of floor area

Exterior Wall	S.F. Area	2000	3000	5000	7000	9000	12000	15000	20000	25000
	L.F. Perimeter	220	260	320	360	420	480	520	640	700
Wood Siding	Wood Truss	209.65	190.00	172.25	163.25	159.35	154.70	151.35	149.15	146.75
Brick Veneer	Wood Truss	258.90	233.35	210.25	198.45	193.40	187.25	182.75	179.95	176.65
Brick on Block	Wood Truss	197.60	170.35	145.55	132.75	127.35	120.70	115.70	112.70	109.05
	Steel Roof Deck	255.85	227.10	200.85	187.30	181.55	174.50	169.25	166.10	162.30
EIFS on Metal Studs	Steel Roof Deck	236.40	212.85	191.55	180.70	176.05	170.45	166.25	163.60	160.65
Tiltup Concrete Panel	Steel Roof Deck	237.55	213.30	191.35	180.25	175.45	169.65	165.40	162.75	159.70
Perimeter Adj., Add or Deduct	Per 100 L.F.	24.45	16.30	9.80	7.05	5.45	4.15	3.25	2.50	1.90
Story Hgt. Adj., Add or Deduct	Per 1 Ft.	2.90	2.25	1.65	1.35	1.30	1.10	0.85	0.90	0.70

For Basement, add $33.65 per square foot of basement area

The above costs were calculated using the basic specifications shown on the facing page. These costs should be adjusted where necessary for design alternatives and owner's requirements. Reported completed project costs, for this type of structure, range from $71.95 to $278.25 per S.F.

Common additives

Description	Unit	$ Cost
Closed circuit surveillance, one station		
Camera and monitor	Ea.	1975
For additional camera stations, add	Ea.	1075
Directory boards, plastic, glass covered		
30" x 20"	Ea.	600
36" x 48"	Ea.	1500
Aluminum, 24" x 18"	Ea.	635
36" x 24"	Ea.	755
48" x 32"	Ea.	1050
48" x 60"	Ea.	2225
Electronic, wall mounted	S.F.	2900
Free standing	S.F.	4100
Pedestal access floor system w/ plastic laminate cover		
Computer room, less than 6000 SF	S.F.	12.10
Greater than 6000 SF	S.F.	11.35
Office, greater than 6000 S.F.	S.F.	8.10

Description	Unit	$ Cost
Security access systems		
Metal detectors, wand type	Ea.	98
Walk-through portal type, single-zone	Ea.	4400
Multi-zone	Ea.	5500
X-ray equipment		
Desk top, for mail, small packages	Ea.	3950
Conveyer type, including monitor, minimum	Ea.	17,900
Maximum	Ea.	31,700
Explosive detection equipment		
Hand held, battery operated	Ea.	28,100
Walk-through portal type	Ea.	48,900
Uninterruptible power supply, 15 kVA/12.75 kW	kW	1810

Important: See the Reference Section for Location Factors.

Model costs calculated for a 1 story building with 12' story height and 7,000 square feet of floor area

				Unit	Unit Cost	Cost Per S.F.	% Of Sub-Total
A. SUBSTRUCTURE							
1010	Standard Foundations	Poured concrete; strip and spread footings		S.F. Ground	2.86	2.86	
1020	Special Foundations	N/A		—	—	—	
1030	Slab on Grade	4" reinforced concrete with vapor barrier and granular base		S.F. Slab	5.44	5.44	9.7%
2010	Basement Excavation	Site preparation for slab and trench for foundation wall and footing		S.F. Ground	.32	.32	
2020	Basement Walls	4' foundation wall		L.F. Wall	159	4.52	
B. SHELL							
	B10 Superstructure						
1010	Floor Construction	N/A		—	—	—	
1020	Roof Construction	Steel joists, girders & deck on columns		S.F. Roof	10.43	10.43	7.7 %
	B20 Exterior Enclosure						
2010	Exterior Walls	E.I.F.S. on metal studs		S.F. Wall	17.56	8.67	
2020	Exterior Windows	Aluminum outward projecting	20% of wall	Each	750	4.02	11.6%
2030	Exterior Doors	Aluminum and glass, hollow metal		Each	3525	3.02	
	B30 Roofing						
3010	Roof Coverings	Single ply membrane, loose laid and ballasted		S.F. Roof	9.69	9.69	7.4%
3020	Roof Openings	Roof hatch		S.F. Roof	.26	.26	
C. INTERIORS							
1010	Partitions	Gypsum board on metal studs	20 S.F. Floor/L.F. Partition	S.F. Partition	9.08	4.54	
1020	Interior Doors	Single leaf hollow metal	200 S.F. Floor/Door	Each	1157	5.79	
1030	Fittings	Toilet partitions		S.F. Floor	.50	.50	
2010	Stair Construction	N/A		—	—	—	18.7%
3010	Wall Finishes	60% vinyl wall covering, 40% paint		S.F. Surface	1.55	1.55	
3020	Floor Finishes	60% carpet, 30% vinyl composition tile, 10% ceramic tile		S.F. Floor	5.82	5.82	
3030	Ceiling Finishes	Mineral fiber tile on concealed zee bars		S.F. Ceiling	7.03	7.03	
D. SERVICES							
	D10 Conveying						
1010	Elevators & Lifts	N/A		—	—	—	0.0 %
1020	Escalators & Moving Walks	N/A		—	—	—	
	D20 Plumbing						
2010	Plumbing Fixtures	Toilet and service fixtures, supply and drainage	1 Fixture/1320 S.F. Floor	Each	8686	6.58	
2020	Domestic Water Distribution	Gas fired water heater		S.F. Floor	1.32	1.32	6.5%
2040	Rain Water Drainage	Roof drains		S.F. Roof	.86	.86	
	D30 HVAC						
3010	Energy Supply	N/A		—	—	—	
3020	Heat Generating Systems	Included in D3050		—	—	—	
3030	Cooling Generating Systems	N/A		—	—	—	16.2 %
3050	Terminal & Package Units	Multizone unit gas heating, electric cooling		S.F. Floor	21.90	21.90	
3090	Other HVAC Sys. & Equipment	N/A		—	—	—	
	D40 Fire Protection						
4010	Sprinklers	Sprinkler system, light hazard		S.F. Floor	3.67	3.67	3.8%
4020	Standpipes	Standpipes and hose systems		S.F. Floor	1.47	1.47	
	D50 Electrical						
5010	Electrical Service/Distribution	400 ampere service, panel board and feeders		S.F. Floor	4.78	4.78	
5020	Lighting & Branch Wiring	High efficiency fluorescent fixtures, receptacles, switches, A.C. and misc. power		S.F. Floor	12.57	12.57	18.4%
5030	Communications & Security	Addressable alarm systems, internet and phone wiring, and emergency lighting		S.F. Floor	7.32	7.32	
5090	Other Electrical Systems	Emergency generator, 7.5 kW		S.F. Floor	.19	.19	
E. EQUIPMENT & FURNISHINGS							
1010	Commercial Equipment	N/A		—	—	—	
1020	Institutional Equipment	N/A		—	—	—	
1030	Vehicular Equipment	N/A		—	—	—	0.0 %
1090	Other Equipment	N/A		—	—	—	
F. SPECIAL CONSTRUCTION							
1020	Integrated Construction	N/A		—	—	—	0.0 %
1040	Special Facilities	N/A		—	—	—	
G. BUILDING SITEWORK	**N/A**						

			Sub-Total	135.12	100%
	CONTRACTOR FEES (General Requirements: 10%, Overhead: 5%, Profit: 10%)		25%	33.76	
	ARCHITECT FEES		7%	11.82	
		Total Building Cost		**180.70**	

For customer support on your Square Foot Cost Data, call 877.694.8329.

175

Costs per square foot of floor area

Exterior Wall	S.F. Area	5000	8000	12000	16000	20000	35000	50000	65000	80000
	L.F. Perimeter	220	260	310	330	360	440	550	600	675
Face Brick with Concrete Block Back-up	Wood Joists	248.00	217.00	199.25	186.65	180.05	166.90	162.85	158.85	156.95
	Steel Joists	250.50	219.50	201.80	189.15	182.55	169.45	165.40	161.30	159.45
Glass and Metal Curtain Wall	Steel Frame	289.80	250.05	227.20	210.60	202.00	184.80	179.45	174.10	171.60
	R/Conc. Frame	289.40	250.15	227.65	211.20	202.65	185.60	180.35	175.00	172.50
Wood Siding	Wood Frame	211.20	189.80	177.70	169.40	165.00	156.45	153.70	151.15	149.90
Brick Veneer	Wood Frame	233.65	206.40	190.90	179.95	174.15	162.80	159.30	155.80	154.15
Perimeter Adj., Add or Deduct	Per 100 L.F.	40.10	25.20	16.75	12.55	10.10	5.75	4.05	3.05	2.55
Story Hgt. Adj., Add or Deduct	Per 1 Ft.	6.55	4.80	3.85	3.10	2.70	1.90	1.60	1.35	1.25

For Basement, add $37.65 per square foot of basement area

The above costs were calculated using the basic specifications shown on the facing page. These costs should be adjusted where necessary for design alternatives and owner's requirements. Reported completed project costs, for this type of structure, range from $75.30 to $291.80 per S.F.

Common additives

Description	Unit	$ Cost
Closed circuit surveillance, one station		
Camera and monitor	Ea.	1975
For additional camera stations, add	Ea.	1075
Directory boards, plastic, glass covered		
30" x 20"	Ea.	600
36" x 48"	Ea.	1500
Aluminum, 24" x 18"	Ea.	635
36" x 24"	Ea.	755
48" x 32"	Ea.	1050
48" x 60"	Ea.	2225
Electronic, wall mounted	S.F.	2900
Free standing	S.F.	4100
Escalators, 10' rise, 32" wide, glass balustrade	Ea.	145,500
Metal balustrade	Ea.	153,500
48" wide, glass balustrade	Ea.	153,000
Metal balustrade	Ea.	162,000
Pedestal access floor system w/ plastic laminate cover		
Computer room, less than 6000 SF	S.F.	12.10
Greater than 6000 SF	S.F.	11.35
Office, greater than 6000 S.F.	S.F.	8.10

Description	Unit	$ Cost
Security access systems		
Metal detectors, wand type	Ea.	98
Walk-through portal type, single-zone	Ea.	4400
Multi-zone	Ea.	5500
X-ray equipment		
Desk top, for mail, small packages	Ea.	3950
Conveyer type, including monitor, minimum	Ea.	17,900
Maximum	Ea.	31,700
Explosive detection equipment		
Hand held, battery operated	Ea.	28,100
Walk-through portal type	Ea.	48,900
Uninterruptible power supply, 15 kVA/12.75 kW	kW	1810

Important: See the Reference Section for Location Factors.

Model costs calculated for a 3 story building with 12' story height and 20,000 square feet of floor area

				Unit	Unit Cost	Cost Per S.F.	% Of Sub-Total
A.	**SUBSTRUCTURE**						
1010	Standard Foundations	Poured concrete; strip and spread footings		S.F. Ground	4.50	1.50	
1020	Special Foundations	N/A		—	—	—	
1030	Slab on Grade	4" reinforced concrete with vapor barrier and granular base		S.F. Slab	5.44	1.81	3.8%
2010	Basement Excavation	Site preparation for slab and trench for foundation wall and footing		S.F. Ground	.32	.11	
2020	Basement Walls	4' foundation wall		L.F. Wall	79	1.77	
B.	**SHELL**						
	B10 Superstructure						
1010	Floor Construction	Open web steel joists, slab form, concrete, columns		S.F. Floor	17.57	11.71	10.4%
1020	Roof Construction	Metal deck, open web steel joists, columns		S.F. Roof	7.41	2.47	
	B20 Exterior Enclosure						
2010	Exterior Walls	Face brick with concrete block backup	80% of wall	S.F. Wall	33.45	17.34	
2020	Exterior Windows	Aluminum outward projecting	20% of wall	Each	750	4.23	16.6%
2030	Exterior Doors	Aluminum and glass, hollow metal		Each	3525	1.06	
	B30 Roofing						
3010	Roof Coverings	Built-up tar and gravel with flashing; perlite/EPS composite		S.F. Roof	7.47	2.49	1.8%
3020	Roof Openings	N/A		—	—	—	
C.	**INTERIORS**						
1010	Partitions	Gypsum board on metal studs	20 S.F. Floor/L.F. Partition	S.F. Partition	10.35	4.14	
1020	Interior Doors	Single leaf hollow metal	200 S.F. Floor/Door	Each	1157	5.79	
1030	Fittings	Toilet partitions		S.F. Floor	1.03	1.03	
2010	Stair Construction	Concrete filled metal pan		Flight	11,600	5.80	22.6%
3010	Wall Finishes	60% vinyl wall covering, 40% paint		S.F. Surface	1.55	1.24	
3020	Floor Finishes	60% carpet, 30% vinyl composition tile, 10% ceramic tile		S.F. Floor	5.82	5.82	
3030	Ceiling Finishes	Mineral fiber tile on concealed zee bars		S.F. Ceiling	7.03	7.03	
D.	**SERVICES**						
	D10 Conveying						
1010	Elevators & Lifts	Two hydraulic passenger elevators		Each	117,300	11.73	8.6%
1020	Escalators & Moving Walks	N/A		—	—	—	
	D20 Plumbing						
2010	Plumbing Fixtures	Toilet and service fixtures, supply and drainage	1 Fixture/1320 S.F. Floor	Each	4963	3.76	
2020	Domestic Water Distribution	Gas fired water heater		S.F. Floor	.46	.46	3.6%
2040	Rain Water Drainage	Roof drains		S.F. Roof	2.10	.70	
	D30 HVAC						
3010	Energy Supply	N/A		—	—	—	
3020	Heat Generating Systems	Included in D3050		—	—	—	
3030	Cooling Generating Systems	N/A		—	—	—	12.9 %
3050	Terminal & Package Units	Multizone unit gas heating, electric cooling		S.F. Floor	17.65	17.65	
3090	Other HVAC Sys. & Equipment	N/A		—	—	—	
	D40 Fire Protection						
4010	Sprinklers	Wet pipe sprinkler system		S.F. Floor	3.84	3.84	3.5%
4020	Standpipes	Standpipes and hose systems		S.F. Floor	.93	.93	
	D50 Electrical						
5010	Electrical Service/Distribution	1000 ampere service, panel board and feeders		S.F. Floor	3.64	3.64	
5020	Lighting & Branch Wiring	High efficiency fluorescent fixtures, receptacles, switches, A.C. and misc. power		S.F. Floor	12.07	12.07	16.2%
5030	Communications & Security	Addressable alarm systems, internet and phone wiring, and emergency lighting		S.F. Floor	6.13	6.13	
5090	Other Electrical Systems	Emergency generator, 7.5 kW, uninterruptible power supply		S.F. Floor	.24	.24	
E.	**EQUIPMENT & FURNISHINGS**						
1010	Commercial Equipment	N/A		—	—	—	
1020	Institutional Equipment	N/A		—	—	—	
1030	Vehicular Equipment	N/A		—	—	—	0.0 %
1090	Other Equipment	N/A		—	—	—	
F.	**SPECIAL CONSTRUCTION**						
1020	Integrated Construction	N/A		—	—	—	0.0 %
1040	Special Facilities	N/A		—	—	—	
G.	**BUILDING SITEWORK**	**N/A**					

		Sub-Total	136.49	100%
CONTRACTOR FEES (General Requirements: 10%, Overhead: 5%, Profit: 10%)		25%	34.12	
ARCHITECT FEES		7%	11.94	
	Total Building Cost		**182.55**	

For customer support on your Square Foot Cost Data, call 877.694.8329.

177

Costs per square foot of floor area

Exterior Wall	S.F. Area	20000	40000	60000	80000	100000	150000	200000	250000	300000
	L.F. Perimeter	260	360	400	420	460	520	650	720	800
Precast Concrete Panel	Steel Frame	233.60	202.70	186.10	176.10	171.40	163.45	161.65	159.10	157.65
	R/Conc. Frame	231.20	200.10	183.40	173.40	168.65	160.70	158.85	156.30	154.85
Face Brick with Concrete Block Back-up	Steel Frame	219.35	191.90	177.80	169.45	165.40	158.65	157.05	154.90	153.65
	R/Conc. Frame	214.30	188.40	174.70	166.60	162.70	156.15	154.65	152.55	151.35
Limestone Panel Concrete Block Back-up	Steel Frame	248.00	212.55	193.40	181.90	176.45	167.25	165.20	162.20	160.50
	R/Conc. Frame	245.15	209.75	190.55	179.05	173.60	164.40	162.35	159.40	157.70
Perimeter Adj., Add or Deduct	Per 100 L.F.	32.05	16.05	10.65	8.00	6.45	4.30	3.20	2.55	2.10
Story Hgt. Adj., Add or Deduct	Per 1 Ft.	6.70	4.60	3.45	2.75	2.40	1.75	1.70	1.50	1.30

For Basement, add $40.60 per square foot of basement area

The above costs were calculated using the basic specifications shown on the facing page. These costs should be adjusted where necessary for design alternatives and owner's requirements. Reported completed project costs, for this type of structure, range from $84.75 to $249.25 per S.F.

Common additives

Description	Unit	$ Cost
Closed circuit surveillance, one station		
Camera and monitor	Ea.	1975
For additional camera stations, add	Ea.	1075
Directory boards, plastic, glass covered		
30" x 20"	Ea.	600
36" x 48"	Ea.	1500
Aluminum, 24" x 18"	Ea.	635
36" x 24"	Ea.	755
48" x 32"	Ea.	1050
48" x 60"	Ea.	2225
Electronic, wall mounted	S.F.	2900
Free standing	S.F.	4100
Escalators, 10' rise, 32" wide, glass balustrade	Ea.	145,500
Metal balustrade	Ea.	153,500
48" wide, glass balustrade	Ea.	153,000
Metal balustrade	Ea.	162,000
Pedestal access floor system w/ plastic laminate cover		
Computer room, less than 6000 SF	S.F.	12.10
Greater than 6000 SF	S.F.	11.35
Office, greater than 6000 S.F.	S.F.	8.10

Description	Unit	$ Cost
Security access systems		
Metal detectors, wand type	Ea.	98
Walk-through portal type, single-zone	Ea.	4400
Multi-zone	Ea.	5500
X-ray equipment		
Desk top, for mail, small packages	Ea.	3950
Conveyer type, including monitor, minimum	Ea.	17,900
Maximum	Ea.	31,700
Explosive detection equipment		
Hand held, battery operated	Ea.	28,100
Walk-through portal type	Ea.	48,900
Uninterruptible power supply, 15 kVA/12.75 kW	kW	1810

Important: See the Reference Section for Location Factors.

Model costs calculated for a 8 story building with 12' story height and 80,000 square feet of floor area

				Unit	Unit Cost	Cost Per S.F.	% Of Sub-Total
A.	**SUBSTRUCTURE**						
1010	Standard Foundations	Poured concrete; strip and spread footings		S.F. Ground	9.04	1.13	
1020	Special Foundations	N/A		—	—	—	
1030	Slab on Grade	4" reinforced concrete with vapor barrier and granular base		S.F. Slab	5.44	.69	1.7%
2010	Basement Excavation	Site preparation for slab and trench for foundation wall and footing		S.F. Ground	.32	.04	
2020	Basement Walls	4' foundation wall		L.F. Wall	84	.44	
B.	**SHELL**						
	B10 Superstructure						
1010	Floor Construction	Concrete slab with metal deck and beams		S.F. Floor	24.93	21.81	17.1%
1020	Roof Construction	Metal deck, open web steel joists, interior columns		S.F. Roof	7.20	.90	
	B20 Exterior Enclosure						
2010	Exterior Walls	Precast concrete panels	80% of wall	S.F. Wall	46.25	18.65	
2020	Exterior Windows	Vertical pivoted steel	20% of wall	Each	588	3.96	17.2%
2030	Exterior Doors	Double aluminum and glass doors and entrance with transoms		Each	4245	.27	
	B30 Roofing						
3010	Roof Coverings	Built-up tar and gravel with flashing; perlite/EPS composite insulation		S.F. Roof	6.48	.81	0.6%
3020	Roof Openings	N/A		—	—	—	
C.	**INTERIORS**						
1010	Partitions	Gypsum board on metal studs	30 S.F. Floor/L.F. Partition	S.F. Partition	10.14	3.38	
1020	Interior Doors	Single leaf hollow metal	400 S.F. Floor/Door	Each	1157	2.90	
1030	Fittings	Toilet Partitions		S.F. Floor	.69	.69	
2010	Stair Construction	Concrete filled metal pan		Flight	11,600	4.35	19.0%
3010	Wall Finishes	60% vinyl wall covering, 40% paint		S.F. Surface	1.56	1.04	
3020	Floor Finishes	60% carpet, 30% vinyl composition tile, 10% ceramic tile		S.F. Floor	5.82	5.82	
3030	Ceiling Finishes	Mineral fiber tile on concealed zee bars		S.F. Ceiling	7.03	7.03	
D.	**SERVICES**						
	D10 Conveying						
1010	Elevators & Lifts	Four geared passenger elevators		Each	296,200	14.81	11.1%
1020	Escalators & Moving Walks	N/A		—	—	—	
	D20 Plumbing						
2010	Plumbing Fixtures	Toilet and service fixtures, supply and drainage	1 Fixture/1370 S.F. Floor	Each	3644	2.66	
2020	Domestic Water Distribution	Gas fired water heater		S.F. Floor	.57	.57	2.7%
2040	Rain Water Drainage	Roof drains		S.F. Roof	2.48	.31	
	D30 HVAC						
3010	Energy Supply	N/A		—	—	—	
3020	Heat Generating Systems	Included in D3050		—	—	—	
3030	Cooling Generating Systems	N/A		—	—	—	13.3 %
3050	Terminal & Package Units	Multizone unit gas heating, electric cooling		S.F. Floor	17.65	17.65	
3090	Other HVAC Sys. & Equipment	N/A		—	—	—	
	D40 Fire Protection						
4010	Sprinklers	Wet pipe sprinkler system		S.F. Floor	2.93	2.93	3.0%
4020	Standpipes	Standpipes and hose systems		S.F. Floor	1.07	1.07	
	D50 Electrical						
5010	Electrical Service/Distribution	1600 ampere service, panel board and feeders		S.F. Floor	1.39	1.39	
5020	Lighting & Branch Wiring	High efficiency fluorescent fixtures, receptacles, switches, A.C. and misc. power		S.F. Floor	11.99	11.99	14.3%
5030	Communications & Security	Addressable alarm systems, internet and phone wiring, emergency lighting		S.F. Floor	4.59	4.59	
5090	Other Electrical Systems	Emergency generator, 100 kW, uninterruptible power supply		S.F. Floor	1.05	1.05	
E.	**EQUIPMENT & FURNISHINGS**						
1010	Commercial Equipment	N/A		—	—	—	
1020	Institutional Equipment	N/A		—	—	—	0.0 %
1030	Vehicular Equipment	N/A		—	—	—	
1090	Other Equipment	N/A		—	—	—	
F.	**SPECIAL CONSTRUCTION**						
1020	Integrated Construction	N/A		—	—	—	0.0 %
1040	Special Facilities	N/A		—	—	—	
G.	**BUILDING SITEWORK**	**N/A**					

		Sub-Total	132.93	100%
	CONTRACTOR FEES (General Requirements: 10%, Overhead: 5%, Profit: 10%)	25%	33.20	
	ARCHITECT FEES	6%	9.97	
	Total Building Cost		**176.10**	

For customer support on your Square Foot Cost Data, call 877.694.8329.

179

Costs per square foot of floor area

Exterior Wall	S.F. Area	120000	145000	170000	200000	230000	260000	400000	600000	800000
	L.F. Perimeter	420	450	470	490	510	530	650	800	900
Double Glazed Heat Absorbing Tinted Plate Glass Panels	Steel Frame	207.00	198.65	191.75	185.45	180.80	177.20	168.70	162.75	158.75
	R/Conc. Frame	197.40	189.25	182.50	176.25	171.70	168.15	159.85	154.00	150.10
Face Brick with Concrete Block Back-up	Steel Frame	179.80	174.55	170.25	166.40	163.60	161.35	156.10	152.40	150.05
	R/Conc. Frame	205.60	200.55	196.45	192.65	189.90	187.75	182.65	179.05	176.70
Precast Concrete Panel With Exposed Aggregate	Steel Frame	186.00	180.00	175.20	170.70	167.45	164.95	158.95	154.75	152.00
	R/Conc. Frame	176.40	170.60	165.90	161.55	158.40	155.90	150.10	146.00	143.35
Perimeter Adj., Add or Deduct	Per 100 L.F.	14.10	11.65	10.00	8.40	7.35	6.45	4.20	2.80	2.10
Story Hgt. Adj., Add or Deduct	Per 1 Ft.	5.70	5.05	4.50	3.95	3.65	3.35	2.65	2.15	1.85
For Basement, add $40.60 per square foot of basement area										

The above costs were calculated using the basic specifications shown on the facing page. These costs should be adjusted where necessary for design alternatives and owner's requirements. Reported completed project costs, for this type of structure, range from $106.30 to $259.40 per S.F.

Common additives

Description	Unit	$ Cost		Description	Unit	$ Cost
Closed circuit surveillance, one station				Security access systems		
Camera and monitor	Ea.	1975		Metal detectors, wand type	Ea.	98
For additional camera stations, add	Ea.	1075		Walk-through portal type, single-zone	Ea.	4400
Directory boards, plastic, glass covered				Multi-zone	Ea.	5500
30" x 20"	Ea.	600		X-ray equipment		
36" x 48"	Ea.	1500		Desk top, for mail, small packages	Ea.	3950
Aluminum, 24" x 18"	Ea.	635		Conveyer type, including monitor, minimum	Ea.	17,900
36" x 24"	Ea.	755		Maximum	Ea.	31,700
48" x 32"	Ea.	1050		Explosive detection equipment		
48" x 60"	Ea.	2225		Hand held, battery operated	Ea.	28,100
Electronic, wall mounted	S.F.	2900		Walk-through portal type	Ea.	48,900
Free standing	S.F.	4100		Uninterruptible power supply, 15 kVA/12.75 kW	kW	1810
Escalators, 10' rise, 32" wide, glass balustrade	Ea.	145,500				
Metal balustrade	Ea.	153,500				
48" wide, glass balustrade	Ea.	153,000				
Metal balustrade	Ea.	162,000				
Pedestal access floor system w/ plastic laminate cover						
Computer room, less than 6000 SF	S.F.	12.10				
Greater than 6000 SF	S.F.	11.35				
Office, greater than 6000 S.F.	S.F.	8.10				

Important: See the Reference Section for Location Factors.

Office, 11-20 Story

Model costs calculated for a 16 story building with 10' story height and 260,000 square feet of floor area

				Unit	Unit Cost	Cost Per S.F.	% Of Sub-Total
A.	**SUBSTRUCTURE**						
1010	Standard Foundations	CIP concrete pile caps		S.F. Ground	10.24	.64	
1020	Special Foundations	Steel H-piles, concrete grade beams		S.F. Ground	61	3.84	
1030	Slab on Grade	4" reinforced concrete with vapor barrier and granular base		S.F. Slab	5.44	.34	3.9%
2010	Basement Excavation	Site preparation for slab, piles and grade beams		S.F. Ground	.32	.02	
2020	Basement Walls	4' foundation wall		L.F. Wall	84	.34	
B.	**SHELL**						
	B10 Superstructure						
1010	Floor Construction	Concrete slab, metal deck, beams		S.F. Floor	29.83	27.97	21.3%
1020	Roof Construction	Metal deck, open web steel joists, beams, columns		S.F. Roof	8.64	.54	
	B20 Exterior Enclosure						
2010	Exterior Walls	N/A		—	—	—	
2020	Exterior Windows	Double glazed heat absorbing, tinted plate glass wall panels	100% of wall	Each	77	25.01	19.2 %
2030	Exterior Doors	Double aluminum & glass doors		Each	6414	.70	
	B30 Roofing						
3010	Roof Coverings	Single ply membrane, fully adhered; perlite/EPS composite insulation		S.F. Roof	5.92	.37	0.3%
3020	Roof Openings	N/A		—	—	—	
C.	**INTERIORS**						
1010	Partitions	Gypsum board on metal studs	30 S.F. Floor/L.F. Partition	S.F. Partition	7.39	1.97	
1020	Interior Doors	Single leaf hollow metal	400 S.F. Floor/Door	Each	1157	2.90	
1030	Fittings	Toilet partitions		S.F. Floor	.39	.39	
2010	Stair Construction	Concrete filled metal pan		Flight	17,300	2.26	15.2%
3010	Wall Finishes	60% vinyl wall covering, 40% paint		S.F. Surface	1.56	.83	
3020	Floor Finishes	60% carpet tile, 30% vinyl composition tile, 10% ceramic tile		S.F. Floor	4.92	4.92	
3030	Ceiling Finishes	Mineral fiber tile on concealed zee bars		S.F. Ceiling	7.03	7.03	
D.	**SERVICES**						
	D10 Conveying						
1010	Elevators & Lifts	Four geared passenger elevators		Each	464,100	7.14	5.3%
1020	Escalators & Moving Walks	N/A		—	—	—	
	D20 Plumbing						
2010	Plumbing Fixtures	Toilet and service fixtures, supply and drainage	1 Fixture/1345 S.F. Floor	Each	5931	4.41	
2020	Domestic Water Distribution	Oil fired water heater		S.F. Floor	.33	.33	3.7%
2040	Rain Water Drainage	Roof drains		S.F. Roof	3.20	.20	
	D30 HVAC						
3010	Energy Supply	N/A		—	—	—	
3020	Heat Generating Systems	Boiler, heat exchanger and fans		Each	497,975	2.65	
3030	Cooling Generating Systems	Chilled water, fan coil units		S.F. Floor	16.72	16.72	14.5 %
3050	Terminal & Package Units	N/A		—	—	—	
3090	Other HVAC Sys. & Equipment	N/A		—	—	—	
	D40 Fire Protection						
4010	Sprinklers	Sprinkler system, light hazard		S.F. Floor	2.83	2.83	2.5%
4020	Standpipes	Standpipes and hose systems		S.F. Floor	.56	.56	
	D50 Electrical						
5010	Electrical Service/Distribution	2400 ampere service, panel board and feeders		S.F. Floor	.89	.89	
5020	Lighting & Branch Wiring	High efficiency fluorescent fixtures, receptacles, switches, A.C. and misc. power		S.F. Floor	11.86	11.86	14.1%
5030	Communications & Security	Addressable alarm systems, internet and phone wiring, emergency lighting		S.F. Floor	5.49	5.49	
5090	Other Electrical Systems	Emergency generator, 200 kW, uninterruptible power supply		S.F. Floor	.56	.56	
E.	**EQUIPMENT & FURNISHINGS**						
1010	Commercial Equipment	N/A		—	—	—	
1020	Institutional Equipment	N/A		—	—	—	
1030	Vehicular Equipment	N/A		—	—	—	0.0 %
1090	Other Equipment	N/A		—	—	—	
F.	**SPECIAL CONSTRUCTION**						
1020	Integrated Construction	N/A		—	—	—	0.0 %
1040	Special Facilities	N/A		—	—	—	
G.	**BUILDING SITEWORK**	**N/A**					

		Sub-Total	133.71	**100%**
CONTRACTOR FEES (General Requirements: 10%, Overhead: 5%, Profit: 10%)			25%	33.46
ARCHITECT FEES			6%	10.03
		Total Building Cost	**177.20**	

For customer support on your Square Foot Cost Data, call 877.694.8329.

Costs per square foot of floor area

Exterior Wall	S.F. Area	5000	7500	10000	12500	15000	17500	20000	22500	25000
	L.F. Perimeter	284	350	410	450	490	530	570	600	632
Brick Veneer	Wood Truss	450.60	437.35	430.10	424.15	420.15	417.35	415.25	413.10	411.50
Brick Veneer	Steel Frame	450.30	437.60	430.70	425.10	421.35	418.60	416.60	414.65	413.10
Decorative Concrete Block	Steel Frame	453.00	439.80	432.70	426.75	422.85	420.05	417.95	415.90	414.25
Decorative Concrete Block	Bearing Wall	443.85	432.30	426.05	420.95	417.60	415.20	413.40	411.60	410.20
Tiltup Concrete Panel	Steel Frame	445.05	433.35	427.00	**421.80**	418.35	415.90	414.05	412.25	410.85
EIFS on Concrete Block	Steel Frame	449.50	435.45	427.90	422.00	418.10	415.30	413.20	411.15	409.60
Perimeter Adj., Add or Deduct	Per 100 L.F.	20.20	13.40	10.10	8.05	6.75	5.75	4.95	4.50	4.00
Story Hgt. Adj., Add or Deduct	Per 1 Ft.	3.05	2.45	2.20	1.85	1.75	1.60	1.45	1.40	1.35

For Basement, add $31.95 per square foot of basement area

The above costs were calculated using the basic specifications shown on the facing page. These costs should be adjusted where necessary for design alternatives and owner's requirements. Reported completed project costs, for this type of structure, range from $186.20 to $965.45 per S.F.

Common additives

Description	Unit	$ Cost
Cabinets, base, door units, metal	L.F.	320
Drawer units	L.F.	630
Tall storage cabinets, open, 7' high	L.F.	605
With glazed doors	L.F.	875
Wall cabinets, metal, 12-1/2" deep, open	L.F.	241
With doors	L.F.	440
Counter top, laminated plastic, no backsplash	L.F.	81
Stainless steel counter top	L.F.	198
For drop-in stainless 43" x 21" sink, add	Ea.	1100
Nurse Call Station		
Single bedside call station	Ea.	335
Ceiling speaker station	Ea.	156
Emergency call station	Ea.	162
Pillow speaker	Ea.	277
Standard call button	Ea.	178
Master control station for 20 stations	Total	6400

Description	Unit	$ Cost
Directory Boards, Plastic, glass covered		
Plastic, glass covered, 30" x 20"	Ea.	600
36" x 48"	Ea.	1500
Building directory, alum., black felt panels, 1 door, 24" x 18"	Ea.	635
36" x 24"	Ea.	755
48" x 32"	Ea.	1050
48" x 60"	Ea.	2225
Tables, Examining, vinyl top		
with base cabinets	Ea.	1500
Utensil washer-sanitizer	Ea.	12,800
X-ray, mobile, minimum	Ea.	18,500

Model costs calculated for a 1 story building with 15'-4" story height and 12,500 square feet of floor area

Outpatient Surgery Center

				Unit	Unit Cost	Cost Per S.F.	% Of Sub-Total
A.	**SUBSTRUCTURE**						
1010	Standard Foundations	Poured concrete; strip and spread footings		S.F. Ground	1.86	1.86	
1020	Special Foundations	N/A		—	—	—	
1030	Slab on Grade	4" reinforced concrete with vapor barrier and granular base		S.F. Slab	6.70	6.70	3.8%
2010	Basement Excavation	Site preparation for slab and trench for foundation wall and footing		S.F. Ground	.32	.32	
2020	Basement Walls	4' Foundation wall		L.F. Wall	84	3	
B.	**SHELL**						
	B10 Superstructure						
1010	Floor Construction	N/A		—	—	—	2.5 %
1020	Roof Construction	Steel Joists and girders on columns		S.F. Roof	7.87	7.87	
	B20 Exterior Enclosure						
2010	Exterior Walls	Brick veneer	85% of wall	S.F. Wall	19.10	8.96	
2020	Exterior Windows	Aluminum frame, fixed	15% of wall	Each	840	4.64	6.1%
2030	Exterior Doors	Aluminum and glass entry; steel flush doors		Each	6790	5.43	
	B30 Roofing						
3010	Roof Coverings	Single-ply membrane, loose laid and ballasted; polyisocyanurate insulation		S.F. Roof	6.78	6.78	2.2%
3020	Roof Openings	Roof hatches		S.F. Roof	.10	.10	
C.	**INTERIORS**						
1010	Partitions	Gypsum board on metal studs	7 S.F. Floor/L.F. Partition	S.F. Partition	7.90	11.29	
1020	Interior Doors	Single and double leaf fire doors	200 S.F. Floor/Door	Each	1391	6.96	
1030	Fittings	Stainless steel toilet partitions, steel lockers		S.F. Floor	1.68	1.68	
2010	Stair Construction	N/A		—	—	—	13.0%
3010	Wall Finishes	80% Paint, 10% vinyl wall covering, 10% ceramic tile		S.F. Surface	2.02	5.78	
3020	Floor Finishes	10% Carpet, 80% vinyl sheet, 10% ceramic tile		S.F. Floor	10.32	10.32	
3030	Ceiling Finishes	Mineral fiber tile and plastic coated tile on suspension system		S.F. Ceiling	4.19	4.19	
D.	**SERVICES**						
	D10 Conveying						
1010	Elevators & Lifts	N/A		—	—	—	0.0 %
1020	Escalators & Moving Walks	N/A		—	—	—	
	D20 Plumbing						
2010	Plumbing Fixtures	Medical, patient & specialty fixtures, supply and drainage	1 fixture/350 S.F. Floor	Each	3483	9.95	
2020	Domestic Water Distribution	Electric water heater		S.F. Floor	1.63	1.63	4.5%
2040	Rain Water Drainage	Roof drains		S.F. Roof	2.32	2.32	
	D30 HVAC						
3010	Energy Supply	Hot water reheat for surgery		S.F. Floor	11.35	11.35	
3020	Heat Generating Systems	Boiler		Each	62,750	6.95	
3030	Cooling Generating Systems	N/A		—	—	—	37.4%
3050	Terminal & Package Units	N/A		—	—	—	
3090	Other HVAC Sys. & Equipment	Rooftop AC, CAV, roof vents, heat exchanger, surgical air curtains		S.F. Floor	98	97.54	
	D40 Fire Protection						
4010	Sprinklers	Wet pipe sprinkler system		S.F. Floor	3.67	3.67	1.7%
4020	Standpipes	Standpipe		S.F. Floor	1.63	1.63	
	D50 Electrical						
5010	Electrical Service/Distribution	1200 ampere service, panel board and feeders		S.F. Floor	6.75	6.75	
5020	Lighting & Branch Wiring	T-8 fluorescent light fixtures, receptacles, switches, A.C. and misc. power		S.F. Floor	17.14	17.14	13.0%
5030	Communications & Security	Addressable alarm system, internet wiring, communications system, emergency lighting		S.F. Floor	10.17	10.17	
5090	Other Electrical Systems	Emergency generator, 300 kW with fuel tank		S.F. Floor	6.11	6.11	
E.	**EQUIPMENT & FURNISHINGS**						
1010	Commercial Equipment	N/A		—	—	—	
1020	Institutional Equipment	Medical gas system, cabinetry, scrub sink, sterilizer		S.F. Floor	47.45	47.45	15.7 %
1030	Vehicular Equipment	N/A		—	—	—	
1090	Other Equipment	Ice maker, washing machine		S.F. Floor	1.02	1.02	
F.	**SPECIAL CONSTRUCTION & DEMOLITION**						
1020	Integrated Construction	N/A		—	—	—	0.0 %
1040	Special Facilities	N/A		—	—	—	
G.	**BUILDING SITEWORK**	**N/A**					

		Sub-Total		309.56	**100%**
CONTRACTOR FEES (General Requirements: 10%, Overhead: 5%, Profit: 5%)			20%	77.41	
ARCHITECT FEES			9%	34.83	
	Total Building Cost			**421.80**	

For customer support on your Square Foot Cost Data, call 877.694.8329.

183

Costs per square foot of floor area

Exterior Wall	S.F. Area	7000	9000	11000	13000	15000	17000	19000	21000	23000
	L.F. Perimeter	240	280	303	325	354	372	397	422	447
Limestone with Concrete Block Back-up	Bearing Walls	278.75	261.35	246.65	236.30	229.85	223.35	219.05	215.65	212.70
	R/Conc. Frame	285.90	269.20	255.40	245.75	239.65	233.60	229.60	226.35	223.65
Face Brick with Concrete Block Back-up	Bearing Walls	249.20	234.35	222.70	214.55	209.30	204.25	200.85	198.05	195.70
	R/Conc. Frame	264.95	250.20	238.60	230.50	225.25	220.20	216.85	214.10	211.75
Decorative Concrete Block	Bearing Walls	237.45	223.70	213.30	206.05	201.25	196.75	193.75	191.25	189.05
	R/Conc. Frame	254.35	240.70	230.35	223.10	218.35	213.90	210.90	208.35	206.30
Perimeter Adj., Add or Deduct	Per 100 L.F.	33.20	25.80	21.10	17.85	15.45	13.65	12.25	11.00	10.15
Story Hgt. Adj., Add or Deduct	Per 1 Ft.	5.95	5.40	4.75	4.30	4.10	3.80	3.65	3.50	3.35

For Basement, add $31.95 per square foot of basement area

The above costs were calculated using the basic specifications shown on the facing page. These costs should be adjusted where necessary for design alternatives and owner's requirements. Reported completed project costs, for this type of structure, range from $122.15 to $318.65 per S.F.

Common additives

Description	Unit	$ Cost
Cells Prefabricated, 5'-6' wide, 7'-8' high, 7'-8' deep	Ea.	12,800
Elevators, Hydraulic passenger, 2 stops		
1500# capacity	Ea.	65,700
2500# capacity	Ea.	68,800
3500# capacity	Ea.	73,300
Emergency Lighting, 25 watt, battery operated		
Lead battery	Ea.	330
Nickel cadmium	Ea.	780
Flagpoles, Complete		
Aluminum, 20' high	Ea.	1850
40' high	Ea.	4200
70' high	Ea.	11,000
Fiberglass, 23' high	Ea.	1325
39'-5" high	Ea.	3450
59' high	Ea.	6850

Description	Unit	$ Cost
Lockers, Steel, Single tier, 60" to 72"	Opng.	221 - 370
2 tier, 60" or 72" total	Opng.	120 - 165
5 tier, box lockers	Opng.	77 - 86.50
Locker bench, lam. maple top only	L.F.	34
Pedestals, steel pipe	Ea.	76
Safe, Office type, 1 hour rating		
30" x 18" x 18"	Ea.	2500
60" x 36" x 18", double door	Ea.	8125
Shooting Range, Incl. bullet traps, target provisions, and contols, not incl. structural shell	Ea.	51,000
Smoke Detectors		
Ceiling type	Ea.	227
Duct type	Ea.	565
Sound System		
Amplifier, 250 watts	Ea.	2050
Speaker, ceiling or wall	Ea.	213
Trumpet	Ea.	410

Important: See the Reference Section for Location Factors.

Model costs calculated for a 2 story building with 12' story height and 11,000 square feet of floor area

Police Station

				Unit	Unit Cost	Cost Per S.F.	% Of Sub-Total
A. SUBSTRUCTURE							
1010	Standard Foundations	Poured concrete; strip and spread footings		S.F. Ground	2.94	1.47	
1020	Special Foundations	N/A		—	—	—	
1030	Slab on Grade	4" reinforced concrete with vapor barrier and granular base		S.F. Slab	5.44	2.72	3.7%
2010	Basement Excavation	Site preparation for slab and trench for foundation wall and footing		S.F. Ground	.32	.16	
2020	Basement Walls	4' foundation wall		L.F. Wall	84	2.30	
B. SHELL							
B10 Superstructure							
1010	Floor Construction	Open web steel joists, slab form, concrete		S.F. Floor	12.46	6.23	
1020	Roof Construction	Metal deck on open web steel joists		S.F. Roof	4.96	2.48	4.8%
B20 Exterior Enclosure							
2010	Exterior Walls	Limestone with concrete block backup	80% of wall	S.F. Wall	57	30.04	
2020	Exterior Windows	Metal horizontal sliding	20% of wall	Each	1159	10.22	23.3%
2030	Exterior Doors	Hollow metal		Each	2723	1.98	
B30 Roofing							
3010	Roof Coverings	Built-up tar and gravel with flashing; perlite/EPS composite insulation		S.F. Roof	7.48	3.74	
3020	Roof Openings	N/A		—	—	—	2.1%
C. INTERIORS							
1010	Partitions	Concrete block	20 S.F. Floor/L.F. Partition	S.F. Partition	14.96	7.48	
1020	Interior Doors	Single leaf kalamein fire door	200 S.F. Floor/Door	Each	1157	5.79	
1030	Fittings	Toilet partitions		S.F. Floor	.94	.94	
2010	Stair Construction	Concrete filled metal pan		Flight	17,300	4.72	18.7%
3010	Wall Finishes	90% paint, 10% ceramic tile		S.F. Surface	3.86	3.86	
3020	Floor Finishes	70% vinyl composition tile, 20% carpet tile, 10% ceramic tile		S.F. Floor	4.02	4.02	
3030	Ceiling Finishes	Mineral fiber tile on concealed zee bars		S.F. Ceiling	7.03	7.03	
D. SERVICES							
D10 Conveying							
1010	Elevators & Lifts	One hydraulic passenger elevator		Each	83,600	7.60	
1020	Escalators & Moving Walks	N/A		—	—	—	4.2%
D20 Plumbing							
2010	Plumbing Fixtures	Toilet and service fixtures, supply and drainage	1 Fixture/580 S.F. Floor	Each	4646	8.01	
2020	Domestic Water Distribution	Oil fired water heater		S.F. Floor	3.74	3.74	7.4%
2040	Rain Water Drainage	Roof drains		S.F. Roof	3.42	1.71	
D30 HVAC							
3010	Energy Supply	N/A		—	—	—	
3020	Heat Generating Systems	N/A		—	—	—	
3030	Cooling Generating Systems	N/A		—	—	—	12.1%
3050	Terminal & Package Units	Multizone HVAC air cooled system		S.F. Floor	21.90	21.90	
3090	Other HVAC Sys. & Equipment	N/A		—	—	—	
D40 Fire Protection							
4010	Sprinklers	Wet pipe sprinkler system		S.F. Floor	3.96	3.96	
4020	Standpipes	Standpipe		S.F. Floor	1.02	1.02	2.8%
D50 Electrical							
5010	Electrical Service/Distribution	400 ampere service, panel board and feeders		S.F. Floor	2.01	2.01	
5020	Lighting & Branch Wiring	T-8 fluorescent fixtures, receptacles, switches, A.C. and misc. power		S.F. Floor	12.24	12.24	
5030	Communications & Security	Addressable alarm systems, internet wiring, intercom and emergency lighting		S.F. Floor	7.33	7.33	12.0%
5090	Other Electrical Systems	Emergency generator, 15 kW		S.F. Floor	.21	.21	
E. EQUIPMENT & FURNISHINGS							
1010	Commercial Equipment	N/A		—	—	—	
1020	Institutional Equipment	Lockers, detention rooms, cells, gasoline dispensers		S.F. Floor	13.62	13.62	
1030	Vehicular Equipment	Gasoline dispenser system		S.F. Floor	2.47	2.47	8.9%
1090	Other Equipment	N/A		—	—	—	
F. SPECIAL CONSTRUCTION							
1020	Integrated Construction	N/A		—	—	—	
1040	Special Facilities	N/A		—	—	—	0.0%
G. BUILDING SITEWORK	**N/A**						

			Sub-Total	181	100%
CONTRACTOR FEES (General Requirements: 10%, Overhead: 5%, Profit: 10%)			25%	45.29	
ARCHITECT FEES			9%	20.36	

Total Building Cost	**246.65**

For customer support on your Square Foot Cost Data, call 877.694.8329.

185

Costs per square foot of floor area

Exterior Wall	S.F. Area	600	750	925	1150	1450	1800	2100	2600	3200
	L.F. Perimeter	100	112	128	150	164	182	200	216	240
Steel Siding on Wood Studs	Wood Truss	75.10	69.70	66.20	63.65	58.60	55.05	53.30	49.90	47.55
Aluminum Siding	Wood Truss	74.55	69.70	66.55	64.20	59.70	56.60	55.00	52.00	49.90
Wood Board Siding	Wood Truss	77.45	72.35	69.05	66.55	61.80	58.50	56.85	53.60	51.45
Plywood Siding	Wood Truss	74.95	70.05	66.90	64.60	60.00	56.90	55.30	52.25	50.15
Perimeter Adj., Add or Deduct	Per 100 L.F.	44.55	35.65	28.95	23.20	18.40	14.85	12.75	10.25	8.30
Story Hgt. Adj., Add or Deduct	Per 1 Ft.	3.20	2.85	2.65	2.45	2.15	1.95	1.85	1.50	1.45
	Basement—Not Applicable									

The above costs were calculated using the basic specifications shown on the facing page. These costs should be adjusted where necessary for design alternatives and owner's requirements. Reported completed project costs, for this type of structure, range from $39.05 to $155.90 per S.F.

Common additives

Description	Unit	$ Cost
Slab on Grade		
4" thick, non-industrial, non-reinforced	S.F.	4.90
Reinforced	S.F.	5.44
Light industrial, non-reinforced	S.F.	6.16
Reinforced	S.F.	6.70
Industrial, non-reinforced	S.F.	10.49
Reinforced	S.F.	11.05
6" thick, non-industrial, non-reinforced	S.F.	5.73
Reinforced	S.F.	6.47
Light industrial, non-reinforced	S.F.	7
Reinforced	S.F.	8.05
Heavy industrial, non-reinforced	S.F.	13.45
Reinforced	S.F.	14.25
Insulation		
Fiberboard, low density, 1" thick, R2.78	S.F.	1.39
Foil on reinforced scrim, single bubble air space, R8.8	C.S.F.	69.50
Vinyl faced fiberglass, 1-1/2" thick, R5	S.F.	0.84
Vinyl/scrim/foil (VSF), 1-1/2" thick, R5	S.F.	1.02

Description	Unit	$ Cost
Roof ventilation, ridge vent, aluminum	L.F.	7.90
Mushroom vent, aluminum	Ea.	90
Gutter, aluminum, 5" K type, enameled	L.F.	8.35
Steel, half round or box, 5", enameled	L.F.	7.75
Wood, clear treated cedar, fir or hemlock, 4" x 5"	L.F.	24.50
Downspout, aluminum, 3" x 4", enameled	L.F.	7.45
Steel, round, 4" diameter	L.F.	7.50

Important: See the Reference Section for Location Factors.

Model costs calculated for a 1 story building with 14' story height and 1,800 square feet of floor area

Post Frame Barn

			Unit	Unit Cost	Cost Per S.F.	% Of Sub-Total
A. SUBSTRUCTURE						
1010	Standard Foundations	Included in B2010 Exterior Walls	–	–	–	
1020	Special Foundations	N/A	–	–	–	
1030	Slab on Grade	N/A	–	–	–	0.0%
2010	Basement Excavation	N/A	–	–	–	
2020	Basement Walls	N/A	–	–	–	
B. SHELL						
	B10 Superstructure					
1010	Floor Construction	N/A	–	–	–	16.3 %
1020	Roof Construction	Wood truss, 4/12 pitch	S.F. Roof	6.65	6.65	
	B20 Exterior Enclosure					
2010	Exterior Walls	2 x 6 Wood framing, 6 x 6 wood posts in concrete; steel siding panels	S.F. Wall	11.78	15.84	
2020	Exterior Windows	Steel windows, fixed	Each	533	4.19	59.3%
2030	Exterior Doors	Steel overhead, hollow metal doors	Each	1870	4.15	
	B30 Roofing					
3010	Roof Coverings	Metal panel roof	S.F. Roof	4.24	4.24	10.4%
3020	Roof Openings	N/A	–	–	–	
C. INTERIORS						
1010	Partitions	N/A	–	–	–	
1020	Interior Doors	N/A	–	–	–	
1030	Fittings	N/A	–	–	–	
2010	Stair Construction	N/A	–	–	–	0.0 %
3010	Wall Finishes	N/A	–	–	–	
3020	Floor Finishes	N/A	–	–	–	
3030	Ceiling Finishes	N/A	–	–	–	
D. SERVICES						
	D10 Conveying					
1010	Elevators & Lifts	N/A	–	–	–	0.0 %
1020	Escalators & Moving Walks	N/A	–	–	–	
	D20 Plumbing					
2010	Plumbing Fixtures	N/A	–	–	–	
2020	Domestic Water Distribution	N/A	–	–	–	0.0 %
2040	Rain Water Drainage	N/A	–	–	–	
	D30 HVAC					
3010	Energy Supply	N/A	–	–	–	
3020	Heat Generating Systems	N/A	–	–	–	
3030	Cooling Generating Systems	N/A	–	–	–	0.0 %
3050	Terminal & Package Units	N/A	–	–	–	
3090	Other HVAC Sys. & Equipment	N/A	–	–	–	
	D40 Fire Protection					
4010	Sprinklers	N/A	–	–	–	0.0 %
4020	Standpipes	N/A	–	–	–	
	D50 Electrical					
5010	Electrical Service/Distribution	60 Ampere service, panel board & feeders	S.F. Floor	1.91	1.91	
5020	Lighting & Branch Wiring	High bay fixtures, receptacles & switches	S.F. Floor	3.81	3.81	14.0%
5030	Communications & Security	N/A	–	–	–	
5090	Other Electrical Systems	N/A	–	–	–	
E. EQUIPMENT & FURNISHINGS						
1010	Commercial Equipment	N/A	–	–	–	
1020	Institutional Equipment	N/A	–	–	–	0.0 %
1030	Vehicular Equipment	N/A	–	–	–	
1090	Other Equipment	N/A	–	–	–	
F. SPECIAL CONSTRUCTION & DEMOLITION						
1020	Integrated Construction	N/A	–	–	–	0.0 %
1040	Special Facilities	N/A	–	–	–	
G. BUILDING SITEWORK	**N/A**					
			Sub-Total		40.79	100%
	CONTRACTOR FEES (General Requirements: 10%, Overhead: 5%, Profit: 5%)			20%	10.18	
	ARCHITECT FEES			8%	4.08	
			Total Building Cost		**55.05**	

For customer support on your Square Foot Cost Data, call 877.694.8329.

187

Costs per square foot of floor area

Exterior Wall	S.F. Area	5000	7000	9000	11000	13000	15000	17000	19000	21000
	L.F. Perimeter	300	380	420	468	486	513	540	580	620
Face Brick with Concrete Block Back-up	Steel Frame	173.10	163.55	154.20	148.85	143.10	139.35	136.60	134.95	133.65
	Bearing Walls	173.60	163.95	154.50	149.10	143.30	139.55	136.70	135.05	133.75
Limestone with Concrete Block Back-up	Steel Frame	194.50	182.85	170.80	164.05	156.40	151.60	147.90	145.85	144.20
	Bearing Walls	194.95	183.30	171.10	164.25	156.60	151.75	148.00	145.95	144.30
Decorative Concrete Block	Steel Frame	161.15	152.70	144.90	140.40	135.65	132.55	130.25	128.85	127.75
	Bearing Walls	161.65	153.15	145.20	140.65	135.85	132.75	130.40	128.95	127.85
Perimeter Adj., Add or Deduct	Per 100 L.F.	18.25	13.00	10.15	8.30	7.05	6.15	5.40	4.80	4.40
Story Hgt. Adj., Add or Deduct	Per 1 Ft.	2.90	2.70	2.25	2.05	1.85	1.70	1.55	1.50	1.45
For Basement, add $31.10 per square foot of basement area										

The above costs were calculated using the basic specifications shown on the facing page. These costs should be adjusted where necessary for design alternatives and owner's requirements. Reported completed project costs, for this type of structure, range from $101.35 to $263.05 per S.F.

Common additives

Description	Unit	$ Cost	Description	Unit	$ Cost
Closed Circuit Surveillance, One station			Mail Boxes, Horizontal, key lock, 15" x 6" x 5"	Ea.	61
Camera and monitor	Ea.	1975	Double 15" x 12" x 5"	Ea.	95.50
For additional camera stations, add	Ea.	1075	Quadruple 15" x 12" x 10"	Ea.	146
Emergency Lighting, 25 watt, battery operated			Vertical, 6" x 5" x 15", aluminum	Ea.	61
Lead battery	Ea.	330	Bronze	Ea.	68
Nickel cadmium	Ea.	780	Steel, enameled	Ea.	61
Flagpoles, Complete			Scales, Dial type, 5 ton cap.		
Aluminum, 20' high	Ea.	1850	8' x 6' platform	Ea.	12,100
40' high	Ea.	4200	9' x 7' platform	Ea.	14,900
70' high	Ea.	11,000	Smoke Detectors		
Fiberglass, 23' high	Ea.	1325	Ceiling type	Ea.	227
39'-5" high	Ea.	3450	Duct type	Ea.	565
59' high	Ea.	6850			

Model costs calculated for a 1 story building with 14' story height and 13,000 square feet of floor area

Post Office

						Unit	Unit Cost	Cost Per S.F.	% Of Sub-Total

A. SUBSTRUCTURE

					Unit	Unit Cost	Cost Per S.F.	% Of Sub-Total
1010	Standard Foundations	Poured concrete; strip and spread footings			S.F. Ground	1.67	1.67	
1020	Special Foundations	N/A			–	–	–	
1030	Slab on Grade	4" reinforced concrete with vapor barrier and granular base			S.F. Slab	5.44	5.44	10.0%
2010	Basement Excavation	Site preparation for slab and trench for foundation wall and footing			S.F. Ground	.32	.32	
2020	Basement Walls	4' foundation wall			L.F. Wall	84	3.12	

B. SHELL

B10 Superstructure

					Unit	Unit Cost	Cost Per S.F.	% Of Sub-Total
1010	Floor Construction	Steel column fireproofing			L.F. Column	35.85	.27	9.2%
1020	Roof Construction	Metal deck, open web steel joists, columns			S.F. Roof	9.42	9.42	

B20 Exterior Enclosure

					Unit	Unit Cost	Cost Per S.F.	% Of Sub-Total
2010	Exterior Walls	Face brick with concrete block backup	80% of wall		S.F. Wall	33.46	14.01	
2020	Exterior Windows	Double strength window glass	20% of wall		Each	750	3.41	17.8%
2030	Exterior Doors	Double aluminum & glass, single aluminum, hollow metal, steel overhead			Each	2700	1.24	

B30 Roofing

					Unit	Unit Cost	Cost Per S.F.	% Of Sub-Total
3010	Roof Coverings	Built-up tar and gravel with flashing; perlite/EPS composite insulation			S.F. Roof	6.72	6.72	6.4%
3020	Roof Openings	N/A			–	–	–	

C. INTERIORS

					Unit	Unit Cost	Cost Per S.F.	% Of Sub-Total
1010	Partitions	Concrete block	15 S.F. Floor/L.F. Partition		S.F. Partition	9.66	7.73	
1020	Interior Doors	Single leaf hollow metal	150 S.F. Floor/Door		Each	1157	7.71	
1030	Fittings	Toilet partitions, cabinets, shelving, lockers			S.F. Floor	1.48	1.48	
2010	Stair Construction	N/A			–	–	–	23.6%
3010	Wall Finishes	Paint			S.F. Surface	2.38	3.81	
3020	Floor Finishes	50% vinyl tile, 50% paint			S.F. Floor	2.29	2.29	
3030	Ceiling Finishes	Mineral fiber tile on concealed zee bars	25% of area		S.F. Ceiling	7.03	1.76	

D. SERVICES

D10 Conveying

					Unit	Unit Cost	Cost Per S.F.	% Of Sub-Total
1010	Elevators & Lifts	N/A			–	–	–	0.0%
1020	Escalators & Moving Walks	N/A			–	–	–	

D20 Plumbing

					Unit	Unit Cost	Cost Per S.F.	% Of Sub-Total
2010	Plumbing Fixtures	Toilet and service fixtures, supply and drainage	1 Fixture/1180 S.F. Floor		Each	2301	1.95	
2020	Domestic Water Distribution	Gas fired water heater			S.F. Floor	.42	.42	3.6%
2040	Rain Water Drainage	Roof drains			S.F. Roof	1.42	1.42	

D30 HVAC

					Unit	Unit Cost	Cost Per S.F.	% Of Sub-Total
3010	Energy Supply	N/A			–	–	–	
3020	Heat Generating Systems	Included in D3050			–	–	–	
3030	Cooling Generating Systems	N/A			–	–	–	8.4%
3050	Terminal & Package Units	Single zone, gas heating, electric cooling			S.F. Floor	8.86	8.86	
3090	Other HVAC Sys. & Equipment	N/A			–	–	–	

D40 Fire Protection

					Unit	Unit Cost	Cost Per S.F.	% Of Sub-Total
4010	Sprinklers	Wet pipe sprinkler system			S.F. Floor	4.73	4.73	5.4%
4020	Standpipes	Standpipe			S.F. Floor	.93	.93	

D50 Electrical

					Unit	Unit Cost	Cost Per S.F.	% Of Sub-Total
5010	Electrical Service/Distribution	600 ampere service, panel board and feeders			S.F. Floor	2.46	2.46	
5020	Lighting & Branch Wiring	T-8 fluorescent fixtures, receptacles, switches, A.C. and misc. power			S.F. Floor	9.09	9.09	15.5%
5030	Communications & Security	Addressable alarm systems, internet wiring and emergency lighting			S.F. Floor	4.65	4.65	
5090	Other Electrical Systems	Emergency generator, 15 kW			S.F. Floor	.12	.12	

E. EQUIPMENT & FURNISHINGS

					Unit	Unit Cost	Cost Per S.F.	% Of Sub-Total
1010	Commercial Equipment	N/A			–	–	–	
1020	Institutional Equipment	N/A			–	–	–	
1030	Vehicular Equipment	N/A			–	–	–	0.0%
1090	Other Equipment	N/A			–	–	–	

F. SPECIAL CONSTRUCTION

					Unit	Unit Cost	Cost Per S.F.	% Of Sub-Total
1020	Integrated Construction	N/A			–	–	–	0.0%
1040	Special Facilities	N/A			–	–	–	

G. BUILDING SITEWORK   N/A

	Sub-Total	105.03	100%
CONTRACTOR FEES (General Requirements: 10%, Overhead: 5%, Profit: 10%)	25%	26.25	
ARCHITECT FEES	9%	11.82	

Total Building Cost	**143.10**

For customer support on your Square Foot Cost Data, call 877.694.8329.

Costs per square foot of floor area

Exterior Wall	S.F. Area	5000	10000	15000	21000	25000	30000	40000	50000	60000
	L.F. Perimeter	287	400	500	600	660	700	834	900	1000
Face Brick with Concrete Block Back-up	Steel Frame	246.80	211.30	198.15	189.15	185.10	179.80	175.30	170.45	168.10
	Bearing Walls	244.65	208.65	195.30	186.20	182.05	176.70	172.10	167.15	164.80
Concrete Block	Steel Frame	212.00	187.00	177.90	171.80	169.10	165.65	162.65	159.55	157.95
Brick Veneer	Steel Frame	231.85	200.90	189.45	181.70	178.25	173.70	169.85	165.75	163.75
Galvanized Steel Siding	Steel Frame	201.85	180.45	172.60	167.45	165.20	162.40	159.85	157.35	156.05
Metal Sandwich Panel	Steel Frame	254.50	216.70	202.60	193.00	188.65	182.95	178.10	172.90	170.30
Perimeter Adj., Add or Deduct	Per 100 L.F.	30.85	15.45	10.25	7.35	6.20	5.15	3.80	3.05	2.55
Story Hgt. Adj., Add or Deduct	Per 1 Ft.	6.35	4.40	3.65	3.15	2.90	2.55	2.30	2.00	1.85
For Basement, add $29.95 per square foot of basement area										

The above costs were calculated using the basic specifications shown on the facing page. These costs should be adjusted where necessary for design alternatives and owner's requirements. Reported completed project costs, for this type of structure, range from $96.00 to $299.35 per S.F.

Common additives

Description	Unit	$ Cost	Description	Unit	$ Cost
Bar, Front Bar	L.F.	425	Lockers, Steel, single tier, 60" or 72"	Opng.	221 - 370
Back Bar	L.F.	340	2 tier, 60" or 72" total	Opng.	120 - 165
Booth, Upholstered, custom straight	L.F.	250 - 460	5 tier, box lockers	Opng.	77 - 86.50
"L" or "U" shaped	L.F.	259 - 440	Locker bench, lam. maple top only	L.F.	34
Bleachers, Telescoping, manual			Pedestals, steel pipe	Ea.	76
To 15 tier	Seat	137 - 190	Sauna, Prefabricated, complete		
21-30 tier	Seat	299 - 390	6' x 4'	Ea.	6575
Courts			6' x 9'	Ea.	10,000
Ceiling	Court	9875	8' x 8'	Ea.	11,500
Floor	Court	16,500	8' x 10'	Ea.	12,800
Walls	Court	28,400	10' x 12'	Ea.	15,500
Emergency Lighting, 25 watt, battery operated			Sound System		
Lead battery	Ea.	330	Amplifier, 250 watts	Ea.	2050
Nickel cadmium	Ea.	780	Speaker, ceiling or wall	Ea.	213
Kitchen Equipment			Trumpet	Ea.	410
Broiler	Ea.	4050	Steam Bath, Complete, to 140 C.F.	Ea.	3000
Cooler, 6 ft. long, reach-in	Ea.	4175	To 300 C.F.	Ea.	3325
Dishwasher, 10-12 racks per hr.	Ea.	4275	To 800 C.F.	Ea.	7100
Food warmer, counter 1.2 KW	Ea.	735	To 2500 C.F.	Ea.	9300
Freezer, reach-in, 44 C.F.	Ea.	5175			
Ice cube maker, 50 lb. per day	Ea.	2000			

Important: See the Reference Section for Location Factors.

Model costs calculated for a 2 story building with 12' story height and 30,000 square feet of floor area

					Unit	Unit Cost	Cost Per S.F.	% Of Sub-Total

A. SUBSTRUCTURE

				Unit	Unit Cost	Cost Per S.F.	% Of Sub-Total
1010	Standard Foundations	Poured concrete; strip and spread footings		S.F. Ground	2.24	1.12	
1020	Special Foundations	N/A		—	—	—	
1030	Slab on Grade	5" reinforced concrete with vapor barrier and granular base		S.F. Slab	7.13	3.57	5.1%
2010	Basement Excavation	Site preparation for slab and trench for foundation wall and footing		S.F. Ground	.32	.21	
2020	Basement Walls	4' foundation wall		L.F. Wall	84	1.95	

B. SHELL

B10 Superstructure

				Unit	Unit Cost	Cost Per S.F.	% Of Sub-Total
1010	Floor Construction	Open web steel joists, slab form, concrete, columns	50% of area	S.F. Floor	18.88	9.44	10.1%
1020	Roof Construction	Metal deck on open web steel joists, columns		S.F. Roof	7.90	3.95	

B20 Exterior Enclosure

				Unit	Unit Cost	Cost Per S.F.	% Of Sub-Total
2010	Exterior Walls	Face brick with concrete block backup	95% of wall	S.F. Wall	32.95	17.53	
2020	Exterior Windows	Storefront	5% of wall	S.F. Window	102	2.85	15.6%
2030	Exterior Doors	Aluminum and glass and hollow metal		Each	3106	.42	

B30 Roofing

				Unit	Unit Cost	Cost Per S.F.	% Of Sub-Total
3010	Roof Coverings	Built-up tar and gravel with flashing; perlite/EPS composite insulation		S.F. Roof	6.88	3.44	2.6%
3020	Roof Openings	Roof hatches		S.F. Roof	.12	.06	

C. INTERIORS

				Unit	Unit Cost	Cost Per S.F.	% Of Sub-Total
1010	Partitions	Concrete block, gypsum board on metal studs	25 S.F. Floor/L.F. Partition	S.F. Partition	14.58	5.83	
1020	Interior Doors	Single leaf hollow metal	810 S.F. Floor/Door	Each	1157	1.43	
1030	Fittings	Toilet partitions		S.F. Floor	.39	.39	
2010	Stair Construction	Concrete filled metal pan		Flight	17,300	1.73	13.4%
3010	Wall Finishes	Paint		S.F. Surface	1.46	1.17	
3020	Floor Finishes	80% carpet tile, 20% ceramic tile	50% of floor area	S.F. Floor	6.26	3.13	
3030	Ceiling Finishes	Mineral fiber tile on concealed zee bars	60% of area	S.F. Ceiling	7.03	4.22	

D. SERVICES

D10 Conveying

				Unit	Unit Cost	Cost Per S.F.	% Of Sub-Total
1010	Elevators & Lifts	N/A		—	—	—	0.0 %
1020	Escalators & Moving Walks	N/A		—	—	—	

D20 Plumbing

				Unit	Unit Cost	Cost Per S.F.	% Of Sub-Total
2010	Plumbing Fixtures	Kitchen, bathroom and service fixtures, supply and drainage	1 Fixture/1000 S.F. Floor	Each	3830	3.83	
2020	Domestic Water Distribution	Gas fired water heater		S.F. Floor	3.19	3.19	5.6%
2040	Rain Water Drainage	Roof drains		S.F. Roof	.78	.39	

D30 HVAC

				Unit	Unit Cost	Cost Per S.F.	% Of Sub-Total
3010	Energy Supply	N/A		—	—	—	
3020	Heat Generating Systems	Included in D3050		—	—	—	
3030	Cooling Generating Systems	N/A		—	—	—	22.3 %
3050	Terminal & Package Units	Multizone unit, gas heating, electric cooling		S.F. Floor	29.70	29.70	
3090	Other HVAC Sys. & Equipment	N/A		—	—	—	

D40 Fire Protection

				Unit	Unit Cost	Cost Per S.F.	% Of Sub-Total
4010	Sprinklers	Sprinklers, light hazard		S.F. Floor	3.43	3.43	2.9%
4020	Standpipes	Standpipe		S.F. Floor	.37	.37	

D50 Electrical

				Unit	Unit Cost	Cost Per S.F.	% Of Sub-Total
5010	Electrical Service/Distribution	400 ampere service, panel board and feeders		S.F. Floor	.90	.90	
5020	Lighting & Branch Wiring	Fluorescent and high intensity discharge fixtures, receptacles, switches, A.C. and misc. power		S.F. Floor	8.24	8.24	7.7%
5030	Communications & Security	Addressable alarm systems and emergency lighting		S.F. Floor	1.10	1.10	
5090	Other Electrical Systems	Emergency generator, 15 kW		S.F. Floor	.06	.06	

E. EQUIPMENT & FURNISHINGS

				Unit	Unit Cost	Cost Per S.F.	% Of Sub-Total
1010	Commercial Equipment	N/A		—	—	—	
1020	Institutional Equipment	N/A		—	—	—	14.7 %
1030	Vehicular Equipment	N/A		—	—	—	
1090	Other Equipment	Courts, sauna baths		S.F. Floor	19.54	19.54	

F. SPECIAL CONSTRUCTION

				Unit	Unit Cost	Cost Per S.F.	% Of Sub-Total
1020	Integrated Construction	N/A		—	—	—	0.0 %
1040	Special Facilities	N/A		—	—	—	

G. BUILDING SITEWORK   N/A

	Sub-Total	133.19	**100%**
CONTRACTOR FEES (General Requirements: 10%, Overhead: 5%, Profit: 10%)	25%	33.29	
ARCHITECT FEES	8%	13.32	
Total Building Cost		**179.80**	

For customer support on your Square Foot Cost Data, call 877.694.8329.

Costs per square foot of floor area

Exterior Wall	S.F. Area	5000	6000	7000	8000	9000	10000	11000	12000	13000
	L.F. Perimeter	286	320	353	386	397	425	454	460	486
Face Brick with Concrete Block Back-up	Steel Joists	192.85	188.35	185.00	182.45	178.40	176.70	175.30	172.55	171.50
	Wood Joists	195.15	190.65	187.25	184.70	180.65	178.90	177.50	174.70	173.65
Stucco on Concrete Block	Steel Joists	191.80	187.90	185.05	182.80	179.45	178.00	176.80	174.55	173.65
	Wood Joists	186.05	182.10	179.20	177.00	173.55	172.10	170.90	168.60	167.70
Limestone with Concrete Block Back-up	Steel Joists	215.45	209.90	205.80	202.60	197.55	195.35	193.65	190.10	188.80
	Wood Joists	210.45	204.90	200.70	197.60	192.45	190.25	188.50	184.95	183.60
Perimeter Adj., Add or Deduct	Per 100 L.F.	17.00	14.15	12.15	10.65	9.50	8.55	7.70	7.10	6.55
Story Hgt. Adj., Add or Deduct	Per 1 Ft.	3.00	2.75	2.65	2.55	2.35	2.25	2.15	2.05	1.95
For Basement, add $31.50 per square foot of basement area										

The above costs were calculated using the basic specifications shown on the facing page. These costs should be adjusted where necessary for design alternatives and owner's requirements. Reported completed project costs, for this type of structure, range from $83.85 to $259.60 per S.F.

Common additives

Description	Unit	$ Cost
Carrels Hardwood	Ea.	975 - 2350
Emergency Lighting, 25 watt, battery operated		
Lead battery	Ea.	330
Nickel cadmium	Ea.	780
Flagpoles, Complete		
Aluminum, 20' high	Ea.	1850
40' high	Ea.	4200
70' high	Ea.	11,000
Fiberglass, 23' high	Ea.	1325
39'-5" high	Ea.	3450
59' high	Ea.	6850
Gym Floor, Incl. sleepers and finish, maple	S.F.	14.55
Intercom System, 25 Station capacity		
Master station	Ea.	3250
Intercom outlets	Ea.	189
Handset	Ea.	520

Description	Unit	$ Cost
Lockers, Steel, single tier, 60" to 72"	Opng.	221 - 370
2 tier, 60" to 72" total	Opng.	120 - 165
5 tier, box lockers	Opng.	77 - 86.50
Locker bench, lam. maple top only	L.F.	34
Pedestals, steel pipe	Ea.	76
Seating		
Auditorium chair, all veneer	Ea.	310
Veneer back, padded seat	Ea.	320
Upholstered, spring seat	Ea.	320
Classroom, movable chair & desk	Set	81 - 171
Lecture hall, pedestal type	Ea.	263 - 620
Smoke Detectors		
Ceiling type	Ea.	227
Duct type	Ea.	565
Sound System		
Amplifier, 250 watts	Ea.	2050
Speaker, ceiling or wall	Ea.	213
Trumpet	Ea.	410
Swimming Pools, Complete, gunite	S.F.	97 - 120

Important: See the Reference Section for Location Factors.

Model costs calculated for a 1 story building with 12′ story height and 10,000 square feet of floor area

				Unit	Unit Cost	Cost Per S.F.	% Of Sub-Total
A. SUBSTRUCTURE							
1010	Standard Foundations	Poured concrete; strip and spread footings		S.F. Ground	2.26	2.26	
1020	Special Foundations	N/A		—	—	—	
1030	Slab on Grade	4″ reinforced concrete with vapor barrier and granular base		S.F. Slab	5.44	5.44	8.9%
2010	Basement Excavation	Site preparation for slab and trench for foundation wall and footing		S.F. Ground	.32	.32	
2020	Basement Walls	4′ foundation wall		L.F. Wall	84	3.55	
B. SHELL							
	B10 Superstructure						
1010	Floor Construction	N/A		—	—	—	7.3 %
1020	Roof Construction	Metal deck, open web steel joists, beams, interior columns		S.F. Roof	9.41	9.41	
	B20 Exterior Enclosure						
2010	Exterior Walls	Face brick with concrete block backup	85% of wall	S.F. Wall	27.54	11.94	
2020	Exterior Windows	Window wall	15% of wall	Each	577	6.47	16.2%
2030	Exterior Doors	Double aluminum and glass		Each	6488	2.61	
	B30 Roofing						
3010	Roof Coverings	Built-up tar and gravel with flashing; perlite/EPS composite insulation		S.F. Roof	6.72	6.72	5.2%
3020	Roof Openings	N/A		—	—	—	
C. INTERIORS							
1010	Partitions	Concrete block	8 S.F. Floor/S.F. Partition	S.F. Partition	9.66	12.08	
1020	Interior Doors	Single leaf hollow metal	700 S.F. Floor/Door	Each	1157	1.65	
1030	Fittings	Toilet partitions		S.F. Floor	1.17	1.17	
2010	Stair Construction	N/A		—	—	—	25.0%
3010	Wall Finishes	Paint		S.F. Surface	2.34	5.85	
3020	Floor Finishes	50% vinyl tile, 50% carpet		S.F. Floor	4.59	4.59	
3030	Ceiling Finishes	Mineral fiber tile on concealed zee bars		S.F. Ceiling	7.03	7.03	
D. SERVICES							
	D10 Conveying						
1010	Elevators & Lifts	N/A		—	—	—	0.0 %
1020	Escalators & Moving Walks	N/A		—	—	—	
	D20 Plumbing						
2010	Plumbing Fixtures	Toilet and service fixtures, supply and drainage	1 Fixture/455 S.F. Floor	Each	3886	8.54	
2020	Domestic Water Distribution	Gas fired water heater		S.F. Floor	1.59	1.59	8.7%
2040	Rain Water Drainage	Roof drains		S.F. Roof	1.10	1.10	
	D30 HVAC						
3010	Energy Supply	N/A		—	—	—	
3020	Heat Generating Systems	N/A		—	—	—	
3030	Cooling Generating Systems	N/A		—	—	—	19.6 %
3050	Terminal & Package Units	Multizone HVAC air cooled system		S.F. Floor	25.45	25.45	
3090	Other HVAC Sys. & Equipment	N/A		—	—	—	
	D40 Fire Protection						
4010	Sprinklers	Sprinkler, light hazard		S.F. Floor	3.67	3.67	3.3%
4020	Standpipes	Standpipe		S.F. Floor	.58	.58	
	D50 Electrical						
5010	Electrical Service/Distribution	200 ampere service, panel board and feeders		S.F. Floor	1.12	1.12	
5020	Lighting & Branch Wiring	T-8 fluorescent fixtures, receptacles, switches, A.C. and misc. power		S.F. Floor	2.91	2.91	5.9%
5030	Communications & Security	Addressable alarm systems, partial internet wiring and emergency lighting		S.F. Floor	3.55	3.55	
5090	Other Electrical Systems	Emergency generator, 15 kW		S.F. Floor	.08	.08	
E. EQUIPMENT & FURNISHINGS							
1010	Commercial Equipment	N/A		—	—	—	
1020	Institutional Equipment	N/A		—	—	—	
1030	Vehicular Equipment	N/A		—	—	—	0.0 %
1090	Other Equipment	N/A		—	—	—	
F. SPECIAL CONSTRUCTION							
1020	Integrated Construction	N/A		—	—	—	0.0 %
1040	Special Facilities	N/A		—	—	—	
G. BUILDING SITEWORK	N/A						

Sub-Total		129.68	**100%**
CONTRACTOR FEES (General Requirements: 10%, Overhead: 5%, Profit: 10%)		25%	32.43	
ARCHITECT FEES		9%	14.59	
Total Building Cost			176.70	

For customer support on your Square Foot Cost Data, call 877.694.8329.

193

Costs per square foot of floor area

Exterior Wall	S.F. Area	2000	2800	3500	4200	5000	5800	6500	7200	8000
	L.F. Perimeter	180	212	240	268	300	314	336	344	368
Wood Siding	Wood Frame	247.45	234.60	228.20	223.95	220.55	216.25	214.25	211.50	210.00
Brick Veneer	Wood Frame	259.50	244.65	237.30	232.35	228.40	223.30	220.90	217.65	215.95
Face Brick with Concrete Block Back-up	Wood Joists	268.05	251.85	243.80	238.45	234.15	228.45	225.85	222.20	220.30
	Steel Joists	263.10	246.50	238.20	232.70	228.30	222.45	219.75	215.95	214.10
Stucco on Concrete Block	Wood Joists	253.95	239.95	233.05	228.40	224.70	219.95	217.80	214.70	213.10
	Steel Joists	249.30	234.90	227.70	222.95	219.10	214.25	211.90	208.80	207.10
Perimeter Adj., Add or Deduct	Per 100 L.F.	29.05	20.85	16.65	13.85	11.65	10.05	8.95	8.05	7.30
Story Hgt. Adj., Add or Deduct	Per 1 Ft.	3.10	2.60	2.35	2.20	2.05	1.80	1.75	1.60	1.60
For Basement, add $36.40 per square foot of basement area										

The above costs were calculated using the basic specifications shown on the facing page. These costs should be adjusted where necessary for design alternatives and owner's requirements. Reported completed project costs, for this type of structure, range from $129.95 to $302.85 per S.F.

Common additives

Description	Unit	$ Cost	Description	Unit	$ Cost
Bar, Front Bar	L.F.	425	Fireplace, Brick, not incl. chimney or foundation		
Back bar	L.F.	340	30" x 29" opening	Ea.	3175
Booth, Upholstered, custom straight	L.F.	250 - 460	Chimney, standard brick		
"L" or "U" shaped	L.F.	259 - 440	Single flue, 16" x 20"	V.L.F.	106
Cupola, Stock unit, redwood			20" x 20"	V.L.F.	125
30" square, 37" high, aluminum roof	Ea.	770	2 Flue, 20" x 24"	V.L.F.	154
Copper roof	Ea.	435	20" x 32"	V.L.F.	191
Fiberglass, 5'-0" base, 63" high	Ea.	4000 - 4725	Kitchen Equipment		
6'-0" base, 63" high	Ea.	5200 - 7525	Broiler	Ea.	4050
Decorative Wood Beams, Non load bearing			Coffee urn, twin 6 gallon	Ea.	3000
Rough sawn, 4" x 6"	L.F.	8.75	Cooler, 6 ft. long	Ea.	4175
4" x 8"	L.F.	9.90	Dishwasher, 10-12 racks per hr.	Ea.	4275
4" x 10"	L.F.	11.70	Food warmer, counter, 1.2 KW	Ea.	735
4" x 12"	L.F.	13.30	Freezer, 44 C.F., reach-in	Ea.	5175
8" x 8"	L.F.	14.15	Ice cube maker, 50 lb. per day	Ea.	2000
Emergency Lighting, 25 watt, battery operated			Range with 1 oven	Ea.	2925
Lead battery	Ea.	330	Refrigerators, Prefabricated, walk-in		
Nickel cadmium	Ea.	780	7'-6" high, 6' x 6'	S.F.	202
			10' x 10'	S.F.	159
			12' x 14'	S.F.	142
			12' x 20'	S.F.	125

Important: See the Reference Section for Location Factors.

Model costs calculated for a 1 story building with 12' story height and 5,000 square feet of floor area

Restaurant

					Unit	Unit Cost	Cost Per S.F.	% Of Sub-Total
A. SUBSTRUCTURE								
1010	Standard Foundations	Poured concrete; strip and spread footings			S.F. Ground	3.34	3.34	
1020	Special Foundations	N/A			—	—	—	
1030	Slab on Grade	4" reinforced concrete with vapor barrier and granular base			S.F. Slab	5.44	5.44	8.7%
2010	Basement Excavation	Site preparation for slab and trench for foundation wall and footing			S.F. Ground	.57	.57	
2020	Basement Walls	4' foundation wall			L.F. Wall	84	5.01	
B. SHELL								
B10 Superstructure								
1010	Floor Construction	Wood columns			S.F. Floor	.45	1.13	5.2%
1020	Roof Construction	Plywood on wood truss (pitched)			S.F. Roof	6.64	7.44	
B20 Exterior Enclosure								
2010	Exterior Walls	Cedar siding on wood studs with insulation		70% of wall	S.F. Wall	12.60	6.35	
2020	Exterior Windows	Storefront windows		30% of wall	Each	45	9.72	13.0%
2030	Exterior Doors	Aluminum and glass doors and entrance with transom			Each	5360	5.38	
B30 Roofing								
3010	Roof Coverings	Cedar shingles with flashing (pitched); rigid fiberglass insulation			S.F. Roof	20.66	8.56	5.2%
3020	Roof Openings	Skylights			S.F. Roof	.09	.09	
C. INTERIORS								
1010	Partitions	Gypsum board on wood studs	25 S.F. Floor/L.F. Partition		S.F. Partition	10.70	4.28	
1020	Interior Doors	Hollow core wood	250 S.F. Floor/Door		Each	537	2.15	
1030	Fittings	Toilet partitions			S.F. Floor	.69	.69	
2010	Stair Construction	N/A			—	—	—	15.2%
3010	Wall Finishes	75% paint, 25% ceramic tile			S.F. Surface	2.45	1.96	
3020	Floor Finishes	65% carpet, 35% quarry tile			S.F. Floor	8.93	8.93	
3030	Ceiling Finishes	Mineral fiber tile on concealed zee bars			S.F. Ceiling	7.03	7.03	
D. SERVICES								
D10 Conveying								
1010	Elevators & Lifts	N/A			—	—	—	0.0 %
1020	Escalators & Moving Walks	N/A			—	—	—	
D20 Plumbing								
2010	Plumbing Fixtures	Kitchen, bathroom and service fixtures, supply and drainage	1 Fixture/355 S.F. Floor		Each	3958	11.15	
2020	Domestic Water Distribution	Gas fired water heater			S.F. Floor	8.42	8.42	11.9%
2040	Rain Water Drainage	Roof drains			S.F. Roof			
D30 HVAC								
3010	Energy Supply	N/A			—	—	—	
3020	Heat Generating Systems	Included in D3050			—	—	—	
3030	Cooling Generating Systems	N/A			—	—	—	23.0 %
3050	Terminal & Package Units	Multizone unit, gas heating, electric cooling; kitchen ventilation			S.F. Floor	37.92	37.92	
3090	Other HVAC Sys. & Equipment	N/A			—	—	—	
D40 Fire Protection								
4010	Sprinklers	Sprinklers, light hazard and ordinary hazard kitchen			S.F. Floor	8.98	8.98	6.7%
4020	Standpipes	Standpipe			S.F. Floor	2.06	2.06	
D50 Electrical								
5010	Electrical Service/Distribution	400 ampere service, panel board and feeders			S.F. Floor	4.63	4.63	
5020	Lighting & Branch Wiring	T-8 fluorescent fixtures, receptacles, switches, A.C. and misc. power			S.F. Floor	9.72	9.72	11.1%
5030	Communications & Security	Addressable alarm systems and emergency lighting			S.F. Floor	3.61	3.61	
5090	Other Electrical Systems	Emergency generator, 15 kW			S.F. Floor	.33	.33	
E. EQUIPMENT & FURNISHINGS								
1010	Commercial Equipment	N/A			—	—	—	
1020	Institutional Equipment	N/A			—	—	—	
1030	Vehicular Equipment	N/A			—	—	—	0.0 %
1090	Other Equipment	N/A			—	—	—	
F. SPECIAL CONSTRUCTION								
1020	Integrated Construction	N/A			—	—	—	
1040	Special Facilities	N/A			—	—	—	0.0 %
G. BUILDING SITEWORK	**N/A**							

		Sub-Total	164.89	100%
CONTRACTOR FEES (General Requirements: 10%, Overhead: 5%, Profit: 10%)		25%	41.23	
ARCHITECT FEES		7%	14.43	

	Total Building Cost	**220.55**

Costs per square foot of floor area

Exterior Wall	S.F. Area	2000	2800	3500	4000	5000	5800	6500	7200	8000
	L.F. Perimeter	180	212	240	260	300	314	336	344	368
Face Brick with Concrete Block Back-up	Bearing Walls	245.20	234.25	228.80	226.10	222.25	217.80	215.90	212.90	211.60
	Steel Frame	246.45	235.50	230.10	227.35	223.55	219.05	217.20	214.20	212.85
Concrete Block With Stucco	Bearing Walls	232.85	223.80	219.30	217.05	213.85	210.20	208.65	206.20	205.05
	Steel Frame	233.50	224.55	220.20	217.95	214.80	211.20	209.75	207.30	206.15
Wood Siding	Wood Frame	225.30	217.20	213.20	211.25	208.40	205.15	203.75	201.55	200.55
Brick Veneer	Steel Frame	238.75	229.05	224.25	221.80	218.40	214.40	212.75	210.10	208.90
Perimeter Adj., Add or Deduct	Per 100 L.F.	38.20	27.35	21.85	19.10	15.30	13.15	11.80	10.65	9.55
Story Hgt. Adj., Add or Deduct	Per 1 Ft.	5.00	4.20	3.80	3.60	3.35	2.95	2.90	2.65	2.55
Basement—Not Applicable										

The above costs were calculated using the basic specifications shown on the facing page. These costs should be adjusted where necessary for design alternatives and owner's requirements. Reported completed project costs, for this type of structure, range from $129.70 to $252.85 per S.F.

Common additives

Description	Unit	$ Cost	Description	Unit	$ Cost
Bar, Front Bar	L.F.	425	Refrigerators, Prefabricated, walk-in		
Back bar	L.F.	340	7'-6" High, 6' x 6'	S.F.	202
Booth, Upholstered, custom straight	L.F.	250 - 460	10' x 10'	S.F.	159
"L" or "U" shaped	L.F.	259 - 440	12' x 14'	S.F.	142
Drive-up Window	Ea.	9075 - 13,000	12' x 20'	S.F.	125
Emergency Lighting, 25 watt, battery operated			Serving		
Lead battery	Ea.	330	Counter top (Stainless steel)	L.F.	198
Nickel cadmium	Ea.	780	Base cabinets	L.F.	415 - 670
Kitchen Equipment			Sound System		
Broiler	Ea.	4050	Amplifier, 250 watts	Ea.	2050
Coffee urn, twin 6 gallon	Ea.	3000	Speaker, ceiling or wall	Ea.	213
Cooler, 6 ft. long	Ea.	4175	Trumpet	Ea.	410
Dishwasher, 10-12 racks per hr.	Ea.	4275	Storage		
Food warmer, counter, 1.2 KW	Ea.	735	Shelving	S.F.	9.65
Freezer, 44 C.F., reach-in	Ea.	5175	Washing		
Ice cube maker, 50 lb. per day	Ea.	2000	Stainless steel counter	L.F.	198
Range with 1 oven	Ea.	2925			

Important: See the Reference Section for Location Factors.

Model costs calculated for a 1 story building with 10' story height and 4,000 square feet of floor area

Restaurant, Fast Food

				Unit	Unit Cost	Cost Per S.F.	% Of Sub-Total
A. SUBSTRUCTURE							
1010	Standard Foundations	Poured concrete; strip and spread footings		S.F. Ground	3.03	3.03	
1020	Special Foundations	N/A		—	—	—	
1030	Slab on Grade	4" reinforced concrete with vapor barrier and granular base		S.F. Slab	5.44	5.44	8.6%
2010	Basement Excavation	Site preparation for slab and trench for foundation wall and footing		S.F. Ground	.57	.57	
2020	Basement Walls	4' foundation wall		L.F. Wall	84	5.43	
B. SHELL							
B10 Superstructure							
1010	Floor Construction	N/A		—	—	—	4.7%
1020	Roof Construction	Metal deck on open web steel joists		S.F. Roof	7.88	7.88	
B20 Exterior Enclosure							
2010	Exterior Walls	Face brick with concrete block backup	70% of wall	S.F. Wall	33.45	15.22	
2020	Exterior Windows	Window wall	30% of wall	Each	51	9.93	19.8%
2030	Exterior Doors	Aluminum and glass		Each	3988	7.98	
B30 Roofing							
3010	Roof Coverings	Built-up tar and gravel with flashing; perlite/EPS composite insulation		S.F. Roof	7.16	7.16	4.6%
3020	Roof Openings	Roof hatches		S.F. Roof	.52	.52	
C. INTERIORS							
1010	Partitions	Gypsum board on metal studs	25 S.F. Floor/L.F. Partition	S.F. Partition	6.31	2.27	
1020	Interior Doors	Hollow core wood	1000 S.F. Floor/Door	Each	537	.54	
1030	Fittings	N/A		—	—	—	
2010	Stair Construction	N/A		—	—	—	16.9%
3010	Wall Finishes	Paint		S.F. Surface	2.97	2.14	
3020	Floor Finishes	Quarry tile		S.F. Floor	16.26	16.26	
3030	Ceiling Finishes	Mineral fiber tile on concealed zee bars		S.F. Ceiling	7.03	7.03	
D. SERVICES							
D10 Conveying							
1010	Elevators & Lifts	N/A		—	—	—	0.0%
1020	Escalators & Moving Walks	N/A		—	—	—	
D20 Plumbing							
2010	Plumbing Fixtures	Kitchen, bathroom and service fixtures, supply and drainage	1 Fixture/400 S.F. Floor	Each	4160	10.40	
2020	Domestic Water Distribution	Gas fired water heater		S.F. Floor	6.07	6.07	10.9%
2040	Rain Water Drainage	Roof drains		S.F. Roof	1.82	1.82	
D30 HVAC							
3010	Energy Supply	N/A		—	—	—	
3020	Heat Generating Systems	Included in D3050		—	—	—	
3030	Cooling Generating Systems	N/A		—	—	—	11.3%
3050	Terminal & Package Units	Multizone unit, gas heating, electric cooling; kitchen ventilation		S.F. Floor	18.96	18.96	
3090	Other HVAC Sys. & Equipment	N/A		—	—	—	
D40 Fire Protection							
4010	Sprinklers	Sprinklers, light hazard		S.F. Floor	8.98	8.98	6.7%
4020	Standpipes	Standpipe		S.F. Floor	2.16	2.16	
D50 Electrical							
5010	Electrical Service/Distribution	400 ampere service, panel board and feeders		S.F. Floor	4.81	4.81	
5020	Lighting & Branch Wiring	T-8 fluorescent fixtures, receptacles, switches, A.C. and misc. power		S.F. Floor	10.54	10.54	12.1%
5030	Communications & Security	Addressable alarm systems and emergency lighting		S.F. Floor	4.49	4.49	
5090	Other Electrical Systems	Emergency generator, 7.5 kW		S.F. Floor	.38	.38	
E. EQUIPMENT & FURNISHINGS							
1010	Commercial Equipment	N/A		—	—	—	
1020	Institutional Equipment	N/A		—	—	—	4.5%
1030	Vehicular Equipment	N/A		—	—	—	
1090	Other Equipment	Walk-in refrigerator		S.F. Floor	7.47	7.47	
F. SPECIAL CONSTRUCTION							
1020	Integrated Construction	N/A		—	—	—	0.0%
1040	Special Facilities	N/A		—	—	—	
G. BUILDING SITEWORK	**N/A**						

	Sub-Total	167.48	**100%**
CONTRACTOR FEES (General Requirements: 10%, Overhead: 5%, Profit: 10%)	25%	41.87	
ARCHITECT FEES	8%	16.75	
	Total Building Cost	**226.10**	

For customer support on your Square Foot Cost Data, call 877.694.8329.

197

Costs per square foot of floor area

Exterior Wall	S.F. Area	10000	15000	20000	25000	30000	35000	40000	45000	50000
	L.F. Perimeter	450	500	600	700	740	822	890	920	966
Face Brick with Concrete Block Back-up	Steel Joists	200.45	182.35	176.85	173.55	168.45	166.55	164.65	162.00	160.25
	Lam. Wood Truss	211.80	194.30	189.25	186.20	181.30	179.50	177.65	175.00	173.35
Concrete Block	Steel Frame	195.80	183.90	180.25	178.00	174.70	173.50	172.25	170.55	169.45
	Lam. Wood Truss	186.00	175.10	171.85	169.95	166.90	165.75	164.60	163.00	161.95
Galvanized Steel Siding	Steel Frame	194.55	183.00	179.45	177.20	174.05	172.85	171.65	169.95	168.85
Metal Sandwich Panel	Steel Frame	197.60	185.30	181.50	179.15	175.75	174.45	173.10	171.35	170.20
Perimeter Adj., Add or Deduct	Per 100 L.F.	14.05	9.40	7.00	5.60	4.75	4.05	3.50	3.15	2.85
Story Hgt. Adj., Add or Deduct	Per 1 Ft.	2.25	1.65	1.50	1.35	1.25	1.20	1.10	1.00	0.95
Basement—Not Applicable										

The above costs were calculated using the basic specifications shown on the facing page. These costs should be adjusted where necessary for design alternatives and owner's requirements. Reported completed project costs, for this type of structure, range from $76.70 to $238.70 per S.F.

Common additives

Description	Unit	$ Cost
Bar, Front Bar	L.F.	425
Back bar	L.F.	340
Booth, Upholstered, custom straight	L.F.	250 - 460
"L" or "U" shaped	L.F.	259 - 440
Bleachers, Telescoping, manual		
To 15 tier	Seat	137 - 190
16-20 tier	Seat	281 - 345
21-30 tier	Seat	299 - 390
For power operation, add	Seat	55 - 86
Emergency Lighting, 25 watt, battery operated		
Lead battery	Ea.	330
Nickel cadmium	Ea.	780
Lockers, Steel, single tier, 60" or 72"	Opng.	221 - 370
2 tier, 60" or 72" total	Opng.	120 - 165
5 tier, box lockers	Opng.	77 - 86.50
Locker bench, lam. maple top only	L.F.	34
Pedestals, steel pipe	Ea.	76

Description	Unit	$ Cost
Rink		
Dasher boards & top guard	Ea.	187,500
Mats, rubber	S.F.	22.50
Score Board	Ea.	13,400 - 18,200

Important: See the Reference Section for Location Factors.

Model costs calculated for a 1 story building with 24' story height and 30,000 square feet of floor area

Rink, Hockey/Indoor Soccer

				Unit	Unit Cost	Cost Per S.F.	% Of Sub-Total
A. SUBSTRUCTURE							
1010	Standard Foundations	Poured concrete; strip and spread footings		S.F. Ground	1.66	1.66	
1020	Special Foundations	N/A		—	—	—	
1030	Slab on Grade	6" reinforced concrete with vapor barrier and granular base		S.F. Slab	6.47	6.47	8.1%
2010	Basement Excavation	Site preparation for slab and trench for foundation wall and footing		S.F. Ground	.18	.18	
2020	Basement Walls	4' foundation wall		L.F. Wall	79	2.24	
B. SHELL							
	B10 Superstructure						
1010	Floor Construction	Wide flange beams and columns		S.F. Floor	43.86	30.70	23.5%
1020	Roof Construction	Metal deck on steel joist		S.F. Roof			
	B20 Exterior Enclosure						
2010	Exterior Walls	Concrete block	95% of wall	S.F. Wall	14.99	8.43	
2020	Exterior Windows	Store front	5% of wall	Each	49.20	1.45	8.2%
2030	Exterior Doors	Aluminum and glass, hollow metal, overhead		Each	3264	.87	
	B30 Roofing						
3010	Roof Coverings	Elastomeric neoprene membrane with flashing; perlite/EPS composite insulation		S.F. Roof	5.09	5.09	4.1%
3020	Roof Openings	Roof hatches		S.F. Roof	.24	.24	
C. INTERIORS							
1010	Partitions	Concrete block	140 S.F. Floor/L.F. Partition	S.F. Partition	9.68	.83	
1020	Interior Doors	Hollow metal	2500 S.F. Floor/Door	Each	1157	.46	
1030	Fittings	N/A		—	—	—	
2010	Stair Construction	N/A		—	—	—	6.7%
3010	Wall Finishes	Paint		S.F. Surface	12.89	2.21	
3020	Floor Finishes	80% rubber mat, 20% paint	50% of floor area	S.F. Floor	9	4.50	
3030	Ceiling Finishes	Mineral fiber tile on concealed zee bar	10% of area	S.F. Ceiling	7.03	.71	
D. SERVICES							
	D10 Conveying						
1010	Elevators & Lifts	N/A		—	—	—	0.0 %
1020	Escalators & Moving Walks	N/A		—	—	—	
	D20 Plumbing						
2010	Plumbing Fixtures	Toilet and service fixtures, supply and drainage	1 Fixture/1070 S.F. Floor	Each	6495	6.07	
2020	Domestic Water Distribution	Oil fired water heater		S.F. Floor	3.19	3.19	7.7%
2040	Rain Water Drainage	N/A		—	—	—	
	D30 HVAC						
3010	Energy Supply	Oil fired hot water, unit heaters	10% of area	S.F. Floor	.93	.93	
3020	Heat Generating Systems	N/A		—	—	—	
3030	Cooling Generating Systems	N/A		—	—	—	11.3%
3050	Terminal & Package Units	Single zone, electric cooling	90% of area	S.F. Floor	13.82	13.82	
3090	Other HVAC Sys. & Equipment	N/A		—	—	—	
	D40 Fire Protection						
4010	Sprinklers	N/A		—	—	—	0.5%
4020	Standpipes	Standpipe		S.F. Floor	.67	.67	
	D50 Electrical						
5010	Electrical Service/Distribution	600 ampere service, panel board and feeders		S.F. Floor	1.50	1.50	
5020	Lighting & Branch Wiring	High intensity discharge and fluorescent fixtures, receptacles, switches, A.C. and misc. power		S.F. Floor	7.96	7.96	9.0%
5030	Communications & Security	Addressable alarm systems, emergency lighting and public address		S.F. Floor	2.10	2.10	
5090	Other Electrical Systems	Emergency generator		S.F. Floor	.19	.19	
E. EQUIPMENT & FURNISHINGS							
1010	Commercial Equipment	N/A		—	—	—	
1020	Institutional Equipment	N/A		—	—	—	
1030	Vehicular Equipment	N/A		—	—	—	0.0 %
1090	Other Equipment	N/A		—	—	—	
F. SPECIAL CONSTRUCTION							
1020	Integrated Construction	N/A		—	—	—	
1040	Special Facilities	Dasher boards and rink (including ice making system)		S.F. Floor	27.33	27.33	20.9 %
G. BUILDING SITEWORK	**N/A**						

						Sub-Total	130.64	**100%**
	CONTRACTOR FEES (General Requirements: 10%, Overhead: 5%, Profit: 10%)					25%	32.63	
	ARCHITECT FEES					7%	11.43	

| | | | | | | **Total Building Cost** | **174.70** | |

For customer support on your Square Foot Cost Data, call 877.694.8329.

199

Costs per square foot of floor area

Exterior Wall	S.F. Area	25000	30000	35000	40000	45000	50000	55000	60000	65000
	L.F. Perimeter	900	1050	1200	1350	1510	1650	1800	1970	2100
Face Brick with Concrete Block Back-up	Steel Frame	182.35	180.90	179.90	179.15	178.70	178.05	177.65	177.60	177.00
	Bearing Walls	179.25	177.70	176.70	175.90	175.50	174.80	174.35	174.35	173.70
Stucco on Concrete Block	Steel Frame	175.00	173.80	172.85	172.25	171.85	171.30	170.95	170.90	170.40
	Bearing Walls	172.15	170.90	169.95	169.30	168.90	168.35	167.95	167.90	167.45
Decorative Concrete Block	Steel Frame	175.25	173.90	173.00	172.40	172.05	171.45	171.05	171.00	170.55
	Bearing Walls	175.90	174.45	173.35	172.60	172.20	171.50	171.10	171.05	170.50
Perimeter Adj., Add or Deduct	Per 100 L.F.	4.05	3.35	2.85	2.55	2.25	2.00	1.80	1.75	1.55
Story Hgt. Adj., Add or Deduct	Per 1 Ft.	1.85	1.85	1.75	1.70	1.75	1.65	1.60	1.65	1.70

For Basement, add $ 27.00 per square foot of basement area

The above costs were calculated using the basic specifications shown on the facing page. These costs should be adjusted where necessary for design alternatives and owner's requirements. Reported completed project costs, for this type of structure, range from $ 89.60 to $ 227.95 per S.F.

Common additives

Description	Unit	$ Cost
Bleachers, Telescoping, manual		
To 15 tier	Seat	137 - 190
16-20 tier	Seat	281 - 345
21-30 tier	Seat	299 - 390
For power operation, add	Seat	55 - 86
Carrels Hardwood	Ea.	975 - 2350
Clock System		
20 room	Ea.	19,900
50 room	Ea.	46,800
Emergency Lighting, 25 watt, battery operated		
Lead battery	Ea.	330
Nickel cadmium	Ea.	780
Flagpoles, Complete		
Aluminum, 20' high	Ea.	1850
40' high	Ea.	4200
Fiberglass, 23' high	Ea.	1325
39'-5" high	Ea.	3450
Kitchen Equipment		
Broiler	Ea.	4050
Cooler, 6 ft. long, reach-in	Ea.	4175

Description	Unit	$ Cost
Kitchen Equipment, cont.		
Dishwasher, 10-12 racks per hr.	Ea.	4275
Food warmer, counter, 1.2 KW	Ea.	735
Freezer, 44 C.F., reach-in	Ea.	5175
Ice cube maker, 50 lb. per day	Ea.	2000
Range with 1 oven	Ea.	2925
Lockers, Steel, single tier, 60" to 72"	Opng.	221 - 370
2 tier, 60" to 72" total	Opng.	120 - 165
5 tier, box lockers	Opng.	77 - 86.50
Locker bench, lam. maple top only	L.F.	34
Pedestals, steel pipe	Ea.	76
Seating		
Auditorium chair, all veneer	Ea.	310
Veneer back, padded seat	Ea.	320
Upholstered, spring seat	Ea.	320
Classroom, movable chair & desk	Set	81 - 171
Lecture hall, pedestal type	Ea.	263 - 620
Sound System		
Amplifier, 250 watts	Ea.	2050
Speaker, ceiling or wall	Ea.	213
Trumpet	Ea.	410

Important: See the Reference Section for Location Factors.

Model costs calculated for a 1 story building with 15' story height and 45,000 square feet of floor area

School, Elementary

				Unit	Unit Cost	Cost Per S.F.	% Of Sub-Total
A. SUBSTRUCTURE							
1010	Standard Foundations	Poured concrete; strip and spread footings		S.F. Ground	2.36	2.36	
1020	Special Foundations	N/A		—	—	—	
1030	Slab on Grade	4" reinforced concrete with vapor barrier and granular base		S.F. Slab	5.44	5.44	8.5%
2010	Basement Excavation	Site preparation for slab and trench for foundation wall and footing		S.F. Ground	.18	.23	
2020	Basement Walls	4' foundation wall		L.F. Wall	84	3.09	
B. SHELL							
	B10 Superstructure						
1010	Floor Construction	N/A		—	—	—	6.2%
1020	Roof Construction	Metal deck on open web steel joists		S.F. Roof	8.08	8.08	
	B20 Exterior Enclosure						
2010	Exterior Walls	Face brick with concrete block backup	70% of wall	S.F. Wall	33.43	11.78	
2020	Exterior Windows	Steel outward projecting	25% of wall	Each	806	6.13	14.2%
2030	Exterior Doors	Metal and glass	5% of wall	Each	3850	.69	
	B30 Roofing						
3010	Roof Coverings	Single-ply membrane with flashing; polyisocyanurate insulation		S.F. Roof	7.60	7.60	5.8%
3020	Roof Openings	N/A		—	—	—	
C. INTERIORS							
1010	Partitions	Concrete block	20 S.F. Floor/L.F. Partition	S.F. Partition	10.54	5.27	
1020	Interior Doors	Single leaf kalamein fire doors	700 S.F. Floor/Door	Each	1157	1.65	
1030	Fittings	Toilet partitions		S.F. Floor	1.84	1.84	
2010	Stair Construction	N/A		—	—	—	20.4%
3010	Wall Finishes	75% paint, 15% glazed coating, 10% ceramic tile		S.F. Surface	4.78	4.78	
3020	Floor Finishes	65% vinyl composition tile, 25% carpet, 10% terrazzo		S.F. Floor	6.16	6.16	
3030	Ceiling Finishes	Mineral fiber tile on concealed zee bars		S.F. Ceiling	7.03	7.03	
D. SERVICES							
	D10 Conveying						
1010	Elevators & Lifts	N/A		—	—	—	0.0%
1020	Escalators & Moving Walks	N/A		—	—	—	
	D20 Plumbing						
2010	Plumbing Fixtures	Kitchen, bathroom and service fixtures, supply and drainage	1 Fixture/625 S.F. Floor	Each	8350	13.36	
2020	Domestic Water Distribution	Gas fired water heater		S.F. Floor	.62	.62	11.7%
2040	Rain Water Drainage	Roof drains		S.F. Roof	1.32	1.32	
	D30 HVAC						
3010	Energy Supply	Oil fired hot water, wall fin radiation		S.F. Floor	10.74	10.74	
3020	Heat Generating Systems	N/A		—	—	—	
3030	Cooling Generating Systems	N/A		—	—	—	18.6%
3050	Terminal & Package Units	Split systems with air cooled condensing units		S.F. Floor	13.70	13.70	
3090	Other HVAC Sys. & Equipment	N/A		—	—	—	
	D40 Fire Protection						
4010	Sprinklers	Sprinklers, light hazard		S.F. Floor	2.94	2.94	2.5%
4020	Standpipes	Standpipe		S.F. Floor	.40	.40	
	D50 Electrical						
5010	Electrical Service/Distribution	800 ampere service, panel board and feeders		S.F. Floor	1.09	1.09	
5020	Lighting & Branch Wiring	High efficiency fluorescent fixtures, receptacles, switches, A.C. and misc. power		S.F. Floor	10.87	10.87	12.1%
5030	Communications & Security	Addressable alarm systems, internet wiring, communications systems and emergency lighting		S.F. Floor	3.79	3.79	
5090	Other Electrical Systems	Emergency generator, 15 kW		S.F. Floor	.07	.07	
E. EQUIPMENT & FURNISHINGS							
1010	Commercial Equipment	N/A		—	—	—	
1020	Institutional Equipment	Chalkboards		S.F. Floor	.18	.18	0.1%
1030	Vehicular Equipment	N/A		—	—	—	
1090	Other Equipment	N/A		—	—	—	
F. SPECIAL CONSTRUCTION							
1020	Integrated Construction	N/A		—	—	—	0.0%
1040	Special Facilities	N/A		—	—	—	
G. BUILDING SITEWORK	N/A						

		Sub-Total		131.21	100%
CONTRACTOR FEES (General Requirements: 10%, Overhead: 5%, Profit: 10%)			25%	32.81	
ARCHITECT FEES			7%	11.48	
		Total Building Cost		**175.50**	

For customer support on your Square Foot Cost Data, call 877.694.8329.

201

Costs per square foot of floor area

Exterior Wall	S.F. Area	50000	70000	90000	110000	130000	150000	170000	190000	210000
	L.F. Perimeter	850	1140	1420	1700	1980	2280	2560	2840	3120
Face Brick with Concrete Block Back-up	Steel Frame	176.75	174.10	172.35	171.25	170.45	170.20	169.80	169.45	169.15
	R/Conc. Frame	182.15	178.40	176.10	174.60	173.55	173.15	172.50	172.10	171.65
Decorative Concrete Block	Steel Frame	172.50	169.05	167.00	165.60	164.65	164.20	163.65	163.25	162.90
	R/Conc. Frame	173.05	169.55	167.55	166.15	165.15	164.75	164.20	163.80	163.40
Limestone with Concrete Block Back-up	Steel Frame	190.95	186.70	184.05	182.35	181.15	180.70	179.95	179.45	178.95
	R/Conc. Frame	198.50	194.25	191.55	189.85	188.65	188.25	187.50	186.95	186.50
Perimeter Adj., Add or Deduct	Per 100 L.F.	4.25	3.00	2.35	1.95	1.70	1.45	1.20	1.10	1.00
Story Hgt. Adj., Add or Deduct	Per 1 Ft.	2.10	2.00	2.00	1.95	1.95	1.95	1.90	1.90	1.90

For Basement, add $35.25 per square foot of basement area

The above costs were calculated using the basic specifications shown on the facing page. These costs should be adjusted where necessary for design alternatives and owner's requirements. Reported completed project costs, for this type of structure, range from $104.15 to $245.35 per S.F.

Common additives

Description	Unit	$ Cost
Bleachers, Telescoping, manual		
To 15 tier	Seat	137 - 190
16-20 tier	Seat	281 - 345
21-30 tier	Seat	299 - 390
For power operation, add	Seat	55 - 86
Carrels Hardwood	Ea.	975 - 2350
Clock System		
20 room	Ea.	19,900
50 room	Ea.	46,800
Elevators, Hydraulic passenger, 2 stops		
1500# capacity	Ea.	65,700
2500# capacity	Ea.	68,800
Emergency Lighting, 25 watt, battery operated		
Lead battery	Ea.	330
Nickel cadmium	Ea.	780
Flagpoles, Complete		
Aluminum, 20' high	Ea.	1850
40' high	Ea.	4200
Fiberglass, 23' high	Ea.	1325
39'-5" high	Ea.	3450

Description	Unit	$ Cost
Kitchen Equipment		
Broiler	Ea.	4050
Cooler, 6 ft. long, reach-in	Ea.	4175
Dishwasher, 10-12 racks per hr.	Ea.	4275
Food warmer, counter, 1.2 KW	Ea.	735
Freezer, 44 C.F., reach-in	Ea.	5175
Lockers, Steel, single tier, 60" or 72"	Opng.	221 - 370
2 tier, 60" or 72" total	Opng.	120 - 165
5 tier, box lockers	Opng.	77 - 86.50
Locker bench, lam. maple top only	L.F.	34
Pedestals, steel pipe	Ea.	76
Seating		
Auditorium chair, all veneer	Ea.	310
Veneer back, padded seat	Ea.	320
Upholstered, spring seat	Ea.	320
Classroom, movable chair & desk	Set	81 - 171
Lecture hall, pedestal type	Ea.	263 - 620
Sound System		
Amplifier, 250 watts	Ea.	2050
Speaker, ceiling or wall	Ea.	213
Trumpet	Ea.	410

Model costs calculated for a 2 story building with 15' story height and 130,000 square feet of floor area

School, High, 2-3 Story

				Unit	Unit Cost	Cost Per S.F.	% Of Sub-Total
A. SUBSTRUCTURE							
1010	Standard Foundations	Poured concrete; strip and spread footings		S.F. Ground	2.62	1.31	
1020	Special Foundations	N/A		—	—	—	
1030	Slab on Grade	4" reinforced concrete with vapor barrier and granular base		S.F. Slab	5.44	2.72	4.2%
2010	Basement Excavation	Site preparation for slab and trench for foundation wall and footing		S.F. Ground	.18	.12	
2020	Basement Walls	4' foundation wall		L.F. Wall	84	1.27	
B. SHELL							
B10 Superstructure							
1010	Floor Construction	Concrete slab without drop panel, concrete columns		S.F. Floor	23.32	11.66	15.0%
1020	Roof Construction	Concrete slab without drop panel		S.F. Roof	15.50	7.75	
B20 Exterior Enclosure							
2010	Exterior Walls	Face brick with concrete block backup	75% of wall	S.F. Wall	33.44	11.46	
2020	Exterior Windows	Window wall	25% of wall	Each	74	8.49	15.8%
2030	Exterior Doors	Metal and glass		Each	2256	.51	
B30 Roofing							
3010	Roof Coverings	Single-ply membrane and standing seam metal; polyisocyanurate insulation		S.F. Roof	11.68	5.84	4.5%
3020	Roof Openings	Roof hatches		S.F. Roof	.10	.05	
C. INTERIORS							
1010	Partitions	Concrete block	25 S.F. Floor/L.F. Partition	S.F. Partition	13.04	6.26	
1020	Interior Doors	Single leaf kalamein fire doors	700 S.F. Floor/Door	Each	1157	1.65	
1030	Fittings	Toilet partitions, chalkboards		S.F. Floor	1.30	1.30	
2010	Stair Construction	Concrete filled metal pan		Flight	11,600	.54	20.2%
3010	Wall Finishes	75% paint, 15% glazed coating, 10% ceramic tile		S.F. Surface	3.58	3.44	
3020	Floor Finishes	70% vinyl composition tile, 20% carpet, 10% terrazzo		S.F. Floor	5.97	5.97	
3030	Ceiling Finishes	Mineral fiber tile on concealed zee bars		S.F. Ceiling	7.03	7.03	
D. SERVICES							
D10 Conveying							
1010	Elevators & Lifts	One hydraulic passenger elevator		Each	83,200	.64	0.5%
1020	Escalators & Moving Walks	N/A		—	—	—	
D20 Plumbing							
2010	Plumbing Fixtures	Kitchen, bathroom and service fixtures, supply and drainage	1 Fixture/860 S.F. Floor	Each	4343	5.05	
2020	Domestic Water Distribution	Gas fired water heater		S.F. Floor	1.09	1.09	5.3%
2040	Rain Water Drainage	Roof drains		S.F. Roof	1.52	.76	
D30 HVAC							
3010	Energy Supply	Oil fired hot water, wall fin radiation		S.F. Floor	5.28	5.28	
3020	Heat Generating Systems	N/A		—	—	—	
3030	Cooling Generating Systems	Chilled water, cooling tower systems		S.F. Floor	18.05	18.05	18.0%
3050	Terminal & Package Units	N/A		—	—	—	
3090	Other HVAC Sys. & Equipment	N/A		—	—	—	
D40 Fire Protection							
4010	Sprinklers	Sprinklers, light hazard		S.F. Floor	2.57	2.57	2.2%
4020	Standpipes	Standpipe, wet, Class III		S.F. Floor	.34	.34	
D50 Electrical							
5010	Electrical Service/Distribution	2000 ampere service, panel board and feeders		S.F. Floor	1.07	1.07	
5020	Lighting & Branch Wiring	High efficiency fluorescent fixtures, receptacles, switches, A.C. and misc. power		S.F. Floor	10.09	10.09	12.1%
5030	Communications & Security	Addressable alarm systems, internet wiring, communications systems and emergency lighting		S.F. Floor	4.09	4.09	
5090	Other Electrical Systems	Emergency generator, 250 kW		S.F. Floor	.44	.44	
E. EQUIPMENT & FURNISHINGS							
1010	Commercial Equipment	N/A		—	—	—	
1020	Institutional Equipment	Laboratory casework and counters		S.F. Floor	1.97	1.97	2.3 %
1030	Vehicular Equipment	N/A		—	—	—	
1090	Other Equipment	Built-in athletic equipment, bleachers and scoreboard		S.F. Floor	.96	.96	
F. SPECIAL CONSTRUCTION							
1020	Integrated Construction	N/A		—	—	—	0.0 %
1040	Special Facilities	N/A		—	—	—	
G. BUILDING SITEWORK	**N/A**						

Sub-Total		129.77	**100%**
CONTRACTOR FEES (General Requirements: 10%, Overhead: 5%, Profit: 10%)	25%	32.43	
ARCHITECT FEES	7%	11.35	
Total Building Cost		**173.55**	

For customer support on your Square Foot Cost Data, call 877.694.8329.

203

Costs per square foot of floor area

Exterior Wall	S.F. Area	50000	65000	80000	95000	110000	125000	140000	155000	170000
	L.F. Perimeter	850	1060	1280	1490	1700	1920	2140	2340	2560
Face Brick with Concrete Block Back-up	Steel Frame	181.30	179.05	177.85	176.90	176.15	175.70	175.40	174.90	174.75
	Bearing Walls	173.75	171.55	170.35	169.35	168.60	168.20	167.85	167.35	167.20
Concrete Block Stucco Face	Steel Frame	174.05	172.10	171.05	170.20	169.60	169.20	168.90	168.50	168.35
	Bearing Walls	166.45	164.55	163.50	162.60	162.05	161.65	161.35	160.95	160.75
Decorative Concrete Block	Steel Frame	174.85	172.95	171.80	170.95	170.30	169.95	169.65	169.25	169.05
	Bearing Walls	167.30	165.35	164.25	163.40	162.75	162.30	162.05	161.60	161.45
Perimeter Adj., Add or Deduct	Per 100 L.F.	3.80	2.95	2.45	1.95	1.75	1.60	1.40	1.25	1.10
Story Hgt. Adj., Add or Deduct	Per 1 Ft.	1.90	1.85	1.80	1.70	1.75	1.75	1.70	1.70	1.70

For Basement, add $36.70 per square foot of basement area

The above costs were calculated using the basic specifications shown on the facing page. These costs should be adjusted where necessary for design alternatives and owner's requirements. Reported completed project costs, for this type of structure, range from $98.35 to $231.05 per S.F.

Common additives

Description	Unit	$ Cost		Description	Unit	$ Cost
Bleachers, Telescoping, manual				Kitchen Equipment		
To 15 tier	Seat	137 - 190		Broiler	Ea.	4050
16-20 tier	Seat	281 - 345		Cooler, 6 ft. long, reach-in	Ea.	4175
21-30 tier	Seat	299 - 390		Dishwasher, 10-12 racks per hr.	Ea.	4275
For power operation, add	Seat	55 - 86		Food warmer, counter, 1.2 KW	Ea.	735
Carrels Hardwood	Ea.	975 - 2350		Freezer, 44 C.F., reach-in	Ea.	5175
Clock System				Lockers, Steel, single tier, 60" to 72"	Opng.	221 - 370
20 room	Ea.	19,900		2 tier, 60" to 72" total	Opng.	120 - 165
50 room	Ea.	46,800		5 tier, box lockers	Opng.	77 - 86.50
Elevators, Hydraulic passenger, 2 stops				Locker bench, lam. maple top only	L.F.	34
1500# capacity	Ea.	65,700		Pedestals, steel pipe	Ea.	76
2500# capacity	Ea.	68,800		Seating		
Emergency Lighting, 25 watt, battery operated				Auditorium chair, all veneer	Ea.	310
Lead battery	Ea.	330		Veneer back, padded seat	Ea.	320
Nickel cadmium	Ea.	780		Upholstered, spring seat	Ea.	320
Flagpoles, Complete				Classroom, movable chair & desk	Set	81 - 171
Aluminum, 20' high	Ea.	1850		Lecture hall, pedestal type	Ea.	263 - 620
40' high	Ea.	4200		Sound System		
Fiberglass, 23' high	Ea.	1325		Amplifier, 250 watts	Ea.	2050
39'-5" high	Ea.	3450		Speaker, ceiling or wall	Ea.	213
				Trumpet	Ea.	410

Important: See the Reference Section for Location Factors.

Model costs calculated for a 2 story building with 15' story height and 110,000 square feet of floor area

				Unit	Unit Cost	Cost Per S.F.	% Of Sub-Total
A.	**SUBSTRUCTURE**						
1010	Standard Foundations	Poured concrete; strip and spread footings		S.F. Ground	2.52	1.26	
1020	Special Foundations	N/A		—	—	—	
1030	Slab on Grade	4" reinforced concrete with vapor barrier and granular base		S.F. Slab	5.44	2.72	4.1%
2010	Basement Excavation	Site preparation for slab and trench for foundation wall and footing		S.F. Ground	.18	.12	
2020	Basement Walls	4' foundation wall		L.F. Wall	84	1.29	
B.	**SHELL**						
	B10 Superstructure						
1010	Floor Construction	Open web steel joists, slab form, concrete, columns		S.F. Floor	29.10	14.55	15.6%
1020	Roof Construction	Metal deck, open web steel joists, columns		S.F. Roof	12.04	6.02	
	B20 Exterior Enclosure						
2010	Exterior Walls	Face brick with concrete block backup	75% of wall	S.F. Wall	33.45	11.63	
2020	Exterior Windows	Window wall	25% of wall	Each	56	6.52	14.3%
2030	Exterior Doors	Double aluminum & glass		Each	2377	.64	
	B30 Roofing						
3010	Roof Coverings	Single-ply membrane and standing seam metal; polyisocyanurate insulation		S.F. Roof	11.86	5.93	4.5%
3020	Roof Openings	Roof hatches		S.F. Roof	.04	.02	
C.	**INTERIORS**						
1010	Partitions	Concrete block	20 S.F. Floor/L.F. Partition	S.F. Partition	9.67	5.80	
1020	Interior Doors	Single leaf kalamein fire doors	750 S.F. Floor/Door	Each	1157	1.54	
1030	Fittings	Toilet partitions, chalkboards		S.F. Floor	1.41	1.41	
2010	Stair Construction	Concrete filled metal pan		Flight	11,600	.63	23.1%
3010	Wall Finishes	50% paint, 40% glazed coatings, 10% ceramic tile		S.F. Surface	4.30	5.16	
3020	Floor Finishes	50% vinyl composition tile, 30% carpet, 20% terrrazzo		S.F. Floor	8.85	8.85	
3030	Ceiling Finishes	Mineral fiberboard on concealed zee bars		S.F. Ceiling	7.03	7.03	
D.	**SERVICES**						
	D10 Conveying						
1010	Elevators & Lifts	One hydraulic passenger elevator		Each	83,600	.76	0.6%
1020	Escalators & Moving Walks	N/A		—	—	—	
	D20 Plumbing						
2010	Plumbing Fixtures	Kitchen, toilet and service fixtures, supply and drainage	1 Fixture/1170 S.F. Floor	Each	6318	5.40	
2020	Domestic Water Distribution	Gas fired water heater		S.F. Floor	.64	.64	5.2%
2040	Rain Water Drainage	Roof drains		S.F. Roof	1.54	.77	
	D30 HVAC						
3010	Energy Supply	N/A		—	—	—	
3020	Heat Generating Systems	Included in D3050		—	—	—	
3030	Cooling Generating Systems	N/A		—	—	—	15.7 %
3050	Terminal & Package Units	Multizone unit, gas heating, electric cooling		S.F. Floor	20.70	20.70	
3090	Other HVAC Sys. & Equipment	N/A		—	—	—	
	D40 Fire Protection						
4010	Sprinklers	Sprinklers, light hazard	10% of area	S.F. Floor	2.57	2.57	2.3%
4020	Standpipes	Standpipe, wet, Class III		S.F. Floor	.41	.41	
	D50 Electrical						
5010	Electrical Service/Distribution	1600 ampere service, panel board and feeders		S.F. Floor	1.02	1.02	
5020	Lighting & Branch Wiring	High efficiency fluorescent fixtures, receptacles, switches, A.C. and misc. power		S.F. Floor	10.09	10.09	12.0%
5030	Communications & Security	Addressable alarm systems, internet wiring, communications systems and emergency lighting		S.F. Floor	4.40	4.40	
5090	Other Electrical Systems	Emergency generator, 100 kW		S.F. Floor	.35	.35	
E.	**EQUIPMENT & FURNISHINGS**						
1010	Commercial Equipment	N/A		—	—	—	
1020	Institutional Equipment	Laboratory casework and counters		S.F. Floor	2.33	2.33	2.6 %
1030	Vehicular Equipment	N/A		—	—	—	
1090	Other Equipment	Bleachers, basketball backstops		S.F. Floor	1.15	1.15	
F.	**SPECIAL CONSTRUCTION**						
1020	Integrated Construction	N/A		—	—	—	0.0 %
1040	Special Facilities	N/A		—	—	—	
G.	**BUILDING SITEWORK**	**N/A**					

			Sub-Total	131.71	**100%**
CONTRACTOR FEES (General Requirements: 10%, Overhead: 5%, Profit: 10%)			25%	32.92	
ARCHITECT FEES			7%	11.52	
			Total Building Cost	176.15	

Costs per square foot of floor area

Exterior Wall	S.F. Area	20000	30000	40000	50000	60000	70000	80000	90000	100000
	L.F. Perimeter	440	590	740	900	1050	1200	1360	1510	1660
Face Brick with Concrete Block Back-up	Steel Frame	189.30	182.95	179.90	178.40	177.10	176.10	175.65	175.10	174.60
	Bearing Walls	184.60	178.35	175.20	173.75	172.40	171.45	171.00	170.40	169.95
Decorative Concrete Block	Steel Frame	175.75	170.90	168.55	167.35	166.30	165.60	165.20	164.75	164.40
	Bearing Walls	170.75	165.95	163.55	162.35	161.30	160.60	160.25	159.80	159.40
Steel Siding on Steel Studs	Steel Frame	169.75	165.55	163.45	162.40	161.50	160.90	160.55	160.15	159.90
Metal Sandwich Panel	Steel Frame	170.10	165.85	163.75	162.70	161.85	161.15	160.80	160.45	160.15
Perimeter Adj., Add or Deduct	Per 100 L.F.	9.50	6.45	4.75	3.85	3.15	2.75	2.40	2.10	1.95
Story Hgt. Adj., Add or Deduct	Per 1 Ft.	2.30	2.10	1.95	1.95	1.85	1.85	1.80	1.80	1.75

For Basement, add $35.60 per square foot of basement area

The above costs were calculated using the basic specifications shown on the facing page. These costs should be adjusted where necessary for design alternatives and owner's requirements. Reported completed project costs, for this type of structure, range from $88.25 to $252.15 per S.F.

Common additives

Description	Unit	$ Cost
Carrels Hardwood	Ea.	975 - 2350
Clock System		
20 room	Ea.	19,900
50 room	Ea.	46,800
Directory Boards, Plastic, glass covered		
30" x 20"	Ea.	600
36" x 48"	Ea.	1500
Aluminum, 24" x 18"	Ea.	635
36" x 24"	Ea.	755
48" x 32"	Ea.	1050
48" x 60"	Ea.	2225
Elevators, Hydraulic passenger, 2 stops		
1500# capacity	Ea.	65,700
2500# capacity	Ea.	68,800
3500# capacity	Ea.	73,300
Emergency Lighting, 25 watt, battery operated		
Lead battery	Ea.	330
Nickel cadmium	Ea.	780

Description	Unit	$ Cost
Flagpoles, Complete		
Aluminum, 20' high	Ea.	1850
40' high	Ea.	4200
Fiberglass, 23' high	Ea.	1325
39'-5" high	Ea.	3450
Seating		
Auditorium chair, all veneer	Ea.	310
Veneer back, padded seat	Ea.	320
Upholstered, spring seat	Ea.	320
Classroom, movable chair & desk	Set	81 - 171
Lecture hall, pedestal type	Ea.	263 - 620
Shops & Workroom:		
Benches, metal	Ea.	835
Parts bins 6'-3" high, 3' wide, 12" deep, 72 bins	Ea.	710
Shelving, metal 1' x 3'	S.F.	12.60
Wide span 6' wide x 24" deep	S.F.	9.65
Sound System		
Amplifier, 250 watts	Ea.	2050
Speaker, ceiling or wall	Ea.	213
Trumpet	Ea.	410

Model costs calculated for a 2 story building with 16' story height and 40,000 square feet of floor area

School, Vocational

				Unit	Unit Cost	Cost Per S.F.	% Of Sub-Total
A. SUBSTRUCTURE							
1010	Standard Foundations	Poured concrete; strip and spread footings		S.F. Ground	3	1.50	
1020	Special Foundations	N/A		—	—	—	
1030	Slab on Grade	5" reinforced concrete with vapor barrier and granular base		S.F. Slab	13.42	6.71	7.5%
2010	Basement Excavation	Site preparation for slab and trench for foundation wall and footing		S.F. Ground	.18	.09	
2020	Basement Walls	4' foundation wall		L.F. Wall	84	1.55	
B. SHELL							
	B10 Superstructure						
1010	Floor Construction	Open web steel joists, slab form, concrete, beams, columns		S.F. Floor	21.34	10.67	10.7%
1020	Roof Construction	Metal deck, open web steel joists, beams, columns		S.F. Roof	6.60	3.30	
	B20 Exterior Enclosure						
2010	Exterior Walls	Face brick with concrete block backup	85% of wall	S.F. Wall	33.45	16.83	
2020	Exterior Windows	Tubular aluminum framing with insulated glass	15% of wall	Each	56	4.99	17.1%
2030	Exterior Doors	Metal and glass doors		Each	2901	.58	
	B30 Roofing						
3010	Roof Coverings	Single-ply membrane with polyisocyanurate insulation		S.F. Roof	6.74	3.37	2.7%
3020	Roof Openings	Roof hatches		S.F. Roof	.34	.17	
C. INTERIORS							
1010	Partitions	Concrete block	20 S.F. Floor/L.F. Partition	S.F. Partition	9.67	5.80	
1020	Interior Doors	Single leaf kalamein fire doors	600 S.F. Floor/Door	Each	1157	1.93	
1030	Fittings	Toilet partitions, chalkboards		S.F. Floor	1.53	1.53	
2010	Stair Construction	Concrete filled metal pan		Flight	14,500	2.18	20.8%
3010	Wall Finishes	50% paint, 40% glazed coating, 10% ceramic tile		S.F. Surface	4.73	5.68	
3020	Floor Finishes	70% vinyl composition tile, 20% carpet, 10% terrazzo		S.F. Floor	5.97	5.97	
3030	Ceiling Finishes	Mineral fiber tile on concealed zee bars	60% of area	S.F. Ceiling	7.03	4.22	
D. SERVICES							
	D10 Conveying						
1010	Elevators & Lifts	One hydraulic passenger elevator		Each	83,600	2.09	1.6%
1020	Escalators & Moving Walks	N/A		—	—	—	
	D20 Plumbing						
2010	Plumbing Fixtures	Toilet and service fixtures, supply and drainage	1 Fixture/700 S.F. Floor	Each	3206	4.58	
2020	Domestic Water Distribution	Gas fired water heater		S.F. Floor	1.05	1.05	5.0%
2040	Rain Water Drainage	Roof drains		S.F. Roof	1.94	.97	
	D30 HVAC						
3010	Energy Supply	Oil fired hot water, wall fin radiation		S.F. Floor	10.14	10.14	
3020	Heat Generating Systems	N/A		—	—	—	
3030	Cooling Generating Systems	Chilled water, cooling tower systems		S.F. Floor	15.35	15.35	19.5%
3050	Terminal & Package Units	N/A		—	—	—	
3090	Other HVAC Sys. & Equipment	N/A		—	—	—	
	D40 Fire Protection						
4010	Sprinklers	Sprinklers, light hazard		S.F. Floor	3.19	3.19	3.2%
4020	Standpipes	Standpipe, wet, Class III		S.F. Floor	.96	.96	
	D50 Electrical						
5010	Electrical Service/Distribution	800 ampere service, panel board and feeders		S.F. Floor	1.24	1.24	
5020	Lighting & Branch Wiring	High efficiency fluorescent fixtures, receptacles, switches, A.C. and misc. power		S.F. Floor	10.90	10.90	11.8%
5030	Communications & Security	Addressable alarm systems, internet wiring, communications systems and emergency lighting		S.F. Floor	3.21	3.21	
5090	Other Electrical Systems	Emergency generator, 11.5 kW		S.F. Floor	.11	.11	
E. EQUIPMENT & FURNISHINGS							
1010	Commercial Equipment	N/A		—	—	—	
1020	Institutional Equipment	Stainless steel countertops		S.F. Floor	.15	.15	0.1%
1030	Vehicular Equipment	N/A		—	—	—	
1090	Other Equipment	N/A		—	—	—	
F. SPECIAL CONSTRUCTION							
1020	Integrated Construction	N/A		—	—	—	0.0%
1040	Special Facilities	N/A		—	—	—	
G. BUILDING SITEWORK	**N/A**						

	Sub-Total	131.01	100%
CONTRACTOR FEES (General Requirements: 10%, Overhead: 5%, Profit: 10%)	25%	32.73	
ARCHITECT FEES	7%	11.46	
	Total Building Cost	**175.20**	

For customer support on your Square Foot Cost Data, call 877.694.8329.

207

Costs per square foot of floor area

Exterior Wall	S.F. Area	1000	2000	3000	4000	6000	8000	10000	12000	15000
	L.F. Perimeter	126	179	219	253	310	358	400	438	490
Wood Siding	Wood Frame	163.15	136.95	126.30	120.30	113.50	109.60	106.95	105.05	102.95
Face Brick Veneer	Wood Frame	205.80	172.75	159.00	151.15	142.15	136.95	133.40	130.90	128.10
Stucco on Concrete Block	Steel Frame	182.30	152.00	139.45	132.40	124.20	119.60	116.45	114.15	111.60
	Bearing Walls	177.30	147.00	134.45	127.35	119.20	114.55	111.35	109.10	106.60
Metal Sandwich Panel	Steel Frame	194.20	160.40	146.30	138.30	129.05	123.75	120.10	117.50	114.65
Precast Concrete	Steel Frame	229.70	185.15	166.20	155.30	142.60	135.25	130.25	126.60	122.60
Perimeter Adj., Add or Deduct	Per 100 L.F.	44.75	22.40	14.90	11.25	7.45	5.60	4.50	3.75	3.05
Story Hgt. Adj., Add or Deduct	Per 1 Ft.	3.25	2.35	1.90	1.65	1.30	1.15	1.05	0.95	0.90

For Basement, add $26.85 per square foot of basement area

The above costs were calculated using the basic specifications shown on the facing page. These costs should be adjusted where necessary for design alternatives and owner's requirements. Reported completed project costs, for this type of structure, range from $73.10 to $253.40 per S.F.

Common additives

Description	Unit	$ Cost
Check Out Counter		
Single belt	Ea.	3600
Double belt	Ea.	5125
Emergency Lighting, 25 watt, battery operated		
Lead battery	Ea.	330
Nickel cadmium	Ea.	780
Refrigerators, Prefabricated, walk-in		
7'-6" high, 6' x 6'	S.F.	202
10' x 10'	S.F.	159
12' x 14'	S.F.	142
12' x 20'	S.F.	125
Refrigerated Food Cases		
Dairy, multi deck, 12' long	Ea.	12,800
Delicatessen case, single deck, 12' long	Ea.	8850
Multi deck, 18 S.F. shelf display	Ea.	8300
Freezer, self-contained chest type, 30 C.F.	Ea.	9725
Glass door upright, 78 C.F.	Ea.	11,600

Description	Unit	$ Cost
Refrigerated Food Cases, cont.		
Frozen food, chest type, 12' long	Ea.	9450
Glass door reach-in, 5 door	Ea.	15,900
Island case 12' long, single deck	Ea.	8425
Multi deck	Ea.	9775
Meat cases, 12' long, single deck	Ea.	8425
Multi deck	Ea.	11,800
Produce, 12' long single deck	Ea.	7950
Multi deck	Ea.	10,000
Safe, Office type, 1 hour rating		
30" x 18" x 18"	Ea.	2500
60" x 36" x 18", double door	Ea.	8125
Smoke Detectors		
Ceiling type	Ea.	227
Duct type	Ea.	565
Sound System		
Amplifier, 250 watts	Ea.	2050
Speaker, ceiling or wall	Ea.	213
Trumpet	Ea.	410

Model costs calculated for a 1 story building with 12' story height and 4,000 square feet of floor area

					Unit	Unit Cost	Cost Per S.F.	% Of Sub-Total

A. SUBSTRUCTURE

				Unit	Unit Cost	Cost Per S.F.	% Of Sub-Total
1010	Standard Foundations	Poured concrete; strip and spread footings		S.F. Ground	1.45	1.45	
1020	Special Foundations	N/A		—	—	—	
1030	Slab on Grade	4" reinforced concrete with vapor barrier and granular base		S.F. Slab	5.44	5.44	13.6%
2010	Basement Excavation	Site preparation for slab and trench for foundation wall and footing		S.F. Ground	.57	.57	
2020	Basement Walls	4' foundation wall		L.F. Wall	75	4.75	

B. SHELL

B10 Superstructure

				Unit	Unit Cost	Cost Per S.F.	% Of Sub-Total
1010	Floor Construction	N/A		—	—	—	7.4 %
1020	Roof Construction	Wood truss with plywood sheathing		S.F. Roof	6.65	6.65	

B20 Exterior Enclosure

				Unit	Unit Cost	Cost Per S.F.	% Of Sub-Total
2010	Exterior Walls	Wood siding on wood studs, insulated	80% of wall	S.F. Wall	10.34	6.28	
2020	Exterior Windows	Storefront	20% of wall	Each	45.85	6.95	17.2%
2030	Exterior Doors	Double aluminum and glass, solid core wood		Each	3010	2.27	

B30 Roofing

				Unit	Unit Cost	Cost Per S.F.	% Of Sub-Total
3010	Roof Coverings	Asphalt shingles; rigid fiberglass insulation		S.F. Roof	5.78	5.78	6.4%
3020	Roof Openings	N/A		—	—	—	

C. INTERIORS

				Unit	Unit Cost	Cost Per S.F.	% Of Sub-Total
1010	Partitions	Gypsum board on wood studs	60 S.F. Floor/L.F. Partition	S.F. Partition	13.62	2.27	
1020	Interior Doors	Single leaf wood, hollow metal	1300 S.F. Floor/Door	Each	1247	.96	
1030	Fittings	N/A		—	—	—	
2010	Stair Construction	N/A		—	—	—	14.1%
3010	Wall Finishes	Paint		S.F. Surface	1.20	.40	
3020	Floor Finishes	Vinyl composition tile		S.F. Floor	2.69	2.69	
3030	Ceiling Finishes	Mineral fiber tile on wood furring		S.F. Ceiling	6.40	6.40	

D. SERVICES

D10 Conveying

				Unit	Unit Cost	Cost Per S.F.	% Of Sub-Total
1010	Elevators & Lifts	N/A		—	—	—	0.0 %
1020	Escalators & Moving Walks	N/A		—	—	—	

D20 Plumbing

				Unit	Unit Cost	Cost Per S.F.	% Of Sub-Total
2010	Plumbing Fixtures	Toilet and service fixtures, supply and drainage	1 Fixture/1000 S.F. Floor	Each	3930	3.93	
2020	Domestic Water Distribution	Gas fired water heater		S.F. Floor	1.36	1.36	5.9%
2040	Rain Water Drainage	N/A		—	—	—	

D30 HVAC

				Unit	Unit Cost	Cost Per S.F.	% Of Sub-Total
3010	Energy Supply	N/A		—	—	—	
3020	Heat Generating Systems	Included in D3050		—	—	—	
3030	Cooling Generating Systems	N/A		—	—	—	8.5 %
3050	Terminal & Package Units	Single zone rooftop unit, gas heating, electric cooling		S.F. Floor	7.64	7.64	
3090	Other HVAC Sys. & Equipment	N/A		—	—	—	

D40 Fire Protection

				Unit	Unit Cost	Cost Per S.F.	% Of Sub-Total
4010	Sprinklers	Sprinkler, ordinary hazard		S.F. Floor	4.98	4.98	7.9%
4020	Standpipes	Standpipe		S.F. floor	2.16	2.16	

D50 Electrical

				Unit	Unit Cost	Cost Per S.F.	% Of Sub-Total
5010	Electrical Service/Distribution	200 ampere service, panel board and feeders		S.F. Floor	3.04	3.04	
5020	Lighting & Branch Wiring	High efficiency fluorescent fixtures, receptacles, switches, A.C. and misc. power		S.F. Floor	10.82	10.82	18.9%
5030	Communications & Security	Addressable alarm systems and emergency lighting		S.F. Floor	2.82	2.82	
5090	Other Electrical Systems	Emergency generator, 7.5 kW		S.F. Floor	.34	.34	

E. EQUIPMENT & FURNISHINGS

				Unit	Unit Cost	Cost Per S.F.	% Of Sub-Total
1010	Commercial Equipment	N/A		—	—	—	
1020	Institutional Equipment	N/A		—	—	—	0.0 %
1030	Vehicular Equipment	N/A		—	—	—	
1090	Other Equipment	N/A		—	—	—	

F. SPECIAL CONSTRUCTION

				Unit	Unit Cost	Cost Per S.F.	% Of Sub-Total
1020	Integrated Construction	N/A		—	—	—	0.0 %
1040	Special Facilities	N/A		—	—	—	

G. BUILDING SITEWORK   N/A

	Sub-Total		89.95	**100%**
CONTRACTOR FEES (General Requirements: 10%, Overhead: 5%, Profit: 10%)	25%	22.48	
ARCHITECT FEES	7%	7.87	
	Total Building Cost	**120.30**	

For customer support on your Square Foot Cost Data, call 877.694.8329.

209

Costs per square foot of floor area

Exterior Wall	S.F. Area	50000	65000	80000	95000	110000	125000	140000	155000	170000
	L.F. Perimeter	920	1065	1167	1303	1333	1433	1533	1633	1733
Face Brick with Concrete Block Back-up	R/Conc. Frame	129.70	127.25	125.10	124.00	122.25	121.55	120.90	120.40	120.05
	Steel Frame	114.30	111.80	109.70	108.60	106.85	106.15	105.50	105.00	104.65
Decorative Concrete Block	R/Conc. Frame	124.15	122.30	120.70	119.85	118.60	118.05	117.60	117.25	117.00
	Steel Joists	109.85	107.85	106.15	105.20	103.90	103.35	102.85	102.45	102.20
Precast Concrete Panels	R/Conc. Frame	128.15	125.85	123.95	122.90	121.35	120.75	120.20	119.75	119.45
	Steel Joists	112.15	109.90	108.00	107.00	105.45	104.80	104.25	103.80	103.50
Perimeter Adj., Add or Deduct	Per 100 L.F.	1.95	1.40	1.20	1.05	0.90	0.75	0.75	0.65	0.55
Story Hgt. Adj., Add or Deduct	Per 1 Ft.	1.00	0.80	0.80	0.70	0.65	0.60	0.60	0.60	0.55

For Basement, add $26.45 per square foot of basement area

The above costs were calculated using the basic specifications shown on the facing page. These costs should be adjusted where necessary for design alternatives and owner's requirements. Reported completed project costs, for this type of structure, range from $63.10 to $156.95 per S.F.

Common additives

Description	Unit	$ Cost
Closed Circuit Surveillance, One station		
Camera and monitor	Ea.	1975
For additional camera stations, add	Ea.	1075
Directory Boards, Plastic, glass covered		
30" x 20"	Ea.	600
36" x 48"	Ea.	1500
Aluminum, 24" x 18"	Ea.	635
36" x 24"	Ea.	755
48" x 32"	Ea.	1050
48" x 60"	Ea.	2225
Emergency Lighting, 25 watt, battery operated		
Lead battery	Ea.	330
Nickel cadmium	Ea.	780
Safe, Office type, 1 hour rating		
30" x 18" x 18"	Ea.	2500
60" x 36" x 18", double door	Ea.	8125
Sound System		
Amplifier, 250 watts	Ea.	2050
Speaker, ceiling or wall	Ea.	213
Trumpet	Ea.	410

Important: See the Reference Section for Location Factors.

Model costs calculated for a 1 story building with 14' story height and 110,000 square feet of floor area

Store, Department, 1 Story

					Unit	Unit Cost	Cost Per S.F.	% Of Sub-Total
A. SUBSTRUCTURE								
1010	Standard Foundations	Poured concrete; strip and spread footings			S.F. Ground	.93	.93	
1020	Special Foundations	N/A			—	—	—	
1030	Slab on Grade	4" reinforced concrete with vapor barrier and granular base			S.F. Slab	5.44	5.44	8.3%
2010	Basement Excavation	Site preparation for slab and trench for foundation wall and footing			S.F. Ground	.32	.32	
2020	Basement Walls	4' foundation wall			L.F. Wall	84	1.01	
B. SHELL								
B10 Superstructure								
1010	Floor Construction	Cast-in-place concrete columns			L.F. Column	74	.94	24.8%
1020	Roof Construction	Precast concrete beam and plank, concrete columns			S.F. Roof	21.95	21.95	
B20 Exterior Enclosure								
2010	Exterior Walls	Face brick with concrete block backup	90% of wall		S.F. Wall	33.01	5.04	
2020	Exterior Windows	Storefront	10% of wall		Each	68	1.15	7.2%
2030	Exterior Doors	Sliding electric operated entrance, hollow metal			Each	7733	.42	
B30 Roofing								
3010	Roof Coverings	Built-up tar and gravel with flashing; perlite/EPS composite insulation			S.F. Roof	5.55	5.55	6.1%
3020	Roof Openings	Roof hatches			S.F. Roof	.05	.05	
C. INTERIORS								
1010	Partitions	Gypsum board on metal studs	60 S.F. Floor/L.F. Partition		S.F. Partition	10.68	1.78	
1020	Interior Doors	Single leaf hollow metal	600 S.F. Floor/Door		Each	1157	1.93	
1030	Fittings	N/A			—	—	—	
2010	Stair Construction	N/A			—	—	—	26.4%
3010	Wall Finishes	Paint			S.F. Surface	.84	.28	
3020	Floor Finishes	50% ceramic tile, 50% carpet tile			S.F. Floor	15.60	15.60	
3030	Ceiling Finishes	Mineral fiber board on exposed grid system, suspended			S.F. Ceiling	4.76	4.76	
D. SERVICES								
D10 Conveying								
1010	Elevators & Lifts	N/A			—	—	—	0.0 %
1020	Escalators & Moving Walks	N/A			—	—	—	
D20 Plumbing								
2010	Plumbing Fixtures	Toilet and service fixtures, supply and drainage	1 Fixture/4075 S.F. Floor		Each	7050	1.73	
2020	Domestic Water Distribution	Gas fired water heater			S.F. Floor	.38	.38	3.2%
2040	Rain Water Drainage	Roof drains			S.F. Roof	.84	.84	
D30 HVAC								
3010	Energy Supply	N/A			—	—	—	
3020	Heat Generating Systems	Included in D3050			—	—	—	
3030	Cooling Generating Systems	N/A			—	—	—	8.8 %
3050	Terminal & Package Units	Single zone unit gas heating, electric cooling			S.F. Floor	8.16	8.16	
3090	Other HVAC Sys. & Equipment	N/A			—	—	—	
D40 Fire Protection								
4010	Sprinklers	Sprinklers, light hazard			S.F. Floor	2.94	2.94	3.5%
4020	Standpipes	Standpipe			S.F. Floor	.28	.28	
D50 Electrical								
5010	Electrical Service/Distribution	1200 ampere service, panel board and feeders			S.F. Floor	.76	.76	
5020	Lighting & Branch Wiring	High efficiency fluorescent fixtures, receptacles, switches, A.C. and misc. power			S.F. Floor	8.73	8.73	11.7%
5030	Communications & Security	Addressable alarm systems, internet wiring and emergency lighting			S.F. Floor	1.26	1.26	
5090	Other Electrical Systems	Emergency generator, 7.5 kW			S.F. Floor	.04	.04	
E. EQUIPMENT & FURNISHINGS								
1010	Commercial Equipment	N/A			—	—	—	
1020	Institutional Equipment	N/A			—	—	—	
1030	Vehicular Equipment	N/A			—	—	—	0.0 %
1090	Other Equipment	N/A			—	—	—	
F. SPECIAL CONSTRUCTION								
1020	Integrated Construction	N/A			—	—	—	0.0 %
1040	Special Facilities	N/A			—	—	—	
G. BUILDING SITEWORK	**N/A**							

		Sub-Total	92.27	100%
CONTRACTOR FEES (General Requirements: 10%, Overhead: 5%, Profit: 10%)		25%	23.06	
ARCHITECT FEES		6%	6.92	
	Total Building Cost		**122.25**	

Costs per square foot of floor area

Exterior Wall	S.F. Area	50000	65000	80000	95000	110000	125000	140000	155000	170000
	L.F. Perimeter	533	593	670	715	778	840	871	923	976
Face Brick with Concrete Block Back-up	Steel Frame	160.80	154.70	151.60	148.50	146.65	145.25	143.55	142.60	141.75
	R/Conc. Frame	158.90	152.90	149.70	146.65	144.80	143.40	141.70	140.70	139.95
Face Brick on Steel Studs	Steel Frame	155.65	150.30	147.50	144.85	143.20	142.00	140.50	139.75	139.00
	R/Conc. Frame	156.55	151.25	148.45	145.75	144.15	142.95	141.50	140.65	139.95
Precast Concrete Panels Exposed Aggregate	Steel Frame	165.65	158.90	155.40	151.90	149.90	148.35	146.45	145.35	144.45
	R/Conc. Frame	165.95	159.20	155.70	152.25	150.25	148.70	146.75	145.70	144.80
Perimeter Adj., Add or Deduct	Per 100 L.F.	5.20	4.00	3.20	2.70	2.35	2.10	1.90	1.65	1.55
Story Hgt. Adj., Add or Deduct	Per 1 Ft.	1.55	1.40	1.20	1.05	1.05	1.00	0.95	0.90	0.90
For Basement, add $44.80 per square foot of basement area										

The above costs were calculated using the basic specifications shown on the facing page. These costs should be adjusted where necessary for design alternatives and owner's requirements. Reported completed project costs, for this type of structure, range from $65.20 to $174.45 per S.F.

Common additives

Description	Unit	$ Cost	Description	Unit	$ Cost
Closed Circuit Surveillance, One station			Escalators, Metal		
Camera and monitor	Ea.	1975	32" wide, 10' story height	Ea.	153,500
For additional camera stations, add	Ea.	1075	20' story height	Ea.	187,500
Directory Boards, Plastic, glass covered			48" wide, 10' story height	Ea.	162,000
30" x 20"	Ea.	600	20' story height	Ea.	197,000
36" x 48"	Ea.	1500	Glass		
Aluminum, 24" x 18"	Ea.	635	32" wide, 10' story height	Ea.	145,500
36" x 24"	Ea.	755	20' story height	Ea.	178,500
48" x 32"	Ea.	1050	48" wide, 10' story height	Ea.	153,000
48" x 60"	Ea.	2225	20' story height	Ea.	187,500
Elevators, Hydraulic passenger, 2 stops			Safe, Office type, 1 hour rating		
1500# capacity	Ea.	65,700	30" x 18" x 18"	Ea.	2500
2500# capacity	Ea.	68,800	60" x 36" x 18", double door	Ea.	8125
3500# capacity	Ea.	73,300	Sound System		
Additional stop, add	Ea.	7825	Amplifier, 250 watts	Ea.	2050
Emergency Lighting, 25 watt, battery operated			Speaker, ceiling or wall	Ea.	213
Lead battery	Ea.	330	Trumpet	Ea.	410
Nickel cadmium	Ea.	780			

Important: See the Reference Section for Location Factors.

Model costs calculated for a 3 story building with 16' story height and 95,000 square feet of floor area

Store, Department, 3 Story

				Unit	Unit Cost	Cost Per S.F.	% Of Sub-Total
A. SUBSTRUCTURE							
1010	Standard Foundations	Poured concrete; strip and spread footings		S.F. Ground	3.81	1.27	
1020	Special Foundations	N/A		—	—	—	
1030	Slab on Grade	4" reinforced concrete with vapor barrier and granular base		S.F. Slab	5.44	1.81	3.4%
2010	Basement Excavation	Site preparation for slab and trench for foundation wall and footing		S.F. Ground	.18	.06	
2020	Basement Walls	4' foundation wall		L.F. Wall	79	.67	
B. SHELL							
	B10 Superstructure						
1010	Floor Construction	Concrete slab with metal deck and beams, steel columns		S.F. Floor	30.78	20.52	21.5%
1020	Roof Construction	Metal deck, open web steel joists, beams, columns		S.F. Roof	10.65	3.55	
	B20 Exterior Enclosure						
2010	Exterior Walls	Face brick with concrete block backup	90% of wall	S.F. Wall	32.97	10.72	
2020	Exterior Windows	Storefront	10% of wall	Each	48.75	1.77	12.7%
2030	Exterior Doors	Revolving and sliding panel, mall-front		Each	12,108	1.79	
	B30 Roofing						
3010	Roof Coverings	Built-up tar and gravel with flashing; perlite/EPS composite insulation		S.F. Roof	6.09	2.03	1.9%
3020	Roof Openings	Roof hatches		S.F. Roof	.21	.07	
C. INTERIORS							
1010	Partitions	Gypsum board on metal studs	60 S.F. Floor/L.F. Partition	S.F. Partition	6.36	1.06	
1020	Interior Doors	Single leaf hollow metal	600 S.F. Floor/Door	Each	1157	1.93	
1030	Fittings	N/A					
2010	Stair Construction	Concrete filled metal pan		Flight	20,200	2.13	26.4%
3010	Wall Finishes	70% paint, 20% vinyl wall covering, 10% ceramic tile		S.F. Surface	5.01	1.67	
3020	Floor Finishes	50% carpet tile, 40% marble tile, 10% terrazzo		S.F. Floor	15.75	15.75	
3030	Ceiling Finishes	Mineral fiber tile on concealed zee bars		S.F. Ceiling	7.03	7.03	
D. SERVICES							
	D10 Conveying						
1010	Elevators & Lifts	One hydraulic passenger, one hydraulic freight		Each	303,050	3.19	8.8%
1020	Escalators & Moving Walks	Four escalators		Each	158,888	6.69	
	D20 Plumbing						
2010	Plumbing Fixtures	Toilet and service fixtures, supply and drainage	1 Fixture/2570 S.F. Floor	Each	4832	1.88	
2020	Domestic Water Distribution	Gas fired water heater		S.F. Floor	.45	.45	2.6%
2040	Rain Water Drainage	Roof drains		S.F. Roof	1.65	.55	
	D30 HVAC						
3010	Energy Supply	N/A		—	—	—	
3020	Heat Generating Systems	Included in D3050		—	—	—	
3030	Cooling Generating Systems	N/A		—	—	—	7.3%
3050	Terminal & Package Units	Multizone rooftop unit, gas heating, electric cooling		S.F. Floor	8.16	8.16	
3090	Other HVAC Sys. & Equipment	N/A		—	—	—	
	D40 Fire Protection						
4010	Sprinklers	Sprinklers, light hazard		S.F. Floor	2.45	2.45	3.7%
4020	Standpipes	Standpipe		S.F. Floor	1.69	1.69	
	D50 Electrical						
5010	Electrical Service/Distribution	1200 ampere service, panel board and feeders		S.F. Floor	1.23	1.23	
5020	Lighting & Branch Wiring	High efficiency fluorescent fixtures, receptacles, switches, A.C. and misc. power		S.F. Floor	9.15	9.15	11.8%
5030	Communications & Security	Addressable alarm systems, internet wiring and emergency lighting		S.F. Floor	2.52	2.52	
5090	Other Electrical Systems	Emergency generator, 50 kW		S.F. Floor	.27	.27	
E. EQUIPMENT & FURNISHINGS							
1010	Commercial Equipment	N/A		—	—	—	
1020	Institutional Equipment	N/A		—	—	—	
1030	Vehicular Equipment	N/A		—	—	—	0.0%
1090	Other Equipment	N/A		—	—	—	
F. SPECIAL CONSTRUCTION							
1020	Integrated Construction	N/A		—	—	—	
1040	Special Facilities	N/A		—	—	—	0.0%
G. BUILDING SITEWORK	**N/A**						
				Sub-Total		112.06	**100%**
	CONTRACTOR FEES (General Requirements: 10%, Overhead: 5%, Profit: 10%)				25%	28.04	
	ARCHITECT FEES				6%	8.40	
				Total Building Cost		**148.50**	

For customer support on your Square Foot Cost Data, call 877.694.8329.

Costs per square foot of floor area

Exterior Wall	S.F. Area	4000	6000	8000	10000	12000	15000	18000	20000	22000
	L.F. Perimeter	260	340	360	410	440	490	540	565	594
Split Face Concrete Block	Steel Joists	152.35	139.10	127.40	122.40	117.90	113.75	110.85	109.20	107.95
Stucco on Concrete Block	Steel Joists	152.45	139.20	127.45	122.45	117.95	113.80	110.95	109.25	107.95
EIFS on Metal Studs	Steel Joists	151.80	138.65	127.05	122.05	117.60	113.50	110.65	109.00	107.70
Face Brick on Concrete Block	Steel Joists	168.55	153.25	138.65	132.60	127.05	121.90	118.35	116.25	114.65
Painted Reinforced Concrete	Steel Joists	160.50	146.20	133.00	127.50	122.50	117.85	114.65	112.75	111.30
Tilt-up Concrete Panels	Steel Joists	146.95	134.40	123.65	119.00	114.90	111.05	108.45	106.90	105.70
Perimeter Adj., Add or Deduct	Per 100 L.F.	16.90	11.35	8.50	6.75	5.70	4.50	3.85	3.45	3.10
Story Hgt. Adj., Add or Deduct	Per 1 Ft.	2.10	1.85	1.45	1.30	1.20	1.10	1.05	0.95	0.90
For Basement, add $37.20 per square foot of basement area										

The above costs were calculated using the basic specifications shown on the facing page. These costs should be adjusted where necessary for design alternatives and owner's requirements. Reported completed project costs, for this type of structure, range from $63.60 to $221.15 per S.F.

Common additives

Description	Unit	$ Cost
Emergency Lighting, 25 watt, battery operated		
Lead battery	Ea.	330
Nickel cadmium	Ea.	780
Safe, Office type, 1 hour rating		
30" x 18" x 18"	Ea.	2500
60" x 36" x 18", double door	Ea.	8125
Smoke Detectors		
Ceiling type	Ea.	227
Duct type	Ea.	565
Sound System		
Amplifier, 250 watts	Ea.	2050
Speaker, ceiling or wall	Ea.	213
Trumpet	Ea.	410

Important: See the Reference Section for Location Factors.

Model costs calculated for a 1 story building with 14' story height and 8,000 square feet of floor area

Store, Retail

				Unit	Unit Cost	Cost Per S.F.	% Of Sub-Total
A.	**SUBSTRUCTURE**						
1010	Standard Foundations	Poured concrete; strip and spread footings		S.F. Ground	1.94	1.94	
1020	Special Foundations	N/A		—	—	—	
1030	Slab on Grade	4" reinforced concrete with vapor barrier and granular base		S.F. Slab	5.44	5.44	12.1%
2010	Basement Excavation	Site preparation for slab and trench for foundation wall and footing		S.F. Ground	.32	.32	
2020	Basement Walls	4' foundation wall		L.F. Wall	84	3.76	
B.	**SHELL**						
	B10 Superstructure						
1010	Floor Construction	N/A		—	—	—	8.0 %
1020	Roof Construction	Metal deck, open web steel joists, beams, interior columns		S.F. Roof	7.55	7.55	
	B20 Exterior Enclosure						
2010	Exterior Walls	Decorative concrete block	90% of wall	S.F. Wall	18.25	10.35	
2020	Exterior Windows	Storefront windows	10% of wall	Each	48.75	3.07	14.8%
2030	Exterior Doors	Sliding entrance door and hollow metal service doors		Each	2288	.57	
	B30 Roofing						
3010	Roof Coverings	Built-up tar and gravel with flashing; perlite/EPS composite insulation		S.F. Roof	6.82	6.82	7.4%
3020	Roof Openings	Roof hatches		S.F. Roof	.12	.12	
C.	**INTERIORS**						
1010	Partitions	Gypsum board on metal studs	60 S.F. Floor/L.F. Partition	S.F. Partition	6.36	1.06	
1020	Interior Doors	Single leaf hollow metal	600 S.F. Floor/Door	Each	1157	1.93	
1030	Fittings	N/A		—	—	—	
2010	Stair Construction	N/A			—	—	15.8%
3010	Wall Finishes	Paint		S.F. Surface	6.54	2.18	
3020	Floor Finishes	Vinyl tile		S.F. Floor	2.69	2.69	
3030	Ceiling Finishes	Mineral fiber tile on concealed zee bars		S.F. Ceiling	7.03	7.03	
D.	**SERVICES**						
	D10 Conveying						
1010	Elevators & Lifts	N/A		—	—	—	0.0 %
1020	Escalators & Moving Walks	N/A		—	—	—	
	D20 Plumbing						
2010	Plumbing Fixtures	Toilet and service fixtures, supply and drainage	1 Fixture/890 S.F. Floor	Each	2901	3.26	
2020	Domestic Water Distribution	Gas fired water heater		S.F. Floor	5.24	5.24	10.8%
2040	Rain Water Drainage	Roof drains		S.F. Roof	1.72	1.72	
	D30 HVAC						
3010	Energy Supply	N/A		—	—	—	
3020	Heat Generating Systems	Included in D3050		—	—	—	
3030	Cooling Generating Systems	N/A		—	—	—	8.6 %
3050	Terminal & Package Units	Single zone unit, gas heating, electric cooling		S.F. Floor	8.16	8.16	
3090	Other HVAC Sys. & Equipment	N/A		—	—	—	
	D40 Fire Protection						
4010	Sprinklers	Wet pipe sprinkler system		S.F. Floor	4.73	4.73	6.2%
4020	Standpipes	Standpipe, wet, Class III		S.F. Floor	1.08	1.08	
	D50 Electrical						
5010	Electrical Service/Distribution	400 ampere service, panel board and feeders		S.F. Floor	2.78	2.78	
5020	Lighting & Branch Wiring	High efficiency fluorescent fixtures, receptacles, switches, A.C. and misc. power		S.F. Floor	10.67	10.67	16.3%
5030	Communications & Security	Addressable alarm systems and emergency lighting		S.F. Floor	1.70	1.70	
5090	Other Electrical Systems	Emergency generator, 15 kW		S.F. Floor	.21	.21	
E.	**EQUIPMENT & FURNISHINGS**						
1010	Commercial Equipment	N/A		—	—	—	
1020	Institutional Equipment	N/A		—	—	—	
1030	Vehicular Equipment	N/A		—	—	—	0.0 %
1090	Other Equipment	N/A		—	—	—	
F.	**SPECIAL CONSTRUCTION**						
1020	Integrated Construction	N/A		—	—	—	0.0 %
1040	Special Facilities	N/A		—	—	—	
G.	**BUILDING SITEWORK**	**N/A**					

		Sub-Total	94.38	100%
CONTRACTOR FEES (General Requirements: 10%, Overhead: 5%, Profit: 10%)		25%	23.58	
ARCHITECT FEES		8%	9.44	

Total Building Cost	**127.40**

For customer support on your Square Foot Cost Data, call 877.694.8329.

215

Costs per square foot of floor area

Exterior Wall	S.F. Area	12000	16000	20000	26000	32000	38000	44000	52000	60000
	L.F. Perimeter	450	510	570	650	716	780	840	920	1020
Face Brick with Concrete Block Back-up	Steel Frame	143.80	133.50	127.35	121.20	116.75	113.70	111.40	109.05	107.80
	Bearing Walls	143.70	133.40	127.30	121.05	116.65	113.60	111.30	109.00	107.75
Stucco on Concrete Block	Steel Frame	133.05	124.40	119.20	114.05	110.35	107.85	105.95	104.00	103.00
	Bearing Walls	133.70	125.00	119.80	114.65	110.95	108.50	106.55	104.60	103.60
Precast Concrete Panels	Steel Frame	151.35	139.90	133.10	126.25	121.25	117.90	115.25	112.60	111.25
Metal Sandwich Panels	Steel Frame	144.35	133.95	127.80	121.60	117.10	114.05	111.70	109.35	108.10
Perimeter Adj., Add or Deduct	Per 100 L.F.	10.40	7.75	6.20	4.75	3.95	3.30	2.80	2.40	2.10
Story Hgt. Adj., Add or Deduct	Per 1 Ft.	2.10	1.80	1.60	1.40	1.30	1.20	1.10	1.05	0.95
For Basement, add $27.05 per square foot of basement area										

The above costs were calculated using the basic specifications shown on the facing page. These costs should be adjusted where necessary for design alternatives and owner's requirements. Reported completed project costs, for this type of structure, range from $74.30 to $212.80 per S.F.

Common additives

Description	Unit	$ Cost	Description	Unit	$ Cost
Check Out Counter			Refrigerated Food Cases, cont.		
Single belt	Ea.	3600	Frozen food, chest type, 12' long	Ea.	9450
Double belt	Ea.	5125	Glass door reach in, 5 door	Ea.	15,900
Scanner, registers, guns & memory 2 lanes	Ea.	18,300	Island case 12' long, single deck	Ea.	8425
10 lanes	Ea.	173,500	Multi deck	Ea.	9775
Power take away	Ea.	6025	Meat case 12' long, single deck	Ea.	8425
Emergency Lighting, 25 watt, battery operated			Multi deck	Ea.	11,800
Lead battery	Ea.	330	Produce, 12' long single deck	Ea.	7950
Nickel cadmium	Ea.	780	Multi deck	Ea.	10,000
Refrigerators, Prefabricated, walk-in			Safe, Office type, 1 hour rating		
7'-6" High, 6' x 6'	S.F.	202	30" x 18" x 18"	Ea.	2500
10' x 10'	S.F.	159	60" x 36" x 18", double door	Ea.	8125
12' x 14'	S.F.	142	Smoke Detectors		
12' x 20'	S.F.	125	Ceiling type	Ea.	227
Refrigerated Food Cases			Duct type	Ea.	565
Dairy, multi deck, 12' long	Ea.	12,800	Sound System		
Delicatessen case, single deck, 12' long	Ea.	8850	Amplifier, 250 watts	Ea.	2050
Multi deck, 18 S.F. shelf display	Ea.	8300	Speaker, ceiling or wall	Ea.	213
Freezer, self-contained chest type, 30 C.F.	Ea.	9725	Trumpet	Ea.	410
Glass door upright, 78 C.F.	Ea.	11,600			

Model costs calculated for a 1 story building with 18' story height and 44,000 square feet of floor area

				Unit	Unit Cost	Cost Per S.F.	% Of Sub-Total
A. SUBSTRUCTURE							
1010	Standard Foundations	Poured concrete; strip and spread footings		S.F. Ground	1.16	1.16	
1020	Special Foundations	N/A		—	—	—	
1030	Slab on Grade	4" reinforced concrete with vapor barrier and granular base		S.F. Slab	5.44	5.44	10.1%
2010	Basement Excavation	Site preparation for slab and trench for foundation wall and footing		S.F. Ground	.18	.18	
2020	Basement Walls	4' foundation wall		L.F. Wall	84	1.60	
B. SHELL							
	B10 Superstructure						
1010	Floor Construction	N/A		—	—	—	9.9 %
1020	Roof Construction	Metal deck, open web steel joists, beams, interior columns		S.F. Roof	8.20	8.20	
	B20 Exterior Enclosure						
2010	Exterior Walls	Face brick with concrete block backup	85% of wall	S.F. Wall	33.45	9.77	
2020	Exterior Windows	Storefront windows	15% of wall	Each	76	3.92	18.1%
2030	Exterior Doors	Sliding entrance doors with electrical operator, hollow metal		Each	5442	1.35	
	B30 Roofing						
3010	Roof Coverings	Built-up tar and gravel with flashing; perlite/EPS composite insulation		S.F. Roof	5.82	5.82	7.2%
3020	Roof Openings	Roof hatches		S.F. Roof	.14	.14	
C. INTERIORS							
1010	Partitions	50% concrete block, 50% gypsum board on metal studs	50 S.F. Floor/L.F. Partition	S.F. Partition	7.96	1.91	
1020	Interior Doors	Single leaf hollow metal	2000 S.F. Floor/Door	Each	1157	.58	
1030	Fittings	N/A		—	—	—	
2010	Stair Construction	N/A		—	—	—	17.8%
3010	Wall Finishes	Paint		S.F. Surface	5.42	2.60	
3020	Floor Finishes	Vinyl composition tile		S.F. Floor	2.69	2.69	
3030	Ceiling Finishes	Mineral fiber tile on concealed zee bars		S.F. Ceiling	7.03	7.03	
D. SERVICES							
	D10 Conveying						
1010	Elevators & Lifts	N/A		—	—	—	0.0 %
1020	Escalators & Moving Walks	N/A		—	—	—	
	D20 Plumbing						
2010	Plumbing Fixtures	Toilet and service fixtures, supply and drainage	1 Fixture/1820 S.F. Floor	Each	2184	1.20	
2020	Domestic Water Distribution	Gas fired water heater		S.F. Floor	.41	.41	3.2%
2040	Rain Water Drainage	Roof drains		S.F. Roof	1.03	1.03	
	D30 HVAC						
3010	Energy Supply	N/A		—	—	—	
3020	Heat Generating Systems	Included in D3050		—	—	—	
3030	Cooling Generating Systems	N/A		—	—	—	6.2 %
3050	Terminal & Package Units	Single zone rooftop unit, gas heating, electric cooling		S.F. Floor	5.17	5.17	
3090	Other HVAC Sys. & Equipment	N/A		—	—	—	
	D40 Fire Protection						
4010	Sprinklers	Sprinklers, light hazard		S.F. Floor	3.67	3.67	5.1%
4020	Standpipes	Standpipe		S.F. Floor	.54	.54	
	D50 Electrical						
5010	Electrical Service/Distribution	1600 ampere service, panel board and feeders		S.F. Floor	2.54	2.54	
5020	Lighting & Branch Wiring	High efficiency fluorescent fixtures, receptacles, switches, A.C. and misc. power		S.F. Floor	13.97	13.97	22.6%
5030	Communications & Security	Addressable alarm systems, partial internet wiring and emergency lighting		S.F. Floor	2.26	2.26	
5090	Other Electrical Systems	Emergency generator, 15 kW		S.F. Floor	.04	.04	
E. EQUIPMENT & FURNISHINGS							
1010	Commercial Equipment	N/A		—	—	—	
1020	Institutional Equipment	N/A		—	—	—	0.0 %
1030	Vehicular Equipment	N/A		—	—	—	
1090	Other Equipment	N/A		—	—	—	
F. SPECIAL CONSTRUCTION							
1020	Integrated Construction	N/A		—	—	—	0.0 %
1040	Special Facilities	N/A		—	—	—	
G. BUILDING SITEWORK	**N/A**						

		Sub-Total	83.22	100%
CONTRACTOR FEES (General Requirements: 10%, Overhead: 5%, Profit: 10%)		25%	20.80	
ARCHITECT FEES		7%	7.28	

	Total Building Cost	**111.30**

For customer support on your Square Foot Cost Data, call 877.694.8329.

217

Costs per square foot of floor area

Exterior Wall	S.F. Area	10000	16000	20000	22000	24000	26000	28000	30000	32000
	L.F. Perimeter	420	510	600	640	660	684	706	740	737
Face Brick with Concrete Block Back-up	Wood Truss	269.60	251.15	247.50	245.80	243.05	240.85	238.95	237.95	235.15
	Precast Conc.	308.30	286.40	282.15	280.10	276.80	274.20	271.90	270.75	267.40
Metal Sandwich Panel	Wood Truss	282.30	262.90	258.45	256.55	253.85	251.65	249.80	248.70	246.25
Precast Concrete Panel	Precast Conc.	309.20	283.10	277.50	274.90	271.20	268.25	265.65	264.25	260.65
Painted Concrete Block	Wood Frame	268.80	255.60	252.90	251.65	249.70	248.10	246.75	246.05	244.05
	Precast Conc.	288.10	267.10	262.40	260.35	257.40	255.05	253.05	251.85	249.10
Perimeter Adj., Add or Deduct	Per 100 L.F.	16.45	10.30	8.20	7.40	6.85	6.40	5.90	5.45	5.15
Story Hgt. Adj., Add or Deduct	Per 1 Ft.	2.25	1.65	1.60	1.50	1.45	1.40	1.35	1.30	1.20

For Basement, add $40.10 per square foot of basement area

The above costs were calculated using the basic specifications shown on the facing page. These costs should be adjusted where necessary for design alternatives and owner's requirements. Reported completed project costs, for this type of structure, range from $117.35 to $345.00 per S.F.

Common additives

Description	Unit	$ Cost
Bleachers, Telescoping, manual		
To 15 tier	Seat	137 - 190
16-20 tier	Seat	281 - 345
21-30 tier	Seat	299 - 390
For power operation, add	Seat	55 - 86
Emergency Lighting, 25 watt, battery operated		
Lead battery	Ea.	330
Nickel cadmium	Ea.	780
Lockers, Steel, single tier, 60" or 72"	Opng.	221 - 370
2 tier, 60" or 72" total	Opng.	120 - 165
5 tier, box lockers	Opng.	77 - 86.50
Locker bench, lam. maple top only	L.F.	34
Pedestal, steel pipe	Ea.	76
Pool Equipment		
Diving stand, 3 meter	Ea.	19,500
1 meter	Ea.	10,500
Diving board, 16' aluminum	Ea.	4775
Fiberglass	Ea.	4000
Lifeguard chair, fixed	Ea.	3975
Portable	Ea.	3050
Lights, underwater, 12 volt, 300 watt	Ea.	1025

Description	Unit	$ Cost
Sauna, Prefabricated, complete		
6' x 4'	Ea.	6575
6' x 6'	Ea.	7925
6' x 9'	Ea.	10,000
8' x 8'	Ea.	11,500
8' x 10'	Ea.	12,800
10' x 12'	Ea.	15,500
Sound System		
Amplifier, 250 watts	Ea.	2050
Speaker, ceiling or wall	Ea.	213
Trumpet	Ea.	410
Steam Bath, Complete, to 140 C.F.	Ea.	3000
To 300 C.F.	Ea.	3325
To 800 C.F.	Ea.	7100
To 2500 C.F.	Ea.	9300

Important: See the Reference Section for Location Factors.

Model costs calculated for a 1 story building with 24' story height and 20,000 square feet of floor area

Swimming Pool, Enclosed

					Unit	Unit Cost	Cost Per S.F.	% Of Sub-Total

A. SUBSTRUCTURE

					Unit	Unit Cost	Cost Per S.F.	% Of Sub-Total
1010	Standard Foundations	Poured concrete; strip and spread footings			S.F. Ground	3.88	3.88	
1020	Special Foundations	N/A			—	—	—	
1030	Slab on Grade	4" reinforced concrete with vapor barrier and granular base			S.F. Slab	5.44	5.44	8.2%
2010	Basement Excavation	Site preparation for slab and trench for foundation wall and footing			S.F. Ground	.32	.64	
2020	Basement Walls	4' foundation wall			L.F. Wall	84	5.05	

B. SHELL

B10 Superstructure

					Unit	Unit Cost	Cost Per S.F.	% Of Sub-Total
1010	Floor Construction	N/A			—	—	—	
1020	Roof Construction	Wood deck on laminated wood truss			S.F. Roof	19.46	19.46	10.6 %

B20 Exterior Enclosure

2010	Exterior Walls	Face brick with concrete block backup	80% of wall		S.F. Wall	32.95	18.98	
2020	Exterior Windows	Outward projecting steel	20% of wall		Each	48.75	7.02	14.6%
2030	Exterior Doors	Double aluminum and glass, hollow metal			Each	2665	.68	

B30 Roofing

| 3010 | Roof Coverings | Asphalt shingles with flashing; perlite/EPS composite insulation | | | S.F. Roof | 4.41 | 4.41 | 2.4% |
| 3020 | Roof Openings | N/A | | | — | — | — | |

C. INTERIORS

					Unit	Unit Cost	Cost Per S.F.	% Of Sub-Total
1010	Partitions	Concrete block	100 S.F. Floor/L.F. Partition		S.F. Partition	9.70	.97	
1020	Interior Doors	Single leaf hollow metal	1000 S.F. Floor/Door		Each	1157	1.16	
1030	Fittings	Toilet partitions			S.F. Floor	.17	.17	
2010	Stair Construction	N/A			—	—	—	15.6%
3010	Wall Finishes	Acrylic glazed coating			S.F. Surface	11.95	2.39	
3020	Floor Finishes	70% terrazzo, 30% ceramic tile			S.F. Floor	22.88	22.88	
3030	Ceiling Finishes	Mineral fiber tile on concealed zee bars	20% of area		S.F. Ceiling	4.76	.96	

D. SERVICES

D10 Conveying

					Unit	Unit Cost	Cost Per S.F.	% Of Sub-Total
1010	Elevators & Lifts	N/A			—	—	—	0.0 %
1020	Escalators & Moving Walks	N/A			—	—	—	

D20 Plumbing

2010	Plumbing Fixtures	Toilet and service fixtures, supply and drainage	1 Fixture/210 S.F. Floor		Each	1262	6.01	
2020	Domestic Water Distribution	Gas fired water heater			S.F. Floor	4.21	4.21	5.6%
2040	Rain Water Drainage	N/A			—	—	—	

D30 HVAC

3010	Energy Supply	Terminal unit heaters	10% of area		S.F. Floor	.93	.93	
3020	Heat Generating Systems	N/A			—	—	—	
3030	Cooling Generating Systems	N/A			—	—	—	8.0%
3050	Terminal & Package Units	Single zone unit gas heating, electric cooling			S.F. Floor	13.82	13.82	
3090	Other HVAC Sys. & Equipment	N/A			—	—	—	

D40 Fire Protection

| 4010 | Sprinklers | Sprinklers, light hazard | | | S.F. Floor | 1.84 | 1.84 | 1.3% |
| 4020 | Standpipes | Standpipe | | | S.F. Floor | .57 | .57 | |

D50 Electrical

5010	Electrical Service/Distribution	400 ampere service, panel board and feeders			S.F. Floor	1.16	1.16	
5020	Lighting & Branch Wiring	High efficiency fluorescent fixtures, receptacles, switches, A.C. and misc. power			S.F. Floor	21.20	21.20	12.8%
5030	Communications & Security	Addressable alarm systems and emergency lighting			S.F. Floor	1.05	1.05	
5090	Other Electrical Systems	Emergency generator, 15 kW			S.F. Floor	.08	.08	

E. EQUIPMENT & FURNISHINGS

					Unit	Unit Cost	Cost Per S.F.	% Of Sub-Total
1010	Commercial Equipment	N/A			—	—	—	
1020	Institutional Equipment	N/A			—	—	—	
1030	Vehicular Equipment	N/A			—	—	—	20.9 %
1090	Other Equipment	Swimming pool			S.F. Floor	38.39	38.39	

F. SPECIAL CONSTRUCTION

| 1020 | Integrated Construction | N/A | | | — | — | — | 0.0 % |
| 1040 | Special Facilities | N/A | | | — | — | — | |

G. BUILDING SITEWORK   N/A

			Sub-Total	183.35	**100%**

| CONTRACTOR FEES (General Requirements: 10%, Overhead: 5%, Profit: 10%) | 25% | 45.81 | |
| ARCHITECT FEES | 8% | 18.34 | |

| **Total Building Cost** | 247.50 |

For customer support on your Square Foot Cost Data, call 877.694.8329.

219

Costs per square foot of floor area

Exterior Wall	S.F. Area	2000	3000	4000	5000	6000	7000	8000	9000	10000
	L.F. Perimeter	180	220	260	286	320	353	368	397	425
Face Brick with Concrete Block Back-up	Steel Frame	227.90	202.35	189.55	179.05	173.40	169.20	163.80	161.20	158.95
	Bearing Walls	217.70	194.90	183.50	173.75	168.65	164.85	159.75	157.35	155.35
Limestone with Concrete Block Back-up	Steel Frame	250.95	222.95	208.95	196.85	190.50	185.85	179.40	176.35	173.85
	Bearing Walls	250.60	219.90	204.50	191.55	184.70	179.60	172.85	169.60	166.90
Decorative Concrete Block	Steel Frame	212.25	189.60	178.30	169.05	164.10	160.40	155.85	153.55	151.60
	Bearing Walls	206.95	184.35	173.00	163.85	158.80	155.15	150.55	148.25	146.30
Perimeter Adj., Add or Deduct	Per 100 L.F.	50.45	33.60	25.25	20.15	16.80	14.40	12.60	11.15	10.10
Story Hgt. Adj., Add or Deduct	Per 1 Ft.	5.90	4.75	4.25	3.70	3.45	3.25	3.00	2.85	2.80

For Basement, add $38.85 per square foot of basement area

The above costs were calculated using the basic specifications shown on the facing page. These costs should be adjusted where necessary for design alternatives and owner's requirements. Reported completed project costs, for this type of structure, range from $103.70 to $374.25 per S.F.

Common additives

Description	Unit	$ Cost
Emergency Lighting, 25 watt, battery operated		
Lead battery	Ea.	330
Nickel cadmium	Ea.	780
Smoke Detectors		
Ceiling type	Ea.	227
Duct type	Ea.	565
Emergency Generators, complete system, gas		
15 kw	Ea.	16,300
85 kw	Ea.	34,300
170 kw	Ea.	85,000
Diesel, 50 kw	Ea.	26,000
150 kw	Ea.	44,100
350 kw	Ea.	70,000

Important: See the Reference Section for Location Factors.

Model costs calculated for a 1 story building with 12' story height and 5,000 square feet of floor area

Telephone Exchange

				Unit	Unit Cost	Cost Per S.F.	% Of Sub-Total
A.	**SUBSTRUCTURE**						
1010	Standard Foundations	Poured concrete; strip and spread footings		S.F. Ground	2.60	2.60	
1020	Special Foundations	N/A		—	—	—	
1030	Slab on Grade	4" reinforced concrete with vapor barrier and granular base		S.F. Slab	5.44	5.44	10.5%
2010	Basement Excavation	Site preparation for slab and trench for foundation wall and footing		S.F. Ground	.57	.71	
2020	Basement Walls	4' foundation wall		L.F. Wall	84	4.77	
B.	**SHELL**						
	B10 Superstructure						
1010	Floor Construction	N/A		—	—	—	8.7 %
1020	Roof Construction	Metal deck, open web steel joists, beams, columns		S.F. Roof	10.39	10.39	
	B20 Exterior Enclosure						
2010	Exterior Walls	Face brick with concrete block backup	80% of wall	S.F. Wall	33.45	18.37	
2020	Exterior Windows	Outward projecting steel	20% of wall	Each	1760	12.08	24.8%
2030	Exterior Doors	Single aluminum glass with transom		Each	3925	1.58	
	B30 Roofing						
3010	Roof Coverings	Built-up tar and gravel with flashing; perlite/EPS composite insulation		S.F. Roof	7.28	7.28	5.6%
3020	Roof Openings						
C.	**INTERIORS**						
1010	Partitions	Double layer gypsum board on metal studs	15 S.F. Floor/L.F. Partition	S.F. Partition	6.32	4.21	
1020	Interior Doors	Single leaf hollow metal	150 S.F. Floor/Door	Each	1157	7.71	
1030	Fittings	Toilet partitions		S.F. Floor	1.17	1.17	
2010	Stair Construction	N/A		—	—	—	21.8%
3010	Wall Finishes	Paint		S.F. Surface	2.23	2.97	
3020	Floor Finishes	90% carpet, 10% terrazzo		S.F. Floor	6.32	6.32	
3030	Ceiling Finishes	Fiberglass board on exposed grid system, suspended		S.F. Ceiling	5.70	5.70	
D.	**SERVICES**						
	D10 Conveying						
1010	Elevators & Lifts	N/A		—	—	—	0.0 %
1020	Escalators & Moving Walks	N/A		—	—	—	
	D20 Plumbing						
2010	Plumbing Fixtures	Kitchen, toilet and service fixtures, supply and drainage	1 Fixture/715 S.F. Floor	Each	3804	5.32	
2020	Domestic Water Distribution	Gas fired water heater		S.F. Floor	2.38	2.38	7.4%
2040	Rain Water Drainage	Roof drains		S.F. Roof	1.79	1.79	
	D30 HVAC						
3010	Energy Supply	N/A		—	—	—	
3020	Heat Generating Systems	Included in D3050		—	—	—	
3030	Cooling Generating Systems	N/A		—	—	—	6.6 %
3050	Terminal & Package Units	Single zone unit, gas heating, electric cooling		S.F. Floor	8.56	8.56	
3090	Other HVAC Sys. & Equipment	N/A		—	—	—	
	D40 Fire Protection						
4010	Sprinklers	Wet pipe sprinkler system		S.F. Floor	5.43	5.43	5.8%
4020	Standpipes	Standpipe		S.F. Floor	2.06	2.06	
	D50 Electrical						
5010	Electrical Service/Distribution	200 ampere service, panel board and feeders		S.F. Floor	1.93	1.93	
5020	Lighting & Branch Wiring	High efficiency fluorescent fixtures, receptacles, switches, A.C. and misc. power		S.F. Floor	6.29	6.29	8.8%
5030	Communications & Security	Addressable alarm systems and emergency lighting		S.F. Floor	2.96	2.96	
5090	Other Electrical Systems	Emergency generator, 15 kW		S.F. Floor	.17	.17	
E.	**EQUIPMENT & FURNISHINGS**						
1010	Commercial Equipment	N/A		—	—	—	
1020	Institutional Equipment	N/A		—	—	—	0.0 %
1030	Vehicular Equipment	N/A		—	—	—	
1090	Other Equipment	N/A		—	—	—	
F.	**SPECIAL CONSTRUCTION**						
1020	Integrated Construction	N/A		—	—	—	0.0 %
1040	Special Facilities	N/A		—	—	—	
G.	**BUILDING SITEWORK**	**N/A**					

		Sub-Total	129.03	100%
CONTRACTOR FEES (General Requirements: 10%, Overhead: 5%, Profit: 10%)		25%	32.28	
ARCHITECT FEES		11%	17.74	
	Total Building Cost		**179.05**	

For customer support on your Square Foot Cost Data, call 877.694.8329.

221

Costs per square foot of floor area

Exterior Wall	S.F. Area	5000	6500	8000	9500	11000	14000	17500	21000	24000
	L.F. Perimeter	300	360	386	396	435	510	550	620	680
Face Brick with Concrete Block Back-up	Steel Joists	166.35	160.10	152.65	146.25	143.70	140.15	135.30	133.15	131.90
	Wood Joists	164.60	158.60	151.60	145.50	143.15	139.75	135.15	133.15	132.00
Stone with Concrete Block Back-up	Steel Joists	174.30	167.45	159.05	151.75	148.95	145.00	139.45	137.10	135.70
	Wood Joists	172.55	165.90	158.05	151.10	148.40	144.60	139.35	137.10	135.75
Brick Veneer	Wood Frame	152.65	147.15	140.85	135.45	133.20	130.15	126.10	124.25	123.15
E.I.F.S.	Wood Frame	140.50	135.65	130.45	126.00	124.10	121.50	118.20	116.70	115.80
Perimeter Adj., Add or Deduct	Per 100 L.F.	16.45	12.70	10.40	8.65	7.50	5.90	4.70	3.95	3.45
Story Hgt. Adj., Add or Deduct	Per 1 Ft.	3.00	2.75	2.45	2.05	1.95	1.80	1.55	1.50	1.45
For Basement, add $32.00 per square foot of basement area										

The above costs were calculated using the basic specifications shown on the facing page. These costs should be adjusted where necessary for design alternatives and owner's requirements. Reported completed project costs, for this type of structure, range from $98.05 to $257.90 per S.F.

Common additives

Description	Unit	$ Cost
Directory Boards, Plastic, glass covered		
30" x 20"	Ea.	600
36" x 48"	Ea.	1500
Aluminum, 24" x 18"	Ea.	635
36" x 24"	Ea.	755
48" x 32"	Ea.	1050
48" x 60"	Ea.	2225
Emergency Lighting, 25 watt, battery operated		
Lead battery	Ea.	330
Nickel cadmium	Ea.	780
Flagpoles, Complete		
Aluminum, 20' high	Ea.	1850
40' high	Ea.	4200
70' high	Ea.	11,000
Fiberglass, 23' high	Ea.	1325
39'-5" high	Ea.	3450
59' high	Ea.	6850
Safe, Office type, 1 hour rating		
30" x 18" x 18"	Ea.	2500
60" x 36" x 18", double door	Ea.	8125

Description	Unit	$ Cost
Smoke Detectors		
Ceiling type	Ea.	227
Duct type	Ea.	565
Vault Front, Door & frame		
1 Hour test, 32" x 78"	Opng.	8275
2 Hour test, 32" door	Opng.	10,100
40" door	Opng.	11,700
4 Hour test, 32" door	Opng.	10,800
40" door	Opng.	12,600
Time lock movement; two movement	Ea.	2300

Important: See the Reference Section for Location Factors.

Model costs calculated for a 1 story building with 12' story height and 11,000 square feet of floor area

Town Hall, 1 Story

				Unit	Unit Cost	Cost Per S.F.	% Of Sub-Total
A. SUBSTRUCTURE							
1010	Standard Foundations	Poured concrete; strip and spread footings		S.F. Ground	1.75	1.75	
1020	Special Foundations	N/A		—	—	—	
1030	Slab on Grade	4" reinforced concrete with vapor barrier and granular base		S.F. Slab	5.44	5.44	10.3%
2010	Basement Excavation	Site preparation for slab and trench for foundation wall and footing		S.F. Ground	.32	.35	
2020	Basement Walls	4' foundation wall		L.F. Wall	84	3.30	
B. SHELL							
B10 Superstructure							
1010	Floor Construction	N/A		—	—	—	7.7 %
1020	Roof Construction	Metal deck, open web steel joists, beams, interior columns		S.F. Roof	7.88	7.88	
B20 Exterior Enclosure							
2010	Exterior Walls	Face brick with concrete block backup	70% of wall	S.F. Wall	33.45	11.11	
2020	Exterior Windows	Metal outward projecting	30% of wall	Each	750	4.64	15.9%
2030	Exterior Doors	Metal and glass with transom		Each	2863	1.04	
B30 Roofing							
3010	Roof Coverings	Built-up tar and gravel with flashing; perlite/EPS composite insulation		S.F. Roof	6.81	6.81	6.5%
3020	Roof Openings	Roof hatches		S.F. Roof			
C. INTERIORS							
1010	Partitions	Gypsum board on metal studs	20 S.F. Floor/L.F. Partition	S.F. Partition	9.48	4.74	
1020	Interior Doors	Wood solid core	200 S.F. Floor/Door	Each	616	3.08	
1030	Fittings	Toilet partitions		S.F. Floor	.47	.47	
2010	Stair Construction	N/A		—	—	—	23.6%
3010	Wall Finishes	90% paint, 10% ceramic tile		S.F. Surface	1.50	1.50	
3020	Floor Finishes	70% carpet tile, 15% terrazzo, 15% vinyl composition tile		S.F. Floor	8.05	8.05	
3030	Ceiling Finishes	Mineral fiber tile on concealed zee bars		S.F. Ceiling	7.03	7.03	
D. SERVICES							
D10 Conveying							
1010	Elevators & Lifts	N/A		—	—	—	0.0 %
1020	Escalators & Moving Walks	N/A		—	—	—	
D20 Plumbing							
2010	Plumbing Fixtures	Kitchen, toilet and service fixtures, supply and drainage	1 Fixture/500 S.F. Floor	Each	2505	5.01	
2020	Domestic Water Distribution	Gas fired water heater		S.F. Floor	1.47	1.47	8.3%
2040	Rain Water Drainage	Roof drains		S.F. Roof	2.28	2.28	
D30 HVAC							
3010	Energy Supply	N/A		—	—	—	
3020	Heat Generating Systems	Included in D3050		—	—	—	
3030	Cooling Generating Systems	N/A		—	—	—	8.4 %
3050	Terminal & Package Units	Multizone unit, gas heating, electric cooling		S.F. Floor	8.86	8.86	
3090	Other HVAC Sys. & Equipment	N/A		—	—	—	
D40 Fire Protection							
4010	Sprinklers	Wet pipe sprinkler system		S.F. Floor	3.67	3.67	4.2%
4020	Standpipes	Standpipe, wet, Class III		S.F. Floor	.78	.78	
D50 Electrical							
5010	Electrical Service/Distribution	400 ampere service, panel board and feeders		S.F. Floor	2.01	2.01	
5020	Lighting & Branch Wiring	High efficiency fluorescent fixtures, receptacles, switches, A.C. and misc. power		S.F. Floor	10.46	10.46	15.1%
5030	Communications & Security	Addressable alarm systems, internet wiring, and emergency lighting		S.F. Floor	3.31	3.31	
5090	Other Electrical Systems	Emergency generator, 15 kW		S.F. Floor	.15	.15	
E. EQUIPMENT & FURNISHINGS							
1010	Commercial Equipment	N/A		—	—	—	
1020	Institutional Equipment	N/A		—	—	—	0.0 %
1030	Vehicular Equipment	N/A		—	—	—	
1090	Other Equipment	N/A		—	—	—	
F. SPECIAL CONSTRUCTION							
1020	Integrated Construction	N/A		—	—	—	0.0 %
1040	Special Facilities	N/A		—	—	—	
G. BUILDING SITEWORK	**N/A**						

		Sub-Total	105.47	100%
CONTRACTOR FEES (General Requirements: 10%, Overhead: 5%, Profit: 10%)		25%	26.36	
ARCHITECT FEES		9%	11.87	
	Total Building Cost		**143.70**	

For customer support on your Square Foot Cost Data, call 877.694.8329.

223

Costs per square foot of floor area

Exterior Wall	S.F. Area	8000	10000	12000	15000	18000	24000	28000	35000	40000
	L.F. Perimeter	206	233	260	300	320	360	393	451	493
Face Brick with Concrete Block Back-up	Steel Frame	221.10	209.45	201.65	193.80	186.30	176.95	173.40	169.15	167.10
	R/Conc. Frame	224.35	212.65	204.85	197.05	189.45	180.15	176.60	172.40	170.30
Stone with Concrete Block Back-up	Steel Frame	232.25	220.10	212.00	203.85	195.40	184.85	180.95	176.25	174.05
	R/Conc. Frame	234.55	222.00	213.50	205.00	196.55	186.15	182.20	177.55	175.25
Limestone with Concrete Block Back-up	Steel Frame	230.95	217.40	208.35	199.20	189.90	178.40	174.10	168.95	166.40
	R/Conc. Frame	234.20	220.60	211.50	202.40	193.15	181.65	177.35	172.25	169.65
Perimeter Adj., Add or Deduct	Per 100 L.F.	25.25	20.25	16.85	13.45	11.25	8.40	7.20	5.75	5.05
Story Hgt. Adj., Add or Deduct	Per 1 Ft.	3.80	3.50	3.15	3.00	2.60	2.20	2.05	1.95	1.85
For Basement, add $31.15 per square foot of basement area										

The above costs were calculated using the basic specifications shown on the facing page. These costs should be adjusted where necessary for design alternatives and owner's requirements. Reported completed project costs, for this type of structure, range from $87.85 to $257.90 per S.F.

Common additives

Description	Unit	$ Cost
Directory Boards, Plastic, glass covered		
30" x 20"	Ea.	600
36" x 48"	Ea.	1500
Aluminum, 24" x 18"	Ea.	635
36" x 24"	Ea.	755
48" x 32"	Ea.	1050
48" x 60"	Ea.	2225
Elevators, Hydraulic passenger, 2 stops		
1500# capacity	Ea.	65,700
2500# capacity	Ea.	68,800
3500# capacity	Ea.	73,300
Additional stop, add	Ea.	7825
Emergency Lighting, 25 watt, battery operated		
Lead battery	Ea.	330
Nickel cadmium	Ea.	780

Description	Unit	$ Cost
Flagpoles, Complete		
Aluminum, 20' high	Ea.	1850
40' high	Ea.	4200
70' high	Ea.	11,000
Fiberglass, 23' high	Ea.	1325
39'-5" high	Ea.	3450
59' high	Ea.	6850
Safe, Office type, 1 hour rating		
30" x 18" x 18"	Ea.	2500
60" x 36" x 18", double door	Ea.	8125
Smoke Detectors		
Ceiling type	Ea.	227
Duct type	Ea.	565
Vault Front, Door & frame		
1 Hour test, 32" x 78"	Opng.	8275
2 Hour test, 32" door	Opng.	10,100
40" door	Opng.	11,700
4 Hour test, 32" door	Opng.	10,800
40" door	Opng.	12,600
Time lock movement; two movement	Ea.	2300

Important: See the Reference Section for Location Factors.

Model costs calculated for a 3 story building with 12' story height and 18,000 square feet of floor area

Town Hall, 2-3 Story

				Unit	Unit Cost	Cost Per S.F.	% Of Sub-Total
A.	**SUBSTRUCTURE**						
1010	Standard Foundations	Poured concrete; strip and spread footings		S.F. Ground	4.50	1.50	
1020	Special Foundations	N/A		—	—	—	
1030	Slab on Grade	4" reinforced concrete with vapor barrier and granular base		S.F. Slab	5.44	1.81	3.5%
2010	Basement Excavation	Site preparation for slab and trench for foundation wall and footing		S.F. Ground	.57	.29	
2020	Basement Walls	4' foundation wall		L.F. Wall	41.75	1.48	
B.	**SHELL**						
	B10 Superstructure						
1010	Floor Construction	Open web steel joists, slab form, concrete, wide flange steel columns		S.F. Floor	28.43	18.95	15.2%
1020	Roof Construction	Metal deck, open web steel joists, beams, interior columns		S.F. Roof	8.64	2.88	
	B20 Exterior Enclosure						
2010	Exterior Walls	Stone with concrete block backup	70% of wall	S.F. Wall	45.04	20.18	
2020	Exterior Windows	Metal outward projecting	10% of wall	Each	1160	9.68	21.3%
2030	Exterior Doors	Metal and glass with transoms		Each	2665	.74	
	B30 Roofing						
3010	Roof Coverings	Built-up tar and gravel with flashing; perlite/EPS composite insulation		S.F. Roof	7.38	2.46	1.7%
3020	Roof Openings	N/A		—	—	—	
C.	**INTERIORS**						
1010	Partitions	Gypsum board on metal studs	20 S.F. Floor/L.F. Partition	S.F. Partition	10.56	5.28	
1020	Interior Doors	Wood solid core	200 S.F. Floor/Door	Each	616	3.08	
1030	Fittings	Toilet partitions		S.F. Floor	.29	.29	
2010	Stair Construction	Concrete filled metal pan		Flight	17,300	5.77	21.6%
3010	Wall Finishes	90% paint, 10% ceramic tile		S.F. Surface	1.50	1.50	
3020	Floor Finishes	70% carpet tile, 15% terrazzo, 15% vinyl composition tile		S.F. Floor	8.05	8.05	
3030	Ceiling Finishes	Mineral fiber tile on concealed zee bars		S.F. Ceiling	7.03	7.03	
D.	**SERVICES**						
	D10 Conveying						
1010	Elevators & Lifts	Two hydraulic elevators		Each	117,270	13.03	9.1%
1020	Escalators & Moving Walks	N/A		—	—	—	
	D20 Plumbing						
2010	Plumbing Fixtures	Toilet and service fixtures, supply and drainage	1 Fixture/1385 S.F. Floor	Each	6606	4.77	
2020	Domestic Water Distribution	Gas fired water heater		S.F. Floor	.64	.64	4.4%
2040	Rain Water Drainage	Roof drains		S.F. Roof	2.49	.83	
	D30 HVAC						
3010	Energy Supply	N/A		—	—	—	
3020	Heat Generating Systems	Included in D3050		—	—	—	
3030	Cooling Generating Systems	N/A		—	—	—	6.2 %
3050	Terminal & Package Units	Multizone unit, gas heating, electric cooling		S.F. Floor	8.86	8.86	
3090	Other HVAC Sys. & Equipment	N/A		—	—	—	
	D40 Fire Protection						
4010	Sprinklers	Sprinklers, light hazard		S.F. Floor	3.38	3.38	4.6%
4020	Standpipes	Standpipe, wet, Class III		S.F. Floor	3.15	3.15	
	D50 Electrical						
5010	Electrical Service/Distribution	400 ampere service, panel board and feeders		S.F. Floor	1.70	1.70	
5020	Lighting & Branch Wiring	High efficiency fluorescent fixtures, receptacles, switches, A.C. and misc. power		S.F. Floor	12.46	12.46	12.4%
5030	Communications & Security	Addressable alarm systems, internet wiring, and emergency lighting		S.F. Floor	3.44	3.44	
5090	Other Electrical Systems	Emergency generator, 15 kW		S.F. Floor	.17	.17	
E.	**EQUIPMENT & FURNISHINGS**						
1010	Commercial Equipment	N/A		—	—	—	
1020	Institutional Equipment	N/A		—	—	—	0.0 %
1030	Vehicular Equipment	N/A		—	—	—	
1090	Other Equipment	N/A		—	—	—	
F.	**SPECIAL CONSTRUCTION**						
1020	Integrated Construction	N/A		—	—	—	0.0 %
1040	Special Facilities	N/A		—	—	—	
G.	**BUILDING SITEWORK**	**N/A**					

	Sub-Total	143.40	**100%**
CONTRACTOR FEES (General Requirements: 10%, Overhead: 5%, Profit: 10%)	25%	35.87	
ARCHITECT FEES	9%	16.13	
	Total Building Cost	**195.40**	

For customer support on your Square Foot Cost Data, call 877.694.8329.

225

Costs per square foot of floor area

Exterior Wall	S.F. Area	3500	5000	7500	9000	11000	12500	15000	20000	25000
	L.F. Perimeter	240	290	350	390	430	460	510	580	650
E.I.F.S.	Wood Truss	174.70	165.35	156.70	154.00	151.00	149.35	147.40	144.25	142.35
	Steel Joists	184.70	174.95	165.90	163.10	159.90	158.20	156.15	152.80	150.85
Wood Siding	Wood Truss	176.75	167.65	159.15	156.50	153.60	152.00	150.05	147.05	145.20
Brick Veneer	Wood Truss	187.70	176.90	166.65	163.45	159.85	157.90	155.50	151.65	149.35
Brick on Block	Wood Truss	190.95	178.95	167.40	163.85	159.75	157.60	154.90	150.45	147.90
Tilt Up Concrete	Steel Joists	176.55	167.85	159.85	157.30	154.55	153.05	151.25	148.35	146.65
Perimeter Adj., Add or Deduct	Per 100 L.F.	13.05	9.15	6.10	5.05	4.15	3.65	3.05	2.30	1.85
Story Hgt. Adj., Add or Deduct	Per 1 Ft.	1.55	1.40	1.10	1.00	0.90	0.85	0.75	0.70	0.60

For Basement, add $33.65 per square foot of basement area

The above costs were calculated using the basic specifications shown on the facing page. These costs should be adjusted where necessary for design alternatives and owner's requirements. Reported completed project costs, for this type of structure, range from $71.95 to $278.25 per S.F.

Common additives

Description	Unit	$ Cost	Description	Unit	$ Cost
Closed circuit surveillance, one station			Kennel doors		
Camera and monitor	Ea.	1975	2 way, swinging type, 13" x 19" opening	Opng.	204
For additional camera stations, add	Ea.	1075	17" x 29" opening	Opng.	226
Directory boards, plastic, glass covered			9" x 9" opening, electronic with accessories	Opng.	263
30" x 20"	Ea.	600			
36" x 48"	Ea.	1500			
Aluminum, 24" x 18"	Ea.	635			
36" x 24"	Ea.	755			
48" x 32"	Ea.	1050			
48" x 60"	Ea.	2225			
Electronic, wall mounted	S.F.	2900			
Free standing	S.F.	4100			
Kennel fencing					
Kennel fencing, 1-1/2" mesh, 6' long, 3'-6" wide, 6'-2" high	Ea.	790			
12' long	Ea.	1025			
Top covers, 1-1/2" mesh, 6' long	Ea.	212			
12' long	Ea.	287			

Important: See the Reference Section for Location Factors.

Model costs calculated for a 1 story building with 12' story height and 7,500 square feet of floor area

Veterinary Hospital

				Unit	Unit Cost	Cost Per S.F.	% Of Sub-Total
A. SUBSTRUCTURE							
1010	Standard Foundations	Poured concrete; strip and spread footings		S.F. Ground	2.07	2.07	
1020	Special Foundations	N/A		—	—	—	
1030	Slab on Grade	4" and 5" reinforced concrete with vapor barrier and granular base		S.F. Slab	5.44	5.44	10.7%
2010	Basement Excavation	Site preparation for slab and trench for foundation wall and footing		S.F. Ground	.57	1.07	
2020	Basement Walls	4' foundation wall		L.F. Wall	84	3.90	
B. SHELL							
	B10 Superstructure						
1010	Floor Construction	6 x 6 wood columns		S.F. Floor	.24	.24	7.5%
1020	Roof Construction	Wood roof truss with plywood sheathing		S.F. Roof	8.48	8.48	
	B20 Exterior Enclosure						
2010	Exterior Walls	Cedar bevel siding on wood studs, insulated	85% of wall	S.F. Wall	13.49	6.42	
2020	Exterior Windows	Wood double hung and awning	15% of wall	Each	1157	3.97	10.8%
2030	Exterior Doors	Aluminum and glass		Each	2708	2.17	
	B30 Roofing						
3010	Roof Coverings	Ashphalt shingles with flashing (pitched); al. gutters and downspouts		S.F. Roof	5	5	4.3%
3020	Roof Openings	N/A		—	—	—	
C. INTERIORS							
1010	Partitions	Gypsum board on wood studs	8 S.F. Floor/L.F. Partition	S.F. Partition	5.20	6.50	
1020	Interior Doors	Solid core wood	180 S.F. Floor/Door	Each	686	3.81	
1030	Fittings	Lockers		Each	.44	.44	
2010	Stair Construction	N/A		—	—	—	19.7%
3010	Wall Finishes	Paint on drywall		S.F. Surface	1.01	2.53	
3020	Floor Finishes	90% vinyl composition tile, 5% ceramic tile, 5% sealed concrete		S.F. Floor	3.27	3.27	
3030	Ceiling Finishes	90% acoustic tile, 10% gypsum board		S.F. Ceiling	6.43	6.42	
D. SERVICES							
	D10 Conveying						
1010	Elevators & Lifts	N/A		—	—	—	0.0 %
1020	Escalators & Moving Walks	N/A		—	—	—	
	D20 Plumbing						
2010	Plumbing Fixtures	Toilet and service fixtures, supply and drainage	1 Fixture/1075 S.F. Floor	Each	1831	3.27	
2020	Domestic Water Distribution	Electric water heater		S.F. Floor	1.14	1.14	3.8%
2040	Rain Water Drainage	N/A		—	—	—	
	D30 HVAC						
3010	Energy Supply	N/A		—	—	—	
3020	Heat Generating Systems	N/A		—	—	—	
3030	Cooling Generating Systems	N/A		—	—	—	5.9 %
3050	Terminal & Package Units	Split system with air cooled condensing units		S.F. Floor	6.85	6.85	
3090	Other HVAC Sys. & Equipment	N/A		—	—	—	
	D40 Fire Protection						
4010	Sprinklers	Wet pipe sprinkler system		S.F. Floor	3.67	3.67	3.1%
4020	Standpipes	N/A		—	—	—	
	D50 Electrical						
5010	Electrical Service/Distribution	800 ampere service, panel board and feeders		S.F. Floor	6.77	6.77	
5020	Lighting & Branch Wiring	High efficiency fluorescent fixtures, receptacles, switches, A.C. and misc. power		S.F. Floor	12.36	12.36	20.6%
5030	Communications & Security	Addressable alarm systems, internet wiring, and emergency lighting		S.F. Floor	4.86	4.86	
5090	Other Electrical Systems	Emergency generator, 100 kW		S.F. Floor	.05	.05	
E. EQUIPMENT & FURNISHINGS							
1010	Commercial Equipment	N/A		—	—	—	
1020	Institutional Equipment	Eye wash station		S.F. Floor	14.17	14.17	
1030	Vehicular Equipment	N/A		—	—	—	13.8 %
1090	Other Equipment	Countertop, plastic laminate		S.F. Floor	1.95	1.95	
F. SPECIAL CONSTRUCTION & DEMOLITION							
1020	Integrated Construction	N/A		—	—	—	0.0 %
1040	Special Facilities	N/A		—	—	—	
G. BUILDING SITEWORK	**N/A**						

		Sub-Total	116.82	100%
CONTRACTOR FEES (General Requirements: 10%, Overhead: 5%, Profit: 10%)		25%	29.19	
ARCHITECT FEES		9%	13.14	
	Total Building Cost		**159.15**	

For customer support on your Square Foot Cost Data, call 877.694.8329.

Costs per square foot of floor area

Exterior Wall	S.F. Area	10000	15000	20000	25000	30000	35000	40000	50000	60000
	L.F. Perimeter	410	500	600	640	700	766	833	966	1000
Tiltup Concrete Panels	Steel Frame	121.55	111.30	106.50	101.90	99.25	97.50	96.20	94.35	91.95
Brick with Block Back-up	Bearing Walls	144.35	128.90	121.85	114.30	110.20	107.45	105.50	102.60	98.45
Concrete Block	Steel Frame	125.75	117.50	113.75	109.75	107.55	106.05	105.00	103.45	101.25
	Bearing Walls	127.80	119.60	115.90	111.85	109.60	108.15	107.15	105.60	103.40
Galvanized Steel Siding	Steel Frame	132.75	124.30	120.50	116.35	114.10	112.60	111.55	109.95	107.70
Metal Sandwich Panels	Steel Frame	134.50	124.65	120.15	115.20	112.50	110.75	109.45	107.60	104.85
Perimeter Adj., Add or Deduct	Per 100 L.F.	7.65	5.05	3.80	2.95	2.50	2.15	1.90	1.50	1.20
Story Hgt. Adj., Add or Deduct	Per 1 Ft.	0.95	0.75	0.70	0.55	0.50	0.50	0.50	0.45	0.35
For Basement, add $29.95 per square foot of basement area										

The above costs were calculated using the basic specifications shown on the facing page. These costs should be adjusted where necessary for design alternatives and owner's requirements. Reported completed project costs, for this type of structure, range from $39.05 to $155.90 per S.F.

Common additives

Description	Unit	$ Cost	Description	Unit	$ Cost
Dock Leveler, 10 ton cap.			Sound System		
6' x 8'	Ea.	7075	Amplifier, 250 watts	Ea.	2050
7' x 8'	Ea.	8100	Speaker, ceiling or wall	Ea.	213
Emergency Lighting, 25 watt, battery operated			Trumpet	Ea.	410
Lead battery	Ea.	330	Yard Lighting, 20' aluminum pole	Ea.	3175
Nickel cadmium	Ea.	780	with 400 watt		
Fence, Chain link, 6' high			high pressure sodium		
9 ga. wire	L.F.	28	fixture.		
6 ga. wire	L.F.	30			
Gate	Ea.	390			
Flagpoles, Complete					
Aluminum, 20' high	Ea.	1850			
40' high	Ea.	4200			
70' high	Ea.	11,000			
Fiberglass, 23' high	Ea.	1325			
39'-5" high	Ea.	3450			
59' high	Ea.	6850			
Paving, Bituminous					
Wearing course plus base course	S.Y.	6.55			
Sidewalks, Concrete 4" thick	S.F.	4.48			

Important: See the Reference Section for Location Factors.

Model costs calculated for a 1 story building with 24' story height and 30,000 square feet of floor area

Warehouse

					Unit	Unit Cost	Cost Per S.F.	% Of Sub-Total
A. SUBSTRUCTURE								
1010	Standard Foundations	Poured concrete; strip and spread footings			S.F. Ground	1.55	1.55	
1020	Special Foundations	N/A			—	—	—	
1030	Slab on Grade	5" reinforced concrete with vapor barrier and granular base			S.F. Slab	13.42	13.42	20.9%
2010	Basement Excavation	Site preparation for slab and trench for foundation wall and footing			S.F. Ground	.18	.18	
2020	Basement Walls	4' foundation wall			L.F. Wall	84	1.95	
B. SHELL								
B10 Superstructure								
1010	Floor Construction	Mezzanine: open web steel joists, slab form, concrete beams, columns	10% of area		S.F. Floor	16.90	1.69	11.1%
1020	Roof Construction	Metal deck, open web steel joists, beams, columns			S.F. Roof	7.43	7.43	
B20 Exterior Enclosure								
2010	Exterior Walls	Concrete block	95% of wall		S.F. Wall	13.95	7.42	
2020	Exterior Windows	N/A			—	—	—	10.5%
2030	Exterior Doors	Steel overhead, hollow metal	5% of wall		Each	3161	1.16	
B30 Roofing								
3010	Roof Coverings	Built-up tar and gravel with flashing; perlite/EPS composite insulation			S.F. Roof	5.98	5.98	7.7%
3020	Roof Openings	Roof hatches and skylight			S.F. Roof	.33	.33	
C. INTERIORS								
1010	Partitions	Concrete block (office and washrooms)	100 S.F. Floor/L.F. Partition		S.F. Partition	9.63	.77	
1020	Interior Doors	Single leaf hollow metal	5000 S.F. Floor/Door		Each	1157	.24	
1030	Fittings	N/A			—	—	—	
2010	Stair Construction	Steel gate with rails			Flight	14,350	.96	8.6%
3010	Wall Finishes	Paint			S.F. Surface	14.56	2.33	
3020	Floor Finishes	90% hardener, 10% vinyl composition tile			S.F. Floor	2.04	2.04	
3030	Ceiling Finishes	Suspended mineral tile on zee channels in office area	10% of area		S.F. Ceiling	7.03	.71	
D. SERVICES								
D10 Conveying								
1010	Elevators & Lifts	N/A			—	—	—	0.0%
1020	Escalators & Moving Walks	N/A			—	—	—	
D20 Plumbing								
2010	Plumbing Fixtures	Toilet and service fixtures, supply and drainage	1 Fixture/2500 S.F. Floor		Each	3775	1.51	
2020	Domestic Water Distribution	Gas fired water heater			S.F. Floor	.40	.40	4.9%
2040	Rain Water Drainage	Roof drains			S.F. Roof	2.14	2.14	
D30 HVAC								
3010	Energy Supply	N/A			—	—	—	
3020	Heat Generating Systems	Ventilation with heat			Each	140,200	5.05	
3030	Cooling Generating Systems	N/A			—	—	—	7.2%
3050	Terminal & Package Units	Single zone unit gas, heating, electric cooling	10% of area		S.F. Floor	.89	.89	
3090	Other HVAC Sys. & Equipment	N/A			—	—	—	
D40 Fire Protection								
4010	Sprinklers	Sprinklers, ordinary hazard			S.F. Floor	4.15	4.15	7.9%
4020	Standpipes	Standpipe			S.F. Floor	2.29	2.29	
D50 Electrical								
5010	Electrical Service/Distribution	200 ampere service, panel board and feeders			S.F. Floor	.56	.56	
5020	Lighting & Branch Wiring	High efficiency fluorescent fixtures, receptacles, switches, A.C. and misc. power			S.F. Floor	6.13	6.13	11.5%
5030	Communications & Security	Addressable alarm systems			S.F. Floor	2.74	2.74	
5090	Other Electrical Systems	N/A			—	—	—	
E. EQUIPMENT & FURNISHINGS								
1010	Commercial Equipment	N/A			—	—	—	
1020	Institutional Equipment	N/A			—	—	—	
1030	Vehicular Equipment	Dock boards, dock levelers			S.F. Floor	2.71	2.71	3.3%
1090	Other Equipment	N/A			—	—	—	
F. SPECIAL CONSTRUCTION								
1020	Integrated Construction	Shipping & receiving air curtain			—	5.23	5.23	6.4%
1040	Special Facilities	N/A			—	—	—	
G. BUILDING SITEWORK	**N/A**							

			Sub-Total	81.96	100%
CONTRACTOR FEES (General Requirements: 10%, Overhead: 5%, Profit: 10%)			25%	20.47	
ARCHITECT FEES			7%	7.17	
		Total Building Cost		**109.60**	

For customer support on your Square Foot Cost Data, call 877.694.8329.

229

Costs per square foot of floor area

Exterior Wall	S.F. Area	2000	3000	5000	8000	12000	20000	30000	50000	100000
	L.F. Perimeter	180	220	300	420	580	900	1300	2100	4100
Concrete Block	Steel Frame	103.35	90.80	80.80	75.10	72.00	69.45	68.20	67.20	66.40
	R/Conc. Frame	105.00	96.40	89.40	85.50	83.40	81.65	80.80	80.05	79.55
Metal Sandwich Panel	Steel Frame	93.75	84.30	76.75	72.45	70.10	68.25	67.25	66.45	65.90
Tilt-up Concrete Panel	Steel and Concrete	95.60	86.95	80.00	76.05	73.90	72.15	71.30	70.55	70.05
Precast Concrete Panel	Steel Frame	112.90	99.60	88.90	82.90	79.60	76.95	75.55	74.50	73.70
	R/Conc. Frame	134.10	121.15	110.80	104.95	101.70	99.15	97.85	96.80	96.05
Perimeter Adj., Add or Deduct	Per 100 L.F.	21.10	14.05	8.40	5.30	3.50	2.15	1.40	0.85	0.40
Story Hgt. Adj., Add or Deduct	Per 1 Ft.	1.40	1.20	0.90	0.85	0.75	0.75	0.65	0.65	0.65
Basement—Not Applicable										

The above costs were calculated using the basic specifications shown on the facing page. These costs should be adjusted where necessary for design alternatives and owner's requirements. Reported completed project costs, for this type of structure, range from $37.40 to $214.45 per S.F.

Common additives

Description	Unit	$ Cost
Dock Leveler, 10 ton cap.		
6' x 8'	Ea.	7075
7' x 8'	Ea.	8100
Emergency Lighting, 25 watt, battery operated		
Lead battery	Ea.	330
Nickel cadmium	Ea.	780
Fence, Chain link, 6' high		
9 ga. wire	L.F.	28
6 ga. wire	L.F.	30
Gate	Ea.	390
Flagpoles, Complete		
Aluminum, 20' high	Ea.	1850
40' high	Ea.	4200
70' high	Ea.	11,000
Fiberglass, 23' high	Ea.	1325
39'-5" high	Ea.	3450
59' high	Ea.	6850
Paving, Bituminous		
Wearing course plus base course	S.Y.	6.55
Sidewalks, Concrete 4" thick	S.F.	4.48

Description	Unit	$ Cost
Sound System		
Amplifier, 250 watts	Ea.	2050
Speaker, ceiling or wall	Ea.	213
Trumpet	Ea.	410
Yard Lighting, 20' aluminum pole with 400 watt high pressure sodium fixture	Ea.	3175

Model costs calculated for a 1 story building with 12' story height and 20,000 square feet of floor area

Warehouse Self Storage

				Unit	Unit Cost	Cost Per S.F.	% Of Sub-Total
A. SUBSTRUCTURE							
1010	Standard Foundations	Poured concrete; strip and spread footings		S.F. Ground	2.32	2.32	
1020	Special Foundations	N/A		–	–	–	
1030	Slab on Grade	4" reinforced concrete with vapor barrier and granular base		S.F. Slab	6.70	6.70	25.8%
2010	Basement Excavation	Site preparation for slab and trench for foundation wall and footing		S.F. Ground	.32	.64	
2020	Basement Walls	4' foundation wall		L.F. Wall	84	3.76	
B. SHELL							
	B10 Superstructure						
1010	Floor Construction	N/A		–	–	–	
1020	Roof Construction	Metal deck, open web steel joists, beams, columns		S.F. Roof	6.72	6.72	14.3 %
	B20 Exterior Enclosure						
2010	Exterior Walls	Concrete block	53% of wall	S.F. Wall	22.68	6.49	
2020	Exterior Windows	Aluminum projecting	1% of wall	Each	1160	.11	24.3%
2030	Exterior Doors	Steel overhead, hollow metal	46% of wall	Each	1485	6.02	
	B30 Roofing						
3010	Roof Coverings	Metal panel roof; perlite/EPS composite insulation		S.F. Roof	6.98	6.98	13.4%
3020	Roof Openings	N/A		–	–	–	
C. INTERIORS							
1010	Partitions	metal panels on metal studs	11S.F. Floor/L.F. Partition	S.F. Partition	1.14	1.24	
1020	Interior Doors	Single leaf hollow metal	20,000 S.F. Floor/Door	Each	600	.06	
1030	Fittings	N/A		–	–	–	
2010	Stair Construction	N/A		–	–	–	2.5%
3010	Wall Finishes	N/A		–	–	–	
3020	Floor Finishes	N/A		–	–	–	
3030	Ceiling Finishes	N/A		–	–	–	
D. SERVICES							
	D10 Conveying						
1010	Elevators & Lifts	N/A		–	–	–	0.0 %
1020	Escalators & Moving Walks	N/A		–	–	–	
	D20 Plumbing						
2010	Plumbing Fixtures	Toilet and service fixtures, supply and drainage	1 Fixture/10,000 S.F. Floor	Each	2800	.28	
2020	Domestic Water Distribution	Gas fired water heater		S.F. Floor	.28	.28	2.3%
2040	Rain Water Drainage	N/A		–	–	–	
	D30 HVAC						
3010	Energy Supply	N/A		–	–	–	
3020	Heat Generating Systems	N/A		–	–	–	
3030	Cooling Generating Systems	N/A		–	–	–	2.8 %
3050	Terminal & Package Units	Single zone rooftop unit		S.F. Floor	1.45	1.45	
3090	Other HVAC Sys. & Equipment	N/A		–	–	–	
	D40 Fire Protection						
4010	Sprinklers	N/A		–	–	–	0.0 %
4020	Standpipes	N/A		–	–	–	
	D50 Electrical						
5010	Electrical Service/Distribution	60 ampere service, panel board and feeders		S.F. Floor	.39	.39	
5020	Lighting & Branch Wiring	High bay fixtures, receptacles, switches and misc. power		S.F. Floor	5.80	5.80	14.5%
5030	Communications & Security	Addressable alarm systems		S.F. Floor	1.35	1.35	
5090	Other Electrical Systems	N/A		–	–	–	
E. EQUIPMENT & FURNISHINGS							
1010	Commercial Equipment	N/A		–	–	–	
1020	Institutional Equipment	N/A		–	–	–	
1030	Vehicular Equipment	N/A		–	–	–	0.0 %
1090	Other Equipment	N/A		–	–	–	
F. SPECIAL CONSTRUCTION							
1020	Integrated Construction	N/A		–	–	–	
1040	Special Facilities	N/A		–	–	–	0.0 %
G. BUILDING SITEWORK	**N/A**						

			Sub-Total	51.94	100%
CONTRACTOR FEES (General Requirements: 10%, Overhead: 5%, Profit: 10%)			25%	12.96	
ARCHITECT FEES			7%	4.55	

Total Building Cost	**69.45**

For customer support on your Square Foot Cost Data, call 877.694.8329.

231

Deterioration occurs within the structure itself and is determined by the observation of both materials and equipment.

Curable deterioration can be remedied either by maintenance, repair or replacement, within prudent economic limits.

Incurable deterioration that has progressed to the point of actually affecting the structural integrity of the structure, making repair or replacement not economically feasible.

Actual Versus Observed Age

The observed age of a structure refers to the age that the structure appears to be. Periodic maintenance, remodeling, and renovation all tend to reduce the amount of deterioration that has taken place, thereby decreasing the observed age. Actual age, on the other hand, relates solely to the year that the structure was built.

The Depreciation Table shown here relates to the observed age.

Obsolescence arises from conditions either occurring within the structure (functional) or caused by factors outside the limits of the structure (economic).

Functional obsolescence is any inadequacy caused by outmoded design, dated construction materials, or oversized or undersized areas, all of which cause excessive operation costs.

Incurable is so costly as not to economically justify the capital expenditure required to correct the deficiency.

Economic obsolescence is caused by factors outside the limits of the structure. The prime causes of economic obsolescence are:

- zoning and environmental laws
- government legislation
- negative neighborhood influences
- business climate
- proximity to transportation facilities

Depreciation Table
Commercial/Industrial/Institutional

Building Material			
Observed Age (Years)	Frame	Masonry On Wood	Masonry On Masonry Or Steel
1	1%	0%	0%
2	2	1	0
3	3	2	1
4	4	3	2
5	6	5	3
10	20	15	8
15	25	20	15
20	30	25	20
25	35	30	25
30	40	35	30
35	45	40	35
40	50	45	40
45	55	50	45
50	60	55	50
55	65	60	55
60	70	65	60

Green Commercial/Industrial/Institutional Section

Table of Contents

Costs per square foot of floor area

Exterior Wall	S.F. Area	2000	2700	3400	4100	4800	5500	6200	6900	7600
	L.F. Perimeter	180	208	236	256	280	303	317	337	357
Face Brick with Concrete Block Back-up	Steel Frame	296.85	277.50	266.10	256.40	250.55	245.95	240.70	237.55	235.00
	R/Conc. Frame	312.40	293.00	281.65	271.95	266.10	261.50	256.25	253.10	250.50
Precast Concrete Panel	Steel Frame	295.65	276.40	265.05	255.45	249.55	245.00	239.80	236.65	234.10
	R/Conc. Frame	314.80	294.40	282.45	272.25	266.15	261.30	255.95	252.65	249.95
Limestone with Concrete Block Back-up	Steel Frame	323.15	300.00	286.40	274.70	267.55	262.05	255.65	251.85	248.75
	R/Conc. Frame	339.45	316.30	302.75	290.95	283.90	278.35	271.95	268.15	265.00
Perimeter Adj., Add or Deduct	Per 100 L.F.	55.35	41.00	32.60	27.05	23.05	20.15	17.90	16.05	14.60
Story Hgt. Adj., Add or Deduct	Per 1 Ft.	5.35	4.55	4.10	3.70	3.45	3.25	3.00	2.90	2.80

For Basement, add $37.65 per square foot of basement area

The above costs were calculated using the basic specifications shown on the facing page. These costs should be adjusted where necessary for design alternatives and owner's requirements. Reported completed project costs, for this type of structure, range from $156.45 to $385.00 per S.F.

Common additives

Description	Unit	$ Cost
Bulletproof Teller Window, 44" x 60"	Ea.	5175
60" x 48"	Ea.	7075
Closed Circuit Surveillance, one station, camera & monitor	Ea.	1975
For additional camera stations, add	Ea.	1075
Counters, complete, dr & frame, 3' x 6'-8", bullet resist stl		
with vision panel	Ea.	7450 - 9625
Drive-up Window, drawer & micr., not incl. glass	Ea.	9075 - 13,000
Night Depository	Ea.	9775 - 14,700
Package Receiver, painted	Ea.	1975
Stainless steel	Ea.	3025
Partitions, bullet resistant to 8' high	L.F.	335 - 535
Pneumatic Tube Systems, 2 station	Ea.	33,800
With TV viewer	Ea.	62,000
Service Windows, pass thru, steel, 24" x 36"	Ea.	3850
48" x 48"	Ea.	4200
Twenty-four Hour Teller, automatic deposit cash & memo	Ea.	54,500
Vault Front, door & frame, 2 hour test, 32" door	Opng.	10,100
4 hour test, 40" door	Opng.	12,600
Time lock, two movement, add	Ea.	2300

Description	Unit	$ Cost
Commissioning Fees, sustainable commercial construction	S.F.	0.24 - 3.04
Energy Modelling Fees, banks to 10,000 SF	Ea.	11,000
Green Bldg Cert Fees for comm construction project reg	Project	900
Photovoltaic Pwr Sys, grid connected, 20 kW (~2400 SF), roof	Ea.	275,000
Green Roofs, 6" soil depth, w/ treated wd edging & sedum mats	S.F.	10.73
10" Soil depth, with treated wood edging & sedum mats	S.F.	12.44
Greywater Recovery Systems, prepackaged comm, 1530 gal	Ea.	52,125
Rainwater Harvest Sys, prepckged comm, 10,000 gal, sys contrller	Ea.	31,500
20,000 gal. w/system controller	Ea.	50,550
30,000 gal. w/system controller	Ea.	80,800
Solar Domestic HW, closed loop, add-on sys, ext heat exchanger	Ea.	10,775
Draindown, hot water system, 120 gal tank	Ea.	13,675

Important: See the Reference Section for Location Factors.

Model costs calculated for a 1 story building with 14' story height and 4,100 square feet of floor area

G Bank

				Unit	Unit Cost	Cost Per S.F.	% Of Sub-Total
A. SUBSTRUCTURE							
1010	Standard Foundations	Poured concrete; strip and spread footings		S.F. Ground	4.33	4.33	
1020	Special Foundations	N/A		—	—	—	
1030	Slab on Grade	4" reinforced concrete with recycled vapor barrier and granular base		S.F. Slab	5.45	5.45	8.6%
2010	Basement Excavation	Site preparation for slab and trench for foundation wall and footing		S.F. Ground	.32	.32	
2020	Basement Walls	4' Foundation wall		L.F. Wall	91	6.82	
B. SHELL							
B10 Superstructure							
1010	Floor Construction	Cast-in-place columns		L.F. Column	6.05	6.05	
1020	Roof Construction	Cast-in-place concrete flat plate		S.F. Roof	16.10	16.10	11.3%
B20 Exterior Enclosure							
2010	Exterior Walls	Face brick with concrete block backup	80% of wall	S.F. Wall	38.69	27.06	
2020	Exterior Windows	Horizontal aluminum sliding	20% of wall	Each	588	6.85	18.1%
2030	Exterior Doors	Double aluminum and glass and hollow metal, low VOC paint		Each	3338	1.63	
B30 Roofing							
3010	Roof Coverings	Single-ply TPO membrane, 60 mils, heat welded seams w/4" thk. R20 insul.		S.F. Roof	7.16	7.16	3.7%
3020	Roof Openings	N/A		—	—	—	
C. INTERIORS							
1010	Partitions	Gypsum board on metal studs w/sound attenuation	20 SF of Flr./LF Part.	S.F. Partition	12.20	6.10	
1020	Interior Doors	Single leaf hollow metal , low VOC paint	200 S.F. Floor/Door	Each	1157	5.79	
1030	Fittings	N/A		—	—	—	
2010	Stair Construction	N/A		—	—	—	13.3%
3010	Wall Finishes	50% vinyl wall covering, 50% paint, low VOC		S.F. Surface	1.51	1.51	
3020	Floor Finishes	50% carpet tile, 40% vinyl composition tile, recycled, 10% quarry tile		S.F. Floor	5.60	5.60	
3030	Ceiling Finishes	Mineral fiber tile on concealed zee bars		S.F. Ceiling	7.03	7.03	
D. SERVICES							
D10 Conveying							
1010	Elevators & Lifts	N/A		—	—	—	
1020	Escalators & Moving Walks	N/A		—	—	—	0.0 %
D20 Plumbing							
2010	Plumbing Fixtures	Toilet low flow, auto sensor and service fixt., supply and drainage	1 Fixt./580 SF Flr.	Each	5928	10.22	
2020	Domestic Water Distribution	Tankless, on demand water heaters, natural gas/propane		S.F. Floor	1	1	6.5%
2040	Rain Water Drainage	Roof drains		S.F. Roof	1.59	1.59	
D30 HVAC							
3010	Energy Supply	N/A		—	—	—	
3020	Heat Generating Systems	Included in D3050		—	—	—	
3040	Distribution Systems	Enthalpy heat recovery packages		Each	14,825	3.62	6.2 %
3050	Terminal & Package Units	Single zone rooftop air conditioner		S.F. Floor	8.59	8.59	
3090	Other HVAC Sys. & Equipment	N/A		—	—	—	
D40 Fire Protection							
4010	Sprinklers	Wet pipe sprinkler system		S.F. Floor	4.98	4.98	
4020	Standpipes	Standpipe		S.F. Floor	2.94	2.94	4.0%
D50 Electrical							
5010	Electrical Service/Distribution	200 ampere service, panel board and feeders		S.F. Floor	1.47	1.47	
5020	Lighting & Branch Wiring	LED fixtures, daylt. dimming & ltg. on/off control, receptacles, switches, A.C.		S.F. Floor	12.93	12.93	
5030	Communications & Security	Alarm systems, internet/phone wiring, and security television		S.F. Floor	10.91	10.91	15.2%
5090	Other Electrical Systems	Emergency generator, 15 kW, UPS, energy monitoring systems		S.F. Floor	4.42	4.42	
E. EQUIPMENT & FURNISHINGS							
1010	Commercial Equipment	Automatic teller, drive up window, night depository		S.F. Floor	8.02	8.02	
1020	Institutional Equipment	Closed circuit TV monitoring system		S.F. Floor	2.93	2.93	
1090	Other Equipment	Waste handling recycling tilt truck		Each	1.38	1.38	6.3%
2020	Moveable Furnishings	No smoking signage		Each	.02	.02	
F. SPECIAL CONSTRUCTION							
1020	Integrated Construction	N/A		—	—	—	
1040	Special Facilities	Security vault door		S.F. Floor	13.17	13.17	6.7 %
G. BUILDING SITEWORK	**N/A**						

Sub-Total		195.99	100%
CONTRACTOR FEES (General Requirements: 10%, Overhead: 5%, Profit: 10%)	25%	49.01	
ARCHITECT FEES	11%	26.95	
Total Building Cost		**271.95**	

For customer support on your Square Foot Cost Data, call 877.694.8329.

235

Costs per square foot of floor area

Exterior Wall	S.F. Area	15000	20000	28000	38000	50000	65000	85000	100000	150000
	L.F. Perimeter	350	400	480	550	630	660	750	825	1035
Face Brick with Concrete Block Back-up	Steel Frame	239.15	229.05	220.30	213.25	208.40	203.05	199.80	198.35	195.20
	Bearing Walls	244.25	231.80	221.20	212.30	206.20	199.20	195.10	193.20	189.20
Decorative Concrete Block	Steel Frame	230.40	221.45	213.80	207.80	203.65	199.20	196.50	195.25	192.65
	Bearing Walls	234.85	223.75	214.20	206.35	200.90	194.85	191.25	189.60	186.10
Stucco on Concrete Block	Steel Frame	244.10	233.20	223.90	216.30	211.05	205.15	201.70	200.05	196.65
	Bearing Walls	249.15	236.00	224.75	215.35	208.85	201.30	196.95	194.95	190.65
Perimeter Adj., Add or Deduct	Per 100 L.F.	11.20	8.30	5.95	4.40	3.35	2.55	2.00	1.70	1.10
Story Hgt. Adj., Add or Deduct	Per 1 Ft.	2.85	2.40	2.05	1.70	1.50	1.15	1.05	1.00	0.80

For Basement, add $ 40.40 per square foot of basement area

The above costs were calculated using the basic specifications shown on the facing page. These costs should be adjusted where necessary for design alternatives and owner's requirements. Reported completed project costs, for this type of structure, range from $ 140.10 to $ 329.30 per S.F.

Common additives

Description	Unit	$ Cost
Clock System, 20 room	Ea.	19,900
50 room	Ea.	46,800
Elevators, hydraulic passenger, 2 stops, 2000# capacity	Ea.	66,600
3500# capacity	Ea.	73,300
Additional stop, add	Ea.	7825
Seating, auditorium chair, all veneer	Ea.	310
Veneer back, padded seat	Ea.	320
Upholstered, spring seat	Ea.	320
Classroom, movable chair & desk	Set	81 - 171
Lecture hall, pedestal type	Ea.	263 - 620
Sound System, amplifier, 250 watts	Ea.	2050
Speaker, ceiling or wall	Ea.	213
Trumpet	Ea.	410
TV Antenna, Master system, 30 outlet	Outlet	221
100 outlet	Outlet	209

Description	Unit	$ Cost
Commissioning Fees, sustainable institutional construction	S.F.	0.58 - 2.47
Energy Modelling Fees, academic buildings to 10,000 SF	Ea.	9000
Greater than 10,000 SF add	S.F.	0.20
Green Bldg Cert Fees for school construction project reg	Project	900
Photovoltaic Pwr Sys, grid connected, 20 kW (~2400 SF), roof	Ea.	275,000
Green Roofs, 6" soil depth, w/treated wd edging & sedum mats	S.F.	10.73
10" Soil depth, with treated wood edging & sedum mats	S.F.	12.44
Greywater Recovery Systems, prepackaged comm, 3060 gal.	Ea.	72,255
4590 gal.	Ea.	89,550
Rainwater Harvest Sys, prepckged comm, 10,000 gal, sys contrller	Ea.	31,500
20,000 gal. w/system controller	Ea.	50,550
30,000 gal. w/system controller	Ea.	80,800
Solar Domestic HW, closed loop, add-on sys, ext heat exchanger	Ea.	10,775
Drainback, hot water system, 120 gal tank	Ea.	13,325
Draindown, hot water system, 120 gal tank	Ea.	13,675

Model costs calculated for a 2 story building with 12' story height and 50,000 square feet of floor area

				Unit	Unit Cost	Cost Per S.F.	% Of Sub-Total
A. SUBSTRUCTURE							
1010	Standard Foundations	Poured concrete; strip and spread footings		S.F. Ground	1.38	.69	
1020	Special Foundations	N/A		—	—	—	
1030	Slab on Grade	4" reinforced concrete with vapor barrier and granular base		S.F. Slab	5.45	2.73	3.6%
2010	Basement Excavation	Site preparation for slab and trench for foundation wall and footing		S.F. Ground	.32	.16	
2020	Basement Walls	4' Foundation wall		L.F. Wall	174	1.76	
B. SHELL							
B10 Superstructure							
1010	Floor Construction	Open web steel joists, slab form, concrete		S.F. Floor	17.36	8.68	9.1%
1020	Roof Construction	Metal deck on open web steel joists, columns		S.F. Roof	9.96	4.98	
B20 Exterior Enclosure							
2010	Exterior Walls	Decorative concrete block	65% of wall	S.F. Wall	20.50	4.03	
2020	Exterior Windows	Window wall	35% of wall	Each	52	5.15	6.6%
2030	Exterior Doors	Double glass and aluminum with transom		Each	5825	.70	
B30 Roofing							
3010	Roof Coverings	Built-up tar and gravel with flashing; perlite/EPS composite insulation		S.F. Roof	5.58	2.79	1.9%
3020	Roof Openings	N/A		—	—	—	
C. INTERIORS							
1010	Partitions	Concrete block	20 SF Flr.LF Part.	S.F. Partition	22.58	11.29	
1020	Interior Doors	Single leaf hollow metal	200 S.F. Floor/Door	Each	1157	5.79	
1030	Fittings	Chalkboards, counters, cabinets		S.F. Floor	6.03	6.03	
2010	Stair Construction	Concrete filled metal pan		Flight	17,300	3.46	27.9%
3010	Wall Finishes	95% paint, 5% ceramic tile		S.F. Surface	4.35	4.35	
3020	Floor Finishes	70% vinyl composition tile, 25% carpet, 5% ceramic tile		S.F. Floor	4.01	4.01	
3030	Ceiling Finishes	Mineral fiber tile on concealed zee bars		S.F. Ceiling	7.03	7.03	
D. SERVICES							
D10 Conveying							
1010	Elevators & Lifts	Two hydraulic passenger elevators		Each	83,500	3.34	2.2%
1020	Escalators & Moving Walks	N/A		—	—	—	
D20 Plumbing							
2010	Plumbing Fixtures	Toilet, low flow, auto sensor & service fixt., supply & drain.	1 Fixt./455 SF Flr.	Each	9400	20.66	
2020	Domestic Water Distribution	Oil fired hot water heater		S.F. Floor	.80	.80	14.8%
2040	Rain Water Drainage	Roof drains		S.F. Roof	1.40	.70	
D30 HVAC							
3010	Energy Supply	N/A		—	—	—	
3020	Heat Generating Systems	Included in D3050		—	—	—	
3040	Distribution Systems	Enthalpy heat recovery packages		Each	36,925	2.21	14.4 %
3050	Terminal & Package Units	Multizone unit, gas heating, electric cooling, SEER 14		S.F. Floor	19.45	19.45	
3090	Other HVAC Sys. & Equipment	N/A		—	—	—	
D40 Fire Protection							
4010	Sprinklers	Sprinklers, light hazard		S.F. Floor	3.19	3.19	2.4%
4020	Standpipes	Standpipe, dry, Class III		S.F. Floor	.36	.36	
D50 Electrical							
5010	Electrical Service/Distribution	2000 ampere service, panel board and feeders		S.F. Floor	3.05	3.05	
5020	Lighting & Branch Wiring	LED fixtures, receptacles, switches, A.C. & misc. power		S.F. Floor	13.94	13.94	17.2%
5030	Communications & Security	Addressable alarm sys., internet wiring, comm. sys. & emerg. ltg.		S.F. Floor	7.22	7.22	
5090	Other Electrical Systems	Emergency generator, 100 kW		S.F. Floor	1.56	1.56	
E. EQUIPMENT & FURNISHINGS							
1010	Commercial Equipment	N/A		—	—	—	
1020	Institutional Equipment	N/A		—	—	—	
1090	Other Equipment	Waste handling recycling tilt truck		Each	.11	.11	0.1 %
2020	Moveable Furnishings	No smoking signage		Each			
F. SPECIAL CONSTRUCTION							
1020	Integrated Construction	N/A		—	—	—	
1040	Special Facilities	N/A		—	—	—	0.0 %
G. BUILDING SITEWORK	**N/A**						

	Sub-Total	150.22	100%
CONTRACTOR FEES (General Requirements: 10%, Overhead: 5%, Profit: 10%)	25%	37.54	
ARCHITECT FEES	7%	13.14	
Total Building Cost		**200.90**	

For customer support on your Square Foot Cost Data, call 877.694.8329.

237

Costs per square foot of floor area

Exterior Wall	S.F. Area	10000	15000	25000	40000	55000	70000	80000	90000	100000
	L.F. Perimeter	260	320	400	476	575	628	684	721	772
Face Brick with Concrete Block Back-up	R/Conc. Frame	245.60	230.90	215.65	204.70	200.65	196.80	195.70	194.35	193.50
	Steel Frame	268.95	254.25	239.00	228.05	224.00	220.15	219.00	217.70	216.85
Decorative Concrete Block	R/Conc. Frame	238.10	226.70	215.15	207.05	204.00	201.20	200.35	199.45	198.80
	Steel Frame	250.70	239.30	227.80	219.70	216.65	213.85	213.00	212.10	211.45
Precast Concrete Panels	R/Conc. Frame	256.55	241.80	226.50	215.55	211.45	207.60	206.45	205.10	204.30
	Steel Frame	269.15	254.45	239.15	228.20	224.05	220.25	219.05	217.75	216.95
Perimeter Adj., Add or Deduct	Per 100 L.F.	21.75	14.55	8.75	5.45	3.95	3.10	2.70	2.35	2.15
Story Hgt. Adj., Add or Deduct	Per 1 Ft.	4.20	3.45	2.65	1.90	1.65	1.45	1.35	1.25	1.30

For Basement, add $39.35 per square foot of basement area

The above costs were calculated using the basic specifications shown on the facing page. These costs should be adjusted where necessary for design alternatives and owner's requirements. Reported completed project costs, for this type of structure, range from $90.50 to $266.70 per S.F.

Common additives

Description	Unit	$ Cost
Closed Circuit Surveillance, one station, camera & monitor	Ea.	1975
For additional camera stations, add	Ea.	1075
Elevators, hydraulic passenger, 2 stops, 3500# capacity	Ea.	73,300
Additional stop, add	Ea.	7825
Furniture	Student	2800 - 5350
Intercom System, 25 station capacity		
Master station	Ea.	3250
Intercom outlets	Ea.	189
Kitchen Equipment		
Broiler	Ea.	4050
Coffee urn, twin 6 gallon	Ea.	3000
Cooler, 6 ft. long	Ea.	4175
Dishwasher, 10-12 racks per hr.	Ea.	4275
Food warmer	Ea.	735
Freezer, 44 C.F., reach-in	Ea.	5175
Ice cube maker, 50 lb. per day	Ea.	2000
Range with 1 oven	Ea.	2925
Laundry Equipment, dryer, 30 lb. capacity	Ea.	3850
Washer, commercial	Ea.	1600
TV Antenna, Master system, 30 outlet	Outlet	221
100 outlet	Outlet	209

Description	Unit	$ Cost
Commissioning Fees, sustainable institutional construction	S.F.	0.58 - 2.47
Energy Modelling Fees, academic buildings to 10,000 SF	Ea.	9000
Greater than 10,000 SF add	S.F.	0.20
Green Bldg Cert Fees for school construction project reg	Project	900
Photovoltaic Pwr Sys, grid connected, 20 kW (~2400 SF), roof	Ea.	275,000
Green Roofs, 6" soil depth, w/treated wd edging & sedum mats	S.F.	10.73
10" Soil depth, with treated wood edging & sedum mats	S.F.	12.44
Greywater Recovery Systems, prepackaged comm, 3060 gal.	Ea.	72,255
4590 gal.	Ea.	89,550
Rainwater Harvest Sys, prepckged comm, 30,000 gal, sys contrller	Ea.	80,800
Solar Domestic HW, closed loop, add-on sys, ext heat exchanger	Ea.	10,775
Drainback, hot water system, 120 gal tank	Ea.	13,325
Draindown, hot water system, 120 gal tank	Ea.	13,675

Important: See the Reference Section for Location Factors.

Model costs calculated for a 3 story building with 12' story height and 25,000 square feet of floor area

G College, Dormitory, 2-3 Story

				Unit	Unit Cost	Cost Per S.F.	% Of Sub-Total
A. SUBSTRUCTURE							
1010	Standard Foundations	Poured concrete; strip and spread footings		S.F. Ground	6.39	2.13	
1020	Special Foundations	N/A		—	—	—	
1030	Slab on Grade	4" reinforced concrete with recycled vapor barrier and granular base		S.F. Slab	5.45	1.81	3.4%
2010	Basement Excavation	Site preparation for slab and trench for foundation wall and footing		S.F. Ground	.18	.06	
2020	Basement Walls	4' Foundation wall		L.F. Wall	91	1.45	
B. SHELL							
B10 Superstructure							
1010	Floor Construction	Concrete flat plate		S.F. Floor	25.59	17.06	13.5%
1020	Roof Construction	Concrete flat plate		S.F. Roof	13.98	4.66	
B20 Exterior Enclosure							
2010	Exterior Walls	Face brick with concrete block backup	80% of wall	S.F. Wall	38.69	17.83	
2020	Exterior Windows	Aluminum horizontal sliding	20% of wall	Each	750	3.75	14.3%
2030	Exterior Doors	Double glass & aluminum doors		Each	5975	1.43	
B30 Roofing							
3010	Roof Coverings	Single-ply TPO membrane, heat welded w/R-20 insul.		S.F. Roof	6.54	2.18	1.4%
3020	Roof Openings	N/A		—	—	—	
C. INTERIORS							
1010	Partitions	Gypsum board on metal studs, CMU w/foamed-in insul.	9 SF Flr./LF Part.	S.F. Partition	7.67	8.52	
1020	Interior Doors	Single leaf wood, low VOC paint	90 S.F. Floor/Door	Each	616	6.84	
1030	Fittings	Closet shelving, mirrors, bathroom accessories		S.F. Floor	1.96	1.96	
2010	Stair Construction	Cast in place concrete		Flight	5375	2.79	19.1%
3010	Wall Finishes	95% paint, low VOC paint, 5% ceramic tile		S.F. Surface	2.07	4.59	
3020	Floor Finishes	80% carpet tile, 10% vinyl comp. tile, recycled content, 10% ceramic tile		S.F. Floor	4.96	4.96	
3030	Ceiling Finishes	90% paint, low VOC, 10% suspended fiberglass board		S.F. Ceiling	1.20	1.20	
D. SERVICES							
D10 Conveying							
1010	Elevators & Lifts	One hydraulic passenger elevator		Each	114,000	4.56	2.8%
1020	Escalators & Moving Walks	N/A		—	—	—	
D20 Plumbing							
2010	Plumbing Fixtures	Toilet, low flow, auto sensor, & service fixt., supply & drain.	1 Fixt./455 SF Flr.	Each	11,825	25.99	
2020	Domestic Water Distribution	Electric, point-of-use water heater		S.F. Floor	1.26	1.26	17.2%
2040	Rain Water Drainage	Roof drains		S.F. Roof	1.47	.49	
D30 HVAC							
3010	Energy Supply	N/A		—	—	—	
3020	Heat Generating Systems	Included in D3050		—	—	—	
3040	Distribution Systems	Enthalpy heat recovery packages		Each	30,400	2.43	9.5 %
3050	Terminal & Package Units	Rooftop multizone unit system, SEER 14		S.F. Floor	12.86	12.86	
3090	Other HVAC Sys. & Equipment	N/A		—	—	—	
D40 Fire Protection							
4010	Sprinklers	Wet pipe sprinkler system		S.F. Floor	3.02	3.02	2.4%
4020	Standpipes	Standpipe, dry, Class III		S.F. Floor	.88	.88	
D50 Electrical							
5010	Electrical Service/Distribution	800 ampere service, panel board and feeders		S.F. Floor	1.60	1.60	
5020	Lighting & Branch Wiring	LED fixtures, daylt. dim., ltg. on/off, recept., switches, and A.C. power		S.F. Floor	12.74	12.74	14.2%
5030	Communications & Security	Alarm sys., internet & phone wiring, comm. sys. and emergency ltg.		S.F. Floor	7.44	7.44	
5090	Other Electrical Systems	Emergency generator, 7.5 kW, energy monitoring systems		S.F. Floor	1.12	1.12	
E. EQUIPMENT & FURNISHINGS							
1010	Commercial Equipment	N/A		—	—	—	
1020	Institutional Equipment	N/A		—	—	—	
1090	Other Equipment	Waste handling recycling tilt truck		Each	.23	.23	2.3 %
2020	Moveable Furnishings	Dormitory furniture, no smoking signage		S.F. Floor	3.40	3.40	
F. SPECIAL CONSTRUCTION							
1020	Integrated Construction	N/A		—	—	—	0.0 %
1040	Special Facilities	N/A		—	—	—	
G. BUILDING SITEWORK	**N/A**						

	Unit	Cost Per S.F.	%
Sub-Total		161.24	**100%**
CONTRACTOR FEES (General Requirements: 10%, Overhead: 5%, Profit: 10%)	25%	40.30	
ARCHITECT FEES	7%	14.11	
Total Building Cost		**215.65**	

For customer support on your Square Foot Cost Data, call 877.694.8329.

239

Costs per square foot of floor area

Exterior Wall	S.F. Area	20000	35000	45000	65000	85000	110000	135000	160000	200000
	L.F. Perimeter	260	340	400	440	500	540	560	590	640
Face Brick with Concrete Block Back-up	R/Conc. Frame	255.05	236.35	231.45	221.00	216.40	211.70	208.15	205.90	203.60
	Steel Frame	269.65	251.00	246.10	235.60	231.05	226.35	222.80	220.55	218.20
Decorative Concrete Block	R/Conc. Frame	238.45	224.00	220.10	212.35	208.90	205.50	202.90	201.20	199.50
	Steel Frame	245.70	231.20	227.35	219.55	216.10	212.70	210.10	208.45	206.75
Precast Concrete Panels With Exposed Aggregate	R/Conc. Frame	254.95	236.30	231.45	220.95	216.40	211.70	208.15	205.90	203.60
	Steel Frame	269.60	250.95	246.10	235.60	231.00	226.35	222.80	220.50	218.15
Perimeter Adj., Add or Deduct	Per 100 L.F.	21.15	12.10	9.40	6.50	5.00	3.85	3.10	2.65	2.10
Story Hgt. Adj., Add or Deduct	Per 1 Ft.	4.20	3.20	2.95	2.20	1.95	1.65	1.35	1.20	1.00
For Basement, add $39.35 per square foot of basement area										

The above costs were calculated using the basic specifications shown on the facing page. These costs should be adjusted where necessary for design alternatives and owner's requirements. Reported completed project costs, for this type of structure, range from $128.65 to $310.25 per S.F.

Common additives

Description	Unit	$ Cost
Closed Circuit Surveillance, one station, camera & monitor	Ea.	1975
For additional camera stations, add	Ea.	1075
Elevators, electric passenger, 5 stops, 3500# capacity	Ea.	171,500
Additional stop, add	Ea.	10,200
Furniture	Student	2800 - 5350
Intercom System, 25 station capacity		
Master station	Ea.	3250
Intercom outlets	Ea.	189
Kitchen Equipment		
Broiler	Ea.	4050
Coffee urn, twin, 6 gallon	Ea.	3000
Cooler, 6 ft. long	Ea.	4175
Dishwasher, 10-12 racks per hr.	Ea.	4275
Food warmer	Ea.	735
Freezer, 44 C.F., reach-in	Ea.	5175
Ice cube maker, 50 lb. per day	Ea.	2000
Range with 1 oven	Ea.	2925
Laundry Equipment, dryer, 30 lb. capacity	Ea.	3850
Washer, commercial	Ea.	1600
TV Antenna, Master system, 30 outlet	Outlet	221
100 outlet	Outlet	209

Description	Unit	$ Cost
Commissioning Fees, sustainable institutional construction	S.F.	0.58 - 2.47
Energy Modelling Fees, academic buildings to 10,000 SF	Ea.	9000
Greater than 10,000 SF add	S.F.	0.20
Green Bldg Cert Fees for school construction project reg	Project	900
Photovoltaic Pwr Sys, grid connected, 20 kW (~2400 SF), roof	Ea.	275,000
Green Roofs, 6" soil depth, w/treated wd edging & sedum mats	S.F.	10.73
10" Soil depth, with treated wood edging & sedum mats	S.F.	12.44
Greywater Recovery Systems, prepackaged comm, 3060 gal.	Ea.	72,255
4590 gal.	Ea.	89,550
Rainwater Harvest Sys, prepckged comm, 30,000 gal, sys contrller	Ea.	80,800
Solar Domestic HW, closed loop, add-on sys, ext heat exchanger	Ea.	10,775
Drainback, hot water system, 120 gal tank	Ea.	13,325
Draindown, hot water system, 120 gal tank	Ea.	13,675

Model costs calculated for a 6 story building with 12' story height and 85,000 square feet of floor area

G College, Dormitory, 4-8 Story

					Unit	Unit Cost	Cost Per S.F.	% Of Sub-Total
A.	**SUBSTRUCTURE**							
1010	Standard Foundations	Poured concrete; strip and spread footings			S.F. Ground	9.36	1.56	
1020	Special Foundations	N/A			—	—	—	
1030	Slab on Grade	4" reinforced concrete with recycled vapor barrier and granular base			S.F. Slab	5.45	.91	2.0%
2010	Basement Excavation	Site preparation for slab and trench for foundation wall and footing			S.F. Ground	.32	.05	
2020	Basement Walls	4' foundation wall			L.F. Wall	174	.74	
B.	**SHELL**							
	B10 Superstructure							
1010	Floor Construction	Concrete slab with metal deck and beams			S.F. Floor	30.26	25.22	18.0%
1020	Roof Construction	Concrete slab with metal deck and beams			S.F. Roof	22.74	3.79	
	B20 Exterior Enclosure							
2010	Exterior Walls	Decorative concrete block	80% of wall		S.F. Wall	22.14	7.50	
2020	Exterior Windows	Aluminum horizontal sliding	20% of wall		Each	528	2.98	6.7%
2030	Exterior Doors	Double glass & aluminum doors			Each	3694	.34	
	B30 Roofing							
3010	Roof Coverings	Single-ply TPO membrane, 60 mils, heat welded seams w/R-20 insul.			S.F. Roof	5.94	.99	0.6%
3020	Roof Openings	N/A			—	—	—	
C.	**INTERIORS**							
1010	Partitions	Concrete block w/foamed-in insulation	9 SF Floor/LF Part.		S.F. Partition	12.35	13.72	
1020	Interior Doors	Single leaf wood, low VOC paint	90 S.F. Floor/Door		Each	621	6.90	
1030	Fittings	Closet shelving, mirrors, bathroom accessories			S.F. Floor	1.60	1.60	
2010	Stair Construction	Concrete filled metal pan			Flight	17,300	3.66	23.7%
3010	Wall Finishes	95% paint, low VOC, 5% ceramic tile			S.F. Surface	2.78	6.18	
3020	Floor Finishes	80% carpet tile, 10% vinyl comp. tile, recycled, 10% ceramic tile			S.F. Floor	4.96	4.96	
3030	Ceiling Finishes	Mineral fiber tile on concealed zee bars, paint			S.F. Ceiling	1.20	1.20	
D.	**SERVICES**							
	D10 Conveying							
1010	Elevators & Lifts	Four geared passenger elevators			Each	243,100	11.44	7.1%
1020	Escalators & Moving Walks	N/A			—	—	—	
	D20 Plumbing							
2010	Plumbing Fixtures	Toilet, low flow, auto sensor, & service fixt., supply & drain.	1 Fixt./390 SF Flr.		Each	10,105	25.91	
2020	Domestic Water Distribution	Electric water heater, point-of-use, energy saver			S.F. Floor	.70	.70	16.7%
2040	Rain Water Drainage	Roof drains			S.F. Roof	2.16	.36	
	D30 HVAC							
3010	Energy Supply	N/A			—	—	—	
3020	Heat Generating Systems	N/A			—	—	—	
3040	Distribution Systems	Enthalpy heat recovery packages			Each	30,400	2.15	9.3 %
3050	Terminal & Package Units	Multizone rooftop air conditioner, SEER 14			S.F. Floor	12.86	12.86	
3090	Other HVAC Sys. & Equipment	N/A						
	D40 Fire Protection							
4010	Sprinklers	Sprinklers, light hazard			S.F. Floor	2.93	2.93	2.2%
4020	Standpipes	Standpipe, dry, Class III			S.F. Floor	.65	.65	
	D50 Electrical							
5010	Electrical Service/Distribution	1200 ampere service, panel board and feeders			S.F. Floor	.85	.85	
5020	Lighting & Branch Wiring	LED fixtures, daylt. dim., ltg. on/off, recept. switches, and A.C. power			S.F. Floor	11.33	11.33	12.0%
5030	Communications & Security	Alarm systems, internet & phone wiring, comm. systems and emerg. ltg.			S.F. Floor	6.41	6.41	
5090	Other Electrical Systems	Emergency generator, 30 kW, energy monitoring systems			S.F. Floor	.86	.86	
E.	**EQUIPMENT & FURNISHINGS**							
1010	Commercial Equipment	N/A			—	—	—	
1020	Institutional Equipment	N/A			—	—	—	
1090	Other Equipment	Waste handling recycling tilt truck			Each	.07	.07	1.8 %
2020	Moveable Furnishings	Built-in dormitory furnishings, no smoking signage			S.F. Floor	2.77	2.77	
F.	**SPECIAL CONSTRUCTION**							
1020	Integrated Construction	N/A			—	—	—	
1040	Special Facilities	N/A			—	—	—	0.0 %
G.	**BUILDING SITEWORK**	**N/A**						

			Sub-Total	161.59	100%
CONTRACTOR FEES (General Requirements: 10%, Overhead: 5%, Profit: 10%)			25%	40.37	
ARCHITECT FEES			7%	14.14	

	Total Building Cost	**216.10**

For customer support on your Square Foot Cost Data, call 877.694.8329.

241

Costs per square foot of floor area

Exterior Wall	S.F. Area	12000	20000	28000	37000	45000	57000	68000	80000	92000
	L.F. Perimeter	470	600	698	793	900	1060	1127	1200	1320
Face Brick with Concrete Brick Back-up	Steel Frame	361.60	298.05	269.05	250.85	241.50	232.45	225.15	219.50	216.10
	Bearing Walls	341.25	277.75	248.75	230.55	221.20	212.15	204.80	199.20	195.75
Decorative Concrete Block	Steel Frame	353.50	291.90	263.95	246.40	237.40	228.60	221.75	216.40	213.15
	Bearing Walls	348.40	286.85	258.85	241.35	232.35	223.55	216.65	211.35	208.05
Stucco on Concrete Block	Steel Frame	366.45	301.80	272.15	253.55	244.00	234.80	227.20	221.45	217.90
	Bearing Walls	361.40	296.70	267.10	248.45	238.95	229.70	222.15	216.35	212.80
Perimeter Adj., Add or Deduct	Per 100 L.F.	12.45	7.50	5.40	4.05	3.30	2.60	2.25	1.90	1.60
Story Hgt. Adj., Add or Deduct	Per 1 Ft.	2.35	1.85	1.50	1.30	1.25	1.10	1.00	0.95	0.90

For Basement, add $25.45 per square foot of basement area

The above costs were calculated using the basic specifications shown on the facing page. These costs should be adjusted where necessary for design alternatives and owner's requirements. Reported completed project costs, for this type of structure, range from $185.25 to $346.00 per S.F.

Common additives

Description	Unit	$ Cost
Cabinets, Base, door units, metal	L.F.	320
Drawer units	L.F.	630
Tall storage cabinets, open	L.F.	605
With doors	L.F.	875
Wall, metal 12-1/2" deep, open	L.F.	241
With doors	L.F.	440
Countertops, not incl. base cabinets, acid proof	S.F.	58.50 - 71.50
Stainless steel	S.F.	158
Fume Hood, Not incl. ductwork	L.F.	765 - 1600
Ductwork	Hood	5850 - 9625
Glassware Washer, Distilled water rinse	Ea.	7700 - 16,000
Seating, auditorium chair, veneer back, padded seat	Ea.	320
Upholstered, spring seat	Ea.	320
Classroom, movable chair & desk	Set	81 - 171
Lecture hall, pedestal type	Ea.	263 - 620
Safety Equipment, Eye wash, hand held	Ea.	450
Deluge shower	Ea.	850
Tables, acid resist. top, drawers	L.F.	219
Titration Unit, Four 2000 ml reservoirs	Ea.	6125

Description	Unit	$ Cost
Commissioning Fees, sustainable institutional construction	S.F.	0.58 - 2.47
Energy Modelling Fees, academic buildings to 10,000 SF	Ea.	9000
Greater than 10,000 SF add	S.F.	0.20
Green Bldg Cert Fees for school construction project reg	Project	900
Photovoltaic Pwr Sys, grid connected, 20 kW (~2400 SF), roof	Ea.	275,000
Green Roofs, 6" soil depth, w/treated wd edging & sedum mats	S.F.	10.73
10" Soil depth, with treated wood edging & sedum mats	S.F.	12.44
Greywater Recovery Systems, prepackaged comm, 3060 gal.	Ea.	72,255
4590 gal.	Ea.	89,550
Rainwater Harvest Sys, prepckged comm, 30,000 gal, sys contrller	Ea.	80,800
Solar Domestic HW, closed loop, add-on sys, ext heat exchanger	Ea.	10,775
Drainback, hot water system, 120 gal tank	Ea.	13,325
Draindown, hot water system, 120 gal tank	Ea.	13,675

Important: See the Reference Section for Location Factors.

Model costs calculated for a 1 story building with 12' story height and 45,000 square feet of floor area

G College, Laboratory

				Unit	Unit Cost	Cost Per S.F.	% Of Sub-Total
A. SUBSTRUCTURE							
1010	Standard Foundations	Poured concrete; strip and spread footings		S.F. Ground	3.22	3.22	
1020	Special Foundations	N/A		—	—	—	
1030	Slab on Grade	4" reinforced concrete with recycled vapor barrier and granular base		S.F. Slab	5.45	5.45	9.9%
2010	Basement Excavation	Site preparation for slab and trench for foundation wall and footing		S.F. Ground	.18	.18	
2020	Basement Walls	4' foundation wall		L.F. Wall	174	7.11	
B. SHELL							
B10 Superstructure							
1010	Floor Construction	Metal deck on open web steel joists	(5680 S.F.)	S.F. Floor	2.26	2.26	4.6%
1020	Roof Construction	Metal deck on open web steel joists		S.F. Roof	5.13	5.13	
B20 Exterior Enclosure							
2010	Exterior Walls	Face brick with concrete block backup	75% of wall	S.F. Wall	38.72	6.97	
2020	Exterior Windows	Window wall	25% of wall	Each	52	2.92	7.5%
2030	Exterior Doors	Glass and metal doors and entrances with transom		Each	5003	2.23	
B30 Roofing							
3010	Roof Coverings	Single-ply TPO membrane, 60 mils, heat welded w/R-20 insul.		S.F. Roof	5.33	5.33	3.5%
3020	Roof Openings	Skylight		S.F. Roof	.35	.35	
C. INTERIORS							
1010	Partitions	Concrete block, foamed-in insulation	10 SF Flr./LF Part.	S.F. Partition	14.16	14.16	
1020	Interior Doors	Single leaf-kalamein fire doors, low VOC paint	820 S.F. Floor/Door	Each	1275	1.55	
1030	Fittings	Lockers		S.F. Floor	.05	.05	
2010	Stair Construction	N/A		—	—	—	21.4%
3010	Wall Finishes	60% paint, low VOC, 40% epoxy coating		S.F. Surface	2.75	5.50	
3020	Floor Finishes	60% epoxy, 20% carpet, 20% vinyl composition tile, recycled content		S.F. Floor	6.12	6.12	
3030	Ceiling Finishes	Mineral fiber tile on concealed zee runners		S.F. Ceiling	7.03	7.03	
D. SERVICES							
D10 Conveying							
1010	Elevators & Lifts	N/A		—	—	—	0.0%
1020	Escalators & Moving Walks	N/A		—	—	—	
D20 Plumbing							
2010	Plumbing Fixtures	Toilet, low flow, auto sensor & service fixt., supply & drain.	1 Fixt./260 SF Flr.	Each	9207	35.41	
2020	Domestic Water Distribution	Electric, point-of-use water heater		S.F. Floor	1.59	1.59	23.5%
2040	Rain Water Drainage	Roof drains		S.F. Roof	.75	.75	
D30 HVAC							
3010	Energy Supply	N/A		—	—	—	
3020	Heat Generating Systems	Included in D3050		—	—	—	
3040	Distribution Systems	Enthalpy heat recovery packages		Each	30,400	4.05	14.6%
3050	Terminal & Package Units	Multizone rooftop air conditioner, SEER 14		S.F. Floor	19.45	19.45	
3090	Other HVAC Sys. & Equipment	N/A		—	—	—	
D40 Fire Protection							
4010	Sprinklers	Sprinklers, light hazard		S.F. Floor	2.94	2.94	2.0%
4020	Standpipes	Standpipe, wet, Class III		S.F. Floor	.33	.33	
D50 Electrical							
5010	Electrical Service/Distribution	1000 ampere service, panel board and feeders		S.F. Floor	1.36	1.36	
5020	Lighting & Branch Wiring	LED fixtures, daylt. dim., ltg. on/off, recept., switches, & A.C. power		S.F. Floor	13.42	13.42	12.0%
5030	Communications & Security	Addressable alarm systems, internet wiring, & emergency lighting		S.F. Floor	3.56	3.56	
5090	Other Electrical Systems	Emergency generator, 11.5 kW, UPS, and energy monitoring systems		S.F. Floor	.90	.90	
E. EQUIPMENT & FURNISHINGS							
1010	Commercial Equipment	N/A		—	—	—	
1020	Institutional Equipment	Cabinets, fume hoods, lockers, glassware washer		S.F. Floor	1.41	1.41	1.0%
1090	Other Equipment	Waste handling recycling tilt truck		Each	.12	.12	
2020	Moveable Furnishings	No smoking signage		Each	.02	.02	
F. SPECIAL CONSTRUCTION							
1020	Integrated Construction	N/A		—	—	—	0.0%
1040	Special Facilities	N/A		—	—	—	
G. BUILDING SITEWORK	**N/A**						

			Sub-Total	160.87	100%
CONTRACTOR FEES (General Requirements: 10%, Overhead: 5%, Profit: 10%)		25%	40.22		
ARCHITECT FEES		10%	20.11		

Total Building Cost	**221.20**

Costs per square foot of floor area

Exterior Wall	S.F. Area	4000	4500	5000	5500	6000	6500	7000	7500	8000
	L.F. Perimeter	260	280	300	310	320	336	353	370	386
Face Brick Concrete Block Back-up	Steel Joists	210.55	206.30	202.95	198.50	194.75	192.40	190.60	189.05	187.60
	Bearing Walls	210.10	205.90	202.50	198.00	194.30	191.95	190.15	188.65	187.10
Decorative Concrete Block	Steel Joists	200.65	197.10	194.30	190.85	187.85	185.95	184.55	183.20	182.05
	Bearing Walls	193.45	189.95	187.20	183.60	180.70	178.75	177.35	176.05	174.85
Limestone with Concrete Block Back-up	Steel Joists	234.75	229.75	225.80	220.40	215.85	213.05	210.95	209.10	207.30
	Bearing Walls	227.55	222.60	218.65	213.25	208.65	205.90	203.75	201.90	200.20
Perimeter Adj., Add or Deduct	Per 100 L.F.	23.75	21.10	19.00	17.25	15.85	14.65	13.60	12.75	11.85
Story Hgt. Adj., Add or Deduct	Per 1 Ft.	3.15	3.00	2.90	2.70	2.60	2.55	2.50	2.40	2.30

For Basement, add $43.10 per square foot of basement area

The above costs were calculated using the basic specifications shown on the facing page. These costs should be adjusted where necessary for design alternatives and owner's requirements. Reported completed project costs, for this type of structure, range from $84.90 to $251.10 per S.F.

Common additives

Description	Unit	$ Cost
Appliances, cooking range, 30" free standing		
1 oven	Ea.	580 - 2450
2 oven	Ea.	1425 - 2900
Microwave oven	Ea.	296 - 840
Compactor, residential, 4-1 compaction	Ea.	825 - 1325
Dishwasher, built-in, 2 cycles	Ea.	600 - 1150
4 cycles	Ea.	755 - 1875
Garbage disposer, sink type	Ea.	232 - 380
Hood for range, 2 speed, vented, 30" wide	Ea.	330 - 1275
Refrigerator, no frost 10-12 C.F.	Ea.	560 - 650
14-16 C.F.	Ea.	660 - 870
18-20 C.F.	Ea.	805 - 1800
Lockers, Steel, single tier, 60" or 72"	Opng.	221 - 370
Locker bench, lam. maple top only	L.F.	34
Pedestals, steel pipe	Ea.	76
Sound System, amplifier, 250 watts	Ea.	2050
Speaker, ceiling or wall	Ea.	213
Trumpet	Ea.	410

Description	Unit	$ Cost
Commissioning Fees, sustainable commercial construction	S.F.	0.24 - 3.04
Energy Modelling Fees, Fire Stations to 10,000 SF	Ea.	11,000
Green Bldg Cert Fees for comm construction project reg	Project	900
Photovoltaic Pwr Sys, grid connected, 20 kW (~2400 SF), roof	Ea.	275,000
Green Roofs, 6" soil depth, w/treated wd edging & sedum mats	S.F.	10.73
10" Soil depth, with treated wood edging & sedum mats	S.F.	12.44
Greywater Recovery Systems, prepackaged comm, 3060 gal.	Ea.	72,255
4590 gal.	Ea.	89,550
Rainwater Harvest Sys, prepckged comm, 10,000 gal, sys contrller	Ea.	31,500
20,000 gal. w/system controller	Ea.	50,550
30,000 gal. w/system controller	Ea.	80,800
Solar Domestic HW, closed loop, add-on sys, ext heat exchanger	Ea.	10,775
Drainback, hot water system, 120 gal tank	Ea.	13,325

Model costs calculated for a 1 story building with 14' story height and 6,000 square feet of floor area

G Fire Station, 1 Story

				Unit	Unit Cost	Cost Per S.F.	% Of Sub-Total
A. SUBSTRUCTURE							
1010	Standard Foundations	Poured concrete; strip and spread footings		S.F. Ground	3.44	3.44	
1020	Special Foundations	N/A		—	—	—	
1030	Slab on Grade	6" reinforced concrete with recycled vapor barrier and granular base		S.F. Slab	7.14	7.14	11.1%
2010	Basement Excavation	Site preparation for slab and trench for foundation wall and footing		S.F. Ground	.57	.57	
2020	Basement Walls	4' foundation wall		L.F. Wall	91	4.86	
B. SHELL							
B10 Superstructure							
1010	Floor Construction	N/A		—	—	—	7.2 %
1020	Roof Construction	Metal deck, open web steel joists, beams on columns		S.F. Roof	10.38	10.38	
B20 Exterior Enclosure							
2010	Exterior Walls	Face brick with concrete block backup	75% of wall	S.F. Wall	38.70	21.67	
2020	Exterior Windows	Aluminum insulated glass	10% of wall	Each	864	2.02	19.6%
2030	Exterior Doors	Single aluminum and glass, overhead, hollow metal, low VOC paint	15% of wall	S.F. Door	2641	4.52	
B30 Roofing							
3010	Roof Coverings	Single-ply TPO membrane, 60 mil, with flashing, R-20 insulation and roof edges		S.F. Roof	6.77	6.77	4.8%
3020	Roof Openings	Skylights, roof hatches		S.F. Roof	.16	.16	
C. INTERIORS							
1010	Partitions	Concrete block w/foamed-in insulation	17 S.F. Floor/LF Part.	S.F. Partition	11.27	6.63	
1020	Interior Doors	Single leaf hollow metal, low VOC paint	500 S.F. Floor/Door	Each	1157	2.31	
1030	Fittings	Toilet partitions		S.F. Floor	.49	.49	
2010	Stair Construction	N/A		—	—	—	12.8%
3010	Wall Finishes	Paint, low VOC		S.F. Surface	2.44	2.87	
3020	Floor Finishes	50% vinyl tile with recycled content, 50% paint, low VOC		S.F. Floor	2.68	2.68	
3030	Ceiling Finishes	Mineral board acoustic ceiling tiles, concealed grid, suspended	50% of area	S.F. Ceiling	3.52	3.52	
D. SERVICES							
D10 Conveying							
1010	Elevators & Lifts	N/A		—	—	—	0.0 %
1020	Escalators & Moving Walks	N/A		—	—	—	
D20 Plumbing							
2010	Plumbing Fixtures	Kitchen, toilet, low flow, auto sensor & service fixt., supply & drain.	1 Fixt./375 SF Flr.	Each	4181	11.15	
2020	Domestic Water Distribution	Water heater, tankless, on-demand, natural gas/propane		S.F. Floor	2.20	2.20	10.2%
2040	Rain Water Drainage	Roof drains		S.F. Roof	1.35	1.35	
D30 HVAC							
3010	Energy Supply	N/A		—	—	—	
3020	Heat Generating Systems	Included in D3050		—	—	—	
3040	Distribution Systems	Enthalpy heat recovery packages		Each	17,525	2.92	17.2 %
3050	Terminal & Package Units	Multizone rooftop air conditioner, SEER 14		S.F. Floor	21.84	21.84	
3090	Other HVAC Sys. & Equipment	N/A		—	—	—	
D40 Fire Protection							
4010	Sprinklers	Wet pipe sprinkler system		S.F. Floor	4.98	4.98	4.4%
4020	Standpipes	Standpipe, wet, Class III		S.F. Floor	1.43	1.43	
D50 Electrical							
5010	Electrical Service/Distribution	200 ampere service, panel board and feeders		S.F. Floor	1.34	1.34	
5020	Lighting & Branch Wiring	LED fixt., daylt. dimming & ltg. on/off control, receptacles, switches, A.C. & misc. pwr.		S.F. Floor	9.87	9.87	12.0%
5030	Communications & Security	Addressable alarm systems		S.F. Floor	1.98	1.98	
5090	Other Electrical Systems	Energy monitoring systems			4.19	4.19	
E. EQUIPMENT & FURNISHINGS							
1010	Commercial Equipment	N/A		—	—	—	
1020	Institutional Equipment	N/A		—	—	—	0.7 %
1090	Other Equipment	Waste handling recycling tilt truck		Each	.93	.93	
2020	Moveable Furnishings	No smoking signage		Each	.04	.04	
F. SPECIAL CONSTRUCTION							
1020	Integrated Construction	N/A		—	—	—	0.0 %
1040	Special Facilities	N/A		—	—	—	
G. BUILDING SITEWORK	**N/A**						

Sub-Total		144.25	**100%**
CONTRACTOR FEES (General Requirements: 10%, Overhead: 5%, Profit: 10%)	25%	36.08	
ARCHITECT FEES	8%	14.42	
Total Building Cost		**194.75**	

For customer support on your Square Foot Cost Data, call 877.694.8329.

Costs per square foot of floor area

Exterior Wall	S.F. Area	6000	7000	8000	9000	10000	11000	12000	13000	14000
	L.F. Perimeter	220	240	260	280	286	303	320	336	353
Face Brick with Concrete Block Back-up	Steel Joists	225.50	218.95	214.10	210.25	205.00	202.20	200.00	197.90	196.25
	Precast Conc.	234.30	227.85	223.00	219.20	213.95	211.15	208.95	206.90	205.25
Decorative Concrete Block	Steel Joists	194.75	189.20	185.05	181.80	177.50	175.15	173.30	171.55	170.15
	Precast Conc.	219.30	213.85	209.70	206.50	202.30	199.90	198.05	196.30	194.95
Limestone with Concrete Block Back-up	Steel Joists	245.05	237.30	231.45	226.85	220.20	216.85	214.15	211.70	209.70
	Precast Conc.	253.60	245.90	240.10	235.55	229.00	225.65	222.95	220.45	218.50
Perimeter Adj., Add or Deduct	Per 100 L.F.	27.60	23.75	20.75	18.50	16.55	15.05	13.80	12.75	11.85
Story Hgt. Adj., Add or Deduct	Per 1 Ft.	3.65	3.45	3.25	3.15	2.85	2.70	2.60	2.60	2.45

For Basement, add $41.85 per square foot of basement area

The above costs were calculated using the basic specifications shown on the facing page. These costs should be adjusted where necessary for design alternatives and owner's requirements. Reported completed project costs, for this type of structure, range from $84.65 to $251.10 per S.F.

Common additives

Description	Unit	$ Cost
Appliances		
Cooking range, 30" free standing		
1 oven	Ea.	580 - 2450
2 oven	Ea.	1425 - 2900
Microwave oven	Ea.	296 - 840
Compactor, residential, 4-1 compaction	Ea.	825 - 1325
Dishwasher, built-in, 2 cycles	Ea.	600 - 1150
4 cycles	Ea.	755 - 1875
Garbage diposer, sink type	Ea.	232 - 380
Hood for range, 2 speed, vented, 30" wide	Ea.	330 - 1275
Refrigerator, no frost 10-12 C.F.	Ea.	560 - 650
14-16 C.F.	Ea.	660 - 870
18-20 C.F.	Ea.	805 - 1800
Elevators, hydraulic passenger, 2 stops, 2500# capacity	Ea.	68,800
3500# capacity	Ea.	73,300
Lockers, Steel, single tier, 60" or 72"	Opng.	221 - 370
Locker bench, lam. maple top only	L.F.	34
Pedestals, steel pipe	Ea.	76
Sound System, amplifier, 250 watts	Ea.	2050
Speaker, ceiling or wall	Ea.	213
Trumpet	Ea.	410

Description	Unit	$ Cost
Commissioning Fees, sustainable commercial construction	S.F.	0.24 - 3.04
Energy Modelling Fees, Fire Stations to 10,000 SF	Ea.	11,000
Greater than 10,000 SF add	S.F.	0.04
Green Bldg Cert Fees for comm construction project reg	Project	900
Photovoltaic Pwr Sys, grid connected, 20 kW (~2400 SF), roof	Ea.	275,000
Green Roofs, 6" soil depth, w/treated wd edging & sedum mats	S.F.	10.73
10" Soil depth, with treated wood edging & sedum mats	S.F.	12.44
Greywater Recovery Systems, prepackaged comm, 3060 gal.	Ea.	72,255
4590 gal.	Ea.	89,550
Rainwater Harvest Sys, prepckged comm, 10,000 gal, sys contrller	Ea.	31,500
20,000 gal. w/system controller	Ea.	50,550
30,000 gal. w/system controller	Ea.	80,800
Solar Domestic HW, closed loop, add-on sys, ext heat exchanger	Ea.	10,775
Drainback, hot water system, 120 gal tank	Ea.	13,325

Important: See the Reference Section for Location Factors.

Model costs calculated for a 2 story building with 14' story height and 10,000 square feet of floor area

G Fire Station, 2 Story

					Unit	Unit Cost	Cost Per S.F.	% Of Sub-Total

A. SUBSTRUCTURE

					Unit	Unit Cost	Cost Per S.F.	% Of Sub-Total
1010	Standard Foundations	Poured concrete; strip and spread footings			S.F. Ground	3.22	1.61	
1020	Special Foundations	N/A			—	—	—	
1030	Slab on Grade	6" reinforced concrete with recycled vapor barrier and granular base			S.F. Slab	7.14	3.58	6.4%
2010	Basement Excavation	Site preparation for slab and trench for foundation wall and footing			S.F. Ground	.57	.29	
2020	Basement Walls	4' foundation wall with insulation			L.F. Wall	101	2.88	

B. SHELL

B10 Superstructure

					Unit	Unit Cost	Cost Per S.F.	% Of Sub-Total
1010	Floor Construction	Open web steel joists, slab form, concrete			S.F. Floor	14.82	7.41	7.6%
1020	Roof Construction	Metal deck on open web steel joists			S.F. Roof	5.14	2.57	

B20 Exterior Enclosure

					Unit	Unit Cost	Cost Per S.F.	% Of Sub-Total
2010	Exterior Walls	Decorative concrete block, insul.	75% of wall		S.F. Wall	24.31	14.60	
2020	Exterior Windows	Aluminum insulated glass	10% of wall		Each	750	2.62	15.9%
2030	Exterior Doors	Single aluminum and glass, steel overhead, hollow metal, low VOC paint	15% of wall		S.F. Door	1936	3.74	

B30 Roofing

					Unit	Unit Cost	Cost Per S.F.	% Of Sub-Total
3010	Roof Coverings	Single-ply TPO, 60 mils, heat welded seams w/R-20 insul.			S.F. Roof	6.94	3.47	2.6%
3020	Roof Openings	N/A			—	—	—	

C. INTERIORS

					Unit	Unit Cost	Cost Per S.F.	% Of Sub-Total
1010	Partitions	Concrete block, foamed-in insul.	10 SF Floor/LF Part.		S.F. Partition	11.27	6.63	
1020	Interior Doors	Single leaf hollow metal, low VOC paint	500 S.F. Floor/Door		Each	1275	2.55	
1030	Fittings	Toilet partitions			S.F. Floor	.44	.44	
2010	Stair Construction	Concrete filled metal pan			Flight	20,200	4.04	18.1%
3010	Wall Finishes	Paint, low VOC			S.F. Surface	3.95	4.65	
3020	Floor Finishes	50% vinyl tile, (recycled content) 50% paint (low VOC)			S.F. Floor	2.68	2.68	
3030	Ceiling Finishes	Fiberglass board on exposed grid, suspended	50% of area		S.F. Ceiling	2.85	2.86	

D. SERVICES

D10 Conveying

					Unit	Unit Cost	Cost Per S.F.	% Of Sub-Total
1010	Elevators & Lifts	One hydraulic passenger elevator			Each	83,600	8.36	6.4%
1020	Escalators & Moving Walks	N/A			—	—	—	

D20 Plumbing

					Unit	Unit Cost	Cost Per S.F.	% Of Sub-Total
2010	Plumbing Fixtures	Kitchen, toilet low flow, auto sensor & service fixt., supply & drain.	1 Fixt./400 SF Flr.		Each	4312	10.78	
2020	Domestic Water Distribution	Gas fired tankless water heater			S.F. Floor	.82	.82	9.5%
2040	Rain Water Drainage	Roof drains			S.F. Roof	1.76	.88	

D30 HVAC

					Unit	Unit Cost	Cost Per S.F.	% Of Sub-Total
3010	Energy Supply	N/A			—	—	—	
3020	Heat Generating Systems	Included in D3050			—	—	—	
3040	Distribution Systems	Enthalpy heat recovery packages			Each	14,825	2.97	18.9 %
3050	Terminal & Package Units	Multizone rooftop air conditioner, SEER 14			S.F. Floor	21.84	21.84	
3090	Other HVAC Sys. & Equipment	N/A						

D40 Fire Protection

					Unit	Unit Cost	Cost Per S.F.	% Of Sub-Total
4010	Sprinklers	Wet pipe sprinkler system			S.F. Floor	3.96	3.96	4.1%
4020	Standpipes	Standpipe, wet, Class III			S.F. Floor	1.46	1.46	

D50 Electrical

					Unit	Unit Cost	Cost Per S.F.	% Of Sub-Total
5010	Electrical Service/Distribution	400 ampere service, panel board and feeders			S.F. Floor	1.12	1.12	
5020	Lighting & Branch Wiring	LED fixt., daylt. dimming & ltg. on/off control., receptacles, switches, A.C. and misc. pwr.			S.F. Floor	8.62	8.62	10.3%
5030	Communications & Security	Addressable alarm systems and emergency lighting			S.F. Floor	1.55	1.55	
5090	Other Electrical Systems	Emergency generator, 15 kW, and energy monitoring systems			S.F. Floor	2.19	2.19	

E. EQUIPMENT & FURNISHINGS

					Unit	Unit Cost	Cost Per S.F.	% Of Sub-Total
1010	Commercial Equipment	N/A			—	—	—	
1020	Institutional Equipment	N/A			—	—	—	
1090	Other Equipment	Waste handling recycling tilt truck			Each	.31	.31	0.3 %
2020	Moveable Furnishings	No smoking signage			Each	.02	.02	

F. SPECIAL CONSTRUCTION

					Unit	Unit Cost	Cost Per S.F.	% Of Sub-Total
1020	Integrated Construction	N/A			—	—	—	
1040	Special Facilities	N/A			—	—	—	0.0 %

G. BUILDING SITEWORK   N/A

		Sub-Total	131.50	100%
CONTRACTOR FEES (General Requirements: 10%, Overhead: 5%, Profit: 10%)		25%	32.85	
ARCHITECT FEES		8%	13.15	

Total Building Cost	**177.50**

For customer support on your Square Foot Cost Data, call 877.694.8329.

247

Costs per square foot of floor area

Exterior Wall	S.F. Area	4000	6000	8000	10000	12000	14000	16000	18000	20000
	L.F. Perimeter	260	320	384	424	460	484	510	540	576
Vertical Redwood Siding	Wood Frame	164.95	153.85	148.50	144.05	140.95	138.25	136.25	134.85	133.85
Brick Veneer	Wood Frame	180.30	167.10	160.85	155.45	151.60	148.20	145.75	144.05	142.80
Aluminum Siding	Wood Frame	166.20	155.55	150.45	146.25	143.25	140.70	138.85	137.55	136.60
Brick on Block	Wood Truss	193.05	177.90	170.75	164.35	159.80	155.80	152.85	150.85	149.40
Limestone on Block	Wood Truss	211.45	192.95	184.20	176.20	170.50	165.35	161.70	159.15	157.35
Stucco on Block	Wood Truss	200.65	183.45	175.30	167.95	162.65	158.00	154.55	152.25	150.60
Perimeter Adj., Add or Deduct	Per 100 L.F.	13.10	8.75	6.55	5.30	4.35	3.75	3.30	2.95	2.65
Story Hgt. Adj., Add or Deduct	Per 1 Ft.	1.55	1.30	1.20	1.10	0.90	0.80	0.75	0.80	0.75

For Basement, add $35.50 per square foot of basement area

The above costs were calculated using the basic specifications shown on the facing page. These costs should be adjusted where necessary for design alternatives and owner's requirements. Reported completed project costs, for this type of structure, range from $120.50 to $370.50 per S.F.

Common additives

Description	Unit	$ Cost
Autopsy Table, standard	Ea.	11,400
Deluxe	Ea.	17,800
Directory Boards, plastic, glass covered, 30" x 20"	Ea.	600
36" x 48"	Ea.	1500
Aluminum, 24" x 18"	Ea.	635
36" x 24"	Ea.	755
48" x 32"	Ea.	1050
Mortuary Refrigerator, end operated		
Two capacity	Ea.	13,500
Six capacity	Ea.	24,400
Planters, precast concrete, 48" diam., 24" high	Ea.	755
7" diam., 36" high	Ea.	1825
Fiberglass, 36" diam., 24" high	Ea.	855
60" diam., 24" high	Ea.	1325

Description	Unit	$ Cost
Commissioning Fees, sustainable commercial construction	S.F.	0.24 - 3.04
Energy Modelling Fees, commercial buildings to 10,000 SF	Ea.	11,000
Greater than 10,000 SF add	S.F.	0.04
Green Bldg Cert Fees for comm construction project reg	Project	900
Photovoltaic Pwr Sys, grid connected, 20 kW (~2400 SF), roof	Ea.	275,000
Green Roofs, 6" soil depth, w/treated wd edging & sedum mats	S.F.	10.73
10" Soil depth, with treated wood edging & sedum mats	S.F.	12.44
Greywater Recovery Systems, prepackaged comm, 3060 gal.	Ea.	72,255
4590 gal.	Ea.	89,550
Rainwater Harvest Sys, prepckged comm, 10,000 gal, sys contrller	Ea.	31,500
20,000 gal. w/system controller	Ea.	50,550
30,000 gal. w/system controller	Ea.	80,800
Solar Domestic HW, closed loop, add-on sys, ext heat exchanger	Ea.	10,775
Drainback, hot water system, 120 gal tank	Ea.	13,325
Draindown, hot water system, 120 gal tank	Ea.	13,675

Important: See the Reference Section for Location Factors.

Model costs calculated for a 1 story building with 12′ story height and 10,000 square feet of floor area

G Funeral Home

				Unit	Unit Cost	Cost Per S.F.	% Of Sub-Total
A. SUBSTRUCTURE							
1010	Standard Foundations	Poured concrete; strip and spread footings		S.F. Ground	1.88	1.88	
1020	Special Foundations	N/A		—	—	—	
1030	Slab on Grade	4″ reinforced concrete with recycled plastic vapor barrier and granular base		S.F. Slab	5.45	5.45	10.9%
2010	Basement Excavation	Site preparation for slab and trench for foundation wall and footing		S.F. Ground	.32	.32	
2020	Basement Walls	4′ foundation wall		L.F. Wall	83	3.85	
B. SHELL							
B10 Superstructure							
1010	Floor Construction	N/A		—	—	—	4.9 %
1020	Roof Construction	Plywood on wood truss		S.F. Roof	5.22	5.22	
B20 Exterior Enclosure							
2010	Exterior Walls	1″ x 4″ vertical T & G redwood siding on 2x6 wood studs with insul.	90% of wall	S.F. Wall	13.15	6.02	
2020	Exterior Windows	Double hung wood	10% of wall	Each	511	2.16	8.9%
2030	Exterior Doors	Wood swinging double doors, single leaf hollow metal, low VOC paint		Each	1985	1.19	
B30 Roofing							
3010	Roof Coverings	Single ply membrane, TPO, 45 mil, fully adhered; polyisocyanurate sheets		S.F. Roof	6.34	6.34	6.0%
3020	Roof Openings	N/A		—	—	—	
C. INTERIORS							
1010	Partitions	Gypsum board on wood studs with sound deadening board	15 SF Flr./LF Part.	S.F. Partition	9.86	5.26	
1020	Interior Doors	Single leaf wood, low VOC paint	150 S.F. Floor/Door	Each	621	4.14	
1030	Fittings	N/A		—	—	—	
2010	Stair Construction	N/A		—	—	—	24.3%
3010	Wall Finishes	50% wallpaper, 25% wood paneling, 25% paint (low VOC)		S.F. Surface	3.50	3.73	
3020	Floor Finishes	70% carpet tile, 30% ceramic tile		S.F. Floor	6.88	6.88	
3030	Ceiling Finishes	Fiberglass board on exposed grid, suspended		S.F. Ceiling	5.70	5.70	
D. SERVICES							
D10 Conveying							
1010	Elevators & Lifts	N/A		—	—	—	0.0 %
1020	Escalators & Moving Walks	N/A		—	—	—	
D20 Plumbing							
2010	Plumbing Fixtures	Toilets, low flow, auto sensor, urinals & service fixt., supply & drain.	1 Fixt./770 SF Flr.	Each	3427	4.45	
2020	Domestic Water Distribution	Electric water heater, point-of-use		S.F. Floor	.45	.45	5.6%
2040	Rain Water Drainage	Roof drain		S.F. Roof	1.04	1.04	
D30 HVAC							
3010	Energy Supply	N/A		—	—	—	
3020	Heat Generating Systems	Included in D3050		—	—	—	
3040	Distribution Systems	Enthalpy heat recovery packages		Each	36,925	3.69	20.3 %
3050	Terminal & Package Units	Multizone rooftop air conditioner, SEER 14		S.F. Floor	17.78	17.78	
3090	Other HVAC Sys. & Equipment	N/A		—	—	—	
D40 Fire Protection							
4010	Sprinklers	Wet pipe sprinkler system		S.F. Floor	3.67	3.67	4.3%
4020	Standpipes	Standpipe, wet, Class III		S.F. Floor	.86	.86	
D50 Electrical							
5010	Electrical Service/Distribution	400 ampere service, panel board and feeders		S.F. Floor	1.54	1.54	
5020	Lighting & Branch Wiring	LED light fixtures, daylt. dimming control & ltg. on/off sys., recept., switches, & A.C. power		S.F. Floor	10.02	10.02	14.2%
5030	Communications & Security	Addressable alarm systems		S.F. Floor	1.55	1.55	
5090	Other Electrical Systems	Emergency generator, 15 kW and energy monitoring systems		S.F. Floor	1.95	1.95	
E. EQUIPMENT & FURNISHINGS							
1010	Commercial Equipment	N/A		—	—	—	
1020	Institutional Equipment	N/A		—	—	—	0.6 %
1090	Other Equipment	Waste handling recycling tilt truck		Each	.56	.56	
2020	Moveable Furnishings	No smoking signage		Each	.04	.04	
F. SPECIAL CONSTRUCTION							
1020	Integrated Construction	N/A		—	—	—	0.0 %
1040	Special Facilities	N/A		—	—	—	
G. BUILDING SITEWORK	**N/A**						

					Sub-Total	105.74	100%
CONTRACTOR FEES (General Requirements: 10%, Overhead: 5%, Profit: 10%)					25%	26.41	
ARCHITECT FEES					9%	11.90	

Total Building Cost — **144.05**

For customer support on your Square Foot Cost Data, call 877.694.8329.

249

Costs per square foot of floor area

Exterior Wall	S.F. Area	7000	10000	13000	16000	19000	22000	25000	28000	31000
	L.F. Perimeter	240	300	336	386	411	435	472	510	524
Face Brick with Concrete Block Back-up	R/Conc. Frame	208.85	197.95	188.70	184.45	179.20	175.25	173.30	171.75	169.10
	Steel Frame	210.65	199.75	190.45	186.25	181.00	177.05	175.05	173.55	170.90
Limestone with Concrete Block	R/Conc. Frame	234.55	220.05	207.50	201.90	194.80	189.50	186.80	184.75	181.15
	Steel Frame	233.55	219.75	207.75	202.35	195.45	190.25	187.65	185.70	182.20
Precast Concrete Panels	R/Conc. Frame	214.10	202.15	192.10	187.50	181.90	177.65	175.55	173.85	171.10
	Steel Frame	214.90	202.90	192.85	188.30	182.65	178.45	176.30	174.60	171.85
Perimeter Adj., Add or Deduct	Per 100 L.F.	26.15	18.35	14.10	11.55	9.65	8.35	7.30	6.55	5.95
Story Hgt. Adj., Add or Deduct	Per 1 Ft.	4.00	3.55	3.00	2.80	2.50	2.30	2.20	2.10	2.00

For Basement, add $52.95 per square foot of basement area

The above costs were calculated using the basic specifications shown on the facing page. These costs should be adjusted where necessary for design alternatives and owner's requirements. Reported completed project costs, for this type of structure, range from $115.30 to $295.85 per S.F.

Common additives

Description	Unit	$ Cost
Carrels Hardwood	Ea.	975 - 2350
Closed Circuit Surveillance, one station, camera and monitor	Ea.	1975
For additional camera stations, add	Ea.	1075
Elevators, hydraulic passenger, 2 stops, 1500# capacity	Ea.	65,700
2500# capacity	Ea.	68,800
3500# capacity	Ea.	73,300
Library Furnishings, bookshelf, 90" high, 10" shelf double face	L.F.	216
Single face	L.F.	185
Charging desk, built-in with counter		
Plastic laminated top	L.F.	420
Reading table, laminated		
Top 60" x 36"	Ea.	775

Description	Unit	$ Cost
Commissioning Fees, sustainable institutional construction	S.F.	0.58 - 2.47
Energy Modelling Fees, commercial buildings to 10,000 SF	Ea.	11,000
Greater than 10,000 SF add	S.F.	0.04
Green Bldg Cert Fees for comm construction project reg	Project	900
Photovoltaic Pwr Sys, grid connected, 20 kW (~2400 SF), roof	Ea.	275,000
Green Roofs, 6" soil depth, w/treated wd edging & sedum mats	S.F.	10.73
10" Soil depth, with treated wood edging & sedum mats	S.F.	12.44
Greywater Recovery Systems, prepackaged comm, 3060 gal.	Ea.	72,255
4590 gal.	Ea.	89,550
Rainwater Harvest Sys, prepckged comm, 10,000 gal, sys contrller	Ea.	31,500
20,000 gal. w/system controller	Ea.	50,550
30,000 gal. w/system controller	Ea.	80,800
Solar Domestic HW, closed loop, add-on sys, ext heat exchanger	Ea.	10,775
Drainback, hot water system, 120 gal tank	Ea.	13,325
Draindown, hot water system, 120 gal tank	Ea.	13,675

Model costs calculated for a 2 story building with 14' story height and 22,000 square feet of floor area

				Unit	Unit Cost	Cost Per S.F.	% Of Sub-Total

A. SUBSTRUCTURE

1010	Standard Foundations	Poured concrete; strip and spread footings		S.F. Ground	3.94	1.97	
1020	Special Foundations	N/A		—	—	—	
1030	Slab on Grade	4" reinforced concrete with recycled vapor barrier and granular base		S.F. Slab	5.45	2.73	5.1%
2010	Basement Excavation	Site preparation for slab and trench for foundation wall and footing		S.F. Ground	.32	.16	
2020	Basement Walls	4' foundation wall		L.F. Wall	91	1.80	

B. SHELL

B10 Superstructure

1010	Floor Construction	Concrete waffle slab		S.F. Floor	27.40	13.70	18.1%
1020	Roof Construction	Concrete waffle slab		S.F. Roof	19.66	9.83	

B20 Exterior Enclosure

2010	Exterior Walls	Face brick with concrete block backup	90% of wall	S.F. Wall	36.81	18.34	
2020	Exterior Windows	Window wall	10% of wall	Each	54	3	16.8%
2030	Exterior Doors	Double aluminum and glass, single leaf hollow metal, low VOC paint		Each	5825	.53	

B30 Roofing

3010	Roof Coverings	Single-ply TPO membrane, 60 mil, heat welded w/R-20 insulation		S.F. Roof	5.96	2.98	2.3%
3020	Roof Openings	Roof hatches		S.F. Roof	.08	.04	

C. INTERIORS

1010	Partitions	Gypsum board on metal studs w/ sound attenuation insul.	30 SF Flr./LF Part.	S.F. Partition	13.53	5.41	
1020	Interior Doors	Single leaf wood, low VOC paint	300 S.F. Floor/Door	Each	621	2.07	
1030	Fittings	N/A		—	—	—	
2010	Stair Construction	Concrete filled metal pan		Flight	9650	.87	16.3%
3010	Wall Finishes	Paint, low VOC		S.F. Surface	1.99	1.59	
3020	Floor Finishes	50% carpet tile, 50% vinyl composition tile, recycled content		S.F. Floor	4.17	4.17	
3030	Ceiling Finishes	Mineral fiber on concealed zee bars		S.F. Ceiling	7.03	7.03	

D. SERVICES

D10 Conveying

1010	Elevators & Lifts	One hydraulic passenger elevator		Each	87,120	3.96	3.1%
1020	Escalators & Moving Walks	N/A		—	—	—	

D20 Plumbing

2010	Plumbing Fixtures	Toilet, low flow, auto sensor & service fixt., supply & drain.	1 Fixt./1835 SF.Flr.	Each	8294	4.52	
2020	Domestic Water Distribution	Tankless, on demand water heater, gas/propane		S.F. Floor	.60	.60	4.4%
2040	Rain Water Drainage	Roof drains		S.F. Roof	1.24	.62	

D30 HVAC

3010	Energy Supply	N/A		—	—	—	
3020	Heat Generating Systems	Included in D3050		—	—	—	
3040	Distribution Systems	Enthalpy heat recovery packages		Each	30,400	2.77	17.5 %
3050	Terminal & Package Units	Multizone rooftop air conditioner, SEER 14		S.F. Floor	20	20	
3090	Other HVAC Sys. & Equipment	N/A		—	—	—	

D40 Fire Protection

4010	Sprinklers	Wet pipe sprinkler system		S.F. Floor	3.19	3.19	3.2%
4020	Standpipes	Standpipe		S.F. Floor	.98	.98	

D50 Electrical

5010	Electrical Service/Distribution	400 ampere service, panel board and feeders		S.F. Floor	.52	.52	
5020	Lighting & Branch Wiring	LED fixtures, daylt. dim., ltg. on/off, recept., switches, and A.C. power		S.F. Floor	12.69	12.69	12.9%
5030	Communications & Security	Addressable alarm systems, internet wiring, and emergency lighting		S.F. Floor	2.10	2.10	
5090	Other Electrical Systems	Emergency generator, 7.5 kW, UPS, energy monitoring systems		S.F. Floor	1.40	1.40	

E. EQUIPMENT & FURNISHINGS

1010	Commercial Equipment	N/A		—	—	—	
1020	Institutional Equipment	N/A		—	—	—	
1090	Other Equipment	Waste handling recycling tilt truck		Each	.25	.25	0.2 %
2020	Moveable Furnishings	No smoking signage		Each			

F. SPECIAL CONSTRUCTION

1020	Integrated Construction	N/A		—	—	—	0.0 %
1040	Special Facilities	N/A		—	—	—	

G. BUILDING SITEWORK   N/A

	Sub-Total	129.82	**100%**
CONTRACTOR FEES (General Requirements: 10%, Overhead: 5%, Profit: 10%)	25%	32.45	
ARCHITECT FEES	8%	12.98	

Total Building Cost	**175.25**

For customer support on your Square Foot Cost Data, call 877.694.8329.

251

Costs per square foot of floor area

Exterior Wall	S.F. Area	4000	5500	7000	8500	10000	11500	13000	14500	16000
	L.F. Perimeter	280	320	380	440	453	503	510	522	560
Face Brick with Concrete Block Back-up	Steel Joists	248.35	237.10	232.90	230.20	224.70	223.10	219.30	216.60	215.65
	Wood Truss	250.35	239.10	234.85	232.15	226.65	225.05	221.30	218.60	217.65
Stucco on Concrete Block	Steel Joists	256.35	243.75	239.10	236.05	229.80	228.05	223.75	220.75	219.70
	Wood Truss	262.75	250.25	245.55	242.50	236.25	234.50	230.20	227.15	226.15
Brick Veneer	Wood Truss	244.55	234.65	230.85	228.40	223.65	222.20	218.90	216.60	215.75
Wood Siding	Wood Frame	237.15	228.90	225.65	223.55	219.70	218.45	215.80	214.00	213.25
Perimeter Adj., Add or Deduct	Per 100 L.F.	18.90	13.75	10.80	8.80	7.55	6.55	5.85	5.20	4.70
Story Hgt. Adj., Add or Deduct	Per 1 Ft.	3.85	3.20	3.00	2.80	2.45	2.40	2.15	2.00	1.95

For Basement, add $34.50 per square foot of basement area

The above costs were calculated using the basic specifications shown on the facing page. These costs should be adjusted where necessary for design alternatives and owner's requirements. Reported completed project costs, for this type of structure, range from $97.15 to $249.90 per S.F.

Common additives

Description	Unit	$ Cost
Cabinets, hospital, base, laminated plastic	L.F.	415
Counter top, laminated plastic	L.F.	81
Nurses station, door type , laminated plastic	L.F.	465
Wall cabinets, laminated plastic	L.F.	305
Directory Boards, plastic, glass covered, 30" x 20"	Ea.	600
36" x 48"	Ea.	1500
Aluminum, 36" x 24"	Ea.	755
48" x 32"	Ea.	1050
Heat Therapy Unit, humidified, 26" x 78" x 28"	Ea.	3975
Tables, examining, vinyl top, with base cabinets	Ea.	1500 - 6750
Utensil Washer, Sanitizer	Ea.	12,800
X-Ray, Mobile	Ea.	18,500 - 85,500

Description	Unit	$ Cost
Commissioning Fees, sustainable commercial construction	S.F.	0.24 - 3.04
Energy Modelling Fees, commercial buildings to 10,000 SF	Ea.	11,000
Greater than 10,000 SF add	S.F.	0.04
Green Bldg Cert Fees for comm construction project reg	Project	900
Photovoltaic Pwr Sys, grid connected, 20 kW (~2400 SF), roof	Ea.	275,000
Green Roofs, 6" soil depth, w/treated wd edging & sedum mats	S.F.	10.73
10" Soil depth, with treated wood edging & sedum mats	S.F.	12.44
Greywater Recovery Systems, prepackaged comm, 3060 gal.	Ea.	72,255
4590 gal.	Ea.	89,550
Rainwater Harvest Sys, prepckged comm, 10,000 gal, sys contrller	Ea.	31,500
20,000 gal. w/system controller	Ea.	50,550
30,000 gal. w/system controller	Ea.	80,800
Solar Domestic HW, closed loop, add-on sys, ext heat exchanger	Ea.	10,775
Drainback, hot water system, 120 gal tank	Ea.	13,325
Draindown, hot water system, 120 gal tank	Ea.	13,675

Important: See the Reference Section for Location Factors.

Model costs calculated for a 1 story building with 10' story height and 7,000 square feet of floor area

				Unit	Unit Cost	Cost Per S.F.	% Of Sub-Total
A.	**SUBSTRUCTURE**						
1010	Standard Foundations	Poured concrete; strip and spread footings		S.F. Ground	2.19	2.19	
1020	Special Foundations	N/A		—	—	—	
1030	Slab on Grade	4" reinforced concrete with recycled vapor barrier and granular base		S.F. Slab	5.45	5.45	7.5%
2010	Basement Excavation	Site preparation for slab and trench for foundation wall and footing		S.F. Ground	.32	.32	
2020	Basement Walls	4' foundation wall		L.F. Wall	91	4.94	
B.	**SHELL**						
	B10	**Superstructure**					
1010	Floor Construction	N/A		—	—	—	
1020	Roof Construction	Plywood on wood trusses		S.F. Roof	6.65	6.65	3.9 %
	B20	**Exterior Enclosure**					
2010	Exterior Walls	Face brick with concrete block backup	70% of wall	S.F. Wall	36.79	13.98	
2020	Exterior Windows	Wood double hung	30% of wall	Each	638	6.11	13.1%
2030	Exterior Doors	Aluminum and glass doors and entrance with transoms		Each	1465	2.51	
	B30	**Roofing**					
3010	Roof Coverings	Asphalt shingles with flashing (Pitched); rigid fiber glass insulation, gutters		S.F. Roof	8.55	8.55	5.0%
3020	Roof Openings	N/A		—	—	—	
C.	**INTERIORS**						
1010	Partitions	Gypsum bd. & sound deadening bd. on wood studs w/insul.	6 SF Flr./LF Part.	S.F. Partition	9.54	12.72	
1020	Interior Doors	Single leaf wood, low VOC paint	60 S.F. Floor/Door	Each	621	10.35	
1030	Fittings	N/A		—	—	—	
2010	Stair Construction	N/A		—	—	—	21.6%
3010	Wall Finishes	50% paint, low VOC, 50% vinyl wall covering		S.F. Surface	1.43	3.82	
3020	Floor Finishes	50% carpet tile, 50% vinyl composition tile, recycled content		S.F. Floor	3.86	3.86	
3030	Ceiling Finishes	Mineral fiber tile on concealed zee bars		S.F. Ceiling	6.40	6.40	
D.	**SERVICES**						
	D10	**Conveying**					
1010	Elevators & Lifts	N/A		—	—	—	0.0 %
1020	Escalators & Moving Walks	N/A		—	—	—	
	D20	**Plumbing**					
2010	Plumbing Fixtures	Toilet, low flow, auto sensor, exam room & service fixt., supply & drain.	1 Fixt./195 SF Flr.	Each	4920	25.23	
2020	Domestic Water Distribution	Gas fired tankless water heater		S.F. Floor	2.84	2.84	16.9%
2040	Rain Water Drainage	Roof drains		S.F. Roof	1.07	1.07	
	D30	**HVAC**					
3010	Energy Supply	N/A		—	—	—	
3020	Heat Generating Systems	Included in D3050		—	—	—	
3040	Distribution Systems	Enthalpy heat recovery packages		Each	17,525	2.50	9.6 %
3050	Terminal & Package Units	Multizone rooftop air conditioner, SEER 14		S.F. Floor	13.98	13.98	
3090	Other HVAC Sys. & Equipment	N/A		—	—	—	
	D40	**Fire Protection**					
4010	Sprinklers	Wet pipe sprinkler system		S.F. Floor	4.98	4.98	3.6%
4020	Standpipes	Standpipe		S.F. Floor	1.23	1.23	
	D50	**Electrical**					
5010	Electrical Service/Distribution	200 ampere service, panel board and feeders		S.F. Floor	1.28	1.28	
5020	Lighting & Branch Wiring	LED fixtures, daylt. dim., ltg. on/off, recept., switches, and A.C. power		S.F. Floor	9.89	9.89	12.9%
5030	Communications & Security	Alarm systems, internet & phone wiring, intercom system, & emerg. ltg.		S.F. Floor	8	8	
5090	Other Electrical Systems	Emergency generator, 7.5 kW, and energy monitoring systems		S.F. Floor	3.05	3.05	
E.	**EQUIPMENT & FURNISHINGS**						
1010	Commercial Equipment	N/A		—	—	—	
1020	Institutional Equipment	Exam room casework and countertops		S.F. Floor	9.58	9.58	6.1 %
1090	Other Equipment	Waste handling recycling tilt truck		Each	.80	.80	
2020	Moveable Furnishings	No smoking signage		Each	.10	.10	
F.	**SPECIAL CONSTRUCTION**						
1020	Integrated Construction	N/A		—	—	—	0.0 %
1040	Special Facilities	N/A		—	—	—	
G.	**BUILDING SITEWORK**	**N/A**					
				Sub-Total		172.38	**100%**
	CONTRACTOR FEES (General Requirements: 10%, Overhead: 5%, Profit: 10%)				25%	43.08	
	ARCHITECT FEES				9%	19.39	
				Total Building Cost		**234.85**	

Costs per square foot of floor area

Exterior Wall	S.F. Area	4000	5500	7000	8500	10000	11500	13000	14500	16000
	L.F. Perimeter	180	210	240	270	286	311	336	361	386
Face Brick with Concrete Block Back-up	Steel Joists	284.80	272.80	265.95	261.55	256.60	253.95	251.90	250.35	249.00
	Wood Joists	289.70	277.70	270.85	266.45	261.50	258.85	256.80	255.25	253.90
Stucco on Concrete Block	Steel Joists	267.75	258.30	252.95	249.40	245.70	243.60	241.95	240.75	239.70
	Wood Joists	278.05	267.85	262.05	258.25	254.15	251.85	250.15	248.80	247.70
Brick Veneer	Wood Frame	278.05	267.85	262.05	258.20	254.10	251.85	250.10	248.80	247.65
Wood Siding	Wood Frame	266.50	258.10	253.25	250.10	246.80	244.95	243.50	242.40	241.50
Perimeter Adj., Add or Deduct	Per 100 L.F.	33.95	24.80	19.50	16.00	13.60	11.85	10.45	9.35	8.50
Story Hgt. Adj., Add or Deduct	Per 1 Ft.	5.05	4.30	3.85	3.55	3.15	3.05	2.90	2.80	2.65

For Basement, add $ 38.25 per square foot of basement area

The above costs were calculated using the basic specifications shown on the facing page. These costs should be adjusted where necessary for design alternatives and owner's requirements. Reported completed project costs, for this type of structure, range from $ 99.70 to $ 340.65 per S.F.

Common additives

Description	Unit	$ Cost
Cabinets, hospital, base, laminated plastic	L.F.	415
Counter top, laminated plastic	L.F.	81
Nurses station, door type , laminated plastic	L.F.	465
Wall cabinets, laminated plastic	L.F.	305
Elevators, hydraulic passenger, 2 stops, 2500# capacity	Ea.	68,800
3500# capacity	Ea.	73,300
Directory Boards, plastic, glass covered, 30" x 20"	Ea.	600
36" x 48"	Ea.	1500
Aluminum, 36" x 24"	Ea.	755
48" x 32"	Ea.	1050
Heat Therapy Unit, humidified, 26" x 78" x 28"	Ea.	3975
Tables, examining, vinyl top, with base cabinets	Ea.	1500 - 6750
Utensil Washer, Sanitizer	Ea.	12,800
X-Ray, Mobile	Ea.	18,500 - 85,500

Description	Unit	$ Cost
Commissioning Fees, sustainable commercial construction	S.F.	0.24 - 3.04
Energy Modelling Fees, commercial buildings to 10,000 SF	Ea.	11,000
Greater than 10,000 SF add	S.F.	0.04
Green Bldg Cert Fees for comm construction project reg	Project	900
Photovoltaic Pwr Sys, grid connected, 20 kW (~2400 SF), roof	Ea.	275,000
Green Roofs, 6" soil depth, w/treated wd edging & sedum mats	S.F.	10.73
10" Soil depth, with treated wood edging & sedum mats	S.F.	12.44
Greywater Recovery Systems, prepackaged comm, 3060 gal.	Ea.	72,255
4590 gal.	Ea.	89,550
Rainwater Harvest Sys, prepckged comm, 10,000 gal, sys contrller	Ea.	31,500
20,000 gal. w/system controller	Ea.	50,550
30,000 gal. w/system controller	Ea.	80,800
Solar Domestic HW, closed loop, add-on sys, ext heat exchanger	Ea.	10,775
Drainback, hot water system, 120 gal tank	Ea.	13,325
Draindown, hot water system, 120 gal tank	Ea.	13,675

Model costs calculated for a 2 story building with 10' story height and 7,000 square feet of floor area

G Medical Office, 2 Story

				Unit	Unit Cost	Cost Per S.F.	% Of Sub-Total
A. SUBSTRUCTURE							
1010	Standard Foundations	Poured concrete; strip and spread footings		S.F. Ground	3.48	1.74	
1020	Special Foundations	N/A		—	—	—	
1030	Slab on Grade	4" reinforced concrete with vapor barrier and granular base		S.F. Slab	5.45	2.73	4.1%
2010	Basement Excavation	Site preparation for slab and trench for foundation wall and footing		S.F. Ground	.18	.09	
2020	Basement Walls	4' foundation wall		L.F. Wall	91	3.12	
B. SHELL							
B10 Superstructure							
1010	Floor Construction	Open web steel joists, slab form, concrete, columns		S.F. Floor	13.76	6.88	5.4%
1020	Roof Construction	Metal deck, open web steel joists, beams, columns		S.F. Roof	6.24	3.12	
B20 Exterior Enclosure							
2010	Exterior Walls	Stucco on concrete block		S.F. Wall	19.35	9.29	
2020	Exterior Windows	Outward projecting metal	30% of wall	Each	528	7.25	9.8%
2030	Exterior Doors	Aluminum and glass doors with transoms		Each	5825	1.67	
B30 Roofing							
3010	Roof Coverings	Built-up tar and gravel with flashing; perlite/EPS composite insulation		S.F. Roof	7.10	3.55	2.0%
3020	Roof Openings	Roof hatches		S.F. Roof	.26	.13	
C. INTERIORS							
1010	Partitions	Gypsum bd. & acous. insul. on metal studs	6 SF Floor/LF Part.	S.F. Partition	7.87	10.49	
1020	Interior Doors	Single leaf wood	60 S.F. Floor/Door	Each	621	10.35	
1030	Fittings	N/A					
2010	Stair Construction	Concrete filled metal pan		Flight	14,500	4.15	21.0%
3010	Wall Finishes	45% paint, 50% vinyl wall coating, 5% ceramic tile		S.F. Surface	1.43	3.82	
3020	Floor Finishes	50% carpet, 50% vinyl composition tile		S.F. Floor	3.86	3.86	
3030	Ceiling Finishes	Mineral fiber tile on concealed zee bars		S.F. Ceiling	6.40	6.40	
D. SERVICES							
D10 Conveying							
1010	Elevators & Lifts	One hydraulic hospital elevator		Each	106,610	15.23	8.2%
1020	Escalators & Moving Walks	N/A		—	—	—	
D20 Plumbing							
2010	Plumbing Fixtures	Toilet, exam room & service fixt., supply & drain.	1 Fixt./160 SF Flr.	Each	4598	28.74	
2020	Domestic Water Distribution	Gas fired water heater		S.F. Floor	2.84	2.84	17.6%
2040	Rain Water Drainage	Roof drains		S.F. Roof	2.28	1.14	
D30 HVAC							
3010	Energy Supply	N/A		—	—	—	
3020	Heat Generating Systems	Included in D3050		—	—	—	
3040	Distribution Systems	Enthalpy heat recovery packages		Each	17,525	2.50	8.9%
3050	Terminal & Package Units	Multizone rooftop air conditioner, SEER 14		S.F. Floor	13.98	13.98	
3090	Other HVAC Sys. & Equipment	N/A		—	—	—	
D40 Fire Protection							
4010	Sprinklers	Wet pipe sprinkler system		S.F. Floor	3.96	3.96	3.1%
4020	Standpipes	Standpipe		S.F. Floor	1.86	1.86	
D50 Electrical							
5010	Electrical Service/Distribution	400 ampere service, panel board and feeders		S.F. Floor	1.61	1.61	
5020	Lighting & Branch Wiring	LED fixtures, receptacles, switches, A.C. and misc. power		S.F. Floor	13.93	13.93	14.4%
5030	Communications & Security	Alarm system, internet & phone wiring, intercom system, & emergency ltg.		S.F. Floor	8	8	
5090	Other Electrical Systems	Emergency generator, 7.5 kW		S.F. Floor	3.17	3.17	
E. EQUIPMENT & FURNISHINGS							
1010	Commercial Equipment	N/A		—	—	—	
1020	Institutional Equipment	Exam room casework and contertops		S.F. Floor	9.58	9.58	
1090	Other Equipment	Waste handling recycling tilt truck		Each	.45	.45	5.4%
2020	Moveable Furnishings	No smoking signage		Each	.02	.02	
F. SPECIAL CONSTRUCTION							
1020	Integrated Construction	N/A		—	—	—	0.0%
1040	Special Facilities	N/A		—	—	—	
G. BUILDING SITEWORK	**N/A**						

				Sub-Total	185.65	100%
CONTRACTOR FEES (General Requirements: 10%, Overhead: 5%, Profit: 10%)				25%	46.41	
ARCHITECT FEES				9%	20.89	

Total Building Cost 252.95

For customer support on your Square Foot Cost Data, call 877.694.8329.

255

Costs per square foot of floor area

Exterior Wall	S.F. Area	2000	3000	4000	6000	8000	10000	12000	14000	16000
	L.F. Perimeter	240	260	280	380	480	560	580	660	740
Brick Veneer	Wood Frame	224.35	196.95	183.30	175.80	172.15	168.70	163.20	162.00	161.10
Aluminum Siding	Wood Frame	207.15	184.50	173.10	166.65	163.40	160.50	156.10	155.05	154.30
Wood Siding	Wood Frame	207.05	184.40	173.10	166.60	163.40	160.45	156.05	155.05	154.25
Wood Shingles	Wood Frame	210.30	186.75	174.90	168.25	164.90	161.90	157.30	156.20	155.40
Precast Concrete Block	Wood Truss	209.90	186.45	174.75	168.05	164.80	161.80	157.25	156.15	155.35
Brick on Concrete Block	Wood Truss	239.80	208.05	192.15	183.90	179.80	175.70	169.30	167.85	166.90
Perimeter Adj., Add or Deduct	Per 100 L.F.	31.45	20.95	15.70	10.55	7.85	6.25	5.15	4.45	3.90
Story Hgt. Adj., Add or Deduct	Per 1 Ft.	5.05	3.60	2.85	2.65	2.45	2.30	2.00	1.90	1.90

For Basement, add $26.35 per square foot of basement area

The above costs were calculated using the basic specifications shown on the facing page. These costs should be adjusted where necessary for design alternatives and owner's requirements. Reported completed project costs, for this type of structure, range from $80.50 to $381.05 per S.F.

Common additives

Description	Unit	$ Cost
Closed Circuit Surveillance, one station, camera & monitor	Ea.	1975
For additional camera stations, add	Ea.	1075
Laundry Equipment, dryer, gas, 30 lb. capacity	Ea.	3850
Washer, commercial	Ea.	1600
Sauna, prefabricated, complete, 6' x 9'	Ea.	10,000
8' x 8'	Ea.	11,500
Swimming Pools, Complete, gunite	S.F.	97 - 120
TV Antenna, Master system, 12 outlet	Outlet	255
30 outlet	Outlet	221
100 outlet	Outlet	209

Description	Unit	$ Cost
Commissioning Fees, sustainable commercial construction	S.F.	0.24 - 3.04
Energy Modelling Fees, commercial buildings to 10,000 SF	Ea.	11,000
Greater than 10,000 SF add	S.F.	0.04
Green Bldg Cert Fees for comm construction project reg	Project	900
Photovoltaic Pwr Sys, grid connected, 20 kW (~2400 SF), roof	Ea.	275,000
Green Roofs, 6" soil depth, w/treated wd edging & sedum mats	S.F.	10.73
10" Soil depth, with treated wood edging & sedum mats	S.F.	12.44
Greywater Recovery Systems, prepackaged comm, 3060 gal.	Ea.	72,255
4590 gal.	Ea.	89,550
Rainwater Harvest Sys, prepckged comm, 10,000 gal, sys contrller	Ea.	31,500
20,000 gal. w/system controller	Ea.	50,550
30,000 gal. w/system controller	Ea.	80,800
Solar Domestic HW, closed loop, add-on sys, ext heat exchanger	Ea.	10,775
Drainback, hot water system, 120 gal tank	Ea.	13,325
Draindown, hot water system, 120 gal tank	Ea.	13,675

Important: See the Reference Section for Location Factors.

Model costs calculated for a 1 story building with 9' story height and 8,000 square feet of floor area

				Unit	Unit Cost	Cost Per S.F.	% Of Sub-Total
A. SUBSTRUCTURE							
1010	Standard Foundations	Poured concrete; strip and spread footings		S.F. Ground	3.31	3.31	
1020	Special Foundations	N/A		—	—	—	
1030	Slab on Grade	4" reinforced concrete with recycled vapor barrier and granular base		S.F. Slab	5.45	5.45	12.9%
2010	Basement Excavation	Site preparation for slab and trench for foundation wall and footing		S.F. Ground	.32	.32	
2020	Basement Walls	4' foundation wall		L.F. Wall	91	7.48	
B. SHELL							
B10 Superstructure							
1010	Floor Construction	N/A		—	—	—	5.2 %
1020	Roof Construction	Plywood on wood trusses		S.F. Roof	6.65	6.65	
B20 Exterior Enclosure							
2010	Exterior Walls	Face brick on wood studs with sheathing, insulation and paper	80% of wall	S.F. Wall	25.14	10.86	
2020	Exterior Windows	Wood double hung	20% of wall	Each	511	3.95	14.6%
2030	Exterior Doors	Wood solid core		Each	1465	4.03	
B30 Roofing							
3010	Roof Coverings	Asphalt strip shingles, 210-235 lbs/Sq., R-10 insul. & weather barrier-recycled		S.F. Roof	8.35	8.35	6.5%
3020	Roof Openings	N/A		—	—	—	
C. INTERIORS							
1010	Partitions	Gypsum bd. and sound deadening bd. on wood studs	9 SF Flr./LF Part.	S.F. Partition	10.22	9.08	
1020	Interior Doors	Single leaf hollow core wood, low VOC paint	300 S.F. Floor/Door	Each	537	1.79	
1030	Fittings	N/A		—	—	—	
2010	Stair Construction	N/A		—	—	—	18.4%
3010	Wall Finishes	90% paint, low VOC, 10% ceramic tile		S.F. Surface	1.60	2.84	
3020	Floor Finishes	85% carpet tile, 15% ceramic tile		S.F. Floor	5.42	5.42	
3030	Ceiling Finishes	Painted gypsum board on furring, low VOC paint		S.F. Ceiling	4.57	4.57	
D. SERVICES							
D10 Conveying							
1010	Elevators & Lifts	N/A		—	—	—	0.0 %
1020	Escalators & Moving Walks	N/A		—	—	—	
D20 Plumbing							
2010	Plumbing Fixtures	Toilet, low flow, auto sensor, & service fixt., supply & drainage	1 Fixt./90 SF Flr.	Each	2214	24.60	
2020	Domestic Water Distribution	Gas fired tankless water heater		S.F. Floor	.40	.40	20.7%
2040	Rain Water Drainage	Roof Drains		S.F. Roof	1.58	1.58	
D30 HVAC							
3010	Energy Supply	N/A		—	—	—	
3020	Heat Generating Systems	Included in D3050		—	—	—	
3030	Cooling Generating Systems	N/A		—	—	—	2.9 %
3050	Terminal & Package Units	Through the wall electric heating and cooling units		S.F. Floor	3.77	3.77	
3090	Other HVAC Sys. & Equipment	N/A		—	—	—	
D40 Fire Protection							
4010	Sprinklers	Wet pipe sprinkler system		S.F. Floor	4.98	4.98	4.7%
4020	Standpipes	Standpipe, wet, Class III		S.F. Floor	1.08	1.08	
D50 Electrical							
5010	Electrical Service/Distribution	200 ampere service, panel board and feeders		S.F. Floor	1.27	1.27	
5020	Lighting & Branch Wiring	LED fixt., daylit. dim., ltg. on/off control, recept., switches and misc. pwr.		S.F. Floor	10.22	10.22	12.7%
5030	Communications & Security	Addressable alarm systems		S.F. Floor	2.73	2.73	
5090	Other Electrical Systems	Emergency generator, 7.5 kW, and energy monitoring systems		S.F. Floor	2.14	2.14	
E. EQUIPMENT & FURNISHINGS							
1010	Commercial Equipment	Laundry equipment		S.F. Floor	.97	.97	
1020	Institutional Equipment	N/A		—	—	—	
1090	Other Equipment	Waste handling recycling tilt truck		Each	.71	.71	1.4%
2020	Moveable Furnishings	No smoking signage		Each	.16	.16	
F. SPECIAL CONSTRUCTION							
1020	Integrated Construction	N/A		—	—	—	0.0 %
1040	Special Facilities	N/A		—	—	—	
G. BUILDING SITEWORK	**N/A**						

	Sub-Total	128.71	**100%**
CONTRACTOR FEES (General Requirements:10%, Overhead: 5%, Profit:10%)	25%	32.18	
ARCHITECT FEES	7%	11.26	
	Total Building Cost	172.15	

For customer support on your Square Foot Cost Data, call 877.694.8329.

257

Costs per square foot of floor area

Exterior Wall		S.F. Area	25000	37000	49000	61000	73000	81000	88000	96000	104000
		L.F. Perimeter	433	593	606	720	835	911	978	1054	1074
Decorative Concrete Block	Wood Joists		180.95	177.20	171.05	169.70	168.70	168.25	167.95	167.55	166.55
	Precast Conc.		193.25	189.25	183.00	181.55	180.45	180.05	179.65	179.25	178.25
Stucco on Concrete Block	Wood Joists		182.75	178.25	171.40	169.80	168.65	168.15	167.75	167.30	166.20
	Precast Conc.		195.50	191.00	184.20	182.60	181.45	180.90	180.50	180.05	178.95
Wood Siding	Wood Frame		174.95	171.60	166.55	165.40	164.50	164.10	163.80	163.45	162.70
Brick Veneer	Wood Frame		182.75	178.25	171.40	169.75	168.65	168.10	167.70	167.25	166.15
Perimeter Adj., Add or Deduct	Per 100 L.F.		5.60	3.75	2.85	2.25	1.95	1.80	1.55	1.45	1.40
Story Hgt. Adj., Add or Deduct	Per 1 Ft.		2.00	1.85	1.40	1.35	1.30	1.30	1.25	1.30	1.20

For Basement, add $34.55 per square foot of basement area

The above costs were calculated using the basic specifications shown on the facing page. These costs should be adjusted where necessary for design alternatives and owner's requirements. Reported completed project costs, for this type of structure, range from $74.10 to $381.05 per S.F.

Common additives

Description	Unit	$ Cost
Closed Circuit Surveillance, one station, camera & monitor	Ea.	1975
For additional camera station, add	Ea.	1075
Elevators, hydraulic passenger, 2 stops, 2500# capacity	Ea.	68,800
3500# capacity	Ea.	73,300
Additional stop, add	Ea.	7825
Laundry Equipment, dryer, gas, 30 lb. capacity	Ea.	3850
Washer, commercial	Ea.	1600
Sauna, prefabricated, complete, 6' x 9'	Ea.	10,000
8' x 8'	Ea.	11,500
Swimming Pools, Complete, gunite	S.F.	97 - 120
TV Antenna, Master system, 12 outlet	Outlet	255
30 outlet	Outlet	221
100 outlet	Outlet	209

Description	Unit	$ Cost
Commissioning Fees, sustainable commercial construction	S.F.	0.24 - 3.04
Energy Modelling Fees, commercial buildings to 10,000 SF	Ea.	11,000
Greater than 10,000 SF add	S.F.	0.04
Green Bldg Cert Fees for comm construction project reg	Project	900
Photovoltaic Pwr Sys, grid connected, 20 kW (~2400 SF), roof	Ea.	275,000
Green Roofs, 6" soil depth, w/treated wd edging & sedum mats	S.F.	10.73
10" Soil depth, with treated wood edging & sedum mats	S.F.	12.44
Greywater Recovery Systems, prepackaged comm, 3060 gal.	Ea.	72,255
4590 gal.	Ea.	89,550
Rainwater Harvest Sys, prepckged comm, 10,000 gal, sys contrller	Ea.	31,500
20,000 gal. w/system controller	Ea.	50,550
30,000 gal. w/system controller	Ea.	80,800
Solar Domestic HW, closed loop, add-on sys, ext heat exchanger	Ea.	10,775
Drainback, hot water system, 120 gal tank	Ea.	13,325
Draindown, hot water system, 120 gal tank	Ea.	13,675

Important: See the Reference Section for Location Factors.

Model costs calculated for a 3 story building with 9' story height and 49,000 square feet of floor area

G Motel, 2-3 Story

			Unit	Unit Cost	Cost Per S.F.	% Of Sub-Total
A. SUBSTRUCTURE						
1010	Standard Foundations	Poured concrete; strip and spread footings	S.F. Ground	1.53	.51	
1020	Special Foundations	N/A	—	—	—	
1030	Slab on Grade	4" reinforced concrete with recycled vapor barrier and granular base	S.F. Slab	5.45	1.81	3.4%
2010	Basement Excavation	Site preparation for slab and trench for foundation wall and footing	S.F. Ground	.18	.06	
2020	Basement Walls	4' foundation wall	L.F. Wall	91	2.25	
B. SHELL						
	B10 Superstructure					
1010	Floor Construction	Precast concrete plank	S.F. Floor	10.40	6.93	7.5%
1020	Roof Construction	Precast concrete plank	S.F. Roof	10.32	3.44	
	B20 Exterior Enclosure					
2010	Exterior Walls	Decorative concrete block 85% of wall	S.F. Wall	22.16	6.29	
2020	Exterior Windows	Aluminum sliding 15% of wall	Each	588	1.96	10.7%
2030	Exterior Doors	Aluminum and glass doors and entrance with transom	Each	1973	6.53	
	B30 Roofing					
3010	Roof Coverings	Single-ply TPO membrane, 60 mils w/R-20 insul.	S.F. Roof	6.39	2.13	1.6%
3020	Roof Openings	Roof hatches	S.F. Roof	.12	.04	
C. INTERIORS						
1010	Partitions	Concrete block w/ foamed-in insul. 7 SF Flr./LF Part.	S.F. Partition	21.73	24.83	
1020	Interior Doors	Wood hollow core, low VOC paint 70 S.F. Floor/Door	Each	537	7.67	
1030	Fittings	N/A	—	—	—	
2010	Stair Construction	Concrete filled metal pan	Flight	14,500	3.55	35.9%
3010	Wall Finishes	90% paint, low VOC, 10% ceramic tile	S.F. Surface	1.61	3.68	
3020	Floor Finishes	85% carpet tile, 5% vinyl composition tile, recycled content, 10% ceramic tile	S.F. Floor	5.42	5.42	
3030	Ceiling Finishes	Textured finish	S.F. Ceiling	4.38	4.38	
D. SERVICES						
	D10 Conveying					
1010	Elevators & Lifts	Two hydraulic passenger elevators	Each	108,045	4.41	3.2%
1020	Escalators & Moving Walks	N/A	—	—	—	
	D20 Plumbing					
2010	Plumbing Fixtures	Toilet, low flow, auto sensor & service fixtures, supply & drain. 1 Fixt./180 SF Flr.	Each	5260	29.22	
2020	Domestic Water Distribution	Gas fired, tankless water heater	S.F. Floor	.30	.30	21.7%
2040	Rain Water Drainage	Roof drains	S.F. Roof	1.41	.47	
	D30 HVAC					
3010	Energy Supply	N/A	—	—	—	
3020	Heat Generating Systems	Included in D3050	—	—	—	
3030	Cooling Generating Systems	N/A	—	—	—	2.5 %
3050	Terminal & Package Units	Through the wall electric heating and cooling units	S.F. Floor	3.40	3.40	
3090	Other HVAC Sys. & Equipment	N/A	—	—	—	
	D40 Fire Protection					
4010	Sprinklers	Sprinklers, wet, light hazard	S.F. Floor	3.03	3.03	2.4%
4020	Standpipes	Standpipe, wet, Class III	S.F. Floor	.23	.23	
	D50 Electrical					
5010	Electrical Service/Distribution	800 ampere service, panel board and feeders	S.F. Floor	1.27	1.27	
5020	Lighting & Branch Wiring	LED fixtures, daylt. dim., ltg. on/off, recept., switches and misc. power	S.F. Floor	10.94	10.94	11.0%
5030	Communications & Security	Addressable alarm systems and emergency lighting	S.F. Floor	1.89	1.89	
5090	Other Electrical Systems	Emergency generator, 7.5 kW, energy monitoring systems	S.F. Floor	1.04	1.04	
E. EQUIPMENT & FURNISHINGS						
1010	Commercial Equipment	Commercial laundry equipment	S.F. Floor	.32	.32	
1020	Institutional Equipment	N/A	—	—	—	
1090	Other Equipment	Waste handling recycling tilt truck	Each	.07	.07	0.3%
2020	Moveable Furnishings	No smoking signage	Each	.02	.02	
F. SPECIAL CONSTRUCTION						
1020	Integrated Construction	N/A	—	—	—	0.0 %
1040	Special Facilities	N/A	—	—	—	
G. BUILDING SITEWORK	**N/A**					

		Sub-Total	138.09	100%
CONTRACTOR FEES (General Requirements: 10%, Overhead: 5%, Profit: 10%)		25%	34.55	
ARCHITECT FEES		6%	10.36	

Total Building Cost	**183**

For customer support on your Square Foot Cost Data, call 877.694.8329.

259

Costs per square foot of floor area

Exterior Wall	S.F. Area	2000	3000	5000	7000	9000	12000	15000	20000	25000
	L.F. Perimeter	220	260	320	360	420	480	520	640	700
Wood Siding	Wood Truss	213.85	194.75	177.50	168.60	164.80	160.25	156.85	154.75	152.30
Brick Veneer	Wood Truss	262.85	238.25	215.75	204.10	199.20	193.20	188.70	186.00	182.65
Brick on Block	Wood Truss	211.85	184.05	158.50	145.20	139.65	132.75	127.55	124.45	120.60
	Steel Roof Deck	265.90	236.60	209.65	195.65	189.75	182.50	176.95	173.70	169.70
EIFS on Metal Studs	Steel Roof Deck	240.95	218.05	197.20	186.45	181.85	176.30	172.10	169.65	166.65
Tiltup Concrete Panel	Steel Roof Deck	249.15	224.05	201.30	189.60	184.60	178.50	174.00	171.25	167.95
Perimeter Adj., Add or Deduct	Per 100 L.F.	25.05	16.70	10.00	7.20	5.55	4.20	3.35	2.55	2.00
Story Hgt. Adj., Add or Deduct	Per 1 Ft.	2.90	2.30	1.65	1.40	1.25	1.05	0.90	0.90	0.75
For Basement, add $36.35 per square foot of basement area										

The above costs were calculated using the basic specifications shown on the facing page. These costs should be adjusted where necessary for design alternatives and owner's requirements. Reported completed project costs, for this type of structure, range from $77.75 to $300.50 per S.F.

Common additives

Description	Unit	$ Cost
Closed Circuit Surveillance, one station, camera & monitor	Ea.	1975
For additional camera stations, add	Ea.	1075
Directory boards, plastic, glass covered, 30" x 20"	Ea.	600
36" x 48"	Ea.	1500
Aluminum, 36" x 24"	Ea.	755
48" x 32"	Ea.	1050
Electronic, wall mounted	S.F.	2900
Pedestal access flr sys w/PLAM cover, comp rm,		
Less than 6000 SF	S.F.	12.10
Greater than 6000 SF	S.F.	11.35
Office, greater than 6000 S.F.	S.F.	8.10
Uninterruptible power supply, 15 kVA/12.75 kW	kW	1810

Description	Unit	$ Cost
Commissioning Fees, sustainable commercial construction	S.F.	0.24 - 3.04
Energy Modelling Fees, commercial buildings to 10,000 SF	Ea.	11,000
Greater than 10,000 SF add	S.F.	0.04
Green Bldg Cert Fees for comm construction project reg	Project	900
Photovoltaic Pwr Sys, grid connected, 20 kW (~2400 SF), roof	Ea.	275,000
Green Roofs, 6" soil depth, w/treated wd edging & sedum mats	S.F.	10.73
10" Soil depth, with treated wood edging & sedum mats	S.F.	12.44
Greywater Recovery Systems, prepackaged comm, 3060 gal.	Ea.	72,255
4590 gal.	Ea.	89,550
Rainwater Harvest Sys, prepckged comm, 10,000 gal, sys contrller	Ea.	31,500
20,000 gal. w/system controller	Ea.	50,550
30,000 gal. w/system controller	Ea.	80,800
Solar Domestic HW, closed loop, add-on sys, ext heat exchanger	Ea.	10,775
Drainback, hot water system, 120 gal tank	Ea.	13,325
Draindown, hot water system, 120 gal tank	Ea.	13,675

Important: See the Reference Section for Location Factors.

Model costs calculated for a 1 story building with 12' story height and 7,000 square feet of floor area

			Unit	Unit Cost	Cost Per S.F.	% Of Sub-Total
A. SUBSTRUCTURE						
1010	Standard Foundations	Poured concrete; strip and spread footings	S.F. Ground	2.86	2.86	
1020	Special Foundations	N/A	—	—	—	
1030	Slab on Grade	4" reinforced concrete with recycled vapor barrier and granular base	S.F. Slab	5.45	5.45	9.7%
2010	Basement Excavation	Site preparation for slab and trench for foundation wall and footing	S.F. Ground	.32	.32	
2020	Basement Walls	4' foundation wall	L.F. Wall	174	4.91	
B. SHELL						
B10 Superstructure						
1010	Floor Construction	N/A	—	—	—	7.5 %
1020	Roof Construction	Steel joists, girders & deck on columns	S.F. Roof	10.43	10.43	
B20 Exterior Enclosure						
2010	Exterior Walls	E.I.F.S. on metal studs	S.F. Wall	17.56	8.67	
2020	Exterior Windows	Aluminum outward projecting	20% of wall	Each	750	4.02	11.3%
2030	Exterior Doors	Aluminum and glass, hollow metal	Each	3525	3.02	
B30 Roofing						
3010	Roof Coverings	Single-ply TPO membrane, 45 mils, heat welded seams w/R-20 insul.	S.F. Roof	9.97	9.97	7.3%
3020	Roof Openings	Roof hatch	S.F. Roof	.26	.26	
C. INTERIORS						
1010	Partitions	Gyp. bd. on mtl. studs w/sound attenuation	20 SF Flr/LF Part.	S.F. Partition	10.24	5.12	
1020	Interior Doors	Single leaf hollow metal, low VOC paint	200 S.F. Floor/Door	Each	1157	5.79	
1030	Fittings	Toilet partitions	S.F. Floor	.84	.84	
2010	Stair Construction	N/A	—	—	—	18.0%
3010	Wall Finishes	60% vinyl wall covering, 40% paint, low VOC	S.F. Surface	1.61	1.61	
3020	Floor Finishes	60% carpet tile, 30% vinyl composition tile, recycled content, 10% ceramic tile	S.F. Floor	4.76	4.76	
3030	Ceiling Finishes	Mineral fiber tile on concealed zee bars	S.F. Ceiling	7.03	7.03	
D. SERVICES						
D10 Conveying						
1010	Elevators & Lifts	N/A	—	—	—	0.0 %
1020	Escalators & Moving Walks	N/A	—	—	—	
D20 Plumbing						
2010	Plumbing Fixtures	Toilet, low flow, auto sensor, & service fixt., supply & drain.	1 Fixt./1320 SF Flr.	Each	9517	7.21	
2020	Domestic Water Distribution	Tankless, on-demand water heater, gas/propane	S.F. Floor	.94	.94	6.5%
2040	Rain Water Drainage	Roof drains	S.F. Roof	.86	.86	
D30 HVAC						
3010	Energy Supply	N/A	—	—	—	
3020	Heat Generating Systems	Included in D3050	—	—	—	
3040	Distribution Systems	Enthalpy heat recovery packages	Each	17,525	2.50	16.0 %
3050	Terminal & Package Units	Multizone rooftop air conditioner, SEER 14	S.F. Floor	19.85	19.85	
3090	Other HVAC Sys. & Equipment	N/A	—	—	—	
D40 Fire Protection						
4010	Sprinklers	Sprinkler system, light hazard	S.F. Floor	3.67	3.67	3.7%
4020	Standpipes	Standpipes and hose systems	S.F. Floor	1.47	1.47	
D50 Electrical						
5010	Electrical Service/Distribution	400 ampere service, panel board and feeders	S.F. Floor	2.03	2.03	
5020	Lighting & Branch Wiring	LED fixtures, daylt. dim., ltg. on/off, recept., switches, and A.C. power	S.F. Floor	15.23	15.23	19.4%
5030	Communications & Security	Addressable alarm systems, internet and phone wiring, and emergency lighting	S.F. Floor	7.32	7.32	
5090	Other Electrical Systems	Emergency generator, 7.5 kW, energy monitoring systems	S.F. Floor	2.42	2.42	
E. EQUIPMENT & FURNISHINGS						
1010	Commercial Equipment	N/A	—	—	—	
1020	Institutional Equipment	N/A	—	—	—	0.6 %
1090	Other Equipment	Waste handling recycling tilt truck	Each	.80	.80	
2020	Moveable Furnishings	No smoking signage	Each	.04	.04	
F. SPECIAL CONSTRUCTION						
1020	Integrated Construction	N/A	—	—	—	0.0 %
1040	Special Facilities	N/A	—	—	—	
G. BUILDING SITEWORK	**N/A**					

				Sub-Total	139.40	100%
CONTRACTOR FEES (General Requirements: 10%, Overhead: 5%, Profit: 10%)				25%	34.85	
ARCHITECT FEES				7%	12.20	

			Total Building Cost	**186.45**	

For customer support on your Square Foot Cost Data, call 877.694.8329.

261

Costs per square foot of floor area

Exterior Wall	S.F. Area	5000	8000	12000	16000	20000	35000	50000	65000	80000
	L.F. Perimeter	220	260	310	330	360	440	490	548	580
Face Brick with Concrete Block Back-up	Wood Joists	269.55	234.95	215.20	201.15	193.85	179.20	172.00	168.40	165.45
	Steel Joists	270.35	235.80	216.05	201.95	194.65	180.00	172.85	169.20	166.25
Glass and Metal Curtain Wall	Steel Frame	320.45	275.00	248.95	229.95	220.15	200.35	190.60	185.65	181.60
	R/Conc. Frame	316.50	271.60	245.80	226.95	217.25	197.65	187.90	183.00	178.90
Wood Siding	Wood Frame	208.55	185.95	173.05	164.45	159.85	150.85	146.55	144.35	142.60
Brick Veneer	Wood Frame	231.30	202.70	186.40	175.05	169.15	157.30	151.60	148.70	146.35
Perimeter Adj., Add or Deduct	Per 100 L.F.	44.45	27.90	18.55	13.85	11.15	6.35	4.50	3.45	2.75
Story Hgt. Adj., Add or Deduct	Per 1 Ft.	7.25	5.45	4.30	3.40	3.00	2.10	1.65	1.45	1.20

For Basement, add $40.65 per square foot of basement area

The above costs were calculated using the basic specifications shown on the facing page. These costs should be adjusted where necessary for design alternatives and owner's requirements. Reported completed project costs, for this type of structure, range from $81.35 to $315.10 per S.F.

Common additives

Description	Unit	$ Cost
Closed Circuit Surveillance, one station, camera and monitor	Ea.	1975
For additional camera stations, add	Ea.	1075
Directory Boards, plastic, glass covered, 30" x 20"	Ea.	600
36" x 48"	Ea.	1500
Aluminum, 36" x 24"	Ea.	755
48" x 32"	Ea.	1050
Electronic, wall mounted	S.F.	2900
Escalators, 10' rise, 32" wide, glass balustrade	Ea.	145,500
48" wide, glass balustrade	Ea.	153,000
Pedestal access flr sys w/PLAM cover, comp rm,		
Less than 6000 SF	S.F.	12.10
Greater than 6000 SF	S.F.	11.35
Office, greater than 6000 S.F.	S.F.	8.10
Uninterruptible power supply, 15 kVA/12.75 kW	kW	1810

Description	Unit	$ Cost
Commissioning Fees, sustainable commercial construction	S.F.	0.24 - 3.04
Energy Modelling Fees, commercial buildings to 10,000 SF	Ea.	11,000
Greater than 10,000 SF add	S.F.	0.04
Green Bldg Cert Fees for comm construction project reg	Project	900
Photovoltaic Pwr Sys, grid connected, 20 kW (~2400 SF), roof	Ea.	275,000
Green Roofs, 6" soil depth, w/treated wd edging & sedum mats	S.F.	10.73
10" Soil depth, with treated wood edging & sedum mats	S.F.	12.44
Greywater Recovery Systems, prepackaged comm, 3060 gal.	Ea.	72,255
4590 gal.	Ea.	89,550
Rainwater Harvest Sys, prepckged comm, 10,000 gal, sys contrller	Ea.	31,500
20,000 gal. w/system controller	Ea.	50,550
30,000 gal. w/system controller	Ea.	80,800
Solar Domestic HW, closed loop, add-on sys, ext heat exchanger	Ea.	10,775
Drainback, hot water system, 120 gal tank	Ea.	13,325
Draindown, hot water system, 120 gal tank	Ea.	13,675

Important: See the Reference Section for Location Factors.

Model costs calculated for a 3 story building with 12' story height and 20,000 square feet of floor area

				Unit	Unit Cost	Cost Per S.F.	% Of Sub-Total
A. SUBSTRUCTURE							
1010	Standard Foundations	Poured concrete; strip and spread footings		S.F. Ground	7.68	2.56	
1020	Special Foundations	N/A		—	—	—	
1030	Slab on Grade	4" reinforced concrete with recycled vapor barrier and granular base		S.F. Slab	5.45	1.81	4.4%
2010	Basement Excavation	Site preparation for slab and trench for foundation wall and footing		S.F. Ground	.18	.06	
2020	Basement Walls	4' foundation wall		L.F. Wall	174	1.93	
B. SHELL							
	B10 Superstructure						
1010	Floor Construction	Open web steel joists, slab form, concrete, columns		S.F. Floor	19.37	12.91	10.6%
1020	Roof Construction	Metal deck, open web steel joists, columns		S.F. Roof	7.41	2.47	
	B20 Exterior Enclosure						
2010	Exterior Walls	Face brick with concrete block backup	80% of wall	S.F. Wall	38.70	20.06	
2020	Exterior Windows	Aluminum outward projecting	20% of wall	Each	750	4.23	17.4%
2030	Exterior Doors	Aluminum and glass, hollow metal		Each	3525	1.06	
	B30 Roofing						
3010	Roof Coverings	Single-ply, TPO membrane, 60 mils, heat welded seams w/ R-20 insul.		S.F. Roof	7.26	2.42	1.7%
3020	Roof Openings	N/A		—	—	—	
C. INTERIORS							
1010	Partitions	Gypsum board on metal studs	20 SF Flr/LF Part.	S.F. Partition	10.53	4.21	
1020	Interior Doors	Single leaf hollow metal, low VOC paint	200 S.F. Floor/Door	Each	1157	5.79	
1030	Fittings	Toilet partitions		S.F. Floor	1.76	1.76	
2010	Stair Construction	Concrete filled metal pan		Flight	14,500	5.08	20.6%
3010	Wall Finishes	60% vinyl wall covering, 40% paint, low VOC		S.F. Surface	1.60	1.28	
3020	Floor Finishes	60% carpet tile, 30% vinyl composition tile, recycled content, 10% ceramic tile		S.F. Floor	4.76	4.76	
3030	Ceiling Finishes	Mineral fiber tile on concealed zee bars		S.F. Ceiling	7.03	7.03	
D. SERVICES							
	D10 Conveying						
1010	Elevators & Lifts	Two hydraulic passenger elevators		Each	117,300	11.73	8.1%
1020	Escalators & Moving Walks	N/A		—	—	—	
	D20 Plumbing						
2010	Plumbing Fixtures	Toilet, low flow, auto sensor, & service fixt., supply & drain.	1 Fixt./1320 SF Flr.	Each	5584	4.23	
2020	Domestic Water Distribution	Gas fired, tankless water heater		S.F. Floor	.20	.20	3.5%
2040	Rain Water Drainage	Roof drains		S.F. Roof	2.10	.70	
	D30 HVAC						
3010	Energy Supply	N/A		—	—	—	
3020	Heat Generating Systems	Included in D3050		—	—	—	
3040	Distribution Systems	Enthalpy heat recovery packages		Each	17,525	2.63	12.8 %
3050	Terminal & Package Units	Multizone rooftop air conditioner. SEER 14		S.F. Floor	16.05	16.05	
3090	Other HVAC Sys. & Equipment	N/A		—	—	—	
	D40 Fire Protection						
4010	Sprinklers	Wet pipe sprinkler system		S.F. Floor	3.84	3.84	3.3%
4020	Standpipes	Standpipes and hose systems		S.F. Floor	.93	.93	
	D50 Electrical						
5010	Electrical Service/Distribution	1000 ampere service, panel board and feeders		S.F. Floor	2.88	2.88	
5020	Lighting & Branch Wiring	LED fixtures, daylt. dim., ltg. on/off, recept., switches, and A.C. power		S.F. Floor	14.94	14.94	17.5%
5030	Communications & Security	Addressable alarm systems, internet and phone wiring, & emergency ltg.		S.F. Floor	6.13	6.13	
5090	Other Electrical Systems	Emergency generator, 7.5 kW, UPS, energy monitoring systems		S.F. Floor	1.55	1.55	
E. EQUIPMENT & FURNISHINGS							
1010	Commercial Equipment	N/A		—	—	—	
1020	Institutional Equipment	N/A		—	—	—	
1090	Other Equipment	Waste handling recycling tilt truck		Each	.29	.29	0.2 %
2020	Moveable Furnishings	No smoking signage		Each	.02	.02	
F. SPECIAL CONSTRUCTION							
1020	Integrated Construction	N/A		—	—	—	
1040	Special Facilities	N/A		—	—	—	0.0 %
G. BUILDING SITEWORK	**N/A**						
				Sub-Total		145.54	**100%**
	CONTRACTOR FEES (General Requirements: 10%, Overhead: 5%, Profit: 10%)				25%	36.37	
	ARCHITECT FEES				7%	12.74	
				Total Building Cost		**194.65**	

COMMERCIAL/INDUSTRIAL/INSTITUTIONAL

G.470 | **G Office, 5-10 Story**

Costs per square foot of floor area

Exterior Wall	S.F. Area	20000	40000	60000	80000	100000	150000	200000	250000	300000
	L.F. Perimeter	260	360	400	420	460	520	600	640	700
Precast Concrete Panel	Steel Frame	242.25	209.20	191.60	181.15	176.20	167.75	164.20	160.95	159.40
	R/Conc. Frame	240.40	207.25	189.60	179.10	174.10	165.65	162.10	158.85	157.25
Face Brick with Concrete Block Back-up	Steel Frame	230.40	201.15	185.75	176.70	172.30	164.95	161.85	159.10	157.70
	R/Conc. Frame	227.65	198.40	183.00	173.95	169.55	162.20	159.15	156.35	154.95
Limestone Panel Concrete Block Back-up	Steel Frame	259.90	221.60	200.90	188.60	182.70	172.85	168.70	164.90	163.00
	R/Conc. Frame	257.15	218.85	198.15	185.85	180.00	170.10	165.95	162.15	160.25
Perimeter Adj., Add or Deduct	Per 100 L.F.	33.45	16.75	11.15	8.35	6.65	4.45	3.35	2.70	2.20
Story Hgt. Adj., Add or Deduct	Per 1 Ft.	7.00	4.85	3.65	2.85	2.45	1.85	1.60	1.40	1.20

For Basement, add $43.85 per square foot of basement area

The above costs were calculated using the basic specifications shown on the facing page. These costs should be adjusted where necessary for design alternatives and owner's requirements. Reported completed project costs, for this type of structure, range from $91.55 to $269.15 per S.F.

Common additives

Description	Unit	$ Cost
Closed Circuit Surveillance, one station, camera and monitor	Ea.	1975
For additional camera stations, add	Ea.	1075
Directory Boards, plastic, glass covered, 30" x 20"	Ea.	600
36" x 48"	Ea.	1500
Aluminum, 36" x 24"	Ea.	755
48" x 32"	Ea.	1050
Electronic, wall mounted	S.F.	2900
Escalators, 10' rise, 32" wide, glass balustrade	Ea.	145,500
48" wide, glass balustrade	Ea.	153,000
Pedestal access flr sys w/PLAM cover, comp rm,		
Less than 6000 SF	S.F.	12.10
Greater than 6000 SF	S.F.	11.35
Office, greater than 6000 S.F.	S.F.	8.10
Uninterruptible power supply, 15 kVA/12.75 kW	kW	1810

Description	Unit	$ Cost
Commissioning Fees, sustainable commercial construction	S.F.	0.24 - 3.04
Energy Modelling Fees, commercial buildings to 10,000 SF	Ea.	11,000
Greater than 10,000 SF add	S.F.	0.04
Green Bldg Cert Fees for comm construction project reg	Project	900
Photovoltaic Pwr Sys, grid connected, 20 kW (~2400 SF), roof	Ea.	275,000
Green Roofs, 6" soil depth, w/treated wd edging & sedum mats	S.F.	10.73
10" Soil depth, with treated wood edging & sedum mats	S.F.	12.44
Greywater Recovery Systems, prepackaged comm, 3060 gal.	Ea.	72,255
4590 gal.	Ea.	89,550
Rainwater Harvest Sys, prepckged comm, 10,000 gal, sys contrller	Ea.	31,500
20,000 gal. w/system controller	Ea.	50,550
30,000 gal. w/system controller	Ea.	80,800
Solar Domestic HW, closed loop, add-on sys, ext heat exchanger	Ea.	10,775
Drainback, hot water system, 120 gal tank	Ea.	13,325
Draindown, hot water system, 120 gal tank	Ea.	13,675

Important: See the Reference Section for Location Factors.

Model costs calculated for a 8 story building with 12' story height and 80,000 square feet of floor area

				Unit	Unit Cost	Cost Per S.F.	% Of Sub-Total
A. SUBSTRUCTURE							
1010	Standard Foundations	Poured concrete; strip and spread footings		S.F. Ground	12.40	1.55	
1020	Special Foundations	N/A		—	—	—	
1030	Slab on Grade	4" reinforced concrete with recycled vapor barrier and granular base		S.F. Slab	5.45	.69	2.1%
2010	Basement Excavation	Site preparation for slab and trench for foundation wall and footing		S.F. Ground	.32	.04	
2020	Basement Walls	4' foundation wall		L.F. Wall	91	.62	
B. SHELL							
B10 Superstructure							
1010	Floor Construction	Concrete slab with metal deck and beams		S.F. Floor	25.33	22.16	16.9%
1020	Roof Construction	Metal deck, open web steel joists, interior columns		S.F. Roof	7.20	.90	
B20 Exterior Enclosure							
2010	Exterior Walls	Precast concrete panels	80% of wall	S.F. Wall	48.34	19.49	
2020	Exterior Windows	Vertical pivoted steel	20% of wall	Each	588	3.96	17.3%
2030	Exterior Doors	Double aluminum and glass doors and entrance with transoms		Each	4245	.27	
B30 Roofing							
3010	Roof Coverings	Single-ply TPO membrane, 60 mils, welded seams w/R-20 insul.		S.F. Roof	5.92	.74	0.5%
3020	Roof Openings	N/A		—	—	—	
C. INTERIORS							
1010	Partitions	Gypsum board on metal studs w/sound attenuation	30 SF Flr./LF Part.	S.F. Partition	11.31	3.77	
1020	Interior Doors	Single leaf hollow metal, low VOC paint	400 S.F. Floor/Door	Each	1157	2.90	
1030	Fittings	Toilet Partitions		S.F. Floor	1.17	1.17	
2010	Stair Construction	Concrete filled metal pan		Flight	14,500	3.08	17.4%
3010	Wall Finishes	60% vinyl wall covering, 40% paint, low VOC paint		S.F. Surface	1.61	1.07	
3020	Floor Finishes	60% carpet tile, 30% vinyl composition tile, recycled content, 10% ceramic tile		S.F. Floor	4.76	4.76	
3030	Ceiling Finishes	Mineral fiber tile on concealed zee bars		S.F. Ceiling	7.03	7.03	
D. SERVICES							
D10 Conveying							
1010	Elevators & Lifts	Four geared passenger elevators		Each	296,200	14.81	10.8%
1020	Escalators & Moving Walks	N/A		—	—	—	
D20 Plumbing							
2010	Plumbing Fixtures	Toilet, low flow, auto sensor, & service fixt., supply & drain.	1 Fixt./1370 SF Flr.	Each	4096	2.99	
2020	Domestic Water Distribution	Gas fired tankless water heater		S.F. Floor	.40	.40	2.7%
2040	Rain Water Drainage	Roof drains		S.F. Roof	2.48	.31	
D30 HVAC							
3010	Energy Supply	N/A		—	—	—	
3020	Heat Generating Systems	Included in D3050		—	—	—	
3040	Distribution Systems	Enthalpy heat recovery packages		Each	36,925	1.85	13.1%
3050	Terminal & Package Units	Multizone rooftop air conditioner, SEER 14		S.F. Floor	16.05	16.05	
3090	Other HVAC Sys. & Equipment	N/A					
D40 Fire Protection							
4010	Sprinklers	Wet pipe sprinkler system		S.F. Floor	2.93	2.93	2.9%
4020	Standpipes	Standpipes and hose systems		S.F. Floor	1.07	1.07	
D50 Electrical							
5010	Electrical Service/Distribution	1600 ampere service, panel board and feeders		S.F. Floor	1.06	1.06	
5020	Lighting & Branch Wiring	LED fixtures, daylt. dim., ltg. on/off, recept., switches, and A.C. power		S.F. Floor	14.65	14.65	16.1%
5030	Communications & Security	Addressable alarm systems, internet and phone wiring, emergency lighting		S.F. Floor	4.59	4.59	
5090	Other Electrical Systems	Emergency generator, 100 kW, UPS, energy monitoring systems		S.F. Floor	1.78	1.78	
E. EQUIPMENT & FURNISHINGS							
1010	Commercial Equipment	N/A		—	—	—	
1020	Institutional Equipment	N/A		—	—	—	
1090	Other Equipment	Waste handling recycling tilt truck		Each	.04	.04	0.0%
2020	Moveable Furnishings	No smoking signage		Each			
F. SPECIAL CONSTRUCTION							
1020	Integrated Construction	N/A		—	—	—	0.0%
1040	Special Facilities	N/A		—	—	—	
G. BUILDING SITEWORK	**N/A**						

	Sub-Total	136.73	**100%**
CONTRACTOR FEES (General Requirements: 10%, Overhead: 5%, Profit: 10%)	25%	34.17	
ARCHITECT FEES	6%	10.25	
	Total Building Cost	**181.15**	

For customer support on your Square Foot Cost Data, call 877.694.8329.

265

Costs per square foot of floor area

Exterior Wall	S.F. Area	120000	145000	170000	200000	230000	260000	400000	600000	800000
	L.F. Perimeter	420	450	470	490	510	530	600	730	820
Double Glazed Heat Absorbing Tinted Plate Glass Panels	Steel Frame	217.05	207.60	199.90	192.90	187.55	183.55	171.60	165.10	160.95
	R/Conc. Frame	207.15	197.95	190.35	183.40	178.25	174.25	162.45	156.10	152.00
Face Brick with Concrete Block Back-up	Steel Frame	185.40	179.55	174.90	170.65	167.50	165.00	157.90	153.95	151.55
	R/Conc. Frame	214.15	208.50	203.95	199.85	196.80	194.35	187.45	183.60	181.20
Precast Concrete Panel With Exposed Aggregate	Steel Frame	189.30	183.10	178.00	173.40	169.95	167.30	159.55	155.35	152.70
	R/Conc. Frame	179.65	173.60	168.70	164.15	160.85	158.25	150.65	146.55	143.95
Perimeter Adj., Add or Deduct	Per 100 L.F.	15.90	13.20	11.30	9.50	8.30	7.30	4.75	3.20	2.40
Story Hgt. Adj., Add or Deduct	Per 1 Ft.	6.45	5.75	5.10	4.45	4.15	3.75	2.75	2.25	1.90

For Basement, add $43.85 per square foot of basement area

The above costs were calculated using the basic specifications shown on the facing page. These costs should be adjusted where necessary for design alternatives and owner's requirements. Reported completed project costs, for this type of structure, range from $114.80 to $280.15 per S.F.

Common additives

Description	Unit	$ Cost
Closed Circuit Surveillance, one station, camera and monitor	Ea.	1975
For additional camera stations, add	Ea.	1075
Directory Boards, plastic, glass covered, 30" x 20"	Ea.	600
36" x 48"	Ea.	1500
Aluminum, 36" x 24"	Ea.	755
48" x 32"	Ea.	1050
Electronic, wall mounted	S.F.	2900
Escalators, 10' rise, 32" wide, glass balustrade	Ea.	145,500
48" wide, glass balustrade	Ea.	153,000
Pedestal access flr sys w/PLAM cover, comp rm,		
Less than 6000 SF	S.F.	12.10
Greater than 6000 SF	S.F.	11.35
Office, greater than 6000 S.F.	S.F.	8.10
Uninterruptible power supply, 15 kVA/12.75 kW	kW	1810

Description	Unit	$ Cost
Commissioning Fees, sustainable commercial construction	S.F.	0.24 - 3.04
Energy Modelling Fees, commercial buildings to 10,000 SF	Ea.	11,000
Greater than 10,000 SF add	S.F.	0.04
Green Bldg Cert Fees for comm construction project reg	Project	900
Photovoltaic Pwr Sys, grid connected, 20 kW (~2400 SF), roof	Ea.	275,000
Green Roofs, 6" soil depth, w/treated wd edging & sedum mats	S.F.	10.73
10" Soil depth, with treated wood edging & sedum mats	S.F.	12.44
Greywater Recovery Systems, prepackaged comm, 3060 gal.	Ea.	72,255
4590 gal.	Ea.	89,550
Rainwater Harvest Sys, prepckged comm, 10,000 gal, sys contrller	Ea.	31,500
20,000 gal. w/system controller	Ea.	50,550
30,000 gal. w/system controller	Ea.	80,800
Solar Domestic HW, closed loop, add-on sys, ext heat exchanger	Ea.	10,775
Drainback, hot water system, 120 gal tank	Ea.	13,325
Draindown, hot water system, 120 gal tank	Ea.	13,675

Important: See the Reference Section for Location Factors.

Model costs calculated for a 16 story building with 10' story height and 260,000 square feet of floor area

				Unit	Unit Cost	Cost Per S.F.	% Of Sub-Total
A. SUBSTRUCTURE							
1010	Standard Foundations	CIP concrete pile caps		S.F. Ground	10.24	.64	
1020	Special Foundations	Steel H-piles, concrete grade beams		S.F. Ground	61	3.84	
1030	Slab on Grade	4" reinforced concrete with recycled vapor barrier and granular base		S.F. Slab	5.45	.34	3.8%
2010	Basement Excavation	Site preparation for slab, piles and grade beams		S.F. Ground	.32	.02	
2020	Basement Walls	4' foundation wall		L.F. Wall	91	.45	
B. SHELL							
	B10 Superstructure						
1010	Floor Construction	Concrete slab, metal deck, beams		S.F. Floor	28.60	26.81	19.7%
1020	Roof Construction	Metal deck, open web steel joists, beams, columns		S.F. Roof	8.64	.54	
	B20 Exterior Enclosure						
2010	Exterior Walls	N/A		—	—	—	
2020	Exterior Windows	Double glazed heat absorbing, tinted plate glass wall panels	100% of wall	Each	83	27.19	20.1 %
2030	Exterior Doors	Double aluminum & glass doors		Each	6414	.70	
	B30 Roofing						
3010	Roof Coverings	Single ply membrane, fully adhered; perlite/EPS R-20 insulation		S.F. Roof	6.72	.42	0.3%
3020	Roof Openings	N/A		—	—	—	
C. INTERIORS							
1010	Partitions	Gypsum board on metal studs, sound attenuation	30 SF Flr./LF Part.	S.F. Partition	12.86	3.43	
1020	Interior Doors	Single leaf hollow metal, low VOC paint	400 S.F. Floor/Door	Each	1157	2.90	
1030	Fittings	Toilet partitions		S.F. Floor	.67	.67	
2010	Stair Construction	Concrete filled metal pan		Flight	17,300	2.33	16.1%
3010	Wall Finishes	60% vinyl wall covering, 40% paint, low VOC		S.F. Surface	1.61	.86	
3020	Floor Finishes	60% carpet tile, 30% vinyl composition tile, recycled content, 10% ceramic tile		S.F. Floor	5.12	5.12	
3030	Ceiling Finishes	Mineral fiber tile on concealed zee bars		S.F. Ceiling	7.03	7.03	
D. SERVICES							
	D10 Conveying						
1010	Elevators & Lifts	Four geared passenger elevators		Each	464,100	7.14	5.2%
1020	Escalators & Moving Walks	N/A		—	—	—	
	D20 Plumbing						
2010	Plumbing Fixtures	Toilet, low flow, auto sensor, & service fixt., supply & drain.	1 Fixt./1345 SF Flr.	Each	6738	5.01	
2020	Domestic Water Distribution	Gas fired, tankless water heater		S.F. Floor	.17	.17	3.9%
2040	Rain Water Drainage	Roof drains		S.F. Roof	3.20	.20	
	D30 HVAC						
3010	Energy Supply	N/A		—	—	—	
3020	Heat Generating Systems	N/A		—	—	—	
3040	Distribution Systems	Enthalpy heat recovery packages		Each	36,925	1.57	12.7 %
3050	Terminal & Package Units	Multizone rooftop air conditioner		S.F. Floor	16.05	16.05	
3090	Other HVAC Sys. & Equipment	N/A		—	—	—	
	D40 Fire Protection						
4010	Sprinklers	Sprinkler system, light hazard		S.F. Floor	2.83	2.83	2.4%
4020	Standpipes	Standpipes and hose systems		S.F. Floor	.56	.56	
	D50 Electrical						
5010	Electrical Service/Distribution	2400 ampere service, panel board and feeders		S.F. Floor	.69	.69	
5020	Lighting & Branch Wiring	LED fixtures, daylt. dim., ltg. on/off, recept., switches, and A.C. power		S.F. Floor	14.52	14.52	15.7%
5030	Communications & Security	Addressable alarm systems, internet & phone wiring, emergency ltg.		S.F. Floor	5.49	5.49	
5090	Other Electrical Systems	Emergency generator, 200 kW, UPS, energy monitoring systems		S.F. Floor	.98	.98	
E. EQUIPMENT & FURNISHINGS							
1010	Commercial Equipment	N/A		—	—	—	
1020	Institutional Equipment	N/A		—	—	—	0.0 %
1090	Other Equipment	Waste handling recycling tilt truck		Each	.01	.01	
2020	Moveable Furnishings	No smoking signage		Each			
F. SPECIAL CONSTRUCTION							
1020	Integrated Construction	N/A		—	—	—	0.0 %
1040	Special Facilities	N/A		—	—	—	
G. BUILDING SITEWORK	**N/A**						

			Sub-Total	138.51	100%
CONTRACTOR FEES (General Requirements: 10%, Overhead: 5%, Profit: 10%)		25%	34.65		
ARCHITECT FEES		6%	10.39		

Total Building Cost	**183.55**

For customer support on your Square Foot Cost Data, call 877.694.8329.

267

Costs per square foot of floor area

Exterior Wall	S.F. Area	7000	9000	11000	13000	15000	17000	19000	21000	23000
	L.F. Perimeter	240	280	303	325	354	372	397	422	447
Limestone with Concrete Block Back-up	Bearing Walls	288.35	272.20	258.20	248.25	242.15	235.90	231.85	228.60	225.75
	R/Conc. Frame	288.70	273.15	260.00	250.70	244.95	239.10	235.30	232.25	229.55
Face Brick with Concrete Block Back-up	Bearing Walls	246.85	233.15	222.10	214.30	209.35	204.55	201.25	198.65	196.40
	R/Conc. Frame	268.70	255.05	243.95	236.10	231.15	226.35	223.05	220.50	218.20
Decorative Concrete Block	Bearing Walls	233.95	221.45	211.75	204.85	200.45	196.30	193.40	191.15	189.10
	R/Conc. Frame	256.90	244.45	234.70	227.80	223.45	219.25	216.35	214.10	212.05
Perimeter Adj., Add or Deduct	Per 100 L.F.	34.70	26.95	22.00	18.65	16.15	14.20	12.70	11.50	10.60
Story Hgt. Adj., Add or Deduct	Per 1 Ft.	6.10	5.50	4.80	4.45	4.20	3.85	3.70	3.55	3.50

For Basement, add $34.50 per square foot of basement area

The above costs were calculated using the basic specifications shown on the facing page. These costs should be adjusted where necessary for design alternatives and owner's requirements. Reported completed project costs, for this type of structure, range from $131.95 to $344.15 per S.F.

Common additives

Description	Unit	$ Cost
Cells Prefabricated, 5'-6' x 7'-8", 7'-8' high	Ea.	12,800
Elevators, hydraulic passenger, 2 stops, 2500# capacity	Ea.	68,800
3500# capacity	Ea.	73,300
Lockers, Steel, Single tier, 60" to 72"	Opng.	221 - 370
Locker bench, lam. maple top only	L.F.	34
Pedestals, steel pipe	Ea.	76
Safe, Office type, 1 hr rating, 60"x36"x18", double door	Ea.	8125
Shooting Range, incl. bullet traps,		
Target provisions & controls, excl. struct shell	Ea.	51,000
Sound System, amplifier, 250 watts	Ea.	2050
Speaker, ceiling or wall	Ea.	213
Trumpet	Ea.	410

Description	Unit	$ Cost
Commissioning Fees, sustainable commercial construction	S.F.	0.24 - 3.04
Energy Modelling Fees, commercial buildings to 10,000 SF	Ea.	11,000
Greater than 10,000 SF add	S.F.	0.04
Green Bldg Cert Fees for comm construction project reg	Project	900
Photovoltaic Pwr Sys, grid connected, 20 kW (~2400 SF), roof	Ea.	275,000
Green Roofs, 6" soil depth, w/treated wd edging & sedum mats	S.F.	10.73
10" Soil depth, with treated wood edging & sedum mats	S.F.	12.44
Greywater Recovery Systems, prepackaged comm, 3060 gal.	Ea.	72,255
4590 gal.	Ea.	89,550
Rainwater Harvest Sys, prepckged comm, 10,000 gal, sys contrller	Ea.	31,500
20,000 gal. w/system controller	Ea.	50,550
30,000 gal. w/system controller	Ea.	80,800
Solar Domestic HW, closed loop, add-on sys, ext heat exchanger	Ea.	10,775
Drainback, hot water system, 120 gal tank	Ea.	13,325
Draindown, hot water system, 120 gal tank	Ea.	13,675

Important: See the Reference Section for Location Factors.

Model costs calculated for a 2 story building with 12' story height and 11,000 square feet of floor area

G Police Station

				Unit	Unit Cost	Cost Per S.F.	% Of Sub-Total
A.	**SUBSTRUCTURE**						
1010	Standard Foundations	Poured concrete; strip and spread footings		S.F. Ground	2.94	1.47	
1020	Special Foundations	N/A		—	—	—	
1030	Slab on Grade	4" reinforced concrete with recycled vapor barrier and granular base		S.F. Slab	5.45	2.73	4.1%
2010	Basement Excavation	Site preparation for slab and trench for foundation wall and footing		S.F. Ground	.32	.16	
2020	Basement Walls	4' foundation wall		L.F. Wall	91	3.33	
B.	**SHELL**						
	B10 Superstructure						
1010	Floor Construction	Open web steel joists, slab form, concrete		S.F. Floor	12.46	6.23	4.6%
1020	Roof Construction	Metal deck on open web steel joists		S.F. Roof	4.96	2.48	
	B20 Exterior Enclosure						
2010	Exterior Walls	Limestone with concrete block backup	80% of wall	S.F. Wall	61	32.27	
2020	Exterior Windows	Metal horizontal sliding	20% of wall	Each	1159	10.22	23.5%
2030	Exterior Doors	Hollow metal		Each	2723	1.98	
	B30 Roofing						
3010	Roof Coverings	Single-ply TPO membrane, 60 mils, heat welded w/R-20 insul.		S.F. Roof	6.84	3.42	1.8%
3020	Roof Openings	N/A		—	—	—	
C.	**INTERIORS**						
1010	Partitions	Concrete block w/foamed-in insul.	20 SF Flr./LF Part.	S.F. Partition	11.28	5.64	
1020	Interior Doors	Single leaf kalamein fire door, low VOC paint	200 S.F. Floor/Door	Each	1157	5.79	
1030	Fittings	Toilet partitions		S.F. Floor	.94	.94	
2010	Stair Construction	Concrete filled metal pan		Flight	17,300	3.14	15.7%
3010	Wall Finishes	90% paint, low VOC, 10% ceramic tile		S.F. Surface	2.76	2.76	
3020	Floor Finishes	70% vinyl composition tile, recycled content, 20% carpet tile, 10% ceramic tile		S.F. Floor	4.47	4.47	
3030	Ceiling Finishes	Mineral fiber tile on concealed zee bars		S.F. Ceiling	7.03	7.03	
D.	**SERVICES**						
	D10 Conveying						
1010	Elevators & Lifts	One hydraulic passenger elevator		Each	83,600	7.60	4.0%
1020	Escalators & Moving Walks	N/A		—	—	—	
	D20 Plumbing						
2010	Plumbing Fixtures	Toilet, low flow, auto sensor, & service fixt., supply & drainage	1 Fixt./580 SF Flr.	Each	5284	9.11	
2020	Domestic Water Distribution	Electric water heater, point-of-use, energy saver		S.F. Floor	1.22	1.22	6.4%
2040	Rain Water Drainage	Roof drains		S.F. Roof	3.42	1.71	
	D30 HVAC						
3010	Energy Supply	N/A		—	—	—	
3020	Heat Generating Systems	N/A		—	—	—	
3040	Distribution Systems	Enthalpy heat recovery packages		Each	30,400	2.77	11.9%
3050	Terminal & Package Units	Multizone rooftop air conditioner, SEER 14		S.F. Floor	19.85	19.85	
3090	Other HVAC Sys. & Equipment	N/A		—	—	—	
	D40 Fire Protection						
4010	Sprinklers	Wet pipe sprinkler system		S.F. Floor	3.96	3.96	2.6%
4020	Standpipes	Standpipe		S.F. Floor	1.02	1.02	
	D50 Electrical						
5010	Electrical Service/Distribution	400 ampere service, panel board and feeders		S.F. Floor	1.01	1.01	
5020	Lighting & Branch Wiring	LED fixtures, daylt. dim., ltg. on/off cntrl., recept., switches, & A.C. pwr.		S.F. Floor	21.50	21.50	16.6%
5030	Communications & Security	Addressable alarm systems, internet wiring, intercom &emergency ltg.		S.F. Floor	7.33	7.33	
5090	Other Electrical Systems	Emergency generator, 15 kW, energy monitoring systems		S.F. Floor	1.69	1.69	
E.	**EQUIPMENT & FURNISHINGS**						
1020	Institutional Equipment	Lockers, detention rooms, cells, gasoline dispensers		S.F. Floor	13.62	13.62	
1030	Vehicular Equipment	Gasoline dispenser system		S.F. Floor	2.47	2.47	8.8%
1090	Other Equipment	Waste handling recycling tilt truck		Each	.51	.51	
2020	Moveable Furnishings	No smoking signage		Each	.04	.04	
F.	**SPECIAL CONSTRUCTION**						
1020	Integrated Construction	N/A		—	—	—	0.0%
1040	Special Facilities	N/A		—	—	—	
G.	**BUILDING SITEWORK**	**N/A**					

		Sub-Total	189.47	**100%**
CONTRACTOR FEES (General Requirements: 10%, Overhead: 5%, Profit: 10%)			25%	47.41
ARCHITECT FEES			9%	21.32

Total Building Cost 258.20

Costs per square foot of floor area

Exterior Wall	S.F. Area	2000	2800	3500	4200	5000	5800	6500	7200	8000
	L.F. Perimeter	180	212	240	268	300	314	336	344	368
Wood Siding	Wood Frame	275.35	260.90	253.65	248.75	244.95	240.10	237.80	234.75	233.10
Brick Veneer	Wood Frame	283.40	266.95	258.80	253.25	248.95	243.25	240.70	237.10	235.20
Face Brick with Concrete Block Back-up	Wood Joists	297.10	278.50	269.20	263.05	258.05	251.55	248.55	244.35	242.25
	Steel Joists	296.45	275.60	265.15	258.20	252.65	245.15	241.85	237.00	234.60
Stucco on Concrete Block	Wood Joists	283.45	267.00	258.80	253.30	248.90	243.30	240.70	237.10	235.25
	Steel Joists	270.60	254.20	245.90	240.50	236.10	230.50	227.90	224.30	222.45
Perimeter Adj., Add or Deduct	Per 100 L.F.	32.30	23.05	18.40	15.45	12.90	11.15	9.95	8.95	8.10
Story Hgt. Adj., Add or Deduct	Per 1 Ft.	3.40	2.85	2.55	2.45	2.25	2.00	2.00	1.85	1.75
For Basement, add $39.35 per square foot of basement area										

The above costs were calculated using the basic specifications shown on the facing page. These costs should be adjusted where necessary for design alternatives and owner's requirements. Reported completed project costs, for this type of structure, range from $140.35 to $327.10 per S.F.

Common additives

Description	Unit	$ Cost
Bar, Front Bar	L.F.	425
Back bar	L.F.	340
Booth, Upholstered, custom straight	L.F.	250 - 460
"L" or "U" shaped	L.F.	259 - 440
Fireplace, brick, excl. chimney or foundation, 30"x29" opening	Ea.	3175
Chimney, standard brick, single flue, 16" x 20"	V.L.F.	106
2 Flue, 20" x 24"	V.L.F.	154
Kitchen Equipment		
Broiler	Ea.	4050
Coffee urn, twin 6 gallon	Ea.	3000
Cooler, 6 ft. long	Ea.	4175
Dishwasher, 10-12 racks per hr.	Ea.	4275
Food warmer, counter, 1.2 KW	Ea.	735
Freezer, 44 C.F., reach-in	Ea.	5175
Ice cube maker, 50 lb. per day	Ea.	2000
Range with 1 oven	Ea.	2925
Refrigerators, prefabricated, walk-in, 7'-6" high, 6' x 6'	S.F.	202
10' x 10'	S.F.	159
12' x 14'	S.F.	142
12' x 20'	S.F.	125

Description	Unit	$ Cost
Commissioning Fees, sustainable commercial construction	S.F.	0.24 - 3.04
Energy Modelling Fees, commercial buildings to 10,000 SF	Ea.	11,000
Greater than 10,000 SF add	S.F.	0.04
Green Bldg Cert Fees for comm construction project reg	Project	900
Photovoltaic Pwr Sys, grid connected, 20 kW (~2400 SF), roof	Ea.	275,000
Green Roofs, 6" soil depth, w/treated wd edging & sedum mats	S.F.	10.73
10" Soil depth, with treated wood edging & sedum mats	S.F.	12.44
Greywater Recovery Systems, prepackaged comm, 3060 gal.	Ea.	72,255
4590 gal.	Ea.	89,550
Rainwater Harvest Sys, prepckged comm, 10,000 gal, sys contrller	Ea.	31,500
20,000 gal. w/system controller	Ea.	50,550
30,000 gal. w/system controller	Ea.	80,800
Solar Domestic HW, closed loop, add-on sys, ext heat exchanger	Ea.	10,775
Drainback, hot water system, 120 gal tank	Ea.	13,325
Draindown, hot water system, 120 gal tank	Ea.	13,675

Model costs calculated for a 1 story building with 12' story height and 5,000 square feet of floor area

G Restaurant

				Unit	Unit Cost	Cost Per S.F.	% Of Sub-Total
A.	**SUBSTRUCTURE**						
1010	Standard Foundations	Poured concrete; strip and spread footings		S.F. Ground	2.72	2.72	
1020	Special Foundations	N/A		—	—	—	
1030	Slab on Grade	4" reinforced concrete with recycled vapor barrier and granular base		S.F. Slab	5.45	5.45	7.8%
2010	Basement Excavation	Site preparation for slab and trench for foundation wall and footing		S.F. Ground	.57	.57	
2020	Basement Walls	4' foundation wall		L.F. Wall	91	5.46	
B.	**SHELL**						
	B10 Superstructure						
1010	Floor Construction	Wood columns		S.F. Floor	.85	.85	
1020	Roof Construction	Plywood on wood truss (pitched)		S.F. Roof	7.44	7.44	4.5%
	B20 Exterior Enclosure						
2010	Exterior Walls	Cedar siding on wood studs with insulation	70% of wall	S.F. Wall	13.19	6.65	
2020	Exterior Windows	Storefront windows	30% of wall	Each	55	11.22	12.7%
2030	Exterior Doors	Aluminum and glass doors and entrance with transom		Each	5360	5.38	
	B30 Roofing						
3010	Roof Coverings	Cedar shingles with flashing (pitched); rigid fiberglass insulation		S.F. Roof	12.58	12.58	
3020	Roof Openings	Skylights		S.F. Roof	.09	.09	6.9%
C.	**INTERIORS**						
1010	Partitions	Gypsum board on wood studs	25 SF Floor/LF Part.	S.F. Partition	12.10	4.84	
1020	Interior Doors	Hollow core wood, low VOC paint	250 S.F. Floor/Door	Each	537	2.15	
1030	Fittings	Toilet partitions		S.F. Floor	1.17	1.17	
2010	Stair Construction	N/A					14.3%
3010	Wall Finishes	75% paint, low VOC, 25% ceramic tile		S.F. Surface	2.54	2.03	
3020	Floor Finishes	65% carpet, 35% quarry tile		S.F. Floor	8.93	8.93	
3030	Ceiling Finishes	Mineral fiber tile on concealed zee bars		S.F. Ceiling	7.03	7.03	
D.	**SERVICES**						
	D10 Conveying						
1010	Elevators & Lifts	N/A		—	—	—	
1020	Escalators & Moving Walks	N/A		—	—	—	0.0 %
	D20 Plumbing						
2010	Plumbing Fixtures	Kitchen, toilet, low flow, auto sensor, & service fixt., supply & drain.	1 Fixt./335 SF Flr.	Each	3894	10.97	
2020	Domestic Water Distribution	Tankless, on-demand water heater gas/propane		S.F. Floor	2.76	2.76	8.6%
2040	Rain Water Drainage	Roof drains		S.F. Roof	1.94	1.94	
	D30 HVAC						
3010	Energy Supply	N/A		—	—	—	
3020	Heat Generating Systems	Included in D3050		—	—	—	
3040	Distribution Systems	Enthalpy heat recovery packages and kitchen exhaust/make-up air system		Each	44,150	8.83	26.3 %
3050	Terminal & Package Units	Multizone rooftop air conditioner, SEER 14		S.F. Floor	39.40	39.40	
3090	Other HVAC Sys. & Equipment	N/A		—	—	—	
	D40 Fire Protection						
4010	Sprinklers	Sprinklers, light hazard and ordinary hazard kitchen		S.F. Floor	8.98	8.98	6.0%
4020	Standpipes	Standpipe		S.F. Floor	2.06	2.06	
	D50 Electrical						
5010	Electrical Service/Distribution	400 ampere service, panel board and feeders		S.F. Floor	2.31	2.31	
5020	Lighting & Branch Wiring	LED fixtures, daylt. dim., ltg. on/off, receptacles, switches, and A.C. power		S.F. Floor	13.06	13.06	12.3%
5030	Communications & Security	Addressable alarm systems and emergency lighting		S.F. Floor	3.61	3.61	
5090	Other Electrical Systems	Emergency generator, 15 kW and energy monitoring systems		S.F. Floor	3.46	3.46	
E.	**EQUIPMENT & FURNISHINGS**						
1010	Commercial Equipment	N/A		—	—	—	
1020	Institutional Equipment	N/A		—	—	—	
1090	Other Equipment	Waste handling recycling tilt truck		Each	1.13	1.13	0.6 %
2020	Moveable Furnishings	No smoking signage		Each	.06	.06	
F.	**SPECIAL CONSTRUCTION**						
1020	Integrated Construction	N/A		—	—	—	
1040	Special Facilities	N/A		—	—	—	0.0 %
G.	**BUILDING SITEWORK**	**N/A**					

	Sub-Total	183.13	**100%**
CONTRACTOR FEES (General Requirements: 10%, Overhead: 5%, Profit: 10%)	25%	45.80	
ARCHITECT FEES	7%	16.02	
Total Building Cost		**244.95**	

For customer support on your Square Foot Cost Data, call 877.694.8329.

271

Costs per square foot of floor area

Exterior Wall	S.F. Area	25000	30000	35000	40000	45000	50000	55000	60000	65000
	L.F. Perimeter	900	1050	1200	1350	1510	1650	1800	1970	2100
Face Brick with Concrete Block Back-up	Steel Frame	199.10	196.85	195.20	194.00	193.30	192.30	191.65	191.55	190.75
	Bearing Walls	191.35	189.30	187.95	186.90	186.35	185.40	184.75	184.75	183.95
Stucco on Concrete Block	Steel Frame	192.30	190.15	188.70	187.65	187.00	186.05	185.45	185.35	184.60
	Bearing Walls	186.05	183.95	182.50	181.35	180.75	179.80	179.20	179.10	178.35
Decorative Concrete Block	Steel Frame	190.40	188.35	186.90	185.80	185.15	184.20	183.60	183.50	182.75
	Bearing Walls	187.30	185.20	183.80	182.70	182.05	181.10	180.50	180.40	179.70
Perimeter Adj., Add or Deduct	Per 100 L.F.	5.00	4.15	3.65	3.20	2.80	2.50	2.30	2.15	1.90
Story Hgt. Adj., Add or Deduct	Per 1 Ft.	1.85	1.80	1.80	1.80	1.80	1.70	1.75	1.75	1.70
For Basement, add $29.15 per square foot of basement area										

The above costs were calculated using the basic specifications shown on the facing page. These costs should be adjusted where necessary for design alternatives and owner's requirements. Reported completed project costs, for this type of structure, range from $96.80 to $246.20 per S.F.

Common additives

Description	Unit	$ Cost
Bleachers, Telescoping, manual, 16-20 tier	Seat	281 - 345
21-30 tier	Seat	299 - 390
For power operation, add	Seat	55 - 86
Carrels Hardwood	Ea.	975 - 2350
Clock System, 20 room	Ea.	19,900
50 room	Ea.	46,800
Kitchen Equipment		
Broiler	Ea.	4050
Cooler, 6 ft. long, reach-in	Ea.	4175
Dishwasher, 10-12 racks per hr.	Ea.	4275
Food warmer, counter, 1.2 KW	Ea.	735
Freezer, 44 C.F., reach-in	Ea.	5175
Ice cube maker, 50 lb. per day	Ea.	2000
Range with 1 oven	Ea.	2925
Lockers, Steel, single tier, 60" to 72"	Opng.	221 - 370
2 tier, 60" to 72" total	Opng.	120 - 165
Locker bench, lam. maple top only	L.F.	34
Pedestals, steel pipe	Ea.	76
Seating, auditorium chair, veneer back, padded seat	Ea.	320
Classroom, movable chair & desk	Set	81 - 171
Lecture hall, pedestal type	Ea.	263 - 620

Description	Unit	$ Cost
Sound System, amplifier, 250 watts	Ea.	2050
Speaker, ceiling or wall	Ea.	213
Commissioning Fees, sustainable institutional construction	S.F.	0.58 - 2.47
Energy Modelling Fees, academic buildings to 10,000 SF	Ea.	9000
Greater than 10,000 SF add	S.F.	0.20
Green Bldg Cert Fees for school construction project reg	Project	900
Photovoltaic Pwr Sys, grid connected, 20 kW (~2400 SF), roof	Ea.	275,000
Green Roofs, 6" soil depth, w/treated wd edging & sedum mats	S.F.	10.73
10" Soil depth, with treated wood edging & sedum mats	S.F.	12.44
Greywater Recovery Systems, prepackaged comm, 3060 gal.	Ea.	72,255
4590 gal.	Ea.	89,550
Rainwater Harvest Sys, prepckged comm, 10,000 gal, sys contrller	Ea.	31,500
20,000 gal. w/system controller	Ea.	50,550
30,000 gal. w/system controller	Ea.	80,800
Solar Domestic HW, closed loop, add-on sys, ext heat exchanger	Ea.	10,775
Drainback, hot water system, 120 gal tank	Ea.	13,325
Draindown, hot water system, 120 gal tank	Ea.	13,675

Important: See the Reference Section for Location Factors.

Model costs calculated for a 1 story building with 15′ story height and 45,000 square feet of floor area

G School, Elementary

				Unit	Unit Cost	Cost Per S.F.	% Of Sub-Total
A. SUBSTRUCTURE							
1010	Standard Foundations	Poured concrete; strip and spread footings		S.F. Ground	5.14	5.14	
1020	Special Foundations	N/A		—	—	—	
1030	Slab on Grade	4″ reinforced concrete with recycled vapor barrier and granular base		S.F. Slab	5.45	5.45	11.5%
2010	Basement Excavation	Site preparation for slab and trench for foundation wall and footing		S.F. Ground	.18	.18	
2020	Basement Walls	4′ foundation wall		L.F. Wall	91	5.20	
B. SHELL							
	B10 Superstructure						
1010	Floor Construction	N/A		—	—	—	3.2 %
1020	Roof Construction	Metal deck on open web steel joists		S.F. Roof	4.52	4.52	
	B20 Exterior Enclosure						
2010	Exterior Walls	Face brick with concrete block backup	70% of wall	S.F. Wall	38.68	13.63	
2020	Exterior Windows	Steel outward projecting	25% of wall	Each	750	4.92	13.8%
2030	Exterior Doors	Metal and glass	5% of wall	Each	3850	.69	
	B30 Roofing						
3010	Roof Coverings	Single-ply TPO membrane, 60 mils, w/ flashing; polyiso. insulation		S.F. Roof	10.45	10.45	7.5%
3020	Roof Openings	N/A		—	—	—	
C. INTERIORS							
1010	Partitions	Concrete block w/ foamed-in insul.	20 SF Flr./LF Part.	S.F. Partition	12.14	6.07	
1020	Interior Doors	Single leaf kalamein fire doors, low VOC paint	700 S.F. Floor/Door	Each	1157	1.65	
1030	Fittings	Toilet partitions		S.F. Floor	1.84	1.84	
2010	Stair Construction	N/A		—	—	—	19.8%
3010	Wall Finishes	75% paint, low VOC, 15% glazed coating, 10% ceramic tile		S.F. Surface	4.92	4.92	
3020	Floor Finishes	65% vinyl composition tile, recycled content, 25% carpet tile, 10% terrazzo		S.F. Floor	6.05	6.05	
3030	Ceiling Finishes	Mineral fiber tile on concealed zee bars		S.F. Ceiling	7.03	7.03	
D. SERVICES							
	D10 Conveying						
1010	Elevators & Lifts	N/A		—	—	—	0.0 %
1020	Escalators & Moving Walks	N/A		—	—	—	
	D20 Plumbing						
2010	Plumbing Fixtures	Kitchen, toilet, low flow, auto sensor, & service fixt., supply & drain.	1 Fixt./625 SF Flr.	Each	9588	15.34	
2020	Domestic Water Distribution	Gas fired, tankless water heater		S.F. Floor	.16	.16	12.1%
2040	Rain Water Drainage	Roof drains		S.F. Roof	1.32	1.32	
	D30 HVAC						
3010	Energy Supply	N/A		—	—	—	
3020	Heat Generating Systems	N/A		—	—	—	
3040	Distribution Systems	Enthalpy heat recovery packages		Each	36,925	2.46	15.7 %
3050	Terminal & Package Units	Multizone rooftop air conditioner		S.F. Floor	19.45	19.45	
3090	Other HVAC Sys. & Equipment	N/A		—	—	—	
	D40 Fire Protection						
4010	Sprinklers	Sprinklers, light hazard		S.F. Floor	2.94	2.94	2.4%
4020	Standpipes	Standpipe		S.F. Floor	.40	.40	
	D50 Electrical						
5010	Electrical Service/Distribution	800 ampere service, panel board and feeders		S.F. Floor	.86	.86	
5020	Lighting & Branch Wiring	LED fixtures, daylt. dim., ltg. on/off, recept., switches, and A.C. power		S.F. Floor	13.53	13.53	13.8%
5030	Communications & Security	Addressable alarm systems, internet wiring, comm. systems & emerg. ltg.		S.F. Floor	3.79	3.79	
5090	Other Electrical Systems	Emergency generator, 15 kW, energy monitoring systems		S.F. Floor	1.05	1.05	
E. EQUIPMENT & FURNISHINGS							
1010	Commercial Equipment	N/A		—	—	—	
1020	Institutional Equipment	Chalkboards		S.F. Floor	.18	.18	0.2 %
1090	Other Equipment	Waste handling recycling tilt truck		Each	.07	.07	
2020	Moveable Furnishings	No smoking signage		Each	.01	.01	
F. SPECIAL CONSTRUCTION							
1020	Integrated Construction	N/A		—	—	—	0.0 %
1040	Special Facilities	N/A		—	—	—	
G. BUILDING SITEWORK	**N/A**						

		Sub-Total	139.30	100%
CONTRACTOR FEES (General Requirements: 10%, Overhead: 5%, Profit: 10%)		25%	34.86	
ARCHITECT FEES		7%	12.19	

Total Building Cost	**186.35**

Costs per square foot of floor area

Exterior Wall	S.F. Area	50000	70000	90000	110000	130000	150000	170000	190000	210000
	L.F. Perimeter	850	1140	1420	1700	1980	2280	2560	2840	3120
Face Brick with Concrete Block Back-up	Steel Frame	185.10	182.15	180.30	179.05	178.20	177.95	177.40	177.05	176.75
	R/Conc. Frame	189.10	186.15	184.30	183.05	182.20	181.95	181.40	181.05	180.75
Decorative Concrete Block	Steel Frame	177.90	175.30	173.60	172.55	171.80	171.55	171.05	170.75	170.50
	R/Conc. Frame	178.75	176.05	174.45	173.35	172.55	172.30	171.85	171.55	171.30
Limestone with Concrete Block Back-up	Steel Frame	196.70	193.25	191.10	189.60	188.60	188.30	187.70	187.30	186.90
	R/Conc. Frame	203.95	200.50	198.35	196.90	195.90	195.60	195.00	194.55	194.20
Perimeter Adj., Add or Deduct	Per 100 L.F.	4.65	3.30	2.55	2.15	1.80	1.55	1.35	1.20	1.10
Story Hgt. Adj., Add or Deduct	Per 1 Ft.	2.35	2.20	2.15	2.15	2.10	2.10	2.10	2.10	2.05

For Basement, add $38.10 per square foot of basement area

The above costs were calculated using the basic specifications shown on the facing page. These costs should be adjusted where necessary for design alternatives and owner's requirements. Reported completed project costs, for this type of structure, range from $112.50 to $264.95 per S.F.

Common additives

Description	Unit	$ Cost
Bleachers, Telescoping, manual, 16-20 tier	Seat	281 - 345
21-30 tier	Seat	299 - 390
For power operation, add	Seat	55 - 86
Carrels Hardwood	Ea.	975 - 2350
Clock System, 20 room	Ea.	19,900
50 room	Ea.	46,800
Elevators, hydraulic passenger, 2 stops, 2500# capacity	Ea.	68,800
3500# capacity	Ea.	73,300
Kitchen Equipment		
Broiler	Ea.	4050
Cooler, 6 ft. long, reach-in	Ea.	4175
Dishwasher, 10-12 racks per hr.	Ea.	4275
Food warmer, counter, 1.2 KW	Ea.	735
Freezer, 44 C.F., reach-in	Ea.	5175
Lockers, Steel, single tier, 60" or 72"	Opng.	221 - 370
2 tier, 60" or 72" total	Opng.	120 - 165
Locker bench, lam. maple top only	L.F.	34
Pedestals, steel pipe	Ea.	76
Seat, auditorium chair, veneer back, padded seat	Ea.	320
Classroom, movable chair & desk	Set	81 - 171
Lecture hall, pedestal type	Ea.	263 - 620

Description	Unit	$ Cost
Sound System, amplifier, 250 watts	Ea.	2050
Speaker, ceiling or wall	Ea.	213
Commissioning Fees, sustainable institutional construction	S.F.	0.58 - 2.47
Energy Modelling Fees, academic buildings to 10,000 SF	Ea.	9000
Greater than 10,000 SF add	S.F.	0.20
Green Bldg Cert Fees for school construction project reg	Project	900
Photovoltaic Pwr Sys, grid connected, 20 kW (~2400 SF), roof	Ea.	275,000
Green Roofs, 6" soil depth, w/treated wd edging & sedum mats	S.F.	10.73
10" Soil depth, with treated wood edging & sedum mats	S.F.	12.44
Greywater Recovery Systems, prepackaged comm, 3060 gal.	Ea.	72,255
4590 gal.	Ea.	89,550
Rainwater Harvest Sys, prepckged comm, 10,000 gal, sys contrller	Ea.	31,500
20,000 gal. w/system controller	Ea.	50,550
30,000 gal. w/system controller	Ea.	80,800
Solar Domestic HW, closed loop, add-on sys, ext heat exchanger	Ea.	10,775
Drainback, hot water system, 120 gal tank	Ea.	13,325
Draindown, hot water system, 120 gal tank	Ea.	13,675

Important: See the Reference Section for Location Factors.

Model costs calculated for a 2 story building with 15' story height and 130,000 square feet of floor area

				Unit	Unit Cost	Cost Per S.F.	% Of Sub-Total
A.	**SUBSTRUCTURE**						
1010	Standard Foundations	Poured concrete; strip and spread footings		S.F. Ground	2.76	1.38	
1020	Special Foundations	N/A		—	—	—	
1030	Slab on Grade	4" reinforced concrete with recycled vapor barrier and granular base		S.F. Slab	5.45	2.73	4.1%
2010	Basement Excavation	Site preparation for slab and trench for foundation wall and footing		S.F. Ground	.18	.09	
2020	Basement Walls	4' foundation wall		L.F. Wall	91	1.39	
B.	**SHELL**						
	B10 Superstructure						
1010	Floor Construction	Concrete slab without drop panel, concrete columns		S.F. Floor	22.42	11.21	13.9%
1020	Roof Construction	Concrete slab without drop panel		S.F. Roof	15.50	7.75	
	B20 Exterior Enclosure						
2010	Exterior Walls	Face brick with concrete block backup	75% of wall	S.F. Wall	38.69	13.26	
2020	Exterior Windows	Window wall	25% of wall	Each	74	8.49	16.3%
2030	Exterior Doors	Metal and glass		Each	2256	.51	
	B30 Roofing						
3010	Roof Coverings	Single-ply TPO membrane and standing seam metal; polyiso. insulation		S.F. Roof	13.94	6.97	5.2%
3020	Roof Openings	Roof hatches		S.F. Roof	.10	.05	
C.	**INTERIORS**						
1010	Partitions	Concrete block w/foamed-in insul.	25 SF Flr./LF Part.	S.F. Partition	14.73	7.07	
1020	Interior Doors	Single leaf kalamein fire doors, low VOC paint	700 S.F. Floor/Door	Each	1157	1.65	
1030	Fittings	Toilet partitions, chalkboards		S.F. Floor	1.30	1.30	
2010	Stair Construction	Concrete filled metal pan		Flight	14,500	.67	19.9%
3010	Wall Finishes	75% paint, low VOC. 15% glazed coating, 10% ceramic tile		S.F. Surface	3.48	3.34	
3020	Floor Finishes	70% vinyl comp. tile, recycled content, 20% carpet tile, 10% terrazzo		S.F. Floor	5.99	5.99	
3030	Ceiling Finishes	Mineral fiber tile on concealed zee bars		S.F. Ceiling	7.03	7.03	
D.	**SERVICES**						
	D10 Conveying						
1010	Elevators & Lifts	One hydraulic passenger elevator		Each	83,200	.64	0.5%
1020	Escalators & Moving Walks	N/A		—	—	—	
	D20 Plumbing						
2010	Plumbing Fixtures	Kitchen, toilet, low flow, auto sensor & service fixt., supply & drain.	1 Fixt./860 SF Flr.	Each	4721	5.49	
2020	Domestic Water Distribution	Gas fired, tankless water heater		S.F. Floor	.30	.30	4.8%
2040	Rain Water Drainage	Roof drains		S.F. Roof	1.52	.76	
	D30 HVAC						
3010	Energy Supply	N/A		—	—	—	
3020	Heat Generating Systems	N/A		—	—	—	
3040	Distribution Systems	Enthalpy heat recovery packages		Each	36,925	2.27	15.9 %
3050	Terminal & Package Units	Multizone rooftop air conditioner, SEER 14		S.F. Floor	19.45	19.45	
3090	Other HVAC Sys. & Equipment	N/A		—	—	—	
	D40 Fire Protection						
4010	Sprinklers	Sprinklers, light hazard		S.F. Floor	2.57	2.57	2.1%
4020	Standpipes	Standpipe, wet, Class III		S.F. Floor	.34	.34	
	D50 Electrical						
5010	Electrical Service/Distribution	2000 ampere service, panel board and feeders		S.F. Floor	.85	.85	
5020	Lighting & Branch Wiring	LED fixtures, daylt. dim., ltg. on/off, recept., switches, and A.C. power		S.F. Floor	14.79	14.79	15.1%
5030	Communications & Security	Addressable alarm systems, internet wiring, comm. systems and emerg. ltg.		S.F. Floor	4.09	4.09	
5090	Other Electrical Systems	Emergency generator, 250 kW, and energy monitoring systems		S.F. Floor	.80	.80	
E.	**EQUIPMENT & FURNISHINGS**						
1010	Commercial Equipment	N/A		—	—	—	
1020	Institutional Equipment	Laboratory casework and counters		S.F. Floor	1.97	1.97	
1090	Other Equipment	Waste handlg. recyc. tilt truck, built-in athletic equip., bleachers & backstp.		Each	.99	.99	2.2 %
2020	Moveable Furnishings	No smoking signage		Each	.02	.02	
F.	**SPECIAL CONSTRUCTION**						
1020	Integrated Construction	N/A		—	—	—	0.0 %
1040	Special Facilities	N/A		—	—	—	
G.	**BUILDING SITEWORK**	**N/A**					

	Sub-Total	136.21	100%
CONTRACTOR FEES (General Requirements: 10%, Overhead: 5%, Profit: 10%)	25%	34.07	
ARCHITECT FEES	7%	11.92	
Total Building Cost		**182.20**	

For customer support on your Square Foot Cost Data, call 877.694.8329.

275

Costs per square foot of floor area

Exterior Wall	S.F. Area	50000	65000	80000	95000	110000	125000	140000	155000	170000
	L.F. Perimeter	850	1060	1280	1490	1700	1920	2140	2340	2560
Face Brick with Concrete Block Back-up	Steel Frame	192.80	190.50	189.35	188.35	187.55	187.25	186.95	186.40	186.20
	Bearing Walls	184.15	181.90	180.75	179.75	179.00	178.65	178.35	177.80	177.60
Concrete Block Stucco Face	Steel Frame	197.95	195.45	194.15	193.00	192.20	191.75	191.45	190.85	190.60
	Bearing Walls	191.15	188.65	187.35	186.20	185.40	185.00	184.65	184.05	183.80
Decorative Concrete Block	Steel Frame	184.65	182.70	181.65	180.75	180.15	179.75	179.55	179.05	178.85
	Bearing Walls	176.35	174.40	173.35	172.45	171.75	171.40	171.20	170.70	170.50
Perimeter Adj., Add or Deduct	Per 100 L.F.	4.25	3.25	2.70	2.25	2.00	1.70	1.50	1.35	1.25
Story Hgt. Adj., Add or Deduct	Per 1 Ft.	2.10	2.05	2.05	1.95	2.00	1.90	1.90	1.90	1.90

For Basement, add $39.65 per square foot of basement area

The above costs were calculated using the basic specifications shown on the facing page. These costs should be adjusted where necessary for design alternatives and owner's requirements. Reported completed project costs, for this type of structure, range from $106.20 to $249.55 per S.F.

Common additives

Description	Unit	$ Cost
Bleachers, Telescoping, manual, 16-20 tier	Seat	281 - 345
21-30 tier	Seat	299 - 390
For power operation, add	Seat	55 - 86
Carrels Hardwood	Ea.	975 - 2350
Clock System, 20 room	Ea.	19,900
50 room	Ea.	46,800
Elevators, hydraulic passenger, 2 stops, 2500# capacity	Ea.	68,800
3500# capacity	Ea.	73,300
Kitchen Equipment		
Broiler	Ea.	4050
Cooler, 6 ft. long, reach-in	Ea.	4175
Dishwasher, 10-12 racks per hr.	Ea.	4275
Food warmer, counter, 1.2 KW	Ea.	735
Freezer, 44 C.F., reach-in	Ea.	5175
Lockers, Steel, single tier, 60" to 72"	Opng.	221 - 370
2 tier, 60" to 72" total	Opng.	120 - 165
Locker bench, lam. maple top only	L.F.	34
Pedestals, steel pipe	Ea.	76

Description	Unit	$ Cost
Seating, auditorium chair, veneer back, padded seat	Ea.	320
Upholstered, spring seat	Ea.	320
Classroom, movable chair & desk	Set	81 - 171
Lecture hall, pedestal type	Ea.	263 - 620
Sound System, amplifier, 250 watts	Ea.	2050
Speaker, ceiling or wall	Ea.	213
Commissioning Fees, sustainable institutional construction	S.F.	0.58 - 2.47
Energy Modelling Fees, academic buildings to 10,000 SF	Ea.	9000
Greater than 10,000 SF add	S.F.	0.20
Green Bldg Cert Fees for school construction project reg	Project	900
Photovoltaic Pwr Sys, grid connected, 20 kW (~2400 SF), roof	Ea.	275,000
Green Roofs, 6" soil depth, w/treated wd edging & sedum mats	S.F.	10.73
10" Soil depth, with treated wood edging & sedum mats	S.F.	12.44
Greywater Recovery Systems, prepackaged comm, 3060 gal.	Ea.	72,255
4590 gal.	Ea.	89,550
Rainwater Harvest Sys, prepckged comm, 10,000 gal, sys contrller	Ea.	31,500
20,000 gal. w/system controller	Ea.	50,550
30,000 gal. w/system controller	Ea.	80,800
Solar Domestic HW, closed loop, add-on sys, ext heat exchanger	Ea.	10,775
Drainback, hot water system, 120 gal tank	Ea.	13,325
Draindown, hot water system, 120 gal tank	Ea.	13,675

Important: See the Reference Section for Location Factors.

G School, Jr High, 2-3 Story

				Unit	Unit Cost	Cost Per S.F.	% Of Sub-Total
A.	**SUBSTRUCTURE**						
1010	Standard Foundations	Poured concrete; strip and spread footings		S.F. Ground	2.52	1.26	
1020	Special Foundations	N/A		—	—	—	
1030	Slab on Grade	4" reinforced concrete with recycled vapor barrier and granular base		S.F. Slab	5.45	2.73	3.9%
2010	Basement Excavation	Site preparation for slab and trench for foundation wall and footing		S.F. Ground	.18	.09	
2020	Basement Walls	4' foundation wall		L.F. Wall	91	1.40	
B.	**SHELL**						
	B10 Superstructure						
1010	Floor Construction	Open web steel joists, slab form, concrete, columns		S.F. Floor	28.86	14.43	14.6%
1020	Roof Construction	Metal deck, open web steel joists, columns		S.F. Roof	12.04	6.02	
	B20 Exterior Enclosure						
2010	Exterior Walls	Face brick with concrete block backup	75% of wall	S.F. Wall	38.71	13.46	
2020	Exterior Windows	Window wall	25% of wall	Each	63	6.95	15.0%
2030	Exterior Doors	Double aluminum & glass		Each	2377	.64	
	B30 Roofing						
3010	Roof Coverings	Single-ply TPO membrane & standing seam metal; polyiso. insulation		S.F. Roof	14.12	7.06	5.0%
3020	Roof Openings	Roof hatches		S.F. Roof	.04	.02	
C.	**INTERIORS**						
1010	Partitions	Concrete block w/foamed-in insul.	20 SF Flr./LF Part.	S.F. Partition	12.35	7.41	
1020	Interior Doors	Single leaf kalamein fire doors, low VOC paint	750 S.F. Floor/Door	Each	1157	1.54	
1030	Fittings	Toilet partitions, chalkboards		S.F. Floor	1.41	1.41	
2010	Stair Construction	Concrete filled metal pan		Flight	14,500	.80	22.3%
3010	Wall Finishes	50% paint, low VOC, 40% glazed coatings, 10% ceramic tile		S.F. Surface	3.83	4.60	
3020	Floor Finishes	50% vinyl comp. tile, recycled content, 30% carpet tile, 20% terrrazzo		S.F. Floor	8.55	8.55	
3030	Ceiling Finishes	Mineral fiberboard on concealed zee bars		S.F. Ceiling	7.03	7.03	
D.	**SERVICES**						
	D10 Conveying						
1010	Elevators & Lifts	One hydraulic passenger elevator		Each	83,600	.76	0.5%
1020	Escalators & Moving Walks	N/A		—	—	—	
	D20 Plumbing						
2010	Plumbing Fixtures	Kitchen, toilet, low flow, auto sensor, & service fixt., supply & drain.	1 Fixt./1170 SF Flr.	Each	6833	5.84	
2020	Domestic Water Distribution	Gas fired, tankless water heater		S.F. Floor	.36	.36	5.0%
2040	Rain Water Drainage	Roof drains		S.F. Roof	1.54	.77	
	D30 HVAC						
3010	Energy Supply	N/A		—	—	—	
3020	Heat Generating Systems	Included in D3050		—	—	—	
3040	Distribution Systems	Enthalpy heat recovery packages		Each	36,925	2.35	15.5 %
3050	Terminal & Package Units	Multizone rooftop air conditioner, SEER 14		S.F. Floor	19.45	19.45	
3090	Other HVAC Sys. & Equipment	N/A		—	—	—	
	D40 Fire Protection						
4010	Sprinklers	Sprinklers, light hazard	10% of area	S.F. Floor	2.57	2.57	2.1%
4020	Standpipes	Standpipe, wet, Class III		S.F. Floor	.41	.41	
	D50 Electrical						
5010	Electrical Service/Distribution	1600 ampere service, panel board and feeders		S.F. Floor	.76	.76	
5020	Lighting & Branch Wiring	LED fixtures, daylt. dim., ltg. on/off, recept., switches, and A.C. power		S.F. Floor	12.89	12.89	13.4%
5030	Communications & Security	Addressable alarm systems, internet wiring, comm. systems and emerg. ltg.		S.F. Floor	4.40	4.40	
5090	Other Electrical Systems	Emergency generator, 100 kW, and energy monitoring systems		S.F. Floor	.77	.77	
E.	**EQUIPMENT & FURNISHINGS**						
1010	Commercial Equipment	N/A		—	—	—	
1020	Institutional Equipment	Laboratory casework and counters		S.F. Floor	2.33	2.33	2.5 %
1090	Other Equipment	Waste handlg. recyc. tilt truck, built-in athletic equip., bleachers & backstps.		Each	1.18	1.18	
2020	Moveable Furnishings	No smoking signage		Each	.02	.02	
F.	**SPECIAL CONSTRUCTION**						
1020	Integrated Construction	N/A		—	—	—	0.0 %
1040	Special Facilities	N/A		—	—	—	
G.	**BUILDING SITEWORK**	N/A					

			Sub-Total	140.26	100%
CONTRACTOR FEES (General Requirements: 10%, Overhead: 5%, Profit: 10%)			25%	35.02	
ARCHITECT FEES			7%	12.27	
			Total Building Cost	**187.55**	

For customer support on your Square Foot Cost Data, call 877.694.8329.

277

Costs per square foot of floor area

Exterior Wall	S.F. Area	20000	30000	40000	50000	60000	70000	80000	90000	100000
	L.F. Perimeter	440	590	740	900	1050	1200	1360	1510	1660
Face Brick with Concrete Block Back-up	Steel Frame	196.20	189.30	185.85	184.20	182.80	181.70	181.15	180.60	180.05
	Bearing Walls	192.35	185.40	182.00	180.40	179.00	177.90	177.40	176.80	176.25
Decorative Concrete Block	Steel Frame	181.10	175.75	173.10	171.80	170.70	169.90	169.50	169.00	168.60
	Bearing Walls	173.75	168.35	165.70	164.40	163.35	162.45	162.05	161.65	161.25
Steel Siding on Steel Studs	Steel Frame	171.25	166.95	164.85	163.75	162.90	162.20	161.90	161.50	161.20
Metal Sandwich Panel	Steel Frame	198.60	191.35	187.85	186.15	184.65	183.50	183.05	182.35	181.85
Perimeter Adj., Add or Deduct	Per 100 L.F.	10.80	7.25	5.40	4.35	3.55	3.05	2.80	2.40	2.20
Story Hgt. Adj., Add or Deduct	Per 1 Ft.	2.70	2.30	2.25	2.15	2.05	2.10	2.10	2.00	2.00

For Basement, add $38.45 per square foot of basement area

The above costs were calculated using the basic specifications shown on the facing page. These costs should be adjusted where necessary for design alternatives and owner's requirements. Reported completed project costs, for this type of structure, range from $95.30 to $272.35 per S.F.

Common additives

Description	Unit	$ Cost
Carrels Hardwood	Ea.	975 - 2350
Clock System, 20 room	Ea.	19,900
50 room	Ea.	46,800
Directory Boards, plastic, glass covered, 36" x 48"	Ea.	1500
Aluminum, 36" x 24"	Ea.	755
48" x 60"	Ea.	2225
Elevators, hydraulic passenger, 2 stops, 2500# capacity	Ea.	68,800
3500# capacity	Ea.	73,300
Seating, auditorium chair, veneer back, padded seat	Ea.	320
Classroom, movable chair & desk	Set	81 - 171
Lecture hall, pedestal type	Ea.	263 - 620
Shops & Workroom, benches, metal	Ea.	835
Parts bins 6'-3" high, 3' wide, 12" deep, 72 bins	Ea.	710
Shelving, metal 1' x 3'	S.F.	12.60
Wide span 6' wide x 24" deep	S.F.	9.65

Description	Unit	$ Cost
Sound System, amplifier, 250 watts	Ea.	2050
Speaker, ceiling or wall	Ea.	213
Commissioning Fees, sustainable institutional construction	S.F.	0.58 - 2.47
Energy Modelling Fees, academic buildings to 10,000 SF	Ea.	9000
Greater than 10,000 SF add	S.F.	0.20
Green Bldg Cert Fees for school construction project reg	Project	900
Photovoltaic Pwr Sys, grid connected, 20 kW (~2400 SF), roof	Ea.	275,000
Green Roofs, 6" soil depth, w/treated wd edging & sedum mats	S.F.	10.73
10" Soil depth, with treated wood edging & sedum mats	S.F.	12.44
Greywater Recovery Systems, prepackaged comm, 3060 gal.	Ea.	72,255
4590 gal.	Ea.	89,550
Rainwater Harvest Sys, prepckged comm, 10,000 gal, sys contrller	Ea.	31,500
20,000 gal. w/system controller	Ea.	50,550
30,000 gal. w/system controller	Ea.	80,800
Solar Domestic HW, closed loop, add-on sys, ext heat exchanger	Ea.	10,775
Drainback, hot water system, 120 gal tank	Ea.	13,325
Draindown, hot water system, 120 gal tank	Ea.	13,675

Important: See the Reference Section for Location Factors.

Model costs calculated for a 2 story building with 16' story height and 40,000 square feet of floor area

G School, Vocational

					Unit	Unit Cost	Cost Per S.F.	% Of Sub-Total
A.	**SUBSTRUCTURE**							
1010	Standard Foundations	Poured concrete; strip and spread footings			S.F. Ground	2.84	1.42	
1020	Special Foundations	N/A			—	—	—	
1030	Slab on Grade	5" reinforced concrete with recycled vapor barrier and granular base			S.F. Slab	13.43	6.72	7.3%
2010	Basement Excavation	Site preparation for slab and trench for foundation wall and footing			S.F. Ground	.18	.09	
2020	Basement Walls	4' foundation wall			L.F. Wall	91	1.68	
B.	**SHELL**							
	B10 Superstructure							
1010	Floor Construction	Open web steel joists, slab form, concrete, beams, columns			S.F. Floor	18.94	9.47	9.4%
1020	Roof Construction	Metal deck, open web steel joists, beams, columns			S.F. Roof	6.60	3.30	
	B20 Exterior Enclosure							
2010	Exterior Walls	Face brick with concrete block backup	85% of wall		S.F. Wall	38.69	19.47	
2020	Exterior Windows	Tubular aluminum framing with insulated glass	15% of wall		Each	63	5.33	18.7%
2030	Exterior Doors	Metal and glass doors, low VOC paint			Each	2901	.58	
	B30 Roofing							
3010	Roof Coverings	Single-ply TPO membrane, 60 mils, heat welded seams w/polyiso. insul.			S.F. Roof	8.98	4.49	3.4%
3020	Roof Openings	Roof hatches			S.F. Roof	.34	.17	
C.	**INTERIORS**							
1010	Partitions	Concrete block w/foamed-in insul.	20 SF Flr./LF Part.		S.F. Partition	11.27	6.76	
1020	Interior Doors	Single leaf kalamein fire doors, low VOC paint	600 S.F. Floor/Door		Each	1157	1.93	
1030	Fittings	Toilet partitions, chalkboards			S.F. Floor	1.53	1.53	
2010	Stair Construction	Concrete filled metal pan			Flight	14,500	1.45	20.2%
3010	Wall Finishes	50% paint, low VOC, 40% glazed coating, 10% ceramic tile			S.F. Surface	4.64	5.57	
3020	Floor Finishes	70% vinyl comp. tile, recycled content, 20% carpet tile, 10% terrazzo			S.F. Floor	5.99	5.99	
3030	Ceiling Finishes	Mineral fiber tile on concealed zee bars	60% of area		S.F. Ceiling	4.22	4.22	
D.	**SERVICES**							
	D10 Conveying							
1010	Elevators & Lifts	One hydraulic passenger elevator			Each	83,600	2.09	1.5%
1020	Escalators & Moving Walks	N/A			—	—	—	
	D20 Plumbing							
2010	Plumbing Fixtures	Toilet, low flow, auto sensor, & service fixt., supply & drain.	1 Fixt./700 SF.Flr.		Each	3514	5.02	
2020	Domestic Water Distribution	Gas fired, tankless water heater			S.F. Floor	.31	.31	4.6%
2040	Rain Water Drainage	Roof drains			S.F. Roof	1.94	.97	
	D30 HVAC							
3010	Energy Supply	N/A			—	—	—	
3020	Heat Generating Systems	N/A			—	—	—	
3040	Distribution Systems	Enthalpy heat recovery packages			Each	30,400	5.32	18.2 %
3050	Terminal & Package Units	Multizone rooftop air conditioner, SEER 14			S.F. Floor	19.45	19.45	
3090	Other HVAC Sys. & Equipment	N/A			—	—	—	
	D40 Fire Protection							
4010	Sprinklers	Sprinklers, light hazard			S.F. Floor	3.19	3.19	3.0%
4020	Standpipes	Standpipe, wet, Class III			S.F. Floor	.96	.96	
	D50 Electrical							
5010	Electrical Service/Distribution	800 ampere service, panel board and feeders			S.F. Floor	.97	.97	
5020	Lighting & Branch Wiring	LED fixtures, daylt. dim., ltg. on/off, recept., switches, and A.C. power			S.F. Floor	12.88	12.88	13.4%
5030	Communications & Security	Addressable alarm systems, internet wiring, comm. systems & emerg. ltg.			S.F. Floor	3.21	3.21	
5090	Other Electrical Systems	Emergency generator, 11.5 kW, and energy monitoring systems			S.F. Floor	1.23	1.23	
E.	**EQUIPMENT & FURNISHINGS**							
1010	Commercial Equipment	N/A			—	—	—	
1020	Institutional Equipment	Stainless steel countertops			S.F. Floor	.15	.15	0.2 %
1090	Other Equipment	Waste handling recycling tilt truck			Each	.14	.14	
2020	Moveable Furnishings	No smoking signage			Each	.02	.02	
F.	**SPECIAL CONSTRUCTION**							
1020	Integrated Construction	N/A			—	—	—	0.0 %
1040	Special Facilities	N/A			—	—	—	
G.	**BUILDING SITEWORK**	**N/A**						

		Sub-Total	136.08	**100%**
CONTRACTOR FEES (General Requirements: 10%, Overhead: 5%, Profit: 10%)		25%	34.01	
ARCHITECT FEES		7%	11.91	
	Total Building Cost		182	

For customer support on your Square Foot Cost Data, call 877.694.8329.

279

Costs per square foot of floor area

Exterior Wall	S.F. Area	12000	16000	20000	26000	32000	38000	44000	52000	60000
	L.F. Perimeter	450	510	570	650	716	780	840	920	1020
Face Brick with Concrete Block Back-up	Steel Frame	176.05	164.90	158.20	151.45	146.70	143.40	140.90	138.35	137.00
	Bearing Walls	162.70	151.50	144.85	138.10	133.35	130.00	127.55	125.00	123.65
Stucco on Concrete Block	Steel Frame	165.70	156.10	150.35	144.60	140.60	137.75	135.65	133.55	132.35
	Bearing Walls	166.30	156.70	150.95	145.20	141.15	138.35	136.25	134.10	132.95
Precast Concrete Panels	Steel Frame	180.35	168.50	161.50	154.35	149.30	145.75	143.10	140.40	138.95
Metal Sandwich Panels	Steel Frame	175.75	164.60	157.95	151.25	146.55	143.20	140.75	138.20	136.85
Perimeter Adj., Add or Deduct	Per 100 L.F.	11.00	8.20	6.60	5.10	4.10	3.45	3.00	2.50	2.20
Story Hgt. Adj., Add or Deduct	Per 1 Ft.	2.30	1.95	1.75	1.60	1.40	1.25	1.20	1.10	1.00
For Basement, add $29.20 per square foot of basement area										

The above costs were calculated using the basic specifications shown on the facing page. These costs should be adjusted where necessary for design alternatives and owner's requirements. Reported completed project costs, for this type of structure, range from $80.25 to $229.80 per S.F.

Common additives

Description	Unit	$ Cost
Check Out Counter, single belt	Ea.	3600
Scanner, registers, guns & memory 10 lanes	Ea.	173,500
Power take away	Ea.	6025
Refrigerators, prefabricated, walk-in, 7'-6" High, 10' x 10'	S.F.	159
12' x 14'	S.F.	142
12' x 20'	S.F.	125
Refrigerated Food Cases, dairy, multi deck, 12' long	Ea.	12,800
Delicatessen case, multi deck, 18 S.F. shelf display	Ea.	8300
Freezer, glass door upright, 78 C.F.	Ea.	11,600
Frozen food, chest type, 12' long	Ea.	9450
Glass door reach in, 5 door	Ea.	15,900
Island case 12' long, multi deck	Ea.	9775
Meat case, 12' long, multi deck	Ea.	11,800
Produce, 12' long multi deck	Ea.	10,000
Safe, Office type, 1 hr rating, 60"x36"x18", double door	Ea.	8125

Description	Unit	$ Cost
Sound System, amplifier, 250 watts	Ea.	2050
Speaker, ceiling or wall	Ea.	213
Commissioning Fees, sustainable commercial construction	S.F.	0.24 - 3.04
Energy Modelling Fees, commercial buildings to 10,000 SF	Ea.	11,000
Greater than 10,000 SF add	S.F.	0.04
Green Bldg Cert Fees for comm construction project reg	Project	900
Photovoltaic Pwr Sys, grid connected, 20 kW (~2400 SF), roof	Ea.	275,000
Green Roofs, 6" soil depth, w/treated wd edging & sedum mats	S.F.	10.73
10" Soil depth, with treated wood edging & sedum mats	S.F.	12.44
Greywater Recovery Systems, prepackaged comm, 3060 gal.	Ea.	72,255
4590 gal.	Ea.	89,550
Rainwater Harvest Sys, prepckged comm, 10,000 gal, sys contrller	Ea.	31,500
20,000 gal. w/system controller	Ea.	50,550
30,000 gal. w/system controller	Ea.	80,800
Solar Domestic HW, closed loop, add-on sys, ext heat exchanger	Ea.	10,775
Drainback, hot water system, 120 gal tank	Ea.	13,325
Draindown, hot water system, 120 gal tank	Ea.	13,675

Model costs calculated for a 1 story building with 18' story height and 44,000 square feet of floor area

					Unit	Unit Cost	Cost Per S.F.	% Of Sub-Total
A. SUBSTRUCTURE								
1010	Standard Foundations	Poured concrete; strip and spread footings			S.F. Ground	1.16	1.16	
1020	Special Foundations	N/A			—	—	—	
1030	Slab on Grade	4" reinforced concrete with recycled vapor barrier and granular base			S.F. Slab	5.45	5.45	8.9%
2010	Basement Excavation	Site preparation for slab and trench for foundation wall and footing			S.F. Ground	.18	.18	
2020	Basement Walls	4' foundation wall			L.F. Wall	91	1.74	
B. SHELL								
B10 Superstructure								
1010	Floor Construction	N/A			—	—	—	
1020	Roof Construction	Metal deck, open web steel joists, beams, interior columns			S.F. Roof	8.20	8.20	8.6 %
B20 Exterior Enclosure								
2010	Exterior Walls	Face brick with concrete block backup		85% of wall	S.F. Wall	38.69	11.30	
2020	Exterior Windows	Storefront windows		15% of wall	Each	63	3.27	16.7%
2030	Exterior Doors	Sliding entrance doors with electrical operator, hollow metal			Each	5442	1.35	
B30 Roofing								
3010	Roof Coverings	Single-ply TPO membrane, 60 mils, heat welded seams w/R-20 insul.			S.F. Roof	5.18	5.18	5.6%
3020	Roof Openings	Roof hatches			S.F. Roof	.14	.14	
C. INTERIORS								
1010	Partitions	50% CMU w/ foamed-in insul., 50% gyp. bd. on mtl. studs	50 SF Flr./LF Part.		S.F. Partition	9.38	2.25	
1020	Interior Doors	Single leaf hollow metal, low VOC paint	2000 S.F. Floor/Door		Each	1157	.58	
1030	Fittings	N/A			—	—	—	
2010	Stair Construction	N/A			—	—	—	16.7%
3010	Wall Finishes	Paint, low VOC			S.F. Surface	5.63	2.70	
3020	Floor Finishes	Vinyl composition tile, recycled content			S.F. Floor	3.34	3.34	
3030	Ceiling Finishes	Mineral fiber tile on concealed zee bars			S.F. Ceiling	7.03	7.03	
D. SERVICES								
D10 Conveying								
1010	Elevators & Lifts	N/A			—	—	—	
1020	Escalators & Moving Walks	N/A			—	—	—	0.0 %
D20 Plumbing								
2010	Plumbing Fixtures	Toilet, low flow, auto sensor, & service fixt., supply & drain.	1 Fixt./1820 SF Flr.		Each	2439	1.34	
2020	Domestic Water Distribution	Point of use water heater, electric, energy saver			S.F. Floor	.20	.20	2.7%
2040	Rain Water Drainage	Roof drains			S.F. Roof	1.03	1.03	
D30 HVAC								
3010	Energy Supply	N/A			—	—	—	
3020	Heat Generating Systems	Included in D3050			—	—	—	
3040	Distribution Systems	Enthalpy heat recovery packages			Each	30,400	2.77	7.2 %
3050	Terminal & Package Units	Single zone rooftop air conditioner			S.F. Floor	4.09	4.09	
3090	Other HVAC Sys. & Equipment	N/A			—	—	—	
D40 Fire Protection								
4010	Sprinklers	Sprinklers, light hazard			S.F. Floor	3.67	3.67	4.4%
4020	Standpipes	Standpipe			S.F. Floor	.54	.54	
D50 Electrical								
5010	Electrical Service/Distribution	1600 ampere service, panel board and feeders			S.F. Floor	1.93	1.93	
5020	Lighting & Branch Wiring	LED fixtures, daylt. dim., ltg. on/off, receptacles, switches, and A.C.power			S.F. Floor	22.46	22.46	29.0%
5030	Communications & Security	Addressable alarm systems, partial internet wiring and emergency lighting			S.F. Floor	2.26	2.26	
5090	Other Electrical Systems	Emergency generator, 15 kW and energy monitoring systems			S.F. Floor	1.05	1.05	
E. EQUIPMENT & FURNISHINGS								
1010	Commercial Equipment	N/A			—	—	—	
1020	Institutional Equipment	N/A			—	—	—	0.2 %
1090	Other Equipment	Waste handling recycling tilt truck			Each	.13	.13	
2020	Moveable Furnishings	No smoking signage			Each	.02	.02	
F. SPECIAL CONSTRUCTION								
1020	Integrated Construction	N/A			—	—	—	0.0 %
1040	Special Facilities	N/A			—	—	—	
G. BUILDING SITEWORK	**N/A**							

	Sub-Total	95.36	100%
CONTRACTOR FEES (General Requirements: 10%, Overhead: 5%, Profit: 10%)	25%	23.85	
ARCHITECT FEES	7%	8.34	
Total Building Cost		**127.55**	

For customer support on your Square Foot Cost Data, call 877.694.8329.

Costs per square foot of floor area

Exterior Wall	S.F. Area	10000	15000	20000	25000	30000	35000	40000	50000	60000
	L.F. Perimeter	410	500	600	640	700	766	833	966	1000
Tiltup Concrete Panels	Steel Frame	153.80	141.35	135.55	129.85	126.80	124.60	123.10	120.85	117.90
Brick with Block Back-up	Bearing Walls	176.50	158.95	151.00	142.45	137.80	134.70	132.50	129.25	124.55
Concrete Block	Steel Frame	158.65	147.95	143.05	137.90	135.05	133.20	131.85	129.85	127.10
	Bearing Walls	142.85	132.15	127.25	122.10	119.30	117.40	116.05	114.05	111.25
Galvanized Steel Siding	Steel Frame	165.45	154.40	149.40	144.05	141.15	139.20	137.80	135.75	132.85
Metal Sandwich Panels	Steel Frame	163.95	152.20	146.95	141.15	138.10	136.00	134.50	132.35	129.25
Perimeter Adj., Add or Deduct	Per 100 L.F.	9.05	5.95	4.50	3.60	3.00	2.60	2.25	1.80	1.50
Story Hgt. Adj., Add or Deduct	Per 1 Ft.	1.05	0.85	0.80	0.70	0.60	0.55	0.55	0.50	0.40
For Basement, add $32.30 per square foot of basement area										

The above costs were calculated using the basic specifications shown on the facing page. These costs should be adjusted where necessary for design alternatives and owner's requirements. Reported completed project costs, for this type of structure, range from $42.15 to $168.35 per S.F.

Common additives

Description	Unit	$ Cost
Dock Leveler, 10 ton cap., 6' x 8'	Ea.	7075
7' x 8'	Ea.	8100
Fence, Chain link, 6' high, 9 ga. wire	L.F.	28
6 ga. wire	L.F.	30
Gate	Ea.	390
Paving, Bituminous, wearing course plus base course	S.Y.	6.55
Sidewalks, Concrete 4" thick	S.F.	4.48
Sound System, amplifier, 250 watts	Ea.	2050
Speaker, ceiling or wall	Ea.	213
Yard Lighting, 20' aluminum pole, 400W, HP sodium fixture	Ea.	3175

Description	Unit	$ Cost
Commissioning Fees, sustainable industrial construction	S.F.	0.20 - 1.06
Energy Modelling Fees, commercial buildings to 10,000 SF	Ea.	11,000
Greater than 10,000 SF add	S.F.	0.04
Green Bldg Cert Fees for comm construction project reg	Project	900
Photovoltaic Pwr Sys, grid connected, 20 kW (~2400 SF), roof	Ea.	275,000
Green Roofs, 6" soil depth, w/treated wd edging & sedum mats	S.F.	10.73
10" Soil depth, with treated wood edging & sedum mats	S.F.	12.44
Greywater Recovery Systems, prepackaged comm, 3060 gal.	Ea.	72,255
4590 gal.	Ea.	89,550
Rainwater Harvest Sys, prepckged comm, 10,000 gal, sys contrller	Ea.	31,500
20,000 gal. w/system controller	Ea.	50,550
30,000 gal. w/system controller	Ea.	80,800
Solar Domestic HW, closed loop, add-on sys, ext heat exchanger	Ea.	10,775
Drainback, hot water system, 120 gal tank	Ea.	13,325
Draindown, hot water system, 120 gal tank	Ea.	13,675

Important: See the Reference Section for Location Factors.

Model costs calculated for a 1 story building with 24' story height and 30,000 square feet of floor area

G Warehouse

					Unit	Unit Cost	Cost Per S.F.	% Of Sub-Total

A. SUBSTRUCTURE

1010	Standard Foundations	Poured concrete; strip and spread footings			S.F. Ground	1.47	1.47	
1020	Special Foundations	N/A			—	—	—	
1030	Slab on Grade	5" reinforced concrete with recycled vapor barrier and granular base			S.F. Slab	13.43	13.43	20.0%
2010	Basement Excavation	Site preparation for slab and trench for foundation wall and footing			S.F. Ground	.18	.18	
2020	Basement Walls	4' foundation wall			L.F. Wall	91	2.76	

B. SHELL

B10 Superstructure

1010	Floor Construction	Mezzanine: open web steel joists, slab form, concrete beams, columns	10% of area		S.F. Floor	2.19	2.19	10.8%
1020	Roof Construction	Metal deck, open web steel joists, beams, columns			S.F. Roof	7.43	7.43	

B20 Exterior Enclosure

2010	Exterior Walls	Concrete block	95% of wall		S.F. Wall	17.20	9.15	
2020	Exterior Windows	N/A			—	—	—	11.6%
2030	Exterior Doors	Steel overhead, hollow metal	5% of wall		Each	3161	1.16	

B30 Roofing

3010	Roof Coverings	Single-ply TPO membrane, 60 mils, heat welded w/R-20 insul.			S.F. Roof	5.34	5.34	6.4%
3020	Roof Openings	Roof hatches and skylight			S.F. Roof	.33	.33	

C. INTERIORS

1010	Partitions	Concrete block w/foamed-in insul. (office and washrooms)	100 SF Flr./LF Part.		S.F. Partition	11.38	.91	
1020	Interior Doors	Single leaf hollow metal, low VOC paint	5000 S.F. Floor/Door		Each	1157	.24	
1030	Fittings	N/A			—	—	—	
2010	Stair Construction	Steel gate with rails			Flight	14,350	.96	8.2%
3010	Wall Finishes	Paint, low VOC			S.F. Surface	14.69	2.35	
3020	Floor Finishes	90% hardener, 10% vinyl composition tile, recycled content			S.F. Floor	2.10	2.10	
3030	Ceiling Finishes	Suspended mineral tile on zee channels in office area	10% of area		S.F. Ceiling	.70	.71	

D. SERVICES

D10 Conveying

1010	Elevators & Lifts	N/A			—	—	—	0.0%
1020	Escalators & Moving Walks	N/A			—	—	—	

D20 Plumbing

2010	Plumbing Fixtures	Toilet, low flow, auto sensor & service fixt., supply & drainage	1 Fixt./2500 SF Flr.		Each	4300	1.72	
2020	Domestic Water Distribution	Tankless, on-demand water heater, gas/propane			S.F. Floor	.44	.44	4.4%
2040	Rain Water Drainage	Roof drains			S.F. Roof	1.77	1.77	

D30 HVAC

3010	Energy Supply	N/A			—	—	—	
3020	Heat Generating Systems	Ventilation with heat			Each	140,200	5.05	
3030	Cooling Generating Systems	N/A			—	—	—	6.5%
3050	Terminal & Package Units	Single zone rooftop air conditioner, SEER 14			S.F. Floor	.74	.74	
3090	Other HVAC Sys. & Equipment	N/A			—	—	—	

D40 Fire Protection

4010	Sprinklers	Sprinklers, ordinary hazard			S.F. Floor	4.15	4.15	7.2%
4020	Standpipes	Standpipe			S.F. Floor	2.29	2.29	

D50 Electrical

5010	Electrical Service/Distribution	200 ampere service, panel board and feeders			S.F. Floor	.31	.31	
5020	Lighting & Branch Wiring	LED fixtures, daylt. dim., ltg. on/off, receptacles, switches, and A.C. power			S.F. Floor	9.83	9.83	15.9%
5030	Communications & Security	Addressable alarm systems			S.F. Floor	2.74	2.74	
5090	Other Electrical Systems	Energy monitoring systems				1.28	1.28	

E. EQUIPMENT & FURNISHINGS

1010	Commercial Equipment	N/A			—	—	—	
1030	Vehicular Equipment	Dock boards, dock levelers			S.F. Floor	2.71	2.71	3.3%
1090	Other Equipment	Waste handling recycling tilt truck			Each	.19	.19	
2020	Moveable Furnishings	No smoking signage			Each	.02	.02	

F. SPECIAL CONSTRUCTION

1020	Integrated Construction	Shipping & receiving air curtain			S.F. Floor	5.23	5.23	5.9%
1040	Special Facilities	N/A			—	—	—	

G. BUILDING SITEWORK   N/A

	Sub-Total	89.18	**100%**
CONTRACTOR FEES (General Requirements: 10%, Overhead: 5%, Profit: 10%)	25%	22.32	
ARCHITECT FEES	7%	7.80	
	Total Building Cost	119.30	

Assemblies Section

Table of Contents

Table of Contents

Table of Contents

Table of Contents

Introduction to the Assemblies Section

This section of the manual contains the technical description and installed cost of all components used in the commercial/industrial/institutional section. In most cases, there is an illustration of the component to aid in identification.

All Costs Include:

Materials purchased in lot sizes typical of normal construction projects with discounts appropriate to established contractors.

Installation work performed by union labor working under standard conditions at a normal pace.

Installing Contractor's Overhead & Profit

Standard installing contractor's overhead & profit are included in the assemblies costs.

These assemblies costs contain no provisions for General Contractor's overhead and profit or Architect's fees, nor do they include premiums for labor or materials, or savings which may be realized under certain economic situations.

The assemblies tables are arranged by physical size (dimensions) of the component wherever possible.

How RSMeans Assemblies Data Works

Assemblies estimating provides a fast and reasonably accurate way to develop construction costs. An assembly is the grouping of individual work items, with appropriate quantities, to provide a cost for a major construction component in a convenient unit of measure.

An assemblies estimate is often used during early stages of design development to compare the cost impact of various design alternatives on total building cost.

Assemblies estimates are also used as an efficient tool to verify construction estimates.

Assemblies estimates do not require a completed design or detailed drawings. Instead, they are based on the general size of the structure and other known parameters of the project. The degree of accuracy of an assemblies estimate is generally within +/- 15%.

NARRATIVE FORMAT

C20 Stairs

C2010 Stair Construction

The table below lists the cost per flight for 4'-0" wide stairs. Side walls are not included. Railings are included in the prices.

RSMeans assemblies are identified by a **unique 12-character identifier**. The assemblies are numbered using UNIFORMAT II, ASTM Standard E1557. The first 6 characters represent this system to Level 4. The last 6 characters represent further breakdown by RSMeans in order to arrange items in understandable groups of similar tasks. Line numbers are consistent across all RSMeans publications, so a line number in any RSMeans assemblies data set will always refer to the same work.

C2010 110	Stairs	COST PER FLIGHT		
		MAT.	INST.	TOTAL
0470	Stairs, C.I.P. concrete, w/o landing, 12 risers, w/o nosing	1,125	2,425	3,550
0480	With nosing	2,050	2,625	4,675
0550	W/landing, 12 risers, w/o nosing	1,275	2,975	4,250
0560	With nosing	2,200	3,175	5,375
0570	16 risers, w/o nosing	1,575	3,725	5,300
0580	With nosing	2,800	4,000	6,800
0590	20 risers, w/o nosing	1,850	4,500	6,350
0600	With nosing	3,400	4,825	8,225
0610	24 risers, w/o nosing	2,150	5,250	7,400
0620	With nosing	4,000	5,650	9,650
0630	Steel, grate type w/nosing & rails, 12 risers, w/o landing	5,700	1,250	6,950
0640	With landing	7,975	1,725	9,700
0660	16 risers, with landing	9,875	2,125	12,000
0680	20 risers, with landing	11,800	2,550	14,350
0700	24 risers, with landing	13,700	2,950	16,650
0701				
0710	Concrete fill metal pan & picket rail, 12 risers, w/o landing	7,375	1,250	8,625
0720	With landing	9,725	1,875	11,600
0740	16 risers, with landing	12,200	2,300	14,500
0760	20 risers, with landing	14,600	2,700	17,300
0780	24 risers, with landing	17,100	3,100	20,200
0790	Cast iron tread & pipe rail, 12 risers, w/o landing	7,375	1,250	8,625
0800	With landing	9,725	1,875	11,600
1120	Wood, prefab box type, oak treads, wood rails 3'-6" wide, 14 risers	2,250	480	2,730
1150	Prefab basement type, oak treads, wood rails 3'-0" wide, 14 risers	1,100	119	1,219

RSMeans assemblies descriptions appear in two formats: narrative and table. **Narrative descriptions** are shown in a hierarchical structure to make them readable. In order to read a complete description, read up through the indents to the top of the section. Include everything that is above and to the left that is not contradicted by information below.

For supplemental customizable square foot estimating forms, visit: **http://www.reedconstructiondata.com/rsmeans/assemblies**

Most assemblies consist of three major elements: a graphic, the system components, and the cost data itself. The **Graphic** is a visual representation showing the typical appearance of the assembly in question, frequently accompanied by additional explanatory technical information describing the class of items. The **Assemblies Data** below lists prices for other similar systems with dimensional and/or size variations.

All RSMeans assemblies costs represent the cost for the installing contractor. An allowance for profit has been added to all material,

labor, and equipment rental costs. A markup for labor burdens, including workers' compensation, fixed overhead, and business overhead, is included with installation costs.

The data published in RSMeans print books represents a "national average" cost. This data should be modified to the project location using the **Location Factors** tables found in the Reference Section in the back of the book.

TABLE FORMAT

A20 Basement Construction

A2020 Basement Walls

The Foundation Bearing Wall System includes: forms up to 16' high (four uses); 3,000 p.s.i. concrete placed and vibrated; and form removal with breaking form ties and patching walls. The wall systems list walls from 6" to 16" thick and are designed with minimum reinforcement.

Excavation and backfill are not included.

Please see the reference section for further design and cost information.

All RSMeans assemblies data includes a typical **Unit of Measure** used for estimating that item. For instance, for continuous footings or foundation walls the unit is linear feet (L.F.). For spread footings the unit is each (Ea.). The estimator needs to take special care that the unit in the data matches the unit in the takeoff. Abbreviations and unit conversions can be found in the Reference Section.

A2020 110						Walls, Cast in Place		
	WALL HEIGHT (FT.)	PLACING METHOD	CONCRETE (C.Y. per L.F.)	REINFORCING (LBS. per L.F.)	WALL THICKNESS (IN.)	COST PER L.F.		
						MAT.	INST.	TOTAL
1500	4'	direct chute	.074	3.3	6	16.80	53	69.80
1520			.099	4.8	8	20.50	54.50	75
1540			.123	6.0	10	23.50	55.50	79
1560			.148	7.2	12	27	56.50	83.50
1580			.173	8.1	14	30.50	57.50	88
1600			.197	9.44	16	34	59	93
3000	6'	direct chute	.111	4.95	6	25	79.50	104.50
3020			.149	7.20	8	30.50	82	112.50
3040			.184	9.00	10	35.50	83	118.50
3060			.222	10.8	12	41	85	126
5000	8'	direct chute	.148	6.6	6	33.50	106	139.50
5020			.199	9.6	8	41	109	150
5040			.250	12	10	48	111	159
5061			.296	14.39	12	54.50	113	167.50
6020	10'	direct chute	.248	12	8	51	136	187
6040			.307	14.99	10	59	139	198
6061			.370	17.99	12	68	142	210
7220	12'	pumped	.298	14.39	8	61.50	169	230.50
7240			.369	17.99	10	71	172	243
7260			.444	21.59	12	81.50	177	258.50
9220	16'	pumped	.397	19.19	8	81.50	224	305.50
9240			.492	23.99	10	95	229	324
9260			.593	28.79	12	109	236	345

Table descriptions work in a similar fashion, except that if there is a blank in the column at a particular line number, read up to the description above in the same column.

How RSMeans Assemblies Data Works (Continued)

Sample Estimate

This sample demonstrates the elements of an estimate, including a tally of the RSMeans data lines. Published assemblies costs include all markups for labor burden and profit for the installing contractor. This estimate adds a summary of the markups applied by a general contractor on the installing contractors' work. These figures represent the total cost to the owner. The RSMeans location factor is added at the bottom of the estimate to adjust the cost of the work to a specific location

Work performed: The body of the estimate shows the RSMeans data selected, including line numbers, a brief description of each item, its takeoff quantity and unit, and the total installed cost, including the installing contractor's overhead and profit.

Location Factor: RSMeans published data is based on national average costs. If necessary, adjust the total cost of the project using a location factor from the "Location Factor" table or the "City Cost Indexes" table, found in the Reference Section. Use location factors if the work is general, covering the work of multiple trades. If the work is by a single trade (e.g. masonry) use the more specific data found in the City Cost Indexes.

To adjust costs by location factors, multiply the base cost by the factor and divide by 100.

Contingency: A factor for contingency may be added to any estimate to represent the cost of unknowns that may occur between the time that the estimate is performed and the time the project is constructed. The amount of the allowance will depend on the stage of design at which the estimate is done, and the contractor's assessment of the risk invloved.

Project Name:	Interior Fit-out, ABC Office
Location:	**Anywhere, USA**
Assembly Number	**Description**
C1010 124 1200	Wood partition, 2 x 4 @ 16" OC w/5/8" FR gypsum board
C1020 114 1800	Metal door & frame, flush hollow core, 3'-0" x 7'-0"
C3010 230 0080	Painting, brushwork, primer & 2 coats
C3020 410 0140	Carpet, tufted, nylon, roll goods, 12' wide, 26 oz
C3030 210 6000	Acoustic ceilings, 24" x 48" tile, tee grid suspension
D5020 125 0560	Receptacles incl plate, box, conduit, wire, 20 A duplex
D5020 125 0720	Light switch incl plate, box, conduit, wire, 20 A single pole
D5020 210 0560	Fluorescent fixtures, recess mounted, 20 per 1000 SF
Assembly Subtotal	
	Sales Tax @
	General Requirements @
Subtotal A	
	GC Overhead @
Subtotal B	
	GC Profit @
Subtotal C	
Adjusted by Location Factor	
	Architects Fee @
	Contingency @
Project Total Cost	

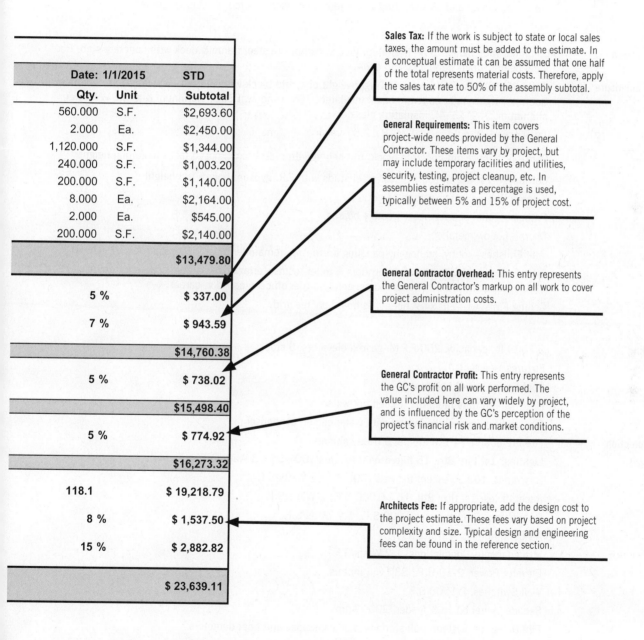

Date: 1/1/2015		STD
Qty.	Unit	Subtotal
560.000	S.F.	$2,693.60
2.000	Ea.	$2,450.00
1,120.000	S.F.	$1,344.00
240.000	S.F.	$1,003.20
200.000	S.F.	$1,140.00
8.000	Ea.	$2,164.00
2.000	Ea.	$545.00
200.000	S.F.	$2,140.00
		$13,479.80
5 %		$ 337.00
7 %		$ 943.59
		$14,760.38
5 %		$ 738.02
		$15,498.40
5 %		$ 774.92
		$16,273.32
118.1		$ 19,218.79
8 %		$ 1,537.50
15 %		$ 2,882.82
		$ 23,639.11

Sales Tax: If the work is subject to state or local sales taxes, the amount must be added to the estimate. In a conceptual estimate it can be assumed that one half of the total represents material costs. Therefore, apply the sales tax rate to 50% of the assembly subtotal.

General Requirements: This item covers project-wide needs provided by the General Contractor. These items vary by project, but may include temporary facilities and utilities, security, testing, project cleanup, etc. In assemblies estimates a percentage is used, typically between 5% and 15% of project cost.

General Contractor Overhead: This entry represents the General Contractor's markup on all work to cover project administration costs.

General Contractor Profit: This entry represents the GC's profit on all work performed. The value included here can vary widely by project, and is influenced by the GC's perception of the project's financial risk and market conditions.

Architects Fee: If appropriate, add the design cost to the project estimate. These fees vary based on project complexity and size. Typical design and engineering fees can be found in the reference section.

Example

Assemblies costs can be used to calculate the cost for each component of a building and, accumulated with appropriate markups, produce a cost for the complete structure. Components from a model building with similar characteristics in conjunction with Assemblies components may be used to calculate costs for a complete structure.

Example:
Outline Specifications

General	Building size—60' x 100', 8 suspended floors, 12' floor to floor, 4' high parapet above roof, full basement 11'-8" floor to floor, bay size 25' x30', ceiling heights -9' in office area and 8' in core area.
A Substructure	Concrete, spread and strip footings, concrete walls waterproofed, 4" slab on grade.
B Shell	
B10 Superstructure	Columns, wide flange; 3 hr. fire rated; Floors, composite steel frame & deck with concrete slab; Roof, steel beams, open web joists and deck.
B20 Exterior Enclosure	Walls; North, East & West, brick & lightweight concrete block with 2" cavity insulation, 25% window; South, 8" lightweight concrete block insulated, 10% window. Doors, aluminum & glass. Windows, aluminum, 3'-0" x5'-4", insulating glass.
B30 Roofing	Tar & gravel, 2" rigid insulation, R 12.5.
C Interiors	Core—6" lightweight concrete block, full height with plaster on exposed faces to ceiling height.
	Corridors—1st & 2nd floor; 3-5/8" steel studs with F.R. gypsum board, full height.
	Exterior Wall—plaster, ceiling height.
	Fittings—Toilet accessories, directory boards.
	Doors—hollow metal.
	Wall Finishes—lobby, mahogany paneling on furring, remainder plaster & gypsum board, paint.
	Floor Finishes—1st floor lobby, corridors & toilet rooms, terrazzo; remainder concrete, tenant developed. 2nd thru 8th, toilet rooms, ceramic tile; office & corridor, carpet.
	Ceiling Finishes—24" x48" fiberglass board on Tee grid.
D Services	
D10 Conveying	2-2500 lb. capacity, 200 F.P.M. geared elevators, 9 stops.
D20 Plumbing	Fixtures, see sketch.
	Roof Drains—2-4" C.I. pipe.
D30 HVAC	Heating—fin tube radiation, forced hot water.
	Air Conditioning—chilled water with water cooled condenser.
D40 Fire Protection	Fire Protection—4" standpipe, 9 hose cabinets.
D50 Electrical	Lighting, 1st thru 8th, 15 fluorescent fixtures/1000 S.F., 3 Watts/S.F.
	Basement, 10 fluorescent fixtures/1000 S.F., 2 Watts/S.F.
	Receptacles, 1st thru 8th, 16.5/1000 S.F., 2 Watts/S.F.
	Basement, 10 receptacles/1000 S.F., 1.2 Watts/S.F.
	Air Conditioning, 4 Watts/S.F.
	Miscellaneous Connections, 1.2 Watts/S.F.
	Elevator Power, 2-10 H.P., 230 volt motors.
	Wall Switches, 2/1000 S.F.
	Service, panel board & feeder, 2000 Amp.
	Fire Detection System, pull stations, signals, smoke and heat detectors.
	Emergency Lighting Generator, 30 KW.
E Equipment & Furnishings	NA.
F Special Construction	NA.
G Building Sitework	NA.

Front Elevation

Basement Plan

100'-0"

60'-0"

Ground Floor Plan

100'-0"

60'-0"

M   W

Lobby

Typical Floor plan

100'-0"

40'-0"

60'-0"

32'-0"

M   W

Example

If spread footing & column sizes are unknown, develop approximate loads as follows. Enter tables with these loads to determine costs.

Approximate loads/S.F. for roof & floors.

Roof. Assume 40 psf superimposed load.

Steel joists, beams & deck.

Table B1020 112—Line 3900

Superimposed Load Ranges

Apartments & Residential Structures	65	to	75 psf
Assembly Areas & Retail Stores	110	to	125 psf
Commercial & Manufacturing	150	to	250 psf
Offices	75	to	100 psf

B1020 112		Steel Joists, Beams, & Deck on Columns						
	BAY SIZE (FT.)	SUPERIMPOSED LOAD (P.S.F.)	DEPTH (IN.)	TOTAL LOAD (P.S.F.)	COLUMN ADD	COST PER S.F.		
						MAT.	INST.	TOTAL
3500	25x30	20	22	40		5.55	1.64	7.19
3600					columns	1.21	.28	1.49
3900		40	25	60		6.70	1.94	8.64
4000					columns	1.45	.34	1.79

B1010 256		Composite Beams, Deck & Slab						
	BAY SIZE (FT.)	SUPERIMPOSED LOAD (P.S.F.)	SLAB THICKNESS (IN.)	TOTAL DEPTH (FT.-IN.)	TOTAL LOAD (P.S.F.)	COST PER S.F.		
						MAT.	INST.	TOTAL
3400	25x30	40	5-1/2	1 - 11-1/2	83	12.90	5.95	18.85
3600		75	5-1/2	1 - 11-1/2	119	13.85	6	19.85
3900		125	5-1/2	1 - 11-1/2	170	16.05	6.80	22.85
4000		200	6-1/4	2 - 6-1/4	252	19.85	7.70	27.55

Floors—Total load, 119 psf.

Interior foundation load.

Roof

[(25' x 30' x 60 psf) + 8 floors x (25' x 30' x 119 psf)] x 1/1000 lb./Kip	=	759 Kips
Approximate Footing Loads, Interior footing	=	759 Kips
Exterior footing (1/2 bay) 759 k x .6	=	455 Kips
Corner footing (1/4 bay) 759 k x .45	=	342 Kips

[Factors to convert Interior load to Exterior & Corner loads]

Approximate average Column load 759 k/2	=	379 Kips

PRELIMINARY ESTIMATE

PROJECT	**Office Building**	TOTAL SITE AREA	
BUILDING TYPE		OWNER	
LOCATION		ARCHITECT	
DATE OF CONSTRUCTION		ESTIMATED CONSTRUCTION PERIOD	

BRIEF DESCRIPTION   **Building size: 60' x 100' - 8 structural floors - 12' floor to floor**

4' high parapet above roof, full basement 11' - 8" floor to floor, bay size

25' x 30', ceiling heights - 9' in office area + 8' in core area

TYPE OF PLAN		TYPE OF CONSTRUCTION	
QUALITY		BUILDING CAPACITY	

Floor			**Wall Area**			
Below Grade Levels			Foundation Walls	L.F.	Ht.	S.F.
Area		S.F.	Frost Walls	L.F.	Ht.	S.F.
Area		S.F.	Exterior Closure		Total	S.F.
Total Area		S.F.	Comment			
Ground Floor			Fenestration		%	S.F.
Area		S.F.			%	S.F.
Area		S.F.	Exterior Wall		%	S.F.
Total Area		S.F.			%	S.F.
Supported Levels			**Site Work**			
Area			Parking		S.F. (For	Cars)
Area		S.F.	Access Roads		L.F. (X	Ft. Wide)
Area		S.F.	Sidewalk		L.F. (X	Ft. Wide)
Area		S.F.	Landscaping		S.F. (	% Unbuilt Site)
Area		S.F.	**Building Codes**			
Total Area		S.F.	City		Country	
Miscellaneous			National		Other	
Area		S.F.	**Loading**			
Area		S.F.	Roof	psf	Ground Floor	psf
Area		S.F.	Supported Floors	psf	Corridor	psf
Area		S.F.	Balcony	psf	Partition, allow	psf
Total Area		S.F.	Miscellaneous			psf
Net Finished Area		S.F.	Live Load Reduction			
Net Floor Area		S.F.	Wind			
Gross Floor Area	**54,000**	S.F.	Earthquake		Zone	
Roof			Comment			
Total Area		S.F.	Soil Type			
Comments			Bearing Capacity			K.S.F.
			Frost Depth			Ft.
Volume			**Frame**			
Depth of Floor System			Type		Bay Spacing	
Minimum		In.	Foundation			
Maximum		In.	Special			
Foundation Wall Height		Ft.	Substructure			
Floor to Floor Height		Ft.	Comment			
Floor to Ceiling Height		Ft.	Superstructure, Vertical			
Subgrade Volume		C.F.	Fireproofing		☐ Columns	Hrs.
Above Grade Volume		C.F.	☐ Girders	Hrs.	☐ Beams	Hrs.
Total Building Volume	**648,000**	C.F.	☐ Floor	Hrs.	☐ None	

NUMBER		QTY.	UNIT	TOTAL COST UNIT	TOTAL	COST PER S.F.
A	**SUBSTRUCTURE**					
A1010 210 7900	Corner Footings 8'-6" SQ. x 27"	4	Ea.	1,945.00	7,780	
8010	Exterior 9' -6" SQ. x 30"	8	"	2,600.00	20,800	
8300	Interior 12' SQ.	3	"	4,750.00	14,250	
A1010 110 2700	Strip 2' Wide x 1' Thick					
	320 L.F. [(4 x 8.5) + 8 x 9.5)] =	320	L.F.	40.25	12,880	
A2020 110 7260	Foundation Wall 12' High, 1' Thick	320		258.50	82,720	
A1010 320 2800	Foundation Waterproofing	320		23.75	7,600	
A1030 120 2240	4", non-industrial, reinf. slab on grade	6000	S.F.	5.44	32,640	
A2010 110 3440	Building Excavation + Backfill	6000	S.F.	7.26	43,560	
3500	(Interpolated ; 12' Between					
4620	8' and 16' ; 6,000 Between					
4680	4,000 and 10,000 S.F.					
	Total				222,230	4.12
B	**SHELL**					
B10	**Superstructure**					
B1010 208 5800	Columns: Load 379K Use 400K					
	Exterior 12 x 96' + Interior 3 x 108'	1476	V.L.F.	150.85	222,655	
B1010 720 3650	Column Fireproofing - Interior 4 sides					
3700	Exterior 1/2 (Interpolated for 12")	900	V.L.F	36.20	32,580	
B1010 256 3600	Floors: Composite steel + lt. wt. conc.	48000	S.F.	19.85	952,800	
B1020 112 3900	Roof: Open web joists, beams + decks	6000	S.F.	8.64	51,840	
	Total				1,259,875	23.33
B20	**Exterior Enclosure**					
B2010 134 1200	4" Brick + 6" Block - Insulated					
	75% NE + W Walls 220' x 100' x .75	16500	S.F.	30.70	506,550	
B2010 110 3410	8" Block - Insulated 90%					
	100' x 100' x .9	9000	S.F.	12.25	110,250	
B2030 110 6950	Double Aluminum + Glass Door	1	Pr.	5,825.00	5,825	
6300	Single	2	Ea.	2,950.00	5,900	
B2020 106 8800	Aluminum windows - Insul. glass					
	(5,500 S.F. + 1,000 S.F.)/(3 x 5.33)	406	Ea.	911.00	369,866	
B2010 105 8480	Precast concrete coping	320	L.F.	41.65	13,328	
	Total				1,011,719	18.74
B30	**Roofing**					
B3010 320 4020	Insulation: Perlite/Polyisocyanurate	6000	S.F.	1.96	11,760	
B3010 105 1400	Roof: 4 Ply T & G composite	6000		3.02	18,120	
B3010 430 0300	Flashing: Aluminum - fabric backed	640		3.35	2,144	
B3020 210 0500	Roof hatch	1	Ea.	1,780.00	1,780	
	Total				33,804	0.63

ASSEMBLY NUMBER		QTY.	UNIT	TOTAL COST UNIT	TOTAL	COST PER S.F.
C	**INTERIORS**					
C1010 104 5500	Core partitions: 6" Lt. Wt. concrete					
	block 288 L.F. x 11.5' x 8 Floors	26,449	S.F.	9.66	255,497	
C1010 144 0920	Plaster: [(196 L.F. x 8') + (144 L.F. x 9')] 8					
	+ Ext. Wall [(320 x 9 x 8 Floors) - 6500]	39,452	S.F.	3.21	126,641	
C1010 126 5400	Corridor partitions 1st + 2nd floors					
	steel studs + F.R. gypsum board	2,530	S.F.	4.24	10,727	
C1020 102 2600	Interior doors	66	Ea.	1,157.00	76,362	
	Wall finishes - Lobby: Mahogany w/furring					
C1010 128 0652	Furring: 150 L.F. x 9' + Ht.	1,350	S.F.	1.63	2,201	
C1030 710 0120	Towel Dispenser	16	Ea.	85.00	1,360	
C1030 710 0140	Grab Bar	32	Ea.	56.00	1,792	
C1030 710 0160	Mirror	31	Ea.	242.00	7,502	
C1030 710 0180	Toilet Tissue Dispenser	31	Ea.	38.85	1,204	
C1030 510 0140	Directory Board	8	Ea.	344.00	2,752	
C2010 110 0780	Stairs: Steel w/conc. fill	18	Flt.	20,200.00	363,600	
C3010 230 1662	Mahogany paneling	1,350	S.F.	5.92	7,992	
C3010 230 0080	Paint, primer + 2 coats finish, on plaster or gyp. board					
	39,452 + [(220' x 9' x 2) - 1350] + 72	42,134	S.F.	1.20	50,561	
	Floor Finishes					
C3020 410 1100	1st floor lobby + terrazzo	2,175	S.F.	20.49	44,566	
	Remainder: Concrete - Tenant finished					
C3020 410 0060	2nd - 8th: Carpet - 4725 S.F. x 7	33,075	S.F.	4.36	144,207	
C3020 410 1720	Ceramic Tile - 300 S.F. x 7	2,100	S.F.	11.39	23,919	
C3030 240 2780	Suspended Ceiling Tiles- 5200 S.F. x 8	41,600	S.F.	2.05	85,280	
C3030 240 3260	Suspended Ceiling Grid- 5200 S.F. x 8	41,600	S.F.	1.53	63,648	
C1030 110 0700	Toilet partitions	31	Ea.	702.00	21,762	
0760	Handicapped addition	16	Ea.	345.00	5,520	
	Total				1,297,093	24.02
D	**SERVICES**					
D10	**Conveying**					
	2 Elevators - 2500# 200'/min. Geared					
D1010 140 1600	5 Floors $171,000					
1800	15 Floors $392,000					
	Diff. $221,000/10 = 22,100/Flr.					
	$171,000 + (4 x22,100) = $259,400/Ea.	2	Ea.	259,400.00	518,800	
	Total				518,800	9.61
D20	**Plumbing**					
D2010 310 1600	Plumbing - Lavatories	31	Ea.	1,200.00	37,200	
D2010 440 4340	-Service Sink	8	Ea.	3,425.00	27,400	
D2010 210 2000	-Urinals	8	Ea.	1,410.00	11,280	

ASSEMBLY NUMBER		QTY.	UNIT	TOTAL COST		COST PER S.F.
				UNIT	TOTAL	
D2010 110 2080	-Water Closets	31	Ea.	3,155.00	97,805	
	Water Control, Waste Vent Piping	45%			78,158	
D2040 210 4200	- Roof Drains, 4" C.I.	2	Ea.	1,915.00	3,830	
4240	- Pipe, 9 Flr. x 12' x 12' Ea.	216	L.F.	43.50	9,396	
	Total				265,069	4.91
D30	HVAC					
D3010 520 2000	10,000 S.F. @ $10.74/S.F. Interpolate					
2040	100,000 S.F. @ $4.59/S.F.	48,000	S.F.	8.14	390,720	
	Cooling - Chilled Water, Air Cool, Cond.					
D3030 115 4000	40,000 S.F. @14.20/S.F. Interpolate	48,000	S.F.	14.60	700,800	
4040	60,000 S.F. @15.20/S.F.					
	Total				1,091,520	20.21
D40	Fire Protection					
D4020 310 0560	Wet Stand Pipe: 4" x 10' 1st floor	12/10	Ea.	8,475.00	10,170	
0580	Additional Floors: (12'/10 x 8)	9.6	Ea.	2,175.00	20,880	
D4020 410 8400	Cabinet Assembly	9	Ea.	1,245.00	11,205	
	Total				42,255	0.78
D50	Electrical					
D5020 210 0280	Office Lighting, 15/1000 S.F. - 3 Watts/S.F.	48,000	S.F.	7.84	376,320	
0240	Basement Lighting, 10/1000 S.F. - 2 Watts/S.F.	6,000	S.F.	5.22	31,320	
D5020 110 0640	Office Receptacles 16.5/1000 S.F. - 2 Watts/S.F.	48,000	S.F.	4.68	224,640	
0560	Basement Receptacles 10/1000 S.F. - 1.2 Watts/S.F.	6,000	S.F.	3.59	21,540	
D5020 140 0280	Central A.C. - 4 Watts/S.F.	48,000	S.F.	0.62	29,760	
D5020 135 0320	Misc. Connections - 1.2 Watts/S.F.	48,000	S.F.	0.34	16,320	
D5020 145 0680	Elevator Motor Power - 10 H.P.	2	Ea.	2,925.00	5,850	
D5020 130 0280	Wall Switches - 2/1000 S.F.	54,000	S.F.	0.48	25,920	
D5010 120 0560	2000 Amp Service	1	Ea.	32,100.00	32,100	
D5010 240 0400	Switchgear	1	Ea.	38,275.00	38,275	
D5010 230 0560	Feeder	50	L.F.	525.00	26,250	
D5030 910 0400	Fire Detection System - 50 Detectors	1	Ea.	34,500.00	34,500	
D5090 210 0320	Emergency Generator - 30 kW	30	kW	553.00	16,590	
	Total				879,385	16.28
E	EQUIPMENT & FURNISHINGS					
G	BUILDING SITEWORK					
	Miscellaneous					

300

Preliminary Estimate Cost Summary

PROJECT: Office Building	TOTAL AREA	54,000 S.F.	SHEET NO.	
LOCATION	TOTAL VOLUME	648,000 C.F.	ESTIMATE NO.	
ARCHITECT	COST PER S.F.		DATE	
OWNER	COST PER C.F.		NO OF STORIES	
QUANTITIES BY	EXTENSIONS BY		CHECKED BY	

DIV	DESCRIPTION	SUBTOTAL COST	COST/S.F.	PERCENTAGE
A	SUBSTRUCTURE	222,230	$ 4.12	
B10	SHELL: SUPERSTRUCTURE	1,259,875	$ 23.33	
B20	SHELL: EXTERIOR ENCLOSURE	1,011,719	$ 18.74	
B30	SHELL: ROOFING	33,804	$ 0.63	
C	INTERIORS	1,297,093	$ 24.02	
D10	SERVICES: CONVEYING	518,800	$ 9.61	
D20	SERVICES: PLUMBING	265,069	$ 4.91	
D30	SERVICES: HVAC	1,091,520	$ 20.21	
D40	SERVICES: FIRE PROTECTION	42,255	$ 0.78	
D50	SERVICES: ELECTRICAL	879,385	$ 16.28	
E	EQUIPMENT & FURNISHINGS			
F	SPECIAL CONSTRUCTION & DEMO.			
G	BUILDING SITEWORK			
	BUILDING SUBTOTAL	$ 6,621,750		$ 6,621,750

Sales Tax   % x Subtotal /2   N/A $ -

General Conditions 10 % x Subtotal $ 662,175

Subtotal "A" $ 7,283,925

Overhead   5 % x Subtotal "A" $ 364,196

Subtotal "B" $ 7,648,121

Profit   10% x Subtotal "B" $ 764,812

Subtotal "C" $ 8,412,934

Location Factor   % x Subtotal "C"   N/A   Localized Cost

Architects Fee 6.5% x Localized Cost = $ 546,841
Contingency   x Localized Cost =   N/A $ -

Project Total Cost $ 8,959,774

Square Foot Cost $8,845,298/ 54,000 S.F. =   S.F. Cost $ 165.92
Cubic Foot Cost $8,845,298/ 648,000 C.F. =   C.F. Cost $ 13.83

A1010 Standard Foundations

The Strip Footing System includes: excavation; hand trim; all forms needed for footing placement; forms for 2″ x 6″ keyway (four uses); dowels; and 3,000 p.s.i. concrete.

The footing size required varies for different soils. Soil bearing capacities are listed for 3 KSF and 6 KSF. Depths of the system range from 8″ and deeper. Widths range from 16″ and wider. Smaller strip footings may not require reinforcement.

Please see the reference section for further design and cost information.

A1010 110	Strip Footings	COST PER L.F.		
		MAT.	INST.	TOTAL
2100	Strip footing, load 2.6 KLF, soil capacity 3 KSF, 16″ wide x 8″ deep, plain	8.65	14.30	22.95
2300	Load 3.9 KLF, soil capacity, 3 KSF, 24″ wide x 8″ deep, plain	10.55	15.70	26.25
2500	Load 5.1 KLF, soil capacity 3 KSF, 24″ wide x 12″ deep, reinf.	16.75	23.50	40.25
2700	Load 11.1 KLF, soil capacity 6 KSF, 24″ wide x 12″ deep, reinf.	16.75	23.50	40.25
2900	Load 6.8 KLF, soil capacity 3 KSF, 32″ wide x 12″ deep, reinf.	20	26	46
3100	Load 14.8 KLF, soil capacity 6 KSF, 32″ wide x 12″ deep, reinf.	20	26	46
3300	Load 9.3 KLF, soil capacity 3 KSF, 40″ wide x 12″ deep, reinf.	23.50	28	51.50
3500	Load 18.4 KLF, soil capacity 6 KSF, 40″ wide x 12″ deep, reinf.	23.50	28	51.50
4500	Load 10 KLF, soil capacity 3 KSF, 48″ wide x 16″ deep, reinf.	33	35.50	68.50
4700	Load 22 KLF, soil capacity 6 KSF, 48″ wide, 16″ deep, reinf.	34	36.50	70.50
5700	Load 15 KLF, soil capacity 3 KSF, 72″ wide x 20″ deep, reinf.	57	50.50	107.50
5900	Load 33 KLF, soil capacity 6 KSF, 72″ wide x 20″ deep, reinf.	60	54	114

A10 Foundations

A1010   Standard Foundations

Dampproofing

Spread Footing

Strip Footing

A1010 210	Spread Footings	COST EACH		
		MAT.	INST.	TOTAL
7090	Spread footings, 3000 psi concrete, chute delivered			
7100	Load 25K, soil capacity 3 KSF, 3'-0" sq. x 12" deep	59.50	124	183.50
7150	Load 50K, soil capacity 3 KSF, 4'-6" sq. x 12" deep	126	213	339
7200	Load 50K, soil capacity 6 KSF, 3'-0" sq. x 12" deep	59.50	124	183.50
7250	Load 75K, soil capacity 3 KSF, 5'-6" sq. x 13" deep	200	300	500
7300	Load 75K, soil capacity 6 KSF, 4'-0" sq. x 12" deep	102	183	285
7350	Load 100K, soil capacity 3 KSF, 6'-0" sq. x 14" deep	253	360	613
7410	Load 100K, soil capacity 6 KSF, 4'-6" sq. x 15" deep	155	251	406
7450	Load 125K, soil capacity 3 KSF, 7'-0" sq. x 17" deep	400	520	920
7500	Load 125K, soil capacity 6 KSF, 5'-0" sq. x 16" deep	200	300	500
7550	Load 150K, soil capacity 3 KSF 7'-6" sq. x 18" deep	485	600	1,085
7610	Load 150K, soil capacity 6 KSF, 5'-6" sq. x 18" deep	266	380	646
7650	Load 200K, soil capacity 3 KSF, 8'-6" sq. x 20" deep	690	795	1,485
7700	Load 200K, soil capacity 6 KSF, 6'-0" sq. x 20" deep	350	465	815
7750	Load 300K, soil capacity 3 KSF, 10'-6" sq. x 25" deep	1,275	1,300	2,575
7810	Load 300K, soil capacity 6 KSF, 7'-6" sq. x 25" deep	660	785	1,445
7850	Load 400K, soil capacity 3 KSF, 12'-6" sq. x 28" deep	2,000	1,925	3,925
7900	Load 400K, soil capacity 6 KSF, 8'-6" sq. x 27" deep	920	1,025	1,945
8010	Load 500K, soil capacity 6 KSF, 9'-6" sq. x 30" deep	1,275	1,325	2,600
8100	Load 600K, soil capacity 6 KSF, 10'-6" sq. x 33" deep	1,700	1,700	3,400
8200	Load 700K, soil capacity 6 KSF, 11'-6" sq. x 36" deep	2,175	2,075	4,250
8300	Load 800K, soil capacity 6 KSF, 12'-0" sq. x 37" deep	2,450	2,300	4,750
8400	Load 900K, soil capacity 6 KSF, 13'-0" sq. x 39" deep	3,025	2,725	5,750
8500	Load 1000K, soil capacity 6 KSF, 13'-6" sq. x 41" deep	3,425	3,050	6,475

A1010 320	Foundation Dampproofing	COST PER L.F.		
		MAT.	INST.	TOTAL
1000	Foundation dampproofing, bituminous, 1 coat, 4' high	1	4.68	5.68
1400	8' high	2	9.35	11.35
1800	12' high	3	14.50	17.50
2000	2 coats, 4' high	1.96	5.80	7.76
2400	8' high	3.92	11.60	15.52
2800	12' high	5.90	17.85	23.75
3000	Asphalt with fibers, 1/16" thick, 4' high	1.60	5.80	7.40
3400	8' high	3.20	11.60	14.80
3800	12' high	4.80	17.85	22.65
4000	1/8" thick, 4' high	2.80	6.90	9.70
4400	8' high	5.60	13.75	19.35
4800	12' high	8.40	21	29.40
5000	Asphalt coated board and mastic, 1/4" thick, 4' high	5.10	6.30	11.40
5400	8' high	10.25	12.55	22.80

305

For customer support on your Square Foot Cost Data, call 877.694.8329.

A1010 Standard Foundations

A1010 320	Foundation Dampproofing	COST PER L.F.		
		MAT.	INST.	TOTAL
5800	12' high	15.35	19.30	34.65
6000	1/2" thick, 4' high	7.70	8.70	16.40
6400	8' high	15.35	17.45	32.80
6800	12' high	23	26.50	49.50
7000	Cementitious coating, on walls, 1/8" thick coating, 4' high	3.20	8.85	12.05
7400	8' high	6.40	17.65	24.05
7800	12' high	9.60	26.50	36.10
8000	Cementitious/metallic slurry, 4 coat, 1/2"thick, 2' high	.70	8.90	9.60
8400	4' high	1.40	17.80	19.20
8800	6' high	2.10	26.50	28.60

A10 Foundations

A1030   Slab on Grade

Fibre Expansion Joint

Wire Mesh Reinforcing

Concrete Slab

Vapor Barrier

Compacted Gravel

Reinforced Slab on Grade

A Slab on Grade system includes fine grading; 6" of compacted gravel; vapor barrier; 3500 p.s.i. concrete; bituminous fiber expansion joint; all necessary edge forms 4 uses; steel trowel finish; and sprayed on membrane curing compound. Wire mesh reinforcing used in all reinforced slabs.

Non-industrial slabs are for foot traffic only with negligible abrasion. Light industrial slabs are for pneumatic wheels and light abrasion. Industrial slabs are for solid rubber wheels and moderate abrasion. Heavy industrial slabs are for steel wheels and severe abrasion.

A1030 120	Plain & Reinforced	COST PER S.F.		
		MAT.	INST.	TOTAL
2220	Slab on grade, 4" thick, non industrial, non reinforced	2.12	2.78	4.90
2240	Reinforced	2.28	3.16	5.44
2260	Light industrial, non reinforced	2.75	3.41	6.16
2280	Reinforced	2.91	3.79	6.70
2300	Industrial, non reinforced	3.39	7.10	10.49
2320	Reinforced	3.55	7.50	11.05
3340	5" thick, non industrial, non reinforced	2.47	2.86	5.33
3360	Reinforced	2.63	3.24	5.87
3380	Light industrial, non reinforced	3.10	3.49	6.59
3400	Reinforced	3.26	3.87	7.13
3420	Heavy industrial, non reinforced	4.42	8.45	12.87
3440	Reinforced	4.52	8.90	13.42
4460	6" thick, non industrial, non reinforced	2.93	2.80	5.73
4480	Reinforced	3.16	3.31	6.47
4500	Light industrial, non reinforced	3.57	3.43	7
4520	Reinforced	3.93	4.12	8.05
4540	Heavy industrial, non reinforced	4.90	8.55	13.45
4560	Reinforced	5.15	9.10	14.25
5580	7" thick, non industrial, non reinforced	3.29	2.89	6.18
5600	Reinforced	3.58	3.44	7.02
5620	Light industrial, non reinforced	3.95	3.52	7.47
5640	Reinforced	4.24	4.07	8.31
5660	Heavy industrial, non reinforced	5.25	8.45	13.70
5680	Reinforced	5.50	8.95	14.45
6700	8" thick, non industrial, non reinforced	3.64	2.95	6.59
6720	Reinforced	3.89	3.41	7.30
6740	Light industrial, non reinforced	4.30	3.58	7.88
6760	Reinforced	4.55	4.04	8.59
6780	Heavy industrial, non reinforced	5.65	8.55	14.20
6800	Reinforced	6	9.05	15.05

A2010 Basement Excavation

Sand & Gravel 1:1

Clay 1/2:1

Line of Excavation

In general, the following items are accounted for in the table below.

Costs:
1) Excavation for building or other structure to depth and extent indicated.
2) Backfill compacted in place
3) Haul of excavated waste
4) Replacement of unsuitable backfill material with bank run gravel.

A2010 110	Building Excavation & Backfill	COST PER S.F.		
		MAT.	INST.	TOTAL
2220	Excav & fill, 1000 S.F., 4' sand, gravel, or common earth, on site storage		1.07	1.07
2240	Off site storage		1.61	1.61
2260	Clay excavation, bank run gravel borrow for backfill	2.45	2.48	4.93
2280	8' deep, sand, gravel, or common earth, on site storage		6.40	6.40
2300	Off site storage		14.15	14.15
2320	Clay excavation, bank run gravel borrow for backfill	9.95	11.60	21.55
2340	16' deep, sand, gravel, or common earth, on site storage		17.55	17.55
2350	Off site storage		35.50	35.50
2360	Clay excavation, bank run gravel borrow for backfill	27.50	29.50	57
3380	4000 S.F., 4' deep, sand, gravel, or common earth, on site storage		.57	.57
3400	Off site storage		1.12	1.12
3420	Clay excavation, bank run gravel borrow for backfill	1.25	1.25	2.50
3440	8' deep, sand, gravel, or common earth, on site storage		4.53	4.53
3460	Off site storage		7.95	7.95
3480	Clay excavation, bank run gravel borrow for backfill	4.53	6.90	11.43
3500	16' deep, sand, gravel, or common earth, on site storage		10.85	10.85
3520	Off site storage		21.50	21.50
3540	Clay, excavation, bank run gravel borrow for backfill	12.15	16.50	28.65
4560	10,000 S.F., 4' deep, sand gravel, or common earth, on site storage		.32	.32
4580	Off site storage		.66	.66
4600	Clay excavation, bank run gravel borrow for backfill	.76	.76	1.52
4620	8' deep, sand, gravel, or common earth, on site storage		3.93	3.93
4640	Off site storage		5.95	5.95
4660	Clay excavation, bank run gravel borrow for backfill	2.74	5.40	8.14
4680	16' deep, sand, gravel, or common earth, on site storage		8.85	8.85
4700	Off site storage		15.15	15.15
4720	Clay excavation, bank run gravel borrow for backfill	7.30	12.30	19.60
5740	30,000 S.F., 4' deep, sand, gravel, or common earth, on site storage		.18	.18
5760	Off site storage		.39	.39
5780	Clay excavation, bank run gravel borrow for backfill	.44	.44	.88
5860	16' deep, sand, gravel, or common earth, on site storage		7.60	7.60
5880	Off site storage		11.05	11.05
5900	Clay excavation, bank run gravel borrow for backfill	4.04	9.55	13.59
6910	100,000 S.F., 4' deep, sand, gravel, or common earth, on site storage		.11	.11
6940	8' deep, sand, gravel, or common earth, on site storage		3.31	3.31
6970	16' deep, sand, gravel, or common earth, on site storage		6.90	6.90
6980	Off site storage		8.70	8.70
6990	Clay excavation, bank run gravel borrow for backfill	2.18	7.95	10.13

A20 Basement Construction

A2020 Basement Walls

The Foundation Bearing Wall System includes: forms up to 16' high (four uses); 3,000 p.s.i. concrete placed and vibrated; and form removal with breaking form ties and patching walls. The wall systems list walls from 6" to 16" thick and are designed with minimum reinforcement.

Excavation and backfill are not included.

Please see the reference section for further design and cost information.

A2020 110		Walls, Cast in Place						
	WALL HEIGHT (FT.)	**PLACING METHOD**	**CONCRETE (C.Y. per L.F.)**	**REINFORCING (LBS. per L.F.)**	**WALL THICKNESS (IN.)**	**COST PER L.F.**		
						MAT.	**INST.**	**TOTAL**
1500	4'	direct chute	.074	3.3	6	16.80	53	69.80
1520			.099	4.8	8	20.50	54.50	75
1540			.123	6.0	10	23.50	55.50	79
1560			.148	7.2	12	27	56.50	83.50
1580			.173	8.1	14	30.50	57.50	88
1600			.197	9.44	16	34	59	93
3000	6'	direct chute	.111	4.95	6	25	79.50	104.50
3020			.149	7.20	8	30.50	82	112.50
3040			.184	9.00	10	35.50	83	118.50
3060			.222	10.8	12	41	85	126
5000	8'	direct chute	.148	6.6	6	33.50	106	139.50
5020			.199	9.6	8	41	109	150
5040			.250	12	10	48	111	159
5061			.296	14.39	12	54.50	113	167.50
6020	10'	direct chute	.248	12	8	51	136	187
6040			.307	14.99	10	59	139	198
6061			.370	17.99	12	68	142	210
7220	12'	pumped	.298	14.39	8	61.50	169	230.50
7240			.369	17.99	10	71	172	243
7260			.444	21.59	12	81.50	177	258.50
9220	16'	pumped	.397	19.19	8	81.50	224	305.50
9240			.492	23.99	10	95	229	324
9260			.593	28.79	12	109	236	345

B10 Superstructure

B1010 Floor Construction

General: It is desirable for purposes of consistency and simplicity to maintain constant column sizes throughout building height. To do this, concrete strength may be varied (higher strength concrete at lower stories and lower strength concrete at upper stories), as well as varying the amount of reinforcing.

The table provides probably minimum column sizes with related costs and weight per lineal foot of story height.

B1010 203		C.I.P. Column, Square Tied						
	LOAD (KIPS)	STORY HEIGHT (FT.)	COLUMN SIZE (IN.)	COLUMN WEIGHT (P.L.F.)	CONCRETE STRENGTH (PSI)	COST PER V.L.F.		
						MAT.	INST.	TOTAL
0640	100	10	10	96	4000	10.80	49	59.80
0680		12	10	97	4000	11	49.50	60.50
0700		14	12	142	4000	14.40	59.50	73.90
0840	200	10	12	140	4000	15.75	62	77.75
0860		12	12	142	4000	16	62.50	78.50
0900		14	14	196	4000	18.65	69.50	88.15
0920	300	10	14	192	4000	19.75	71	90.75
0960		12	14	194	4000	20	71.50	91.50
0980		14	16	253	4000	23.50	79	102.50
1020	400	10	16	248	4000	24.50	81.50	106
1060		12	16	251	4000	25	82.50	107.50
1080		14	16	253	4000	25.50	83	108.50
1200	500	10	18	315	4000	29	91	120
1250		12	20	394	4000	35	103	138
1300		14	20	397	4000	36	105	141
1350	600	10	20	388	4000	36.50	106	142.50
1400		12	20	394	4000	37.50	107	144.50
1600		14	20	397	4000	38	108	146
3400	900	10	24	560	4000	49.50	131	180.50
3800		12	24	567	4000	50.50	133	183.50
4000		14	24	571	4000	52	135	187
7300	300	10	14	192	6000	19.90	70.50	90.40
7500		12	14	194	6000	20.50	71	91.50
7600		14	14	196	6000	20.50	71.50	92
8000	500	10	16	248	6000	25.50	81.50	107
8050		12	16	251	6000	26	82.50	108.50
8100		14	16	253	6000	26.50	83	109.50
8200	600	10	18	315	6000	29.50	91.50	121
8300		12	18	319	6000	30	92.50	122.50
8400		14	18	321	6000	30.50	93.50	124
8800	800	10	20	388	6000	36	104	140
8900		12	20	394	6000	37	105	142
9000		14	20	397	6000	37.50	106	143.50

B1010 Floor Construction

B1010 203 | C.I.P. Column, Square Tied

	LOAD (KIPS)	STORY HEIGHT (FT.)	COLUMN SIZE (IN.)	COLUMN WEIGHT (P.L.F.)	CONCRETE STRENGTH (PSI)	COST PER V.L.F.		
						MAT.	INST.	TOTAL
9100	900	10	20	388	6000	39	108	147
9300		12	20	394	6000	40	110	150
9600		14	20	397	6000	40.50	111	151.50

B1010 204 | C.I.P. Column, Square Tied-Minimum Reinforcing

	LOAD (KIPS)	STORY HEIGHT (FT.)	COLUMN SIZE (IN.)	COLUMN WEIGHT (P.L.F.)	CONCRETE STRENGTH (PSI)	COST PER V.L.F.		
						MAT.	INST.	TOTAL
9913	150	10-14	12	135	4000	13.05	57.50	70.55
9918	300	10-14	16	240	4000	21	75.50	96.50
9924	500	10-14	20	375	4000	31.50	97.50	129
9930	700	10-14	24	540	4000	43	121	164
9936	1000	10-14	28	740	4000	57	148	205
9942	1400	10-14	32	965	4000	70	166	236
9948	1800	10-14	36	1220	4000	87	194	281

For customer support on your Square Foot Cost Data, call 877.694.8329.

313

B1010 Floor Construction

Concentric Load

Eccentric Load

General: Data presented here is for plant produced members transported 50 miles to 100 miles to the site and erected.

Design and pricing assumptions:
Normal wt. concrete, f'c = 5 KSI

Main reinforcement, fy = 60 KSI
Ties, fy = 40 KSI

Minimum design eccentricity, 0.1t.

Concrete encased structural steel haunches are assumed where practical; otherwise galvanized rebar haunches are assumed.

Base plates are integral with columns.

Foundation anchor bolts, nuts and washers are included in price.

B1010 206 — Tied, Concentric Loaded Precast Concrete Columns

	LOAD (KIPS)	STORY HEIGHT (FT.)	COLUMN SIZE (IN.)	COLUMN WEIGHT (P.L.F.)	LOAD LEVELS	COST PER V.L.F.		
						MAT.	INST.	TOTAL
0560	100	10	12x12	164	2	166	14.20	180.20
0570		12	12x12	162	2	160	11.85	171.85
0580		14	12x12	161	2	160	11.60	171.60
0590	150	10	12x12	166	3	160	11.50	171.50
0600		12	12x12	169	3	159	10.30	169.30
0610		14	12x12	162	3	159	10.20	169.20
0620	200	10	12x12	168	4	160	12.30	172.30
0630		12	12x12	170	4	159	11.10	170.10
0640		14	12x12	220	4	159	11	170

B1010 207 — Tied, Eccentric Loaded Precast Concrete Columns

	LOAD (KIPS)	STORY HEIGHT (FT.)	COLUMN SIZE (IN.)	COLUMN WEIGHT (P.L.F.)	LOAD LEVELS	COST PER V.L.F.		
						MAT.	INST.	TOTAL
1130	100	10	12x12	161	2	157	14.20	171.20
1140		12	12x12	159	2	157	11.85	168.85
1150		14	12x12	159	2	157	11.60	168.60
1390	600	10	18x18	385	4	230	12.30	242.30
1400		12	18x18	380	4	230	11.10	241.10
1410		14	18x18	375	4	229	11	240
1480	800	10	20x20	490	4	210	12.30	222.30
1490		12	20x20	480	4	209	11.10	220.10
1500		14	20x20	475	4	210	11	221

B10 Superstructure

B1010 Floor Construction

  (A) Wide Flange

  (B) Pipe

  (C) Pipe, Concrete Filled

  (G) Rectangular Tube, Concrete Filled

B1010 208						Steel Columns			
	LOAD (KIPS)	UNSUPPORTED HEIGHT (FT.)	WEIGHT (P.L.F.)	SIZE (IN.)	TYPE	COST PER V.L.F.			
						MAT.	INST.	TOTAL	
1000	25	10	13	4	A	23	10.85	33.85	
1020			7.58	3	B	13.40	10.85	24.25	
1040			15	3-1/2	C	16.30	10.85	27.15	
1120			20	4x3	G	18.85	10.85	29.70	
1200		16	16	5	A	26	8.10	34.10	
1220			10.79	4	B	17.65	8.10	25.75	
1240			36	5-1/2	C	24.50	8.10	32.60	
1320			64	8x6	G	38	8.10	46.10	
1600	50	10	16	5	A	28.50	10.85	39.35	
1620			14.62	5	B	26	10.85	36.85	
1640			24	4-1/2	C	19.45	10.85	30.30	
1720			28	6x3	G	25	10.85	35.85	
1800		16	24	8	A	39	8.10	47.10	
1840			36	5-1/2	C	24.50	8.10	32.60	
1920			64	8x6	G	38	8.10	46.10	
2000		20	28	8	A	43.50	8.10	51.60	
2040			49	6-5/8	C	30.50	8.10	38.60	
2120			64	8x6	G	36	8.10	44.10	
2200	75	10	20	6	A	35.50	10.85	46.35	
2240			36	4-1/2	C	49	10.85	59.85	
2320			35	6x4	G	28	10.85	38.85	
2400		16	31	8	A	50.50	8.10	58.60	
2440			49	6-5/8	C	32	8.10	40.10	
2520			64	8x6	G	38	8.10	46.10	
2600		20	31	8	A	48	8.10	56.10	
2640			81	8-5/8	C	46	8.10	54.10	
2720			64	8x6	G	36	8.10	44.10	
2800	100	10	24	8	A	42.50	10.85	53.35	
2840			35	4-1/2	C	49	10.85	59.85	
2920			46	8x4	G	34.50	10.85	45.35	

B1010 Floor Construction

B1010 208	Steel Columns

	LOAD (KIPS)	UNSUPPORTED HEIGHT (FT.)	WEIGHT (P.L.F.)	SIZE (IN.)	TYPE	COST PER V.L.F.		
						MAT.	INST.	TOTAL
3000	100	16	31	8	A	50.50	8.10	58.60
3040			56	6-5/8	C	47.50	8.10	55.60
3120			64	8x6	G	38	8.10	46.10
3200		20	40	8	A	62	8.10	70.10
3240			81	8-5/8	C	46	8.10	54.10
3320			70	8x6	G	51.50	8.10	59.60
3400	125	10	31	8	A	55	10.85	65.85
3440			81	8	C	52	10.85	62.85
3520			64	8x6	G	41	10.85	51.85
3600		16	40	8	A	65.50	8.10	73.60
3640			81	8	C	48.50	8.10	56.60
3720			64	8x6	G	38	8.10	46.10
3800		20	48	8	A	74.50	8.10	82.60
3840			81	8	C	46	8.10	54.10
3920			60	8x6	G	51.50	8.10	59.60
4000	150	10	35	8	A	62	10.85	72.85
4040			81	8-5/8	C	52	10.85	62.85
4120			64	8x6	G	41	10.85	51.85
4200		16	45	10	A	73.50	8.10	81.60
4240			81	8-5/8	C	48.50	8.10	56.60
4320			70	8x6	G	54.50	8.10	62.60
4400		20	49	10	A	76	8.10	84.10
4440			123	10-3/4	C	65	8.10	73.10
4520			86	10x6	G	51	8.10	59.10
4600	200	10	45	10	A	79.50	10.85	90.35
4640			81	8-5/8	C	52	10.85	62.85
4720			70	8x6	G	59	10.85	69.85
4800		16	49	10	A	80	8.10	88.10
4840			123	10-3/4	C	68.50	8.10	76.60
4920			85	10x6	G	63	8.10	71.10
5200	300	10	61	14	A	108	10.85	118.85
5240			169	12-3/4	C	91	10.85	101.85
5320			86	10x6	G	88	10.85	98.85
5400		16	72	12	A	118	8.10	126.10
5440			169	12-3/4	C	84.50	8.10	92.60
5600		20	79	12	A	122	8.10	130.10
5640			169	12-3/4	C	80.50	8.10	88.60
5800	400	10	79	12	A	140	10.85	150.85
5840			178	12-3/4	C	119	10.85	129.85
6000		16	87	12	A	142	8.10	150.10
6040			178	12-3/4	C	110	8.10	118.10
6400	500	10	99	14	A	175	10.85	185.85
6600		16	109	14	A	178	8.10	186.10
6800		20	120	12	A	186	8.10	194.10
7000	600	10	120	12	A	212	10.85	222.85
7200		16	132	14	A	216	8.10	224.10
7400		20	132	14	A	204	8.10	212.10
7600	700	10	136	12	A	240	10.85	250.85
7800		16	145	14	A	237	8.10	245.10
8000		20	145	14	A	224	8.10	232.10
8200	800	10	145	14	A	256	10.85	266.85
8300		16	159	14	A	260	8.10	268.10
8400		20	176	14	A	272	8.10	280.10

B10 Superstructure

B1010 Floor Construction

B1010 208			Steel Columns					
	LOAD (KIPS)	UNSUPPORTED HEIGHT (FT.)	WEIGHT (P.L.F.)	SIZE (IN.)	TYPE	COST PER V.L.F.		
						MAT.	INST.	TOTAL
8800	900	10	159	14	A	281	10.85	291.85
8900		16	176	14	A	288	8.10	296.10
9000		20	193	14	A	299	8.10	307.10
9100	1000	10	176	14	A	310	10.85	320.85
9200		16	193	14	A	315	8.10	323.10
9300		20	211	14	A	325	8.10	333.10

B1010 Floor Construction

Columns

Interior Bay

Beams

Exterior Bay

Joists

Corner Bay

Description: Table below lists costs per S.F. of bay size for wood columns of various sizes and unsupported heights and the maximum allowable total load per S.F. per bay size.

B1010 210	Wood Columns							
	NOMINAL COLUMN SIZE (IN.)	BAY SIZE (FT.)	UNSUPPORTED HEIGHT (FT.)	MATERIAL (BF per M.S.F.)	TOTAL LOAD (P.S.F.)	COST PER S.F.		
						MAT.	INST.	TOTAL
1000	4 x 4	10 x 8	8	133	100	.19	.26	.45
1050			10	167	60	.23	.32	.55
1200		10 x 10	8	106	80	.15	.20	.35
1250			10	133	50	.19	.26	.45
1400		10 x 15	8	71	50	.10	.14	.24
1450			10	88	30	.12	.17	.29
1600		15 x 15	8	47	30	.07	.09	.16
1650			10	59	15	.08	.11	.19
2000	6 x 6	10 x 15	8	160	230	.25	.28	.53
2050			10	200	210	.31	.36	.67
2200		15 x 15	8	107	150	.17	.19	.36
2250			10	133	140	.21	.24	.45
2400		15 x 20	8	80	110	.12	.14	.26
2450			10	100	100	.16	.18	.34
2600		20 x 20	8	60	80	.09	.11	.20
2650			10	75	70	.12	.13	.25
2800		20 x 25	8	48	60	.07	.09	.16
2850			10	60	50	.09	.11	.20
3400	8 x 8	20 x 20	8	107	160	.16	.18	.34
3450			10	133	160	.20	.22	.42
3600		20 x 25	8	85	130	.12	.16	.28
3650			10	107	130	.15	.21	.36
3800		25 x 25	8	68	100	.11	.11	.22
3850			10	85	100	.14	.13	.27
4200	10 x 10	20 x 25	8	133	210	.21	.21	.42
4250			10	167	210	.27	.26	.53
4400		25 x 25	8	107	160	.17	.17	.34
4450			10	133	160	.21	.21	.42
4700	12 x 12	20 x 25	8	192	310	.32	.28	.60
4750			10	240	310	.40	.35	.75
4900		25 x 25	8	154	240	.26	.22	.48
4950			10	192	240	.32	.28	.60

B10 Superstructure

B1010 Floor Construction

"T" Shaped Precast Beams

"L" Shaped Precast Beams

B1010 214					"T" Shaped Precast Beams			
	SPAN (FT.)	SUPERIMPOSED LOAD (K.L.F.)	SIZE W X D (IN.)	BEAM WEIGHT (P.L.F.)	TOTAL LOAD (K.L.F.)	COST PER L.F.		
						MAT.	INST.	TOTAL
2300	15	2.8	12x16	260	3.06	176	20.50	196.50
2500		8.37	12x28	515	8.89	209	22	231
8900	45	3.34	12x60	1165	4.51	335	13.20	348.20
9900		6.5	24x60	1915	8.42	365	18.50	383.50

B1010 215					"L" Shaped Precast Beams			
	SPAN (FT.)	SUPERIMPOSED LOAD (K.L.F.)	SIZE W X D (IN.)	BEAM WEIGHT (P.L.F.)	TOTAL LOAD (K.L.F.)	COST PER L.F.		
						MAT.	INST.	TOTAL
2250	15	2.58	12x16	230	2.81	166	20.50	186.50
2400		5.92	12x24	370	6.29	182	20.50	202.50
4000	25	2.64	12x28	435	3.08	183	13.30	196.30
4450		6.44	18x36	790	7.23	226	15.85	241.85
5300	30	2.80	12x36	565	3.37	204	12.05	216.05
6400		8.66	24x44	1245	9.90	271	17.35	288.35

B10 Superstructure

B1010 Floor Construction

Cast in Place Floor Slab, One Way
General: Solid concrete slabs of uniform depth reinforced for flexure in one direction and for temperature and shrinkage in the other direction.

Cast in Place Beam & Slab, One Way
General: Solid concrete one way slab cast monolithically with reinforced concrete beams and girders.

B1010 217 — Cast in Place Slabs, One Way

	SLAB DESIGN & SPAN (FT.)	SUPERIMPOSED LOAD (P.S.F.)	THICKNESS (IN.)	TOTAL LOAD (P.S.F.)	COST PER S.F.		
					MAT.	INST.	TOTAL
2500	Single 8	40	4	90	4.09	8.40	12.49
2600		75	4	125	4.13	9.15	13.28
2700		125	4-1/2	181	4.31	9.20	13.51
2800		200	5	262	4.77	9.50	14.27
3000	Single 10	40	4	90	4.27	9.05	13.32
3100		75	4	125	4.27	9.05	13.32
3200		125	5	188	4.74	9.25	13.99
3300		200	7-1/2	293	6.25	10.40	16.65
3500	Single 15	40	5-1/2	90	5.10	9.15	14.25
3600		75	6-1/2	156	5.65	9.65	15.30
3700		125	7-1/2	219	6.15	9.90	16.05
3800		200	8-1/2	306	6.65	10.10	16.75
4000	Single 20	40	7-1/2	115	6.10	9.65	15.75
4100		75	9	200	6.65	9.65	16.30
4200		125	10	250	7.35	10.05	17.40
4300		200	10	324	7.85	10.45	18.30

B1010 219 — Cast in Place Beam & Slab, One Way

	BAY SIZE (FT.)	SUPERIMPOSED LOAD (P.S.F.)	MINIMUM COL. SIZE (IN.)	SLAB THICKNESS (IN.)	TOTAL LOAD (P.S.F.)	COST PER S.F.		
						MAT.	INST.	TOTAL
5000	20x25	40	12	5-1/2	121	5.20	11.55	16.75
5200		125	16	5-1/2	215	6.10	13.20	19.30
5500	25x25	40	12	6	129	5.45	11.30	16.75
5700		125	18	6	227	6.95	13.90	20.85
7500	30x35	40	16	8	158	6.85	12.80	19.65
7600		75	18	8	196	7.30	13.20	20.50
7700		125	22	8	254	8.15	14.65	22.80
7800		200	26	8	332	8.90	15.15	24.05
8000	35x35	40	16	9	169	7.65	13.20	20.85
8200		75	20	9	213	8.25	14.35	22.60
8400		125	24	9	272	9.10	14.90	24
8600		200	26	9	355	10	15.90	25.90

B10 Superstructure

B1010 Floor Construction

General: Solid concrete two way slab cast monolithically with reinforced concrete support beams and girders.

General: Flat Slab: Solid uniform depth concrete two way slabs with drop panels at columns and no column capitals.

B1010 220			Cast in Place Beam & Slab, Two Way					
	BAY SIZE (FT.)	SUPERIMPOSED LOAD (P.S.F.)	MINIMUM COL. SIZE (IN.)	SLAB THICKNESS (IN.)	TOTAL LOAD (P.S.F.)	COST PER S.F.		
						MAT.	INST.	TOTAL
4000	20 x 25	40	12	7	141	6	11.60	17.60
4300		75	14	7	181	6.90	12.70	19.60
4500		125	16	7	236	7	13.05	20.05
5100	25 x 25	40	12	7-1/2	149	6.25	11.75	18
5200		75	16	7-1/2	185	6.85	12.60	19.45
5300		125	18	7-1/2	250	7.40	13.65	21.05
7600	30 x 35	40	16	10	188	8.30	13.30	21.60
7700		75	18	10	225	8.85	13.80	22.65
8000		125	22	10	282	9.80	14.85	24.65
8500	35 x 35	40	16	10-1/2	193	8.90	13.55	22.45
8600		75	20	10-1/2	233	9.35	14.05	23.40
9000		125	24	10-1/2	287	10.45	15.10	25.55

B1010 222			Cast in Place Flat Slab with Drop Panels					
	BAY SIZE (FT.)	SUPERIMPOSED LOAD (P.S.F.)	MINIMUM COL. SIZE (IN.)	SLAB & DROP (IN.)	TOTAL LOAD (P.S.F.)	COST PER S.F.		
						MAT.	INST.	TOTAL
1960	20 x 20	40	12	7 - 3	132	5.35	9.25	14.60
1980		75	16	7 - 4	168	5.65	9.45	15.10
2000		125	18	7 - 6	221	6.30	9.80	16.10
3200	25 x 25	40	12	8-1/2 - 5-1/2	154	6.30	9.70	16
4000		125	20	8-1/2 - 8-1/2	243	7.10	10.40	17.50
4400		200	24	9 - 8-1/2	329	7.50	10.60	18.10
5000	25 x 30	40	14	9-1/2 - 7	168	6.85	10.05	16.90
5200		75	18	9-1/2 - 7	203	7.30	10.40	17.70
5600		125	22	9-1/2 - 8	256	7.65	10.65	18.30
6400	30 x 30	40	14	10-1/2 - 7-1/2	182	7.40	10.30	17.70
6600		75	18	10-1/2 - 7-1/2	217	7.90	10.70	18.60
6800		125	22	10-1/2 - 9	269	8.25	10.95	19.20
7400	30 x 35	40	16	11-1/2 - 9	196	8.10	10.70	18.80
7900		75	20	11-1/2 - 9	231	8.65	11.15	19.80
8000		125	24	11-1/2 - 11	284	9	11.40	20.40
9000	35 x 35	40	16	12 - 9	202	8.30	10.80	19.10
9400		75	20	12 - 11	240	8.95	11.30	20.25
9600		125	24	12 - 11	290	9.25	11.50	20.75

B10 Superstructure

B1010 Floor Construction

Multispan Joist Slab

General: Flat Plates: Solid uniform depth concrete two way slab without drops or interior beams. Primary design limit is shear at columns.

General: Combination of thin concrete slab and monolithic ribs at uniform spacing to reduce dead weight and increase rigidity.

B1010 223	**Cast in Place Flat Plate**							
	BAY SIZE (FT.)	SUPERIMPOSED LOAD (P.S.F.)	MINIMUM COL. SIZE (IN.)	SLAB THICKNESS (IN.)	TOTAL LOAD (P.S.F.)	COST PER S.F.		
						MAT.	INST.	TOTAL
3000	15 x 20	40	14	7	127	5	8.95	13.95
3400		75	16	7-1/2	169	5.35	9.10	14.45
3600		125	22	8-1/2	231	5.90	9.35	15.25
3800		175	24	8-1/2	281	5.95	9.35	15.30
4200	20 x 20	40	16	7	127	5.05	8.95	14
4400		75	20	7-1/2	175	5.40	9.10	14.50
4600		125	24	8-1/2	231	5.95	9.30	15.25
5000		175	24	8-1/2	281	5.95	9.35	15.30
5600	20 x 25	40	18	8-1/2	146	5.85	9.35	15.20
6000		75	20	9	188	6.10	9.40	15.50
6400		125	26	9-1/2	244	6.60	9.75	16.35
6600		175	30	10	300	6.85	9.85	16.70
7000	25 x 25	40	20	9	152	6.10	9.40	15.50
7400		75	24	9-1/2	194	6.45	9.65	16.10
7600		125	30	10	250	6.85	9.85	16.70

B1010 226	**Cast in Place Multispan Joist Slab**							
	BAY SIZE (FT.)	SUPERIMPOSED LOAD (P.S.F.)	MINIMUM COL. SIZE (IN.)	RIB DEPTH (IN.)	TOTAL LOAD (P.S.F.)	COST PER S.F.		
						MAT.	INST.	TOTAL
2000	15 x 15	40	12	8	115	6.55	10.90	17.45
2100		75	12	8	150	6.60	10.95	17.55
2200		125	12	8	200	6.70	11.05	17.75
2300		200	14	8	275	6.90	11.50	18.40
2600	15 x 20	40	12	8	115	6.70	10.95	17.65
2800		75	12	8	150	6.80	11.55	18.35
3000		125	14	8	200	7.05	11.70	18.75
3300		200	16	8	275	7.35	11.95	19.30
3600	20 x 20	40	12	10	120	6.85	10.75	17.60
3900		75	14	10	155	7.10	11.40	18.50
4000		125	16	10	205	7.15	11.65	18.80
4100		200	18	10	280	7.50	12.15	19.65
6200	30 x 30	40	14	14	131	7.55	11.30	18.85
6400		75	18	14	166	7.75	11.65	19.40
6600		125	20	14	216	8.20	12.25	20.45
6700		200	24	16	297	8.75	12.70	21.45

For customer support on your Square Foot Cost Data, call 877.694.8329.

Waffle Slab

B1010 227	Cast in Place Waffle Slab							

	BAY SIZE (FT.)	SUPERIMPOSED LOAD (P.S.F.)	MINIMUM COL. SIZE (IN.)	RIB DEPTH (IN.)	TOTAL LOAD (P.S.F.)	COST PER S.F.		
						MAT.	INST.	TOTAL
3900	20 x 20	40	12	8	144	8.30	10.85	19.15
4000		75	12	8	179	8.45	11	19.45
4100		125	16	8	229	8.60	11.10	19.70
4200		200	18	8	304	9.05	11.50	20.55
4400	20 x 25	40	12	8	146	8.45	10.90	19.35
4500		75	14	8	181	8.65	11.05	19.70
4600		125	16	8	231	8.80	11.20	20
4700		200	18	8	306	9.20	11.55	20.75
4900	25 x 25	40	12	10	150	8.65	11	19.65
5000		75	16	10	185	8.95	11.25	20.20
5300		125	18	10	235	9.20	11.45	20.65
5500		200	20	10	310	9.45	11.65	21.10
5700	25 x 30	40	14	10	154	8.85	11.10	19.95
5800		75	16	10	189	9.10	11.30	20.40
5900		125	18	10	239	9.35	11.50	20.85
6000		200	20	12	329	10.25	12	22.25
6400	30 x 30	40	14	12	169	9.45	11.35	20.80
6500		75	18	12	204	9.65	11.50	21.15
6600		125	20	12	254	9.80	11.65	21.45
6700		200	24	12	329	10.65	12.30	22.95
6900	30 x 35	40	16	12	169	9.60	11.45	21.05
7000		75	18	12	204	9.60	11.45	21.05
7100		125	22	12	254	10.10	11.85	21.95
7200		200	26	14	334	11.20	12.60	23.80
7400	35 x 35	40	16	14	174	10.15	11.75	21.90
7500		75	20	14	209	10.40	11.95	22.35
7600		125	24	14	259	10.60	12.15	22.75
7700		200	26	16	346	11.40	12.75	24.15
8000	35 x 40	40	18	14	176	10.45	11.95	22.40
8300		75	22	14	211	10.75	12.25	23
8500		125	26	16	271	11.30	12.50	23.80
8750		200	30	20	372	12.35	13.20	25.55
9200	40 x 40	40	18	14	176	10.75	12.25	23
9400		75	24	14	211	11.20	12.60	23.80
9500		125	26	16	271	11.50	12.65	24.15
9700	40 x 45	40	20	16	186	11.20	12.40	23.60
9800		75	24	16	221	11.60	12.75	24.35
9900		125	28	16	271	11.80	12.90	24.70

B1010   Floor Construction

Precast Plank with No Topping

Precast Plank with 2" Concrete Topping

B1010 229					Precast Plank with No Topping			
	SPAN (FT.)	SUPERIMPOSED LOAD (P.S.F.)	TOTAL DEPTH (IN.)	DEAD LOAD (P.S.F.)	TOTAL LOAD (P.S.F.)	COST PER S.F.		
						MAT.	INST.	TOTAL
0720	10	40	4	50	90	6.45	3.47	9.92
0750		75	6	50	125	7.35	2.97	10.32
0770		100	6	50	150	7.35	2.97	10.32
0800	15	40	6	50	90	7.35	2.97	10.32
0820		75	6	50	125	7.35	2.97	10.32
0850		100	6	50	150	7.35	2.97	10.32
0950	25	40	6	50	90	7.35	2.97	10.32
0970		75	8	55	130	7.80	2.60	10.40
1000		100	8	55	155	7.80	2.60	10.40
1200	30	40	8	55	95	7.80	2.60	10.40
1300		75	8	55	130	7.80	2.60	10.40
1400		100	10	70	170	8.50	2.31	10.81
1500	40	40	10	70	110	8.50	2.31	10.81
1600		75	12	70	145	8.65	2.08	10.73
1700	45	40	12	70	110	8.65	2.08	10.73

B1010 230					Precast Plank with 2" Concrete Topping			
	SPAN (FT.)	SUPERIMPOSED LOAD (P.S.F.)	TOTAL DEPTH (IN.)	DEAD LOAD (P.S.F.)	TOTAL LOAD (P.S.F.)	COST PER S.F.		
						MAT.	INST.	TOTAL
2000	10	40	6	75	115	7.45	5.90	13.35
2100		75	8	75	150	8.35	5.40	13.75
2200		100	8	75	175	8.35	5.40	13.75
2500	15	40	8	75	115	8.35	5.40	13.75
2600		75	8	75	150	8.35	5.40	13.75
2700		100	8	75	175	8.35	5.40	13.75
3100	25	40	8	75	115	8.35	5.40	13.75
3200		75	8	75	150	8.35	5.40	13.75
3300		100	10	80	180	8.80	5	13.80
3400	30	40	10	80	120	8.80	5	13.80
3500		75	10	80	155	8.80	5	13.80
3600		100	10	80	180	8.80	5	13.80
4000	40	40	12	95	135	9.50	4.72	14.22
4500		75	14	95	170	9.65	4.49	14.14
5000	45	40	14	95	135	9.65	4.49	14.14

B10 Superstructure

B1010 Floor Construction

Precast Double "T" Beams with No Topping

Precast Double "T" Beams with 2" Topping

Most widely used for moderate span floors and moderate and long span roofs. At shorter spans, they tend to be competitive with hollow core slabs. They are also used as wall panels.

B1010 234		Precast Double "T" Beams with No Topping						
	SPAN (FT.)	SUPERIMPOSED LOAD (P.S.F.)	DBL. "T" SIZE D (IN.) W (FT.)	CONCRETE "T" TYPE	TOTAL LOAD (P.S.F.)	COST PER S.F.		
						MAT.	INST.	TOTAL
4300	50	30	20x8	Lt. Wt.	66	9.30	1.58	10.88
4400		40	20x8	Lt. Wt.	76	9.35	1.45	10.80
4500		50	20x8	Lt. Wt.	86	9.45	1.74	11.19
4600		75	20x8	Lt. Wt.	111	9.50	2	11.50
5600	70	30	32x10	Lt. Wt.	78	9.35	1.10	10.45
5750		40	32x10	Lt. Wt.	88	9.40	1.33	10.73
5900		50	32x10	Lt. Wt.	98	9.45	1.53	10.98
6000		75	32x10	Lt. Wt.	123	9.55	1.86	11.41
6100		100	32x10	Lt. Wt.	148	9.80	2.52	12.32
6200	80	30	32x10	Lt. Wt.	78	9.45	1.53	10.98
6300		40	32x10	Lt. Wt.	88	9.55	1.86	11.41
6400		50	32x10	Lt. Wt.	98	9.70	2.19	11.89

B1010 235		Precast Double "T" Beams With 2" Topping						
	SPAN (FT.)	SUPERIMPOSED LOAD (P.S.F.)	DBL. "T" SIZE D (IN.) W (FT.)	CONCRETE "T" TYPE	TOTAL LOAD (P.S.F.)	COST PER S.F.		
						MAT.	INST.	TOTAL
7100	40	30	18x8	Reg. Wt.	120	8.20	3.22	11.42
7200		40	20x8	Reg. Wt.	130	9.60	3.08	12.68
7300		50	20x8	Reg. Wt.	140	9.70	3.37	13.07
7400		75	20x8	Reg. Wt.	165	9.75	3.51	13.26
7500		100	20x8	Reg. Wt.	190	9.90	3.92	13.82
7550	50	30	24x8	Reg. Wt.	120	9.65	3.23	12.88
7600		40	24x8	Reg. Wt.	130	9.70	3.34	13.04
7750		50	24x8	Reg. Wt.	140	9.70	3.36	13.06
7800		75	24x8	Reg. Wt.	165	9.80	3.78	13.58
7900		100	32x10	Reg. Wt.	189	9.75	3.10	12.85

325

B1010 Floor Construction

Precast Beam and Plank with No Topping

Precast Beam and Plank with 2″ Topping

B1010 236	Precast Beam & Plank with No Topping							
	BAY SIZE (FT.)	SUPERIMPOSED LOAD (P.S.F.)	PLANK THICKNESS (IN.)	TOTAL DEPTH (IN.)	TOTAL LOAD (P.S.F.)	COST PER S.F.		
						MAT.	INST.	TOTAL
6200	25x25	40	8	28	118	20.50	3.45	23.95
6400		75	8	36	158	20.50	3.46	23.96
6500		100	8	36	183	20.50	3.46	23.96
7000	25x30	40	8	36	110	18.85	3.35	22.20
7200		75	10	36	159	19.95	3.07	23.02
7400		100	12	36	188	19.75	2.80	22.55
7600	30x30	40	8	36	121	19.25	3.24	22.49
8000		75	10	44	140	19.15	3.26	22.41
8250		100	12	52	206	21.50	2.75	24.25
8500	30x35	40	12	44	135	18.60	2.65	21.25
8750		75	12	52	176	19.55	2.64	22.19
9000	35x35	40	12	52	141	19.55	2.60	22.15
9250		75	12	60	181	20	2.59	22.59
9500	35x40	40	12	52	137	19.55	2.60	22.15

B1010 238	Precast Beam & Plank with 2″ Topping							
	BAY SIZE (FT.)	SUPERIMPOSED LOAD (P.S.F.)	PLANK THICKNESS (IN.)	TOTAL DEPTH (IN.)	TOTAL LOAD (P.S.F.)	COST PER S.F.		
						MAT.	INST.	TOTAL
4300	20x20	40	6	22	135	22.50	6.20	28.70
4400		75	6	24	173	23	6.25	29.25
4500		100	6	28	200	23.50	6.25	29.75
4600	20x25	40	6	26	134	21	6.05	27.05
5000		75	8	30	177	21.50	5.60	27.10
5200		100	8	30	202	21.50	5.60	27.10
5400	25x25	40	6	38	143	21.50	5.80	27.30
5600		75	8	38	183	17.20	5.50	22.70
6000		100	8	46	216	17.05	5.05	22.10
6200	25x30	40	8	38	144	19.65	5.20	24.85
6400		75	10	46	200	20.50	4.92	25.42
6600		100	10	46	225	21	4.94	25.94
7000	30x30	40	8	46	150	20.50	5.15	25.65
7200		75	10	54	181	21.50	5.15	26.65
7600		100	10	54	231	21.50	5.15	26.65
7800	30x35	40	10	54	166	20.50	4.75	25.25
8000		75	12	54	200	20.50	4.51	25.01

326

For customer support on your Square Foot Cost Data, call 877.694.8329.

B1010 Floor Construction

General: Beams and double tees priced here are for plant produced prestressed members transported to the site and erected.

The 2″ structural topping is applied after the beams and double tees are in place and is reinforced with W.W.F.

With Topping

B1010 239	Precast Double "T" & 2″ Topping on Precast Beams							

	BAY SIZE (FT.)	SUPERIMPOSED LOAD (P.S.F.)	DEPTH (IN.)		TOTAL LOAD (P.S.F.)	COST PER S.F.		
						MAT.	INST.	TOTAL
3000	25x30	40	38		130	21	4.37	25.37
3100		75	38		168	21	4.37	25.37
3300		100	46		196	20.50	4.34	24.84
3600	30x30	40	46		150	21	4.24	25.24
3750		75	46		174	21	4.24	25.24
4000		100	54		203	22	4.23	26.23
4100	30x40	40	46		136	18.65	3.83	22.48
4300		75	54		173	19.25	3.81	23.06
4400		100	62		204	19.85	3.81	23.66
4600	30x50	40	54		138	17.40	3.54	20.94
4800		75	54		181	17.25	3.57	20.82
5000		100	54		219	17.30	3.26	20.56
5200	30x60	40	62		151	16.95	3.21	20.16
5400		75	62		192	16.55	3.26	19.81
5600		100	62		215	16.70	3.26	19.96
5800	35x40	40	54		139	19.50	3.76	23.26
6000		75	62		179	19.55	3.60	23.15
6250		100	62		212	19.75	3.74	23.49
6500	35x50	40	62		142	18	3.51	21.51
6750		75	62		186	18.05	3.64	21.69
7300		100	62		231	18.20	3.32	21.52
7600	35x60	40	54		154	17.45	3.08	20.53
7750		75	54		179	17.55	3.09	20.64
8000		100	62		224	18	3.15	21.15
8250	40x40	40	62		145	21	3.77	24.77
8400		75	62		187	20.50	3.78	24.28
8750		100	62		223	21	3.87	24.87
9000	40x50	40	62		151	18.75	3.54	22.29
9300		75	62		193	18.30	2.63	20.93
9800	40x60	40	62		164	21	2.52	23.52

327

B10 Superstructure

B1010 Floor Construction

General: The following table is based upon structural wide flange (WF) beam and girder framing. Non-composite action is assumed between WF framing and decking. Deck costs not included.

The deck spans the short direction. The steel beams and girders are fireproofed with sprayed fiber fireproofing.

No columns included in price.

B1010 241		W Shape Beams & Girders						
	BAY SIZE (FT.) BEAM X GIRD	SUPERIMPOSED LOAD (P.S.F.)	STEEL FRAMING DEPTH (IN.)	FIREPROOFING (S.F. PER S.F.)	TOTAL LOAD (P.S.F.)	COST PER S.F.		
						MAT.	INST.	TOTAL
1350	15x20	40	12	.535	50	5.10	1.97	7.07
1400		40	16	.65	90	6.65	2.54	9.19
1450		75	18	.694	125	8.75	3.20	11.95
1500		125	24	.796	175	12.10	4.43	16.53
2000	20x20	40	12	.55	50	5.70	2.17	7.87
2050		40	14	.579	90	7.80	2.82	10.62
2100		75	16	.672	125	9.35	3.36	12.71
2150		125	16	.714	175	11.15	4.07	15.22
2900	20x25	40	16	.53	50	6.25	2.32	8.57
2950		40	18	.621	96	9.90	3.48	13.38
3000		75	18	.651	131	11.40	3.95	15.35
3050		125	24	.77	200	15.05	5.35	20.40
5200	25x25	40	18	.486	50	6.85	2.46	9.31
5300		40	18	.592	96	10.15	3.53	13.68
5400		75	21	.668	131	12.30	4.23	16.53
5450		125	24	.738	191	16.20	5.65	21.85
6300	25x30	40	21	.568	50	7.80	2.81	10.61
6350		40	21	.694	90	10.55	3.73	14.28
6400		75	24	.776	125	13.55	4.69	18.24
6450		125	30	.904	175	16.30	5.85	22.15
7700	30x30	40	21	.52	50	8.35	2.93	11.28
7750		40	24	.629	103	12.90	4.36	17.26
7800		75	30	.715	138	15.30	5.15	20.45
7850		125	36	.822	206	20	6.95	26.95
8250	30x35	40	21	.508	50	9.55	3.27	12.82
8300		40	24	.651	109	14.10	4.74	18.84
8350		75	33	.732	150	17.10	5.70	22.80
8400		125	36	.802	225	21.50	7.30	28.80
9550	35x35	40	27	.560	50	9.85	3.42	13.27
9600		40	36	.706	109	18	5.95	23.95
9650		75	36	.750	150	18.60	6.15	24.75
9820		125	36	.797	225	25	8.40	33.40

B1010   Floor Construction

Description: Table below lists costs for light gage CEE or PUNCHED DOUBLE joists to suit the span and loading with the minimum thickness subfloor required by the joist spacing.

B1010 244	Light Gauge Steel Floor Systems							
	SPAN (FT.)	SUPERIMPOSED LOAD (P.S.F.)	FRAMING DEPTH (IN.)	FRAMING SPAC. (IN.)	TOTAL LOAD (P.S.F.)	COST PER S.F.		
						MAT.	INST.	TOTAL
1500	15	40	8	16	54	2.77	2.26	5.03
1550			8	24	54	2.91	2.31	5.22
1600		65	10	16	80	3.56	2.75	6.31
1650			10	24	80	3.54	2.68	6.22
1700		75	10	16	90	3.56	2.75	6.31
1750			10	24	90	3.54	2.68	6.22
1800		100	10	16	116	4.61	3.49	8.10
1850			10	24	116	3.54	2.68	6.22
1900		125	10	16	141	4.61	3.49	8.10
1950			10	24	141	3.54	2.68	6.22
2500	20	40	8	16	55	3.17	2.55	5.72
2550			8	24	55	2.91	2.31	5.22
2600		65	8	16	80	3.66	2.92	6.58
2650			10	24	80	3.55	2.70	6.25
2700		75	10	16	90	3.37	2.61	5.98
2750			12	24	90	3.97	2.37	6.34
2800		100	10	16	115	3.37	2.61	5.98
2850			10	24	116	4.53	3.05	7.58
2900		125	12	16	142	5.25	2.99	8.24
2950			12	24	141	4.91	3.23	8.14
3500	25	40	10	16	55	3.56	2.75	6.31
3550			10	24	55	3.54	2.68	6.22
3600		65	10	16	81	4.61	3.49	8.10
3650			12	24	81	4.39	2.56	6.95
3700		75	12	16	92	5.25	2.99	8.24
3750			12	24	91	4.91	3.23	8.14
3800		100	12	16	117	5.90	3.30	9.20
3850		125	12	16	143	6.70	4.33	11.03
4500	30	40	12	16	57	5.25	2.99	8.24
4550			12	24	56	4.37	2.56	6.93
4600		65	12	16	82	5.85	3.29	9.14
4650		75	12	16	92	5.85	3.29	9.14

Description: The table below lists costs per S.F. for a floor system on bearing walls using open web steel joists, galvanized steel slab form and 2-1/2″ concrete slab reinforced with welded wire fabric. Costs of the bearing walls are not included.

B1010 246	Deck & Joists on Bearing Walls						
	SPAN (FT.)	SUPERIMPOSED LOAD (P.S.F.)	JOIST SPACING FT. - IN.	DEPTH (IN.)	TOTAL LOAD (P.S.F.)	COST PER S.F.		
						MAT.	INST.	TOTAL
1050	20	40	2-0	14-1/2	83	5.95	3.94	9.89
1070		65	2-0	16-1/2	109	6.45	4.17	10.62
1100		75	2-0	16-1/2	119	6.45	4.17	10.62
1120		100	2-0	18-1/2	145	6.50	4.18	10.68
1150		125	1-9	18-1/2	170	7.10	4.43	11.53
1170	25	40	2-0	18-1/2	84	6.50	4.18	10.68
1200		65	2-0	20-1/2	109	6.70	4.27	10.97
1220		75	2-0	20-1/2	119	7.15	4.45	11.60
1250		100	2-0	22-1/2	145	7.25	4.50	11.75
1270		125	1-9	22-1/2	170	8.25	4.90	13.15
1300	30	40	2-0	22-1/2	84	7.25	4.10	11.35
1320		65	2-0	24-1/2	110	7.85	4.28	12.13
1350		75	2-0	26-1/2	121	8.10	4.36	12.46
1370		100	2-0	26-1/2	146	8.65	4.52	13.17
1400		125	2-0	24-1/2	172	10.15	4.97	15.12
1420	35	40	2-0	26-1/2	85	9.60	4.80	14.40
1450		65	2-0	28-1/2	111	9.90	4.91	14.81
1470		75	2-0	28-1/2	121	9.90	4.91	14.81
1500		100	1-11	28-1/2	147	10.75	5.15	15.90
1520		125	1-8	28-1/2	172	11.85	5.50	17.35
1550	5/8″ gyp. fireproof.							
1560	On metal furring, add					.82	3.06	3.88

B10 Superstructure

B1010 Floor Construction

The table below lists costs for a floor system on exterior bearing walls and interior columns and beams using open web steel joists, galvanized steel slab form, 2-1/2″ concrete slab reinforced with welded wire fabric.

Columns are sized to accommodate one floor plus roof loading but costs for columns are from suspended floor to slab on grade only. Costs of the bearing walls are not included.

Steel Joists and Concrete Slab on Walls and Beams

B1010 248		Steel Joists on Beam & Wall						
	BAY SIZE (FT.)	SUPERIMPOSED LOAD (P.S.F.)	DEPTH (IN.)	TOTAL LOAD (P.S.F.)	COLUMN ADD	COST PER S.F.		
						MAT.	INST.	TOTAL
1720	25x25	40	23	84		8.10	4.49	12.59
1730					columns	.42	.13	.55
1750	25x25	65	29	110		8.55	4.64	13.19
1760					columns	.42	.13	.55
1770	25x25	75	26	120		9.50	4.98	14.48
1780					columns	.48	.14	.62
1800	25x25	100	29	145		10.70	5.45	16.15
1810					columns	.48	.14	.62
1820	25x25	125	29	170		11.25	5.60	16.85
1830					columns	.54	.16	.70
2020	30x30	40	29	84		9.30	4.54	13.84
2030					columns	.37	.11	.48
2050	30x30	65	29	110		10.55	4.90	15.45
2060					columns	.37	.11	.48
2070	30x30	75	32	120		10.85	4.99	15.84
2080					columns	.43	.13	.56
2100	30x30	100	35	145		12	5.35	17.35
2110					columns	.50	.15	.65
2120	30x30	125	35	172		13.85	5.90	19.75
2130					columns	.60	.18	.78
2170	30x35	40	29	85		10.50	4.90	15.40
2180					columns	.36	.10	.46
2200	30x35	65	29	111		12.50	5.50	18
2210					columns	.41	.13	.54
2220	30x35	75	32	121		12.50	5.50	18
2230					columns	.42	.13	.55
2250	30x35	100	35	148		12.70	5.55	18.25
2260					columns	.52	.15	.67
2270	30x35	125	38	173		14.25	6.05	20.30
2280					columns	.53	.15	.68
2320	35x35	40	32	85		10.80	4.99	15.79
2330					columns	.37	.11	.48
2350	35x35	65	35	111		13.20	5.70	18.90
2360					columns	.44	.13	.57
2370	35x35	75	35	121		13.45	5.80	19.25
2380					columns	.44	.13	.57
2400	35x35	100	38	148		14.05	5.95	20
2410					columns	.55	.16	.71

B1010 Floor Construction

B1010 248	Steel Joists on Beam & Wall							
	BAY SIZE (FT.)	SUPERIMPOSED LOAD (P.S.F.)	DEPTH (IN.)	TOTAL LOAD (P.S.F.)	COLUMN ADD	COST PER S.F.		
						MAT.	INST.	TOTAL
2460	5/8 gyp. fireproof.							
2475	On metal furring, add					.82	3.06	3.88

B1010 Floor Construction

The table below lists costs per S.F. for a floor system on steel columns and beams using open web steel joists, galvanized steel slab form, 2-1/2″ concrete slab reinforced with welded wire fabric.

Columns are sized to accommodate one floor plus roof loading but costs for columns are from floor to grade only.

Steel Joists and Concrete Slab on Steel Columns and Beams

B1010 250		Steel Joists, Beams & Slab on Columns						
	BAY SIZE (FT.)	SUPERIMPOSED LOAD (P.S.F.)	DEPTH (IN.)	TOTAL LOAD (P.S.F.)	COLUMN ADD	COST PER S.F.		
						MAT.	INST.	TOTAL
2350	15x20	40	17	83		8.45	4.54	12.99
2400					column	1.38	.41	1.79
2450	15x20	65	19	108		9.35	4.81	14.16
2500					column	1.38	.41	1.79
2550	15x20	75	19	119		9.70	4.96	14.66
2600					column	1.51	.44	1.95
2650	15x20	100	19	144		10.35	5.15	15.50
2700					column	1.51	.44	1.95
2750	15x20	125	19	170		10.85	5.30	16.15
2800					column	2.01	.59	2.60
2850	20x20	40	19	83		9.20	4.75	13.95
2900					column	1.13	.33	1.46
2950	20x20	65	23	109		10.10	5.05	15.15
3000					column	1.51	.44	1.95
3100	20x20	75	26	119		10.70	5.25	15.95
3200					column	1.51	.44	1.95
3400	20x20	100	23	144		11.10	5.35	16.45
3450					column	1.51	.44	1.95
3500	20x20	125	23	170		12.35	5.80	18.15
3600					column	1.81	.53	2.34
3700	20x25	40	44	83		9.80	4.98	14.78
3800					column	1.21	.35	1.56
3900	20x25	65	26	110		10.75	5.25	16
4000					column	1.21	.35	1.56
4100	20x25	75	26	120		11.25	5.45	16.70
4200					column	1.45	.43	1.88
4300	20x25	100	26	145		11.90	5.70	17.60
4400					column	1.45	.43	1.88
4500	20x25	125	29	170		13.25	6.15	19.40
4600					column	1.69	.49	2.18
4700	25x25	40	23	84		10.65	5.20	15.85
4800					column	1.16	.34	1.50
4900	25x25	65	29	.110		11.30	5.40	16.70
5000					column	1.16	.34	1.50
5100	25x25	75	26	120		12.45	5.85	18.30
5200					column	1.35	.39	1.74
5300	25x25	100	29	145		13.85	6.35	20.20
5400					column	1.35	.39	1.74

B1010   Floor Construction

B1010 250	Steel Joists, Beams & Slab on Columns

	BAY SIZE (FT.)	SUPERIMPOSED LOAD (P.S.F.)	DEPTH (IN.)	TOTAL LOAD (P.S.F.)	COLUMN ADD	COST PER S.F.		
						MAT.	INST.	TOTAL
5500	25x25	125	32	170		14.65	6.60	21.25
5600					column	1.50	.44	1.94
5700	25x30	40	29	84		12.20	5.35	17.55
5800					column	1.12	.33	1.45
5900	25x30	65	29	110		12.70	5.50	18.20
6000					column	1.12	.33	1.45
6050	25x30	75	29	120		13.05	5.65	18.70
6100					column	1.24	.37	1.61
6150	25x30	100	29	145		14.15	5.95	20.10
6200					column	1.24	.37	1.61
6250	25x30	125	32	170		15.90	6.50	22.40
6300					column	1.43	.42	1.85
6350	30x30	40	29	84		12.20	5.35	17.55
6400					column	1.04	.30	1.34
6500	30x30	65	29	110		13.85	5.85	19.70
6600					column	1.04	.30	1.34
6700	30x30	75	32	120		14.15	5.95	20.10
6800					column	1.19	.35	1.54
6900	30x30	100	35	145		15.70	6.40	22.10
7000					column	1.39	.41	1.80
7100	30x30	125	35	172		17.90	7.05	24.95
7200					column	1.55	.46	2.01
7300	30x35	40	29	85		13.70	5.85	19.55
7400					column	.89	.26	1.15
7500	30x35	65	29	111		16	6.50	22.50
7600					column	1.15	.34	1.49
7700	30x35	75	32	121		16	6.50	22.50
7800					column	1.17	.34	1.51
7900	30x35	100	35	148		16.50	6.65	23.15
8000					column	1.43	.42	1.85
8100	30x35	125	38	173		18.30	7.20	25.50
8200					column	1.46	.43	1.89
8300	35x35	40	32	85		14.10	5.90	20
8400					column	1.02	.30	1.32
8500	35x35	65	35	111		16.80	6.75	23.55
8600					column	1.23	.37	1.60
9300	35x35	75	38	121		17.25	6.85	24.10
9400					column	1.23	.37	1.60
9500	35x35	100	38	148		18.60	7.30	25.90
9600					column	1.52	.44	1.96
9750	35x35	125	41	173		19.20	7.45	26.65
9800					column	1.54	.46	2
9810	5/8 gyp. fireproof.							
9815	On metal furring, add					.82	3.06	3.88

334

For customer support on your Square Foot Cost Data, call 877.694.8329.

General: Composite construction of wide flange beams and concrete slabs is most efficiently used when loads are heavy and spans are moderately long. It is stiffer with less deflection than non-composite construction of similar depth and spans.

Composite Beam & Cast in Place Slab

B1010 252	Composite Beam & Cast in Place Slab

	BAY SIZE (FT.)	SUPERIMPOSED LOAD (P.S.F.)	SLAB THICKNESS (IN.)	TOTAL DEPTH (FT.-IN.)	TOTAL LOAD (P.S.F.)	COST PER S.F.		
						MAT.	INST.	TOTAL
3800	20x25	40	4	1 - 8	94	10.25	11.10	21.35
3900		75	4	1 - 8	130	11.95	11.80	23.75
4000		125	4	1 - 10	181	13.95	12.60	26.55
4100		200	5	2 - 2	272	18.20	14.25	32.45
4200	25x25	40	4-1/2	1 - 8-1/2	99	10.65	11.25	21.90
4300		75	4-1/2	1 - 10-1/2	136	12.95	12.15	25.10
4400		125	5-1/2	2 - 0-1/2	200	15.30	13.15	28.45
4500		200	5-1/2	2 - 2-1/2	278	19.10	14.50	33.60
4600	25x30	40	4-1/2	1 - 8-1/2	100	12	11.75	23.75
4700		75	4-1/2	1 - 10-1/2	136	14.10	12.50	26.60
4800		125	5-1/2	2 - 2-1/2	202	17.30	13.75	31.05
4900		200	5-1/2	2 - 5-1/2	279	20.50	15	35.50
5000	30x30	40	4	1 - 8	95	12	11.75	23.75
5200		75	4	2 - 1	131	14	12.40	26.40
5400		125	4	2 - 4	183	17.30	13.65	30.95
5600		200	5	2 - 10	274	22	15.45	37.45
5800	30x35	40	4	2 - 1	95	12.75	11.95	24.70
6000		75	4	2 - 4	139	15.45	12.90	28.35
6250		125	4	2 - 4	185	20	14.50	34.50
6500		200	5	2 - 11	276	24.50	16.20	40.70
8000	35x35	40	4	2 - 1	96	13.45	12.15	25.60
8250		75	4	2 - 4	133	16.05	13.15	29.20
8500		125	4	2 - 10	185	19.30	14.35	33.65
8750		200	5	3 - 5	276	25	16.40	41.40
9000	35x40	40	4	2 - 4	97	15	12.70	27.70
9250		75	4	2 - 4	134	17.55	13.50	31.05
9500		125	4	2 - 7	186	21.50	14.95	36.45
9750		200	5	3 - 5	278	28	17.20	45.20

The table below lists costs per S.F. for floors using steel beams and girders, composite steel deck, concrete slab reinforced with W.W.F. and sprayed fiber fireproofing (non- asbestos) on the steel beams and girders and on the steel deck.

Wide Flange, Composite Deck & Slab

B1010 254		W Shape, Composite Deck, & Slab						
	BAY SIZE (FT.) BEAM X GIRD	SUPERIMPOSED LOAD (P.S.F.)	SLAB THICKNESS (IN.)	TOTAL DEPTH (FT.-IN.)	TOTAL LOAD (P.S.F.)	COST PER S.F.		
						MAT.	INST.	TOTAL
0540	15x20	40	5	1-7	89	11.85	6.75	18.60
0560		75	5	1-9	125	13.05	7.15	20.20
0580		125	5	1-11	176	15.30	7.85	23.15
0600		200	5	2-2	254	18.80	8.95	27.75
0700	20x20	40	5	1-9	90	12.90	7.05	19.95
0720		75	5	1-9	126	14.60	7.65	22.25
0740		125	5	1-11	177	16.60	8.25	24.85
0760		200	5	2-5	255	20	9.25	29.25
0780	20x25	40	5	1-9	90	13.15	6.95	20.10
0800		75	5	1-9	127	16.45	8.10	24.55
0820		125	5	1-11	180	19.70	8.80	28.50
0840		200	5	2-5	256	22	9.95	31.95
0940	25x25	40	5	1-11	91	14.90	7.55	22.45
0960		75	5	2-5	178	17.20	8.30	25.50
0980		125	5	2-5	181	21	9.20	30.20
1000		200	5-1/2	2-8-1/2	263	24	10.65	34.65
1400	25x30	40	5	2-5	91	14.80	7.70	22.50
1500		75	5	2-5	128	18.05	8.60	26.65
1600		125	5	2-8	180	20.50	9.50	30
1700		200	5	2-11	259	25.50	10.95	36.45
2200	30x30	40	5	2-2	92	16.45	8.20	24.65
2300		75	5	2-5	129	19.30	9.10	28.40
2400		125	5	2-11	182	23	10.20	33.20
2500		200	5	3-2	263	31	12.80	43.80
2600	30x35	40	5	2-5	94	17.95	8.65	26.60
2700		75	5	2-11	131	21	9.60	30.60
2800		125	5	3-2	183	24.50	10.70	35.20
2900		200	5-1/2	3-5-1/2	268	30	12.30	42.30
3800	35x35	40	5	2-8	94	18.50	8.65	27.15
3900		75	5	2-11	131	21.50	9.60	31.10
4000		125	5	3-5	184	26	10.95	36.95
4100		200	5-1/2	3-5-1/2	270	33.50	13	46.50

B1010 Floor Construction

Composite Beam, Deck & Slab

The table below lists costs per S.F. for floors using composite steel beams with welded shear studs, composite steel deck and light weight concrete slab reinforced with W.W.F.

B1010 256				Composite Beams, Deck & Slab				
	BAY SIZE (FT.)	SUPERIMPOSED LOAD (P.S.F.)	SLAB THICKNESS (IN.)	TOTAL DEPTH (FT.-IN.)	TOTAL LOAD (P.S.F.)	COST PER S.F.		
						MAT.	INST.	TOTAL
2400	20x25	40	5-1/2	1 - 5-1/2	80	12.80	6.30	19.10
2500		75	5-1/2	1 - 9-1/2	115	13.30	6.30	19.60
2750		125	5-1/2	1 - 9-1/2	167	16.20	7.35	23.55
2900		200	6-1/4	1 - 11-1/2	251	18.30	7.90	26.20
3000	25x25	40	5-1/2	1 - 9-1/2	82	12.60	6	18.60
3100		75	5-1/2	1 - 11-1/2	118	14	6.10	20.10
3200		125	5-1/2	2 - 2-1/2	169	14.65	6.55	21.20
3300		200	6-1/4	2 - 6-1/4	252	19.75	7.65	27.40
3400	25x30	40	5-1/2	1 - 11-1/2	83	12.90	5.95	18.85
3600		75	5-1/2	1 - 11-1/2	119	13.85	6	19.85
3900		125	5-1/2	1 - 11-1/2	170	16.05	6.80	22.85
4000		200	6-1/4	2 - 6-1/4	252	19.85	7.70	27.55
4200	30x30	40	5-1/2	1 - 11-1/2	81	12.95	6.10	19.05
4400		75	5-1/2	2 - 2-1/2	116	13.95	6.40	20.35
4500		125	5-1/2	2 - 5-1/2	168	16.85	7.15	24
4700		200	6-1/4	2 - 9-1/4	252	20	8.30	28.30
4900	30x35	40	5-1/2	2 - 2-1/2	82	13.50	6.30	19.80
5100		75	5-1/2	2 - 5-1/2	117	14.75	6.50	21.25
5300		125	5-1/2	2 - 5-1/2	169	17.35	7.30	24.65
5500		200	6-1/4	2 - 9-1/4	254	20.50	8.30	28.80
5750	35x35	40	5-1/2	2 - 5-1/2	84	14.40	6.35	20.75
6000		75	5-1/2	2 - 5-1/2	121	16.40	6.80	23.20
7000	35x35	125	5-1/2	2 - 8-1/2	170	19.25	7.80	27.05
7200		200	5-1/2	2 - 11-1/2	254	22	8.55	30.55
7400	35x40	40	5-1/2	2 - 5-1/2	85	15.90	6.85	22.75
7600		75	5-1/2	2 - 5-1/2	121	17.25	7.05	24.30
8000		125	5-1/2	2 - 5-1/2	171	19.75	7.90	27.65
9000		200	5-1/2	2 - 11-1/2	255	24	8.95	32.95

B1010 Floor Construction

Description: Table B1010 258 lists S.F. costs for steel deck and concrete slabs for various spans.

Description: Table B1010 261 lists the S.F. costs for wood joists and a minimum thickness plywood subfloor.

Description: Table B1010 264 lists the S.F. costs, total load, and member sizes, for various bay sizes and loading conditions.

Metal Deck/Concrete Fill

Wood Joist

Wood Beam and Joist

B1010 258 — Metal Deck/Concrete Fill

	SUPERIMPOSED LOAD (P.S.F.)	DECK SPAN (FT.)	DECK GAGE DEPTH	SLAB THICKNESS (IN.)	TOTAL LOAD (P.S.F.)	COST PER S.F.		
						MAT.	INST.	TOTAL
0900	125	6	22 1-1/2	4	164	3	2.57	5.57
0920		7	20 1-1/2	4	164	3.32	2.72	6.04
0950		8	20 1-1/2	4	165	3.32	2.72	6.04
0970		9	18 1-1/2	4	165	3.89	2.72	6.61
1000		10	18 2	4	165	3.78	2.86	6.64
1020		11	18 3	5	169	4.67	3.11	7.78

B1010 261 — Wood Joist

		COST PER S.F.		
		MAT.	INST.	TOTAL
2900	2"x8", 12" O.C.	1.87	1.96	3.83
2950	16" O.C.	1.60	1.68	3.28
3000	24" O.C.	1.63	1.55	3.18
3300	2"x10", 12" O.C.	2.39	2.24	4.63
3350	16" O.C.	1.98	1.88	3.86
3400	24" O.C.	1.88	1.68	3.56
3700	2"x12", 12" O.C.	2.85	2.26	5.11
3750	16" O.C.	2.33	1.90	4.23
3800	24" O.C.	2.12	1.70	3.82
4100	2"x14", 12" O.C.	3.19	2.46	5.65
4150	16" O.C.	2.59	2.05	4.64
4200	24" O.C.	2.30	1.80	4.10
7101	Note: Subfloor cost is included in these prices.			

B1010 264 — Wood Beam & Joist

	BAY SIZE (FT.)	SUPERIMPOSED LOAD (P.S.F.)	GIRDER BEAM (IN.)	JOISTS (IN.)	TOTAL LOAD (P.S.F.)	COST PER S.F.		
						MAT.	INST.	TOTAL
2000	15x15	40	8 x 12 4 x 12	2 x 6 @ 16	53	9.20	4.55	13.75
2050		75	8 x 16 4 x 16	2 x 8 @ 16	90	12.70	4.97	17.67
2100		125	12 x 16 6 x 16	2 x 8 @ 12	144	19.50	6.40	25.90
2150		200	14 x 22 12 x 16	2 x 10 @ 12	227	39	9.75	48.75
3000	20x20	40	10 x 14 10 x 12	2 x 8 @ 16	63	10.95	4.57	15.52
3050		75	12 x 16 8 x 16	2 x 10 @ 16	102	16.75	5.10	21.85
3100		125	14 x 22 12 x 16	2 x 10 @ 12	163	33	8.20	41.20

338

For customer support on your Square Foot Cost Data, call 877.694.8329.

B1010   Floor Construction

Steel Column Fireproofing

"T" Shaped
Precast Beams

"L" Shaped
Precast Beams

B1010 720						Steel Column Fireproofing		
	ENCASEMENT SYSTEM	COLUMN SIZE (IN.)	THICKNESS (IN.)	FIRE RATING (HRS.)	WEIGHT (P.L.F.)	COST PER V.L.F.		
						MAT.	INST.	TOTAL
3000	Concrete	8	1	1	110	6.60	37	43.60
3300		14	1	1	258	10.90	53.50	64.40
3400			2	3	325	13	60.50	73.50
3450	Gypsum board	8	1/2	2	8	3.61	22.50	26.11
3550	1 layer	14	1/2	2	18	3.91	24	27.91
3600	Gypsum board	8	1	3	14	5	28.50	33.50
3650	1/2" fire rated	10	1	3	17	5.35	30.50	35.85
3700	2 layers	14	1	3	22	5.55	31	36.55
3750	Gypsum board	8	1-1/2	3	23	6.65	36	42.65
3800	1/2" fire rated	10	1-1/2	3	27	7.50	40	47.50
3850	3 layers	14	1-1/2	3	35	8.30	43.50	51.80
3900	Sprayed fiber	8	1-1/2	2	6.3	3.93	7.20	11.13
3950	Direct application		2	3	8.3	5.40	9.95	15.35
4050		10	1-1/2	2	7.9	4.75	8.65	13.40
4200	Sprayed fiber	14	1-1/2	2	10.8	5.90	10.75	16.65

B1020 Roof Construction

The table below lists prices per S.F. for roof rafters and sheathing by nominal size and spacing. Sheathing is 5/16″ CDX for 12″ and 16″ spacing and 3/8″ CDX for 24″ spacing.

Factors for Converting Inclined to Horizontal

Roof Slope	Approx. Angle	Factor	Roof Slope	Approx. Angle	Factor
Flat	0°	1.000	12 in 12	45.0°	1.414
1 in 12	4.8°	1.003	13 in 12	47.3°	1.474
2 in 12	9.5°	1.014	14 in 12	49.4°	1.537
3 in 12	14.0°	1.031	15 in 12	51.3°	1.601
4 in 12	18.4°	1.054	16 in 12	53.1°	1.667
5 in 12	22.6°	1.083	17 in 12	54.8°	1.734
6 in 12	26.6°	1.118	18 in 12	56.3°	1.803
7 in 12	30.3°	1.158	19 in 12	57.7°	1.873
8 in 12	33.7°	1.202	20 in 12	59.0°	1.943
9 in 12	36.9°	1.250	21 in 12	60.3°	2.015
10 in 12	39.8°	1.302	22 in 12	61.4°	2.088
11 in 12	42.5°	1.357	23 in 12	62.4°	2.162

B1020 102	Wood/Flat or Pitched	COST PER S.F.		
		MAT.	INST.	TOTAL
2500	Flat rafter, 2″x4″, 12″ O.C.	1.15	1.78	2.93
2550	16″ O.C.	1.03	1.53	2.56
2600	24″ O.C.	.95	1.32	2.27
2900	2″x6″, 12″ O.C.	1.45	1.78	3.23
2950	16″ O.C.	1.25	1.53	2.78
3000	24″ O.C.	1.09	1.31	2.40
3300	2″x8″, 12″ O.C.	1.75	1.91	3.66
3350	16″ O.C.	1.48	1.63	3.11
3400	24″ O.C.	1.25	1.38	2.63
3700	2″x10″, 12″ O.C.	2.27	2.19	4.46
3750	16″ O.C.	1.86	1.83	3.69
3800	24″ O.C.	1.50	1.51	3.01
4100	2″x12″, 12″ O.C.	2.73	2.21	4.94
4150	16″ O.C.	2.21	1.85	4.06
4200	24″ O.C.	1.74	1.53	3.27
4500	2″x14″, 12″ O.C.	3.08	3.23	6.31
4550	16″ O.C.	2.48	2.62	5.10
4600	24″ O.C.	1.92	2.04	3.96
4900	3″x6″, 12″ O.C.	2.91	1.88	4.79
4950	16″ O.C.	2.35	1.60	3.95
5000	24″ O.C.	1.83	1.36	3.19
5300	3″x8″, 12″ O.C.	3.72	2.10	5.82
5350	16″ O.C.	2.95	1.77	4.72
5400	24″ O.C.	2.23	1.47	3.70
5700	3″x10″, 12″ O.C.	4.49	2.40	6.89
5750	16″ O.C.	3.54	1.99	5.53
5800	24″ O.C.	2.62	1.62	4.24
6100	3″x12″, 12″ O.C.	5.25	2.87	8.12
6150	16″ O.C.	4.11	2.35	6.46
6200	24″ O.C.	3	1.86	4.86
7001	Wood truss, 4 in 12 slope, 24″ O.C., 24′ to 29′ span	3.77	2.66	6.43
7100	30′ to 43′ span	4.06	2.59	6.65
7200	44′ to 60′ span	4.42	2.66	7.08

The table below lists the cost per S.F. for a roof system with steel columns, beams and deck using open web steel joists and 1-1/2″ galvanized metal deck. Perimeter of system is supported on bearing walls.

Fireproofing is not included. Costs/S.F. are based on a building 4 bays long and 4 bays wide.

Column costs are additive. Costs for the bearing walls are not included.

Steel Joists, Beams and Deck
on Bearing Walls

B1020 108		Steel Joists, Beams & Deck on Columns & Walls						
	BAY SIZE (FT.)	SUPERIMPOSED LOAD (P.S.F.)	DEPTH (IN.)	TOTAL LOAD (P.S.F.)	COLUMN ADD	COST PER S.F.		
						MAT.	INST.	TOTAL
3000	25x25	20	18	40		4.29	1.44	5.73
3100					columns	.39	.09	.48
3200		30	22	50		4.60	1.53	6.13
3300					columns	.52	.12	.64
3400		40	20	60		5.10	1.69	6.79
3500					columns	.52	.12	.64
3600	25x30	20	22	40		4.52	1.41	5.93
3700					columns	.44	.10	.54
3800		30	20	50		5.15	1.57	6.72
3900					columns	.44	.10	.54
4000		40	25	60		5.30	1.61	6.91
4100					columns	.52	.12	.64
4200	30x30	20	25	42		4.98	1.51	6.49
4300					columns	.36	.09	.45
4400		30	22	52		5.45	1.64	7.09
4500					columns	.44	.10	.54
4600		40	28	62		5.65	1.69	7.34
4700					columns	.44	.10	.54
4800	30x35	20	22	42		5.15	1.57	6.72
4900					columns	.37	.09	.46
5000		30	28	52		5.45	1.64	7.09
5100					columns	.37	.09	.46
5200		40	25	62		5.90	1.76	7.66
5300					columns	.44	.10	.54
5400	35x35	20	28	42		5.15	1.58	6.73
5500					columns	.32	.08	.40

341

Description: The table below lists the cost per S.F. for a roof system with steel columns, beams, and deck, using open web steel joists and 1-1/2″ galvanized metal deck.

Column costs are additive. Fireproofing is not included.

Steel Joists, Beams and Deck on Columns

B1020 112	Steel Joists, Beams, & Deck on Columns							
	BAY SIZE (FT.)	SUPERIMPOSED LOAD (P.S.F.)	DEPTH (IN.)	TOTAL LOAD (P.S.F.)	COLUMN ADD	COST PER S.F.		
						MAT.	INST.	TOTAL
1100	15x20	20	16	40		4.24	1.44	5.68
1200					columns	2.26	.53	2.79
1500		40	18	60		4.73	1.59	6.32
1600					columns	2.26	.53	2.79
2300	20x25	20	18	40		4.82	1.60	6.42
2400					columns	1.35	.31	1.66
2700		40	20	60		5.40	1.78	7.18
2800					columns	1.81	.42	2.23
2900	25x25	20	18	40		5.65	1.81	7.46
3000					columns	1.08	.25	1.33
3100		30	22	50		5.85	1.85	7.70
3200					columns	1.45	.34	1.79
3300		40	20	60		6.55	2.05	8.60
3400					columns	1.45	.34	1.79
3500	25x30	20	22	40		5.55	1.64	7.19
3600					columns	1.21	.28	1.49
3900		40	25	60		6.70	1.94	8.64
4000					columns	1.45	.34	1.79
4100	30x30	20	25	42		6.30	1.82	8.12
4200					columns	1	.24	1.24
4300		30	22	52		6.90	1.99	8.89
4400					columns	1.21	.28	1.49
4500		40	28	62		7.25	2.08	9.33
4600					columns	1.21	.28	1.49
5300	35x35	20	28	42		6.85	1.97	8.82
5400					columns	.88	.20	1.08
5500		30	25	52		7.65	2.18	9.83
5600					columns	1.03	.24	1.27
5700		40	28	62		8.30	2.33	10.63
5800					columns	1.14	.26	1.40

Description: The table below lists the cost per S.F. for a roof system using open web steel joists and 1-1/2″ galvanized metal deck. The system is assumed supported on bearing walls or other suitable support. Costs for the supports are not included.

B1020 116	Steel Joists & Deck on Bearing Walls							
	BAY SIZE (FT.)	SUPERIMPOSED LOAD (P.S.F.)	DEPTH (IN.)	TOTAL LOAD (P.S.F.)		COST PER S.F.		
						MAT.	INST.	TOTAL
1100	20	20	13-1/2	40		2.71	1.05	3.76
1200		30	15-1/2	50		2.77	1.07	3.84
1300		40	15-1/2	60		2.97	1.15	4.12
1400	25	20	17-1/2	40		2.97	1.16	4.13
1500		30	17-1/2	50		3.20	1.25	4.45
1600		40	19-1/2	60		3.25	1.27	4.52
1700	30	20	19-1/2	40		3.24	1.27	4.51
1800		30	21-1/2	50		3.31	1.29	4.60
1900		40	23-1/2	60		3.56	1.39	4.95
2000	35	20	23-1/2	40		3.52	1.19	4.71
2100		30	25-1/2	50		3.62	1.22	4.84
2200		40	25-1/2	60		3.84	1.29	5.13
2300	40	20	25-1/2	41		3.90	1.31	5.21
2400		30	25-1/2	51		4.14	1.38	5.52
2500		40	25-1/2	61		4.28	1.42	5.70
2600	45	20	27-1/2	41		4.69	1.55	6.24
2700		30	31-1/2	51		4.96	1.63	6.59
2800		40	31-1/2	61		5.25	1.71	6.96
2900	50	20	29-1/2	42		5.25	1.72	6.97
3000		30	31-1/2	52		5.75	1.86	7.61
3100		40	31-1/2	62		6.10	1.98	8.08
3200	60	20	37-1/2	42		6.20	2.21	8.41
3300		30	37-1/2	52		6.90	2.44	9.34
3400		40	37-1/2	62		6.90	2.44	9.34
3500	70	20	41-1/2	42		6.90	2.44	9.34
3600		30	41-1/2	52		7.35	2.60	9.95
3700		40	41-1/2	64		8.90	3.14	12.04
3800	80	20	45-1/2	44		8.65	3.05	11.70
3900		30	45-1/2	54		8.65	3.05	11.70
4000		40	45-1/2	64		9.60	3.37	12.97
4400	100	20	57-1/2	44		8.55	2.90	11.45
4500		30	57-1/2	54		10.20	3.44	13.64
4600		40	57-1/2	65		11.25	3.79	15.04
4700	125	20	69-1/2	44		10.15	3.43	13.58
4800		30	69-1/2	56		11.80	3.98	15.78
4900		40	69-1/2	67		13.45	4.52	17.97

343

Description: The table below lists costs for a roof system supported on exterior bearing walls and interior columns. Costs include bracing, joist girders, open web steel joists and 1-1/2″ galvanized metal deck.

Column costs are additive. Fireproofing is not included.

Costs/S.F. are based on a building 4 bays long and 4 bays wide. Costs for bearing walls are not included.

B1020 120		Steel Joists, Joist Girders & Deck on Columns & Walls						
	BAY SIZE (FT.) GIRD X JOISTS	SUPERIMPOSED LOAD (P.S.F.)	DEPTH (IN.)	TOTAL LOAD (P.S.F.)	COLUMN ADD	COST PER S.F.		
						MAT.	INST.	TOTAL
2350	30x35	20	32-1/2	40		3.96	1.47	5.43
2400					columns	.37	.09	.46
2550		40	36-1/2	60		4.44	1.62	6.06
2600					columns	.43	.10	.53
3000	35x35	20	36-1/2	40		4.32	1.82	6.14
3050					columns	.32	.08	.40
3200		40	36-1/2	60		4.84	1.99	6.83
3250					columns	.41	.10	.51
3300	35x40	20	36-1/2	40		4.38	1.66	6.04
3350					columns	.33	.08	.41
3500		40	36-1/2	60		4.93	1.83	6.76
3550					columns	.36	.09	.45
3900	40x40	20	40-1/2	41		5.10	1.85	6.95
3950					columns	.32	.08	.40
4100		40	40-1/2	61		5.70	2.02	7.72
4150					columns	.32	.08	.40
5100	45x50	20	52-1/2	41		5.75	2.09	7.84
5150					columns	.23	.05	.28
5300		40	52-1/2	61		6.95	2.48	9.43
5350					columns	.33	.08	.41
5400	50x45	20	56-1/2	41		5.55	2.04	7.59
5450					columns	.23	.05	.28
5600		40	56-1/2	61		7.15	2.54	9.69
5650					columns	.31	.07	.38
5700	50x50	20	56-1/2	42		6.20	2.28	8.48
5750					columns	.23	.05	.28
5900		40	59	64		7.15	2.78	9.93
5950					columns	.32	.08	.40
6300	60x50	20	62-1/2	43		6.25	2.31	8.56
6350					columns	.36	.09	.45
6500		40	71	65		7.30	2.87	10.17
6550					columns	.48	.11	.59

Description: The table below lists the cost per S.F. for a roof system supported on columns. Costs include joist girders, open web steel joists and 1-1/2" galvanized metal deck. Column costs are additive.

Fireproofing is not included. Costs/S.F. are based on a building 4 bays long and 4 bays wide.

Costs for wind columns are not included.

B1020 124	Steel Joists & Joist Girders on Columns							
	BAY SIZE (FT.) GIRD X JOISTS	SUPERIMPOSED LOAD (P.S.F.)	DEPTH (IN.)	TOTAL LOAD (P.S.F.)	COLUMN ADD	COST PER S.F.		
						MAT.	INST.	TOTAL
2000	30x30	20	17-1/2	40		3.97	1.77	5.74
2050					columns	1.21	.29	1.50
2100		30	17-1/2	50		4.30	1.90	6.20
2150					columns	1.21	.29	1.50
2200		40	21-1/2	60		4.42	1.95	6.37
2250					columns	1.21	.29	1.50
2300	30x35	20	32-1/2	40		4.25	1.64	5.89
2350					columns	1.03	.24	1.27
2500		40	36-1/2	60		4.78	1.81	6.59
2550					columns	1.21	.29	1.50
3200	35x40	20	36-1/2	40		5.50	2.21	7.71
3250					columns	.90	.21	1.11
3400		40	36-1/2	60		6.20	2.44	8.64
3450					columns	1	.24	1.24
3800	40x40	20	40-1/2	41		5.60	2.22	7.82
3850					columns	.88	.20	1.08
4000		40	40-1/2	61		6.30	2.45	8.75
4050					columns	.88	.20	1.08
4100	40x45	20	40-1/2	41		6	2.52	8.52
4150					columns	.78	.19	.97
4200		30	40-1/2	51		6.55	2.68	9.23
4250					columns	.78	.19	.97
4300		40	40-1/2	61		7.25	2.91	10.16
4350					columns	.88	.21	1.09
5000	45x50	20	52-1/2	41		6.35	2.55	8.90
5050					columns	.62	.15	.77
5200		40	52-1/2	61		7.75	2.99	10.74
5250					columns	.80	.19	.99
5300	50x45	20	56-1/2	41		6.15	2.50	8.65
5350					columns	.62	.15	.77
5500		40	56-1/2	61		7.95	3.06	11.01
5550					columns	.80	.19	.99
5600	50x50	20	56-1/2	42		6.90	2.73	9.63
5650					columns	.63	.15	.78
5800		40	59	64		6.15	2.62	8.77
5850					columns	.89	.21	1.10

B2010 Exterior Walls

Cast in Place Concrete Wall

Flat Precast Concrete Wall

Table B2010 101 describes a concrete wall system for exterior closure. There are several types of wall finishes priced from plain finish to a finish with 3/4" rustication strip.

In Table B2010 103, precast concrete wall panels are either solid or insulated. Prices below are based on delivery within 50 miles of a plant. Small jobs can double the prices below. For large, highly repetitive jobs, deduct up to 15%.

B2010 101	Cast In Place Concrete	COST PER S.F.		
		MAT.	INST.	TOTAL
2100	Conc wall reinforced, 8' high, 6" thick, plain finish, 3000 PSI	4.79	16.55	21.34
2700	Aged wood liner, 3000 PSI	5.95	18.85	24.80
3000	Sand blast light 1 side, 3000 PSI	5.30	18.85	24.15
3700	3/4" bevel rustication strip, 3000 PSI	4.91	17.55	22.46
4000	8" thick, plain finish, 3000 PSI	5.65	17	22.65
4100	4000 PSI	5.80	17	22.80
4300	Rub concrete 1 side, 3000 PSI	5.65	20	25.65
4550	8" thick, aged wood liner, 3000 PSI	6.80	19.25	26.05
4750	Sand blast light 1 side, 3000 PSI	6.15	19.25	25.40
5300	3/4" bevel rustication strip, 3000 PSI	5.75	18	23.75
5600	10" thick, plain finish, 3000 PSI	6.45	17.40	23.85
5900	Rub concrete 1 side, 3000 PSI	6.45	20.50	26.95
6200	Aged wood liner, 3000 PSI	7.65	19.65	27.30
6500	Sand blast light 1 side, 3000 PSI	6.95	19.70	26.65
7100	3/4" bevel rustication strip, 3000 PSI	6.55	18.40	24.95
7400	12" thick, plain finish, 3000 PSI	7.40	17.85	25.25
7700	Rub concrete 1 side, 3000 PSI	7.40	21	28.40
8000	Aged wood liner, 3000 PSI	8.60	20	28.60
8300	Sand blast light 1 side, 3000 PSI	7.90	20	27.90
8900	3/4" bevel rustication strip, 3000 PSI	7.55	18.85	26.40

B2010 102		Flat Precast Concrete				COST PER S.F.		
	THICKNESS (IN.)	PANEL SIZE (FT.)	FINISHES	RIGID INSULATION (IN)	TYPE	MAT.	INST.	TOTAL
3000	4	5x18	smooth gray	none	low rise	12.40	7.95	20.35
3050		6x18				10.35	6.65	17
3100		8x20				20.50	3.21	23.71
3150		12x20				19.40	3.03	22.43
3200	6	5x18	smooth gray	2	low rise	13.40	8.55	21.95
3250		6x18				11.40	7.20	18.60
3300		8x20				22	3.88	25.88
3350		12x20				19.80	3.56	23.36
3400	8	5x18	smooth gray	2	low rise	28.50	4.90	33.40
3450		6x18				27	4.67	31.67
3500		8x20				24.50	4.30	28.80
3550		12x20				22.50	3.95	26.45

346

For customer support on your Square Foot Cost Data, call 877.694.8329.

B20 Exterior Enclosure

B2010 Exterior Walls

B2010 102	Flat Precast Concrete

	THICKNESS (IN.)	PANEL SIZE (FT.)	FINISHES	RIGID INSULATION (IN)	TYPE	COST PER S.F. MAT.	COST PER S.F. INST.	COST PER S.F. TOTAL
3600	4	4x8	white face	none	low rise	47.50	4.59	52.09
3650		8x8				36	5.15	41.15
3700		10x10				31.50	3.02	34.52
3750		20x10				28.50	2.74	31.24
3800	5	4x8	white face	none	low rise	48.50	4.68	53.18
3850		8x8				36.50	3.54	40.04
3900		10x10				32.50	3.13	35.63
3950		20x20				29.50	2.87	32.37
4000	6	4x8	white face	none	low rise	50.50	4.86	55.36
4050		8x8				38	3.70	41.70
4100		10x10				33.50	3.25	36.75
4150		20x10				31	2.99	33.99
4200	6	4x8	white face	2	low rise	51.50	5.50	57
4250		8x8				39.50	4.35	43.85
4300		10x10				35	3.90	38.90
4350		20x10				31	2.99	33.99
4400	7	4x8	white face	none	low rise	51.50	4.97	56.47
4450		8x8				39.50	3.82	43.32
4500		10x10				35.50	3.42	38.92
4550		20x10				32.50	3.13	35.63
4600	7	4x8	white face	2	low rise	52.50	5.60	58.10
4650		8x8				40.50	4.47	44.97
4700		10x10				36.50	4.07	40.57
4750		20x10				33.50	3.78	37.28
4800	8	4x8	white face	none	low rise	52.50	5.10	57.60
4850		8x8				40.50	3.92	44.42
4900		10x10				36.50	3.53	40.03
4950		20x10				33.50	3.25	36.75
5000	8	4x8	white face	2	low rise	54	5.75	59.75
5050		8x8				42	4.57	46.57
5100		10x10				37.50	4.18	41.68
5150		20x10				35	3.90	38.90

B2010 103	Fluted Window or Mullion Precast Concrete

	THICKNESS (IN.)	PANEL SIZE (FT.)	FINISHES	RIGID INSULATION (IN)	TYPE	COST PER S.F. MAT.	COST PER S.F. INST.	COST PER S.F. TOTAL
5200	4	4x8	smooth gray	none	high rise	28.50	18.20	46.70
5250		8x8				20.50	13	33.50
5300		10x10				39	6.10	45.10
5350		20x10				34.50	5.35	39.85
5400	5	4x8	smooth gray	none	high rise	29	18.55	47.55
5450		8x8				21	13.45	34.45
5500		10x10				41	6.40	47.40
5550		20x10				36	5.65	41.65
5600	6	4x8	smooth gray	none	high rise	29.50	19	48.50
5650		8x8				21.50	13.95	35.45
5700		10x10				42	6.55	48.55
5750		20x10				37.50	5.85	43.35
5800	6	4x8	smooth gray	2	high rise	31	19.65	50.65
5850		8x8				23	14.60	37.60
5900		10x10				43	7.20	50.20
5950		20x10				38.50	6.50	45

B2010 Exterior Walls

B2010 103	Fluted Window or Mullion Precast Concrete

	THICKNESS (IN.)	PANEL SIZE (FT.)	FINISHES	RIGID INSULATION (IN)	TYPE	COST PER S.F.		
						MAT.	INST.	TOTAL
6000	7	4x8	smooth gray	none	high rise	30	19.40	49.40
6050		8x8				22.50	14.35	36.85
6100		10x10				44	6.85	50.85
6150		20x10				38.50	6	44.50
6200	7	4x8	smooth gray	2	high rise	31.50	20	51.50
6250		8x8				23.50	15	38.50
6300		10x10				45	7.50	52.50
6350		20x10				39.50	6.70	46.20
6400	8	4x8	smooth gray	none	high rise	30.50	19.70	50.20
6450		8x8				23	14.65	37.65
6500		10x10				45	7.10	52.10
6550		20x10				40.50	6.30	46.80
6600	8	4x8	smooth gray	2	high rise	32	20.50	52.50
6650		8x8				24	15.30	39.30
6700		10x10				46.50	7.75	54.25
6750		20x10				41.50	6.95	48.45

B2010 104	Ribbed Precast Concrete

	THICKNESS (IN.)	PANEL SIZE (FT.)	FINISHES	RIGID INSULATION(IN.)	TYPE	COST PER S.F.		
						MAT.	INST.	TOTAL
6800	4	4x8	aggregate	none	high rise	33.50	18.20	51.70
6850		8x8				24.50	13.30	37.80
6900		10x10				41.50	4.80	46.30
6950		20x10				37	4.25	41.25
7000	5	4x8	aggregate	none	high rise	34	18.55	52.55
7050		8x8				25	13.65	38.65
7100		10x10				43	4.99	47.99
7150		20x10				38.50	4.45	42.95
7200	6	4x8	aggregate	none	high rise	34.50	18.85	53.35
7250		8x8				26	14.05	40.05
7300		10x10				44.50	5.15	49.65
7350		20x10				40	4.62	44.62
7400	6	4x8	aggregate	2	high rise	36	19.50	55.50
7450		8x8				27	14.70	41.70
7500		10x10				46	5.80	51.80
7550		20x10				41	5.25	46.25
7600	7	4x8	aggregate	none	high rise	35.50	19.30	54.80
7650		8x8				26.50	14.40	40.90
7700		10x10				46	5.30	51.30
7750		20x10				41	4.75	45.75
7800	7	4x8	aggregate	2	high rise	36.50	19.95	56.45
7850		8x8				27.50	15.05	42.55
7900		10x10				47	5.95	52.95
7950		20x10				42.50	5.40	47.90
8000	8	4x8	aggregate	none	high rise	36	19.65	55.65
8050		8x8				27	14.85	41.85
8100		10x10				47.50	5.50	53
8150		20x10				43	4.96	47.96
8200	8	4x8	aggregate	2	high rise	37	20.50	57.50
8250		8x8				28.50	15.50	44
8300		10x10				49	7.90	56.90
8350		20x10				44	5.60	49.60

B2010 105		Precast Concrete Specialties							
	TYPE	SIZE					COST PER L.F.		
							MAT.	INST.	TOTAL
8400	Coping, precast	6" wide					18.80	13.70	32.50
8450	Stock units	10" wide					20.50	14.70	35.20
8460		12" wide					22.50	15.80	38.30
8480		14" wide					24.50	17.15	41.65
8500	Window sills	6" wide					13.10	14.70	27.80
8550	Precast	10" wide					17.35	17.15	34.50
8600		14" wide					17.05	20.50	37.55
8610									

For customer support on your Square Foot Cost Data, call 877.694.8329.

349

B2010 Exterior Walls

The advantage of tilt up construction is in the low cost of forms and placing of concrete and reinforcing. Tilt up has been used for several types of buildings, including warehouses, stores, offices, and schools. The panels are cast in forms on the ground, or floor slab. Most jobs use 5-1/2" thick solid reinforced concrete panels.

Design Assumptions:
Conc. f'c = 3000 psi
Reinf. fy = 60,000

B2010 106	Tilt-Up Concrete Panel	COST PER S.F.		
		MAT.	INST.	TOTAL
3200	Tilt-up conc panels, broom finish, 5-1/2" thick, 3000 PSI	4.25	6.30	10.55
3250	5000 PSI	4.35	6.20	10.55
3300	6" thick, 3000 PSI	4.64	6.50	11.14
3350	5000 PSI	4.78	6.35	11.13
3400	7-1/2" thick, 3000 PSI	5.95	6.70	12.65
3450	5000 PSI	6.10	6.60	12.70
3500	8" thick, 3000 PSI	6.35	6.85	13.20
3550	5000 PSI	6.60	6.75	13.35
3700	Steel trowel finish, 5-1/2" thick, 3000 PSI	4.23	6.40	10.63
3750	5000 PSI	4.35	6.25	10.60
3800	6" thick, 3000 PSI	4.64	6.55	11.19
3850	5000 PSI	4.78	6.45	11.23
3900	7-1/2" thick, 3000 PSI	5.95	6.80	12.75
3950	5000 PSI	6.10	6.65	12.75
4000	8" thick, 3000 PSI	6.35	7	13.35
4050	5000 PSI	6.60	6.85	13.45
4200	Exp. aggregate finish, 5-1/2" thick, 3000 PSI	4.85	6.45	11.30
4250	5000 PSI	4.96	6.35	11.31
4300	6" thick, 3000 PSI	5.25	6.60	11.85
4350	5000 PSI	5.40	6.50	11.90
4400	7-1/2" thick, 3000 PSI	6.55	6.85	13.40
4450	5000 PSI	6.75	6.75	13.50
4500	8" thick, 3000 PSI	7	7	14
4550	5000 PSI	7.20	6.90	14.10
4600	Exposed aggregate & vert. rustication 5-1/2" thick, 3000 PSI	7.10	7.95	15.05
4650	5000 PSI	7.20	7.85	15.05
4700	6" thick, 3000 PSI	7.50	8.15	15.65
4750	5000 PSI	7.65	8	15.65
4800	7-1/2" thick, 3000 PSI	8.80	8.35	17.15
4850	5000 PSI	9	8.25	17.25
4900	8" thick, 3000 PSI	9.20	8.50	17.70
4950	5000 PSI	9.45	8.40	17.85
5000	Vertical rib & light sandblast, 5-1/2" thick, 3000 PSI	6.55	10.45	17
5100	6" thick, 3000 PSI	6.95	10.60	17.55
5200	7-1/2" thick, 3000 PSI	8.25	10.85	19.10
5300	8" thick, 3000 PSI	8.70	11	19.70
6000	Broom finish w/2" polystyrene insulation, 6" thick, 3000 PSI	4.03	7.80	11.83
6100	Broom finish 2" fiberplank insulation 6" thick, 3000 PSI	4.61	7.70	12.31
6200	Exposed aggregate w/2" polystyrene insulation, 6" thick, 3000 PSI	4.54	7.75	12.29
6300	Exposed aggregate 2" fiberplank insulation, 6" thick, 3000 PSI	5.10	7.70	12.80

350

For customer support on your Square Foot Cost Data, call 877.694.8329.

B2010 Exterior Walls

Exterior concrete block walls are defined in the following terms; structural reinforcement, weight, percent solid, size, strength and insulation. Within each of these categories, two to four variations are shown. No costs are included for brick shelf or relieving angles.

B2010 109 — Concrete Block Wall - Regular Weight

	TYPE	SIZE (IN.)	STRENGTH (P.S.I.)	CORE FILL		COST PER S.F. MAT.	INST.	TOTAL
1200	Hollow	4x8x16	2,000	none		2.13	6.30	8.43
1250			4,500	none		2.55	6.30	8.85
1400		8x8x16	2,000	perlite		4.75	7.70	12.45
1410				styrofoam		4.76	7.25	12.01
1440				none		3.27	7.25	10.52
1450			4,500	perlite		5.25	7.70	12.95
1460				styrofoam		5.25	7.25	12.50
1490				none		3.78	7.25	11.03
2000	75% solid	4x8x16	2,000	none		2.56	6.35	8.91
2100		6x8x16	2,000	perlite		3.75	7.05	10.80
2500	Solid	4x8x16	2,000	none		2.41	6.50	8.91
2700		8x8x16	2,000	none		4.04	7.55	11.59

B2010 110 — Concrete Block Wall - Lightweight

	TYPE	SIZE (IN.)	WEIGHT (P.C.F.)	CORE FILL		COST PER S.F. MAT.	INST.	TOTAL
3100	Hollow	8x4x16	105	perlite		4.29	7.25	11.54
3110				styrofoam		4.30	6.80	11.10
3200		4x8x16	105	none		2.24	6.15	8.39
3250			85	none		2.86	6	8.86
3300		6x8x16	105	perlite		4.06	6.95	11.01
3310				styrofoam		4.38	6.60	10.98
3340				none		3.06	6.60	9.66
3400		8x8x16	105	perlite		5.15	7.50	12.65
3410				styrofoam		5.20	7.05	12.25
3440				none		3.69	7.05	10.74
3450			85	perlite		5.90	7.35	13.25
4000	75% solid	4x8x16	105	none		3.22	6.20	9.42
4050			85	none		3.52	6.10	9.62
4100		6x8x16	105	perlite		4.97	6.85	11.82
4500	Solid	4x8x16	105	none		2.05	6.45	8.50
4700		8x8x16	105	none		4.43	7.45	11.88

For customer support on your Square Foot Cost Data, call 877.694.8329.

351

B2010 Exterior Walls

B2010 111	Reinforced Concrete Block Wall - Regular Weight

	TYPE	SIZE (IN.)	STRENGTH (P.S.I.)	VERT. REINF & GROUT SPACING		COST PER S.F.		
						MAT.	INST.	TOTAL
5200	Hollow	4x8x16	2,000	#4 @ 48"		2.29	7	9.29
5300		6x8x16	2,000	#4 @ 48"		2.95	7.50	10.45
5330				#5 @ 32"		3.16	7.75	10.91
5340				#5 @ 16"		3.61	8.75	12.36
5350			4,500	#4 @ 28"		3.37	7.50	10.87
5390				#5 @ 16"		4.03	8.75	12.78
5400		8x8x16	2,000	#4 @ 48"		3.65	8	11.65
5430				#5 @ 32"		3.81	8.45	12.26
5440				#5 @ 16"		4.35	9.60	13.95
5450		8x8x16	4,500	#4 @ 48"		4.09	8.10	12.19
5490				#5 @ 16"		4.86	9.60	14.46
5500		12x8x16	2,000	#4 @ 48"		5.05	10.25	15.30
5540				#5 @ 16"		6.10	11.85	17.95
6100	75% solid	6x8x16	2,000	#4 @ 48"		3.39	7.45	10.84
6140				#5 @ 16"		3.82	8.50	12.32
6150			4,500	#4 @ 48"		3.95	7.45	11.40
6190				#5 @ 16"		4.38	8.50	12.88
6200		8x8x16	2,000	#4 @ 48"		4.12	8.05	12.17
6230				#5 @ 32"		4.27	8.30	12.57
6240				#5 @ 16"		4.58	9.30	13.88
6250			4,500	#4 @ 48"		4.79	8.05	12.84
6280				#5 @ 32"		4.94	8.30	13.24
6290				#5 @ 16"		5.25	9.30	14.55
6500	Solid-double	2-4x8x16	2,000	#4 @ 48" E.W.		5.60	14.75	20.35
6530	Wythe			#5 @ 16" E.W.		6.30	15.65	21.95
6550			4,500	#4 @ 48" E.W.		7.80	14.65	22.45
6580				#5 @ 16" E.W.		8.50	15.55	24.05

B2010 112	Reinforced Concrete Block Wall - Lightweight

	TYPE	SIZE (IN.)	WEIGHT (P.C.F.)	VERT REINF. & GROUT SPACING		COST PER S.F.		
						MAT.	INST.	TOTAL
7100	Hollow	8x4x16	105	#4 @ 48"		3.12	7.65	10.77
7140				#5 @ 16"		3.89	9.15	13.04
7150			85	#4 @ 48"		4.65	7.50	12.15
7190				#5 @ 16"		5.40	9	14.40
7400		8x8x16	105	#4 @ 48"		4	7.90	11.90
7440				#5 @ 16"		4.77	9.40	14.17
7450		8x8x16	85	#4 @ 48"		4.73	7.75	12.48
7490				#5 @ 16"		5.50	9.25	14.75
7800		8x8x24	105	#4 @ 48"		3.07	8.55	11.62
7840				#5 @ 16"		3.84	10.05	13.89
7850			85	#4 @ 48"		6.95	7.25	14.20
7890				#5 @ 16"		7.75	8.75	16.50
8100	75% solid	6x8x16	105	#4 @ 48"		4.61	7.25	11.86
8130				#5 @ 32"		4.76	7.50	12.26
8150			85	#4 @ 48"		4.63	7.10	11.73
8180				#5 @ 32"		4.78	7.35	12.13
8200		8x8x16	105	#4 @ 48"		5.55	7.85	13.40
8230				#5 @ 32"		5.70	8.10	13.80
8250			85	#4 @ 48"		5.60	7.65	13.25
8280				#5 @ 32"		5.75	7.90	13.65

B2010 Exterior Walls

| B2010 112 | | Reinforced Concrete Block Wall - Lightweight | | | | | | |

	TYPE	SIZE (IN.)	WEIGHT (P.C.F.)	VERT REINF. & GROUT SPACING		COST PER S.F.		
						MAT.	INST.	TOTAL
8500	Solid-double	2-4x8x16	105	#4 @ 48"		4.89	14.65	19.54
8530	Wythe		105	#5 @ 16"		5.60	15.55	21.15
8600		2-6x8x16	105	#4 @ 48"		7.80	15.60	23.40
8630				#5 @ 16"		8.50	16.50	25
8650			85	#4 @ 48"		11.80	15	26.80

353

For customer support on your Square Foot Cost Data, call 877.694.8329.

B2010 Exterior Walls

Exterior split ribbed block walls are defined in the following terms; structural reinforcement, weight, percent solid, size, number of ribs and insulation. Within each of these categories two to four variations are shown. No costs are included for brick shelf or relieving angles. Costs include control joints every 20′ and horizontal reinforcing.

B2010 113		Split Ribbed Block Wall - Regular Weight				COST PER S.F.		
	TYPE	SIZE (IN.)	RIBS	CORE FILL		MAT.	INST.	TOTAL
1220	Hollow	4x8x16	4	none		4.27	7.80	12.07
1250			8	none		4.64	7.80	12.44
1280			16	none		5	7.90	12.90
1430		8x8x16	8	perlite		7.45	9.30	16.75
1440				styrofoam		7.25	8.85	16.10
1450				none		5.95	8.85	14.80
1530		12x8x16	8	perlite		9.35	12.35	21.70
1540				styrofoam		8.65	11.50	20.15
1550				none		6.95	11.50	18.45
2120	75% solid	4x8x16	4	none		5.35	7.90	13.25
2150			8	none		5.80	7.90	13.70
2180			16	none		6.30	8	14.30
2520	Solid	4x8x16	4	none		6.10	8	14.10
2550			8	none		7.15	8	15.15
2580			16	none		7.15	8.15	15.30

B2010 115		Reinforced Split Ribbed Block Wall - Regular Weight				COST PER S.F.		
	TYPE	SIZE (IN.)	RIBS	VERT. REINF. & GROUT SPACING		MAT.	INST.	TOTAL
5200	Hollow	4x8x16	4	#4 @ 48″		4.43	8.50	12.93
5230			8	#4 @ 48″		4.80	8.50	13.30
5260			16	#4 @ 48″		5.15	8.60	13.75
5430		8x8x16	8	#4 @ 48″		6.25	9.70	15.95
5440				#5 @ 32″		6.50	10.05	16.55
5450				#5 @ 16″		7.05	11.20	18.25
5530		12x8x16	8	#4 @ 48″		7.35	12.35	19.70
5540				#5 @ 32″		7.65	12.75	20.40
5550				#5 @ 16″		8.40	13.95	22.35
6230	75% solid	6x8x16	8	#4 @ 48″		6.80	9	15.80
6240				#5 @ 32″		6.95	9.25	16.20
6250				#5 @ 16″		7.20	10.05	17.25
6330		8x8x16	8	#4 @ 48″		7.60	9.70	17.30
6340				#5 @ 32″		7.75	9.95	17.70
6350				#5 @ 16″		8.05	10.95	19

Exterior split face block walls are defined in the following terms; structural reinforcement, weight, percent solid, size, scores and insulation. Within each of these categories two to four variations are shown. No costs are included for brick shelf or relieving angles. Costs include control joints every 20' and horizontal reinforcing.

B2010 117			Split Face Block Wall - Regular Weight					
	TYPE	SIZE (IN.)	SCORES	CORE FILL		COST PER S.F.		
						MAT.	INST.	TOTAL
1200	Hollow	8x4x16	0	perlite		7.15	10.25	17.40
1210				styrofoam		7.15	9.80	16.95
1240				none		5.65	9.80	15.45
1250			1	perlite		7.60	10.25	17.85
1260				styrofoam		7.60	9.80	17.40
1290				none		6.10	9.80	15.90
1300		12x4x16	0	perlite		9.40	11.65	21.05
1310				styrofoam		8.70	10.80	19.50
1340				none		7	10.80	17.80
1350			1	perlite		9.85	11.65	21.50
1360				styrofoam		9.15	10.80	19.95
1390				none		7.45	10.80	18.25
1400		4x8x16	0	none		3.91	7.70	11.61
1450			1	none		4.24	7.80	12.04
1500		6x8x16	0	perlite		5.45	8.85	14.30
1510				styrofoam		5.75	8.50	14.25
1540				none		4.45	8.50	12.95
1550			1	perlite		5.80	9	14.80
1560				styrofoam		6.15	8.65	14.80
1590				none		4.81	8.65	13.46
1600		8x8x16	0	perlite		6.45	9.60	16.05
1610				styrofoam		6.50	9.15	15.65
1640				none		4.99	9.15	14.14
1650		8x8x16	1	perlite		6.90	9.75	16.65
1660				styrofoam		6.90	9.30	16.20
1690				none		5.40	9.30	14.70
1700		12x8x16	0	perlite		8.40	12.60	21
1710				styrofoam		7.70	11.75	19.45
1740				none		6	11.75	17.75
1750			1	perlite		8.90	12.80	21.70
1760				styrofoam		8.20	11.95	20.15
1790				none		6.50	11.95	18.45

355

B2010 Exterior Walls

B2010 117	Split Face Block Wall - Regular Weight

	TYPE	SIZE (IN.)	SCORES	CORE FILL		COST PER S.F. MAT.	INST.	TOTAL
1800	75% solid	8x4x16	0	perlite		7.65	10.15	17.80
1840				none		6.90	9.95	16.85
1850			1	perlite		8.15	10.15	18.30
1890				none		7.40	9.95	17.35
2000		4x8x16	0	none		4.89	7.80	12.69
2050			1	none		5.30	7.90	13.20
2400	Solid	8x4x16	0	none		7.70	10.15	17.85
2450			1	none		8.30	10.15	18.45
2900		12x8x16	0	none		8.40	12.15	20.55
2950			1	none		9.10	12.40	21.50

B2010 119	Reinforced Split Face Block Wall - Regular Weight

	TYPE	SIZE (IN.)	SCORES	VERT. REINF. & GROUT SPACING		COST PER S.F. MAT.	INST.	TOTAL
5200	Hollow	8x4x16	0	#4 @ 48"		5.95	10.65	16.60
5210				#5 @ 32"		6.20	11	17.20
5240				#5 @ 16"		6.75	12.15	18.90
5250			1	#4 @ 48"		6.40	10.65	17.05
5260				#5 @ 32"		6.65	11	17.65
5290				#5 @ 16"		7.20	12.15	19.35
5700		12x8x16	0	#4 @ 48"		6.40	12.60	19
5710				#5 @ 32"		6.70	13	19.70
5740				#5 @ 16"		7.45	14.20	21.65
5750			1	#4 @ 48"		6.90	12.80	19.70
5760				#5 @ 32"		7.20	13.20	20.40
5790				#5 @ 16"		7.95	14.40	22.35
6000	75% solid	8x4x16	0	#4 @ 48"		7.05	10.65	17.70
6010				#5 @ 32"		7.20	10.90	18.10
6040				#5 @ 16"		7.50	11.90	19.40
6050			1	#4 @ 48"		7.55	10.65	18.20
6060				#5 @ 32"		7.70	10.90	18.60
6090				#5 @ 16"		8	11.90	19.90
6100		12x4x16	0	#4 @ 48"		8.65	11.70	20.35
6110				#5 @ 32"		8.85	12	20.85
6140				#5 @ 16"		9.25	12.75	22
6150			1	#4 @ 48"		9.30	11.70	21
6160				#5 @ 32"		9.50	12	21.50
6190				#5 @ 16"		9.95	13	22.95
6700	Solid-double Wythe	2-4x8x16	0	#4 @ 48" E.W.		11.90	17.55	29.45
6710				#5 @ 32" E.W.		12.15	17.70	29.85
6740				#5 @ 16" E.W.		12.60	18.45	31.05
6750			1	#4 @ 48" E.W.		12.90	17.75	30.65
6760				#5 @ 32" E.W.		13.15	17.90	31.05
6790				#5 @ 16" E.W.		13.60	18.65	32.25
6800		2-6x8x16	0	#4 @ 48" E.W.		13.40	19.40	32.80
6810				#5 @ 32" E.W.		13.65	19.55	33.20
6840				#5 @ 16" E.W.		14.10	20	34.10
6850			1	#4 @ 48" E.W.		14.50	19.70	34.20
6860				#5 @ 32" E.W.		14.75	19.85	34.60
6890				#5 @ 16" E.W.		15.20	20.50	35.70

B2010 Exterior Walls

Exterior solid brick walls are defined in the following terms; structural reinforcement, type of brick, thickness, and bond. Nine different types of face bricks are presented, with single wythes shown in four different bonds. Shelf angles are included in system components. These walls do not include ties and as such are not tied to a backup wall.

B2010 125			Solid Brick Walls - Single Wythe			COST PER S.F.		
	TYPE	THICKNESS (IN.)	BOND			MAT.	INST.	TOTAL
1000	Common	4	running			5.65	14.20	19.85
1010			common			6.85	16.25	23.10
1050			Flemish			7.55	19.70	27.25
1100			English			8.50	21	29.50
1150	Standard	4	running			5.35	14.70	20.05
1160			common			6.30	16.95	23.25
1200			Flemish			7	20	27
1250			English			7.90	21.50	29.40
1300	Glazed	4	running			20.50	15.30	35.80
1310			common			24.50	17.75	42.25
1350			Flemish			27	21.50	48.50
1400			English			30.50	23	53.50
1600	Economy	4	running			8.10	11.25	19.35
1610			common			9.55	12.90	22.45
1650			Flemish			10.55	15.30	25.85
1700			English			22	16.25	38.25
1900	Fire	4-1/2	running			13.10	12.90	26
1910			common			15.60	14.70	30.30
1950			Flemish			17.25	17.75	35
2000			English			19.45	18.70	38.15
2050	King	3-1/2	running			4.44	11.55	15.99
2060			common			5.20	13.30	18.50
2100			Flemish			6.25	15.90	22.15
2150			English			6.95	16.60	23.55
2200	Roman	4	running			8.65	13.30	21.95
2210			common			10.30	15.30	25.60
2250			Flemish			11.35	18.20	29.55
2300			English			12.85	19.70	32.55
2350	Norman	4	running			6.75	11	17.75
2360			common			7.95	12.50	20.45
2400			Flemish			18.60	15	33.60
2450			English			9.85	15.90	25.75

B2010 Exterior Walls

B2010 125 — Solid Brick Walls - Single Wythe

	TYPE	THICKNESS (IN.)	BOND		COST PER S.F. MAT.	INST.	TOTAL
2500	Norwegian	4	running		5.90	9.80	15.70
2510			common		7	11.10	18.10
2550			Flemish		7.70	13.30	21
2600			English		8.65	13.95	22.60

B2010 126 — Solid Brick Walls - Double Wythe

	TYPE	THICKNESS (IN.)	COLLAR JOINT THICKNESS (IN.)		COST PER S.F. MAT.	INST.	TOTAL
4100	Common	8	3/4		10.90	23.50	34.40
4150	Standard	8	1/2		10.10	24.50	34.60
4200	Glazed	8	1/2		40	25.50	65.50
4250	Engineer	8	1/2		10.15	21.50	31.65
4300	Economy	8	3/4		15.50	18.65	34.15
4350	Double	8	3/4		14.30	15.10	29.40
4400	Fire	8	3/4		25.50	21.50	47
4450	King	7	3/4		8.25	19.10	27.35
4500	Roman	8	1		16.60	22.50	39.10
4550	Norman	8	3/4		12.80	18.25	31.05
4600	Norwegian	8	3/4		11.25	16.20	27.45
4650	Utility	8	3/4		12	14	26
4700	Triple	8	3/4		9.30	12.60	21.90
4750	SCR	12	3/4		15.30	18.70	34
4800	Norwegian	12	3/4		13.25	16.55	29.80
4850	Jumbo	12	3/4		10.95	14.45	25.40

B2010 127 — Reinforced Brick Walls

	TYPE	THICKNESS (IN.)	WYTHE	REINF. & SPACING	COST PER S.F. MAT.	INST.	TOTAL
7200	Common	4"	1	#4 @ 48"vert	5.85	14.75	20.60
7220				#5 @ 32"vert	6	14.95	20.95
7230				#5 @ 16"vert	6.25	15.60	21.85
7700	King	9"	2	#4 @ 48" E.W.	8.60	20	28.60
7720				#5 @ 32" E.W.	8.90	20	28.90
7730				#5 @ 16" E.W.	9.30	20.50	29.80
7800	Common	10"	2	#4 @ 48" E.W.	11.25	24.50	35.75
7820				#5 @ 32" E.W.	11.55	24.50	36.05
7830				#5 @ 16" E.W.	11.95	25.50	37.45
7900	Standard	10"	2	#4 @ 48" E.W.	10.45	25.50	35.95
7920				#5 @ 32" E.W.	10.75	25.50	36.25
7930				#5 @ 16" E.W.	11.15	26.50	37.65
8100	Economy	10"	2	#4 @ 48" E.W.	15.85	19.60	35.45
8120				#5 @ 32" E.W.	16.15	19.75	35.90
8130				#5 @ 16" E.W.	16.55	20	36.55
8200	Double	10"	2	#4 @ 48" E.W.	14.65	16.05	30.70
8220				#5 @ 32" E.W.	14.95	16.20	31.15
8230				#5 @ 16" E.W.	15.35	16.80	32.15
8300	Fire	10"	2	#4 @ 48" E.W.	26	22.50	48.50
8320				#5 @ 32" E.W.	26	22.50	48.50
8330				#5 @ 16" E.W.	26.50	23.50	50
8400	Roman	10"	2	#4 @ 48" E.W.	16.95	23	39.95
8420				#5 @ 32" E.W.	17.25	23.50	40.75
8430				#5 @ 16" E.W.	17.65	24	41.65

B2010 Exterior Walls

B2010 127	Reinforced Brick Walls

	TYPE	THICKNESS (IN.)	WYTHE	REINF. & SPACING		COST PER S.F.		
						MAT.	INST.	TOTAL
8500	Norman	10″	2	#4 @ 48″ E.W.		13.15	19.20	32.35
8520				#5 @ 32″ E.W.		13.45	19.35	32.80
8530				#5 @ 16″ E.W.		13.85	19.95	33.80
8600	Norwegian	10″	2	#4 @ 48″ E.W.		11.60	17.15	28.75
8620				#5 @ 32″ E.W.		11.90	17.30	29.20
8630				#5 @ 16″ E.W.		12.30	17.90	30.20
8800	Triple	10″	2	#4 @ 48″ E.W.		9.65	13.55	23.20
8820				#5 @ 32″ E.W.		9.95	13.70	23.65
8830				#5 @ 16″ E.W.		10.35	14.30	24.65

359

For customer support on your Square Foot Cost Data, call 877.694.8329.

B2010 Exterior Walls

The table below lists costs per S.F. for stone veneer walls on various backup using different stone.

Stone Veneer

B2010 128	Stone Veneer	COST PER S.F.		
		MAT.	INST.	TOTAL
2000	Ashlar veneer, 4", 2" x 4" stud backup, 16" O.C., 8' high, low priced stone	14.95	22.50	37.45
2100	Metal stud backup, 8' high, 16" O.C.	15.70	22.50	38.20
2150	24" O.C.	15.35	22	37.35
2200	Conc. block backup, 4" thick	16.40	26.50	42.90
2300	6" thick	17	27	44
2350	8" thick	17.55	27.50	45.05
2400	10" thick	17.70	29	46.70
2500	12" thick	18.90	31	49.90
3100	High priced stone, wood stud backup, 10' high, 16" O.C.	22	25.50	47.50
3200	Metal stud backup, 10' high, 16" O.C.	22.50	26	48.50
3250	24" O.C.	22.50	25.50	48
3300	Conc. block backup, 10' high, 4" thick	23.50	30	53.50
3350	6" thick	24	30	54
3400	8" thick	25	32	57
3450	10" thick	24.50	32	56.50
3500	12" thick	26	34	60
4000	Indiana limestone 2" thick, sawn finish, wood stud backup, 10' high, 16" O.	33.50	13.70	47.20
4100	Metal stud backup, 10' high, 16" O.C.	34.50	14	48.50
4150	24" O.C.	34	13.50	47.50
4200	Conc. block backup, 4" thick	35	17.95	52.95
4250	6" thick	35.50	18.20	53.70
4300	8" thick	36.50	18.80	55.30
4350	10" thick	36.50	20	56.50
4400	12" thick	37.50	22.50	60
4450	2" thick, smooth finish, wood stud backup, 8' high, 16" O.C.	33.50	13.70	47.20
4550	Metal stud backup, 8' high, 16" O.C.	34.50	13.80	48.30
4600	24" O.C.	34	13.30	47.30
4650	Conc. block backup, 4" thick	35	17.75	52.75
4700	6" thick	35.50	18	53.50
4750	8" thick	36.50	18.60	55.10
4800	10" thick	36.50	20	56.50
4850	12" thick	37.50	22.50	60
5350	4" thick, smooth finish, wood stud backup, 8' high, 16" O.C.	35	13.70	48.70
5450	Metal stud backup, 8' high, 16" O.C.	36	14	50
5500	24" O.C.	35.50	13.50	49
5550	Conc. block backup, 4" thick	36.50	17.95	54.45
5600	6" thick	37	18.20	55.20
5650	8" thick	38	18.80	56.80

B2010 Exterior Walls

B2010 128	Stone Veneer	COST PER S.F.		
		MAT.	INST.	TOTAL
5700	10" thick	38	20	58
5750	12" thick	39	22.50	61.50
6000	Granite, gray or pink, 2" thick, wood stud backup, 8' high, 16" O.C.	33.50	25	58.50
6100	Metal studs, 8' high, 16" O.C.	34.50	25	59.50
6150	24" O.C.	34	24.50	58.50
6200	Conc. block backup, 4" thick	35	29	64
6250	6" thick	35.50	29.50	65
6300	8" thick	36.50	30	66.50
6350	10" thick	36.50	31.50	68
6400	12" thick	37.50	33.50	71
6900	4" thick, wood stud backup, 8' high, 16" O.C.	44.50	28.50	73
7000	Metal studs, 8' high, 16" O.C.	45.50	29	74.50
7050	24" O.C.	45	28.50	73.50
7100	Conc. block backup, 4" thick	46	33	79
7150	6" thick	46.50	33	79.50
7200	8" thick	47.50	34	81.50
7250	10" thick	47.50	35	82.50
7300	12" thick	48.50	37	85.50

361

For customer support on your Square Foot Cost Data, call 877.694.8329.

B2010 Exterior Walls

Exterior brick veneer/stud backup walls are defined in the following terms; type of brick and studs, stud spacing and bond. All systems include a brick shelf, ties to the backup and all necessary dampproofing and insulation.

B2010 129		Brick Veneer/Wood Stud Backup						
	FACE BRICK	STUD BACKUP	STUD SPACING (IN.)	BOND	FACE	COST PER S.F.		
						MAT.	INST.	TOTAL
1100	Standard	2x4-wood	16	running		7.15	17.45	24.60
1120				common		8.15	19.70	27.85
1140				Flemish		8.80	23	31.80
1160				English		9.70	24	33.70
1400		2x6-wood	16	running		7.45	17.55	25
1420				common		8.40	19.80	28.20
1440				Flemish		9.10	23	32.10
1460				English		10	24.50	34.50
1700	Glazed	2x4-wood	16	running		22	18.05	40.05
1720				common		26	20.50	46.50
1740				Flemish		28.50	24	52.50
1760				English		32	25.50	57.50
2300	Engineer	2x4-wood	16	running		7.20	15.65	22.85
2320				common		8.15	17.45	25.60
2340				Flemish		8.80	20.50	29.30
2360				English		9.65	21.50	31.15
2900	Roman	2x4-wood	16	running		10.45	16.05	26.50
2920				common		12.10	18.05	30.15
2940				Flemish		13.15	21	34.15
2960				English		14.65	22.50	37.15
4100	Norwegian	2x4-wood	16	running		7.75	12.55	20.30
4120				common		8.80	13.85	22.65
4140				Flemish		9.50	16.05	25.55
4160				English		10.45	16.70	27.15

B2010 130		Brick Veneer/Metal Stud Backup						
	FACE BRICK	STUD BACKUP	STUD SPACING (IN.)	BOND	FACE	COST PER S.F.		
						MAT.	INST.	TOTAL
5100	Standard	25ga.x6"NLB	24	running		6.65	17.05	23.70
5120				common		7.60	19.30	26.90
5140				Flemish		7.95	22	29.95
5160				English		9.20	24	33.20
5200		20ga.x3-5/8"NLB	16	running		6.75	17.75	24.50
5220				common		7.70	20	27.70
5240				Flemish		8.40	23.50	31.90
5260				English		9.30	24.50	33.80

B2010 Exterior Walls

B2010 130	Brick Veneer/Metal Stud Backup

	FACE BRICK	STUD BACKUP	STUD SPACING (IN.)	BOND	FACE	COST PER S.F.		
						MAT.	INST.	TOTAL
5400	Standard	16ga.x3-5/8"LB	16	running		7.50	18	25.50
5420				common		8.50	20	28.50
5440				Flemish		9.15	23.50	32.65
5460				English		10.05	25	35.05
5500			24	running		7.15	17.50	24.65
5520				common		8.15	19.75	27.90
5540				Flemish		8.80	23	31.80
5560				English		9.70	24.50	34.20
5700	Glazed	25ga.x6"NLB	24	running		21.50	17.65	39.15
5720				common		25.50	20	45.50
5740				Flemish		28	24	52
5760				English		31.50	25	56.50
5800		20ga.x3-5/8"NLB	24	running		21.50	17.80	39.30
5820				common		25.50	20.50	46
5840				Flemish		28	24	52
5860				English		31.50	25.50	57
6000		16ga.x3-5/8"LB	16	running		22.50	18.60	41.10
6020				common		26.50	21	47.50
6040				Flemish		29	25	54
6060				English		32.50	26	58.50
6300	Engineer	25ga.x6"NLB	24	running		6.70	15.25	21.95
6320				common		7.65	17.05	24.70
6340				Flemish		8.30	20	28.30
6360				English		9.15	21	30.15
6400		20ga.x3-5/8"NLB	16	running		6.80	15.95	22.75
6420				common		7.75	17.75	25.50
6440				Flemish		8.40	21	29.40
6460				English		9.25	21.50	30.75
6900	Roman	25ga.x6"NLB	24	running		9.95	15.65	25.60
6920				common		11.60	17.55	29.15
6940				Flemish		12.65	20.50	33.15
6960				English		14.15	22	36.15
7000		20ga.x3-5/8"NLB	16	running		10.05	16.35	26.40
7020				common		11.70	18.35	30.05
7040				Flemish		12.75	21	33.75
7060				English		14.25	22.50	36.75
7500	Norman	25ga.x6"NLB	24	running		8.05	13.35	21.40
7520				common		9.25	14.85	24.10
7540				Flemish		19.90	17.35	37.25
7560				English		11.15	18.25	29.40
7600		20ga.x3-5/8"NLB	24	running		8	13.50	21.50
7620				common		9.20	15	24.20
7640				Flemish		19.85	17.50	37.35
7660				English		11.10	18.40	29.50
8100	Norwegian	25ga.x6"NLB	24	running		7.25	12.15	19.40
8120				common		8.30	13.45	21.75
8140				Flemish		9	15.65	24.65
8160				English		9.95	16.30	26.25
8400		16ga.x3-5/8"LB	16	running		8.10	13.10	21.20
8420				common		9.15	14.40	23.55
8440				Flemish		9.85	16.60	26.45
8460				English		10.80	17.25	28.05

Brick Face Composite Wall

B2010 132				Brick Face Composite Wall - Double Wythe				
	FACE BRICK	BACKUP MASONRY	BACKUP THICKNESS (IN.)	BACKUP CORE FILL		COST PER S.F.		
						MAT.	INST.	TOTAL
1000	Standard	common brick	4	none		10.25	27	37.25
1040		SCR brick	6	none		12.60	24	36.60
1080		conc. block	4	none		7.80	21.50	29.30
1120			6	perlite		9.40	22.50	31.90
1160				styrofoam		9.70	22	31.70
1200			8	perlite		10.45	23	33.45
1240				styrofoam		10.45	22.50	32.95
1520		glazed block	4	none		16.85	23	39.85
1560			6	perlite		18.70	24	42.70
1600				styrofoam		19	23.50	42.50
1640			8	perlite		20	24.50	44.50
1680				styrofoam		20	24	44
1720		clay tile	4	none		11.95	21	32.95
1760			6	none		14.05	21.50	35.55
1800			8	none		16	22	38
1840		glazed tile	4	none		19.35	27.50	46.85
1880								
2000	Glazed	common brick	4	none		25.50	27.50	53
2040		SCR brick	6	none		27.50	24.50	52
2080		conc. block	4	none		23	22	45
2200			8	perlite		25.50	23.50	49
2240				styrofoam		25.50	23	48.50
2280		L.W. block	4	none		23	22	45
2400			8	perlite		26	23.50	49.50
2520		glazed block	4	none		32	23.50	55.50
2560			6	perlite		33.50	24.50	58
2600				styrofoam		34	24	58
2640			8	perlite		35	25	60
2680				styrofoam		35	24.50	59.50

B2010 Exterior Walls

B2010 132	Brick Face Composite Wall - Double Wythe

	FACE BRICK	BACKUP MASONRY	BACKUP THICKNESS (IN.)	BACKUP CORE FILL		COST PER S.F.		
						MAT.	INST.	TOTAL
2720	Glazed	clay tile	4	none		27	21.50	48.50
2760			6	none		29	22	51
2800			8	none		31	22.50	53.50
2840		glazed tile	4	none		34.50	28	62.50
3000	Engineer	common brick	4	none		10.30	25	35.30
3040		SCR brick	6	none		12.60	22	34.60
3080		conc. block	4	none		7.85	19.90	27.75
3200			8	perlite		10.45	21	31.45
3280		L.W. block	4	none		6.05	14.10	20.15
3320			6	perlite		9.80	20.50	30.30
3520		glazed block	4	none		16.85	21.50	38.35
3560			6	perlite		18.70	22	40.70
3600				styrofoam		19.05	21.50	40.55
3640			8	perlite		20	22.50	42.50
3680				styrofoam		20	22	42
3720		clay tile	4	none		11.95	19.05	31
3800			8	none		16	20.50	36.50
3840		glazed tile	4	none		19.35	25.50	44.85
4000	Roman	common brick	4	none		13.55	25.50	39.05
4040		SCR brick	6	none		15.85	22.50	38.35
4080		conc. block	4	none		11.10	20	31.10
4120			6	perlite		12.65	21	33.65
4200			8	perlite		13.70	21.50	35.20
5000	Norman	common brick	4	none		11.65	23	34.65
5040		SCR brick	6	none		13.95	20.50	34.45
5280		L.W. block	4	none		9.30	17.85	27.15
5320			6	perlite		11.15	18.75	29.90
5720		clay tile	4	none		13.30	17.15	30.45
5760			6	none		15.40	17.70	33.10
5840		glazed tile	4	none		20.50	23.50	44
6000	Norwegian	common brick	4	none		10.80	22	32.80
6040		SCR brick	6	none		13.15	19.15	32.30
6080		conc. block	4	none		8.35	16.80	25.15
6120			6	perlite		9.95	17.60	27.55
6160				styrofoam		10.25	17.25	27.50
6200			8	perlite		11	18.10	29.10
6520		glazed block	4	none		17.40	18.30	35.70
6560			6	perlite		19.25	19	38.25
7000	Utility	common brick	4	none		11.20	21	32.20
7040		SCR brick	6	none		13.55	17.95	31.50
7080		conc. block	4	none		8.75	15.60	24.35
7120			6	perlite		10.35	16.40	26.75
7160				styrofoam		10.65	16.05	26.70
7200			8	perlite		11.40	16.90	28.30
7240				styrofoam		11.40	16.50	27.90
7280		L.W. block	4	none		8.85	15.45	24.30
7320			6	perlite		10.70	16.25	26.95
7520		glazed block	4	none		17.80	17.10	34.90
7560			6	perlite		19.65	17.80	37.45
7720		clay tile	4	none		12.90	14.75	27.65
7760			6	none		15	15.30	30.30
7840		glazed tile	4	none		20.50	21.50	42

B2010 Exterior Walls

B2010 133	Brick Face Composite Wall - Triple Wythe

	FACE BRICK	MIDDLE WYTHE	INSIDE MASONRY	TOTAL THICKNESS (IN.)		COST PER S.F.		
						MAT.	INST.	TOTAL
8000	Standard	common brick	standard brick	12		14.80	38.50	53.30
8100		4" conc. brick	standard brick	12		14.15	39.50	53.65
8120		4" conc. brick	common brick	12		14.45	39	53.45
8200	Glazed	common brick	standard brick	12		30	39	69
8300		4" conc. brick	standard brick	12		29	40.50	69.50
8320		4" conc. brick	glazed brick	12		29.50	40	69.50

B2010 Exterior Walls

Brick Face Cavity Wall

B2010 134			Brick Face Cavity Wall					
	FACE BRICK	BACKUP MASONRY	TOTAL THICKNESS (IN.)	CAVITY INSULATION		COST PER S.F.		
						MAT.	INST.	TOTAL
1000	Standard	4" common brick	10	polystyrene		10.15	27	37.15
1020				none		9.90	26.50	36.40
1040		6" SCR brick	12	polystyrene		12.50	24	36.50
1060				none		12.25	23.50	35.75
1080		4" conc. block	10	polystyrene		7.70	21.50	29.20
1100				none		7.45	21	28.45
1120		6" conc. block	12	polystyrene		8.30	22	30.30
1140				none		8.05	21.50	29.55
1160		4" L.W. block	10	polystyrene		7.80	21.50	29.30
1180				none		7.55	21	28.55
1200		6" L.W. block	12	polystyrene		8.70	22	30.70
1220				none		8.40	21.50	29.90
1240		4" glazed block	10	polystyrene		16.75	23	39.75
1260				none		16.45	22.50	38.95
1280		6" glazed block	12	polystyrene		16.75	23	39.75
1300				none		16.50	22.50	39
1320		4" clay tile	10	polystyrene		11.85	21	32.85
1340				none		11.55	20	31.55
1360		4" glazed tile	10	polystyrene		19.25	27.50	46.75
1380				none		18.95	27	45.95
1500	Glazed	4" common brick	10	polystyrene		25	27.50	52.50
1520				none		25	27	52
1580		4" conc. block	10	polystyrene		22.50	22.50	45
1600				none		22.50	21.50	44
1660		4" L.W. block	10	polystyrene		23	22	45
1680				none		22.50	21.50	44
1740		4" glazed block	10	polystyrene		32	24	56
1760				none		31.50	23	54.50
1820		4" clay tile	10	polystyrene		27	21.50	48.50
1840				none		26.50	21	47.50
1860		4" glazed tile	10	polystyrene		29	22	51
1880				none		28.50	21.50	50
2000	Engineer	4" common brick	10	polystyrene		10.20	25	35.20
2020				none		9.90	24.50	34.40
2080		4" conc. block	10	polystyrene		7.75	19.95	27.70
2100				none		7.45	19.20	26.65

B2010 134	Brick Face Cavity Wall

	FACE BRICK	BACKUP MASONRY	TOTAL THICKNESS (IN.)	CAVITY INSULATION		COST PER S.F. MAT.	COST PER S.F. INST.	COST PER S.F. TOTAL
2160	Engineer	4" L.W. block	10	polystyrene		7.85	19.80	27.65
2180				none		7.60	19.05	26.65
2240		4" glazed block	10	polystyrene		16.80	21.50	38.30
2260				none		16.50	20.50	37
2320		4" clay tile	10	polystyrene		11.90	19.10	31
2340				none		11.60	18.35	29.95
2360		4" glazed tile	10	polystyrene		19.30	25.50	44.80
2380				none		19	25	44
2500	Roman	4" common brick	10	polystyrene		13.45	25.50	38.95
2520				none		13.15	25	38.15
2580		4" conc. block	10	polystyrene		11	20.50	31.50
2600				none		10.70	19.60	30.30
2660		4" L.W. block	10	polystyrene		11.10	20	31.10
2680				none		10.80	19.45	30.25
2740		4" glazed block	10	polystyrene		20	22	42
2760				none		19.75	21	40.75
2820		4" clay tile	10	polystyrene		15.10	19.50	34.60
2840				none		14.85	18.75	33.60
2860		4" glazed tile	10	polystyrene		22.50	26	48.50
2880				none		22	25.50	47.50
3000	Norman	4" common brick	10	polystyrene		11.55	23.50	35.05
3020				none		11.25	22.50	33.75
3080		4" conc. block	10	polystyrene		9.10	18.05	27.15
3100				none		8.80	17.30	26.10
3160		4" L.W. block	10	polystyrene		9.20	17.90	27.10
3180				none		8.90	17.15	26.05
3240		4" glazed block	10	polystyrene		18.10	19.55	37.65
3260				none		17.85	18.80	36.65
3320		4" clay tile	10	polystyrene		13.20	17.20	30.40
3340				none		12.95	16.45	29.40
3360		4" glazed tile	10	polystyrene		20.50	24	44.50
3380				none		20.50	23	43.50
3500	Norwegian	4" common brick	10	polystyrene		10.75	22	32.75
3520				none		10.45	21.50	31.95
3580		4" conc. block	10	polystyrene		8.30	16.85	25.15
3600				none		8	16.10	24.10
3660		4" L.W. block	10	polystyrene		8.40	16.70	25.10
3680				none		8.10	15.95	24.05
3740		4" glazed block	10	polystyrene		17.30	18.35	35.65
3760				none		17.05	17.60	34.65
3820		4" clay tile	10	polystyrene		12.40	16	28.40
3840				none		12.15	15.25	27.40
3860		4" glazed tile	10	polystyrene		19.80	22.50	42.30
3880				none		19.55	22	41.55
4000	Utility	4" common brick	10	polystyrene		11.10	21	32.10
4020				none		10.85	20	30.85
4080		4" conc. block	10	polystyrene		8.65	15.65	24.30
4100				none		8.40	14.90	23.30
4160		4" L.W. block	10	polystyrene		8.75	15.50	24.25
4180				none		8.50	14.75	23.25
4240		4" glazed block	10	polystyrene		17.70	17.15	34.85
4260				none		17.40	16.40	33.80

B20 Exterior Enclosure

B2010 Exterior Walls

B2010 134 — Brick Face Cavity Wall

	FACE BRICK	BACKUP MASONRY	TOTAL THICKNESS (IN.)	CAVITY INSULATION		COST PER S.F.		
						MAT.	INST.	TOTAL
4320	Utility	4" clay tile	10	polystyrene		12.80	14.80	27.60
4340				none		12.50	14.05	26.55
4360		4" glazed tile	10	polystyrene		20	21.50	41.50
4380				none		19.90	20.50	40.40

B2010 135 — Brick Face Cavity Wall - Insulated Backup

	FACE BRICK	BACKUP MASONRY	TOTAL THICKNESS (IN.)	BACKUP CORE FILL		COST PER S.F.		
						MAT.	INST.	TOTAL
5100	Standard	6" conc. block	10	perlite		9	22	31
5120				styrofoam		9.35	21.50	30.85
5180		6" L.W. block	10	perlite		9.40	21.50	30.90
5200				styrofoam		9.70	21.50	31.20
5260		6" glazed block	10	perlite		18.30	23	41.30
5280				styrofoam		18.65	23	41.65
5340		6" clay tile	10	none		13.65	20.50	34.15
5360		8" clay tile	12	none		15.65	21.50	37.15
5600	Glazed	6" conc. block	10	perlite		24	22.50	46.50
5620				styrofoam		24.50	22	46.50
5680		6" L.W. block	10	perlite		24.50	22.50	47
5700				styrofoam		24.50	22	46.50
5760		6" glazed block	10	perlite		33.50	24	57.50
5780				styrofoam		33.50	23.50	57
5840		6" clay tile	10	none		28.50	21.50	50
5860		8" clay tile	8	none		30.50	22	52.50
6100	Engineer	6" conc. block	10	perlite		9.05	20	29.05
6120				styrofoam		9.35	19.65	29
6180		6" L.W. block	10	perlite		9.40	19.85	29.25
6200				styrofoam		9.75	19.50	29.25
6260		6" glazed block	10	perlite		18.35	21.50	39.85
6280				styrofoam		18.65	21	39.65
6340		6" clay tile	10	none		13.70	18.90	32.60
6360		8" clay tile	12	none		15.70	19.65	35.35
6600	Roman	6" conc. block	10	perlite		12.30	20.50	32.80
6620				styrofoam		12.60	20	32.60
6680		6" L.W. block	10	perlite		12.65	20.50	33.15
6700				styrofoam		13	19.90	32.90
6760		6" glazed block	10	perlite		21.50	22	43.50
6780				styrofoam		22	21.50	43.50
6840		6" clay tile	10	none		16.95	19.30	36.25
6860		8" clay tile	12	none		18.90	20	38.90
7100	Norman	6" conc. block	10	perlite		10.40	18.10	28.50
7120				styrofoam		10.70	17.75	28.45
7180		6" L.W. block	10	perlite		10.75	17.95	28.70
7200				styrofoam		11.10	17.60	28.70
7260		6" glazed block	10	perlite		19.70	19.50	39.20
7280				styrofoam		20	19.15	39.15
7340		6" clay tile	10	none		15.05	17	32.05
7360		8" clay tile	12	none		17	17.75	34.75
7600	Norwegian	6" conc. block	10	perlite		9.60	16.90	26.50
7620				styrofoam		9.90	16.55	26.45
7680		6" L.W. block	10	perlite		9.95	16.75	26.70
7700				styrofoam		10.25	16.40	26.65

For customer support on your Square Foot Cost Data, call 877.694.8329.

369

B2010 Exterior Walls

B2010 135	Brick Face Cavity Wall - Insulated Backup

	FACE BRICK	BACKUP MASONRY	TOTAL THICKNESS (IN.)	BACKUP CORE FILL		COST PER S.F.		
						MAT.	INST.	TOTAL
7760		6" glazed block	10	perlite		18.90	18.30	37.20
7780				styrofoam		19.20	17.95	37.15
7840		6" clay tile	10	none		14.25	15.80	30.05
7860		8" clay tile	12	none		16.20	16.55	32.75
8100	Utility	6" conc. block	10	perlite		9.95	15.70	25.65
8120				styrofoam		10.30	15.35	25.65
8180		6" L.W. block	10	perlite		10.35	15.55	25.90
8200				styrofoam		10.65	15.20	25.85
8220		8" L.W. block	12	perlite		11.45	16.05	27.50
8240				styrofoam		11.45	15.60	27.05
8260		6" glazed block	10	perlite		19.25	17.10	36.35
8280				styrofoam		19.60	16.75	36.35
8300		8" glazed block	12	perlite		20.50	17.70	38.20
8340		6" clay tile	10	none		14.60	14.60	29.20
8360		8" clay tile	12	none		16.60	15.35	31.95

B2010 Exterior Walls

Glass Block

B2010 140	Glass Block	COST PER S.F.		
		MAT.	**INST.**	**TOTAL**
2300	Glass block 4" thick, 6"x6" plain, under 1,000 S.F.	29	25.50	54.50
2400	1,000 to 5,000 S.F.	28.50	22	50.50
2500	Over 5,000 S.F.	27.50	20.50	48
2600	Solar reflective, under 1,000 S.F.	41	35	76
2700	1,000 to 5,000 S.F.	40	30	70
2800	Over 5,000 S.F.	38.50	28	66.50
3500	8"x8" plain, under 1,000 S.F.	17.10	19	36.10
3600	1,000 to 5,000 S.F.	16.75	16.40	33.15
3700	Over 5,000 S.F.	16.25	14.80	31.05
3800	Solar reflective, under 1,000 S.F.	24	25.50	49.50
3900	1,000 to 5,000 S.F.	23.50	22	45.50
4000	Over 5,000 S.F.	22.50	19.70	42.20
5000	12"x12" plain, under 1,000 S.F.	24.50	17.55	42.05
5100	1,000 to 5,000 S.F.	24	14.80	38.80
5200	Over 5,000 S.F.	23	13.50	36.50
5300	Solar reflective, under 1,000 S.F.	34.50	23.50	58
5400	1,000 to 5,000 S.F.	34	19.70	53.70
5600	Over 5,000 S.F.	32.50	17.90	50.40
5800	3" thinline, 6"x6" plain, under 1,000 S.F.	23	25.50	48.50
5900	Over 5,000 S.F.	22	20.50	42.50
6000	Solar reflective, under 1,000 S.F.	32.50	35	67.50
6100	Over 5,000 S.F.	31	28	59
6200	8"x8" plain, under 1,000 S.F.	12.95	19	31.95
6300	Over 5,000 S.F.	12.35	14.80	27.15
6400	Solar reflective, under 1,000 S.F.	18.05	25.50	43.55
6500	Over 5,000 S.F.	17.20	19.70	36.90

371

For customer support on your Square Foot Cost Data, call 877.694.8329.

B2010 Exterior Walls

Metal Siding

The table below lists costs for metal siding of various descriptions, not including the steel frame, or the structural steel, of a building. Costs are per S.F. including all accessories and insulation.

B2010 146	Metal Siding Panel	COST PER S.F.		
		MAT.	INST.	TOTAL
1400	Metal siding aluminum panel, corrugated, .024" thick, natural	3.06	4.20	7.26
1450	Painted	3.23	4.20	7.43
1500	.032" thick, natural	3.39	4.20	7.59
1550	Painted	3.99	4.20	8.19
1600	Ribbed 4" pitch, .032" thick, natural	3.33	4.20	7.53
1650	Painted	4.05	4.20	8.25
1700	.040" thick, natural	3.83	4.20	8.03
1750	Painted	4.36	4.20	8.56
1800	.050" thick, natural	4.29	4.20	8.49
1850	Painted	4.85	4.20	9.05
1900	Ribbed 8" pitch, .032" thick, natural	3.17	4.03	7.19
1950	Painted	3.88	4.05	7.93
2000	.040" thick, natural	3.68	4.05	7.73
2050	Painted	4.19	4.09	8.28
2100	.050" thick, natural	4.14	4.07	8.21
2150	Painted	4.72	4.10	8.82
3000	Steel, corrugated or ribbed, 29 Ga., .0135" thick, galvanized	2.10	3.84	5.94
3050	Colored	3.02	3.88	6.90
3100	26 Ga., .0179" thick, galvanized	2.19	3.86	6.05
3150	Colored	3.17	3.90	7.07
3200	24 Ga., .0239" thick, galvanized	2.76	3.87	6.63
3250	Colored	3.48	3.91	7.39
3300	22 Ga., .0299" thick, galvanized	2.82	3.89	6.71
3350	Colored	4.17	3.93	8.10
3400	20 Ga., .0359" thick, galvanized	2.82	3.89	6.71
3450	Colored	4.43	4.16	8.59
4100	Sandwich panels, factory fab., 1" polystyrene, steel core, 26 Ga., galv.	5.85	6.05	11.90
4200	Colored, 1 side	7.25	6.05	13.30
4300	2 sides	9.05	6.05	15.10
4400	2" polystyrene core, 26 Ga., galvanized	6.95	6.05	13
4500	Colored, 1 side	8.35	6.05	14.40
4600	2 sides	10.15	6.05	16.20
4700	22 Ga., baked enamel exterior	13.40	6.35	19.75
4800	Polyvinyl chloride exterior	14.10	6.35	20.45
5100	Textured aluminum, 4' x 8' x 5/16" plywood backing, single face	4.48	3.65	8.13
5200	Double face	5.80	3.65	9.45

B2010 Exterior Walls

The table below lists costs per S.F. for exterior walls with wood siding. A variety of systems are presented using both wood and metal studs at 16" and 24" O.C.

Exterior Wall Siding

B2010 148	Panel, Shingle & Lap Siding	COST PER S.F.		
		MAT.	INST.	TOTAL
1400	Wood siding w/2" x 4" studs, 16" O.C., insul. wall, 5/8" text 1-11 fir ply.	3.78	5.91	9.69
1450	5/8" text 1-11 cedar plywood	4.88	5.70	10.58
1500	1" x 4" vert T.&G. redwood	4.84	6.85	11.69
1600	1" x 8" vert T.&G. redwood	6.55	5.95	12.50
1650	1" x 5" rabbetted cedar bev. siding	6.60	6.15	12.75
1700	1" x 6" cedar drop siding	6.70	6.20	12.90
1750	1" x 12" rough sawn cedar	3.47	5.95	9.42
1800	1" x 12" sawn cedar, 1" x 4" battens	7	5.60	12.60
1850	1" x 10" redwood shiplap siding	6.60	5.75	12.35
1900	18" no. 1 red cedar shingles, 5-1/2" exposed	4.84	7.80	12.64
1950	6" exposed	4.64	7.55	12.19
2000	6-1/2" exposed	4.44	7.30	11.74
2100	7" exposed	4.24	7.05	11.29
2150	7-1/2" exposed	4.04	6.75	10.79
3000	8" wide aluminum siding	4.50	5.30	9.80
3150	8" plain vinyl siding	3.19	5.40	8.59
3250	8" insulated vinyl siding	3.53	6	9.53
3300				
3400	2" x 6" studs, 16" O.C., insul. wall, w/ 5/8" text 1-11 fir plywood	4.25	6.10	10.35
3500	5/8" text 1-11 cedar plywood	5.35	6.10	11.45
3600	1" x 4" vert T.&G. redwood	5.30	7	12.30
3700	1" x 8" vert T.&G. redwood	7.05	6.10	13.15
3800	1" x 5" rabbetted cedar bev siding	7.05	6.35	13.40
3900	1" x 6" cedar drop siding	7.15	6.35	13.50
4000	1" x 12" rough sawn cedar	3.94	6.15	10.09
4200	1" x 12" sawn cedar, 1" x 4" battens	7.45	5.75	13.20
4500	1" x 10" redwood shiplap siding	7.05	5.90	12.95
4550	18" no. 1 red cedar shingles, 5-1/2" exposed	5.30	7.95	13.25
4600	6" exposed	5.10	7.70	12.80
4650	6-1/2" exposed	4.91	7.45	12.36
4700	7" exposed	4.71	7.20	11.91
4750	7-1/2" exposed	4.51	6.95	11.46
4800	8" wide aluminum siding	4.97	5.45	10.42
4850	8" plain vinyl siding	3.66	5.55	9.21
4900	8" insulated vinyl siding	4	6.15	10.15
4910				
5000	2" x 6" studs, 24" O.C., insul. wall, 5/8" text 1-11, fir plywood	4.06	5.75	9.81
5050	5/8" text 1-11 cedar plywood	5.15	5.75	10.90
5100	1" x 4" vert T.&G. redwood	5.10	6.65	11.75
5150	1" x 8" vert T.&G. redwood	6.85	5.75	12.60

B2010 148	Panel, Shingle & Lap Siding	COST PER S.F.		
		MAT.	INST.	TOTAL
5200	1" x 5" rabbetted cedar bev siding	6.90	6	12.90
5250	1" x 6" cedar drop siding	6.95	6	12.95
5300	1" x 12" rough sawn cedar	3.75	5.80	9.55
5400	1" x 12" sawn cedar, 1" x 4" battens	7.25	5.40	12.65
5450	1" x 10" redwood shiplap siding	6.85	5.55	12.40
5500	18" no. 1 red cedar shingles, 5-1/2" exposed	5.10	7.60	12.70
5550	6" exposed	4.92	7.35	12.27
5650	7" exposed	4.52	6.85	11.37
5700	7-1/2" exposed	4.32	6.60	10.92
5750	8" wide aluminum siding	4.78	5.10	9.88
5800	8" plain vinyl siding	3.47	5.20	8.67
5850	8" insulated vinyl siding	3.81	5.80	9.61
5900	3-5/8" metal studs, 16 Ga. 16" O.C. insul. wall, 5/8" text 1-11 fir plywood	4.52	6.20	10.72
5950	5/8" text 1-11 cedar plywood	5.60	6.20	11.80
6000	1" x 4" vert T.&G. redwood	5.60	7.15	12.75
6050	1" x 8" vert T.&G. redwood	7.30	6.25	13.55
6100	1" x 5" rabbetted cedar bev siding	7.35	6.45	13.80
6150	1" x 6" cedar drop siding	7.45	6.50	13.95
6200	1" x 12" rough sawn cedar	4.21	6.25	10.46
6250	1" x 12" sawn cedar, 1" x 4" battens	7.70	5.90	13.60
6300	1" x 10" redwood shiplap siding	7.30	6.05	13.35
6350	18" no. 1 red cedar shingles, 5-1/2" exposed	5.60	8.10	13.70
6500	6" exposed	5.40	7.85	13.25
6550	6-1/2" exposed	5.20	7.60	12.80
6600	7" exposed	4.98	7.35	12.33
6650	7-1/2" exposed	4.98	7.15	12.13
6700	8" wide aluminum siding	5.25	5.60	10.85
6750	8" plain vinyl siding	3.93	5.50	9.43
6800	8" insulated vinyl siding	4.27	6.10	10.37
7000	3-5/8" metal studs, 16 Ga. 24" O.C. insul wall, 5/8" text 1-11 fir plywood	4.17	5.55	9.72
7050	5/8" text 1-11 cedar plywood	5.25	5.55	10.80
7100	1" x 4" vert T.&G. redwood	5.25	6.65	11.90
7150	1" x 8" vert T.&G. redwood	6.95	5.75	12.70
7200	1" x 5" rabbetted cedar bev siding	7	6	13
7250	1" x 6" cedar drop siding	7.10	6	13.10
7300	1" x 12" rough sawn cedar	3.86	5.80	9.66
7350	1" x 12" sawn cedar 1" x 4" battens	7.35	5.40	12.75
7400	1" x 10" redwood shiplap siding	6.95	5.55	12.50
7450	18" no. 1 red cedar shingles, 5-1/2" exposed	5.45	7.90	13.35
7500	6" exposed	5.25	7.60	12.85
7550	6-1/2" exposed	5.05	7.35	12.40
7600	7" exposed	4.83	7.10	11.93
7650	7-1/2" exposed	4.43	6.60	11.03
7700	8" wide aluminum siding	4.89	5.10	9.99
7750	8" plain vinyl siding	3.58	5.20	8.78
7800	8" insul. vinyl siding	3.92	5.80	9.72

B2010 Exterior Walls

The table below lists costs for some typical stucco walls including all the components as demonstrated in the component block below. Prices are presented for backup walls using wood studs, metal studs and CMU.

Exterior Stucco Wall

B2010 151	Stucco Wall	COST PER S.F.		
		MAT.	INST.	TOTAL
2100	Cement stucco, 7/8" th., plywood sheathing, stud wall, 2" x 4", 16" O.C.	2.93	9.55	12.48
2200	24" O.C.	2.82	9.30	12.12
2300	2" x 6", 16" O.C.	3.40	9.75	13.15
2400	24" O.C.	3.21	9.40	12.61
2500	No sheathing, metal lath on stud wall, 2" x 4", 16" O.C.	2.01	8.30	10.31
2600	24" O.C.	1.90	8.05	9.95
2700	2" x 6", 16" O.C.	2.48	8.50	10.98
2800	24" O.C.	2.29	8.15	10.44
2900	1/2" gypsum sheathing, 3-5/8" metal studs, 16" O.C.	3.02	8.70	11.72
2950	24" O.C.	2.67	8.20	10.87
3000	Cement stucco, 5/8" th., 2 coats on std. CMU block, 8"x 16", 8" thick	3.73	11.70	15.43
3100	10" thick	4.87	12	16.87
3200	12" thick	5	13.80	18.80
3300	Std. light Wt. block 8" x 16", 8" thick	4.09	11.50	15.59
3400	10" thick	4.84	11.80	16.64
3500	12" thick	5.10	13.55	18.65
3600	3 coat stucco, self furring metal lath 3.4 Lb/SY, on 8" x 16", 8" thick	4.57	13.80	18.37
3700	10" thick	4.92	12.65	17.57
3800	12" thick	5.05	14.45	19.50
3900	Lt. Wt. block, 8" thick	4.14	12.15	16.29
4000	10" thick	4.89	12.45	17.34
4100	12" thick	5.15	14.20	19.35

B2010 152	E.I.F.S.	COST PER S.F.		
		MAT.	INST.	TOTAL
5100	E.I.F.S., plywood sheathing, stud wall, 2" x 4", 16" O.C., 1" EPS	3.99	9.65	13.64
5110	2" EPS	4.26	9.65	13.91
5120	3" EPS	4.54	9.65	14.19
5130	4" EPS	4.81	9.65	14.46
5140	2" x 6", 16" O.C., 1" EPS	4.46	9.85	14.31
5150	2" EPS	4.73	9.85	14.58
5160	3" EPS	5	9.85	14.85
5170	4" EPS	5.30	9.85	15.15
5180	Cement board sheathing, 3-5/8" metal studs, 16" O.C., 1" EPS	4.68	12	16.68
5190	2" EPS	4.95	12	16.95
5200	3" EPS	5.25	12	17.25
5210	4" EPS	6.30	12	18.30
5220	6" metal studs, 16" O.C., 1" EPS	5.25	12.05	17.30
5230	2" EPS	5.50	12.05	17.55

B20 Exterior Enclosure

B2010 Exterior Walls

B2010 152	E.I.F.S.	COST PER S.F.		
		MAT.	INST.	TOTAL
5240	3" EPS	5.80	12.05	17.85
5250	4" EPS	6.85	12.05	18.90
5260	CMU block, 8" x 8" x 16", 1" EPS	5.35	13.80	19.15
5270	2" EPS	5.60	13.80	19.40
5280	3" EPS	5.90	13.80	19.70
5290	4" EPS	6.95	15.25	22.20
5300	8" x 10" x 16", 1" EPS	6.50	14.10	20.60
5310	2" EPS	6.75	14.10	20.85
5320	3" EPS	7.05	14.10	21.15
5330	4" EPS	7.30	14.10	21.40
5340	8" x 12" x 16", 1" EPS	6.65	15.90	22.55
5350	2" EPS	6.90	15.90	22.80
5360	3" EPS	7.20	15.90	23.10
5370	4" EPS	7.45	15.90	23.35

B2010 Exterior Walls

Description: The table below lists costs, $/S.F., for channel girts with sag rods and connector angles top and bottom for various column spacings, building heights and wind loads. Additive costs are shown for wind columns.

How to Use this Table: Add the cost of girts, sag rods, and angles to the framing costs of steel buildings clad in metal or composition siding. If the column spacing is in excess of the column spacing shown, use intermediate wind columns. Additive costs are shown under "Wind Columns".

B2010 154	BLDG. HEIGHT (FT.)	WIND LOAD (P.S.F.)	COL. SPACING (FT.)		INTERMEDIATE COLUMNS	COST PER S.F. MAT.	COST PER S.F. INST.	COST PER S.F. TOTAL
Metal Siding Support								
3000	18	20	20			2.42	4.08	6.50
3100					wind cols.	1.25	.29	1.54
3200		20	25			2.65	4.13	6.78
3300					wind cols.	1	.23	1.23
3400		20	30			2.92	4.20	7.12
3500					wind cols.	.83	.19	1.02
3600		20	35			3.21	4.27	7.48
3700					wind cols.	.71	.16	.87
3800		30	20			2.68	4.14	6.82
3900					wind cols.	1.25	.29	1.54
4000		30	25			2.92	4.20	7.12
4100					wind cols.	1	.23	1.23
4200		30	30			3.23	4.27	7.50
4300					wind cols.	1.13	.26	1.39
4600	30	20	20			2.35	2.87	5.22
4700					wind cols.	1.77	.41	2.18
4800		20	25			2.64	2.94	5.58
4900					wind cols.	1.41	.33	1.74
5000		20	30			2.97	3.02	5.99
5100					wind cols.	1.39	.33	1.72
5200		20	35			3.32	3.10	6.42
5300					wind cols.	1.38	.32	1.70
5400		30	20			2.67	2.95	5.62
5500					wind cols.	2.09	.49	2.58
5600		30	25			2.96	3.01	5.97
5700					wind cols.	1.99	.47	2.46
5800		30	30			3.34	3.10	6.44
5900					wind cols.	1.87	.44	2.31

B2010 Exterior Walls

B2010 160	Pole Barn Exterior Wall	COST PER S.F.		
		MAT.	INST.	TOTAL
2000	Pole barn exterior wall, pressure treated pole in concrete, 8' O.C.			
3000	Steel siding, 8' eave	4.49	9.80	14.29
3050	10' eave	4.30	9.05	13.35
3100	12' eave	4.09	8.25	12.34
3150	14' eave	3.98	7.80	11.78
3200	16' eave	3.88	7.45	11.33
4000	Aluminum siding, 8' eave	4.16	9.80	13.96
4050	10' eave	3.91	8.90	12.81
4100	12' eave	3.76	8.25	12.01
4150	14' eave	3.65	7.80	11.45
4200	16' eave	3.55	7.45	11
5000	Wood siding, 8' eave	4.75	8.60	13.35
5050	10' eave	4.50	7.65	12.15
5100	12' eave	4.35	7.05	11.40
5150	14' eave	4.24	6.60	10.84
5200	16' eave	4.14	6.25	10.39
6000	Plywood siding, 8' eave	4.07	8.55	12.62
6050	10' eave	3.82	7.65	11.47
6100	12' eave	3.67	7	10.67
6150	14' eave	3.56	6.55	10.11
6200	16' eave	3.46	6.20	9.66

B2020 Exterior Windows

The table below lists window systems by material, type and size. Prices between sizes listed can be interpolated with reasonable accuracy. Prices include frame, hardware, and casing as illustrated in the component block below.

B2020 102			**Wood Windows**				
	MATERIAL	TYPE	GLAZING	SIZE	DETAIL	COST PER UNIT		
						MAT.	INST.	TOTAL
3000	Wood	double hung	std. glass	2'-8" x 4'-6"		245	251	496
3050				3'-0" x 5'-6"		320	288	608
3100			insul. glass	2'-8" x 4'-6"		260	251	511
3150				3'-0" x 5'-6"		350	288	638
3200		sliding	std. glass	3'-4" x 2'-7"		325	209	534
3250				4'-4" x 3'-3"		340	228	568
3300				5'-4" x 6'-0"		430	275	705
3350			insul. glass	3'-4" x 2'-7"		400	244	644
3400				4'-4" x 3'-3"		415	265	680
3450				5'-4" x 6'-0"		520	315	835
3500		awning	std. glass	2'-10" x 1'-9"		229	126	355
3600				4'-4" x 2'-8"		365	121	486
3700			insul. glass	2'-10" x 1'-9"		279	145	424
3800				4'-4" x 2'-8"		450	133	583
3900		casement	std. glass	1'-10" x 3'-2"	1 lite	330	165	495
3950				4'-2" x 4'-2"	2 lite	595	219	814
4000				5'-11" x 5'-2"	3 lite	895	282	1,177
4050				7'-11" x 6'-3"	4 lite	1,250	340	1,590
4100				9'-11" x 6'-3"	5 lite	1,650	395	2,045
4150			insul. glass	1'-10" x 3'-2"	1 lite	330	165	495
4200				4'-2" x 4'-2"	2 lite	610	219	829
4250				5'-11" x 5'-2"	3 lite	940	282	1,222
4300				7'-11" x 6'-3"	4 lite	1,325	340	1,665
4350				9'-11" x 6'-3"	5 lite	1,750	395	2,145
4400		picture	std. glass	4'-6" x 4'-6"		465	283	748
4450				5'-8" x 4'-6"		525	315	840
4500		picture	insul. glass	4'-6" x 4'-6"		565	330	895
4550				5'-8" x 4'-6"		645	365	1,010
4600		fixed bay	std. glass	8' x 5'		1,625	520	2,145
4650				9'-9" x 5'-4"		1,150	730	1,880
4700			insul. glass	8' x 5'		2,150	520	2,670
4750				9'-9" x 5'-4"		1,250	730	1,980
4800		casement bay	std. glass	8' x 5'		1,625	595	2,220
4850			insul. glass	8' x 5'		1,825	595	2,420
4900		vert. bay	std. glass	8' x 5'		1,825	595	2,420
4950			insul. glass	8' x 5'		1,925	595	2,520

B2020 Exterior Windows

B2020 104 — Steel Windows

	MATERIAL	TYPE	GLAZING	SIZE	DETAIL	COST PER UNIT		
						MAT.	INST.	TOTAL
5000	Steel	double hung	1/4" tempered	2'-8" x 4'-6"		905	197	1,102
5050				3'-4" x 5'-6"		1,375	300	1,675
5100			insul. glass	2'-8" x 4'-6"		920	226	1,146
5150				3'-4" x 5'-6"		1,400	345	1,745
5200		horz. pivoted	std. glass	2' x 2'		271	66	337
5250				3' x 3'		610	148	758
5300				4' x 4'		1,075	263	1,338
5350				6' x 4'		1,625	395	2,020
5400			insul. glass	2' x 2'		276	75.50	351.50
5450				3' x 3'		620	170	790
5500				4' x 4'		1,100	300	1,400
5550				6' x 4'		1,650	450	2,100
5600		picture window	std. glass	3' x 3'		385	148	533
5650				6' x 4'		1,025	395	1,420
5700			insul. glass	3' x 3'		395	170	565
5750				6' x 4'		1,050	450	1,500
5800		industrial security	std. glass	2'-9" x 4'-1"		815	185	1,000
5850				4'-1" x 5'-5"		1,600	365	1,965
5900			insul. glass	2'-9" x 4'-1"		830	212	1,042
5950				4'-1" x 5'-5"		1,650	415	2,065
6000		comm. projected	std. glass	3'-9" x 5'-5"		1,350	335	1,685
6050				6'-9" x 4'-1"		1,825	455	2,280
6100			insul. glass	3'-9" x 5'-5"		1,375	385	1,760
6150				6'-9" x 4'-1"		1,850	520	2,370
6200		casement	std. glass	4'-2" x 4'-2"	2 lite	1,200	285	1,485
6250			insul. glass	4'-2" x 4'-2"		1,225	325	1,550
6300			std. glass	5'-11" x 5'-2"	3 lite	2,525	505	3,030
6350			insul. glass	5'-11" x 5'-2"		2,575	575	3,150

B2020 106 — Aluminum Windows

	MATERIAL	TYPE	GLAZING	SIZE	DETAIL	COST PER UNIT		
						MAT.	INST.	TOTAL
6400	Aluminum	double hung	insul. glass	3'-0" x 4'-0"		530	146	676
6450				4'-5" x 5'-3"		440	183	623
6500				3'-1" x 3'-2"		470	175	645
6550				4'-5" x 5'-3"		530	220	750
6600		sliding	std. glass	3' x 2'		239	146	385
6650				5' x 3'		365	163	528
6700				8' x 4'		385	244	629
6750				9' x 5'		585	365	950
6800			insul. glass	3' x 2'		255	146	401
6850				5' x 3'		425	163	588
6900				8' x 4'		620	244	864
6950				9' x 5'		935	365	1,300
7000		single hung	std. glass	2' x 3'		229	146	375
7050				2'-8" x 6'-8"		400	183	583
7100				3'-4" x 5'-0"		330	163	493
7150			insul. glass	2' x 3'		277	146	423
7200				2'-8" x 6'-8"		520	183	703
7250				3'-4" x 5'		365	163	528
7300		double hung	std. glass	2' x 3'		320	98.50	418.50
7350				2'-8" x 6'-8"		950	293	1,243

For customer support on your Square Foot Cost Data, call 877.694.8329.

B2020 Exterior Windows

B2020 106	Aluminum Windows							
	MATERIAL	TYPE	GLAZING	SIZE	DETAIL	COST PER UNIT		
						MAT.	INST.	TOTAL
7400	Aluminum	Double hinge		3'-4" x 5'		885	274	1,159
7450			insul. glass	2' x 3'		330	113	443
7500				2'-8" x 6'-8"		970	335	1,305
7550				3'-4" x 5'-0"		910	315	1,225
7600		casement	std. glass	3'-1" x 3'-2"		281	160	441
7650				4'-5" x 5'-3"		680	390	1,070
7700			insul. glass	3'-1" x 3'-2"		294	184	478
7750				4'-5" x 5'-3"		715	445	1,160
7800		hinged swing	std. glass	3' x 4'		585	197	782
7850				4' x 5'		975	330	1,305
7900			insul. glass	3' x 4'		600	226	826
7950				4' x 5'		1,000	375	1,375
8200		picture unit	std. glass	2'-0" x 3'-0"		182	98.50	280.50
8250				2'-8" x 6'-8"		540	293	833
8300				3'-4" x 5'-0"		505	274	779
8350			insul. glass	2'-0" x 3'-0"		190	113	303
8400				2'-8" x 6'-8"		560	335	895
8450				3'-4" x 5'-0"		525	315	840
8500		awning type	std. glass	3'-0" x 3'-0"	2 lite	470	105	575
8550				3'-0" x 4'-0"	3 lite	540	146	686
8600				3'-0" x 5'-4"	4 lite	650	146	796
8650				4'-0" x 5'-4"	4 lite	715	163	878
8700			insul. glass	3'-0" x 3'-0"	2 lite	495	105	600
8750				3'-0" x 4'-0"	3 lite	620	146	766
8800				3'-0" x 5'-4"	4 lite	765	146	911
8850				4'-0" x 5'-4"	4 lite	855	163	1,018
9051								
9052								

B2020 210	Tubular Aluminum Framing	COST/S.F. OPNG.		
		MAT.	INST.	TOTAL
1100	Alum flush tube frame, for 1/4" glass, 1-3/4"x4", 5'x6'opng, no inter horiz	14.60	11.50	26.10
1150	One intermediate horizontal	19.95	13.75	33.70
1200	Two intermediate horizontals	25.50	15.95	41.45
1250	5' x 20' opening, three intermediate horizontals	13.45	9.85	23.30
1400	1-3/4" x 4-1/2", 5' x 6' opening, no intermediate horizontals	16.80	11.50	28.30
1450	One intermediate horizontal	22.50	13.75	36.25
1500	Two intermediate horizontals	28	15.95	43.95
1550	5' x 20' opening, three intermediate horizontals	15.10	9.85	24.95
1700	For insulating glass, 2"x4-1/2", 5'x6' opening, no intermediate horizontals	17.05	12.15	29.20
1750	One intermediate horizontal	22	14.55	36.55
1800	Two intermediate horizontals	26.50	16.85	43.35
1850	5' x 20' opening, three intermediate horizontals	14.60	10.40	25
2000	Thermal break frame, 2-1/4"x4-1/2", 5'x6'opng, no intermediate horizontals	17.80	12.35	30.15
2050	One intermediate horizontal	23.50	15.15	38.65
2100	Two intermediate horizontals	29.50	17.90	47.40
2150	5' x 20' opening, three intermediate horizontals	16.10	10.90	27

B2020 Exterior Windows

The table below lists costs of curtain wall and spandrel panels per S.F. Costs do not include structural framing used to hang the panels.

Glazing Panel   Plate Glass

Spandrel Glass   Polycarbonate   Sandwich Panel

B2020 220	Curtain Wall Panels	MAT.	INST.	TOTAL
		COST PER S.F.		
1000	Glazing panel, insulating, 1/2" thick, 2 lites 1/8" float, clear	11.10	11.55	22.65
1100	Tinted	15.45	11.55	27
1200	5/8" thick units, 2 lites 3/16" float, clear	15.05	12.20	27.25
1300	Tinted	15.15	12.20	27.35
1400	1" thick units, 2 lites, 1/4" float, clear	18.50	14.65	33.15
1500	Tinted	26	14.65	40.65
1600	Heat reflective film inside	30.50	12.95	43.45
1700	Light and heat reflective glass, tinted	34.50	12.95	47.45
2000	Plate glass, 1/4" thick, clear	6.55	9.15	15.70
2050	Tempered	9.75	9.15	18.90
2100	Tinted	10	9.15	19.15
2200	3/8" thick, clear	11.25	14.65	25.90
2250	Tempered	18.40	14.65	33.05
2300	Tinted	17.25	14.65	31.90
2400	1/2" thick, clear	19	20	39
2450	Tempered	27.50	20	47.50
2500	Tinted	30	20	50
2600	3/4" thick, clear	39	31.50	70.50
2650	Tempered	45.50	31.50	77
3000	Spandrel glass, panels, 1/4" plate glass insul w/fiberglass, 1" thick	18.55	9.15	27.70
3100	2" thick	22	9.15	31.15
3200	Galvanized steel backing, add	6.45		6.45
3300	3/8" plate glass, 1" thick	31.50	9.15	40.65
3400	2" thick	35	9.15	44.15
4000	Polycarbonate, masked, clear or colored, 1/8" thick	15.40	6.45	21.85
4100	3/16" thick	17.60	6.65	24.25
4200	1/4" thick	18.70	7.10	25.80
4300	3/8" thick	28.50	7.35	35.85
5000	Facing panel, textured alum., 4' x 8' x 5/16" plywood backing, sgl face	4.48	3.65	8.13
5100	Double face	5.80	3.65	9.45
5200	4' x 10' x 5/16" plywood backing, single face	4.69	3.65	8.34
5300	Double face	6.40	3.65	10.05
5400	4' x 12' x 5/16" plywood backing, single face	4.69	3.65	8.34
5500	Sandwich panel, 22 Ga. galv., both sides 2" insulation, enamel exterior	13.40	6.35	19.75
5600	Polyvinylidene fluoride exterior finish	14.10	6.35	20.45
5700	26 Ga., galv. both sides, 1" insulation, colored 1 side	7.25	6.05	13.30
5800	Colored 2 sides	9.05	6.05	15.10

B2030 Exterior Doors

Costs are listed for exterior door systems by material, type and size. Prices between sizes listed can be interpolated with reasonable accuracy. Prices are per opening for a complete door system including frame.

B2030 110		Glazed Doors, Steel or Aluminum						
	MATERIAL	TYPE	DOORS	SPECIFICATION	OPENING	COST PER OPNG.		
						MAT.	INST.	TOTAL
5600	St. Stl. & glass	revolving	stock unit	manual oper.	6'-0" x 7'-0"	45,100	9,700	54,800
5650				auto Cntrls.	6'-10" x 7'-0"	61,500	10,400	71,900
5700	Bronze		stock unit	manual oper.	6'-10" x 7'-0"	53,000	19,500	72,500
5750				auto Cntrls.	6'-10" x 7'-0"	69,500	20,200	89,700
5800	St. Stl. & glass	balanced	standard	economy	3'-0" x 7'-0"	10,100	1,625	11,725
5850				premium	3'-0" x 7'-0"	17,300	2,150	19,450
6300	Alum. & glass	w/o transom	narrow stile	w/panic Hrdwre.	3'-0" x 7'-0"	1,875	1,075	2,950
6350				dbl. door, Hrdwre.	6'-0" x 7'-0"	3,025	1,775	4,800
6400			wide stile	hdwre.	3'-0" x 7'-0"	2,075	1,050	3,125
6450				dbl. door, Hdwre.	6'-0" x 7'-0"	4,000	2,100	6,100
6500			full vision	hdwre.	3'-0" x 7'-0"	2,775	1,675	4,450
6550				dbl. door, Hdwre.	6'-0" x 7'-0"	3,600	2,375	5,975
6600			non-standard	hdwre.	3'-0" x 7'-0"	2,275	1,050	3,325
6650				dbl. door, Hdwre.	6'-0" x 7'-0"	4,550	2,100	6,650
6700			bronze fin.	hdwre.	3'-0" x 7'-0"	1,650	1,050	2,700
6750				dbl. door, Hrdwre.	6'-0" x 7'-0"	3,325	2,100	5,425
6800			black fin.	hdwre.	3'-0" x 7'-0"	2,350	1,050	3,400
6850				dbl. door, Hdwre.	6'-0" x 7'-0"	4,700	2,100	6,800
6900		w/transom	narrow stile	hdwre.	3'-0" x 10'-0"	2,350	1,225	3,575
6950				dbl. door, Hdwre.	6'-0" x 10'-0"	3,700	2,125	5,825
7000			wide stile	hdwre.	3'-0" x 10'-0"	2,700	1,450	4,150
7050				dbl. door, Hdwre.	6'-0" x 10'-0"	4,050	2,525	6,575
7100			full vision	hdwre.	3'-0" x 10'-0"	2,975	1,625	4,600
7150				dbl. door, Hdwre.	6'-0" x 10'-0"	4,400	2,800	7,200
7200			non-standard	hdwre.	3'-0" x 10'-0"	2,350	1,125	3,475
7250				dbl. door, Hdwre.	6'-0" x 10'-0"	4,725	2,275	7,000
7300			bronze fin.	hdwre.	3'-0" x 10'-0"	1,750	1,125	2,875
7350				dbl. door, Hdwre.	6'-0" x 10'-0"	3,500	2,275	5,775
7400			black fin.	hdwre.	3'-0" x 10'-0"	2,425	1,125	3,550
7450				dbl. door, Hdwre.	6'-0" x 10'-0"	4,875	2,275	7,150
7500		revolving	stock design	minimum	6'-10" x 7'-0"	24,500	3,900	28,400
7550				average	6'-0" x 7'-0"	28,900	4,875	33,775

B2030 Exterior Doors

B2030 110			Glazed Doors, Steel or Aluminum				

	MATERIAL	TYPE	DOORS	SPECIFICATION	OPENING	COST PER OPNG.		
						MAT.	INST.	TOTAL
7600	Alum. & glass	w/transom	non-standard	maximum	6'-10" x 7'-0"	36,000	6,500	42,500
7650				min., automatic	6'-10" x 7'-0"	41,100	4,550	45,650
7700				avg., automatic	6'-10" x 7'-0"	45,500	5,525	51,025
7750				max., automatic	6'-10" x 7'-0"	52,500	7,150	59,650
7800		balanced	standard	economy	3'-0" x 7'-0"	7,275	1,625	8,900
7850				premium	3'-0" x 7'-0"	8,725	2,100	10,825
7900		mall front	sliding panels	alum. fin.	16'-0" x 9'-0"	3,825	845	4,670
7950					24'-0" x 9'-0"	5,425	1,575	7,000
8000				bronze fin.	16'-0" x 9'-0"	4,450	985	5,435
8050					24'-0" x 9'-0"	6,325	1,825	8,150
8100			fixed panels	alum. fin.	48'-0" x 9'-0"	10,000	1,225	11,225
8150				bronze fin.	48'-0" x 9'-0"	11,700	1,425	13,125
8200		sliding entrance	5' x 7' door	electric oper.	12'-0" x 7'-6"	9,400	1,575	10,975
8250		sliding patio	temp. glass	economy	6'-0" x 7'-0"	1,825	289	2,114
8300				economy	12'-0" x 7'-0"	4,275	385	4,660
8350				premium	6'-0" x 7'-0"	2,750	435	3,185
8400					12'-0" x 7'-0"	6,425	580	7,005

B2030 Exterior Doors

Exterior Door System

The table below lists exterior door systems by material, type and size. Prices between sizes listed can be interpolated with reasonable accuracy. Prices are per opening for a complete door system including frame and required hardware.

Wood doors in this table are designed with wood frames and metal doors with hollow metal frames. Depending upon quality the total material and installation cost of a wood door frame is about the same as a hollow metal frame.

B2030 210 — Wood Doors

	MATERIAL	TYPE	DOORS	SPECIFICATION	OPENING	COST PER OPNG. MAT.	INST.	TOTAL
2350	Birch	solid core	single door	hinged	2'-6" x 6'-8"	1,150	315	1,465
2400					2'-6" x 7'-0"	1,150	315	1,465
2450					2'-8" x 7'-0"	1,125	315	1,440
2500					3'-0" x 7'-0"	1,150	315	1,465
2550			double door		2'-6" x 6'-8"	2,150	575	2,725
2600					2'-6" x 7'-0"	2,150	575	2,725
2650					2'-8" x 7'-0"	2,125	575	2,700
2700					3'-0" x 7'-0"	2,000	505	2,505
2750	Wood	combination	storm & screen		3'-0" x 6'-8"	335	77	412
2800					3'-0" x 7'-0"	370	84.50	454.50
2850		overhead	panels, H.D.	manual oper.	8'-0" x 8'-0"	1,150	580	1,730
2900					10'-0" x 10'-0"	1,650	640	2,290
2950					12'-0" x 12'-0"	2,250	770	3,020
3000					14'-0" x 14'-0"	3,625	890	4,515
3050					20'-0" x 16'-0"	6,025	1,775	7,800
3100				electric oper.	8'-0" x 8'-0"	2,200	870	3,070
3150					10'-0" x 10'-0"	2,700	930	3,630
3200					12'-0" x 12'-0"	3,300	1,050	4,350
3250					14'-0" x 14'-0"	4,675	1,175	5,850
3300					20'-0" x 16'-0"	7,275	2,350	9,625

B2030 220 — Steel Doors

	MATERIAL	TYPE	DOORS	SPECIFICATION	OPENING	COST PER OPNG. MAT.	INST.	TOTAL
3350	Steel 18 Ga.	hollow metal	1 door w/frame	no label	2'-6" x 7'-0"	1,550	325	1,875
3400					2'-8" x 7'-0"	1,550	325	1,875
3450					3'-0" x 7'-0"	1,550	325	1,875
3500					3'-6" x 7'-0"	1,675	345	2,020
3550					4'-0" x 8'-0"	1,950	345	2,295
3600			2 doors w/frame		5'-0" x 7'-0"	3,000	605	3,605
3650					5'-4" x 7'-0"	3,000	605	3,605
3700					6'-0" x 7'-0"	3,000	605	3,605
3750					7'-0" x 7'-0"	3,250	635	3,885
3800					8'-0" x 8'-0"	3,825	640	4,465

For customer support on your Square Foot Cost Data, call 877.694.8329.

385

B2030 Exterior Doors

B2030 220 — Steel Doors

	MATERIAL	TYPE	DOORS	SPECIFICATION	OPENING	COST PER OPNG.		
						MAT.	INST.	TOTAL
3850	Steel 18 Ga.	hollow metal	1 door w/frame	"A" label	2'-6" x 7'-0"	1,950	395	2,345
3900					2'-8" x 7'-0"	1,950	400	2,350
3950					3'-0" x 7'-0"	1,950	400	2,350
4000					3'-6" x 7'-0"	2,100	405	2,505
4050					4'-0" x 8'-0"	2,300	425	2,725
4100			2 doors w/frame		5'-0" x 7'-0"	3,725	730	4,455
4150					5'-4" x 7'-0"	3,750	740	4,490
4200					6'-0" x 7'-0"	3,750	740	4,490
4250					7'-0" x 7'-0"	4,075	755	4,830
4300					8'-0" x 8'-0"	4,075	755	4,830
4350	Steel 24 Ga.	overhead	sectional	manual oper.	8'-0" x 8'-0"	900	580	1,480
4400					10'-0" x 10'-0"	1,225	640	1,865
4450					12'-0" x 12'-0"	1,475	770	2,245
4500					20'-0" x 14'-0"	4,100	1,650	5,750
4550				electric oper.	8'-0" x 8'-0"	1,950	870	2,820
4600					10'-0" x 10'-0"	2,275	930	3,205
4650					12'-0" x 12'-0"	2,525	1,050	3,575
4700					20'-0" x 14'-0"	5,350	2,225	7,575
4750	Steel		rolling	manual oper.	8'-0" x 8'-0"	1,275	915	2,190
4800					10'-0" x 10'-0"	2,100	1,050	3,150
4850					12'-0" x 12'-0"	2,125	1,225	3,350
4900					14'-0" x 14'-0"	3,425	1,825	5,250
4950					20'-0" x 12'-0"	2,400	1,625	4,025
5000					20'-0" x 16'-0"	4,100	2,450	6,550
5050				electric oper.	8'-0" x 8'-0"	2,550	1,200	3,750
5100					10'-0" x 10'-0"	3,375	1,350	4,725
5150					12'-0" x 12'-0"	3,400	1,525	4,925
5200					14'-0" x 14'-0"	4,700	2,125	6,825
5250					20'-0" x 12'-0"	3,575	1,925	5,500
5300					20'-0" x 16'-0"	5,275	2,750	8,025
5350				fire rated	10'-0" x 10'-0"	2,400	1,325	3,725
5400			rolling grille	manual oper.	10'-0" x 10'-0"	2,750	1,475	4,225
5450					15'-0" x 8'-0"	3,175	1,825	5,000
5500		vertical lift	1 door w/frame	motor operator	16'-0" x 16'-0"	23,700	5,675	29,375
5550					32'-0" x 24'-0"	53,500	3,775	57,275

B2030 230 — Aluminum Doors

	MATERIAL	TYPE	DOORS	SPECIFICATION	OPENING	COST PER OPNG.		
						MAT.	INST.	TOTAL
6000	Aluminum	combination	storm & screen	hinged	3'-0" x 6'-8"	340	82.50	422.50
6050					3'-0" x 7'-0"	375	91	466
6100		overhead	rolling grille	manual oper.	12'-0" x 12'-0"	4,475	2,575	7,050
6150				motor oper.	12'-0" x 12'-0"	5,850	2,875	8,725
6200	Alum. & Fbrgls.		heavy duty	manual oper.	12'-0" x 12'-0"	2,975	770	3,745
6250				electric oper.	12'-0" x 12'-0"	4,025	1,050	5,075

B30 Roofing

B3010 Roof Coverings

Built Up Ply

Multiple ply roofing is the most popular covering for minimum pitch roofs.

B3010 105	Built-Up	COST PER S.F.		
		MAT.	INST.	TOTAL
1200	Asphalt flood coat w/gravel; not incl. insul, flash., nailers			
1300				
1400	Asphalt base sheets & 3 plies #15 asphalt felt, mopped	1.10	1.92	3.02
1500	On nailable deck	1.12	2.01	3.13
1600	4 plies #15 asphalt felt, mopped	1.49	2.11	3.60
1700	On nailable deck	1.32	2.22	3.54
1800	Coated glass base sheet, 2 plies glass (type IV), mopped	1.18	1.92	3.10
1900	For 3 plies	1.43	2.11	3.54
2000	On nailable deck	1.34	2.22	3.56
2300	4 plies glass fiber felt (type IV), mopped	1.76	2.11	3.87
2400	On nailable deck	1.59	2.22	3.81
2500	Organic base sheet & 3 plies #15 organic felt, mopped	1.17	2.11	3.28
2600	On nailable deck	1.14	2.22	3.36
2700	4 plies #15 organic felt, mopped	1.43	1.92	3.35
2750				
2800	Asphalt flood coat, smooth surface, not incl. insul, flash., nailers			
2900	Asphalt base sheet & 3 plies #15 asphalt felt, mopped	1.15	1.75	2.90
3000	On nailable deck	1.06	1.83	2.89
3100	Coated glass fiber base sheet & 2 plies glass fiber felt, mopped	1.13	1.68	2.81
3200	On nailable deck	1.06	1.75	2.81
3300	For 3 plies, mopped	1.37	1.83	3.20
3400	On nailable deck	1.29	1.92	3.21
3700	4 plies glass fiber felt (type IV), mopped	1.62	1.83	3.45
3800	On nailable deck	1.53	1.92	3.45
3900	Organic base sheet & 3 plies #15 organic felt, mopped	1.18	1.75	2.93
4000	On nailable decks	1.09	1.83	2.92
4100	4 plies #15 organic felt, mopped	1.38	1.92	3.30
4200	Coal tar pitch with gravel surfacing			
4300	4 plies #15 tarred felt, mopped	2.57	2.01	4.58
4400	3 plies glass fiber felt (type IV), mopped	2.12	2.22	4.34
4500	Coated glass fiber base sheets 2 plies glass fiber felt, mopped	2.16	2.22	4.38
4600	On nailable decks	1.87	2.34	4.21
4800	3 plies glass fiber felt (type IV), mopped	2.90	2.01	4.91
4900	On nailable decks	2.61	2.11	4.72

B3010   Roof Coverings

Full Adhered
Single Ply Membrane

Preformed Metal

Flat Seam

Batten Seam

Standing Seam

B3010 120	Single Ply Membrane	COST PER S.F.		
		MAT.	INST.	TOTAL
1000	CSPE (Chlorosulfonated polyethylene), 45 mils, plate attached	3.40	.77	4.17
2000	EPDM (Ethylene propylene diene monomer), 45 mils, fully adhered	1.22	1.04	2.26
4000	Modified bit., SBS modified, granule surface cap sheet, mopped, 150 mils	1.38	2.10	3.48
4500	APP modified, granule surface cap sheet, torched, 180 mils	.98	1.36	2.34
6000	Reinforced PVC, 48 mils, loose laid and ballasted with stone	1.30	.53	1.83
6200	Fully adhered with adhesive	1.68	1.04	2.72

B3010 130	Preformed Metal Roofing	COST PER S.F.		
		MAT.	INST.	TOTAL
0200	Corrugated roofing, aluminum, mill finish, .0175" thick, .272 P.S.F.	1.09	1.91	3
0250	.0215" thick, .334 P.S.F.	1.43	1.91	3.34

B3010 135	Formed Metal	COST PER S.F.		
		MAT.	INST.	TOTAL
1000	Batten seam, formed copper roofing, 3" min slope, 16 oz., 1.2 P.S.F.	13.35	6.30	19.65
1100	18 oz., 1.35 P.S.F.	15.10	6.95	22.05
2000	Zinc copper alloy, 3" min slope, .020" thick, .88 P.S.F.	13.60	5.80	19.40
3000	Flat seam, copper, 1/4" min. slope, 16 oz., 1.2 P.S.F.	9.90	5.80	15.70
3100	18 oz., 1.35 P.S.F.	11.10	6.05	17.15
5000	Standing seam, copper, 2-1/2" min. slope, 16 oz., 1.25 P.S.F.	10.60	5.35	15.95
5100	18 oz., 1.40 P.S.F.	11.85	5.80	17.65
6000	Zinc copper alloy, 2-1/2" min. slope, .020" thick, .87 P.S.F.	13.10	5.80	18.90
6100	.032" thick, 1.39 P.S.F.	21	6.30	27.30

B3010 Roof Coverings

Shingle and Tile

Shingles and tiles are practical in applications where the roof slope is more than 3-1/2″ per foot of rise. Lines 1100 through 6000 list the various materials and the weight per square foot.

Roof Edges

The table below lists the costs for various types of roof perimeter edge treatments.

Roof edge systems include the cost per L.F. for a 2″ x 8″ treated wood nailer fastened at 4′-0″ O.C. and a diagonally cut 4″ x 6″ treated wood cant.

Roof edge and base flashing are assumed to be made from the same material.

B3010 140	Shingle & Tile	COST PER S.F.		
		MAT.	INST.	TOTAL
1095	Asphalt roofing			
1100	Strip shingles, 4″ slope, inorganic class A 210-235 lb./sq.	.94	1.16	2.10
1150	Organic, class C, 235-240 lb./sq.	1.09	1.26	2.35
1200	Premium laminated multi-layered, class A, 260-300 lb./sq.	1.74	1.72	3.46
1545	Metal roofing			
1550	Alum., shingles, colors, 3″ min slope, .019″ thick, 0.4 PSF	2.78	1.25	4.03
1850	Steel, colors, 3″ min slope, 26 gauge, 1.0 PSF	4.44	2.58	7.02
2795	Slate roofing			
2800	Slate roofing, 4″ min. slope, shingles, 3/16″ thick, 8.0 PSF	5.25	3.24	8.49
3495	Wood roofing			
3500	4″ min slope, cedar shingles, 16″ x 5″, 5″ exposure 1.6 PSF	3.27	2.49	5.76
4000	Shakes, 18″, 8-1/2″ exposure, 2.8 PSF	2.10	3.07	5.17
5095	Tile roofing			
5100	Aluminum, mission, 3″ min slope, .019″ thick, 0.65 PSF	9.20	2.40	11.60
6000	Clay tile, flat shingle, interlocking, 15″, 166 pcs/sq, fireflashed blend	5.65	2.83	8.48

B3010   Roof Coverings

B3010 320	Roof Deck Rigid Insulation	COST PER S.F.		
		MAT.	INST.	TOTAL
0100	Fiberboard low density			
0150	1" thick R2.78	.68	.71	1.39
0300	1 1/2" thick R4.17	.99	.71	1.70
0350	2" thick R5.56	1.28	.71	1.99
0370	Fiberboard high density, 1/2" thick R1.3	.44	.60	1.04
0380	1" thick R2.5	.75	.71	1.46
0390	1 1/2" thick R3.8	1.08	.71	1.79
0410	Fiberglass, 3/4" thick R2.78	.78	.60	1.38
0450	15/16" thick R3.70	1	.60	1.60
0550	1-5/16" thick R5.26	1.62	.60	2.22
0650	2 7/16" thick R10	1.96	.71	2.67
1510	Polyisocyanurate 2#/CF density, 1" thick	.64	.48	1.12
1550	1 1/2" thick	.81	.52	1.33
1600	2" thick	1.02	.56	1.58
1650	2 1/2" thick	1.28	.58	1.86
1700	3" thick	1.51	.60	2.11
1750	3 1/2" thick	2.26	.60	2.86
1800	Tapered for drainage	.81	.48	1.29
1810	Expanded polystyrene, 1#/CF density, 3/4" thick R2.89	.32	.46	.78
1820	2" thick R7.69	.66	.52	1.18
1830	Extruded polystyrene, 15 PSI compressive strength, 1" thick, R5	.72	.46	1.18
1835	2" thick R10	.90	.52	1.42
1840	3" thick R15	1.68	.60	2.28
2550	40 PSI compressive strength, 1" thick R5	.69	.46	1.15
2600	2" thick R10	1.22	.52	1.74
2650	3" thick R15	1.71	.60	2.31
2700	4" thick R20	2.21	.60	2.81
2750	Tapered for drainage	.95	.48	1.43
2810	60 PSI compressive strength, 1" thick R5	.92	.47	1.39
2850	2" thick R10	1.66	.53	2.19
2900	Tapered for drainage	1.17	.48	1.65
3050	Composites with 2" EPS			
3060	1" Fiberboard	1.65	.59	2.24
3070	7/16" oriented strand board	1.35	.71	2.06
3080	1/2" plywood	1.64	.71	2.35
3090	1" perlite	1.36	.71	2.07
4000	Composites with 1-1/2" polyisocyanurate			
4010	1" fiberboard	1.42	.71	2.13
4020	1" perlite	1.28	.68	1.96
4030	7/16" oriented strand board	1.13	.71	1.84

B3010 420			Roof Edges					

	EDGE TYPE	DESCRIPTION	SPECIFICATION	FACE HEIGHT		COST PER L.F.		
						MAT.	INST.	TOTAL
1000	Aluminum	mill finish	.050" thick	4"		14.80	10.55	25.35
1100				6"		15	10.85	25.85
1300		duranodic	.050" thick	4"		15.35	10.55	25.90
1400				6"		15.90	10.85	26.75
1600		painted	.050" thick	4"		15.55	10.55	26.10
1700				6"		16.90	10.85	27.75
2000	Copper	plain	16 oz.	4"		34.50	10.55	45.05
2100				6"		44.50	10.85	55.35
2300			20 oz.	4"		43	12.10	55.10
2400				6"		50	11.65	61.65
2700	Sheet Metal	galvanized	20 Ga.	4"		17.80	12.55	30.35
2800				6"		18.15	12.55	30.70
3000			24 Ga.	4"		14.80	10.55	25.35

B30 Roofing

B3010 Roof Coverings

B3010 420	Roof Edges							

	EDGE TYPE	DESCRIPTION	SPECIFICATION	FACE HEIGHT		COST PER L.F.		
						MAT.	INST.	TOTAL
3100	Sheet Metal	galvanized	20 Ga.	6"		15.05	10.55	25.60

B3010 430	Flashing							

	MATERIAL	BACKING	SIDES	SPECIFICATION	QUANTITY	COST PER S.F.		
						MAT.	INST.	TOTAL
0040	Aluminum	none		.019"		1.27	3.78	5.05
0050				.032"		1.49	3.78	5.27
0300		fabric	2	.004"		1.69	1.66	3.35
0400		mastic		.004"		1.69	1.66	3.35
0700	Copper	none		16 oz.	<500 lbs.	8.45	4.76	13.21
0800				24 oz.		14.65	5.20	19.85
3500	PVC black			.010"		.23	1.92	2.15
3700				.030"		.39	1.92	2.31
4200	Neoprene			1/16"		2.45	1.92	4.37
4500	Stainless steel			.015"	<500 lbs.	7.05	4.76	11.81
4600	Copper clad				>2000 lbs.	6.70	3.53	10.23
5000	Plain			32 ga.		3.30	3.53	6.83

B3010   Roof Coverings

B3010 610 — Gutters

	SECTION	MATERIAL	THICKNESS	SIZE	FINISH	COST PER L.F. MAT.	INST.	TOTAL
0050	Box	aluminum	.027"	5"	enameled	3.02	5.60	8.62
0100					mill	3.09	5.60	8.69
0200			.032"	5"	enameled	3.32	5.45	8.77
0500		copper	16 Oz.	5"	lead coated	20	5.45	25.45
0600					mill	8.85	5.45	14.30
1000		steel galv.	28 Ga.	5"	enameled	2.31	5.45	7.76
1200			26 Ga.	5"	mill	2.40	5.45	7.85
1800		vinyl		4"	colors	1.38	5.05	6.43
1900				5"	colors	1.71	5.05	6.76
2300		hemlock or fir		4"x5"	treated	18.45	5.80	24.25
3000	Half round	copper	16 Oz.	4"	lead coated	15.20	5.45	20.65
3100					mill	9.80	5.45	15.25
3600		steel galv.	28 Ga.	5"	enameled	2.31	5.45	7.76
4102		stainless steel		5"	mill	10.40	5.45	15.85
5000		vinyl		4"	white	1.43	5.05	6.48

B3010 620 — Downspouts

	MATERIALS	SECTION	SIZE	FINISH	THICKNESS	COST PER V.L.F. MAT.	INST.	TOTAL
0100	Aluminum	rectangular	2"x3"	embossed mill	.020"	1.08	3.60	4.68
0150				enameled	.020"	1.51	3.60	5.11
0250			3"x4"	enameled	.024"	2.56	4.89	7.45
0300		round corrugated	3"	enameled	.020"	2.15	3.60	5.75
0350			4"	enameled	.025"	2.78	4.89	7.67
0500	Copper	rectangular corr.	2"x3"	mill	16 Oz.	9.40	3.60	13
0600		smooth		mill	16 Oz.	12.15	3.60	15.75
0700		rectangular corr.	3"x4"	mill	16 Oz.	11.45	4.72	16.17
1300	Steel	rectangular corr.	2"x3"	galvanized	28 Ga.	2.13	3.60	5.73
1350				epoxy coated	24 Ga.	2.48	3.60	6.08
1400		smooth		galvanized	28 Ga.	4.02	3.60	7.62
1450		rectangular corr.	3"x4"	galvanized	28 Ga.	2	4.72	6.72
1500				epoxy coated	24 Ga.	3.08	4.72	7.80
1550		smooth		galvanized	28 Ga.	4.47	4.72	9.19
1600		round corrugated	2"	galvanized	28 Ga.	2.29	3.60	5.89
1650			3"	galvanized	28 Ga.	2.29	3.60	5.89
1700			4"	galvanized	28 Ga.	2.77	4.72	7.49
1750			5"	galvanized	28 Ga.	4.40	5.25	9.65
2552	S.S. tubing sch.5	rectangular	3"x4"	mill		54	4.72	58.72

B3010 630 — Gravel Stop

	MATERIALS	SECTION	SIZE	FINISH	THICKNESS	COST PER L.F. MAT.	INST.	TOTAL
5100	Aluminum	extruded	4"	mill	.050"	6.90	4.72	11.62
5200			4"	duranodic	.050"	7.45	4.72	12.17
5300			8"	mill	.050"	8.10	5.45	13.55
5400			8"	duranodic	.050"	9.25	5.45	14.70
6000			12"-2 pc.	duranodic	.050"	11.30	6.85	18.15
6100	Stainless	formed	6"	mill	24 Ga.	16.85	5.05	21.90

B3020 Roof Openings

Roof Hatch

Smoke Hatch

Skylight

B3020 110	Skylights	COST PER S.F.		
		MAT.	INST.	TOTAL
5100	Skylights, plastic domes, insul curbs, nom. size to 10 S.F., single glaze	31	14.35	45.35
5200	Double glazing	35.50	17.65	53.15
5300	10 S.F. to 20 S.F., single glazing	31.50	5.80	37.30
5400	Double glazing	30.50	7.30	37.80
5500	20 S.F. to 30 S.F., single glazing	24	4.93	28.93
5600	Double glazing	28.50	5.80	34.30
5700	30 S.F. to 65 S.F., single glazing	22.50	3.76	26.26
5800	Double glazing	29	4.93	33.93
6000	Sandwich panels fiberglass, 1-9/16" thick, 2 S.F. to 10 S.F.	19.70	11.45	31.15
6100	10 S.F. to 18 S.F.	17.75	8.65	26.40
6200	2-3/4" thick, 25 S.F. to 40 S.F.	28.50	7.80	36.30
6300	40 S.F. to 70 S.F.	23.50	6.95	30.45

B3020 210	Hatches	COST PER OPNG.		
		MAT.	INST.	TOTAL
0200	Roof hatches with curb and 1" fiberglass insulation, 2'-6"x3'-0", aluminum	1,125	229	1,354
0300	Galvanized steel 165 lbs.	705	229	934
0400	Primed steel 164 lbs.	785	229	1,014
0500	2'-6"x4'-6" aluminum curb and cover, 150 lbs.	1,525	255	1,780
0600	Galvanized steel 220 lbs.	975	255	1,230
0650	Primed steel 218 lbs.	1,050	255	1,305
0800	2'x6"x8'-0" aluminum curb and cover, 260 lbs.	2,100	350	2,450
0900	Galvanized steel, 360 lbs.	1,975	350	2,325
0950	Primed steel 358 lbs.	1,675	350	2,025
1200	For plexiglass panels, add to the above	485		485
2100	Smoke hatches, unlabeled not incl. hand winch operator, 2'-6"x3', galv	855	277	1,132
2200	Plain steel, 160 lbs.	950	277	1,227
2400	2'-6"x8'-0", galvanized steel, 360 lbs.	2,150	385	2,535
2500	Plain steel, 350 lbs.	1,825	385	2,210
3000	4'-0"x8'-0", double leaf low profile, aluminum cover, 359 lb.	3,075	287	3,362
3100	Galvanized steel 475 lbs.	2,625	287	2,912
3200	High profile, aluminum cover, galvanized curb, 361 lbs.	2,950	287	3,237

C1010 Partitions

The Concrete Block Partition Systems are defined by weight and type of block, thickness, type of finish and number of sides finished. System components include joint reinforcing on alternate courses and vertical control joints.

C1010 102	Concrete Block Partitions - Regular Weight						
					COST PER S.F.		
	TYPE	THICKNESS (IN.)	TYPE FINISH	SIDES FINISHED	MAT.	INST.	TOTAL
1000	Hollow	4	none	0	2.13	6.30	8.43
1010			gyp. plaster 2 coat	1	2.52	9.20	11.72
1020				2	2.90	12.10	15
1200			portland - 3 coat	1	2.40	9.70	12.10
1400			5/8" drywall	1	2.74	8.35	11.09
1500		6	none	0	2.69	6.75	9.44
1510			gyp. plaster 2 coat	1	3.08	9.65	12.73
1520				2	3.46	12.55	16.01
1700			portland - 3 coat	1	2.96	10.15	13.11
1900			5/8" drywall	1	3.30	8.80	12.10
1910				2	3.91	10.85	14.76
2000		8	none	0	3.27	7.25	10.52
2010			gyp. plaster 2 coat	1	3.66	10.10	13.76
2020				2	4.04	13.05	17.09
2200			portland - 3 coat	1	3.54	10.60	14.14
2400			5/8" drywall	1	3.88	9.30	13.18
2410				2	4.49	11.30	15.79
2500	Hollow	10	none	0	3.42	7.60	11.02
2510			gyp. plaster 2 coat	1	3.81	10.45	14.26
2520				2	4.19	13.40	17.59
2700			portland - 3 coat	1	3.69	10.95	14.64
2900			5/8" drywall	1	4.03	9.60	13.63
2910				2	4.64	11.65	16.29
3000	Solid	2	none	0	1.85	6.20	8.05
3010			gyp. plaster	1	2.24	9.10	11.34
3020				2	2.62	12	14.62
3200			portland - 3 coat	1	2.12	9.60	11.72
3400			5/8" drywall	1	2.46	8.25	10.71
3410				2	3.07	10.30	13.37
3500		4	none	0	2.41	6.50	8.91
3510			gyp. plaster	1	2.91	9.40	12.31
3520				2	3.18	12.30	15.48
3700			portland - 3 coat	1	2.68	9.90	12.58
3900			5/8" drywall	1	3.02	8.55	11.57
3910				2	3.63	10.60	14.23

C1010 Partitions

C1010 102		Concrete Block Partitions - Regular Weight						

	TYPE	THICKNESS (IN.)	TYPE FINISH	SIDES FINISHED		COST PER S.F.		
						MAT.	INST.	TOTAL
4000	Solid	6	none	0		2.87	7	9.87
4010			gyp. plaster	1		3.26	9.90	13.16
4020				2		3.64	12.80	16.44
4200			portland - 3 coat	1		3.14	10.40	13.54
4400			5/8" drywall	1		3.48	9.05	12.53
4410				2		4.09	11.10	15.19

C1010 104		Concrete Block Partitions - Lightweight						

	TYPE	THICKNESS (IN.)	TYPE FINISH	SIDES FINISHED		COST PER S.F.		
						MAT.	INST.	TOTAL
5000	Hollow	4	none	0		2.24	6.15	8.39
5010			gyp. plaster	1		2.63	9.05	11.68
5020				2		3.01	11.95	14.96
5200			portland - 3 coat	1		2.51	9.55	12.06
5400			5/8" drywall	1		2.85	8.20	11.05
5410				2		3.46	10.25	13.71
5500		6	none	0		3.06	6.60	9.66
5520			gyp. plaster	2		3.83	12.40	16.23
5700			portland - 3 coat	1		3.33	10	13.33
5900			5/8" drywall	1		3.67	8.65	12.32
5910				2		4.28	10.70	14.98
6000		8	none	0		3.69	7.05	10.74
6010			gyp. plaster	1		4.08	9.90	13.98
6020				2		4.46	12.85	17.31
6200			portland - 3 coat	1		3.96	10.40	14.36
6400			5/8" drywall	1		4.30	9.10	13.40
6410				2		4.91	11.10	16.01
6500		10	none	0		4.44	7.40	11.84
6510			gyp. plaster	1		4.83	10.25	15.08
6520				2		5.20	13.20	18.40
6700			portland - 3 coat	1		4.71	10.75	15.46
6900			5/8" drywall	1		5.05	9.40	14.45
6910				2		5.65	11.45	17.10
7000	Solid	4	none	0		2.05	6.45	8.50
7010			gyp. plaster	1		2.44	9.35	11.79
7020				2		2.82	12.25	15.07
7200			portland - 3 coat	1		2.32	9.85	12.17
7400			5/8" drywall	1		2.66	8.50	11.16
7410				2		3.27	10.55	13.82
7500		6	none	0		3.49	6.90	10.39
7510			gyp. plaster	1		4.02	9.85	13.87
7520				2		4.26	12.70	16.96
7700			portland - 3 coat	1		3.76	10.30	14.06
7900			5/8" drywall	1		4.10	8.95	13.05
7910				2		4.71	11	15.71
8000		8	none	0		4.43	7.45	11.88
8010			gyp. plaster	1		4.82	10.30	15.12
8020				2		5.20	13.25	18.45
8200			portland - 3 coat	1		4.70	10.80	15.50
8400			5/8" drywall	1		5.05	9.50	14.55
8410				2		5.65	11.50	17.15

For customer support on your Square Foot Cost Data, call 877.694.8329.

397

C1010 Partitions

Single Wythe

Double Wythe

C1010 120	Tile Partitions	COST PER S.F.		
		MAT.	INST.	TOTAL
1000	8W series 8"x16", 4" thick wall, reinf every 2 courses, glazed 1 side	15.80	7.60	23.40
1100	Glazed 2 sides	18.85	8.05	26.90
1200	Glazed 2 sides, using 2 wythes of 2" thick tile	23	14.60	37.60
1300	6" thick wall, horizontal reinf every 2 courses, glazed 1 side	27	7.95	34.95
1400	Glazed 2 sides, each face different color, 2" and 4" tile	27.50	14.90	42.40
1500	8" thick wall, glazed 2 sides using 2 wythes of 4" thick tile	31.50	15.20	46.70
1600	10" thick wall, glazed 2 sides using 1 wythe of 4" & 1 wythe of 6" tile	43	15.55	58.55
1700	Glazed 2 sides cavity wall, using 2 wythes of 4" thick tile	31.50	15.20	46.70
1800	12" thick wall, glazed 2 sides using 2 wythes of 6" thick tile	54	15.90	69.90
1900	Glazed 2 sides cavity wall, using 2 wythes of 4" thick tile	31.50	15.20	46.70
2100	6T series 5-1/3"x12" tile, 4" thick, non load bearing glazed one side	13.40	11.90	25.30
2200	Glazed two sides	17.60	13.45	31.05
2300	Glazed two sides, using two wythes of 2" thick tile	19.70	23.50	43.20
2400	6" thick, glazed one side	20	12.50	32.50
2500	Glazed two sides	24	14.15	38.15
2600	Glazed two sides using 2" thick tile and 4" thick tile	23.50	23.50	47
2700	8" thick, glazed one side	26.50	14.55	41.05
2800	Glazed two sides using two wythes of 4" thick tile	27	24	51
2900	Glazed two sides using 6" thick tile and 2" thick tile	30	24	54
3000	10" thick cavity wall, glazed two sides using two wythes of 4" tile	27	24	51
3100	12" thick, glazed two sides using 4" thick tile and 8" thick tile	40	26.50	66.50
3200	2" thick facing tile, glazed one side, on 6" concrete block	13.95	13.85	27.80
3300	On 8" concrete block	14.50	14.30	28.80
3400	On 10" concrete block	14.65	14.60	29.25

C1010 Partitions

Wood Stud Framing

Metal Stud Framing

The Drywall Partitions/Stud Framing Systems are defined by type of drywall and number of layers, type and spacing of stud framing, and treatment on the opposite face. Components include taping and finishing.

Cost differences between regular and fire resistant drywall are negligible, and terminology is interchangeable. In some cases fiberglass insulation is included for additional sound deadening.

C1010 124	Drywall Partitions/Wood Stud Framing							
	FACE LAYER	BASE LAYER	FRAMING	OPPOSITE FACE	INSULATION	COST PER S.F.		
						MAT.	INST.	TOTAL
1200	5/8" FR drywall	none	2 x 4, @ 16" O.C.	same	0	1.33	3.48	4.81
1250				5/8" reg. drywall	0	1.32	3.48	4.80
1300				nothing	0	.91	2.32	3.23
1400		1/4" SD gypsum	2 x 4 @ 16" O.C.	same	1-1/2" fiberglass	2.63	5.35	7.98
1450				5/8" FR drywall	1-1/2" fiberglass	2.20	4.70	6.90
1500				nothing	1-1/2" fiberglass	1.78	3.54	5.32
1600		resil. channels	2 x 4 @ 16", O.C.	same	1-1/2" fiberglass	2.11	6.80	8.91
1650				5/8" FR drywall	1-1/2" fiberglass	1.94	5.40	7.34
1700				nothing	1-1/2" fiberglass	1.52	4.26	5.78
1800		5/8" FR drywall	2 x 4 @ 24" O.C.	same	0	1.95	4.41	6.36
1850				5/8" FR drywall	0	1.58	3.83	5.41
1900				nothing	0	1.16	2.67	3.83
2200			2 rows-2 x 4 16"O.C.	same	2" fiberglass	3.06	6.35	9.41
2250				5/8" FR drywall	2" fiberglass	2.69	5.80	8.49
2300				nothing	2" fiberglass	2.27	4.63	6.90
2400	5/8" WR drywall	none	2 x 4, @ 16" O.C.	same	0	1.51	3.48	4.99
2450				5/8" FR drywall	0	1.42	3.48	4.90
2500				nothing	0	1	2.32	3.32
2600		5/8" FR drywall	2 x 4, @ 24" O.C.	same	0	2.13	4.41	6.54
2650				5/8" FR drywall	0	1.67	3.83	5.50
2700				nothing	0	1.25	2.67	3.92
2800	5/8" VF drywall	none	2 x 4, @ 16" O.C.	same	0	2.17	3.72	5.89
2850				5/8" FR drywall	0	1.75	3.60	5.35
2900				nothing	0	1.33	2.44	3.77
3000		5/8" FR drywall	2 x 4 , 24" O.C.	same	0	2.79	4.65	7.44
3050				5/8" FR drywall	0	2	3.95	5.95
3100				nothing	0	1.58	2.79	4.37
3200	1/2" reg drywall	3/8" reg drywall	2 x 4, @ 16" O.C.	same	0	1.99	4.64	6.63
3250				5/8" FR drywall	0	1.66	4.06	5.72
3300				nothing	0	1.24	2.90	4.14

For customer support on your Square Foot Cost Data, call 877.694.8329.

399

C10 Interior Construction

C1010 Partitions

| C1010 126 | Drywall Partitions/Metal Stud Framing |

	FACE LAYER	BASE LAYER	FRAMING	OPPOSITE FACE	INSULATION	MAT.	INST.	TOTAL
5000	5/8" FR drywall	none	1-5/8" @ 16" O.C.	same	0	1.12	3.49	4.61
5010				5/8" reg. drywall	0	1.11	3.49	4.60
5020				nothing	0	.70	2.33	3.03
5030			2-1/2"" @ 16" O.C.	same	0	1.18	3.50	4.68
5040				5/8" reg. drywall	0	1.17	3.50	4.67
5050				nothing	0	.76	2.34	3.10
5060			3-5/8" @ 16" O.C.	same	0	1.24	3.52	4.76
5070				5/8" reg. drywall	0	1.23	3.52	4.75
5080				nothing	0	.82	2.36	3.18
5200			1-5/8" @ 24" O.C.	same	0	1.04	3.08	4.12
5250				5/8" reg. drywall	0	1.03	3.08	4.11
5300				nothing	0	.62	1.92	2.54
5310			2-1/2" @ 24" O.C.	same	0	1.09	3.09	4.18
5320				5/8" reg. drywall	0	1.08	3.09	4.17
5330				nothing	0	.67	1.93	2.60
5400			3-5/8" @ 24" O.C.	same	0	1.14	3.10	4.24
5450				5/8" reg. drywall	0	1.13	3.10	4.23
5500				nothing	0	.72	1.94	2.66
5530		1/4" SD gypsum	1-5/8" @ 16" O.C.	same	0	1.98	4.77	6.75
5535				5/8" FR drywall	0	1.12	3.49	4.61
5540				nothing	0	1.13	2.97	4.10
5545			2-1/2" @ 16" O.C.	same	0	2.04	4.78	6.82
5550				5/8" FR drywall	0	1.18	3.50	4.68
5555				nothing	0	1.19	2.98	4.17
5560			3-5/8" @ 16" O.C.	same	0	2.10	4.80	6.90
5565				5/8" FR drywall	0	1.24	3.52	4.76
5570				nothing	0	1.25	3	4.25
5600			1-5/8" @ 24" O.C.	same	0	1.90	4.36	6.26
5650				5/8" FR drywall	0	1.47	3.72	5.19
5700				nothing	0	1.05	2.56	3.61
5800			2-1/2" @ 24" O.C.	same	0	1.95	4.37	6.32
5850				5/8" FR drywall	0	1.52	3.73	5.25
5900				nothing	0	1.10	2.57	3.67
5910			3-5/8" @ 24" O.C.	same	0	2	4.38	6.38
5920				5/8" FR drywall	0	1.57	3.74	5.31
5930				nothing	0	1.15	2.58	3.73
6000		5/8" FR drywall	2-1/2" @ 16" O.C.	same	0	2.01	4.96	6.97
6050				5/8" FR drywall	0	1.64	4.38	6.02
6100				nothing	0	1.22	3.22	4.44
6110			3-5/8" @ 16" O.C.	same	0	1.98	4.68	6.66
6120				5/8" FR drywall	0	1.61	4.10	5.71
6130				nothing	0	1.19	2.94	4.13
6170			2-1/2" @ 24" O.C.	same	0	1.83	4.25	6.08
6180				5/8" FR drywall	0	1.46	3.67	5.13
6190				nothing	0	1.04	2.51	3.55
6200			3-5/8" @ 24" O.C.	same	0	1.88	4.26	6.14
6250				5/8"FR drywall	3-1/2" fiberglass	2.01	4.11	6.12
6300				nothing	0	1.09	2.52	3.61

400

For customer support on your Square Foot Cost Data, call 877.694.8329.

C1010 Partitions

C1010 126	Drywall Partitions/Metal Stud Framing

	FACE LAYER	BASE LAYER	FRAMING	OPPOSITE FACE	INSULATION	COST PER S.F.		
						MAT.	INST.	TOTAL
6310	5/8" WR drywall	none	1-5/8" @ 16" O.C.	same	0	1.30	3.49	4.79
6320				5/8" FR drywall	0	1.49	4.07	5.56
6330				nothing	0	1.07	2.91	3.98
6340			2-1/2" @ 16" O.C.	same	0	1.92	4.66	6.58
6350				5/8" FR drywall	0	1.55	4.08	5.63
6360				nothing	0	1.13	2.92	4.05
6370			3-5/8" @ 16" O.C.	same	0	1.98	4.68	6.66
6380				5/8" FR drywall	0	1.61	4.10	5.71
6390				nothing	0	1.19	2.94	4.13
6400	5/8" WR drywall	none	1-5/8" @ 24" O.C.	same	0	1.22	3.08	4.30
6450				5/8" FR drywall	0	1.13	3.08	4.21
6500				nothing	0	.71	1.92	2.63
6510			2-1/2" @ 24" O.C.	same	0	1.83	4.25	6.08
6520				5/8" FR drywall	0	1.18	3.09	4.27
6530				nothing	0	.76	1.93	2.69
6600			3-5/8" @ 24" O.C.	same	0	1.32	3.10	4.42
6650				5/8" FR drywall	0	1.23	3.10	4.33
6700				nothing	0	.81	1.94	2.75
6800		5/8" FR drywall	2-1/2" @ 16" O.C.	same	0	2.19	4.96	7.15
6850				5/8" FR drywall	0	1.73	4.38	6.11
6900				nothing	0	1.31	3.22	4.53
6910			3-5/8" @ 16" O.C.	same	0	2.16	4.68	6.84
6920				5/8" FR drywall	0	1.70	4.10	5.80
6930				nothing	0	1.28	2.94	4.22
6940			2-1/2" @ 24" O.C.	same	0	1.83	4.25	6.08
6950				5/8" FR drywall	0	1.55	3.67	5.22
6960				nothing	0	1.13	2.51	3.64
7000			3-5/8" @ 24" O.C.	same	0	2.06	4.26	6.32
7050				5/8"FR drywall	3-1/2" fiberglass	2.10	4.11	6.21
7100				nothing	0	1.18	2.52	3.70
7110	5/8" VF drywall	none	1-5/8" @ 16" O.C.	same	0	1.96	3.73	5.69
7120				5/8" FR drywall	0	1.58	4.07	5.65
7130				nothing	0	1.16	2.91	4.07
7140			3-5/8" @ 16" O.C.	same	0	2.16	4.68	6.84
7150				5/8" FR drywall	0	1.70	4.10	5.80
7160				nothing	0	1.28	2.94	4.22
7200	5/8" VF drywall	none	1-5/8" @ 24" O.C.	same	0	1.88	3.32	5.20
7250				5/8" FR drywall	0	1.46	3.20	4.66
7300				nothing	0	1.04	2.04	3.08
7400			3-5/8" @ 24" O.C.	same	0	1.98	3.34	5.32
7450				5/8" FR drywall	0	1.56	3.22	4.78
7500				nothing	0	1.14	2.06	3.20

For customer support on your Square Foot Cost Data, call 877.694.8329.

401

C1010 Partitions

C1010 126	Drywall Partitions/Metal Stud Framing

	FACE LAYER	BASE LAYER	FRAMING	OPPOSITE FACE	INSULATION	COST PER S.F.		
						MAT.	INST.	TOTAL
7600		5/8" FR drywall	2-1/2" @ 16" O.C.	same	0	2.85	5.20	8.05
7650				5/8" FR drywall	0	2.06	4.50	6.56
7700				nothing	0	1.64	3.34	4.98
7710			3-5/8" @ 16" O.C.	same	0	2.82	4.92	7.74
7720				5/8" FR drywall	0	2.03	4.22	6.25
7730				nothing	0	1.61	3.06	4.67
7740			2-1/2" @ 24" O.C.	same	0	2.67	4.49	7.16
7750				5/8" FR drywall	0	1.88	3.79	5.67
7760				nothing	0	1.46	2.63	4.09
7800			3-5/8" @ 24" O.C.	same	0	2.72	4.50	7.22
7850				5/8"FR drywall	3-1/2" fiberglass	2.43	4.23	6.66
7900				nothing	0	1.51	2.64	4.15

C1010 Partitions

C1010 128	Drywall Components	COST PER S.F.		
		MAT.	INST.	TOTAL
0060	Metal studs, 24" O.C. including track, load bearing, 20 gage, 2-1/2"	.66	1.08	1.74
0080	3-5/8"	.78	1.10	1.88
0100	4"	.70	1.12	1.82
0120	6"	1.04	1.15	2.19
0140	Metal studs, 24" O.C. including track, load bearing, 18 gage, 2-1/2"	.66	1.08	1.74
0160	3-5/8"	.78	1.10	1.88
0180	4"	.70	1.12	1.82
0200	6"	1.04	1.15	2.19
0220	16 gage, 2-1/2"	.77	1.23	2
0240	3-5/8"	.91	1.26	2.17
0260	4"	.96	1.29	2.25
0280	6"	1.20	1.32	2.52
0300	Non load bearing, 25 gage, 1-5/8"	.20	.76	.96
0340	3-5/8"	.30	.78	1.08
0360	4"	.33	.78	1.11
0380	6"	.40	.80	1.20
0400	20 gage, 2-1/2"	.32	.96	1.28
0420	3-5/8"	.37	.98	1.35
0440	4"	.44	.98	1.42
0460	6"	.51	1	1.51
0540	Wood studs including blocking, shoe and double top plate, 2"x4", 12" O.C.	.61	1.45	2.06
0560	16" O.C.	.49	1.16	1.65
0580	24" O.C.	.37	.93	1.30
0600	2"x6", 12" O.C.	.96	1.65	2.61
0620	16" O.C.	.77	1.29	2.06
0640	24" O.C.	.59	1.01	1.60
0642	Furring one side only, steel channels, 3/4", 12" O.C.	.40	2.16	2.56
0644	16" O.C.	.36	1.91	2.27
0646	24" O.C.	.24	1.45	1.69
0647	1-1/2", 12" O.C.	.54	2.42	2.96
0648	16" O.C.	.48	2.11	2.59
0649	24" O.C.	.32	1.66	1.98
0650	Wood strips, 1" x 3", on wood, 12" O.C.	.42	1.05	1.47
0651	16" O.C.	.32	.79	1.11
0652	On masonry, 12" O.C.	.46	1.17	1.63
0653	16" O.C.	.35	.88	1.23
0654	On concrete, 12" O.C.	.46	2.22	2.68
0655	16" O.C.	.35	1.67	2.02
0665	Gypsum board, one face only, exterior sheathing, 1/2"	.50	1.03	1.53
0680	Interior, fire resistant, 1/2"	.39	.58	.97
0700	5/8"	.37	.58	.95
0720	Sound deadening board 1/4"	.43	.64	1.07
0740	Standard drywall 3/8"	.37	.58	.95
0760	1/2"	.33	.58	.91
0780	5/8"	.36	.58	.94
0800	Tongue & groove coreboard 1"	.86	2.41	3.27
0820	Water resistant, 1/2"	.43	.58	1.01
0840	5/8"	.46	.58	1.04
0860	Add for the following:, foil backing	.17		.17
0880	Fiberglass insulation, 3-1/2"	.50	.43	.93
0900	6"	.66	.43	1.09
0920	Rigid insulation 1"	.57	.58	1.15
0940	Resilient furring @ 16" O.C.	.22	1.82	2.04
0960	Taping and finishing	.05	.58	.63
0980	Texture spray	.04	.63	.67
1000	Thin coat plaster	.12	.72	.84
1050	Sound wall framing, 2x6 plates, 2x4 staggered studs, 12" O.C.	.75	1.41	2.16

For customer support on your Square Foot Cost Data, call 877.694.8329.

403

Gypsum plaster and gypsum lath on metal studs.

Gypsum plaster and diamond metal lath on metal studs.

Gypsum plaster and gypsum lath on wood studs.

Gypsum plaster and diamond metal lath on wood studs.

C1010 140				Plaster Partitions/Metal Stud Framing				
	TYPE	FRAMING	LATH	OPPOSITE FACE		COST PER S.F.		
						MAT.	INST.	TOTAL
1000	2 coat gypsum	2-1/2" @ 16"O.C.	3/8" gypsum	same		2.28	8.45	10.73
1010				nothing		1.35	4.89	6.24
1100		3-1/4" @ 24"O.C.	1/2" gypsum	same		2.08	8.15	10.23
1110				nothing		1.24	4.54	5.78
1500	2 coat vermiculite	2-1/2" @ 16"O.C.	3/8" gypsum	same		2.60	9.10	11.70
1510				nothing		1.51	5.25	6.76
1600		3-1/4" @ 24"O.C.	1/2" gypsum	same		2.40	8.80	11.20
1610				nothing		1.40	4.90	6.30
2000	3 coat gypsum	2-1/2" @ 16"O.C.	3/8" gypsum	same		2.11	9.60	11.71
2010				nothing		1.26	5.50	6.76
2020			3.4lb. diamond	same		2.39	9.60	11.99
2030				nothing		1.41	5.50	6.91
2040			2.75lb. ribbed	same		2.19	9.60	11.79
2050				nothing		1.31	5.50	6.81
2100		3-1/4" @ 24"O.C.	1/2" gypsum	same		1.91	9.30	11.21
2110				nothing		1.15	5.15	6.30
2120			3.4lb. ribbed	same		2.33	9.30	11.63
2130				nothing		1.35	5.15	6.50
3500	3 coat gypsum	2-1/2" @ 16"O.C.	3/8" gypsum	same		2.59	12.30	14.89
3510	W/med. Keenes			nothing		1.50	6.85	8.35
3520			3.4lb. diamond	same		2.87	12.30	15.17
3530				nothing		1.65	6.85	8.50
3540			2.75lb. ribbed	same		2.67	12.30	14.97
3550				nothing		1.55	6.85	8.40
3600		3-1/4" @ 24"O.C.	1/2" gypsum	same		2.39	12	14.39
3610				nothing		1.39	6.50	7.89
3620			3.4lb. ribbed	same		2.81	12	14.81

C1010 Partitions

C1010 140	Plaster Partitions/Metal Stud Framing							

	TYPE	FRAMING	LATH	OPPOSITE FACE		COST PER S.F.		
						MAT.	INST.	TOTAL
3630				nothing		1.59	6.50	8.09
4000	3 coat gypsum	2-1/2" @ 16"O.C.	3/8" gypsum	same		2.59	13.65	16.24
4010	W/hard Keenes			nothing		1.50	7.50	9
4020			3.4lb. diamond	same		2.87	13.65	16.52
4030				nothing		1.65	7.50	9.15
4040			2.75lb. ribbed	same		2.67	13.65	16.32
4050				nothing		1.55	7.50	9.05
4100		3-1/4" @ 24"O.C.	1/2" gypsum	same		2.39	13.35	15.74
4110				nothing		1.39	7.15	8.54
4120			3.4lb. ribbed	same		2.81	13.35	16.16
4130				nothing		1.59	7.15	8.74

C1010 142	Plaster Partitions/Wood Stud Framing							

	TYPE	FRAMING	LATH	OPPOSITE FACE		COST PER S.F.		
						MAT.	INST.	TOTAL
5000	2 coat gypsum	2"x4" @ 16"O.C.	3/8" gypsum	same		2.43	8.25	10.68
5010				nothing		1.50	4.78	6.28
5100		2"x4" @ 24"O.C.	1/2" gypsum	same		2.15	8.10	10.25
5110				nothing		1.31	4.59	5.90
5500	2 coat vermiculite	2"x4" @ 16"O.C.	3/8" gypsum	same		2.75	8.90	11.65
5510				nothing		1.66	5.15	6.81
5600		2"x4" @ 24"O.C.	1/2" gypsum	same		2.47	8.75	11.22
5610				nothing		1.47	4.95	6.42
6000	3 coat gypsum	2"x4" @ 16"O.C.	3/8" gypsum	same		2.26	9.40	11.66
6010				nothing		1.41	5.35	6.76
6020			3.4lb. diamond	same		2.54	9.50	12.04
6030				nothing		1.56	5.40	6.96
6040			2.75lb. ribbed	same		2.31	9.55	11.86
6050				nothing		1.44	5.45	6.89
6100		2"x4" @ 24"O.C.	1/2" gypsum	same		1.98	9.25	11.23
6110				nothing		1.22	5.20	6.42
6120			3.4lb. ribbed	same		2.22	9.35	11.57
6130				nothing		1.34	5.25	6.59
7500	3 coat gypsum	2"x4" @ 16"O.C.	3/8" gypsum	same		2.74	12.10	14.84
7510	W/med Keenes			nothing		1.65	6.75	8.40
7520			3.4lb. diamond	same		3.02	12.20	15.22
7530				nothing		1.80	6.80	8.60
7540			2.75lb. ribbed	same		2.79	12.20	14.99
7550				nothing		1.68	6.80	8.48
7600		2"x4" @ 24"O.C.	1/2" gypsum	same		2.46	11.95	14.41
7610				nothing		1.46	6.55	8.01
7620			3.4lb. ribbed	same		2.88	12.15	15.03
7630				nothing		1.66	6.65	8.31
8000	3 coat gypsum	2"x4" @ 16"O.C.	3/8" gypsum	same		2.74	13.45	16.19
8010	W/hard Keenes			nothing		1.65	7.35	9
8020			3.4lb. diamond	same		3.02	13.55	16.57
8030				nothing		1.80	7.40	9.20
8040			2.75lb. ribbed	same		2.79	13.60	16.39
8050				nothing		1.68	7.45	9.13
8100		2"x4" @ 24"O.C.	1/2" gypsum	same		2.46	13.30	15.76
8110				nothing		1.46	7.20	8.66
8120			3.4lb. ribbed	same		2.88	13.50	16.38
8130				nothing		1.66	7.30	8.96

C1010 Partitions

C1010 144	Plaster Partition Components	COST PER S.F.		
		MAT.	INST.	TOTAL
0060	Metal studs, 16" O.C., including track, non load bearing, 25 gage, 1-5/8"	.34	1.18	1.52
0080	2-1/2"	.34	1.18	1.52
0100	3-1/4"	.40	1.20	1.60
0120	3-5/8"	.40	1.20	1.60
0140	4"	.45	1.22	1.67
0160	6"	.54	1.23	1.77
0180	Load bearing, 20 gage, 2-1/2"	.89	1.50	2.39
0200	3-5/8"	1.05	1.52	2.57
0220	4"	1.11	1.56	2.67
0240	6"	1.40	1.59	2.99
0260	16 gage 2-1/2"	1.05	1.70	2.75
0280	3-5/8"	1.26	1.75	3.01
0300	4"	1.32	1.78	3.10
0320	6"	1.65	1.81	3.46
0340	Wood studs, including blocking, shoe and double plate, 2"x4" , 12" O.C.	.61	1.45	2.06
0360	16" O.C.	.49	1.16	1.65
0380	24" O.C.	.37	.93	1.30
0400	2"x6" , 12" O.C.	.96	1.65	2.61
0420	16" O.C.	.77	1.29	2.06
0440	24" O.C.	.59	1.01	1.60
0460	Furring one face only, steel channels, 3/4" , 12" O.C.	.40	2.16	2.56
0480	16" O.C.	.36	1.91	2.27
0500	24" O.C.	.24	1.45	1.69
0520	1-1/2" , 12" O.C.	.54	2.42	2.96
0540	16" O.C.	.48	2.11	2.59
0560	24"O.C.	.32	1.66	1.98
0580	Wood strips 1"x3", on wood., 12" O.C.	.42	1.05	1.47
0600	16"O.C.	.32	.79	1.11
0620	On masonry, 12" O.C.	.46	1.17	1.63
0640	16" O.C.	.35	.88	1.23
0660	On concrete, 12" O.C.	.46	2.22	2.68
0680	16" O.C.	.35	1.67	2.02
0700	Gypsum lath. plain or perforated, nailed to studs, 3/8" thick	.37	.66	1.03
0720	1/2" thick	.30	.70	1
0740	Clipped to studs, 3/8" thick	.37	.75	1.12
0760	1/2" thick	.30	.80	1.10
0780	Metal lath, diamond painted, nailed to wood studs, 2.5 lb.	.44	.66	1.10
0800	3.4 lb.	.52	.70	1.22
0820	Screwed to steel studs, 2.5 lb.	.44	.70	1.14
0840	3.4 lb.	.50	.75	1.25
0860	Rib painted, wired to steel, 2.75 lb	.42	.75	1.17
0880	3.4 lb	.50	.80	1.30
0900	4.0 lb	.69	.87	1.56
0910				
0920	Gypsum plaster, 2 coats	.44	2.77	3.21
0940	3 coats	.63	3.36	3.99
0960	Perlite or vermiculite plaster, 2 coats	.72	3.13	3.85
0980	3 coats	.78	3.89	4.67
1000	Stucco, 3 coats, 1" thick, on wood framing	.76	5.45	6.21
1020	On masonry	.31	4.31	4.62
1100	Metal base galvanized and painted 2-1/2" high	.71	2.11	2.82

C1010 Partitions

Folding Accordion

Folding Leaf

Movable and Borrow Lites

C1010 205	Partitions	COST PER S.F.		
		MAT.	INST.	TOTAL
0360	Folding accordion, vinyl covered, acoustical, 3 lb. S.F., 17 ft max. hgt	35	11.55	46.55
0380	5 lb. per S.F. 27 ft max height	48.50	12.15	60.65
0400	5.5 lb. per S.F., 17 ft. max height	56.50	12.85	69.35
0420	Commercial, 1.75 lb per S.F., 8 ft. max height	31.50	5.15	36.65
0440	2.0 Lb per S.F., 17 ft. max height	32.50	7.70	40.20
0460	Industrial, 4.0 lb. per S.F. 27 ft. max height	49	15.40	64.40
0480	Vinyl clad wood or steel, electric operation 6 psf	69	7.25	76.25
0500	Wood, non acoustic, birch or mahogany	37	3.85	40.85
0560	Folding leaf, alum framed acoustical 12' high., 5.5 lb/S.F. standard trim	45	19.25	64.25
0580	Premium trim	54	38.50	92.50
0600	6.5 lb. per S.F., standard trim	47	19.25	66.25
0620	Premium trim	58.50	38.50	97
0640	Steel acoustical, 7.5 per S.F., vinyl faced, standard trim	66.50	19.25	85.75
0660	Premium trim	81.50	38.50	120
0680	Wood acoustic type, vinyl faced to 18' high 6 psf, economy trim	62.50	19.25	81.75
0700	Standard trim	74.50	25.50	100
0720	Premium trim	96.50	38.50	135
0740	Plastic lam. or hardwood faced, standard trim	64.50	19.25	84
0760	Premium trim	68.50	38.50	107
0780	Wood, low acoustical type to 12 ft. high 4.5 psf	47	23	70
0840	Demountable, trackless wall, cork finish, semi acous, 1-5/8" th, unsealed	42.50	3.56	46.06
0860	Sealed	47	6.10	53.10
0880	Acoustic, 2" thick, unsealed	40	3.79	43.79
0900	Sealed	61.50	5.15	66.65
0920	In-plant modular office system, w/prehung steel door			
0940	3" thick honeycomb core panels			
0960	12' x 12', 2 wall	13.35	.73	14.08
0970	4 wall	13.50	.97	14.47
0980	16' x 16', 2 wall	13.50	.51	14.01
0990	4 wall	9.15	.52	9.67
1000	Gypsum, demountable, 3" to 3-3/4" thick x 9' high, vinyl clad	7.35	2.66	10.01
1020	Fabric clad	18.30	2.94	21.24
1040	1.75 system, vinyl clad hardboard, paper honeycomb core panel			
1060	1-3/4" to 2-1/2" thick x 9' high	12.30	2.66	14.96
1080	Unitized gypsum panel system, 2" to 2-1/2" thick x 9' high			
1100	Vinyl clad gypsum	15.85	2.66	18.51
1120	Fabric clad gypsum	26	2.94	28.94
1140	Movable steel walls, modular system			
1160	Unitized panels, 48" wide x 9' high			
1180	Baked enamel, pre-finished	17.85	2.14	19.99
1200	Fabric clad	26	2.28	28.28

C10 Interior Construction

C1020 Interior Doors

Single Leaf | Sliding Entrance | Rolling Overhead

C1020 102	Special Doors	MAT.	INST.	TOTAL
2500	Single leaf, wood, 3'-0"x7'-0"x1 3/8", birch, solid core	400	216	616
2510	Hollow core	330	207	537
2530	Hollow core, lauan	320	207	527
2540	Louvered pine	465	207	672
2550	Paneled pine	495	207	702
2600	Hollow metal, comm. quality, flush, 3'-0"x7'-0"x1-3/8"	930	227	1,157
2650	3'-0"x10'-0" openings with panel	1,475	294	1,769
2700	Metal fire, comm. quality, 3'-0"x7'-0"x1-3/8"	1,150	236	1,386
2800	Kalamein fire, comm. quality, 3'-0"x7'-0"x1-3/4"	840	435	1,275
3200	Double leaf, wood, hollow core, 2 - 3'-0"x7'-0"x1-3/8"	585	395	980
3300	Hollow metal, comm. quality, B label, 2'-3'-0"x7'-0"x1-3/8"	2,225	495	2,720
3400	6'-0"x10'-0" opening, with panel	2,825	615	3,440
3500	Double swing door system, 12'-0"x7'-0", mill finish	9,250	3,225	12,475
3700	Black finish	9,375	3,325	12,700
3800	Sliding entrance door and system mill finish	10,800	2,775	13,575
3900	Bronze finish	11,800	3,000	14,800
4000	Black finish	12,300	3,125	15,425
4100	Sliding panel mall front, 16'x9' opening, mill finish	3,825	845	4,670
4200	Bronze finish	4,975	1,100	6,075
4300	Black finish	6,125	1,350	7,475
4400	24'x9' opening mill finish	5,425	1,575	7,000
4500	Bronze finish	7,050	2,050	9,100
4600	Black finish	8,675	2,525	11,200
4700	48'x9' opening mill finish	10,000	1,225	11,225
4800	Bronze finish	13,000	1,600	14,600
4900	Black finish	16,000	1,950	17,950
5000	Rolling overhead steel door, manual, 8' x 8' high	1,275	915	2,190
5100	10' x 10' high	2,100	1,050	3,150
5200	20' x 10' high	3,475	1,475	4,950
5300	12' x 12' high	2,125	1,225	3,350
5400	Motor operated, 8' x 8' high	2,550	1,200	3,750
5500	10' x 10' high	3,375	1,350	4,725
5600	20' x 10' high	4,750	1,775	6,525
5700	12' x 12' high	3,400	1,525	4,925
5800	Roll up grille, aluminum, manual, 10' x 10' high, mill finish	3,100	1,775	4,875
5900	Bronze anodized	4,900	1,775	6,675
6000	Motor operated, 10' x 10' high, mill finish	4,475	2,075	6,550
6100	Bronze anodized	6,275	2,075	8,350
6200	Steel, manual, 10' x 10' high	2,750	1,475	4,225
6300	15' x 8' high	3,175	1,825	5,000
6400	Motor operated, 10' x 10' high	4,125	1,775	5,900
6500	15' x 8' high	4,550	2,125	6,675
8970	Counter door, rolling, 6' high, 14' wide, aluminum	2,950	825	3,775

C1030 Fittings

Toilet Units **Entrance Screens** **Urinal Screens**

C1030 110	Toilet Partitions	COST PER UNIT		
		MAT.	INST.	TOTAL
0380	Toilet partitions, cubicles, ceiling hung, marble	1,975	530	2,505
0400	Painted metal	575	289	864
0420	Plastic laminate	570	289	859
0460	Stainless steel	1,175	289	1,464
0480	Handicap addition	495		495
0520	Floor and ceiling anchored, marble	2,175	425	2,600
0540	Painted metal	645	231	876
0560	Plastic laminate	865	231	1,096
0600	Stainless steel	1,400	231	1,631
0620	Handicap addition	345		345
0660	Floor mounted marble	1,350	355	1,705
0680	Painted metal	650	165	815
0700	Plastic laminate	585	165	750
0740	Stainless steel	1,550	165	1,715
0760	Handicap addition	345		345
0780	Juvenile deduction	45.50		45.50
0820	Floor mounted with headrail marble	1,275	355	1,630
0840	Painted metal	430	193	623
0860	Plastic laminate	825	193	1,018
0900	Stainless steel	1,125	193	1,318
0920	Handicap addition	410		410
0960	Wall hung, painted metal	700	165	865
1020	Stainless steel	1,825	165	1,990
1040	Handicap addition	410		410
1080	Entrance screens, floor mounted, 54" high, marble	870	118	988
1100	Painted metal	266	77	343
1140	Stainless steel	1,000	77	1,077
1300	Urinal screens, floor mounted, 24" wide, laminated plastic	220	145	365
1320	Marble	710	171	881
1340	Painted metal	253	145	398
1380	Stainless steel	625	145	770
1428	Wall mounted wedge type, painted metal	148	116	264
1460	Stainless steel	635	116	751
1500	Partitions, shower stall, single wall, painted steel, 2'-8" x 2'-8"	1,125	274	1,399
1510	Fiberglass, 2'-8" x 2'-8"	875	305	1,180
1520	Double wall, enameled steel, 2'-8" x 2'-8"	1,125	274	1,399
1530	Stainless steel, 2'-8" x 2'-8"	2,525	274	2,799
1560	Tub enclosure, sliding panels, tempered glass, aluminum frame	450	340	790
1570	Chrome/brass frame, clear glass	1,325	455	1,780

C10 Interior Construction

C1030 Fittings

C1030 310 — Storage Specialties, EACH

		COST EACH		
		MAT.	INST.	TOTAL
0200	Lockers, steel, single tier, 5' to 6' high, per opening, 1 wide	172	49	221
0210	3 wide	315	57	372
0600	Shelving, metal industrial, braced, 3' wide, 1' deep	25	12.55	37.55
0610	2' deep	33.50	13.30	46.80

C1030 510 — Identifying/Visual Aid Specialties, EACH

		COST EACH		
		MAT.	INST.	TOTAL
0100	Control boards, magnetic, porcelain finish, framed, 24" x 18"	213	145	358
0110	96" x 48"	1,175	231	1,406
0120	Directory boards, outdoor, black plastic, 36" x 24"	840	580	1,420
0130	36" x 36"	970	770	1,740
0140	Indoor, economy, open faced, 18" x 24"	179	165	344
0500	Signs, interior electric exit sign, wall mounted, 6"	101	82	183
0510	Street, reflective alum., dbl. face, 4 way, w/bracket	254	38.50	292.50
0520	Letters, cast aluminum, 1/2" deep, 4" high	27.50	32	59.50
0530	1" deep, 10" high	58	32	90
0540	Plaques, cast aluminum, 20" x 30"	2,025	289	2,314
0550	Cast bronze, 36" x 48"	5,125	580	5,705

C1030 520 — Identifying/Visual Aid Specialties, S.F.

		COST PER S.F.		
		MAT.	INST.	TOTAL
0100	Bulletin board, cork sheets, no frame, 1/4" thick	1.69	3.99	5.68
0120	Aluminum frame, 1/4" thick, 3' x 5'	9.70	4.86	14.56
0200	Chalkboards, wall hung, alum, frame & chalktrough	14.75	2.56	17.31
0210	Wood frame & chalktrough	11.80	2.76	14.56
0220	Sliding board, one board with back panel	57	2.46	59.46
0230	Two boards with back panel	89	2.46	91.46
0240	Liquid chalk type, alum. frame & chalktrough	13.35	2.56	15.91
0250	Wood frame & chalktrough	33.50	2.56	36.06

C1030 710 — Bath and Toilet Accessories, EACH

		COST EACH		
		MAT.	INST.	TOTAL
0100	Specialties, bathroom accessories, st. steel, curtain rod, 5' long, 1" diam	31	44.50	75.50
0120	Dispenser, towel, surface mounted	49	36	85
0140	Grab bar, 1-1/4" diam., 12" long	32	24	56
0160	Mirror, framed with shelf, 18" x 24"	213	29	242
0170	72" x 24"	330	96.50	426.50
0180	Toilet tissue dispenser, surface mounted, single roll	19.60	19.25	38.85
0200	Towel bar, 18" long	46	25	71
0300	Medicine cabinets, sliding mirror doors, 20" x 16" x 4-3/4", unlighted	136	82.50	218.50
0310	24" x 19" x 8-1/2", lighted	197	116	313

C1030 730 — Bath and Toilet Accessories, L.F.

		COST PER L.F.		
		MAT.	INST.	TOTAL
0100	Partitions, hospital curtain, ceiling hung, polyester oxford cloth	24.50	5.65	30.15
0110	Designer oxford cloth	17.20	7.15	24.35

C1030 830 — Fabricated Cabinets & Counters, L.F.

		COST PER L.F.		
		MAT.	INST.	TOTAL
0110	Household, base, hardwood, one top drawer & one door below x 12" wide	292	46.50	338.50
0120	Four drawer x 24" wide	380	52	432
0130	Wall, hardwood, 30" high with one door x 12" wide	238	52.50	290.50
0140	Two doors x 48" wide	550	63	613
0150	Counter top-laminated plastic, stock, economy	18.70	19.25	37.95
0160	Custom-square edge, 7/8" thick	16.55	43.50	60.05

410

C10 Interior Construction

C1030 Fittings

C1030 830	Fabricated Cabinets & Counters, L.F.	COST PER L.F.		
		MAT.	INST.	TOTAL
0170	School, counter, wood, 32" high	264	58	322
0180	Metal, 84" high	470	77	547

C1030 910	Other Fittings, EACH	COST EACH		
		MAT.	INST.	TOTAL
0500	Mail boxes, horizontal, rear loaded, aluminum, 5" x 6" x 15" deep	44	17	61
0510	Front loaded, aluminum, 10" x 12" x 15" deep	97	29	126
0520	Vertical, front loaded, aluminum, 15" x 5" x 6" deep	44	17	61
0530	Bronze, duranodic finish	51	17	68
0540	Letter slot, post office	114	72.50	186.50
0550	Mail counter, window, post office, with grille	605	289	894
0700	Turnstiles, one way, 4' arm, 46" diam., manual	1,875	231	2,106
0710	Electric	2,275	965	3,240
0720	3 arm, 5'-5" diam. & 7' high, manual	6,475	1,150	7,625
0730	Electric	9,025	1,925	10,950

C2010 Stair Construction

The table below lists the cost per flight for 4'-0" wide stairs. Side walls are not included. Railings are included in the prices.

C2010 110	Stairs	MAT.	INST.	TOTAL
		COST PER FLIGHT		
0470	Stairs, C.I.P. concrete, w/o landing, 12 risers, w/o nosing	1,125	2,425	3,550
0480	With nosing	2,050	2,625	4,675
0550	W/landing, 12 risers, w/o nosing	1,275	2,975	4,250
0560	With nosing	2,200	3,175	5,375
0570	16 risers, w/o nosing	1,575	3,725	5,300
0580	With nosing	2,800	4,000	6,800
0590	20 risers, w/o nosing	1,850	4,500	6,350
0600	With nosing	3,400	4,825	8,225
0610	24 risers, w/o nosing	2,150	5,250	7,400
0620	With nosing	4,000	5,650	9,650
0630	Steel, grate type w/nosing & rails, 12 risers, w/o landing	5,700	1,250	6,950
0640	With landing	7,975	1,725	9,700
0660	16 risers, with landing	9,875	2,125	12,000
0680	20 risers, with landing	11,800	2,550	14,350
0700	24 risers, with landing	13,700	2,950	16,650
0701				
0710	Concrete fill metal pan & picket rail, 12 risers, w/o landing	7,375	1,250	8,625
0720	With landing	9,725	1,875	11,600
0740	16 risers, with landing	12,200	2,300	14,500
0760	20 risers, with landing	14,600	2,700	17,300
0780	24 risers, with landing	17,100	3,100	20,200
0790	Cast iron tread & pipe rail, 12 risers, w/o landing	7,375	1,250	8,625
0800	With landing	9,725	1,875	11,600
1120	Wood, prefab box type, oak treads, wood rails 3'-6" wide, 14 risers	2,250	480	2,730
1150	Prefab basement type, oak treads, wood rails 3'-0" wide, 14 risers	1,100	119	1,219

C30 Interior Finishes

C3010 Wall Finishes

C3010 230	Paint & Covering	COST PER S.F.		
		MAT.	INST.	TOTAL
0060	Painting, interior on plaster and drywall, brushwork, primer & 1 coat	.14	.75	.89
0080	Primer & 2 coats	.21	.99	1.20
0100	Primer & 3 coats	.29	1.22	1.51
0120	Walls & ceilings, roller work, primer & 1 coat	.14	.50	.64
0140	Primer & 2 coats	.21	.64	.85
0160	Woodwork incl. puttying, brushwork, primer & 1 coat	.14	1.08	1.22
0180	Primer & 2 coats	.21	1.43	1.64
0200	Primer & 3 coats	.29	1.95	2.24
0260	Cabinets and casework, enamel, primer & 1 coat	.14	1.22	1.36
0280	Primer & 2 coats	.22	1.50	1.72
0300	Masonry or concrete, latex, brushwork, primer & 1 coat	.31	1.01	1.32
0320	Primer & 2 coats	.41	1.45	1.86
0340	Addition for block filler	.24	1.25	1.49
0380	Fireproof paints, intumescent, 1/8" thick 3/4 hour	2.43	.99	3.42
0400	3/16" thick 1 hour	5.40	1.50	6.90
0420	7/16" thick 2 hour	7	3.47	10.47
0440	1-1/16" thick 3 hour	11.40	6.95	18.35
0500	Gratings, primer & 1 coat	.36	1.52	1.88
0600	Pipes over 12" diameter	.83	4.86	5.69
0700	Structural steel, brushwork, light framing 300-500 S.F./Ton	.11	1.82	1.93
0720	Heavy framing 50-100 S.F./Ton	.11	.91	1.02
0740	Spraywork, light framing 300-500 S.F./Ton	.12	.41	.53
0760	Heavy framing 50-100 S.F./Ton	.12	.45	.57
0800	Varnish, interior wood trim, no sanding sealer & 1 coat	.08	1.22	1.30
0820	Hardwood floor, no sanding 2 coats	.17	.26	.43
0840	Wall coatings, acrylic glazed coatings, minimum	.34	.93	1.27
0860	Maximum	.72	1.59	2.31
0880	Epoxy coatings, solvent based	.44	.93	1.37
0900	Water based	1.32	2.86	4.18
0940	Exposed epoxy aggregate, troweled on, 1/16" to 1/4" aggregate, topping mix	.68	2.07	2.75
0960	Integral mix	1.46	3.74	5.20
0980	1/2" to 5/8" aggregate, topping mix	1.32	3.74	5.06
1000	Integral mix	2.29	6.10	8.39
1020	1" aggregate, topping mix	2.33	5.40	7.73
1040	Integral mix	3.55	8.85	12.40
1060	Sprayed on, topping mix	.63	1.65	2.28
1080	Water based	1.16	3.35	4.51
1100	High build epoxy 50 mil, solvent based	.75	1.25	2
1120	Water based	1.27	5.10	6.35
1140	Laminated epoxy with fiberglass solvent based	.80	1.65	2.45
1160	Water based	1.45	3.35	4.80
1180	Sprayed perlite or vermiculite 1/16" thick, solvent based	.30	.17	.47
1200	Water based	.81	.76	1.57
1260	Wall coatings, vinyl plastic, solvent based	.36	.66	1.02
1280	Water based	.90	2.03	2.93
1300	Urethane on smooth surface, 2 coats, solvent based	.30	.43	.73
1320	Water based	.65	.73	1.38
1340	3 coats, solvent based	.39	.58	.97
1360	Water based	.87	1.03	1.90
1380	Ceramic-like glazed coating, cementitious, solvent based	.53	1.11	1.64
1400	Water based	.89	1.41	2.30
1420	Resin base, solvent based	.36	.76	1.12
1440	Water based	.59	1.47	2.06
1460	Wall coverings, aluminum foil	1.14	1.79	2.93
1500	Vinyl backing	6.10	2.05	8.15
1520	Cork tiles, 12"x12", light or dark, 3/16" thick	4.96	2.05	7.01

C3010 Wall Finishes

C3010 230	Paint & Covering	COST PER S.F.		
		MAT.	INST.	TOTAL
1540	5/16" thick	3.44	2.10	5.54
1560	Basketweave, 1/4" thick	3.85	2.05	5.90
1580	Natural, non-directional, 1/2" thick	7.60	2.05	9.65
1600	12"x36", granular, 3/16" thick	1.44	1.28	2.72
1620	1" thick	1.85	1.33	3.18
1640	12"x12", polyurethane coated, 3/16" thick	4.49	2.05	6.54
1660	5/16" thick	7.25	2.10	9.35
1661	Paneling, prefinished plywood, birch	1.32	2.75	4.07
1662	Mahogany, African	3.03	2.89	5.92
1663	Philippine (lauan)	.72	2.31	3.03
1664	Oak or cherry	2.26	2.89	5.15
1665	Rosewood	3.19	3.61	6.80
1666	Teak	3.19	2.89	6.08
1667	Chestnut	5.30	3.08	8.38
1668	Pecan	2.75	2.89	5.64
1669	Walnut	5.80	2.89	8.69
1670	Wood board, knotty pine, finished	2.44	4.78	7.22
1671	Rough sawn cedar	3.86	4.78	8.64
1672	Redwood	5.80	4.78	10.58
1673	Aromatic cedar	2.89	5.15	8.04
1680	Cork wallpaper, paper backed, natural	2.27	1.03	3.30
1700	Color	3.16	1.03	4.19
1720	Gypsum based, fabric backed, minimum	.94	.62	1.56
1740	Average	1.40	.68	2.08
1760	Small quantities	.84	.77	1.61
1780	Vinyl wall covering, fabric back, light weight	1.07	.77	1.84
1800	Medium weight	.99	1.03	2.02
1820	Heavy weight	1.51	1.13	2.64
1840	Wall paper, double roll, solid pattern, avg. workmanship	.67	.77	1.44
1860	Basic pattern, avg. workmanship	1.22	.92	2.14
1880	Basic pattern, quality workmanship	2.42	1.13	3.55
1900	Grass cloths with lining paper, minimum	1	1.23	2.23
1920	Maximum	3.20	1.41	4.61
1940	Ceramic tile, thin set, 4-1/4" x 4-1/4"	2.48	4.75	7.23
1942				
1960	12" x 12"	4.78	5.65	10.43

C3010 235	Paint Trim	COST PER L.F.		
		MAT.	INST.	TOTAL
2040	Painting, wood trim, to 6" wide, enamel, primer & 1 coat	.14	.61	.75
2060	Primer & 2 coats	.22	.77	.99
2080	Misc. metal brushwork, ladders	.72	6.10	6.82
2100	Pipes, to 4" dia.	.11	1.28	1.39
2120	6" to 8" dia.	.22	2.56	2.78
2140	10" to 12" dia.	.65	3.83	4.48
2160	Railings, 2" pipe	.26	3.04	3.30
2180	Handrail, single	.18	1.22	1.40
2185	Caulking & Sealants, Polyurethane, In place, 1 or 2 component, 1/2" X 1/4"	.35	1.96	2.31

C3020 Floor Finishes

C3020 410	Tile & Covering	COST PER S.F.		
		MAT.	INST.	TOTAL
0060	Carpet tile, nylon, fusion bonded, 18" x 18" or 24" x 24", 24 oz.	3.66	.70	4.36
0080	35 oz.	4.27	.70	4.97
0100	42 oz.	5.65	.70	6.35
0140	Carpet, tufted, nylon, roll goods, 12' wide, 26 oz.	3.04	1.14	4.18
0160	36 oz.	4.22	1.14	5.36
0180	Woven, wool, 36 oz.	11.10	.80	11.90
0200	42 oz.	12.20	.80	13
0220	Padding, add to above, 2.7 density	.73	.38	1.11
0240	13.0 density	.98	.38	1.36
0260	Composition flooring, acrylic, 1/4" thick	1.83	5.65	7.48
0280	3/8" thick	2.31	6.55	8.86
0300	Epoxy, 3/8" thick	3.09	4.35	7.44
0320	1/2" thick	4.46	6	10.46
0340	Epoxy terrazzo, granite chips	6.55	6.30	12.85
0360	Recycled porcelain	9.95	8.45	18.40
0380	Mastic, hot laid, 1-1/2" thick, minimum	4.57	4.26	8.83
0400	Maximum	5.85	5.65	11.50
0420	Neoprene 1/4" thick, minimum	4.50	5.40	9.90
0440	Maximum	6.15	6.85	13
0460	Polyacrylate with ground granite 1/4", granite chips	3.85	4	7.85
0480	Recycled porcelain	7.05	6.15	13.20
0500	Polyester with colored quartz chips 1/16", minimum	3.48	2.76	6.25
0520	Maximum	5.20	4.35	9.55
0540	Polyurethane with vinyl chips, clear	8.35	2.76	11.11
0560	Pigmented	12.15	3.42	15.57
0600	Concrete topping, granolithic concrete, 1/2" thick	.34	4.67	5.01
0620	1" thick	.68	4.79	5.47
0640	2" thick	1.35	5.50	6.85
0660	Heavy duty 3/4" thick, minimum	.46	7.25	7.71
0680	Maximum	.88	8.60	9.48
0700	For colors, add to above, minimum	.48	1.67	2.15
0720	Maximum	.81	1.84	2.65
0740	Exposed aggregate finish, minimum	.51	.85	1.36
0760	Maximum	.96	1.15	2.11
0780	Abrasives, .25 P.S.F. add to above, minimum	.58	.63	1.21
0800	Maximum	.87	.63	1.50
0820	Dust on coloring, add, minimum	.48	.41	.89
0840	Maximum	.81	.85	1.66
0860	Floor coloring using 0.6 psf powdered color, 1/2" integral, minimum	5.45	4.67	10.12
0880	Maximum	5.80	4.67	10.47
0900	Dustproofing, add, minimum	.19	.28	.47
0920	Maximum	.70	.41	1.11
0930	Paint	.41	1.45	1.86
0940	Hardeners, metallic add, minimum	.59	.63	1.22
0960	Maximum	1.77	.93	2.70
0980	Non-metallic, minimum	.24	.63	.87
1000	Maximum	.71	.93	1.64
1020	Integral topping and finish, 1:1:2 mix, 3/16" thick	.11	2.75	2.86
1040	1/2" thick	.31	2.90	3.21
1060	3/4" thick	.46	3.24	3.70
1080	1" thick	.61	3.67	4.28
1100	Terrazzo, minimum	3.69	16.80	20.49
1120	Maximum	6.80	21	27.80
1340	Cork tile, minimum	7.65	1.61	9.26
1360	Maximum	13.35	1.61	14.96
1380	Polyethylene, in rolls, minimum	4.50	1.84	6.34
1400	Maximum	7.50	1.84	9.34
1420	Polyurethane, thermoset, minimum	5.95	5.05	11

C3020 Floor Finishes

C3020 410	Tile & Covering	COST PER S.F.		
		MAT.	INST.	TOTAL
1440	Maximum	7.45	10.15	17.60
1460	Rubber, sheet goods, minimum	8.55	4.22	12.77
1480	Maximum	13.65	5.65	19.30
1500	Tile, minimum	6.25	1.27	7.52
1520	Maximum	10.95	1.84	12.79
1580	Vinyl, composition tile, minimum	1.30	1.01	2.31
1600	Maximum	1.68	1.01	2.69
1620	Vinyl tile, 3/32"	3.95	1.01	4.96
1640	Maximum	3.58	1.01	4.59
1660	Sheet goods, plain pattern/colors	4.50	2.03	6.53
1680	Intricate pattern/colors	8.25	2.53	10.78
1720	Tile, ceramic natural clay	6.45	4.94	11.39
1730	Marble, synthetic 12"x12"x5/8"	11.15	15.05	26.20
1740	Porcelain type, minimum	5.70	4.94	10.64
1760	Maximum	6.90	5.85	12.75
1800	Quarry tile, mud set, minimum	8.30	6.45	14.75
1820	Maximum	9.55	8.20	17.75
1840	Thin set, deduct		1.29	1.29
1850	Tile, natural stone, marble, in mortar bed, 12" x 12" x 3/8" thick	17.60	25	42.60
1860	Terrazzo precast, minimum	5.25	6.75	12
1880	Maximum	10.25	6.75	17
1900	Non-slip, minimum	22	17.15	39.15
1920	Maximum	22.50	23	45.50
1960	Stone flooring, polished marble in mortar bed	16.25	25	41.25
2020	Wood block, end grain factory type, natural finish, 2" thick	4.88	4.62	9.50
2040	Fir, vertical grain, 1"x4", no finish, minimum	2.89	2.27	5.16
2060	Maximum	3.07	2.27	5.34
2080	Prefinished white oak, prime grade, 2-1/4" wide	5.15	3.40	8.55
2100	3-1/4" wide	5.60	3.12	8.72
2120	Maple strip, sanded and finished, minimum	5.70	4.97	10.67
2140	Maximum	6	4.97	10.97
2160	Oak strip, sanded and finished, minimum	4.50	4.97	9.47
2180	Maximum	5.40	4.97	10.37
2200	Parquetry, sanded and finished, plain pattern	6.75	5.20	11.95
2220	Intricate pattern	11.55	7.35	18.90
2260	Add for sleepers on concrete, treated, 24" O.C., 1"x2"	1.78	2.98	4.76
2280	1"x3"	2	2.30	4.30
2300	2"x4"	.98	1.18	2.16
2340	Underlayment, plywood, 3/8" thick	1.01	.77	1.78
2350	1/2" thick	1.21	.80	2.01
2360	5/8" thick	1.32	.83	2.15
2370	3/4" thick	1.57	.89	2.46
2380	Particle board, 3/8" thick	.44	.77	1.21
2390	1/2" thick	.47	.80	1.27
2400	5/8" thick	.57	.83	1.40
2410	3/4" thick	.69	.89	1.58
2420	Hardboard, 4' x 4', .215" thick	.64	.77	1.41
9200	Vinyl, composition tile, 12" x 12" x 1/8" thick, recycled content	2.33	1.01	3.34

C3030 Ceiling Finishes

2 Coats of Plaster on Gypsum Lath on Wood Furring

Fiberglass Board on Exposed Suspended Grid System

Plaster and Metal Lath on Metal Furring

C3030 105 — Plaster Ceilings

	TYPE	LATH	FURRING	SUPPORT		MAT.	INST.	TOTAL
2400	2 coat gypsum	3/8" gypsum	1"x3" wood, 16" O.C.	wood		1.27	5.80	7.07
2500	Painted			masonry		1.30	5.90	7.20
2600				concrete		1.30	6.60	7.90
2700	3 coat gypsum	3.4# metal		wood		1.61	6.25	7.86
2800	Painted			masonry		1.64	6.35	7.99
2900				concrete		1.64	7.05	8.69
3000	2 coat perlite	3/8" gypsum		wood		1.55	6.05	7.60
3100	Painted			masonry		1.58	6.15	7.73
3200				concrete		1.58	6.85	8.43
3300	3 coat perlite	3.4# metal		wood		1.70	6.20	7.90
3400	Painted			masonry		1.73	6.30	8.03
3500				concrete		1.73	7	8.73
3600	2 coat gypsum	3/8" gypsum	3/4" CRC, 12" O.C.	1-1/2" CRC, 48"O.C.		1.35	7	8.35
3700	Painted		3/4" CRC, 16" O.C.	1-1/2" CRC, 48"O.C.		1.31	6.30	7.61
3800			3/4" CRC, 24" O.C.	1-1/2" CRC, 48"O.C.		1.19	5.75	6.94
3900	2 coat perlite		3/4" CRC, 12" O.C.	1-1/2" CRC, 48"O.C		1.63	7.50	9.13
4000	Painted		3/4" CRC, 16" O.C.	1-1/2" CRC, 48"O.C.		1.59	6.80	8.39
4100			3/4" CRC, 24" O.C.	1-1/2" CRC, 48"O.C.		1.47	6.25	7.72
4200	3 coat gypsum	3.4# metal	3/4" CRC, 12" O.C.	1-1/2" CRC, 36" O.C.		1.85	9.15	11
4300	Painted		3/4" CRC, 16" O.C.	1-1/2" CRC, 36" O.C.		1.81	8.50	10.31
4400			3/4" CRC, 24" O.C.	1-1/2" CRC, 36" O.C.		1.69	7.95	9.64
4500	3 coat perlite		3/4" CRC, 12" O.C.	1-1/2" CRC,36" O.C.		2	10.05	12.05
4600	Painted		3/4" CRC, 16" O.C.	1-1/2" CRC, 36" O.C.		1.96	9.35	11.31
4700			3/4" CRC, 24" O.C.	1-1/2" CRC, 36" O.C.		1.84	8.80	10.64

C3030 110 — Drywall Ceilings

	TYPE	FINISH	FURRING	SUPPORT		MAT.	INST.	TOTAL
4800	1/2" F.R. drywall	painted and textured	1"x3" wood, 16" O.C.	wood		.94	3.57	4.51
4900				masonry		.97	3.69	4.66
5000				concrete		.97	4.39	5.36
5100	5/8" F.R. drywall			wood		.93	3.57	4.50
5200				masonry		.96	3.69	4.65
5300				concrete		.96	4.39	5.35
5400	1/2" F.R. drywall	painted and textured	7/8" resil. channels	24" O.C.		.76	3.47	4.23

For customer support on your Square Foot Cost Data, call 877.694.8329.

417

C30 Interior Finishes

C3030 Ceiling Finishes

C3030 110 — Drywall Ceilings

	TYPE	FINISH	FURRING	SUPPORT		COST PER S.F.		
						MAT.	INST.	TOTAL
5500	1/2" F.R. drywall		1"x2" wood	stud clips		.94	3.34	4.28
5602		painted	1-5/8" metal studs	24" O.C.		.82	3.09	3.91
5700	5/8" F.R. drywall	painted and textured	1-5/8"metal studs			.81	3.09	3.90
5702								

C3030 140 — Plaster Ceiling Components

		COST PER S.F.		
		MAT.	INST.	TOTAL
0060	Plaster, gypsum incl. finish			
0080	3 coats	.63	3.72	4.35
0100	Perlite, incl. finish, 2 coats	.72	3.66	4.38
0120	3 coats	.78	4.60	5.38
0140	Thin coat on drywall	.12	.72	.84
0200	Lath, gypsum, 3/8" thick	.37	.92	1.29
0220	1/2" thick	.30	.96	1.26
0240	5/8" thick	.34	1.13	1.47
0260	Metal, diamond, 2.5 lb.	.44	.75	1.19
0280	3.4 lb.	.52	.80	1.32
0300	Flat rib, 2.75 lb.	.42	.75	1.17
0320	3.4 lb.	.50	.80	1.30
0440	Furring, steel channels, 3/4" galvanized , 12" O.C.	.40	2.42	2.82
0460	16" O.C.	.36	1.75	2.11
0480	24" O.C.	.24	1.21	1.45
0500	1-1/2" galvanized , 12" O.C.	.54	2.67	3.21
0520	16" O.C.	.48	1.95	2.43
0540	24" O.C.	.32	1.30	1.62
0560	Wood strips, 1"x3", on wood, 12" O.C.	.42	1.65	2.07
0580	16" O.C.	.32	1.24	1.56
0600	24" O.C.	.21	.83	1.04
0620	On masonry, 12" O.C.	.46	1.81	2.27
0640	16" O.C.	.35	1.36	1.71
0660	24" O.C.	.23	.91	1.14
0680	On concrete, 12" O.C.	.46	2.75	3.21
0700	16" O.C.	.35	2.06	2.41
0720	24" O.C.	.23	1.38	1.61
0940	Paint on plaster or drywall, roller work, primer + 1 coat	.14	.50	.64
0960	Primer + 2 coats	.21	.64	.85

C3030 210 — Acoustical Ceilings

	TYPE	TILE	GRID	SUPPORT		COST PER S.F.		
						MAT.	INST.	TOTAL
5800	5/8" fiberglass board	24" x 48"	tee	suspended		2.40	1.73	4.13
5900		24" x 24"	tee	suspended		2.63	1.90	4.53
6000	3/4" fiberglass board	24" x 48"	tee	suspended		3.93	1.77	5.70
6100		24" x 24"	tee	suspended		4.16	1.94	6.10
6500	5/8" mineral fiber	12" x 12"	1"x3" wood, 12" O.C.	wood		2.82	3.58	6.40
6600				masonry		2.86	3.74	6.60
6700				concrete		2.86	4.68	7.54
6800	3/4" mineral fiber			wood		3.06	3.58	6.64
6900				masonry		3.06	3.58	6.64
7000				concrete		3.06	3.58	6.64
7100	3/4"mineral fiber on		25 ga. channels	runners		3.48	4.44	7.92
7102	5/8" F.R. drywall							

C30 Interior Finishes

C3030 Ceiling Finishes

C3030 210 — Acoustical Ceilings

	TYPE	TILE	GRID	SUPPORT		COST PER S.F. MAT.	COST PER S.F. INST.	COST PER S.F. TOTAL
7200 7201 7202	5/8" plastic coated Mineral fiber	12" x 12"		adhesive backed		2.83	1.93	4.76
7300 7301 7302	3/4" plastic coated Mineral fiber					3.07	1.93	5
7400 7401 7402	3/4" mineral fiber		conceal 2" bar & channels	suspended		2.67	4.36	7.03

C3030 240 — Acoustical Ceiling Components

		COST PER S.F. MAT.	COST PER S.F. INST.	COST PER S.F. TOTAL
2480	Ceiling boards, eggcrate, acrylic, 1/2" x 1/2" x 1/2" cubes	1.99	1.16	3.15
2500	Polystyrene, 3/8" x 3/8" x 1/2" cubes	1.68	1.13	2.81
2520	1/2" x 1/2" x 1/2" cubes	2	1.16	3.16
2540	Fiberglass boards, plain, 5/8" thick	1.36	.92	2.28
2560	3/4" thick	2.89	.96	3.85
2580	Grass cloth faced, 3/4" thick	2.92	1.16	4.08
2600	1" thick	3.44	1.19	4.63
2620	Luminous panels, prismatic, acrylic	2.90	1.45	4.35
2640	Polystyrene	1.65	1.45	3.10
2660	Flat or ribbed, acrylic	4.24	1.45	5.69
2680	Polystyrene	2.52	1.45	3.97
2700	Drop pan, white, acrylic	6.20	1.45	7.65
2720	Polystyrene	4.84	1.45	6.29
2740	Mineral fiber boards, 5/8" thick, standard	.75	.86	1.61
2760	Plastic faced	2.56	1.45	4.01
2780	2 hour rating	1.19	.86	2.05
2800	Perforated aluminum sheets, .024 thick, corrugated painted	2.57	1.18	3.75
2820	Plain	5	1.16	6.16
3080	Mineral fiber, plastic coated, 12" x 12" or 12" x 24", 5/8" thick	2.40	1.93	4.33
3100	3/4" thick	2.64	1.93	4.57
3120	Fire rated, 3/4" thick, plain faced	1.40	1.93	3.33
3140	Mylar faced	2.02	1.93	3.95
3160	Add for ceiling primer	.14		.14
3180	Add for ceiling cement	.43		.43
3240	Suspension system, furring, 1" x 3" wood 12" O.C.	.42	1.65	2.07
3260	T bar suspension system, 2' x 4' grid	.81	.72	1.53
3280	2' x 2' grid	1.04	.89	1.93
3300	Concealed Z bar suspension system 12" module	.92	1.11	2.03
3320	Add to above for 1-1/2" carrier channels 4' O.C.	.12	1.23	1.35
3340	Add to above for carrier channels for recessed lighting	.22	1.26	1.48

D10 Conveying

D1010 Elevators and Lifts

Hydraulic   Traction Geared   Traction Gearless

D1010 110	Hydraulic	COST EACH		
		MAT.	INST.	TOTAL
1300	Pass. elev., 1500 lb., 2 Floors, 100 FPM	46,900	18,800	65,700
1400	5 Floors, 100 FPM	82,500	49,900	132,400
1600	2000 lb., 2 Floors, 100 FPM	47,800	18,800	66,600
1700	5 floors, 100 FPM	83,500	49,900	133,400
1900	2500 lb., 2 Floors, 100 FPM	50,000	18,800	68,800
2000	5 floors, 100 FPM	85,500	49,900	135,400
2200	3000 lb., 2 Floors, 100 FPM	51,500	18,800	70,300
2300	5 floors, 100 FPM	87,000	49,900	136,900
2500	3500 lb., 2 Floors, 100 FPM	54,500	18,800	73,300
2600	5 floors, 100 FPM	90,000	49,900	139,900
2800	4000 lb., 2 Floors, 100 FPM	56,000	18,800	74,800
2900	5 floors, 100 FPM	91,500	49,900	141,400
3100	4500 lb., 2 Floors, 100 FPM	59,000	18,800	77,800
3200	5 floors, 100 FPM	94,500	49,900	144,400
4000	Hospital elevators, 3500 lb., 2 Floors, 100 FPM	79,000	18,800	97,800
4100	5 floors, 100 FPM	127,500	49,900	177,400
4300	4000 lb., 2 Floors, 100 FPM	79,000	18,800	97,800
4400	5 floors, 100 FPM	127,500	49,900	177,400
4600	4500 lb., 2 Floors, 100 FPM	86,500	18,800	105,300
4800	5 floors, 100 FPM	135,000	49,900	184,900
4900	5000 lb., 2 Floors, 100 FPM	90,000	18,800	108,800
5000	5 floors, 100 FPM	138,500	49,900	188,400
6700	Freight elevators (Class "B"), 3000 lb., 2 Floors, 50 FPM	118,500	23,900	142,400
6800	5 floors, 100 FPM	189,500	62,500	252,000
7000	4000 lb., 2 Floors, 50 FPM	123,000	23,900	146,900
7100	5 floors, 100 FPM	194,500	62,500	257,000
7500	10,000 lb., 2 Floors, 50 FPM	154,000	23,900	177,900
7600	5 floors, 100 FPM	225,500	62,500	288,000
8100	20,000 lb., 2 Floors, 50 FPM	201,500	23,900	225,400
8200	5 Floors, 100 FPM	273,000	62,500	335,500

D1010 140	Traction Geared Elevators	COST EACH		
		MAT.	INST.	TOTAL
1300	Passenger, 2000 Lb., 5 floors, 200 FPM	120,000	47,000	167,000
1500	15 floors, 350 FPM	245,500	142,000	387,500
1600	2500 Lb., 5 floors, 200 FPM	124,000	47,000	171,000
1800	15 floors, 350 FPM	250,000	142,000	392,000
2200	3500 Lb., 5 floors, 200 FPM	126,000	47,000	173,000
2400	15 floors, 350 FPM	252,000	142,000	394,000
2500	4000 Lb., 5 floors, 200 FPM	127,500	47,000	174,500
2700	15 floors, 350 FPM	253,000	142,000	395,000

D10 Conveying

D1010 Elevators and Lifts

D1010 140	Traction Geared Elevators	COST EACH		
		MAT.	INST.	TOTAL
2800	4500 Lb., 5 floors, 200 FPM	130,000	47,000	177,000
3000	15 floors, 350 FPM	255,500	142,000	397,500
3100	5000 Lb., 5 floors, 200 FPM	132,500	47,000	179,500
3300	15 floors, 350 FPM	258,000	142,000	400,000
4000	Hospital, 3500 Lb., 5 floors, 200 FPM	125,000	47,000	172,000
4200	15 floors, 350 FPM	296,500	142,000	438,500
4300	4000 Lb., 5 floors, 200 FPM	125,000	47,000	172,000
4500	15 floors, 350 FPM	296,500	142,000	438,500
4600	4500 Lb., 5 floors, 200 FPM	131,000	47,000	178,000
4800	15 floors, 350 FPM	302,500	142,000	444,500
4900	5000 Lb., 5 floors, 200 FPM	133,000	47,000	180,000
5100	15 floors, 350 FPM	304,500	142,000	446,500
6000	Freight, 4000 Lb., 5 floors, 50 FPM class 'B'	150,500	48,600	199,100
6200	15 floors, 200 FPM class 'B'	351,000	169,500	520,500
6300	8000 Lb., 5 floors, 50 FPM class 'B'	175,000	48,600	223,600
6500	15 floors, 200 FPM class 'B'	375,500	169,500	545,000
7000	10,000 Lb., 5 floors, 50 FPM class 'B'	202,500	48,600	251,100
7200	15 floors, 200 FPM class 'B'	587,000	169,500	756,500
8000	20,000 Lb., 5 floors, 50 FPM class 'B'	224,000	48,600	272,600
8200	15 floors, 200 FPM class 'B'	608,500	169,500	778,000

D1010 150	Traction Gearless Elevators	COST EACH		
		MAT.	INST.	TOTAL
1700	Passenger, 2500 Lb., 10 floors, 200 FPM	280,000	122,500	402,500
1900	30 floors, 600 FPM	560,500	313,000	873,500
2000	3000 Lb., 10 floors, 200 FPM	281,000	122,500	403,500
2200	30 floors, 600 FPM	561,000	313,000	874,000
2300	3500 Lb., 10 floors, 200 FPM	282,000	122,500	404,500
2500	30 floors, 600 FPM	562,500	313,000	875,500
2700	50 floors, 800 FPM	808,000	503,500	1,311,500
2800	4000 Lb., 10 floors, 200 FPM	283,500	122,500	406,000
3000	30 floors, 600 FPM	564,000	313,000	877,000
3200	50 floors, 800 FPM	809,500	503,500	1,313,000
3300	4500 Lb., 10 floors, 200 FPM	286,000	122,500	408,500
3500	30 floors, 600 FPM	566,500	313,000	879,500
3700	50 floors, 800 FPM	812,000	503,500	1,315,500
3800	5000 Lb., 10 floors, 200 FPM	288,500	122,500	411,000
4000	30 floors, 600 FPM	569,000	313,000	882,000
4200	50 floors, 800 FPM	814,500	503,500	1,318,000
6000	Hospital, 3500 Lb., 10 floors, 200 FPM	304,000	122,500	426,500
6200	30 floors, 600 FPM	680,000	313,000	993,000
6400	4000 Lb., 10 floors, 200 FPM	304,000	122,500	426,500
6600	30 floors, 600 FPM	680,000	313,000	993,000
6800	4500 Lb., 10 floors, 200 FPM	310,000	122,500	432,500
7000	30 floors, 600 FPM	686,000	313,000	999,000
7200	5000 Lb., 10 floors, 200 FPM	312,000	122,500	434,500
7400	30 floors, 600 FPM	688,000	313,000	1,001,000

D20 Plumbing

D2010 Plumbing Fixtures

Minimum Plumbing Fixture Requirements

Classification	Occupancy	Description	Water		Lavatories		Bathtubs/Showers	Drinking Fountains	Other
			Male	Female	Male	Female			
Assembly	A-1	Theaters and other buildings for the performing arts and motion pictures	1:125	1:65	1:200			1:500	1 Service Sink
	A-2	Nightclubs, bars, taverns dance halls	1:40		1:75			1:500	1 Service Sink
		Restaurants, banquet halls, food courts	1:75		1:200			1:500	1 Service Sink
	A-3	Auditorium w/o permanent seating, art galleries, exhibition halls, museums, lecture halls, libraries, arcades & gymnasiums	1:125	1:65	1:200			1:500	1 Service Sink
		Passenger terminals and transportation facilities	1:500		1:750			1:1000	1 Service Sink
		Places of worship and other religious services	1:150	1:75	1:200			1:1000	1 Service Sink
	A-4	Indoor sporting events and activities, coliseums, arenas, skating rinks, pools, and tennis courts	1:75 for the first 1500, then 1:120 for the remainder	1:40 for the first 1520, then 1:60 for the remainder	1:200	1:150		1:1000	1 Service Sink
	A-5	Outdoor sporting events and activities, stadiums, amusement parks, bleachers, grandstands	1:75 for the first 1500, then 1:120 for the remainder	1:40 for the first 1520, then 1:60 for the remainder	1:200	1:150		1:1000	1 Service Sink
Business	B	Buildings for the transaction of business, professional services, other services involving merchandise, office buildings, banks, light industrial	1:25 for the first 50, then 1:50 for the remainder		1:40 for the first 80, then 1:80 for the remainder			1:100	1 Service Sink
Educational	E	Educational Facilities	1:50		1:50			1:100	1 Service Sink
Factory and industrial	F-1 and F-2	Structures in which occupants are engaged in work fabricating, assembly or processing of products or materials	1:100		1:100		See International Plumbing Code	1:400	1 Service Sink
Institutional	I-1	Residential Care	1:10		1:10		1:8	1:100	1 Service Sink
	I-2	Hospitals, ambulatory nursing home care recipient	1 per room		1 per room		1:15	1:100	1 Service Sink
		Employees, other than residential care	1:25		1:35			1:100	
		Visitors, other than residential care	1:75		1:100			1:500	
	I-3	Prisons	1 per cell		1 per cell		1:15	1:100	1 Service Sink
		Reformatories, detention and correction centers	1:15		1:15		1:15	1:100	1 Service Sink
		Employees	1:25		1:35			1:100	
	I-4	Adult and child day care	1:15		1:15		1	1:100	1 Service Sink
Mercantile	M	Retail stores, service stations, shops, salesrooms, markets and shopping centers	1:500		1:750			1:1000	1 Service Sink
Residential	R-1	Hotels, Motels, boarding houses (transient)	1 per sleeping unit		1 per sleeping unit		1 per sleeping unit		1 Service Sink
	R-2	Dormitories, fraternities, sororities and boarding houses (not transient)	1:10		1:10		1:8	1:100	1 Service Sink
		Apartment House	1 per dwelling unit		1 per dwelling unit		1 per dwelling unit		1 Kitchen sink per dwelling; 1 clothes washer connection per 20 dwellings
	R-3	1 and 2 Family Dwellings	1 per dwelling unit		1:10		1 per dwelling unit		1 Kitchen sink per dwelling; 1 clothes washer connection per dwelling
	R-3	Congregate living facilities w/<16 people	1:10		1:10		1:8	1:100	1 Service Sink
	R-4	Congregate living facilities w/<16 people	1:10		1:10		1:8	1:100	1 Service Sink
Storage	S-1 and S-2	Structures for the storage of good, warehouses, storehouses and freight depots, low and moderate hazard	1:100		1:100		See International Plumbing Code	1;1000	1 Service Sink

Table 2902.1

Excerpted from the 2012 *International Building Code*, Copyright 2011. Washington, D.C.: International Code Council. Reproduced with permission. All rights reserved. www.ICCSAFE.org

D2010 Plumbing Fixtures

One Piece Wall Hung Water Closet

Wall Hung Urinal

Side By Side Water Closet Group

Floor Mount Water Closet

Stall Type Urinal

Back to Back Water Closet Group

D2010 110	Water Closet Systems	COST EACH		
		MAT.	INST.	TOTAL
1800	Water closet, vitreous china			
1840	Tank type, wall hung			
1880	Close coupled two piece	1,825	755	2,580
1920	Floor mount, one piece	1,600	800	2,400
1960	One piece low profile	1,625	800	2,425
2000	Two piece close coupled	670	800	1,470
2040	Bowl only with flush valve			
2080	Wall hung	2,300	855	3,155
2120	Floor mount	795	815	1,610
2122				
2160	Floor mount, ADA compliant with 18" high bowl	790	835	1,625

D2010 120	Water Closets, Group	COST EACH		
		MAT.	INST.	TOTAL
1760	Water closets, battery mount, wall hung, side by side, first closet	2,425	880	3,305
1800	Each additional water closet, add	2,300	830	3,130
3000	Back to back, first pair of closets	4,250	1,175	5,425
3100	Each additional pair of closets, back to back	4,150	1,150	5,300
9000	Back to back, first pair of closets, auto sensor flush valve, 1.28 gpf	4,725	1,250	5,975
9100	Ea additional pair of cls, back to back, auto sensor flush valve, 1.28 gpf	4,575	1,175	5,750

D2010 210	Urinal Systems	COST EACH		
		MAT.	INST.	TOTAL
2000	Urinal, vitreous china, wall hung	565	845	1,410
2040	Stall type	1,200	1,000	2,200

425

For customer support on your Square Foot Cost Data, call 877.694.8329.

D20 Plumbing

D2010 Plumbing Fixtures

Systems are complete with trim and rough-in (supply, waste and vent) to connect to supply branches and waste mains.

Vanity Top

Wall Hung

Counter Top Single Bowl

Counter Top Double Bowl

D2010 310	Lavatory Systems	COST EACH		
		MAT.	INST.	TOTAL
1560	Lavatory w/trim, vanity top, PE on CI, 20" x 18", Vanity top by others.	625	755	1,380
1600	19" x 16" oval	445	755	1,200
1640	18" round	735	755	1,490
1680	Cultured marble, 19" x 17"	450	755	1,205
1720	25" x 19"	485	755	1,240
1760	Stainless, self-rimming, 25" x 22"	660	755	1,415
1800	17" x 22"	645	755	1,400
1840	Steel enameled, 20" x 17"	435	775	1,210
1880	19" round	445	775	1,220
1920	Vitreous china, 20" x 16"	540	790	1,330
1960	19" x 16"	550	790	1,340
2000	22" x 13"	550	790	1,340
2040	Wall hung, PE on CI, 18" x 15"	945	835	1,780
2080	19" x 17"	970	835	1,805
2120	20" x 18"	805	835	1,640
2160	Vitreous china, 18" x 15"	755	855	1,610
2200	19" x 17"	730	855	1,585
2240	24" x 20"	810	855	1,665
2300	20" x 27", handicap	1,475	920	2,395

D2010 410	Kitchen Sink Systems	COST EACH		
		MAT.	INST.	TOTAL
1720	Kitchen sink w/trim, countertop, PE on CI, 24"x21", single bowl	605	825	1,430
1760	30" x 21" single bowl	970	825	1,795
1800	32" x 21" double bowl	700	890	1,590
1880	Stainless steel, 19" x 18" single bowl	940	825	1,765
1920	25" x 22" single bowl	1,025	825	1,850
1960	33" x 22" double bowl	1,350	890	2,240
2000	43" x 22" double bowl	1,550	905	2,455
2040	44" x 22" triple bowl	1,800	940	2,740
2080	44" x 24" corner double bowl	1,250	905	2,155
2120	Steel, enameled, 24" x 21" single bowl	850	825	1,675
2160	32" x 21" double bowl	855	890	1,745
2240	Raised deck, PE on CI, 32" x 21", dual level, double bowl	785	1,125	1,910
2280	42" x 21" dual level, triple bowl	1,625	1,225	2,850

D2010 Plumbing Fixtures

Laboratory Sink

Service Sink

Single Compartment Sink

Double Compartment Sink

Systems are complete with trim and rough-in (supply, waste and vent) to connect to supply branches and waste mains.

D2010 420	Laundry Sink Systems	COST EACH		
		MAT.	INST.	TOTAL
1740	Laundry sink w/trim, PE on CI, black iron frame			
1760	24" x 20", single compartment	925	810	1,735
1840	48" x 21" double compartment	1,225	880	2,105
1920	Molded stone, on wall, 22" x 21" single compartment	470	810	1,280
1960	45"x 21" double compartment	675	880	1,555
2040	Plastic, on wall or legs, 18" x 23" single compartment	435	795	1,230
2080	20" x 24" single compartment	460	795	1,255
2120	36" x 23" double compartment	515	855	1,370
2160	40" x 24" double compartment	615	855	1,470

D2010 430	Laboratory Sink Systems	COST EACH		
		MAT.	INST.	TOTAL
1580	Laboratory sink w/trim, stainless steel, single bowl,			
1590	Stainless steel, single bowl,			
1600	Double drainboard, 54" x 24" O.D.	2,125	1,050	3,175
1640	Single drainboard, 47" x 24"O.D.	1,550	1,050	2,600
1670	Stainless steel, double bowl,			
1680	70" x 24" O.D.	2,325	1,050	3,375
1750	Polyethylene, single bowl,			
1760	Flanged, 14-1/2" x 14-1/2" O.D.	500	950	1,450
1800	18-1/2" x 18-1/2" O.D.	615	950	1,565
1840	23-1/2" x 20-1/2" O.D.	635	950	1,585
1920	Polypropylene, cup sink, oval, 7" x 4" O.D.	370	840	1,210
1960	10" x 4-1/2" O.D.	395	840	1,235
1961				

427

D2010 Plumbing Fixtures

D2010 440	Service Sink Systems	COST EACH		
		MAT.	INST.	TOTAL
4260	Service sink w/trim, PE on CI, corner floor, 28" x 28", w/rim guard	1,875	1,075	2,950
4300	Wall hung w/rim guard, 22" x 18"	2,125	1,225	3,350
4340	24" x 20"	2,200	1,225	3,425
4380	Vitreous china, wall hung 22" x 20"	1,950	1,225	3,175
4383				

For customer support on your Square Foot Cost Data, call 877.694.8329.

D2010 Plumbing Fixtures

Recessed Bathtub   **Corner Bathtub**

Systems are complete with trim and rough-in (supply, waste and vent) to connect to supply branches and waste mains.

Circular Wash Fountain

Semi-Circular Wash Fountain

D2010 510	Bathtub Systems	COST EACH		
		MAT.	INST.	TOTAL
2000	Bathtub, recessed, P.E. on Cl., 48" x 42"	3,325	935	4,260
2040	72" x 36"	3,425	1,050	4,475
2080	Mat bottom, 5' long	1,675	905	2,580
2120	5'-6" long	2,375	935	3,310
2160	Corner, 48" x 42"	3,325	905	4,230
2200	Formed steel, enameled, 4'-6" long	965	835	1,800

D2010 610	Group Wash Fountain Systems	COST EACH		
		MAT.	INST.	TOTAL
1740	Group wash fountain, precast terrazzo			
1760	Circular, 36" diameter	8,400	1,350	9,750
1800	54" diameter	10,400	1,500	11,900
1840	Semi-circular, 36" diameter	7,425	1,350	8,775
1880	54" diameter	9,825	1,500	11,325
1960	Stainless steel, circular, 36" diameter	7,600	1,275	8,875
2000	54" diameter	9,250	1,400	10,650
2040	Semi-circular, 36" diameter	5,950	1,275	7,225
2080	54" diameter	8,125	1,400	9,525
2160	Thermoplastic, circular, 36" diameter	5,300	1,025	6,325
2200	54" diameter	6,100	1,200	7,300
2240	Semi-circular, 36" diameter	4,900	1,025	5,925
2280	54" diameter	5,900	1,200	7,100

D2010 Plumbing Fixtures

Square Shower Stall

Corner Angle Shower Stall

Systems are complete with trim and rough-in (supply, waste and vent) to connect to supply branches and waste mains.

Wall Mounted, Low Back

Wall Mounted, No Back

D2010 710	Shower Systems	COST EACH		
		MAT.	INST.	TOTAL
1560	Shower, stall, baked enamel, molded stone receptor, 30" square	1,550	870	2,420
1600	32" square	1,550	880	2,430
1640	Terrazzo receptor, 32" square	1,750	880	2,630
1680	36" square	1,900	890	2,790
1720	36" corner angle	2,175	890	3,065
1800	Fiberglass one piece, three walls, 32" square	820	855	1,675
1840	36" square	835	855	1,690
1880	Polypropylene, molded stone receptor, 30" square	980	1,275	2,255
1920	32" square	995	1,275	2,270
1960	Built-in head, arm, bypass, stops and handles	112	325	437
2050	Shower, stainless steel panels, handicap			
2100	w/fixed and handheld head, control valves, grab bar, and seat	5,400	3,925	9,325
2500	Shower, group with six heads, thermostatic mix valves & balancing valve	12,400	955	13,355
2520	Five heads	8,875	870	9,745

D2010 810	Drinking Fountain Systems	COST EACH		
		MAT.	INST.	TOTAL
1740	Drinking fountain, one bubbler, wall mounted			
1760	Non recessed			
1800	Bronze, no back	1,300	495	1,795
1840	Cast iron, enameled, low back	1,250	495	1,745
1880	Fiberglass, 12" back	2,325	495	2,820
1920	Stainless steel, no back	1,350	495	1,845
1960	Semi-recessed, poly marble	1,250	495	1,745
2040	Stainless steel	1,475	495	1,970
2080	Vitreous china	1,175	495	1,670
2120	Full recessed, poly marble	1,950	495	2,445
2200	Stainless steel	1,900	495	2,395
2240	Floor mounted, pedestal type, aluminum	2,675	675	3,350
2320	Bronze	2,325	675	3,000
2360	Stainless steel	2,125	675	2,800

D2010   Plumbing Fixtures

Wall Hung Water Cooler

Floor Mounted Water Cooler

Systems are complete with trim and rough-in (supply, waste and vent) to connect to supply branches and waste mains.

D2010 820	Water Cooler Systems	COST EACH		
		MAT.	INST.	TOTAL
1840	Water cooler, electric, wall hung, 8.2 G.P.H.	1,025	640	1,665
1880	Dual height, 14.3 G.P.H.	1,275	655	1,930
1920	Wheelchair type, 7.5 G.P.H.	1,200	640	1,840
1960	Semi recessed, 8.1 G.P.H.	1,025	640	1,665
2000	Full recessed, 8 G.P.H.	2,425	685	3,110
2040	Floor mounted, 14.3 G.P.H.	1,175	555	1,730
2080	Dual height, 14.3 G.P.H.	1,525	675	2,200
2120	Refrigerated compartment type, 1.5 G.P.H.	1,850	555	2,405

431

For customer support on your Square Foot Cost Data, call 877.694.8329.

D2010 Plumbing Fixtures

Two Fixture

Three Fixture

*Common wall is with adjacent bathrom

D2010 920	Two Fixture Bathroom, Two Wall Plumbing	COST EACH		
		MAT.	INST.	TOTAL
1180	Bathroom, lavatory & water closet, 2 wall plumbing, stand alone	1,750	2,075	3,825
1200	Share common plumbing wall*	1,600	1,800	3,400

D2010 922	Two Fixture Bathroom, One Wall Plumbing	COST EACH		
		MAT.	INST.	TOTAL
2220	Bathroom, lavatory & water closet, one wall plumbing, stand alone	1,650	1,875	3,525
2240	Share common plumbing wall*	1,425	1,600	3,025

D2010 924	Three Fixture Bathroom, One Wall Plumbing	COST EACH		
		MAT.	INST.	TOTAL
1150	Bathroom, three fixture, one wall plumbing			
1160	Lavatory, water closet & bathtub			
1170	Stand alone	3,025	2,425	5,450
1180	Share common plumbing wall *	2,575	1,750	4,325

D2010 926	Three Fixture Bathroom, Two Wall Plumbing	COST EACH		
		MAT.	INST.	TOTAL
2130	Bathroom, three fixture, two wall plumbing			
2140	Lavatory, water closet & bathtub			
2160	Stand alone	3,050	2,475	5,525
2180	Long plumbing wall common *	2,725	1,975	4,700
3610	Lavatory, bathtub & water closet			
3620	Stand alone	3,250	2,800	6,050
3640	Long plumbing wall common *	3,075	2,525	5,600
4660	Water closet, corner bathtub & lavatory			
4680	Stand alone	4,700	2,500	7,200
4700	Long plumbing wall common *	4,325	1,875	6,200
6100	Water closet, stall shower & lavatory			
6120	Stand alone	3,325	2,800	6,125
6140	Long plumbing wall common *	3,175	2,600	5,775
7060	Lavatory, corner stall shower & water closet			
7080	Stand alone	3,700	2,475	6,175
7100	Short plumbing wall common *	3,200	1,675	4,875

D2010 Plumbing Fixtures

Four Fixture   Five Fixture

*Common wall is with adjacent bathroom

D2010 928		Four Fixture Bathroom, Two Wall Plumbing	COST EACH		
			MAT.	INST.	TOTAL
1140	Bathroom, four fixture, two wall plumbing				
1150		Bathtub, water closet, stall shower & lavatory			
1160		Stand alone	4,250	2,675	6,925
1180		Long plumbing wall common *	3,875	2,075	5,950
2260		Bathtub, lavatory, corner stall shower & water closet			
2280		Stand alone	4,875	2,700	7,575
2320		Long plumbing wall common *	4,500	2,075	6,575
3620		Bathtub, stall shower, lavatory & water closet			
3640		Stand alone	4,700	3,375	8,075
3660		Long plumbing wall (opp. door) common *	4,325	2,750	7,075
4680	Bathroom, four fixture, three wall plumbing				
4700		Bathtub, stall shower, lavatory & water closet			
4720		Stand alone	5,525	3,725	9,250
4760		Long plumbing wall (opposite door) common *	5,350	3,425	8,775

D2010 932		Five Fixture Bathroom, Two Wall Plumbing	COST EACH		
			MAT.	INST.	TOTAL
1320	Bathroom, five fixture, two wall plumbing				
1340		Bathtub, water closet, stall shower & two lavatories			
1360		Stand alone	5,525	4,300	9,825
1400		One short plumbing wall common *	5,125	3,700	8,825
2360	Bathroom, five fixture, three wall plumbing				
2380		Water closet, bathtub, two lavatories & stall shower			
2400		Stand alone	6,150	4,325	10,475
2440		One short plumbing wall common *	5,750	3,700	9,450
4080	Bathroom, five fixture, one wall plumbing				
4100		Bathtub, two lavatories, corner stall shower & water closet			
4120		Stand alone	5,850	3,850	9,700
4160		Share common wall *	4,900	2,500	7,400

433

For customer support on your Square Foot Cost Data, call 877.694.8329.

D2020 Domestic Water Distribution

Installation includes piping and fittings within 10' of heater. Gas and oil fired heaters require vent piping (not included with these units).

Gas Fired

Oil Fired

D2020 220	Gas Fired Water Heaters - Residential Systems	COST EACH		
		MAT.	INST.	TOTAL
2200	Gas fired water heater, residential, 100°F rise			
2260	30 gallon tank, 32 GPH	1,725	1,475	3,200
2300	40 gallon tank, 32 GPH	1,925	1,650	3,575
2340	50 gallon tank, 63 GPH	1,975	1,650	3,625
2380	75 gallon tank, 63 GPH	2,725	1,850	4,575
2420	100 gallon tank, 63 GPH	3,000	1,950	4,950
2422				

D2020 230	Oil Fired Water Heaters - Residential Systems	COST EACH		
		MAT.	INST.	TOTAL
2200	Oil fired water heater, residential, 100°F rise			
2220	30 gallon tank, 103 GPH	2,000	1,375	3,375
2260	50 gallon tank, 145 GPH	2,450	1,525	3,975
2300	70 gallon tank, 164 GPH	3,350	1,725	5,075
2340	85 gallon tank, 181 GPH	10,900	1,775	12,675

434

For customer support on your Square Foot Cost Data, call 877.694.8329.

D2020 Domestic Water Distribution

Systems below include piping and fittings within 10' of heater. Electric water heaters do not require venting. Gas fired heaters require vent piping (not included in these prices).

Electric

Gas Fired

D2020 240	Electric Water Heaters - Commercial Systems	COST EACH		
		MAT.	INST.	TOTAL
1800	Electric water heater, commercial, 100°F rise			
1820	50 gallon tank, 9 KW 37 GPH	5,450	1,225	6,675
1860	80 gal, 12 KW 49 GPH	7,025	1,525	8,550
1900	36 KW 147 GPH	9,475	1,650	11,125
1940	120 gal, 36 KW 147 GPH	10,200	1,775	11,975
1980	150 gal, 120 KW 490 GPH	30,500	1,900	32,400
2020	200 gal, 120 KW 490 GPH	32,100	1,950	34,050
2060	250 gal, 150 KW 615 GPH	35,900	2,275	38,175
2100	300 gal, 180 KW 738 GPH	47,900	2,400	50,300
2140	350 gal, 30 KW 123 GPH	29,000	2,575	31,575
2180	180 KW 738 GPH	40,100	2,575	42,675
2220	500 gal, 30 KW 123 GPH	37,700	3,050	40,750
2260	240 KW 984 GPH	54,500	3,050	57,550
2300	700 gal, 30 KW 123 GPH	31,200	3,475	34,675
2340	300 KW 1230 GPH	45,600	3,475	49,075
2380	1000 gal, 60 KW 245 GPH	38,400	4,825	43,225
2420	480 KW 1970 GPH	61,000	4,850	65,850
2460	1500 gal, 60 KW 245 GPH	80,500	6,000	86,500
2500	480 KW 1970 GPH	110,000	6,000	116,000

D2020 250	Gas Fired Water Heaters - Commercial Systems	COST EACH		
		MAT.	INST.	TOTAL
1760	Gas fired water heater, commercial, 100°F rise			
1780	75.5 MBH input, 63 GPH	5,075	1,900	6,975
1820	95 MBH input, 86 GPH	7,050	1,900	8,950
1860	100 MBH input, 91 GPH	7,250	1,975	9,225
1900	115 MBH input, 110 GPH	8,525	2,050	10,575
1980	155 MBH input, 150 GPH	10,300	2,300	12,600
2020	175 MBH input, 168 GPH	11,000	2,475	13,475
2060	200 MBH input, 192 GPH	11,500	2,775	14,275
2100	240 MBH input, 230 GPH	12,000	3,025	15,025
2140	300 MBH input, 278 GPH	13,100	3,450	16,550
2180	390 MBH input, 374 GPH	15,700	3,500	19,200
2220	500 MBH input, 480 GPH	21,000	3,750	24,750
2260	600 MBH input, 576 GPH	24,100	4,050	28,150

435

For customer support on your Square Foot Cost Data, call 877.694.8329.

D2020 Domestic Water Distribution

Oil fired water heater systems include piping and fittings within 10' of heater. Oil fired heaters require vent piping (not included in these systems).

Oil Fired

D2020 260	Oil Fired Water Heaters - Commercial Systems	COST EACH		
		MAT.	INST.	TOTAL
1800	Oil fired water heater, commercial, 100°F rise			
1820	140 gal., 140 MBH input, 134 GPH	22,600	1,625	24,225
1900	140 gal., 255 MBH input, 247 GPH	24,500	2,025	26,525
1940	140 gal., 270 MBH input, 259 GPH	30,000	2,325	32,325
1980	140 gal., 400 MBH input, 384 GPH	31,400	2,675	34,075
2060	140 gal., 720 MBH input, 691 GPH	33,100	2,775	35,875
2100	221 gal., 300 MBH input, 288 GPH	43,700	3,050	46,750
2140	221 gal., 600 MBH input, 576 GPH	48,500	3,075	51,575
2180	221 gal., 800 MBH input, 768 GPH	49,400	3,200	52,600
2220	201 gal., 1000 MBH input, 960 GPH	50,500	3,225	53,725
2260	201 gal., 1250 MBH input, 1200 GPH	51,500	3,325	54,825
2300	201 gal., 1500 MBH input, 1441 GPH	55,500	3,400	58,900
2340	411 gal., 600 MBH input, 576 GPH	56,000	3,475	59,475
2380	411 gal., 800 MBH input, 768 GPH	56,500	3,575	60,075
2420	411 gal., 1000 MBH input, 960 GPH	61,000	4,100	65,100
2460	411 gal., 1250 MBH input, 1200 GPH	62,500	4,225	66,725
2500	397 gal., 1500 MBH input, 1441 GPH	66,000	4,350	70,350
2540	397 gal., 1750 MBH input, 1681 GPH	68,000	4,500	72,500

D2040   Rain Water Drainage

Design Assumptions: Vertical conductor size is based on a maximum rate of rainfall of 4" per hour. To convert roof area to other rates multiply "Max. S.F. Roof Area" shown by four and divide the result by desired local rate. The answer is the local roof area that may be handled by the indicated pipe diameter.

Basic cost is for roof drain, 10' of vertical leader and 10' of horizontal, plus connection to the main.

Pipe Dia.	Max. S.F. Roof Area	Gallons per Min.
2"	544	23
3"	1610	67
4"	3460	144
5"	6280	261
6"	10,200	424
8"	22,000	913

D2040 210	Roof Drain Systems	COST EACH		
		MAT.	INST.	TOTAL
1880	Roof drain, DWV PVC, 2" diam., piping, 10' high	310	710	1,020
1920	For each additional foot add	6.20	21.50	27.70
1960	3" diam., 10' high	385	825	1,210
2000	For each additional foot add	8.85	24	32.85
2040	4" diam., 10' high	455	930	1,385
2080	For each additional foot add	11	26.50	37.50
2120	5" diam., 10' high	1,400	1,075	2,475
2160	For each additional foot add	29	29.50	58.50
2200	6" diam., 10' high	1,375	1,200	2,575
2240	For each additional foot add	18.05	32.50	50.55
2280	8" diam., 10' high	3,000	2,025	5,025
2320	For each additional foot add	40	41.50	81.50
3940	C.I., soil, single hub, service wt., 2" diam. piping, 10' high	635	785	1,420
3980	For each additional foot add	12.50	20.50	33
4120	3" diam., 10' high	875	845	1,720
4160	For each additional foot add	16.50	21.50	38
4200	4" diam., 10' high	1,000	915	1,915
4240	For each additional foot add	20.50	23	43.50
4280	5" diam., 10' high	1,400	1,000	2,400
4320	For each additional foot add	28	26	54
4360	6" diam., 10' high	1,750	1,075	2,825
4400	For each additional foot add	34.50	27	61.50
4440	8" diam., 10' high	3,400	2,225	5,625
4480	For each additional foot add	52	46	98
6040	Steel galv. sch 40 threaded, 2" diam. piping, 10' high	715	755	1,470
6080	For each additional foot add	11.35	19.95	31.30
6120	3" diam., 10' high	1,400	1,075	2,475
6160	For each additional foot add	23	29.50	52.50
6200	4" diam., 10' high	1,975	1,400	3,375
6240	For each additional foot add	33	35.50	68.50
6280	5" diam., 10' high	2,500	1,150	3,650
6320	For each additional foot add	65.50	34.50	100
6360	6" diam, 10' high	2,925	1,500	4,425
6400	For each additional foot add	71	47.50	118.50
6440	8" diam., 10' high	4,675	2,175	6,850
6480	For each additional foot add	83.50	53.50	137

D2090 Other Plumbing Systems

D2090 810	Piping - Installed - Unit Costs	COST PER L.F.		
		MAT.	INST.	TOTAL
0840	Cast iron, soil, B & S, service weight, 2" diameter	12.50	20.50	33
0860	3" diameter	16.50	21.50	38
0880	4" diameter	20.50	23	43.50
0900	5" diameter	28	26	54
0920	6" diameter	34.50	27	61.50
0940	8" diameter	52	46	98
0960	10" diameter	82	50	132
0980	12" diameter	116	56.50	172.50
1040	No hub, 1-1/2" diameter	12.15	18	30.15
1060	2" diameter	12.40	19.05	31.45
1080	3" diameter	16.70	19.95	36.65
1100	4" diameter	21	22	43
1120	5" diameter	28	24	52
1140	6" diameter	35.50	25	60.50
1160	8" diameter	60	39	99
1180	10" diameter	97	44.50	141.50
1220	Copper tubing, hard temper, solder, type K, 1/2" diameter	5.25	9.10	14.35
1260	3/4" diameter	8.55	9.60	18.15
1280	1" diameter	11.40	10.75	22.15
1300	1-1/4" diameter	13.45	12.65	26.10
1320	1-1/2" diameter	17.15	14.20	31.35
1340	2" diameter	25.50	17.75	43.25
1360	2-1/2" diameter	38.50	21.50	60
1380	3" diameter	52.50	23.50	76
1400	4" diameter	88	33.50	121.50
1480	5" diameter	167	40	207
1500	6" diameter	244	52.50	296.50
1520	8" diameter	425	58.50	483.50
1560	Type L, 1/2" diameter	4.07	8.75	12.82
1600	3/4" diameter	5.70	9.35	15.05
1620	1" diameter	8.60	10.45	19.05
1640	1-1/4" diameter	11.35	12.25	23.60
1660	1-1/2" diameter	14.25	13.65	27.90
1680	2" diameter	20.50	16.90	37.40
1700	2-1/2" diameter	32.50	20.50	53
1720	3" diameter	40.50	23	63.50
1740	4" diameter	72	32.50	104.50
1760	5" diameter	132	37.50	169.50
1780	6" diameter	182	49.50	231.50
1800	8" diameter	315	55	370
1840	Type M, 1/2" diameter	3.36	8.45	11.81
1880	3/4" diameter	4.54	9.10	13.64
1900	1" diameter	7.25	10.15	17.40
1920	1-1/4" diameter	9.65	11.80	21.45
1940	1-1/2" diameter	12.70	13.15	25.85
1960	2" diameter	19.10	16.10	35.20
1980	2-1/2" diameter	28	19.95	47.95
2000	3" diameter	37.50	22	59.50
2020	4" diameter	66	32	98
2060	6" diameter	182	47.50	229.50
2080	8" diameter	315	52.50	367.50
2120	Type DWV, 1-1/4" diameter	10.05	11.80	21.85
2160	1-1/2" diameter	12.30	13.15	25.45
2180	2" diameter	16.10	16.10	32.20
2200	3" diameter	29	22	51
2220	4" diameter	48	32	80
2240	5" diameter	118	35.50	153.50
2260	6" diameter	168	47.50	215.50

D2090 Other Plumbing Systems

D2090 810	Piping - Installed - Unit Costs	COST PER L.F.		
		MAT.	INST.	TOTAL
2280	8" diameter	340	52.50	392.50
2800	Plastic,PVC, DWV, schedule 40, 1-1/4" diameter	5.90	16.90	22.80
2820	1-1/2" diameter	5.80	19.70	25.50
2830	2" diameter	6.20	21.50	27.70
2840	3" diameter	8.85	24	32.85
2850	4" diameter	11	26.50	37.50
2890	6" diameter	18.05	32.50	50.55
3010	Pressure pipe 200 PSI, 1/2" diameter	5.25	13.15	18.40
3030	3/4" diameter	5.60	13.90	19.50
3040	1" diameter	6.15	15.40	21.55
3050	1-1/4" diameter	7.05	16.90	23.95
3060	1-1/2" diameter	7.45	19.70	27.15
3070	2" diameter	8.45	21.50	29.95
3080	2-1/2" diameter	13.05	23	36.05
3090	3" diameter	14.45	24	38.45
3100	4" diameter	22.50	26.50	49
3110	6" diameter	37	32.50	69.50
3120	8" diameter	55	41.50	96.50
4000	Steel, schedule 40, black, threaded, 1/2" diameter	3.66	11.25	14.91
4020	3/4" diameter	4.30	11.65	15.95
4030	1" diameter	5.55	13.40	18.95
4040	1-1/4" diameter	6.80	14.35	21.15
4050	1-1/2" diameter	7.80	15.95	23.75
4060	2" diameter	9.90	19.95	29.85
4070	2-1/2" diameter	15.40	25.50	40.90
4080	3" diameter	19.60	29.50	49.10
4090	4" diameter	29.50	35.50	65
4100	Grooved, 5" diameter	42	34.50	76.50
4110	6" diameter	53	47.50	100.50
4120	8" diameter	86.50	53.50	140
4130	10" diameter	117	64	181
4140	12" diameter	130	73.50	203.50
4200	Galvanized, threaded, 1/2" diameter	4.06	11.25	15.31
4220	3/4" diameter	4.67	11.65	16.32
4230	1" diameter	6.30	13.40	19.70
4240	1-1/4" diameter	7.70	14.35	22.05
4250	1-1/2" diameter	8.85	15.95	24.80
4260	2" diameter	11.35	19.95	31.30
4270	2-1/2" diameter	17.95	25.50	43.45
4280	3" diameter	23	29.50	52.50
4290	4" diameter	33	35.50	68.50
4300	Grooved, 5" diameter	65.50	34.50	100
4310	6" diameter	71	47.50	118.50
4320	8" diameter	83.50	53.50	137
4330	10" diameter	134	64	198
4340	12" diameter	163	73.50	236.50
5010	Flanged, black, 1" diameter	10.90	19.15	30.05
5020	1-1/4" diameter	12.25	21	33.25
5030	1-1/2" diameter	12.85	23	35.85
5040	2" diameter	15.30	30	45.30
5050	2-1/2" diameter	21.50	37.50	59
5060	3" diameter	25	42	67
5070	4" diameter	34	51.50	85.50
5080	5" diameter	54.50	64	118.50
5090	6" diameter	65	82	147
5100	8" diameter	106	108	214
5110	10" diameter	151	128	279
5120	12" diameter	182	147	329

D20 Plumbing

D2090 Other Plumbing Systems

D2090 810	Piping - Installed - Unit Costs	COST PER L.F.		
		MAT.	INST.	TOTAL
5720	Grooved joints, black, 3/4" diameter	6	10	16
5730	1" diameter	5.80	11.25	17.05
5740	1-1/4" diameter	7.25	12.25	19.50
5750	1-1/2" diameter	8.10	13.90	22
5760	2" diameter	9.60	17.75	27.35
5770	2-1/2" diameter	15	22.50	37.50
5900	3" diameter	18.40	25.50	43.90
5910	4" diameter	26	28.50	54.50
5920	5" diameter	42	34.50	76.50
5930	6" diameter	53	47.50	100.50
5940	8" diameter	86.50	53.50	140
5950	10" diameter	117	64	181
5960	12" diameter	130	73.50	203.50

D3010 Energy Supply

Forced Hot Water Heating System

Fin Tube Radiation

Terminal Unit Heater

D3010 510	Apartment Building Heating - Fin Tube Radiation	COST PER S.F.		
		MAT.	INST.	TOTAL
1740	Heating systems, fin tube radiation, forced hot water			
1760	1,000 S.F. area, 10,000 C.F. volume	7.45	6.90	14.35
1800	10,000 S.F. area, 100,000 C.F. volume	3.13	4.13	7.26
1840	20,000 S.F. area, 200,000 C.F. volume	3.45	4.64	8.09
1880	30,000 S.F. area, 300,000 C.F. volume	3.34	4.49	7.83

D3010 520	Commercial Building Heating - Fin Tube Radiation	COST PER S.F.		
		MAT.	INST.	TOTAL
1940	Heating systems, fin tube radiation, forced hot water			
1960	1,000 S.F. bldg, one floor	18.30	15.50	33.80
2000	10,000 S.F., 100,000 C.F., total two floors	4.89	5.85	10.74
2040	100,000 S.F., 1,000,000 C.F., total three floors	1.97	2.62	4.59
2080	1,000,000 S.F., 10,000,000 C.F., total five floors	1.07	1.37	2.44

D3010 530	Commercial Bldg. Heating - Terminal Unit Heaters	COST PER S.F.		
		MAT.	INST.	TOTAL
1860	Heating systems, terminal unit heaters, forced hot water			
1880	1,000 S.F. bldg., one floor	17.90	14.20	32.10
1920	10,000 S.F. bldg., 100,000 C.F. total two floors	4.41	4.85	9.26
1960	100,000 S.F. bldg., 1,000,000 C.F. total three floors	2.01	2.39	4.40
2000	1,000,000 S.F. bldg., 10,000,000 C.F. total five floors	1.36	1.54	2.90

441

For customer support on your Square Foot Cost Data, call 877.694.8329.

D3020 Heat Generating Systems

Small Electric Boiler Systems Considerations:

1. Terminal units are fin tube baseboard radiation rated at 720 BTU/hr with 200° water temperature or 820 BTU/hr steam.
2. Primary use being for residential or smaller supplementary areas, the floor levels are based on 7-1/2" ceiling heights.
3. All distribution piping is copper for boilers through 205 MBH. All piping for larger systems is steel pipe.

Large Electric Boiler System Considerations:

1. Terminal units are all unit heaters of the same size. Quantities are varied to accommodate total requirements.
2. All air is circulated through the heaters a minimum of three times per hour.
3. As the capacities are adequate for commercial use, gross output rating by floor levels are based on 10' ceiling height.
4. All distribution piping is black steel pipe.

D3020 102	Small Heating Systems, Hydronic, Electric Boilers	COST PER S.F.		
		MAT.	INST.	TOTAL
1100	Small heating systems, hydronic, electric boilers			
1120	Steam, 1 floor, 1480 S.F., 61 M.B.H.	7.43	8.85	16.28
1160	3,000 S.F., 123 M.B.H.	5.65	7.75	13.40
1200	5,000 S.F., 205 M.B.H.	4.93	7.15	12.08
1240	2 floors, 12,400 S.F., 512 M.B.H.	3.92	7.10	11.02
1280	3 floors, 24,800 S.F., 1023 M.B.H.	4.41	7.05	11.46
1320	34,750 S.F., 1,433 M.B.H.	4.04	6.85	10.89
1360	Hot water, 1 floor, 1,000 S.F., 41 M.B.H.	12.55	4.91	17.46
1400	2,500 S.F., 103 M.B.H.	8.80	8.85	17.65
1440	2 floors, 4,850 S.F., 205 M.B.H.	8.50	10.60	19.10
1480	3 floors, 9,700 S.F., 410 M.B.H.	8.80	10.95	19.75

D3020 104	Large Heating Systems, Hydronic, Electric Boilers	COST PER S.F.		
		MAT.	INST.	TOTAL
1230	Large heating systems, hydronic, electric boilers			
1240	9,280 S.F., 135 K.W., 461 M.B.H., 1 floor	4.14	3.28	7.42
1280	14,900 S.F., 240 K.W., 820 M.B.H., 2 floors	5.55	5.35	10.90
1320	18,600 S.F., 296 K.W., 1,010 M.B.H., 3 floors	5.45	5.75	11.20
1360	26,100 S.F., 420 K.W., 1,432 M.B.H., 4 floors	5.35	5.65	11
1400	39,100 S.F., 666 K.W., 2,273 M.B.H., 4 floors	4.59	4.73	9.32
1440	57,700 S.F., 900 K.W., 3,071 M.B.H., 5 floors	4.39	4.67	9.06
1480	111,700 S.F., 1,800 K.W., 6,148 M.B.H., 6 floors	3.89	3.99	7.88
1520	149,000 S.F., 2,400 K.W., 8,191 M.B.H., 8 floors	3.84	4	7.84
1560	223,300 S.F., 3,600 K.W., 12,283 M.B.H., 14 floors	4.10	4.53	8.63

D3020 Heat Generating Systems

Boiler Selection: The maximum allowable working pressures are limited by ASME "Code for Heating Boilers" to 15 PSI for steam and 160 PSI for hot water heating boilers, with a maximum temperature limitation of 250° F. Hot water boilers are generally rated for a working pressure of 30 PSI. High pressure boilers are governed by the ASME "Code for Power Boilers" which is used almost universally for boilers operating over 15 PSIG. High pressure boilers used for a combination of heating/process loads are usually designed for 150 PSIG.

Boiler ratings are usually indicated as either Gross or Net Output. The Gross Load is equal to the Net Load plus a piping and pickup allowance. When this allowance cannot be determined, divide the gross output rating by 1.25 for a value equal to or greater than the next heat loss requirement of the building.

Table below lists installed cost per boiler and includes insulating jacket, standard controls, burner and safety controls. Costs do not include piping or boiler base pad. Outputs are Gross.

D3020 106	Boilers, Hot Water & Steam	COST EACH		
		MAT.	INST.	TOTAL
0600	Boiler, electric, steel, hot water, 12 K.W., 41 M.B.H.	5,475	1,500	6,975
0620	30 K.W., 103 M.B.H.	5,950	1,625	7,575
0640	60 K.W., 205 M.B.H.	6,600	1,775	8,375
0660	120 K.W., 410 M.B.H.	7,025	2,175	9,200
0680	210 K.W., 716 M.B.H.	8,375	3,250	11,625
0700	510 K.W., 1,739 M.B.H.	25,400	6,075	31,475
0720	720 K.W., 2,452 M.B.H.	31,300	6,875	38,175
0740	1,200 K.W., 4,095 M.B.H.	39,100	7,875	46,975
0760	2,100 K.W., 7,167 M.B.H.	65,500	9,925	75,425
0780	3,600 K.W., 12,283 M.B.H.	103,500	16,700	120,200
0820	Steam, 6 K.W., 20.5 M.B.H.	4,350	1,625	5,975
0840	24 K.W., 81.8 M.B.H.	5,300	1,775	7,075
0860	60 K.W., 205 M.B.H.	7,325	1,950	9,275
0880	150 K.W., 512 M.B.H.	10,600	3,000	13,600
0900	510 K.W., 1,740 M.B.H.	35,900	7,425	43,325
0920	1,080 K.W., 3,685 M.B.H.	44,700	10,700	55,400
0940	2,340 K.W., 7,984 M.B.H.	95,000	16,700	111,700
0980	Gas, cast iron, hot water, 80 M.B.H.	2,175	1,875	4,050
1000	100 M.B.H.	2,825	2,025	4,850
1020	163 M.B.H.	3,475	2,750	6,225
1040	280 M.B.H.	4,750	3,050	7,800
1060	544 M.B.H.	10,800	5,400	16,200
1080	1,088 M.B.H.	14,900	6,875	21,775
1100	2,000 M.B.H.	23,500	10,700	34,200
1120	2,856 M.B.H.	33,900	13,700	47,600
1140	4,720 M.B.H.	81,500	19,000	100,500
1160	6,970 M.B.H.	130,500	30,900	161,400
1180	For steam systems under 2,856 M.B.H., add 8%			
1520	Oil, cast iron, hot water, 109 M.B.H.	2,550	2,300	4,850
1540	173 M.B.H.	3,225	2,750	5,975
1560	236 M.B.H.	4,100	3,225	7,325
1580	1,084 M.B.H.	12,000	7,300	19,300
1600	1,600 M.B.H.	15,400	10,500	25,900
1620	2,480 M.B.H.	22,600	13,400	36,000
1640	3,550 M.B.H.	30,100	16,100	46,200
1660	Steam systems same price as hot water			

443

D3020 Heat Generating Systems

Unit Heater

Fossil Fuel Boiler System Considerations:

1. Terminal units are horizontal unit heaters. Quantities are varied to accommodate total heat loss per building.
2. Unit heater selection was determined by their capacity to circulate the building volume a minimum of three times per hour in addition to the BTU output.
3. Systems shown are forced hot water. Steam boilers cost slightly more than hot water boilers. However, this is compensated for by the smaller size or fewer terminal units required with steam.
4. Floor levels are based on 10' story heights.
5. MBH requirements are gross boiler output.

D3020 108	Heating Systems, Unit Heaters	COST PER S.F.		
		MAT.	INST.	TOTAL
1260	Heating systems, hydronic, fossil fuel, terminal unit heaters,			
1280	Cast iron boiler, gas, 80 M.B.H., 1,070 S.F. bldg.	10.93	10.56	21.49
1320	163 M.B.H., 2,140 S.F. bldg.	7.40	7.15	14.55
1360	544 M.B.H., 7,250 S.F. bldg.	5.50	5.05	10.55
1400	1,088 M.B.H., 14,500 S.F. bldg.	4.57	4.76	9.33
1440	3,264 M.B.H., 43,500 S.F. bldg.	3.74	3.60	7.34
1480	5,032 M.B.H., 67,100 S.F. bldg.	4.32	3.71	8.03
1520	Oil, 109 M.B.H., 1,420 S.F. bldg.	12.05	9.40	21.45
1560	235 M.B.H., 3,150 S.F. bldg.	7.60	6.65	14.25
1600	940 M.B.H., 12,500 S.F. bldg.	5.45	4.37	9.82
1640	1,600 M.B.H., 21,300 S.F. bldg.	5.05	4.18	9.23
1680	2,480 M.B.H., 33,100 S.F. bldg.	5.05	3.78	8.83
1720	3,350 M.B.H., 44,500 S.F. bldg.	4.46	3.85	8.31
1760	Coal, 148 M.B.H., 1,975 S.F. bldg.	10.35	6.10	16.45
1800	300 M.B.H., 4,000 S.F. bldg.	8.20	4.80	13
1840	2,360 M.B.H., 31,500 S.F. bldg.	5.50	3.82	9.32

D3020 Heat Generating Systems

Fin Tube Radiator

Fossil Fuel Boiler System Considerations:

1. Terminal units are commercial steel fin tube radiation. Quantities are varied to accommodate total heat loss per building.
2. Systems shown are forced hot water. Steam boilers cost slightly more than hot water boilers. However, this is compensated for by the smaller size or fewer terminal units required with steam.
3. Floor levels are based on 10′ story heights.
4. MBH requirements are gross boiler output.

D3020 110	Heating System, Fin Tube Radiation	COST PER S.F.		
		MAT.	INST.	TOTAL
3230	Heating systems, hydronic, fossil fuel, fin tube radiation			
3240	Cast iron boiler, gas, 80 MBH, 1,070 S.F. bldg.	13.55	15.89	29.44
3280	169 M.B.H., 2,140 S.F. bldg.	8.60	10.05	18.65
3320	544 M.B.H., 7,250 S.F. bldg.	7.35	8.60	15.95
3360	1,088 M.B.H., 14,500 S.F. bldg.	6.55	8.45	15
3400	3,264 M.B.H., 43,500 S.F. bldg.	5.90	7.45	13.35
3440	5,032 M.B.H., 67,100 S.F. bldg.	6.50	7.55	14.05
3480	Oil, 109 M.B.H., 1,420 S.F. bldg.	16.55	17.35	33.90
3520	235 M.B.H., 3,150 S.F. bldg.	9.45	10.15	19.60
3560	940 M.B.H., 12,500 S.F. bldg.	7.50	8.05	15.55
3600	1,600 M.B.H., 21,300 S.F. bldg.	7.20	8	15.20
3640	2,480 M.B.H., 33,100 S.F. bldg.	7.20	7.60	14.80
3680	3,350 M.B.H., 44,500 S.F. bldg.	6.60	7.70	14.30
3720	Coal, 148 M.B.H., 1,975 S.F. bldg.	12.15	9.60	21.75
3760	300 M.B.H., 4,000 S.F. bldg.	10	8.25	18.25
3800	2,360 M.B.H., 31,500 S.F. bldg.	7.60	7.60	15.20
4080	Steel boiler, oil, 97 M.B.H., 1,300 S.F. bldg.	12.30	14.25	26.55
4120	315 M.B.H., 4,550 S.F. bldg.	6.45	7.25	13.70
4160	525 M.B.H., 7,000 S.F. bldg.	9.10	8.30	17.40
4200	1,050 M.B.H., 14,000 S.F. bldg.	8.20	8.30	16.50
4240	2,310 M.B.H., 30,800 S.F. bldg.	7.20	7.65	14.85
4280	3,150 M.B.H., 42,000 S.F. bldg.	6.70	7.75	14.45

D3030 Cooling Generating Systems

Chilled Water Supply & Return Piping — Air Cooled Water Chiller Unit — Roof — Insulate — Return — Fan Coil Unit — Supply — Finish Ceiling

*Cooling requirements would lead to a
 choice of multiple chillers.

D3030 110	Chilled Water, Air Cooled Condenser Systems	COST PER S.F.		
		MAT.	INST.	TOTAL
1180	Packaged chiller, air cooled, with fan coil unit			
1200	Apartment corridors, 3,000 S.F., 5.50 ton	7.43	8.80	16.23
1240	6,000 S.F., 11.00 ton	5.95	7	12.95
1280	10,000 S.F., 18.33 ton	5.05	5.30	10.35
1320	20,000 S.F., 36.66 ton	3.97	3.83	7.80
1360	40,000 S.F., 73.33 ton	5.05	3.83	8.88
1440	Banks and libraries, 3,000 S.F., 12.50 ton	10.75	10.20	20.95
1480	6,000 S.F., 25.00 ton	9.80	8.30	18.10
1520	10,000 S.F., 41.66 ton	7.70	5.75	13.45
1560	20,000 S.F., 83.33 ton	8.55	5.15	13.70
1600	40,000 S.F., 167 ton*			
1680	Bars and taverns, 3,000 S.F., 33.25 ton	18.80	11.95	30.75
1720	6,000 S.F., 66.50 ton	16.90	9.10	26
1760	10,000 S.F., 110.83 ton	13.80	2.72	16.52
1800	20,000 S.F., 220 ton*			
1840	40,000 S.F., 440 ton*			
1920	Bowling alleys, 3,000 S.F., 17.00 ton	13.45	11.45	24.90
1960	6,000 S.F., 34.00 ton	10.50	8	18.50
2000	10,000 S.F., 56.66 ton	9.10	5.70	14.80
2040	20,000 S.F., 113.33 ton	9.80	5.20	15
2080	40,000 S.F., 227 ton*			
2160	Department stores, 3,000 S.F., 8.75 ton	10.30	9.80	20.10
2200	6,000 S.F., 17.50 ton	7.55	7.50	15.05
2240	10,000 S.F., 29.17 ton	6.20	5.40	11.60
2280	20,000 S.F., 58.33 ton	5.20	3.94	9.14
2320	40,000 S.F., 116.66 ton	6.40	3.94	10.34
2400	Drug stores, 3,000 S.F., 20.00 ton	15.20	11.80	27
2440	6,000 S.F., 40.00 ton	12.40	8.75	21.15
2480	10,000 S.F., 66.66 ton	13.05	7.40	20.45
2520	20,000 S.F., 133.33 ton	11.65	5.45	17.10
2560	40,000 S.F., 267 ton*			
2640	Factories, 2,000 S.F., 10.00 ton	9.50	9.85	19.35
2680	6,000 S.F., 20.00 ton	8.55	7.90	16.45
2720	10,000 S.F., 33.33 ton	6.70	5.50	12.20
2760	20,000 S.F., 66.66 ton	7.50	4.93	12.43
2800	40,000 S.F., 133.33 ton	6.90	4.04	10.94
2880	Food supermarkets, 3,000 S.F., 8.50 ton	10.10	9.75	19.85
2920	6,000 S.F., 17.00 ton	7.45	7.45	14.90
2960	10,000 S.F., 28.33 ton	5.85	5.25	11.10
3000	20,000 S.F., 56.66 ton	5	3.87	8.87

D3030 Cooling Generating Systems

D3030 110	Chilled Water, Air Cooled Condenser Systems	COST PER S.F.		
		MAT.	INST.	TOTAL
3040	40,000 S.F., 113.33 ton	6.20	3.95	10.15
3120	Medical centers, 3,000 S.F., 7.00 ton	9.10	9.50	18.60
3160	6,000 S.F., 14.00 ton	6.70	7.25	13.95
3200	10,000 S.F., 23.33 ton	5.90	5.50	11.40
3240	20,000 S.F., 46.66 ton	4.73	4.02	8.75
3280	40,000 S.F., 93.33 ton	5.65	3.89	9.54
3360	Offices, 3,000 S.F., 9.50 ton	9	9.70	18.70
3400	6,000 S.F., 19.00 ton	8.30	7.85	16.15
3440	10,000 S.F., 31.66 ton	6.50	5.45	11.95
3480	20,000 S.F., 63.33 ton	7.55	5.05	12.60
3520	40,000 S.F., 126.66 ton	6.80	4.08	10.88
3600	Restaurants, 3,000 S.F., 15.00 ton	12	10.60	22.60
3640	6,000 S.F., 30.00 ton	10	8.05	18.05
3680	10,000 S.F., 50.00 ton	8.75	5.85	14.60
3720	20,000 S.F., 100.00 ton	10	5.50	15.50
3760	40,000 S.F., 200 ton*			
3840	Schools and colleges, 3,000 S.F., 11.50 ton	10.25	10.05	20.30
3880	6,000 S.F., 23.00 ton	9.30	8.15	17.45
3920	10,000 S.F., 38.33 ton	7.30	5.65	12.95
3960	20,000 S.F., 76.66 ton	8.40	5.20	13.60
4000	40,000 S.F., 153 ton*			

D3030 Cooling Generating Systems

*Cooling requirements would lead to a choice of multiple chillers

D3030 115	Chilled Water, Cooling Tower Systems	COST PER S.F.		
		MAT.	INST.	TOTAL
1300	Packaged chiller, water cooled, with fan coil unit			
1320	Apartment corridors, 4,000 S.F., 7.33 ton	7.20	8.63	15.83
1360	6,000 S.F., 11.00 ton	5.90	7.40	13.30
1400	10,000 S.F., 18.33 ton	5.75	5.45	11.20
1440	20,000 S.F., 26.66 ton	4.65	4.06	8.71
1480	40,000 S.F., 73.33 ton	4.96	4.14	9.10
1520	60,000 S.F., 110.00 ton	5.25	4.25	9.50
1600	Banks and libraries, 4,000 S.F., 16.66 ton	12.45	9.35	21.80
1640	6,000 S.F., 25.00 ton	10.50	8.20	18.70
1680	10,000 S.F., 41.66 ton	8.40	6.05	14.45
1720	20,000 S.F., 83.33 ton	8.20	5.80	14
1760	40,000 S.F., 166.66 ton	9.05	7	16.05
1800	60,000 S.F., 250.00 ton	9.25	7.45	16.70
1880	Bars and taverns, 4,000 S.F., 44.33 ton	19.80	11.60	31.40
1920	6,000 S.F., 66.50 ton	22.50	12	34.50
1960	10,000 S.F., 110.83 ton	18.20	9.45	27.65
2000	20,000 S.F., 221.66 ton	21.50	9.55	31.05
2040	40,000 S.F., 440 ton*			
2080	60,000 S.F., 660 ton*			
2160	Bowling alleys, 4,000 S.F., 22.66 ton	14.30	10.20	24.50
2200	6,000 S.F., 34.00 ton	12.75	9	21.75
2240	10,000 S.F., 56.66 ton	11.10	6.55	17.65
2280	20,000 S.F., 113.33 ton	10.25	6.05	16.30
2320	40,000 S.F., 226.66 ton	12.20	6.95	19.15
2360	60,000 S.F., 340 ton			
2440	Department stores, 4,000 S.F., 11.66 ton	7.95	9.30	17.25
2480	6,000 S.F., 17.50 ton	9.20	7.65	16.85
2520	10,000 S.F., 29.17 ton	6.95	5.65	12.60
2560	20,000 S.F., 58.33 ton	5.70	4.19	9.89
2600	40,000 S.F., 116.66 ton	6.25	4.32	10.57
2640	60,000 S.F., 175.00 ton	8.20	6.80	15
2720	Drug stores, 4,000 S.F., 26.66 ton	15.55	10.50	26.05
2760	6,000 S.F., 40.00 ton	13.30	9	22.30
2800	10,000 S.F., 66.66 ton	14.50	8.10	22.60
2840	20,000 S.F., 133.33 ton	11.75	6.35	18.10
2880	40,000 S.F., 266.67 ton	12.35	7.85	20.20
2920	60,000 S.F., 400 ton*			
3000	Factories, 4,000 S.F., 13.33 ton	10.70	8.90	19.60
3040	6,000 S.F., 20.00 ton	9.15	7.70	16.85
3080	10,000 S.F., 33.33 ton	7.60	5.85	13.45
3120	20,000 S.F., 66.66 ton	8.10	5.25	13.35
3160	40,000 S.F., 133.33 ton	6.95	4.47	11.42

D3030 Cooling Generating Systems

D3030 115	Chilled Water, Cooling Tower Systems	COST PER S.F.		
		MAT.	INST.	TOTAL
3200	60,000 S.F., 200.00 ton	8.35	7.15	15.50
3280	Food supermarkets, 4,000 S.F., 11.33 ton	7.80	9.20	17
3320	6,000 S.F., 17.00 ton	8.15	7.50	15.65
3360	10,000 S.F., 28.33 ton	6.85	5.65	12.50
3400	20,000 S.F., 56.66 ton	5.90	4.21	10.11
3440	40,000 S.F., 113.33 ton	6.20	4.34	10.54
3480	60,000 S.F., 170.00 ton	8.10	6.85	14.95
3560	Medical centers, 4.000 S.F., 9.33 ton	6.80	8.45	15.25
3600	6,000 S.F., 14.00 ton	8	7.35	15.35
3640	10,000 S.F., 23.33 ton	6.30	5.45	11.75
3680	20,000 S.F., 46.66 ton	5.30	4.11	9.41
3720	40,000 S.F., 93.33 ton	5.65	4.26	9.91
3760	60,000 S.F., 140.00 ton	6.90	6.90	13.80
3840	Offices, 4,000 S.F., 12.66 ton	10.40	8.85	19.25
3880	6,000 S.F., 19.00 ton	9.05	7.85	16.90
3920	10,000 S.F., 31.66 ton	7.55	5.90	13.45
3960	20,000 S.F., 63.33 ton	8	5.25	13.25
4000	40,000 S.F., 126.66 ton	7.50	6.70	14.20
4040	60,000 S.F., 190.00 ton	8.10	7.10	15.20
4120	Restaurants, 4,000 S.F., 20.00 ton	12.80	9.50	22.30
4160	6,000 S.F., 30.00 ton	11.50	8.45	19.95
4200	10,000 S.F., 50.00 ton	10.20	6.30	16.50
4240	20,000 S.F., 100.00 ton	9.65	6.05	15.70
4280	40,000 S.F., 200.00 ton	9.60	6.75	16.35
4320	60,000 S.F., 300.00 ton	10.30	7.70	18
4400	Schools and colleges, 4,000 S.F., 15.33 ton	11.75	9.15	20.90
4440	6,000 S.F., 23.00 ton	9.95	8.05	18
4480	10,000 S.F., 38.33 ton	7.95	5.95	13.90
4520	20,000 S.F., 76.66 ton	7.95	5.75	13.70
4560	40,000 S.F., 153.33 ton	8.50	6.85	15.35
4600	60,000 S.F., 230.00 ton	8.55	7.15	15.70

449

For customer support on your Square Foot Cost Data, call 877.694.8329.

D30 HVAC

D3050 Terminal & Package Units

Rooftop, Single Zone System

*Above normal capacity

D3050 150	Rooftop Single Zone Unit Systems	COST PER S.F.		
		MAT.	INST.	TOTAL
1260	Rooftop, single zone, air conditioner			
1280	Apartment corridors, 500 S.F., .92 ton	4.35	4.05	8.40
1320	1,000 S.F., 1.83 ton	4.32	4.02	8.34
1360	1500 S.F., 2.75 ton	2.67	3.32	5.99
1400	3,000 S.F., 5.50 ton	2.42	3.19	5.61
1440	5,000 S.F., 9.17 ton	2.27	2.89	5.16
1480	10,000 S.F., 18.33 ton	2.76	2.72	5.48
1560	Banks or libraries, 500 S.F., 2.08 ton	9.80	9.15	18.95
1600	1,000 S.F., 4.17 ton	6.05	7.55	13.60
1640	1,500 S.F., 6.25 ton	5.50	7.25	12.75
1680	3,000 S.F., 12.50 ton	5.15	6.55	11.70
1720	5,000 S.F., 20.80 ton	6.25	6.15	12.40
1760	10,000 S.F., 41.67 ton	5.40	6.10	11.50
1840	Bars and taverns, 500 S.F. 5.54 ton	13.55	12.20	25.75
1880	1,000 S.F., 11.08 ton	12.60	10.40	23
1920	1,500 S.F., 16.62 ton	12.10	9.70	21.80
1960	3,000 S.F., 33.25 ton	13.60	9.25	22.85
2000	5,000 S.F., 55.42 ton	12.60	9.25	21.85
2040	10,000 S.F., 110.83 ton*			
2080	Bowling alleys, 500 S.F., 2.83 ton	8.25	10.25	18.50
2120	1,000 S.F., 5.67 ton	7.50	9.85	17.35
2160	1,500 S.F., 8.50 ton	7	8.95	15.95
2200	3,000 S.F., 17.00 ton	6.75	8.55	15.30
2240	5,000 S.F., 28.33 ton	7.50	8.35	15.85
2280	10,000 S.F., 56.67 ton	7	8.35	15.35
2360	Department stores, 500 S.F., 1.46 ton	6.90	6.40	13.30
2400	1,000 S.F., 2.92 ton	4.25	5.30	9.55
2480	3,000 S.F., 8.75 ton	3.60	4.60	8.20
2520	5,000 S.F., 14.58 ton	3.47	4.41	7.88
2560	10,000 S.F., 29.17 ton	3.87	4.29	8.16
2640	Drug stores, 500 S.F., 3.33 ton	9.70	12.05	21.75
2680	1,000 S.F., 6.67 ton	8.80	11.60	20.40
2720	1,500 S.F., 10.00 ton	8.25	10.50	18.75
2760	3,000 S.F., 20.00 ton	10.05	9.90	19.95
2800	5,000 S.F., 33.33 ton	8.85	9.80	18.65
2840	10,000 S.F., 66.67 ton	8.25	9.80	18.05
2920	Factories, 500 S.F., 1.67 ton	7.85	7.35	15.20
3000	1,500 S.F., 5.00 ton	4.40	5.80	10.20
3040	3,000 S.F., 10.00 ton	4.12	5.25	9.37
3080	5,000 S.F., 16.67 ton	3.96	5.05	9.01
3120	10,000 S.F., 33.33 ton	4.42	4.90	9.32
3200	Food supermarkets, 500 S.F., 1.42 ton	6.65	6.25	12.90

450

D3050 Terminal & Package Units

D3050 150	Rooftop Single Zone Unit Systems	COST PER S.F.		
		MAT.	INST.	TOTAL
3240	1,000 S.F., 2.83 ton	4.09	5.10	9.19
3280	1,500 S.F., 4.25 ton	3.74	4.93	8.67
3320	3,000 S.F., 8.50 ton	3.50	4.47	7.97
3360	5,000 S.F., 14.17 ton	3.36	4.28	7.64
3400	10,000 S.F., 28.33 ton	3.75	4.16	7.91
3480	Medical centers, 500 S.F., 1.17 ton	5.50	5.15	10.65
3520	1,000 S.F., 2.33 ton	5.50	5.15	10.65
3560	1,500 S.F., 3.50 ton	3.41	4.23	7.64
3640	5,000 S.F., 11.67 ton	2.89	3.69	6.58
3680	10,000 S.F., 23.33 ton	3.51	3.45	6.96
3760	Offices, 500 S.F., 1.58 ton	7.45	6.95	14.40
3800	1,000 S.F., 3.17 ton	4.62	5.75	10.37
3840	1,500 S.F., 4.75 ton	4.18	5.50	9.68
3880	3,000 S.F., 9.50 ton	3.91	5	8.91
3920	5,000 S.F., 15.83 ton	3.77	4.79	8.56
3960	10,000 S.F., 31.67 ton	4.20	4.66	8.86
4000	Restaurants, 500 S.F., 2.50 ton	11.80	11	22.80
4040	1,000 S.F., 5.00 ton	6.60	8.70	15.30
4080	1,500 S.F., 7.50 ton	6.20	7.90	14.10
4120	3,000 S.F., 15.00 ton	5.95	7.55	13.50
4160	5,000 S.F., 25.00 ton	7.55	7.40	14.95
4200	10,000 S.F., 50.00 ton	6.20	7.35	13.55
4240	Schools and colleges, 500 S.F., 1.92 ton	9.05	8.45	17.50
4280	1,000 S.F., 3.83 ton	5.60	6.95	12.55
4360	3,000 S.F., 11.50 ton	4.74	6.05	10.79
4400	5,000 S.F., 19.17 ton	5.75	5.70	11.45

451

For customer support on your Square Foot Cost Data, call 877.694.8329.

*Note A: Small single zone unit recommended.

*Note B: A combination of multizone units recommended.

D3050 155	Rooftop Multizone Unit Systems	COST PER S.F.		
		MAT.	INST.	TOTAL
1240	Rooftop, multizone, air conditioner			
1260	Apartment corridors, 1,500 S.F., 2.75 ton. See Note A.			
1280	3,000 S.F., 5.50 ton	11.03	5.17	16.20
1320	10,000 S.F., 18.30 ton	7.60	4.95	12.55
1360	15,000 S.F., 27.50 ton	7.40	4.93	12.33
1400	20,000 S.F., 36.70 ton	7.70	4.95	12.65
1440	25,000 S.F., 45.80 ton	7.20	4.97	12.17
1520	Banks or libraries, 1,500 S.F., 6.25 ton	25	11.70	36.70
1560	3,000 S.F., 12.50 ton	21	11.45	32.45
1600	10,000 S.F., 41.67 ton	16.35	11.30	27.65
1640	15,000 S.F., 62.50 ton	12	11.25	23.25
1680	20,000 S.F., 83.33 ton	12	11.25	23.25
1720	25,000 S.F., 104.00 ton	10.70	11.20	21.90
1800	Bars and taverns, 1,500 S.F., 16.62 ton	54.50	16.90	71.40
1840	3,000 S.F., 33.24 ton	45	16.30	61.30
1880	10,000 S.F., 110.83 ton	27	16.20	43.20
1920	15,000 S.F., 165 ton, See Note B.			
1960	20,000 S.F., 220 ton, See Note B.			
2000	25,000 S.F., 275 ton, See Note B.			
2080	Bowling alleys, 1,500 S.F., 8.50 ton	34	15.95	49.95
2120	3,000 S.F., 17.00 ton	28.50	15.60	44.10
2160	10,000 S.F., 56.70 ton	22.50	15.40	37.90
2200	15,000 S.F., 85.00 ton	16.35	15.25	31.60
2240	20,000 S.F., 113.00 ton	14.50	15.20	29.70
2280	25,000 S.F., 140.00 ton see Note B.			
2360	Department stores, 1,500 S.F., 4.37 ton, See Note A.			
2400	3,000 S.F., 8.75 ton	17.55	8.20	25.75
2440	10,000 S.F., 29.17 ton	12.30	7.85	20.15
2520	20,000 S.F., 58.33 ton	8.40	7.85	16.25
2560	25,000 S.F., 72.92 ton	8.40	7.85	16.25
2640	Drug stores, 1,500 S.F., 10.00 ton	40	18.75	58.75
2680	3,000 S.F., 20.00 ton	27.50	18.05	45.55
2720	10,000 S.F., 66.66 ton	19.20	17.95	37.15
2760	15,000 S.F., 100.00 ton	17.15	17.90	35.05
2800	20,000 S.F., 135 ton, See Note B.			
2840	25,000 S.F., 165 ton, See Note B.			
2920	Factories, 1,500 S.F., 5 ton, See Note A.			
3000	10,000 S.F., 33.33 ton	14.05	9	23.05

D3050 Terminal & Package Units

D3050 155	Rooftop Multizone Unit Systems	COST PER S.F.		
		MAT.	INST.	TOTAL
3040	15,000 S.F., 50.00 ton	13.10	9.05	22.15
3080	20,000 S.F., 66.66 ton	9.60	9	18.60
3120	25,000 S.F., 83.33 ton	9.60	9	18.60
3200	Food supermarkets, 1,500 S.F., 4.25 ton, See Note A.			
3240	3,000 S.F., 8.50 ton	17.05	7.95	25
3280	10,000 S.F., 28.33 ton	11.45	7.65	19.10
3320	15,000 S.F., 42.50 ton	11.15	7.70	18.85
3360	20,000 S.F., 56.67 ton	8.15	7.65	15.80
3400	25,000 S.F., 70.83 ton	8.15	7.65	15.80
3480	Medical centers, 1,500 S.F., 3.5 ton, See Note A.			
3520	3,000 S.F., 7.00 ton	14.05	6.55	20.60
3560	10,000 S.F., 23.33 ton	9.25	6.30	15.55
3600	15,000 S.F., 35.00 ton	9.80	6.30	16.10
3640	20,000 S.F., 46.66 ton	9.15	6.35	15.50
3680	25,000 S.F., 58.33 ton	6.70	6.30	13
3760	Offices, 1,500 S.F., 4.75 ton, See Note A.			
3800	3,000 S.F., 9.50 ton	19.05	8.90	27.95
3840	10,000 S.F., 31.66 ton	13.35	8.55	21.90
3920	20,000 S.F., 63.33 ton	9.10	8.55	17.65
3960	25,000 S.F., 79.16 ton	9.10	8.55	17.65
4000	Restaurants, 1,500 S.F., 7.50 ton	30	14.05	44.05
4040	3,000 S.F., 15.00 ton	25.50	13.75	39.25
4080	10,000 S.F., 50.00 ton	19.65	13.55	33.20
4120	15,000 S.F., 75.00 ton	14.40	13.45	27.85
4160	20,000 S.F., 100.00 ton	12.85	13.45	26.30
4200	25,000 S.F., 125 ton, See Note B.			
4240	Schools and colleges, 1,500 S.F., 5.75 ton	23	10.80	33.80
4320	10,000 S.F., 38.33 ton	15.05	10.40	25.45
4360	15,000 S.F., 57.50 ton	11.05	10.35	21.40
4400	20,000 S.F.,76.66 ton	11.05	10.35	21.40
4440	25,000 S.F., 95.83 ton	10.35	10.35	20.70

Self-Contained Water Cooled System

D3050 160	Self-contained, Water Cooled Unit Systems	COST PER S.F.		
		MAT.	INST.	TOTAL
1280	Self-contained, water cooled unit	3.56	3.29	6.85
1300	Apartment corridors, 500 S.F., .92 ton	3.55	3.30	6.85
1320	1,000 S.F., 1.83 ton	3.54	3.28	6.82
1360	3,000 S.F., 5.50 ton	2.94	2.75	5.69
1400	5,000 S.F., 9.17 ton	2.89	2.55	5.44
1440	10,000 S.F., 18.33 ton	3.96	2.28	6.24
1520	Banks or libraries, 500 S.F., 2.08 ton	7.75	3.40	11.15
1560	1,000 S.F., 4.17 ton	6.70	6.20	12.90
1600	3,000 S.F., 12.50 ton	6.55	5.75	12.30
1640	5,000 S.F., 20.80 ton	9	5.15	14.15
1680	10,000 S.F., 41.66 ton	7.85	5.15	13
1760	Bars and taverns, 500 S.F., 5.54 ton	16.85	5.75	22.60
1800	1,000 S.F., 11.08 ton	16.95	9.95	26.90
1840	3,000 S.F., 33.25 ton	21	7.80	28.80
1880	5,000 S.F., 55.42 ton	19.90	8.15	28.05
1920	10,000 S.F., 110.00 ton	19.35	8.05	27.40
2000	Bowling alleys, 500 S.F., 2.83 ton	10.50	4.63	15.13
2040	1,000 S.F., 5.66 ton	9.05	8.45	17.50
2080	3,000 S.F., 17.00 ton	12.25	7.05	19.30
2120	5,000 S.F., 28.33 ton	11.10	6.75	17.85
2160	10,000 S.F., 56.66 ton	10.40	6.90	17.30
2200	Department stores, 500 S.F., 1.46 ton	5.40	2.39	7.79
2240	1,000 S.F., 2.92 ton	4.67	4.36	9.03
2280	3,000 S.F., 8.75 ton	4.59	4.04	8.63
2320	5,000 S.F., 14.58 ton	6.30	3.61	9.91
2360	10,000 S.F., 29.17 ton	5.70	3.48	9.18
2440	Drug stores, 500 S.F., 3.33 ton	12.40	5.45	17.85
2480	1,000 S.F., 6.66 ton	10.70	9.95	20.65
2520	3,000 S.F., 20.00 ton	14.40	8.25	22.65
2560	5,000 S.F., 33.33 ton	15.10	8.15	23.25
2600	10,000 S.F., 66.66 ton	12.25	8.15	20.40
2680	Factories, 500 S.F., 1.66 ton	6.20	2.72	8.92
2720	1,000 S.F. 3.37 ton	5.40	5.05	10.45
2760	3,000 S.F., 10.00 ton	5.25	4.63	9.88
2800	5,000 S.F., 16.66 ton	7.20	4.13	11.33
2840	10,000 S.F., 33.33 ton	6.55	3.98	10.53

454

For customer support on your Square Foot Cost Data, call 877.694.8329.

D3050 Terminal & Package Units

D3050 160	Self-contained, Water Cooled Unit Systems	COST PER S.F.		
		MAT.	INST.	TOTAL
2920	Food supermarkets, 500 S.F., 1.42 ton	5.25	2.31	7.56
2960	1,000 S.F., 2.83 ton	5.50	5.05	10.55
3000	3,000 S.F., 8.50 ton	4.54	4.23	8.77
3040	5,000 S.F., 14.17 ton	4.46	3.92	8.38
3080	10,000 S.F., 28.33 ton	5.55	3.38	8.93
3160	Medical centers, 500 S.F., 1.17 ton	4.33	1.90	6.23
3200	1,000 S.F., 2.33 ton	4.52	4.17	8.69
3240	3,000 S.F., 7.00 ton	3.74	3.49	7.23
3280	5,000 S.F., 11.66 ton	3.68	3.22	6.90
3320	10,000 S.F., 23.33 ton	5.05	2.88	7.93
3400	Offices, 500 S.F., 1.58 ton	5.90	2.59	8.49
3440	1,000 S.F., 3.17 ton	6.10	5.70	11.80
3480	3,000 S.F., 9.50 ton	4.98	4.39	9.37
3520	5,000 S.F., 15.83 ton	6.80	3.93	10.73
3560	10,000 S.F., 31.67 ton	6.20	3.79	9.99
3640	Restaurants, 500 S.F., 2.50 ton	9.30	4.09	13.39
3680	1,000 S.F., 5.00 ton	8	7.50	15.50
3720	3,000 S.F., 15.00 ton	7.90	6.95	14.85
3760	5,000 S.F., 25.00 ton	10.80	6.20	17
3800	10,000 S.F., 50.00 ton	6.75	5.70	12.45
3880	Schools and colleges, 500 S.F., 1.92 ton	7.10	3.13	10.23
3920	1,000 S.F., 3.83 ton	6.15	5.75	11.90
3960	3,000 S.F., 11.50 ton	6.05	5.30	11.35
4000	5,000 S.F., 19.17 ton	8.25	4.75	13
4040	10,000 S.F., 38.33 ton	7.25	4.75	12

D3050 Terminal & Package Units

Self-Contained Air Cooled System

D3050 165	Self-contained, Air Cooled Unit Systems	COST PER S.F.		
		MAT.	INST.	TOTAL
1300	Self-contained, air cooled unit			
1320	Apartment corridors, 500 S.F., .92 ton	5.35	4.20	9.55
1360	1,000 S.F., 1.83 ton	5.30	4.17	9.47
1400	3,000 S.F., 5.50 ton	5.05	3.90	8.95
1440	5,000 S.F., 9.17 ton	3.75	3.71	7.46
1480	10,000 S.F., 18.33 ton	3.38	3.40	6.78
1560	Banks or libraries, 500 S.F., 2.08 ton	11.60	5.40	17
1600	1,000 S.F., 4.17 ton	11.45	8.85	20.30
1640	3,000 S.F., 12.50 ton	8.55	8.45	17
1680	5,000 S.F., 20.80 ton	7.80	7.75	15.55
1720	10,000 S.F., 41.66 ton	8.55	7.55	16.10
1800	Bars and taverns, 500 S.F., 5.54 ton	27.50	11.80	39.30
1840	1,000 S.F., 11.08 ton	22.50	17	39.50
1880	3,000 S.F., 33.25 ton	23	14.65	37.65
1920	5,000 S.F., 55.42 ton	23	14.70	37.70
1960	10,000 S.F., 110.00 ton	23.50	14.60	38.10
2040	Bowling alleys, 500 S.F., 2.83 ton	15.90	7.40	23.30
2080	1,000 S.F., 5.66 ton	15.60	12.05	27.65
2120	3,000 S.F., 17.00 ton	10.50	10.50	21
2160	5,000 S.F., 28.33 ton	11.90	10.25	22.15
2200	10,000 S.F., 56.66 ton	11.90	10.25	22.15
2240	Department stores, 500 S.F., 1.46 ton	8.20	3.81	12.01
2280	1,000 S.F., 2.92 ton	8.05	6.20	14.25
2320	3,000 S.F., 8.75 ton	5.95	5.90	11.85
2360	5,000 S.F., 14.58 ton	5.95	5.90	11.85
2400	10,000 S.F., 29.17 ton	6.10	5.30	11.40
2480	Drug stores, 500 S.F., 3.33 ton	18.70	8.70	27.40
2520	1,000 S.F., 6.66 ton	18.35	14.15	32.50
2560	3,000 S.F., 20.00 ton	12.50	12.40	24.90
2600	5,000 S.F., 33.33 ton	13.95	12.10	26.05
2640	10,000 S.F., 66.66 ton	14.15	12.10	26.25
2720	Factories, 500 S.F., 1.66 ton	9.50	4.39	13.89
2760	1,000 S.F., 3.33 ton	9.25	7.10	16.35
2800	3,000 S.F., 10.00 ton	6.80	6.75	13.55
2840	5,000 S.F., 16.66 ton	6.15	6.20	12.35
2880	10,000 S.F., 33.33 ton	6.95	6.05	13
2960	Food supermarkets, 500 S.F., 1.42 ton	7.95	3.70	11.65
3000	1,000 S.F., 2.83 ton	8.15	6.45	14.60
3040	3,000 S.F., 8.50 ton	7.80	6	13.80
3080	5,000 S.F., 14.17 ton	5.80	5.70	11.50

D30 HVAC

D3050 Terminal & Package Units

D3050 165	Self-contained, Air Cooled Unit Systems	COST PER S.F.		
		MAT.	INST.	TOTAL
3120	10,000 S.F., 28.33 ton	5.95	5.15	11.10
3200	Medical centers, 500 S.F., 1.17 ton	6.60	3.06	9.66
3240	1,000 S.F., 2.33 ton	6.75	5.35	12.10
3280	3,000 S.F., 7.00 ton	6.45	4.96	11.41
3320	5,000 S.F., 16.66 ton	4.80	4.72	9.52
3360	10,000 S.F., 23.33 ton	4.32	4.32	8.64
3440	Offices, 500 S.F., 1.58 ton	8.90	4.14	13.04
3480	1,000 S.F., 3.16 ton	9.20	7.25	16.45
3520	3,000 S.F., 9.50 ton	6.50	6.40	12.90
3560	5,000 S.F., 15.83 ton	5.95	5.90	11.85
3600	10,000 S.F., 31.66 ton	6.65	5.75	12.40
3680	Restaurants, 500 S.F., 2.50 ton	14	6.50	20.50
3720	1,000 S.F., 5.00 ton	13.80	10.60	24.40
3760	3,000 S.F., 15.00 ton	10.25	10.10	20.35
3800	5,000 S.F., 25.00 ton	10.45	9.05	19.50
3840	10,000 S.F., 50.00 ton	10.85	9.05	19.90
3920	Schools and colleges, 500 S.F., 1.92 ton	10.75	5	15.75
3960	1,000 S.F., 3.83 ton	10.60	8.15	18.75
4000	3,000 S.F., 11.50 ton	7.85	7.75	15.60
4040	5,000 S.F., 19.17 ton	7.10	7.10	14.20
4080	10,000 S.F., 38.33 ton	7.90	6.95	14.85

457

For customer support on your Square Foot Cost Data, call 877.694.8329.

D3050 Terminal & Package Units

*Cooling requirements would lead to more
than one system.

D3050 170	Split Systems With Air Cooled Condensing Units	COST PER S.F.		
		MAT.	INST.	TOTAL
1260	Split system, air cooled condensing unit			
1280	Apartment corridors, 1,000 S.F., 1.83 ton	2.63	2.63	5.26
1320	2,000 S.F., 3.66 ton	1.95	2.67	4.62
1360	5,000 S.F., 9.17 ton	2.05	3.32	5.37
1400	10,000 S.F., 18.33 ton	2.36	3.62	5.98
1440	20,000 S.F., 36.66 ton	2.70	3.69	6.39
1520	Banks and libraries, 1,000 S.F., 4.17 ton	4.43	6.05	10.48
1560	2,000 S.F., 8.33 ton	4.67	7.55	12.22
1600	5,000 S.F., 20.80 ton	5.35	8.25	13.60
1640	10,000 S.F., 41.66 ton	6.10	8.40	14.50
1680	20,000 S.F., 83.32 ton	6.75	8.75	15.50
1760	Bars and taverns, 1,000 S.F., 11.08 ton	10.95	11.90	22.85
1800	2,000 S.F., 22.16 ton	14.90	14.25	29.15
1840	5,000 S.F., 55.42 ton	13.90	13.35	27.25
1880	10,000 S.F., 110.84 ton	14.85	13.95	28.80
1920	20,000 S.F., 220 ton*			
2000	Bowling alleys, 1,000 S.F., 5.66 ton	6.10	11.30	17.40
2040	2,000 S.F., 11.33 ton	6.35	10.25	16.60
2080	5,000 S.F., 28.33 ton	7.30	11.20	18.50
2120	10,000 S.F., 56.66 ton	8.35	11.40	19.75
2160	20,000 S.F., 113.32 ton	9.70	12.40	22.10
2320	Department stores, 1,000 S.F., 2.92 ton	3.11	4.16	7.27
2360	2,000 S.F., 5.83 ton	3.14	5.85	8.99
2400	5,000 S.F., 14.58 ton	3.27	5.30	8.57
2440	10,000 S.F., 29.17 ton	3.76	5.80	9.56
2480	20,000 S.F., 58.33 ton	4.29	5.90	10.19
2560	Drug stores, 1,000 S.F., 6.66 ton	7.15	13.30	20.45
2600	2,000 S.F., 13.32 ton	7.50	12.10	19.60
2640	5,000 S.F., 33.33 ton	9.80	13.45	23.25
2680	10,000 S.F., 66.66 ton	9.85	14	23.85
2720	20,000 S.F., 133.32 ton*			
2800	Factories, 1,000 S.F., 3.33 ton	3.56	4.75	8.31
2840	2,000 S.F., 6.66 ton	3.58	6.65	10.23
2880	5,000 S.F., 16.66 ton	4.30	6.60	10.90
2920	10,000 S.F., 33.33 ton	4.90	6.70	11.60
2960	20,000 S.F., 66.66 ton	4.90	7	11.90
3040	Food supermarkets, 1,000 S.F., 2.83 ton	3.03	4.04	7.07
3080	2,000 S.F., 5.66 ton	3.04	5.65	8.69
3120	5,000 S.F., 14.66 ton	3.18	5.15	8.33
3160	10,000 S.F., 28.33 ton	3.64	5.60	9.24
3200	20,000 S.F., 56.66 ton	4.16	5.70	9.86
3280	Medical centers, 1,000 S.F., 2.33 ton	2.68	3.26	5.94

D3050 Terminal & Package Units

D3050 170	Split Systems With Air Cooled Condensing Units	COST PER S.F.		
		MAT.	INST.	TOTAL
3320	2,000 S.F., 4.66 ton	2.51	4.65	7.16
3360	5,000 S.F., 11.66 ton	2.62	4.23	6.85
3400	10,000 S.F., 23.33 ton	3	4.63	7.63
3440	20,000 S.F., 46.66 ton	3.43	4.71	8.14
3520	Offices, 1,000 S.F., 3.17 ton	3.38	4.53	7.91
3560	2,000 S.F., 6.33 ton	3.41	6.35	9.76
3600	5,000 S.F., 15.83 ton	3.55	5.75	9.30
3640	10,000 S.F., 31.66 ton	4.08	6.30	10.38
3680	20,000 S.F., 63.32 ton	4.66	6.65	11.31
3760	Restaurants, 1,000 S.F., 5.00 ton	5.40	10	15.40
3800	2,000 S.F., 10.00 ton	5.60	9.05	14.65
3840	5,000 S.F., 25.00 ton	6.45	9.90	16.35
3880	10,000 S.F., 50.00 ton	7.35	10.10	17.45
3920	20,000 S.F., 100.00 ton	8.55	10.95	19.50
4000	Schools and colleges, 1,000 S.F., 3.83 ton	4.07	5.60	9.67
4040	2,000 S.F., 7.66 ton	4.30	6.95	11.25
4080	5,000 S.F., 19.17 ton	4.93	7.60	12.53
4120	10,000 S.F., 38.33 ton	5.65	7.75	13.40

For customer support on your Square Foot Cost Data, call 877.694.8329.

459

D3090 Other HVAC Systems/Equip

Cast Iron Garage Exhaust System

Dual Exhaust System

D3090 320	Garage Exhaust Systems	COST PER BAY		
		MAT.	INST.	TOTAL
1040	Garage, single 3" exhaust outlet, cars & light trucks, one bay	4,400	1,950	6,350
1060	Additional bays up to seven bays	1,025	525	1,550
1500	4" outlet, trucks, one bay	4,425	1,950	6,375
1520	Additional bays up to six bays	1,075	525	1,600
1600	5" outlet, diesel trucks, one bay	4,750	1,950	6,700
1650	Additional single bays up to six	1,500	620	2,120
1700	Two adjoining bays	4,750	1,950	6,700
2000	Dual exhaust, 3" outlets, pair of adjoining bays	5,500	2,450	7,950
2100	Additional pairs of adjoining bays	1,675	620	2,295

Reference - System Classification

System Classification
Rules for installation of sprinkler systems vary depending on the classification of occupancy falling into one of three categories as follows:

Light Hazard Occupancy
The protection area allotted per sprinkler should not exceed 225 S.F., with the maximum distance between lines and sprinklers on lines being 15'. The sprinklers do not need to be staggered. Branch lines should not exceed eight sprinklers on either side of a cross main. Each large area requiring more than 100 sprinklers and without a sub-dividing partition should be supplied by feed mains or risers sized for ordinary hazard occupancy.
Maximum system area = 52,000 S.F.

Included in this group are:
Churches	Nursing Homes
Clubs	Offices
Educational	Residential
Hospitals	Restaurants
Institutional	Theaters and Auditoriums
Libraries	(except stages and prosceniums)
(except large stack rooms)	Unused Attics
Museums	

Ordinary Hazard Occupancy
The protection area allotted per sprinkler shall not exceed 130 S.F. of noncombustible ceiling and 130 S.F. of combustible ceiling. The maximum allowable distance between sprinkler lines and sprinklers on line is 15'. Sprinklers shall be staggered if the distance between heads exceeds 12'. Branch lines should not exceed eight sprinklers on either side of a cross main.
Maximum system area = 52,000 S.F.

Included in this group are:
Group 1	Group 2
Automotive Parking and Showrooms	Cereal Mills
Bakeries	Chemical Plants—Ordinary
Beverage manufacturing	Confectionery Products
Canneries	Distilleries
Dairy Products Manufacturing/Processing	Dry Cleaners
Electronic Plans	Feed Mills
Glass and Glass Products Manufacturing	Horse Stables
Laundries	Leather Goods Manufacturing
Restaurant Service Areas	Libraries—Large Stack Room Areas
	Machine Shops
	Metal Working
	Mercantile
	Paper and Pulp Mills
	Paper Process Plants
	Piers and Wharves
	Post Offices
	Printing and Publishing
	Repair Garages
	Stages
	Textile Manufacturing
	Tire Manufacturing
	Tobacco Products Manufacturing
	Wood Machining
	Wood Product Assembly

Extra Hazard Occupancy
The protection area allotted per sprinkler shall not exceed 100 S.F. of noncombustible ceiling and 100 S.F. of combustible ceiling. The maximum allowable distance between lines and between sprinklers on lines is 12'. Sprinklers on alternate lines shall be staggered if the distance between sprinklers on lines exceeds 8'. Branch lines should not exceed six sprinklers on either side of a cross main.
Maximum system area:
   Design by pipe schedule = 25,000 S.F.
   Design by hydraulic calculation = 40,000 S.F.

Included in this group are:
Group 1	Group 2
Aircraft hangars	Asphalt Saturating
Combustible Hydraulic Fluid Use Area	Flammable Liquids Spraying
Die Casting	Flow Coating
Metal Extruding	Manufactured/Modular Home
Plywood/Particle Board Manufacturing	Building Assemblies (where
Printing (inks with flash points < 100	finished enclosure is present
degrees F	and has combustible interiors)
Rubber Reclaiming, Compounding,	Open Oil Quenching
Drying, Milling, Vulcanizing	Plastics Processing
Saw Mills	Solvent Cleaning
Textile Picking, Opening, Blending,	Varnish and Paint Dipping
Garnetting, Carding, Combing of	
Cotton, Synthetics, Wood Shoddy,	
or Burlap	
Upholstering with Plastic Foams	

461

For customer support on your Square Foot Cost Data, call 877.694.8329.

Reference - Sprinkler Systems (Automatic)

Sprinkler systems may be classified by type as follows:

1. **Wet Pipe System.** A system employing automatic sprinklers attached to a piping system containing water and connected to a water supply so that water discharges immediately from sprinklers opened by a fire.

2. **Dry Pipe System.** A system employing automatic sprinklers attached to a piping system containing air under pressure, the release of which as from the opening of sprinklers permits the water pressure to open a valve known as a "dry pipe valve". The water then flows into the piping system and out the opened sprinklers.

3. **Pre-Action System.** A system employing automatic sprinklers attached to a piping system containing air that may or may not be under pressure, with a supplemental heat responsive system of generally more sensitive characteristics than the automatic sprinklers themselves, installed in the same areas as the sprinklers; actuation of the heat responsive system, as from a fire, opens a valve which permits water to flow into the sprinkler piping system and to be discharged from any sprinklers which may be open.

4. **Deluge System.** A system employing open sprinklers attached to a piping system connected to a water supply through a valve which is opened by the operation of a heat responsive system installed in the same areas as the sprinklers. When this valve opens, water flows into the piping system and discharges from all sprinklers attached thereto.

5. **Combined Dry Pipe and Pre-Action Sprinkler System.** A system employing automatic sprinklers attached to a piping system containing air under pressure with a supplemental heat responsive system of generally more sensitive characteristics than the automatic sprinklers themselves, installed in the same areas as the sprinklers; operation of the heat responsive system, as from a fire, actuates tripping devices which open dry pipe valves simultaneously and without loss of air pressure in the system. Operation of the heat responsive system also opens approved air exhaust valves at the end of the feed main which facilitates the filling of the system with water which usually precedes the opening of sprinklers. The heat responsive system also serves as an automatic fire alarm system.

6. **Limited Water Supply System.** A system employing automatic sprinklers and conforming to these standards but supplied by a pressure tank of limited capacity.

7. **Chemical Systems.** Systems using halon, carbon dioxide, dry chemical or high expansion foam as selected for special requirements. Agent may extinguish flames by chemically inhibiting flame propagation, suffocate flames by excluding oxygen, interrupting chemical action of oxygen uniting with fuel or sealing and cooling the combustion center.

8. **Firecycle System.** Firecycle is a fixed fire protection sprinkler system utilizing water as its extinguishing agent. It is a time delayed, recycling, preaction type which automatically shuts the water off when heat is reduced below the detector operating temperature and turns the water back on when that temperature is exceeded. The system senses a fire condition through a closed circuit electrical detector system which controls water flow to the fire automatically. Batteries supply up to 90 hour emergency power supply for system operation. The piping system is dry (until water is required) and is monitored with pressurized air. Should any leak in the system piping occur, an alarm will sound, but water will not enter the system until heat is sensed by a firecycle detector.

Area coverage sprinkler systems may be laid out and fed from the supply in any one of several patterns as shown below. It is desirable, if possible, to utilize a central feed and achieve a shorter flow path from the riser to the furthest sprinkler. This permits use of the smallest sizes of pipe possible with resulting savings.

Reprinted with permission from NFPA 13-2013, *Installation of Sprinkler Systems*, Copyright © 2012, National Fire Protection Association, Quincy, MA. This reprinted material is not the complete and official position of the NFPA on the referenced subject, which is represented only by the standard in its entirety.

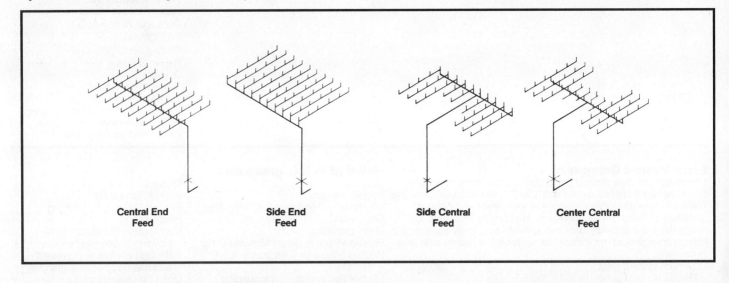

Central End Feed   Side End Feed   Side Central Feed   Center Central Feed

D40 Fire Protection

D4010 Sprinklers

Dry Pipe System: A system employing automatic sprinklers attached to a piping system containing air under pressure, the release of which from the opening of sprinklers permits the water pressure to open a valve known as a "dry pipe valve". The water then flows into the piping system and out the opened sprinklers.

All areas are assumed to be open.

D4010 310	Dry Pipe Sprinkler Systems	COST PER S.F.		
		MAT.	INST.	TOTAL
0520	Dry pipe sprinkler systems, steel, black, sch. 40 pipe			
0530	Light hazard, one floor, 500 S.F.	9.75	6.25	16
0560	1000 S.F.	5.55	3.63	9.18
0580	2000 S.F.	5.35	3.69	9.04
0600	5000 S.F.	2.85	2.50	5.35
0620	10,000 S.F.	2.02	2.05	4.07
0640	50,000 S.F.	1.55	1.85	3.40
0660	Each additional floor, 500 S.F.	2.13	3.07	5.20
0680	1000 S.F.	1.86	2.50	4.36
0700	2000 S.F.	1.85	2.32	4.17
0720	5000 S.F.	1.51	2	3.51
0740	10,000 S.F.	1.39	1.82	3.21
0760	50,000 S.F.	1.26	1.64	2.90
1000	Ordinary hazard, one floor, 500 S.F.	9.95	6.30	16.25
1020	1000 S.F.	5.70	3.69	9.39
1040	2000 S.F.	5.50	3.82	9.32
1060	5000 S.F.	3.28	2.68	5.96
1080	10,000 S.F.	2.63	2.73	5.36
1100	50,000 S.F.	2.42	2.59	5.01
1140	Each additional floor, 500 S.F.	2.34	3.14	5.48
1160	1000 S.F.	2.17	2.83	5
1180	2000 S.F.	2.19	2.56	4.75
1200	5000 S.F.	2.02	2.22	4.24
1220	10,000 S.F.	1.84	2.16	4
1240	50,000 S.F.	1.79	1.88	3.67
1500	Extra hazard, one floor, 500 S.F.	13.85	7.85	21.70
1520	1000 S.F.	8.50	5.65	14.15
1540	2000 S.F.	6.15	4.93	11.08
1560	5000 S.F.	3.79	3.76	7.55
1580	10,000 S.F.	4	3.56	7.56
1600	50,000 S.F.	4.34	3.46	7.80
1660	Each additional floor, 500 S.F.	3.31	3.90	7.21
1680	1000 S.F.	3.24	3.69	6.93
1700	2000 S.F.	3	3.69	6.69
1720	5000 S.F.	2.55	3.24	5.79
1740	10,000 S.F.	3.04	2.93	5.97
1760	50,000 S.F.	3.10	2.82	5.92
2020	Grooved steel, black, sch. 40 pipe, light hazard, one floor, 2000 S.F.	5.15	3.15	8.30

463

D40 Fire Protection

D4010 Sprinklers

D4010 310	Dry Pipe Sprinkler Systems	COST PER S.F.		
		MAT.	INST.	TOTAL
2060	10,000 S.F.	2.01	1.77	3.78
2100	Each additional floor, 2000 S.F.	1.83	1.89	3.72
2150	10,000 S.F.	1.38	1.54	2.92
2200	Ordinary hazard, one floor, 2000 S.F.	5.45	3.34	8.79
2250	10,000 S.F.	2.48	2.30	4.78
2300	Each additional floor, 2000 S.F.	2.10	2.08	4.18
2350	10,000 S.F.	1.86	2.07	3.93
2400	Extra hazard, one floor, 2000 S.F.	6.15	4.22	10.37
2450	10,000 S.F.	3.50	2.96	6.46
2500	Each additional floor, 2000 S.F.	3.01	3.05	6.06
2550	10,000 S.F.	2.71	2.58	5.29
3050	Grooved steel, black, sch. 10 pipe, light hazard, one floor, 2000 S.F.	5.10	3.13	8.23
3100	10,000 S.F.	1.96	1.74	3.70
3150	Each additional floor, 2000 S.F.	1.75	1.87	3.62
3200	10,000 S.F.	1.33	1.51	2.84
3250	Ordinary hazard, one floor, 2000 S.F.	5.35	3.31	8.66
3300	10,000 S.F.	2.38	2.25	4.63
3350	Each additional floor, 2000 S.F.	2.02	2.05	4.07
3400	10,000 S.F.	1.76	2.02	3.78
3450	Extra hazard, one floor, 2000 S.F.	6.10	4.19	10.29
3500	10,000 S.F.	3.31	2.90	6.21
3550	Each additional floor, 2000 S.F.	2.95	3.02	5.97
3600	10,000 S.F.	2.63	2.53	5.16
4050	Copper tubing, type M, light hazard, one floor, 2000 S.F.	5.45	3.12	8.57
4100	10,000 S.F.	2.38	1.78	4.16
4150	Each additional floor, 2000 S.F.	2.14	1.90	4.04
4200	10,000 S.F.	1.76	1.56	3.32
4250	Ordinary hazard, one floor, 2000 S.F.	5.85	3.47	9.32
4300	10,000 S.F.	2.94	2.11	5.05
4350	Each additional floor, 2000 S.F.	2.79	2.18	4.97
4400	10,000 S.F.	2.26	1.84	4.10
4450	Extra hazard, one floor, 2000 S.F.	6.70	4.25	10.95
4500	10,000 S.F.	5.10	3.22	8.32
4550	Each additional floor, 2000 S.F.	3.57	3.08	6.65
4600	10,000 S.F.	3.76	2.81	6.57
5050	Copper tubing, type M, T-drill system, light hazard, one floor			
5060	2000 S.F.	5.45	2.90	8.35
5100	10,000 S.F.	2.26	1.48	3.74
5150	Each additional floor, 2000 S.F.	2.13	1.68	3.81
5200	10,000 S.F.	1.64	1.26	2.90
5250	Ordinary hazard, one floor, 2000 S.F.	5.65	2.96	8.61
5300	10,000 S.F.	2.84	1.89	4.73
5350	Each additional floor, 2000 S.F.	2.30	1.70	4
5400	10,000 S.F.	2.12	1.58	3.70
5450	Extra hazard, one floor, 2000 S.F.	6.20	3.51	9.71
5500	10,000 S.F.	4.32	2.37	6.69
5550	Each additional floor, 2000 S.F.	3.06	2.34	5.40
5600	10,000 S.F.	2.99	1.96	4.95

D4010 Sprinklers

Preaction System: A system employing automatic sprinklers attached to a piping system containing air that may or may not be under pressure, with a supplemental heat responsive system of generally more sensitive characteristics than the automatic sprinklers themselves, installed in the same areas as the sprinklers. Actuation of the heat responsive system, as from a fire, opens a valve which permits water to flow into the sprinkler piping system and to be discharged from those sprinklers which were opened by heat from the fire.

All areas are assumed to be opened.

D4010 350	Preaction Sprinkler Systems	COST PER S.F.		
		MAT.	INST.	TOTAL
0520	Preaction sprinkler systems, steel, black, sch. 40 pipe			
0530	Light hazard, one floor, 500 S.F.	9.60	4.97	14.57
0560	1000 S.F.	5.60	3.72	9.32
0580	2000 S.F.	5.20	3.68	8.88
0600	5000 S.F.	2.72	2.49	5.21
0620	10,000 S.F.	1.91	2.05	3.96
0640	50,000 S.F.	1.45	1.85	3.30
0660	Each additional floor, 500 S.F.	2.42	2.73	5.15
0680	1000 S.F.	1.89	2.50	4.39
0700	2000 S.F.	1.87	2.32	4.19
0720	5000 S.F.	1.38	1.99	3.37
0740	10,000 S.F.	1.28	1.82	3.10
0760	50,000 S.F.	1.26	1.70	2.96
1000	Ordinary hazard, one floor, 500 S.F.	9.90	5.40	15.30
1020	1000 S.F.	5.55	3.69	9.24
1040	2000 S.F.	5.65	3.82	9.47
1060	5000 S.F.	3	2.66	5.66
1080	10,000 S.F.	2.30	2.72	5.02
1100	50,000 S.F.	2.10	2.57	4.67
1140	Each additional floor, 500 S.F.	2.73	3.15	5.88
1160	1000 S.F.	1.88	2.56	4.44
1180	2000 S.F.	1.75	2.53	4.28
1200	5000 S.F.	1.86	2.39	4.25
1220	10,000 S.F.	1.68	2.49	4.17
1240	50,000 S.F.	1.63	2.23	3.86
1500	Extra hazard, one floor, 500 S.F.	13.40	6.90	20.30
1520	1000 S.F.	7.75	5.20	12.95
1540	2000 S.F.	5.60	4.90	10.50

D4010 Sprinklers

D4010 350	Preaction Sprinkler Systems	COST PER S.F.		
		MAT.	INST.	TOTAL
1560	5000 S.F.	3.52	4.06	7.58
1580	10,000 S.F.	3.51	3.95	7.46
1600	50,000 S.F.	3.76	3.84	7.60
1660	Each additional floor, 500 S.F.	3.35	3.89	7.24
1680	1000 S.F.	2.73	3.66	6.39
1700	2000 S.F.	2.48	3.66	6.14
1720	5000 S.F.	2.05	3.26	5.31
1740	10,000 S.F.	2.39	2.97	5.36
1760	50,000 S.F.	2.33	2.78	5.11
2020	Grooved steel, black, sch. 40 pipe, light hazard, one floor, 2000 S.F.	5.20	3.15	8.35
2060	10,000 S.F.	1.90	1.77	3.67
2100	Each additional floor of 2000 S.F.	1.85	1.89	3.74
2150	10,000 S.F.	1.27	1.54	2.81
2200	Ordinary hazard, one floor, 2000 S.F.	5.25	3.33	8.58
2250	10,000 S.F.	2.15	2.29	4.44
2300	Each additional floor, 2000 S.F.	1.93	2.07	4
2350	10,000 S.F.	1.53	2.06	3.59
2400	Extra hazard, one floor, 2000 S.F.	5.60	4.19	9.79
2450	10,000 S.F.	2.85	2.94	5.79
2500	Each additional floor, 2000 S.F.	2.49	3.02	5.51
2550	10,000 S.F.	2.02	2.55	4.57
3050	Grooved steel, black, sch. 10 pipe light hazard, one floor, 2000 S.F.	5.10	3.13	8.23
3100	10,000 S.F.	1.85	1.74	3.59
3150	Each additional floor, 2000 S.F.	1.77	1.87	3.64
3200	10,000 S.F.	1.22	1.51	2.73
3250	Ordinary hazard, one floor, 2000 S.F.	5.10	3.10	8.20
3300	10,000 S.F.	1.77	2.23	4
3350	Each additional floor, 2000 S.F.	1.85	2.04	3.89
3400	10,000 S.F.	1.43	2.01	3.44
3450	Extra hazard, one floor, 2000 S.F.	5.55	4.16	9.71
3500	10,000 S.F.	2.62	2.87	5.49
3550	Each additional floor, 2000 S.F.	2.43	2.99	5.42
3600	10,000 S.F.	1.94	2.50	4.44
4050	Copper tubing, type M, light hazard, one floor, 2000 S.F.	5.45	3.12	8.57
4100	10,000 S.F.	2.27	1.78	4.05
4150	Each additional floor, 2000 S.F.	2.17	1.90	4.07
4200	10,000 S.F.	1.37	1.55	2.92
4250	Ordinary hazard, one floor, 2000 S.F.	5.65	3.46	9.11
4300	10,000 S.F.	2.61	2.10	4.71
4350	Each additional floor, 2000 S.F.	2.13	1.93	4.06
4400	10,000 S.F.	1.76	1.69	3.45
4450	Extra hazard, one floor, 2000 S.F.	6.20	4.22	10.42
4500	10,000 S.F.	4.36	3.18	7.54
4550	Each additional floor, 2000 S.F.	3.05	3.05	6.10
4600	10,000 S.F.	3.07	2.78	5.85
5050	Copper tubing, type M, T-drill system, light hazard, one floor			
5060	2000 S.F.	5.45	2.90	8.35
5100	10,000 S.F.	2.15	1.48	3.63
5150	Each additional floor, 2000 S.F.	2.15	1.68	3.83
5200	10,000 S.F.	1.53	1.26	2.79
5250	Ordinary hazard, one floor, 2000 S.F.	5.45	2.95	8.40
5300	10,000 S.F.	2.51	1.88	4.39
5350	Each additional floor, 2000 S.F.	2.14	1.70	3.84
5400	10,000 S.F.	1.89	1.65	3.54
5450	Extra hazard, one floor, 2000 S.F.	5.65	3.48	9.13
5500	10,000 S.F.	3.59	2.33	5.92
5550	Each additional floor, 2000 S.F.	2.54	2.31	4.85
5600	10,000 S.F.	2.30	1.93	4.23

Deluge System: A system employing open sprinklers attached to a piping system connected to a water supply through a valve which is opened by the operation of a heat responsive system installed in the same areas as the sprinklers. When this valve opens, water flows into the piping system and discharges from all sprinklers attached thereto.

D4010 370	Deluge Sprinkler Systems	COST PER S.F.		
		MAT.	INST.	TOTAL
0520	Deluge sprinkler systems, steel, black, sch. 40 pipe			
0530	Light hazard, one floor, 500 S.F.	29	5.05	34.05
0560	1000 S.F.	15.35	3.58	18.93
0580	2000 S.F.	10.40	3.69	14.09
0600	5000 S.F.	4.80	2.50	7.30
0620	10,000 S.F.	2.95	2.05	5
0640	50,000 S.F.	1.66	1.85	3.51
0660	Each additional floor, 500 S.F.	2.42	2.73	5.15
0680	1000 S.F.	1.89	2.50	4.39
0700	2000 S.F.	1.87	2.32	4.19
0720	5000 S.F.	1.38	1.99	3.37
0740	10,000 S.F.	1.28	1.82	3.10
0760	50,000 S.F.	1.26	1.70	2.96
1000	Ordinary hazard, one floor, 500 S.F.	30.50	5.75	36.25
1020	1000 S.F.	15.35	3.72	19.07
1040	2000 S.F.	10.85	3.83	14.68
1060	5000 S.F.	5.10	2.67	7.77
1080	10,000 S.F.	3.34	2.72	6.06
1100	50,000 S.F.	2.36	2.61	4.97
1140	Each additional floor, 500 S.F.	2.73	3.15	5.88
1160	1000 S.F.	1.88	2.56	4.44
1180	2000 S.F.	1.75	2.53	4.28
1200	5000 S.F.	1.74	2.20	3.94
1220	10,000 S.F.	1.64	2.20	3.84
1240	50,000 S.F.	1.54	2.05	3.59
1500	Extra hazard, one floor, 500 S.F.	33	7	40
1520	1000 S.F.	18	5.40	23.40
1540	2000 S.F.	10.85	4.91	15.76
1560	5000 S.F.	5.35	3.74	9.09
1580	10,000 S.F.	4.41	3.61	8.02
1600	50,000 S.F.	4.46	3.53	7.99
1660	Each additional floor, 500 S.F.	3.35	3.89	7.24

For customer support on your Square Foot Cost Data, call 877.694.8329.

467

D4010 Sprinklers

D4010 370	Deluge Sprinkler Systems	COST PER S.F.		
		MAT.	INST.	TOTAL
1680	1000 S.F.	2.73	3.66	6.39
1700	2000 S.F.	2.48	3.66	6.14
1720	5000 S.F.	2.05	3.26	5.31
1740	10,000 S.F.	2.44	3.08	5.52
1760	50,000 S.F.	2.45	2.97	5.42
2000	Grooved steel, black, sch. 40 pipe, light hazard, one floor			
2020	2000 S.F.	10.40	3.16	13.56
2060	10,000 S.F.	2.96	1.79	4.75
2100	Each additional floor, 2,000 S.F.	1.85	1.89	3.74
2150	10,000 S.F.	1.27	1.54	2.81
2200	Ordinary hazard, one floor, 2000 S.F.	5.25	3.33	8.58
2250	10,000 S.F.	3.19	2.29	5.48
2300	Each additional floor, 2000 S.F.	1.93	2.07	4
2350	10,000 S.F.	1.53	2.06	3.59
2400	Extra hazard, one floor, 2000 S.F.	10.85	4.20	15.05
2450	10,000 S.F.	3.92	2.95	6.87
2500	Each additional floor, 2000 S.F.	2.49	3.02	5.51
2550	10,000 S.F.	2.02	2.55	4.57
3000	Grooved steel, black, sch. 10 pipe, light hazard, one floor			
3050	2000 S.F.	9.70	3.01	12.71
3100	10,000 S.F.	2.89	1.74	4.63
3150	Each additional floor, 2000 S.F.	1.77	1.87	3.64
3200	10,000 S.F.	1.22	1.51	2.73
3250	Ordinary hazard, one floor, 2000 S.F.	10.40	3.31	13.71
3300	10,000 S.F.	2.81	2.23	5.04
3350	Each additional floor, 2000 S.F.	1.85	2.04	3.89
3400	10,000 S.F.	1.43	2.01	3.44
3450	Extra hazard, one floor, 2000 S.F.	10.75	4.17	14.92
3500	10,000 S.F.	3.66	2.87	6.53
3550	Each additional floor, 2000 S.F.	2.43	2.99	5.42
3600	10,000 S.F.	1.94	2.50	4.44
4000	Copper tubing, type M, light hazard, one floor			
4050	2000 S.F.	10.65	3.13	13.78
4100	10,000 S.F.	3.31	1.78	5.09
4150	Each additional floor, 2000 S.F.	2.15	1.90	4.05
4200	10,000 S.F.	1.37	1.55	2.92
4250	Ordinary hazard, one floor, 2000 S.F.	10.85	3.47	14.32
4300	10,000 S.F.	3.65	2.10	5.75
4350	Each additional floor, 2000 S.F.	2.13	1.93	4.06
4400	10,000 S.F.	1.76	1.69	3.45
4450	Extra hazard, one floor, 2000 S.F.	11.40	4.23	15.63
4500	10,000 S.F.	5.45	3.20	8.65
4550	Each additional floor, 2000 S.F.	3.05	3.05	6.10
4600	10,000 S.F.	3.07	2.78	5.85
5000	Copper tubing, type M, T-drill system, light hazard, one floor			
5050	2000 S.F.	10.65	2.91	13.56
5100	10,000 S.F.	3.19	1.48	4.67
5150	Each additional floor, 2000 S.F.	2.17	1.70	3.87
5200	10,000 S.F.	1.53	1.26	2.79
5250	Ordinary hazard, one floor, 2000 S.F.	10.65	2.96	13.61
5300	10,000 S.F.	3.55	1.88	5.43
5350	Each additional floor, 2000 S.F.	2.13	1.69	3.82
5400	10,000 S.F.	1.89	1.65	3.54
5450	Extra hazard, one floor, 2000 S.F.	10.90	3.49	14.39
5500	10,000 S.F.	4.63	2.33	6.96
5550	Each additional floor, 2000 S.F.	2.54	2.31	4.85
5600	10,000 S.F.	2.30	1.93	4.23

Firecycle is a fixed fire protection sprinkler system utilizing water as its extinguishing agent. It is a time delayed, recycling, preaction type which automatically shuts the water off when heat is reduced below the detector operating temperature and turns the water back on when that temperature is exceeded.

The system senses a fire condition through a closed circuit electrical detector system which controls water flow to the fire automatically. Batteries supply up to 90 hours emergency power supply for system operation. The piping system is dry (until water is required) and is monitored with pressurized air. Should any leak in the system piping occur, an alarm will sound, but water will not enter the system until heat is sensed by a Firecycle detector.

D4010 390	Firecycle Sprinkler Systems	COST PER S.F.		
		MAT.	INST.	TOTAL
0520	Firecycle sprinkler systems, steel black sch. 40 pipe			
0530	Light hazard, one floor, 500 S.F.			
0560	1000 S.F.	49	9.75	58.75
0580	2000 S.F.	25	6.10	31.10
0600	5000 S.F.	14.85	4.50	19.35
0620	10,000 S.F.	6.60	2.83	9.43
0640	50,000 S.F.	3.92	2.23	6.15
0660	Each additional floor of 500 S.F.	1.88	1.88	3.76
0680	1000 S.F.	2.49	2.74	5.23
0700	2000 S.F.	1.92	2.50	4.42
0720	5000 S.F.	1.61	2.31	3.92
0740	10,000 S.F.	1.41	2	3.41
0760	50,000 S.F.	1.36	1.83	3.19
1000	Ordinary hazard, one floor, 500 S.F.	1.30	1.69	2.99
1020	1000 S.F.	49	10.15	59.15
1040	2000 S.F.	25	6.05	31.05
1060	5000 S.F.	15	4.63	19.63
1080	10,000 S.F.	6.90	3	9.90
1100	50,000 S.F.	4.31	2.90	7.21
1140	Each additional floor, 500 S.F.	2.79	2.91	5.70
1160	1000 S.F.	2.80	3.16	5.96
1180	2000 S.F.	1.91	2.56	4.47
1200	5000 S.F.	1.96	2.33	4.29
1220	10,000 S.F.	1.77	2.21	3.98
1240	50,000 S.F.	1.59	2.16	3.75
1500	Extra hazard, one floor, 500 S.F.	1.55	1.94	3.49
1520	1000 S.F.	52.50	11.65	64.15
1540	2000 S.F.	27	7.55	34.55
1560	5000 S.F.	15.30	5.70	21
1580	10,000 S.F.	7.15	4.07	11.22
		5.45	4.10	9.55

For customer support on your Square Foot Cost Data, call 877.694.8329.

469

D4010 Sprinklers

D4010 390	Firecycle Sprinkler Systems	COST PER S.F.		
		MAT.	INST.	TOTAL
1600	50,000 S.F.	4.59	4.53	9.12
1660	Each additional floor, 500 S.F.	3.42	3.90	7.32
1680	1000 S.F.	2.76	3.66	6.42
1700	2000 S.F.	2.52	3.66	6.18
1720	5000 S.F.	2.08	3.27	5.35
1740	10,000 S.F.	2.47	2.98	5.45
1760	50,000 S.F.	2.46	2.86	5.32
2020	Grooved steel, black, sch. 40 pipe, light hazard, one floor			
2030	2000 S.F.	14.85	3.97	18.82
2060	10,000 S.F.	4.23	2.69	6.92
2100	Each additional floor, 2000 S.F.	1.89	1.89	3.78
2150	10,000 S.F.	1.35	1.55	2.90
2200	Ordinary hazard, one floor, 2000 S.F.	14.95	4.15	19.10
2250	10,000 S.F.	4.46	2.60	7.06
2300	Each additional floor, 2000 S.F.	1.97	2.07	4.04
2350	10,000 S.F.	1.61	2.07	3.68
2400	Extra hazard, one floor, 2000 S.F.	15.30	5	20.30
2450	10,000 S.F.	4.80	3.09	7.89
2500	Each additional floor, 2000 S.F.	2.53	3.02	5.55
2550	10,000 S.F.	2.10	2.56	4.66
3050	Grooved steel, black, sch. 10 pipe light hazard, one floor,			
3060	2000 S.F.	14.75	3.95	18.70
3100	10,000 S.F.	3.86	1.92	5.78
3150	Each additional floor, 2000 S.F.	1.81	1.87	3.68
3200	10,000 S.F.	1.30	1.52	2.82
3250	Ordinary hazard, one floor, 2000 S.F.	14.85	4.12	18.97
3300	10,000 S.F.	4.06	2.42	6.48
3350	Each additional floor, 2000 S.F.	1.89	2.04	3.93
3400	10,000 S.F.	1.51	2.02	3.53
3450	Extra hazard, one floor, 2000 S.F.	15.25	4.98	20.23
3500	10,000 S.F.	4.61	3.03	7.64
3550	Each additional floor, 2000 S.F.	2.47	2.99	5.46
3600	10,000 S.F.	2.02	2.51	4.53
4060	Copper tubing, type M, light hazard, one floor, 2000 S.F.	15.15	3.94	19.09
4100	10,000 S.F.	4.28	1.96	6.24
4150	Each additional floor, 2000 S.F.	2.20	1.90	4.10
4200	10,000 S.F.	1.73	1.57	3.30
4250	Ordinary hazard, one floor, 2000 S.F.	15.35	4.28	19.63
4300	10,000 S.F.	4.62	2.28	6.90
4350	Each additional floor, 2000 S.F.	2.17	1.93	4.10
4400	10,000 S.F.	1.81	1.68	3.49
4450	Extra hazard, one floor, 2000 S.F.	15.85	5.05	20.90
4500	10,000 S.F.	6.40	3.39	9.79
4550	Each additional floor, 2000 S.F.	3.09	3.05	6.14
4600	10,000 S.F.	3.15	2.79	5.94
5060	Copper tubing, type M, T-drill system, light hazard, one floor 2000 S.F.	15.10	3.72	18.82
5100	10,000 S.F.	4.16	1.66	5.82
5150	Each additional floor, 2000 S.F.	2.39	1.77	4.16
5200	10,000 S.F.	1.61	1.27	2.88
5250	Ordinary hazard, one floor, 2000 S.F.	15.15	3.77	18.92
5300	10,000 S.F.	4.52	2.06	6.58
5350	Each additional floor, 2000 S.F.	2.17	1.69	3.86
5400	10,000 S.F.	1.97	1.66	3.63
5450	Extra hazard, one floor, 2000 S.F.	15.35	4.30	19.65
5500	10,000 S.F.	5.60	2.49	8.09
5550	Each additional floor, 2000 S.F.	2.58	2.31	4.89
5600	10,000 S.F.	2.38	1.94	4.32

D4010 Sprinklers

Wet Pipe System. A system employing automatic sprinklers attached to a piping system containing water and connected to a water supply so that water discharges immediately from sprinklers opened by heat from a fire.

All areas are assumed to be open.

D4010 410	Wet Pipe Sprinkler Systems	COST PER S.F.		
		MAT.	INST.	TOTAL
0520	Wet pipe sprinkler systems, steel, black, sch. 40 pipe			
0530	Light hazard, one floor, 500 S.F.	2.95	3.22	6.17
0560	1000 S.F.	6.15	3.30	9.45
0580	2000 S.F.	5.55	3.34	8.89
0600	5000 S.F.	2.63	2.35	4.98
0620	10,000 S.F.	1.70	1.97	3.67
0640	50,000 S.F.	1.13	1.81	2.94
0660	Each additional floor, 500 S.F.	1.33	2.74	4.07
0680	1000 S.F.	1.35	2.54	3.89
0700	2000 S.F.	1.32	2.29	3.61
0720	5000 S.F.	.94	1.97	2.91
0740	10,000 S.F.	.89	1.80	2.69
0760	50,000 S.F.	.74	1.43	2.17
1000	Ordinary hazard, one floor, 500 S.F.	3.23	3.45	6.68
1020	1000 S.F.	6.10	3.27	9.37
1040	2000 S.F.	5.70	3.47	9.17
1060	5000 S.F.	2.91	2.52	5.43
1080	10,000 S.F.	2.09	2.64	4.73
1100	50,000 S.F.	1.75	2.50	4.25
1140	Each additional floor, 500 S.F.	1.62	3.10	4.72
1160	1000 S.F.	1.32	2.53	3.85
1180	2000 S.F.	1.47	2.52	3.99
1200	5000 S.F.	1.45	2.41	3.86
1220	10,000 S.F.	1.29	2.47	3.76
1240	50,000 S.F.	1.27	2.21	3.48
1500	Extra hazard, one floor, 500 S.F.	11.80	5.30	17.10
1520	1000 S.F.	7.35	4.60	11.95
1540	2000 S.F.	6	4.70	10.70
1560	5000 S.F.	3.63	4.15	7.78
1580	10,000 S.F.	3.23	3.91	7.14
1600	50,000 S.F.	3.57	3.78	7.35
1660	Each additional floor, 500 S.F.	2.24	3.84	6.08
1680	1000 S.F.	2.17	3.63	5.80
1700	2000 S.F.	1.93	3.63	5.56
1720	5000 S.F.	1.61	3.24	4.85
1740	10,000 S.F.	2	2.95	4.95
1760	50,000 S.F.	1.99	2.81	4.80
2020	Grooved steel, black sch. 40 pipe, light hazard, one floor, 2000 S.F.	5.50	2.81	8.31

D4010 Sprinklers

D4010 410	Wet Pipe Sprinkler Systems	COST PER S.F.		
		MAT.	INST.	TOTAL
2060	10,000 S.F.	2.16	1.76	3.92
2100	Each additional floor, 2000 S.F.	1.30	1.86	3.16
2150	10,000 S.F.	.88	1.52	2.40
2200	Ordinary hazard, one floor, 2000 S.F.	5.60	2.99	8.59
2250	10,000 S.F.	1.94	2.21	4.15
2300	Each additional floor, 2000 S.F.	1.38	2.04	3.42
2350	10,000 S.F.	1.14	2.04	3.18
2400	Extra hazard, one floor, 2000 S.F.	5.95	3.85	9.80
2450	10,000 S.F.	2.60	2.85	5.45
2500	Each additional floor, 2000 S.F.	1.94	2.99	4.93
2550	10,000 S.F.	1.63	2.53	4.16
3050	Grooved steel, black sch. 10 pipe, light hazard, one floor, 2000 S.F.	5.45	2.79	8.24
3100	10,000 S.F.	1.64	1.66	3.30
3150	Each additional floor, 2000 S.F.	1.22	1.84	3.06
3200	10,000 S.F.	.83	1.49	2.32
3250	Ordinary hazard, one floor, 2000 S.F.	5.50	2.96	8.46
3300	10,000 S.F.	1.84	2.16	4
3350	Each additional floor, 2000 S.F.	1.30	2.01	3.31
3400	10,000 S.F.	1.04	1.99	3.03
3450	Extra hazard, one floor, 2000 S.F.	5.90	3.82	9.72
3500	10,000 S.F.	2.41	2.79	5.20
3550	Each additional floor, 2000 S.F.	1.88	2.96	4.84
3600	10,000 S.F.	1.55	2.48	4.03
4050	Copper tubing, type M, light hazard, one floor, 2000 S.F.	5.80	2.78	8.58
4100	10,000 S.F.	2.06	1.70	3.76
4150	Each additional floor, 2000 S.F.	1.61	1.87	3.48
4200	10,000 S.F.	1.26	1.54	2.80
4250	Ordinary hazard, one floor, 2000 S.F.	6	3.12	9.12
4300	10,000 S.F.	2.40	2.02	4.42
4350	Each additional floor, 2000 S.F.	1.81	2.07	3.88
4400	10,000 S.F.	1.54	1.81	3.35
4450	Extra hazard, one floor, 2000 S.F.	6.50	3.88	10.38
4500	10,000 S.F.	4.15	3.10	7.25
4550	Each additional floor, 2000 S.F.	2.50	3.02	5.52
4600	10,000 S.F.	2.68	2.76	5.44
5050	Copper tubing, type M, T-drill system, light hazard, one floor			
5060	2000 S.F.	5.80	2.56	8.36
5100	10,000 S.F.	1.94	1.40	3.34
5150	Each additional floor, 2000 S.F.	1.60	1.65	3.25
5200	10,000 S.F.	1.14	1.24	2.38
5250	Ordinary hazard, one floor, 2000 S.F.	5.80	2.61	8.41
5300	10,000 S.F.	2.30	1.80	4.10
5350	Each additional floor, 2000 S.F.	1.58	1.66	3.24
5400	10,000 S.F.	1.50	1.63	3.13
5450	Extra hazard, one floor, 2000 S.F.	6	3.14	9.14
5500	10,000 S.F.	3.38	2.25	5.63
5550	Each additional floor, 2000 S.F.	2.09	2.33	4.42
5600	10,000 S.F.	1.91	1.91	3.82

D4020 Standpipes

Roof

Roof connections with hose gate valves (for combustible roof)

Hose connections on each floor (size based on class of service)

Check Valve

Siamese inlet connections (for fire department use)

D4020 310	Wet Standpipe Risers, Class I	COST PER FLOOR		
		MAT.	INST.	TOTAL
0550	Wet standpipe risers, Class I, steel, black sch. 40, 10' height			
0560	4" diameter pipe, one floor	4,975	3,500	8,475
0580	Additional floors	1,100	1,075	2,175
0600	6" diameter pipe, one floor	8,550	6,125	14,675
0620	Additional floors	2,075	1,725	3,800
0640	8" diameter pipe, one floor	13,200	7,425	20,625
0660	Additional floors	3,100	2,075	5,175

D4020 310	Wet Standpipe Risers, Class II	COST PER FLOOR		
		MAT.	INST.	TOTAL
1030	Wet standpipe risers, Class II, steel, black sch. 40, 10' height			
1040	2" diameter pipe, one floor	1,950	1,250	3,200
1060	Additional floors	570	485	1,055
1080	2-1/2" diameter pipe, one floor	2,875	1,850	4,725
1100	Additional floors	660	565	1,225

D4020 310	Wet Standpipe Risers, Class III	COST PER FLOOR		
		MAT.	INST.	TOTAL
1530	Wet standpipe risers, Class III, steel, black sch. 40, 10' height			
1540	4" diameter pipe, one floor	5,100	3,500	8,600
1560	Additional floors	950	910	1,860
1580	6" diameter pipe, one floor	8,675	6,125	14,800
1600	Additional floors	2,125	1,725	3,850
1620	8" diameter pipe, one floor	13,300	7,425	20,725
1640	Additional floors	3,150	2,075	5,225

Roof

Check valve

Roof connections with hose gate valves (for combustible roof)

Hose connections on each floor (size based on class of service)

Siamese inlet connections (for fire department use)

D4020 330	Dry Standpipe Risers, Class I	COST PER FLOOR		
		MAT.	INST.	TOTAL
0530	Dry standpipe riser, Class I, steel, black sch. 40, 10' height			
0540	4" diameter pipe, one floor	2,975	2,825	5,800
0560	Additional floors	1,025	1,025	2,050
0580	6" diameter pipe, one floor	6,100	4,850	10,950
0600	Additional floors	1,975	1,650	3,625
0620	8" diameter pipe, one floor	9,700	5,900	15,600
0640	Additional floors	3,000	2,025	5,025

D4020 330	Dry Standpipe Risers, Class II	COST PER FLOOR		
		MAT.	INST.	TOTAL
1030	Dry standpipe risers, Class II, steel, black sch. 40, 10' height			
1040	2" diameter pipe, one floor	1,575	1,300	2,875
1060	Additional floor	480	430	910
1080	2-1/2" diameter pipe, one floor	2,250	1,525	3,775
1100	Additional floors	570	510	1,080

D4020 330	Dry Standpipe Risers, Class III	COST PER FLOOR		
		MAT.	INST.	TOTAL
1530	Dry standpipe risers, Class III, steel, black sch. 40, 10' height			
1540	4" diameter pipe, one floor	3,025	2,775	5,800
1560	Additional floors	870	930	1,800
1580	6" diameter pipe, one floor	6,150	4,850	11,000
1600	Additional floors	2,025	1,650	3,675
1620	8" diameter pipe, one floor	9,775	5,900	15,675
1640	Additional floor	3,075	2,025	5,100

474

For customer support on your Square Foot Cost Data, call 877.694.8329.

D40 Fire Protection

D4020 Standpipes

D4020 410	Fire Hose Equipment	COST EACH		
		MAT.	INST.	TOTAL
0100	Adapters, reducing, 1 piece, FxM, hexagon, cast brass, 2-1/2" x 1-1/2"	61		61
0200	Pin lug, 1-1/2" x 1"	52.50		52.50
0250	3" x 2-1/2"	137		137
0300	For polished chrome, add 75% mat.			
0400	Cabinets, D.S. glass in door, recessed, steel box, not equipped			
0500	Single extinguisher, steel door & frame	128	153	281
0550	Stainless steel door & frame	226	153	379
0600	Valve, 2-1/2" angle, steel door & frame	149	102	251
0650	Aluminum door & frame	180	102	282
0700	Stainless steel door & frame	243	102	345
0750	Hose rack assy, 2-1/2" x 1-1/2" valve & 100' hose, steel door & frame	278	204	482
0800	Aluminum door & frame	410	204	614
0850	Stainless steel door & frame	545	204	749
0900	Hose rack assy & extinguisher,2-1/2"x1-1/2" valve & hose,steel door & frame	291	245	536
0950	Aluminum	525	245	770
1000	Stainless steel	590	245	835
1550	Compressor, air, dry pipe system, automatic, 200 gal., 3/4 H.P.	840	525	1,365
1600	520 gal., 1 H.P.	870	525	1,395
1650	Alarm, electric pressure switch (circuit closer)	97.50	26	123.50
2500	Couplings, hose, rocker lug, cast brass, 1-1/2"	58.50		58.50
2550	2-1/2"	77		77
3000	Escutcheon plate, for angle valves, polished brass, 1-1/2"	16.70		16.70
3050	2-1/2"	27		27
3500	Fire pump, electric, w/controller, fittings, relief valve			
3550	4" pump, 30 H.P., 500 G.P.M.	16,400	3,800	20,200
3600	5" pump, 40 H.P., 1000 G.P.M.	18,200	4,325	22,525
3650	5" pump, 100 H.P., 1000 G.P.M.	24,800	4,800	29,600
3700	For jockey pump system, add	2,850	610	3,460
5000	Hose, per linear foot, synthetic jacket, lined,			
5100	300 lb. test, 1-1/2" diameter	3.56	.47	4.03
5150	2-1/2" diameter	6.15	.56	6.71
5200	500 lb. test, 1-1/2" diameter	3.69	.47	4.16
5250	2-1/2" diameter	6.50	.56	7.06
5500	Nozzle, plain stream, polished brass, 1-1/2" x 10"	52.50		52.50
5550	2-1/2" x 15" x 13/16" or 1-1/2"	110		110
5600	Heavy duty combination adjustable fog and straight stream w/handle 1-1/2"	495		495
5650	2-1/2" direct connection	615		615
6000	Rack, for 1-1/2" diameter hose 100 ft. long, steel	67	61	128
6050	Brass	115	61	176
6500	Reel, steel, for 50 ft. long 1-1/2" diameter hose	144	87.50	231.50
6550	For 75 ft. long 2-1/2" diameter hose	234	87.50	321.50
7050	Siamese, w/plugs & chains, polished brass, sidewalk, 4" x 2-1/2" x 2-1/2"	695	490	1,185
7100	6" x 2-1/2" x 2-1/2"	860	610	1,470
7200	Wall type, flush, 4" x 2-1/2" x 2-1/2"	570	245	815
7250	6" x 2-1/2" x 2-1/2"	795	266	1,061
7300	Projecting, 4" x 2-1/2" x 2-1/2"	520	245	765
7350	6" x 2-1/2" x 2-1/2"	880	266	1,146
7400	For chrome plate, add 15% mat.			
8000	Valves, angle, wheel handle, 300 Lb., rough brass, 1-1/2"	104	56.50	160.50
8050	2-1/2"	192	97	289
8100	Combination pressure restricting, 1-1/2"	94	56.50	150.50
8150	2-1/2"	192	97	289
8200	Pressure restricting, adjustable, satin brass, 1-1/2"	117	56.50	173.50
8250	2-1/2"	176	97	273
8300	Hydrolator, vent and drain, rough brass, 1-1/2"	92.50	56.50	149
8350	2-1/2"	92.50	56.50	149
8400	Cabinet assy, incls. adapter, rack, hose, and nozzle	875	370	1,245

For customer support on your Square Foot Cost Data, call 877.694.8329.

475

General: Automatic fire protection (suppression) systems other than water sprinklers may be desired for special environments, high risk areas, isolated locations or unusual hazards. Some typical applications would include:

1. Paint dip tanks
2. Securities vaults
3. Electronic data processing
4. Tape and data storage
5. Transformer rooms
6. Spray booths
7. Petroleum storage
8. High rack storage

Piping and wiring costs are highly variable and are not included.

D4090 910	Fire Suppression Unit Components	COST EACH		
		MAT.	INST.	TOTAL
0020	Detectors with brackets			
0040	Fixed temperature heat detector	56	82	138
0060	Rate of temperature rise detector	56	82	138
0080	Ion detector (smoke) detector	121	106	227
0200	Extinguisher agent			
0240	200 lb FM200, container	8,550	320	8,870
0280	75 lb carbon dioxide cylinder	1,400	213	1,613
0320	Dispersion nozzle			
0340	FM200 1-1/2″ dispersion nozzle	73.50	50.50	124
0380	Carbon dioxide 3″ x 5″ dispersion nozzle	73.50	39.50	113
0420	Control station			
0440	Single zone control station with batteries	1,900	655	2,555
0470	Multizone (4) control station with batteries	3,600	1,300	4,900
0500	Electric mechanical release	184	340	524
0550	Manual pull station	66.50	118	184.50
0640	Battery standby power 10″ x 10″ x 17″	455	164	619
0740	Bell signalling device	86.50	82	168.50
0900	Standard low-rise sprinkler accessory package, 3 story	2,750	1,825	4,575
1000	Standard mid-rise sprinkler accessory package, 8 story	5,850	3,475	9,325
1100	Standard high-rise sprinkler accessory package, 16 story	12,600	6,825	19,425

D4090 920	FM200 Systems	COST PER C.F.		
		MAT.	INST.	TOTAL
0820	Average FM200 system, minimum			1.94
0840	Maximum			3.85

476

D5010 Electrical Service/Distribution

D5010 120	Overhead Electric Service, 3 Phase - 4 Wire	COST EACH		
		MAT.	INST.	TOTAL
0200	Service installation, includes breakers, metering, 20' conduit & wire			
0220	3 phase, 4 wire, 120/208 volts, 60 A	595	1,075	1,670
0240	3 phase, 4 wire, 120/208 volts, 100 A	790	1,200	1,990
0245	100 A w/circuit breaker	1,575	1,500	3,075
0280	200 A	1,475	1,650	3,125
0285	200 A w/circuit breaker	3,100	2,100	5,200
0320	400 A	2,925	3,325	6,250
0325	400 A w/circuit breaker	6,000	4,150	10,150
0360	600 A	5,025	4,900	9,925
0365	600 A, w/switchboard	9,675	6,200	15,875
0400	800 A	6,950	5,875	12,825
0405	800 A, w/switchboard	11,600	7,375	18,975
0440	1000 A	8,625	7,250	15,875
0445	1000 A, w/switchboard	14,200	8,900	23,100
0480	1200 A	11,500	8,275	19,775
0485	1200 A, w/groundfault switchboard	35,600	10,200	45,800
0520	1600 A	14,200	10,800	25,000
0525	1600 A, w/groundfault switchboard	40,300	12,900	53,200
0560	2000 A	19,000	13,100	32,100
0565	2000 A, w/groundfault switchboard	48,700	16,200	64,900
0610	1 phase, 3 wire, 120/240 volts, 100 A (no safety switch)	190	590	780
0615	100 A w/load center	555	1,125	1,680
0620	200 A	445	815	1,260
0625	200 A w/load center	1,250	1,650	2,900

477

For customer support on your Square Foot Cost Data, call 877.694.8329.

D5010 130	Underground Electric Service	COST EACH		
		MAT.	INST.	TOTAL
0950	Underground electric service including excavation, backfill, and compaction			
1000	3 phase, 4 wire, 277/480 volts, 2000 A	40,200	20,900	61,100
1050	2000 A w/groundfault switch	68,500	23,400	91,900
1100	1600 A	31,900	16,900	48,800
1150	1600 A w/groundfault switchboard	58,000	19,000	77,000
1200	1200 A	24,400	13,400	37,800
1250	1200 A w/groundfault switchboard	48,500	15,400	63,900
1400	800 A	17,700	11,200	28,900
1450	800 A w/ switchboard	22,300	12,700	35,000
1500	600 A	11,400	10,300	21,700
1550	600 A w/ switchboard	16,000	11,500	27,500
1600	1 phase, 3 wire, 120/240 volts, 200 A	3,225	3,550	6,775
1650	200 A w/load center	4,050	4,400	8,450
1700	100 A	2,200	2,900	5,100
1750	100 A w/load center	3,000	3,750	6,750

D5010 Electrical Service/Distribution

D5010 230	Feeder Installation	COST PER L.F.		
		MAT.	INST.	TOTAL
0200	Feeder installation 600 V, including RGS conduit and XHHW wire, 60 A	5.55	12.05	17.60
0240	100 A	10.05	15.70	25.75
0280	200 A	23.50	23.50	47
0320	400 A	46.50	47.50	94
0360	600 A	87.50	77	164.50
0400	800 A	113	91.50	204.50
0440	1000 A	142	120	262
0480	1200 A	175	154	329
0520	1600 A	227	183	410
0560	2000 A	284	241	525
1200	Branch installation 600 V, including EMT conduit and THW wire, 15 A	1.15	5.90	7.05
1240	20 A	1.37	6.25	7.62
1280	30 A	1.74	6.50	8.24
1320	50 A	3.40	9	12.40
1360	65 A	4.27	9.55	13.82
1400	85 A	6.60	11.10	17.70
1440	100 A	7.40	11.30	18.70
1480	130 A	11	13.25	24.25
1520	150 A	13.55	15.20	28.75
1560	200 A	18.75	17.30	36.05

D5010 240	Switchgear	COST EACH		
		MAT.	INST.	TOTAL
0190	Switchgear installation, including switchboard, panels, & circuit breaker			
0200	120/208 V, 1 phase, 400 A	8,375	2,775	11,150
0240	600 A	10,500	3,100	13,600
0280	800 A	13,900	3,625	17,525
0300	1000 A	17,000	4,000	21,000
0320	1200 A	17,800	4,500	22,300
0360	1600 A	27,400	5,175	32,575
0400	2000 A	32,600	5,675	38,275
0500	277/480 V, 3 phase, 400 A	11,300	5,175	16,475
0520	600 A	15,500	5,900	21,400
0540	800 A	18,500	6,625	25,125
0560	1000 A	23,100	7,300	30,400
0580	1200 A	25,000	8,150	33,150
0600	1600 A	34,200	8,925	43,125
0620	2000 A	41,800	9,925	51,725

D5010 Electrical Service/Distribution

D5010 250	Panelboard	COST EACH		
		MAT.	INST.	TOTAL
0900	Panelboards, NQOD, 4 wire, 120/208 volts w/conductor & conduit			
1000	100 A, 0 stories, 0' horizontal	1,300	1,400	2,700
1020	1 stories, 25' horizontal	1,725	2,000	3,725
1040	5 stories, 50' horizontal	2,525	3,175	5,700
1060	10 stories, 75' horizontal	3,475	4,550	8,025
1080	225A, 0 stories, 0' horizontal	2,425	1,825	4,250
2000	1 stories, 25' horizontal	3,650	2,900	6,550
2020	5 stories, 50' horizontal	6,100	5,025	11,125
2040	10 stories, 75' horizontal	8,925	7,500	16,425
2060	400A, 0 stories, 0' horizontal	3,575	2,725	6,300
2080	1 stories, 25' horizontal	5,275	4,475	9,750
3000	5 stories, 50' horizontal	8,600	7,925	16,525
3020	10 stories, 75' horizontal	18,000	15,500	33,500
3040	600 A, 0 stories, 0' horizontal	5,300	3,275	8,575
3060	1 stories, 25' horizontal	8,800	6,125	14,925
3080	5 stories, 50' horizontal	15,700	11,700	27,400
4000	10 stories, 75' horizontal	35,500	23,100	58,600
4010	Panelboards, NEHB, 4 wire, 277/480 volts w/conductor, conduit, & safety switch			
4020	100 A, 0 stories, 0' horizontal, includes safety switch	2,975	1,900	4,875
4040	1 stories, 25' horizontal	3,375	2,500	5,875
4060	5 stories, 50' horizontal	4,200	3,675	7,875
4080	10 stories, 75' horizontal	5,150	5,050	10,200
5000	225 A, 0 stories, 0' horizontal	4,450	2,325	6,775
5020	1 stories, 25' horizontal	5,675	3,400	9,075
5040	5 stories, 50' horizontal	8,100	5,525	13,625
5060	10 stories, 75' horizontal	10,900	8,025	18,925
5080	400 A, 0 stories, 0' horizontal	7,225	3,575	10,800
6000	1 stories, 25' horizontal	8,925	5,325	14,250
6020	5 stories, 50' horizontal	12,300	8,775	21,075
6040	10 stories, 75' horizontal	21,700	16,400	38,100
6060	600 A, 0 stories, 0' horizontal	10,900	4,550	15,450
6080	1 stories, 25' horizontal	14,400	7,400	21,800
7000	5 stories, 50' horizontal	21,300	13,000	34,300
7020	10 stories, 75' horizontal	41,100	24,400	65,500

Duplex Wall Receptacle

Undercarpet Receptacle System

D5020 110	Receptacle (by Wattage)	COST PER S.F.		
		MAT.	INST.	TOTAL
0190	Receptacles include plate, box, conduit, wire & transformer when required			
0200	2.5 per 1000 S.F., .3 watts per S.F.	.39	1.46	1.85
0240	With transformer	.47	1.53	2
0280	4 per 1000 S.F., .5 watts per S.F.	.44	1.70	2.14
0320	With transformer	.57	1.82	2.39
0360	5 per 1000 S.F., .6 watts per S.F.	.52	2.01	2.53
0400	With transformer	.68	2.16	2.84
0440	8 per 1000 S.F., .9 watts per S.F.	.54	2.23	2.77
0480	With transformer	.80	2.46	3.26
0520	10 per 1000 S.F., 1.2 watts per S.F.	.57	2.41	2.98
0560	With transformer	.89	2.70	3.59
0600	16.5 per 1000 S.F., 2.0 watts per S.F.	.65	3.02	3.67
0640	With transformer	1.18	3.50	4.68
0680	20 per 1000 S.F., 2.4 watts per S.F.	.70	3.30	4
0720	With transformer	1.34	3.89	5.23

D5020 115	Receptacles, Floor	COST PER S.F.		
		MAT.	INST.	TOTAL
0200	Receptacle systems, underfloor duct, 5' on center, low density	7.15	3.22	10.37
0240	High density	7.70	4.14	11.84
0280	7' on center, low density	5.65	2.78	8.43
0320	High density	6.20	3.70	9.90
0400	Poke thru fittings, low density	1.26	1.58	2.84
0440	High density	2.50	3.14	5.64
0520	Telepoles, using Romex, low density	1.19	.97	2.16
0560	High density	2.38	1.94	4.32
0600	Using EMT, low density	1.26	1.28	2.54
0640	High density	2.50	2.56	5.06
0720	Conduit system with floor boxes, low density	1.17	1.11	2.28
0760	High density	2.34	2.21	4.55
0840	Undercarpet power system, 3 conductor with 5 conductor feeder, low density	1.55	.39	1.94
0880	High density	3.04	.81	3.85

D5020 Lighting and Branch Wiring

Description: Table D5020 130 includes the cost for switch, plate, box, conduit in slab or EMT exposed and copper wire. Add 20% for exposed conduit.

No power required for switches.

Federal energy guidelines recommend the maximum lighting area controlled per switch shall not exceed 1000 S.F. and that areas over 500 S.F. shall be so controlled that total illumination can be reduced by at least 50%.

D5020 130	Wall Switch by Sq. Ft.	COST PER S.F.		
		MAT.	INST.	TOTAL
0200	Wall switches, 1.0 per 1000 S.F.	.06	.23	.29
0240	1.2 per 1000 S.F.	.07	.26	.33
0280	2.0 per 1000 S.F.	.10	.38	.48
0320	2.5 per 1000 S.F.	.13	.48	.61
0360	5.0 per 1000 S.F.	.25	1.02	1.27
0400	10.0 per 1000 S.F.	.53	2.05	2.58

D5020 135	Miscellaneous Power	COST PER S.F.		
		MAT.	INST.	TOTAL
0200	Miscellaneous power, to .5 watts	.03	.12	.15
0240	.8 watts	.05	.17	.22
0280	1 watt	.07	.22	.29
0320	1.2 watts	.08	.26	.34
0360	1.5 watts	.09	.30	.39
0400	1.8 watts	.10	.35	.45
0440	2 watts	.12	.42	.54
0480	2.5 watts	.15	.52	.67
0520	3 watts	.18	.61	.79

D5020 140	Central A. C. Power (by Wattage)	COST PER S.F.		
		MAT.	INST.	TOTAL
0200	Central air conditioning power, 1 watt	.07	.26	.33
0220	2 watts	.08	.30	.38
0240	3 watts	.11	.33	.44
0280	4 watts	.18	.44	.62
0320	6 watts	.34	.59	.93
0360	8 watts	.43	.63	1.06
0400	10 watts	.63	.74	1.37

D5020 Lighting and Branch Wiring

Power System   Switch   Motor Starter   Switch   Motor Connection   Motor

Motor Installation

D5020 145	Motor Installation	COST EACH		
		MAT.	INST.	TOTAL
0200	Motor installation, single phase, 115V, 1/3 HP motor size	550	995	1,545
0240	1 HP motor size	570	995	1,565
0280	2 HP motor size	620	1,050	1,670
0320	3 HP motor size	710	1,075	1,785
0360	230V, 1 HP motor size	550	1,000	1,550
0400	2 HP motor size	590	1,000	1,590
0440	3 HP motor size	660	1,075	1,735
0520	Three phase, 200V, 1-1/2 HP motor size	640	1,100	1,740
0560	3 HP motor size	720	1,200	1,920
0600	5 HP motor size	760	1,350	2,110
0640	7-1/2 HP motor size	790	1,375	2,165
0680	10 HP motor size	1,200	1,725	2,925
0720	15 HP motor size	1,625	1,925	3,550
0760	20 HP motor size	2,075	2,200	4,275
0800	25 HP motor size	2,125	2,225	4,350
0840	30 HP motor size	3,300	2,625	5,925
0880	40 HP motor size	4,125	3,100	7,225
0920	50 HP motor size	7,250	3,600	10,850
0960	60 HP motor size	7,550	3,800	11,350
1000	75 HP motor size	9,500	4,350	13,850
1040	100 HP motor size	27,200	5,125	32,325
1080	125 HP motor size	27,500	5,650	33,150
1120	150 HP motor size	31,800	6,650	38,450
1160	200 HP motor size	32,800	7,850	40,650
1240	230V, 1-1/2 HP motor size	615	1,100	1,715
1280	3 HP motor size	690	1,200	1,890
1320	5 HP motor size	730	1,325	2,055
1360	7-1/2 HP motor size	730	1,325	2,055
1400	10 HP motor size	1,125	1,625	2,750
1440	15 HP motor size	1,275	1,775	3,050
1480	20 HP motor size	1,975	2,150	4,125
1520	25 HP motor size	2,075	2,200	4,275
1560	30 HP motor size	2,100	2,225	4,325
1600	40 HP motor size	3,975	3,025	7,000
1640	50 HP motor size	4,300	3,200	7,500
1680	60 HP motor size	7,275	3,600	10,875
1720	75 HP motor size	8,750	4,075	12,825
1760	100 HP motor size	10,300	4,550	14,850
1800	125 HP motor size	27,700	5,300	33,000
1840	150 HP motor size	29,300	6,025	35,325
1880	200 HP motor size	31,500	6,700	38,200
1960	460V, 2 HP motor size	760	1,100	1,860
2000	5 HP motor size	835	1,200	2,035
2040	10 HP motor size	840	1,325	2,165
2080	15 HP motor size	1,125	1,525	2,650
2120	20 HP motor size	1,200	1,625	2,825

D5020 Lighting and Branch Wiring

D5020 145	Motor Installation	COST EACH		
		MAT.	INST.	TOTAL
2160	25 HP motor size	1,275	1,725	3,000
2200	30 HP motor size	1,625	1,850	3,475
2240	40 HP motor size	2,100	2,000	4,100
2280	50 HP motor size	2,350	2,200	4,550
2320	60 HP motor size	3,500	2,600	6,100
2360	75 HP motor size	4,150	2,875	7,025
2400	100 HP motor size	4,625	3,200	7,825
2440	125 HP motor size	7,600	3,625	11,225
2480	150 HP motor size	9,425	4,050	13,475
2520	200 HP motor size	11,100	4,575	15,675
2600	575V, 2 HP motor size	760	1,100	1,860
2640	5 HP motor size	835	1,200	2,035
2680	10 HP motor size	840	1,325	2,165
2720	20 HP motor size	1,125	1,525	2,650
2760	25 HP motor size	1,200	1,625	2,825
2800	30 HP motor size	1,625	1,850	3,475
2840	50 HP motor size	1,750	1,925	3,675
2880	60 HP motor size	3,475	2,575	6,050
2920	75 HP motor size	3,500	2,600	6,100
2960	100 HP motor size	4,150	2,875	7,025
3000	125 HP motor size	7,400	3,550	10,950
3040	150 HP motor size	7,600	3,625	11,225
3080	200 HP motor size	9,575	4,100	13,675

D5020 155	Motor Feeder	COST PER L.F.		
		MAT.	INST.	TOTAL
0200	Motor feeder systems, single phase, feed up to 115V 1HP or 230V 2 HP	2.23	7.75	9.98
0240	115V 2HP, 230V 3HP	2.38	7.85	10.23
0280	115V 3HP	2.71	8.20	10.91
0360	Three phase, feed to 200V 3HP, 230V 5HP, 460V 10HP, 575V 10HP	2.36	8.35	10.71
0440	200V 5HP, 230V 7.5HP, 460V 15HP, 575V 20HP	2.59	8.50	11.09
0520	200V 10HP, 230V 10HP, 460V 30HP, 575V 30HP	3.08	9	12.08
0600	200V 15HP, 230V 15HP, 460V 40HP, 575V 50HP	4.35	10.35	14.70
0680	200V 20HP, 230V 25HP, 460V 50HP, 575V 60HP	6.50	13.05	19.55
0760	200V 25HP, 230V 30HP, 460V 60HP, 575V 75HP	7.30	13.30	20.60
0840	200V 30HP	8.25	13.75	22
0920	230V 40HP, 460V 75HP, 575V 100HP	10.40	15	25.40
1000	200V 40HP	11.55	16.05	27.60
1080	230V 50HP, 460V 100HP, 575V 125HP	14.85	17.75	32.60
1160	200V 50HP, 230V 60HP, 460V 125HP, 575V 150HP	17.15	18.80	35.95
1240	200V 60HP, 460V 150HP	21	22	43
1320	230V 75HP, 575V 200HP	24	23	47
1400	200V 75HP	28	23.50	51.50
1480	230V 100HP, 460V 200HP	36.50	27.50	64
1560	200V 100HP	50.50	34.50	85
1640	230V 125HP	50.50	34.50	85
1720	200V 125HP, 230V 150HP	58.50	42.50	101
1800	200V 150HP	71	46.50	117.50
1880	200V 200HP	90	49	139
1960	230V 200HP	92	50.50	142.50

For customer support on your Square Foot Cost Data, call 877.694.8329.

485

D5020   Lighting and Branch Wiring

Fluorescent Fixture

Incandescent Fixture

D5020 210	Fluorescent Fixtures (by Wattage)	COST PER S.F.		
		MAT.	INST.	TOTAL
0190	Fluorescent fixtures recess mounted in ceiling			
0195	T12, standard 40 watt lamps			
0200	1 watt per S.F., 20 FC, 5 fixtures @40 watts per 1000 S.F.	.72	1.93	2.65
0240	2 watt per S.F., 40 FC, 10 fixtures @40 watt per 1000 S.F.	1.44	3.78	5.22
0280	3 watt per S.F., 60 FC, 15 fixtures @40 watt per 1000 S.F	2.14	5.70	7.84
0320	4 watt per S.F., 80 FC, 20 fixtures @40 watt per 1000 S.F.	2.86	7.55	10.41
0400	5 watt per S.F., 100 FC, 25 fixtures @40 watt per 1000 S.F.	3.58	9.50	13.08
0450	T8, energy saver 32 watt lamps			
0500	0.8 watt per S.F., 20 FC, 5 fixtures @32 watt per 1000 S.F.	.79	1.93	2.72
0520	1.6 watt per S.F., 40 FC, 10 fixtures @32 watt per 1000 S.F.	1.58	3.78	5.36
0540	2.4 watt per S.F., 60 FC, 15 fixtures @ 32 watt per 1000 S.F	2.36	5.70	8.06
0560	3.2 watt per S.F., 80 FC, 20 fixtures @32 watt per 1000 S.F.	3.15	7.55	10.70
0580	4 watt per S.F., 100 FC, 25 fixtures @32 watt per 1000 S.F.	3.94	9.50	13.44

D5020 216	Incandescent Fixture (by Wattage)	COST PER S.F.		
		MAT.	INST.	TOTAL
0190	Incandescent fixture recess mounted, type A			
0200	1 watt per S.F., 8 FC, 6 fixtures per 1000 S.F.	.98	1.56	2.54
0240	2 watt per S.F., 16 FC, 12 fixtures per 1000 S.F.	1.96	3.11	5.07
0280	3 watt per S.F., 24 FC, 18 fixtures, per 1000 S.F.	2.94	4.59	7.53
0320	4 watt per S.F., 32 FC, 24 fixtures per 1000 S.F.	3.91	6.15	10.06
0400	5 watt per S.F., 40 FC, 30 fixtures per 1000 S.F.	4.91	7.70	12.61

D50 Electrical

D5020 Lighting and Branch Wiring

High Bay Fixture

Low Bay Fixture

D5020 226	H.I.D. Fixture, High Bay, 16' (by Wattage)	COST PER S.F.		
		MAT.	INST.	TOTAL
0190	High intensity discharge fixture, 16' above work plane			
0240	1 watt/S.F., type E, 42 FC, 1 fixture/1000 S.F.	1.05	1.62	2.67
0280	Type G, 52 FC, 1 fixture/1000 S.F.	1.05	1.62	2.67
0320	Type C, 54 FC, 2 fixture/1000 S.F.	1.23	1.82	3.05
0440	2 watt/S.F., type E, 84 FC, 2 fixture/1000 S.F.	2.12	3.31	5.43
0480	Type G, 105 FC, 2 fixture/1000 S.F.	2.12	3.31	5.43
0520	Type C, 108 FC, 4 fixture/1000 S.F.	2.45	3.60	6.05
0640	3 watt/S.F., type E, 126 FC, 3 fixture/1000 S.F.	3.18	4.91	8.09
0680	Type G, 157 FC, 3 fixture/1000 S.F.	3.18	4.91	8.09
0720	Type C, 162 FC, 6 fixture/1000 S.F.	3.66	5.40	9.06
0840	4 watt/S.F., type E, 168 FC, 4 fixture/1000 S.F.	4.24	6.60	10.84
0880	Type G, 210 FC, 4 fixture/1000 S.F.	4.24	6.60	10.84
0920	Type C, 243 FC, 9 fixture/1000 S.F.	5.35	7.55	12.90
1040	5 watt/S.F., type E, 210 FC, 5 fixture/1000 S.F.	5.30	8.20	13.50
1080	Type G, 262 FC, 5 fixture/1000 S.F.	5.30	8.20	13.50
1120	Type C, 297 FC, 11 fixture/1000 S.F.	6.55	9.35	15.90

D5020 234	H.I.D. Fixture, High Bay, 30' (by Wattage)	COST PER S.F.		
		MAT.	INST.	TOTAL
0190	High intensity discharge fixture, 30' above work plane			
0240	1 watt/S.F., type E, 37 FC, 1 fixture/1000 S.F.	1.18	2.04	3.22
0280	Type G, 45 FC., 1 fixture/1000 S.F.	1.18	2.04	3.22
0320	Type F, 50 FC, 1 fixture/1000 S.F.	.99	1.60	2.59
0440	2 watt/S.F., type E, 74 FC, 2 fixtures/1000 S.F.	2.37	4.11	6.48
0480	Type G, 92 FC, 2 fixtures/1000 S.F.	2.37	4.11	6.48
0520	Type F, 100 FC, 2 fixtures/1000 S.F.	2.01	3.28	5.29
0640	3 watt/S.F., type E, 110 FC, 3 fixtures/1000 S.F.	3.57	6.20	9.77
0680	Type G, 138 FC, 3 fixtures/1000 S.F.	3.57	6.20	9.77
0720	Type F, 150 FC, 3 fixtures/1000 S.F.	3.01	4.87	7.88
0840	4 watt/S.F., type E, 148 FC, 4 fixtures/1000 S.F.	4.74	8.25	12.99
0880	Type G, 185 FC, 4 fixtures/1000 S.F.	4.74	8.25	12.99
0920	Type F, 200 FC, 4 fixtures/1000 S.F.	4.03	6.55	10.58
1040	5 watt/S.F., type E, 185 FC, 5 fixtures/1000 S.F.	5.95	10.30	16.25
1080	Type G, 230 FC, 5 fixtures/1000 S.F.	5.95	10.30	16.25
1120	Type F, 250 FC, 5 fixtures/1000 S.F.	5	8.15	13.15

D5020 238	H.I.D. Fixture, Low Bay, 8'-10' (by Wattage)	COST PER S.F.		
		MAT.	INST.	TOTAL
0190	High intensity discharge fixture, 8'-10' above work plane			
0240	1 watt/S.F., type J, 30 FC, 4 fixtures/1000 S.F.	2.40	3.33	5.73
0280	Type K, 29 FC, 5 fixtures/1000 S.F.	2.44	3.01	5.45
0400	2 watt/S.F., type J, 52 FC, 7 fixtures/1000 S.F.	4.29	6.15	10.44
0440	Type K, 63 FC, 11 fixtures/1000 S.F.	5.30	6.30	11.60
0560	3 watt/S.F., type J, 81 FC, 11 fixtures/1000 S.F.	6.65	9.25	15.90

D50 Electrical

D5020 Lighting and Branch Wiring

D5020 238	H.I.D. Fixture, Low Bay, 8'-10' (by Wattage)	COST PER S.F.		
		MAT.	INST.	TOTAL
0600	Type K, 92 FC, 16 fixtures/1000 S.F.	7.75	9.30	17.05
0720	4 watt/S.F., type J, 103 FC, 14 fixtures/1000 S.F.	8.55	12.20	20.75
0760	Type K, 127 FC, 22 fixtures/1000 S.F.	10.60	12.60	23.20
0880	5 watt/S.F., type J, 133 FC, 18 fixtures/1000 S.F.	10.90	15.35	26.25
0920	Type K, 155 FC, 27 fixtures/1000 S.F.	13.05	15.65	28.70

D5020 242	H.I.D. Fixture, Low Bay, 16' (by Wattage)	COST PER S.F.		
		MAT.	INST.	TOTAL
0190	High intensity discharge fixture, mounted 16' above work plane			
0240	1 watt/S.F., type J, 28 FC, 4 fixt./1000 S.F.	2.55	3.79	6.34
0280	Type K, 27 FC, 5 fixt./1000 S.F.	2.83	4.25	7.08
0400	2 watt/S.F., type J, 48 FC, 7 fixt/1000 S.F.	4.65	7.35	12
0440	Type K, 58 FC, 11 fixt/1000 S.F.	6	8.70	14.70
0560	3 watt/S.F., type J, 75 FC, 11 fixt/1000 S.F.	7.20	11.15	18.35
0600	Type K, 85 FC, 16 fixt/1000 S.F.	8.85	12.90	21.75
0720	4 watt/S.F., type J, 95 FC, 14 fixt/1000 S.F.	9.30	14.65	23.95
0760	Type K, 117 FC, 22 fixt/1000 S.F.	12.05	17.35	29.40
0880	5 watt/S.F., type J, 122 FC, 18 fixt/1000 S.F.	11.85	18.45	30.30
0920	Type K, 143 FC, 27 fixt/1000 S.F.	14.85	21.50	36.35

D5030 Communications and Security

D5030 910	Communication & Alarm Systems	COST EACH		
		MAT.	INST.	TOTAL
0200	Communication & alarm systems, includes outlets, boxes, conduit & wire			
0210	Sound system, 6 outlets	5,800	8,300	14,100
0220	12 outlets	8,350	13,300	21,650
0240	30 outlets	14,900	25,100	40,000
0280	100 outlets	45,300	84,000	129,300
0320	Fire detection systems, non-addressable, 12 detectors	3,300	6,875	10,175
0360	25 detectors	5,800	11,600	17,400
0400	50 detectors	11,400	23,100	34,500
0440	100 detectors	21,200	41,900	63,100
0450	Addressable type, 12 detectors	4,975	6,925	11,900
0452	25 detectors	9,075	11,700	20,775
0454	50 detectors	17,500	23,000	40,500
0456	100 detectors	33,900	42,300	76,200
0458	Fire alarm control panel, 8 zone, excluding wire and conduit	860	1,300	2,160
0459	12 zone	2,625	1,975	4,600
0460	Fire alarm command center, addressable without voice, excl. wire & conduit	4,925	1,150	6,075
0462	Addressable with voice	10,400	1,800	12,200
0480	Intercom systems, 6 stations	4,000	5,675	9,675
0560	25 stations	12,900	21,700	34,600
0640	100 stations	49,200	79,500	128,700
0680	Master clock systems, 6 rooms	5,175	9,250	14,425
0720	12 rooms	8,100	15,800	23,900
0760	20 rooms	11,200	22,300	33,500
0800	30 rooms	18,300	41,100	59,400
0840	50 rooms	29,700	69,500	99,200
0920	Master TV antenna systems, 6 outlets	2,425	5,850	8,275
0960	12 outlets	4,500	10,900	15,400
1000	30 outlets	11,200	25,200	36,400
1040	100 outlets	37,300	82,000	119,300

D5090   Other Electrical Systems

Generator System

A: Engine
B: Battery
C: Charger
D: Transfer Switch
E: Muffler

D5090 210	Generators (by kW)	COST PER kW		
		MAT.	INST.	TOTAL
0190	Generator sets, include battery, charger, muffler & transfer switch			
0200	Gas/gasoline operated, 3 phase, 4 wire, 277/480 volt, 7.5 kW	1,075	288	1,363
0240	11.5 kW	1,000	219	1,219
0280	20 kW	675	142	817
0320	35 kW	460	93	553
0360	80 kW	330	56	386
0400	100 kW	289	54	343
0440	125 kW	470	51	521
0480	185 kW	420	39	459
0560	Diesel engine with fuel tank, 30 kW	515	108	623
0600	50 kW	435	84.50	519.50
0640	75 kW	310	68	378
0680	100 kW	330	57.50	387.50
0760	150 kW	248	46	294
0840	200 kW	233	37	270
0880	250 kW	199	31	230
0960	350 kW	174	25.50	199.50
1040	500 kW	190	19.80	209.80

E1010 Commercial Equipment

E1010 110	Security/Vault, EACH	COST EACH		
		MAT.	INST.	TOTAL
0100	Bank equipment, drive up window, drawer & mike, no glazing, economy	7,600	1,475	9,075
0110	Deluxe	10,100	2,925	13,025
0120	Night depository, economy	8,300	1,475	9,775
0130	Deluxe	11,800	2,925	14,725
0140	Pneumatic tube systems, 2 station, standard	28,800	5,000	33,800
0150	Teller, automated, 24 hour, single unit	49,700	5,000	54,700
0160	Teller window, bullet proof glazing, 44" x 60"	4,250	915	5,165
0170	Pass through, painted steel, 72" x 40"	4,675	1,825	6,500
0300	Safe, office type, 1 hr. rating, 34" x 20" x 20"	2,500		2,500
0310	4 hr. rating, 62" x 33" x 20"	9,750		9,750
0320	Data storage, 4 hr. rating, 63" x 44" x 16"	14,800		14,800
0330	Jewelers, 63" x 44" x 16"	15,800		15,800
0340	Money, "B" label, 9" x 14" x 14"	595		595
0350	Tool and torch resistive, 24" x 24" x 20"	8,675	231	8,906
0500	Security gates-scissors type, painted steel, single, 6' high, 5-1/2' wide	248	365	613
0510	Double gate, 7-1/2' high, 14' wide	695	730	1,425

E1010 510	Mercantile Equipment, EACH	COST EACH		
		MAT.	INST.	TOTAL
0015	Barber equipment, chair, hydraulic, economy	615	24	639
0020	Deluxe	3,825	36	3,861
0100	Checkout counter, single belt	3,500	92.50	3,592.50
0110	Double belt, power take-away	5,025	103	5,128
0200	Display cases, freestanding, glass and aluminum, 3'-6" x 3' x 1'-0" deep	1,350	145	1,495
0220	Wall mounted, glass and aluminum, 3' x 4' x 1'-4" deep	2,400	231	2,631
0320	Frozen food, chest type, 12 ft. long	9,050	395	9,445

E1010 610	Laundry/Dry Cleaning, EACH	COST EACH		
		MAT.	INST.	TOTAL
0100	Laundry equipment, dryers, gas fired, residential, 16 lb. capacity	740	236	976
0110	Commercial, 30 lb. capacity, single	3,625	236	3,861
0120	Dry cleaners, electric, 20 lb. capacity	37,000	6,825	43,825
0130	30 lb. capacity	56,500	9,100	65,600
0140	Ironers, commercial, 120" with canopy, 8 roll	194,000	19,500	213,500
0150	Institutional, 110", single roll	35,300	3,275	38,575
0160	Washers, residential, 4 cycle	965	236	1,201
0170	Commercial, coin operated, deluxe	3,700	236	3,936

E10 Equipment

E1020 Institutional Equipment

E1020 110	Ecclesiastical Equipment, EACH	COST EACH		
		MAT.	INST.	TOTAL
0090	Church equipment, altar, wood, custom, plain	2,825	415	3,240
0100	Granite, custom, deluxe	41,800	5,300	47,100
0110	Baptistry, fiberglass, economy	6,300	1,500	7,800
0120	Bells & carillons, keyboard operation	20,000	8,700	28,700
0130	Confessional, wood, single, economy	3,500	965	4,465
0140	Double, deluxe	20,900	2,900	23,800
0150	Steeples, translucent fiberglas, 30" square, 15' high	9,200	1,825	11,025

E1020 130	Ecclesiastical Equipment, L.F.	COST PER L.F.		
		MAT.	INST.	TOTAL
0100	Arch. equip., church equip. pews, bench type, hardwood, economy	104	29	133
0110	Deluxe	206	38.50	244.50

E1020 210	Library Equipment, EACH	COST EACH		
		MAT.	INST.	TOTAL
0110	Library equipment, carrels, metal, economy	286	116	402
0120	Hardwood, deluxe	2,200	145	2,345

E1020 230	Library Equipment, L.F.	COST PER L.F.		
		MAT.	INST.	TOTAL
0100	Library equipment, book shelf, metal, single face, 90" high x 10" shelf	137	48	185
0110	Double face, 90" high x 10" shelf	345	105	450
0120	Charging desk, built-in, with counter, plastic laminate	335	82.50	417.50

E1020 310	Theater and Stage Equipment, EACH	COST EACH		
		MAT.	INST.	TOTAL
0200	Movie equipment, changeover, economy	520		520
0210	Film transport, incl. platters and autowind, economy	5,600		5,600
0220	Lamphouses, incl. rectifiers, xenon, 1000W	7,400	330	7,730
0230	4000W	10,500	435	10,935
0240	Projector mechanisms, 35 mm, economy	12,300		12,300
0250	Deluxe	16,900		16,900
0260	Sound systems, incl. amplifier, single, economy	3,675	730	4,405
0270	Dual, Dolby/super sound	18,400	1,650	20,050
0280	Projection screens, wall hung, manual operation, 50 S.F., economy	325	116	441
0290	Electric operation, 100 S.F., deluxe	2,200	580	2,780

E1020 320	Theater and Stage Equipment, S.F.	COST PER S.F.		
		MAT.	INST.	TOTAL
0090	Movie equipment, projection screens, rigid in wall, acrylic, 1/4" thick	46.50	5.65	52.15
0100	1/2" thick	54	8.45	62.45
0110	Stage equipment, curtains, velour, medium weight	8.80	1.93	10.73
0120	Silica based yarn, fireproof	16.90	23	39.90
0130	Stages, portable with steps, folding legs, 8" high	36		36
0140	Telescoping platforms, aluminum, deluxe	52.50	30	82.50

E1020 330	Theater and Stage Equipment, L.F.	COST PER L.F.		
		MAT.	INST.	TOTAL
0100	Stage equipment, curtain track, heavy duty	68.50	64	132.50
0110	Lights, border, quartz, colored	192	33	225

E1020 610	Detention Equipment, EACH	COST EACH		
		MAT.	INST.	TOTAL
0110	Detention equipment, cell front rolling door, 7/8" bars, 5' x 7' high	5,875	1,550	7,425
0120	Cells, prefab., including front, 5' x 7' x 7' deep	10,700	2,075	12,775

E10 Equipment

E1020 Institutional Equipment

E1020 610 — Detention Equipment, EACH

		COST EACH		
		MAT.	INST.	TOTAL
0130	Doors and frames, 3' x 7', single plate	5,050	780	5,830
0140	Double plate	6,150	780	6,930
0150	Toilet apparatus, incl wash basin	3,750	1,000	4,750
0160	Visitor cubicle, vision panel, no intercom	3,525	1,550	5,075

E1020 710 — Laboratory Equipment, EACH

		COST EACH		
		MAT.	INST.	TOTAL
0110	Laboratory equipment, glassware washer, distilled water, economy	6,950	760	7,710
0120	Deluxe	14,600	1,375	15,975
0140	Radio isotope	18,800		18,800

E1020 720 — Laboratory Equipment, S.F.

		COST PER S.F.		
		MAT.	INST.	TOTAL
0100	Arch. equip., lab equip., counter tops, acid proof, economy	44.50	14.10	58.60
0110	Stainless steel	144	14.10	158.10

E1020 730 — Laboratory Equipment, L.F.

		COST PER L.F.		
		MAT.	INST.	TOTAL
0110	Laboratory equipment, cabinets, wall, open	183	58	241
0120	Base, drawer units	565	64	629
0130	Fume hoods, not incl. HVAC, economy	550	214	764
0140	Deluxe incl. fixtures	885	480	1,365

E1020 810 — Medical Equipment, EACH

		COST EACH		
		MAT.	INST.	TOTAL
0100	Dental equipment, central suction system, economy	1,725	590	2,315
0110	Compressor-air, deluxe	9,850	1,200	11,050
0120	Chair, hydraulic, economy	2,475	1,200	3,675
0130	Deluxe	8,600	2,400	11,000
0140	Drill console with accessories, economy	2,300	375	2,675
0150	Deluxe	5,400	375	5,775
0160	X-ray unit, portable	2,575	150	2,725
0170	Panoramic unit	17,300	1,000	18,300
0300	Medical equipment, autopsy table, standard	10,700	710	11,410
0310	Deluxe	16,600	1,175	17,775
0320	Incubators, economy	3,350		3,350
0330	Deluxe	13,000		13,000
0700	Station, scrub-surgical, single, economy	4,725	236	4,961
0710	Dietary, medium, with ice	18,400		18,400
0720	Sterilizers, general purpose, single door, 20" x 20" x 28"	14,400		14,400
0730	Floor loading, double door, 28" x 67" x 52"	227,000		227,000
0740	Surgery tables, standard	13,400	1,050	14,450
0750	Deluxe	28,900	1,475	30,375
0770	Tables, standard, with base cabinets, economy	1,125	385	1,510
0780	Deluxe	6,175	580	6,755
0790	X-ray, mobile, economy	18,500		18,500
0800	Stationary, deluxe	247,000		247,000

E10 Equipment

E1030 Vehicular Equipment

E1030 110 — Vehicular Service Equipment, EACH

		COST EACH		
		MAT.	INST.	TOTAL
0110	Automotive equipment, compressors, electric, 1-1/2 H.P., std. controls	500	1,100	1,600
0120	5 H.P., dual controls	3,725	1,650	5,375
0130	Hoists, single post, 4 ton capacity, swivel arms	6,125	4,100	10,225
0140	Dual post, 12 ton capacity, adjustable frame	10,500	865	11,365
0150	Lube equipment, 3 reel type, with pumps	9,675	3,275	12,950
0160	Product dispenser, 6 nozzles, w/vapor recovery, not incl. piping, installed	27,200		27,200
0800	Scales, dial type, built in floor, 5 ton capacity, 8' x 6' platform	8,575	3,475	12,050
0810	10 ton capacity, 9' x 7' platform	12,100	4,950	17,050
0820	Truck (including weigh bridge), 20 ton capacity, 24' x 10'	13,100	5,775	18,875

E1030 210 — Parking Control Equipment, EACH

		COST EACH		
		MAT.	INST.	TOTAL
0110	Parking equipment, automatic gates, 8 ft. arm, one way	4,750	1,200	5,950
0120	Traffic detectors, single treadle	2,025	545	2,570
0130	Booth for attendant, economy	8,325		8,325
0140	Deluxe	31,200		31,200
0150	Ticket printer/dispenser, rate computing	9,850	935	10,785
0160	Key station on pedestal	890	320	1,210

E1030 310 — Loading Dock Equipment, EACH

		COST EACH		
		MAT.	INST.	TOTAL
0110	Dock bumpers, rubber blocks, 4-1/2" thick, 10" high, 14" long	70	22	92
0120	6" thick, 20" high, 11" long	163	44.50	207.50
0130	Dock boards, H.D., 5' x 5', aluminum, 5000 lb. capacity	1,600		1,600
0140	16,000 lb. capacity	1,925		1,925
0150	Dock levelers, hydraulic, 7' x 8', 10 ton capacity	10,400	1,525	11,925
0160	Dock lifters, platform, 6' x 6', portable, 3000 lb. capacity	10,200		10,200
0170	Dock shelters, truck, scissor arms, economy	1,950	580	2,530
0180	Deluxe	2,675	1,150	3,825

For customer support on your Square Foot Cost Data, call 877.694.8329.

495

E10 Equipment

E1090 Other Equipment

E1090 210	Solid Waste Handling Equipment, EACH	COST EACH		
		MAT.	INST.	TOTAL
0110	Waste handling, compactors, single bag, 250 lbs./hr., hand fed	16,600	685	17,285
0120	Heavy duty industrial, 5 C.Y. capacity	35,800	3,275	39,075
0130	Incinerator, electric, 100 lbs./hr., economy	71,500	3,275	74,775
0140	Gas, 2000 lbs./hr., deluxe	445,500	27,000	472,500
0150	Shredder, no baling, 35 tons/hr.	335,000		335,000
0160	Incl. baling, 50 tons/day	669,000		669,000

E1090 350	Food Service Equipment, EACH	COST EACH		
		MAT.	INST.	TOTAL
0110	Kitchen equipment, bake oven, single deck	6,000	160	6,160
0120	Broiler, without oven	3,900	160	4,060
0130	Commercial dish washer, semiautomatic, 50 racks/hr.	8,450	980	9,430
0140	Automatic, 275 racks/hr.	34,600	4,150	38,750
0150	Cooler, beverage, reach-in, 6 ft. long	3,950	213	4,163
0160	Food warmer, counter, 1.65 kw	735		735
0170	Fryers, with submerger, single	1,375	182	1,557
0180	Double	2,675	255	2,930
0185	Ice maker, 1000 lb. per day, with bin	5,425	1,275	6,700
0190	Kettles, steam jacketed, 20 gallons	9,050	278	9,328
0200	Range, restaurant type, burners, 2 ovens and 24" griddle	5,025	213	5,238
0210	Range hood, incl. carbon dioxide system, economy	2,075	425	2,500
0220	Deluxe	2,075	425	2,500

E1090 360	Food Service Equipment, S.F.	COST PER S.F.		
		MAT.	INST.	TOTAL
0110	Refrigerators, prefab, walk-in, 7'-6" high, 6' x 6'	181	21	202
0120	12' x 20'	114	10.55	124.55

E1090 410	Residential Equipment, EACH	COST EACH		
		MAT.	INST.	TOTAL
0110	Arch. equip., appliances, range, cook top, 4 burner, economy	325	109	434
0120	Built in, single oven 30" wide, economy	750	109	859
0130	Standing, single oven-21" wide, economy	500	92.50	592.50
0140	Double oven-30" wide, deluxe	2,800	92.50	2,892.50
0150	Compactor, residential, economy	710	116	826
0160	Deluxe	1,125	193	1,318
0170	Dish washer, built-in, 2 cycles, economy	261	340	601
0180	4 or more cycles, deluxe	1,200	680	1,880
0190	Garbage disposer, sink type, economy	96	136	232
0200	Deluxe	242	136	378
0210	Refrigerator, no frost, 10 to 12 C.F., economy	465	92.50	557.50
0220	21 to 29 C.F., deluxe	3,700	310	4,010
0300	Washing machine, automatic	1,675	710	2,385

E1090 610	School Equipment, EACH	COST EACH		
		MAT.	INST.	TOTAL
0110	School equipment, basketball backstops, wall mounted, wood, fixed	1,525	1,025	2,550
0120	Suspended type, electrically operated	7,025	1,950	8,975
0130	Bleachers-telescoping, manual operation, 15 tier, economy (per seat)	101	36	137
0140	Power operation, 30 tier, deluxe (per seat)	410	64	474
0150	Weight lifting gym, universal, economy	365	925	1,290
0160	Deluxe	16,400	1,850	18,250
0170	Scoreboards, basketball, 1 side, economy	2,650	865	3,515
0180	4 sides, deluxe	6,325	11,900	18,225

E10 Equipment

E1090 Other Equipment

E1090 610	School Equipment, EACH	COST EACH		
		MAT.	INST.	TOTAL
0800	Vocational shop equipment, benches, metal	605	231	836
0810	Wood	700	231	931
0820	Dust collector, not incl. ductwork, 6' diam.	5,125	620	5,745
0830	Planer, 13" x 6"	1,200	289	1,489

E1090 620	School Equipment, S.F.	COST PER S.F.		
		MAT.	INST.	TOTAL
0110	School equipment, gym mats, naugahyde cover, 2" thick	4.08		4.08
0120	Wrestling, 1" thick, heavy duty	6.15		6.15

E1090 810	Athletic, Recreational, and Therapeutic Equipment, EACH	COST EACH		
		MAT.	INST.	TOTAL
0050	Bowling alley with gutters	51,000	12,200	63,200
0060	Combo. table and ball rack	1,275		1,275
0070	Bowling alley automatic scorer	11,000		11,000
0110	Sauna, prefabricated, incl. heater and controls, 7' high, 6' x 4'	5,700	885	6,585
0120	10' x 12'	13,500	1,950	15,450
0130	Heaters, wall mounted, to 200 C.F.	755		755
0140	Floor standing, to 1000 C.F., 12500 W	4,200	219	4,419
0610	Shooting range incl. bullet traps, controls, separators, ceilings, economy	39,000	7,675	46,675
0620	Deluxe	58,000	12,900	70,900
0650	Sport court, squash, regulation, in existing building, economy			40,400
0660	Deluxe			45,000
0670	Racketball, regulation, in existing building, economy	43,800	11,000	54,800
0680	Deluxe	47,400	22,000	69,400
0700	Swimming pool equipment, diving stand, stainless steel, 1 meter	10,100	430	10,530
0710	3 meter	16,600	2,900	19,500
0720	Diving boards, 16 ft. long, aluminum	4,350	430	4,780
0730	Fiberglass	3,575	430	4,005
0740	Filter system, sand, incl. pump, 6000 gal./hr.	2,175	790	2,965
0750	Lights, underwater, 12 volt with transformer, 300W	365	655	1,020
0760	Slides, fiberglass with aluminum handrails & ladder, 6' high, straight	3,950	725	4,675
0780	12' high, straight with platform	15,200	965	16,165

E1090 820	Athletic, Recreational, and Therapeutic Equipment, S.F.	COST PER S.F.		
		MAT.	INST.	TOTAL
0110	Swimming pools, residential, vinyl liner, metal sides	21.50	8.15	29.65
0120	Concrete sides	25.50	14.65	40.15
0130	Gunite shell, plaster finish, 350 S.F.	48	30.50	78.50
0140	800 S.F.	38.50	17.55	56.05
0150	Motel, gunite shell, plaster finish	59	38	97
0160	Municipal, gunite shell, tile finish, formed gutters	226	66.50	292.50

For customer support on your Square Foot Cost Data, call 877.694.8329.

497

E2010 Fixed Furnishings

E2010 310	Window Treatment, EACH	COST EACH		
		MAT.	INST.	TOTAL
0110	Furnishings, blinds, exterior, aluminum, louvered, 1'-4" wide x 3'-0" long	220	58	278
0120	1'-4" wide x 6'-8" long	390	64	454
0130	Hemlock, solid raised, 1'-4" wide x 3'-0" long	78.50	58	136.50
0140	1'-4" wide x 6'-9" long	167	64	231
0150	Polystyrene, louvered, 1'-3" wide x 3'-3" long	39.50	58	97.50
0160	1'-3" wide x 6'-8" long	75	64	139
0200	Interior, wood folding panels, louvered, 7" x 20" (per pair)	87.50	34	121.50
0210	18" x 40" (per pair)	148	34	182

E2010 320	Window Treatment, S.F.	COST PER S.F.		
		MAT.	INST.	TOTAL
0110	Furnishings, blinds-interior, venetian-aluminum, stock, 2" slats	5.95	.98	6.93
0120	Custom, 1" slats, deluxe	5.40	.98	6.38
0130	Vertical, PVC or cloth, T&B track, economy	8.75	1.26	10.01
0140	Deluxe	26	1.45	27.45
0150	Draperies, unlined, economy	3.44		3.44
0160	Lightproof, deluxe	9.90		9.90
0510	Shades, mylar, wood roller, single layer, non-reflective	5.55	.84	6.39
0520	Metal roller, triple layer, heat reflective	7.35	.84	8.19
0530	Vinyl, light weight, 4 ga.	.67	.84	1.51
0540	Heavyweight, 6 ga.	2.06	.84	2.90
0550	Vinyl coated cotton, lightproof decorator shades	3.30	.84	4.14
0560	Woven aluminum, 3/8" thick, light and fireproof	6.45	1.65	8.10

E2010 420	Fixed Floor Grilles and Mats, S.F.	COST PER S.F.		
		MAT.	INST.	TOTAL
0110	Floor mats, recessed, inlaid black rubber, 3/8" thick, solid	28.50	2.99	31.49
0120	Colors, 1/2" thick, perforated	30.50	2.99	33.49
0130	Link-including nosings, steel-galvanized, 3/8" thick	26	2.99	28.99
0140	Vinyl, in colors	27	2.99	29.99

E2010 510	Fixed Multiple Seating, EACH	COST EACH		
		MAT.	INST.	TOTAL
0110	Seating, painted steel, upholstered, economy	147	33	180
0120	Deluxe	470	41.50	511.50
0400	Seating, lecture hall, pedestal type, economy	210	52.50	262.50
0410	Deluxe	540	79.50	619.50
0500	Auditorium chair, veneer construction	255	52.50	307.50
0510	Fully upholstered, spring seat	266	52.50	318.50

E2020 Moveable Furnishings

E2020 210	Furnishings/EACH	COST EACH		
		MAT.	INST.	TOTAL
0200	Hospital furniture, beds, manual, economy	880		880
0210	Deluxe	2,800		2,800
0220	All electric, economy	1,800		1,800
0230	Deluxe	4,925		4,925
0240	Patient wall systems, no utilities, economy, per room	1,575		1,575
0250	Deluxe, per room	2,125		2,125
0300	Hotel furnishings, standard room set, economy, per room	2,650		2,650
0310	Deluxe, per room	9,450		9,450
0500	Office furniture, standard employee set, economy, per person	680		680
0510	Deluxe, per person	2,650		2,650
0550	Posts, portable, pedestrian traffic control, economy	147		147
0560	Deluxe	465		465
0700	Restaurant furniture, booth, molded plastic, stub wall and 2 seats, economy	405	289	694
0710	Deluxe	1,600	385	1,985
0720	Upholstered seats, foursome, single-economy	860	116	976
0730	Foursome, double-deluxe	1,925	193	2,118

E2020 220	Furniture and Accessories, L.F.	COST PER L.F.		
		MAT.	INST.	TOTAL
0210	Dormitory furniture, desk top (built-in),laminated plastc, 24"deep, economy	48.50	23	71.50
0220	30" deep, deluxe	272	29	301
0230	Dressing unit, built-in, economy	197	96.50	293.50
0240	Deluxe	590	145	735
0310	Furnishings, cabinets, hospital, base, laminated plastic	300	116	416
0320	Stainless steel	555	116	671
0330	Countertop, laminated plastic, no backsplash	52	29	81
0340	Stainless steel	169	29	198
0350	Nurses station, door type, laminated plastic	350	116	466
0360	Stainless steel	670	116	786
0710	Restaurant furniture, bars, built-in, back bar	224	116	340
0720	Front bar	310	116	426
0910	Wardrobes & coatracks, standing, steel, single pedestal, 30" x 18" x 63"	116		116
0920	Double face rack, 39" x 26" x 70"	132		132
0930	Wall mounted rack, steel frame & shelves, 12" x 15" x 26"	77.50	8.20	85.70
0940	12" x 15" x 50"	47.50	4.30	51.80

For customer support on your Square Foot Cost Data, call 877.694.8329.

499

F1010 Special Structures

F1010 120	Air-Supported Structures, S.F.	COST PER S.F.		
		MAT.	INST.	TOTAL
0110	Air supported struc., polyester vinyl fabric, 24oz., warehouse, 5000 S.F.	29	.37	29.37
0120	50,000 S.F.	13.55	.30	13.85
0130	Tennis, 7,200 S.F.	26	.31	26.31
0140	24,000 S.F.	18.05	.31	18.36
0150	Woven polyethylene, 6 oz., shelter, 3,000 S.F.	17.60	.62	18.22
0160	24,000 S.F.	13.20	.31	13.51
0170	Teflon coated fiberglass, stadium cover, economy	63	.16	63.16
0180	Deluxe	75	.22	75.22
0190	Air supported storage tank covers, reinf. vinyl fabric, 12 oz., 400 S.F.	25.50	.52	26.02
0200	18,000 S.F.	9.30	.47	9.77

F1010 210	Pre-Engineered Structures, EACH	COST EACH		
		MAT.	INST.	TOTAL
0600	Radio towers, guyed, 40 lb. section, 50' high, 70 MPH basic wind speed	3,025	1,475	4,500
0610	90 lb. section, 400' high, wind load 70 MPH basic wind speed	41,900	16,400	58,300
0620	Self supporting, 60' high, 70 MPH basic wind speed	4,950	2,875	7,825
0630	190' high, wind load 90 MPH basic wind speed	31,100	11,500	42,600
0700	Shelters, aluminum frame, acrylic glazing, 8' high, 3' x 9'	3,375	1,275	4,650
0710	9' x 12'	8,025	2,000	10,025

F1010 320	Other Special Structures, S.F.	COST PER S.F.		
		MAT.	INST.	TOTAL
0110	Swimming pool enclosure, transluscent, freestanding, economy	41	5.80	46.80
0120	Deluxe	102	16.50	118.50
0510	Tension structures, steel frame, polyester vinyl fabric, 12,000 S.F.	14.55	2.68	17.23
0520	20,800 S.F.	14.30	2.42	16.72

F1010 330	Special Structures, EACH	COST EACH		
		MAT.	INST.	TOTAL
0110	Kiosks, round, 5' diam., 7' high, aluminum wall, illuminated	25,800		25,800
0120	Rectangular, 5' x 9', 1" insulated dbl. wall fiberglass, 7'-6" high	26,400		26,400
0220	Silos, steel prefab, 30,000 gal., painted, economy	24,300	5,850	30,150
0230	Epoxy-lined, deluxe	50,500	11,700	62,200

F1010 340	Special Structures, S.F.	COST PER S.F.		
		MAT.	INST.	TOTAL
0110	Comfort stations, prefab, mobile on steel frame, economy	180		180
0120	Permanent on concrete slab, deluxe	211	48	259
0210	Domes, bulk storage, wood framing, wood decking, 50' diam.	79.50	1.83	81.33
0220	116' diam.	38.50	2.12	40.62
0230	Steel framing, metal decking, 150' diam.	38	11.85	49.85
0240	400' diam.	31	9.05	40.05
0250	Geodesic, wood framing, wood panels, 30' diam.	36	2.07	38.07
0260	60' diam.	23.50	1.29	24.79
0270	Aluminum framing, acrylic panels, 40' diam.			83
0280	Aluminum panels, 400' diam.			26
0310	Garden house, prefab, wood, shell only, 48 S.F.	56.50	5.80	62.30
0320	200 S.F.	42	24	66
0410	Greenhouse, shell-stock, residential, lean-to, 8'-6" long x 3'-10" wide	46	34	80
0420	Freestanding, 8'-6" long x 13'-6" wide	42.50	10.70	53.20
0430	Commercial-truss frame, under 2000 S.F., deluxe			14.30
0440	Over 5,000 S.F., economy			13.15
0450	Institutional-rigid frame, under 500 S.F., deluxe			28
0460	Over 2,000 S.F., economy			11.75
0510	Hangar, prefab, galv. roof and walls, bottom rolling doors, economy	12.90	4.71	17.61
0520	Electric bifolding doors, deluxe	12.75	6.15	18.90

F10 Special Construction

F1020 Integrated Construction

F1020 110	Integrated Construction, EACH	COST EACH		
		MAT.	INST.	TOTAL
0110	Integrated ceilings, radiant electric, 2' x 4' panel, manila finish	101	26	127
0120	ABS plastic finish	114	50.50	164.50

F1020 120	Integrated Construction, S.F.	COST PER S.F.		
		MAT.	INST.	TOTAL
0110	Integrated ceilings, Luminaire, suspended, 5' x 5' modules, 50% lighted	4.41	13.85	18.26
0120	100% lighted	6.55	25	31.55
0130	Dimensionaire, 2' x 4' module tile system, no air bar	2.66	2.31	4.97
0140	With air bar, deluxe	3.90	2.31	6.21
0220	Pedestal access floors, stl pnls, no stringers, vinyl covering, >6000 S.F.	13.50	1.81	15.31
0230	With stringers, high pressure laminate covering, under 6000 S.F.	12.40	1.93	14.33
0240	Carpet covering, under 6000 S.F.	13.80	1.93	15.73
0250	Aluminum panels, no stringers, no covering	36	2.31	38.31

F1020 250	Special Purpose Room, EACH	COST EACH		
		MAT.	INST.	TOTAL
0110	Portable booth, acoustical, 27 db 1000 hz., 15 S.F. floor	3,925		3,925
0120	55 S.F. flr.	8,050		8,050

F1020 260	Special Purpose Room, S.F.	COST PER S.F.		
		MAT.	INST.	TOTAL
0110	Anechoic chambers, 7' high, 100 cps cutoff, 25 S.F.			2,600
0120	200 cps cutoff, 100 S.F.			1,350
0130	Audiometric rooms, under 500 S.F.	57.50	23.50	81
0140	Over 500 S.F.	55	19.25	74.25
0300	Darkrooms, shell, not including door, 240 S.F., 8' high	29.50	9.65	39.15
0310	64 S.F., 12' high	73	18.05	91.05
0510	Music practice room, modular, perforated steel, under 500 S.F.	35.50	16.50	52
0520	Over 500 S.F.	30	14.45	44.45

F1020 330	Special Construction, L.F.	COST PER L.F.		
		MAT.	INST.	TOTAL
0110	Spec. const., air curtains, shipping & receiving, 8'high x 5'wide, economy	750	137	887
0120	20' high x 8' wide, heated, deluxe	3,200	340	3,540
0130	Customer entrance, 10' high x 5' wide, economy	1,125	137	1,262
0140	12' high x 4' wide, heated, deluxe	2,025	171	2,196

For customer support on your Square Foot Cost Data, call 877.694.8329.

503

F10 Special Construction

F1030 Special Construction Systems

F1030 120 — Sound, Vibration, and Seismic Construction, S.F.

		COST PER S.F.		
		MAT.	INST.	TOTAL
0020	Special construction, acoustical, enclosure, 4" thick, 8 psf panels	36	24	60
0030	Reverb chamber, 4" thick, parallel walls	51	29	80
0110	Sound absorbing panels, 2'-6" x 8', painted metal	13.55	8.05	21.60
0120	Vinyl faced	10.60	7.25	17.85
0130	Flexible transparent curtain, clear	8.20	9.55	17.75
0140	With absorbing foam, 75% coverage	11.40	9.55	20.95
0150	Strip entrance, 2/3 overlap	8.35	15.20	23.55
0160	Full overlap	10.55	17.85	28.40
0200	Audio masking system, plenum mounted, over 10,000 S.F.	.80	.30	1.10
0210	Ceiling mounted, under 5,000 S.F.	1.39	.55	1.94

F1030 210 — Radiation Protection, EACH

		COST EACH		
		MAT.	INST.	TOTAL
0110	Shielding, lead x-ray protection, radiography room, 1/16" lead, economy	11,100	4,050	15,150
0120	Deluxe	13,400	6,775	20,175
0210	Deep therapy x-ray, 1/4" lead, economy	31,100	12,700	43,800
0220	Deluxe	38,400	16,900	55,300

F1030 220 — Radiation Protection, S.F.

		COST PER S.F.		
		MAT.	INST.	TOTAL
0110	Shielding, lead, gypsum board, 5/8" thick, 1/16" lead	12.05	6.85	18.90
0120	1/8" lead	25.50	7.85	33.35
0130	Lath, 1/16" thick	11.55	7.50	19.05
0140	1/8" thick	34	8.45	42.45
0150	Radio frequency, galvanized steel, prefab type, economy	4.91	3.08	7.99
0160	Radio frequency, door, copper/wood laminate, 4" x 7'	8.65	8.25	16.90

F1030 910 — Other Special Construction Systems, EACH

		COST EACH		
		MAT.	INST.	TOTAL
0110	Disappearing stairways, folding, pine, 8'-6" ceiling	194	145	339
0220	Automatic electric, wood, 8' to 9' ceiling	9,400	1,150	10,550
0300	Fireplace prefabricated, freestanding or wall hung, painted	1,525	445	1,970
0310	Stainless steel	3,375	640	4,015
0320	Woodburning stoves, cast iron, economy, less than 1500 s.f. htg. area	1,300	890	2,200
0330	Greater than 2000 sf htg. area	3,050	1,450	4,500

F10 Special Construction

F1040 Special Facilities

F1040 210	Ice Rinks, EACH	COAT EACH		
		MAT.	INST.	TOTAL
0100	Ice skating rink, 85' x 200', 55° system, 5 mos., 100 ton			632,500
0110	90° system, 12 mos., 135 ton			715,000
0120	Dash boards, acrylic screens, polyethylene coated plywood	148,500	38,900	187,400
0130	Fiberglass and aluminum construction	170,500	38,900	209,400

F1040 510	Liquid & Gas Storage Tanks, EACH	COST EACH		
		MAT.	INST.	TOTAL
0100	Tanks, steel, ground level, 100,000 gal.			226,500
0110	10,000,000 gal.			5,151,500
0120	Elevated water, 50,000 gal.		193,500	465,500
0130	1,000,000 gal.		903,000	2,063,000
0150	Cypress wood, ground level, 3,000 gal.	9,900	11,600	21,500
0160	Redwood, ground level, 45,000 gal.	77,500	31,400	108,900

F10 Special Construction

F1040 Special Facilities

F1040 910	Special Construction, EACH	COST EACH		
		MAT.	INST.	TOTAL
0110	Special construction, bowling alley incl. pinsetter, scorer etc., economy	46,400	11,600	58,000
0120	Deluxe	58,500	12,800	71,300
0130	For automatic scorer, economy, add	9,150		9,150
0140	Deluxe, add	11,000		11,000
0300	Control tower, modular, 12' x 10', incl. instrumentation, economy			832,500
0310	Deluxe			1,310,000
0400	Garage costs, residential, prefab, wood, single car economy	5,925	1,150	7,075
0410	Two car deluxe	13,300	2,300	15,600
0500	Hangars, prefab, galv. steel, bottom rolling doors, economy (per plane)	17,100	5,225	22,325
0510	Electrical bi-folding doors, deluxe (per plane)	15,200	7,175	22,375

G1030 Site Earthwork

G1030 805	Trenching Common Earth	COST PER L.F.		
		MAT.	INST.	TOTAL
1310	Trenching, common earth, no slope, 2' wide, 2' deep, 3/8 C.Y. bucket		2.65	2.65
1330	4' deep, 3/8 C.Y. bucket		5.19	5.19
1360	10' deep, 1 C.Y. bucket		11.41	11.41
1400	4' wide, 2' deep, 3/8 C.Y. bucket		5.43	5.43
1420	4' deep, 1/2 C.Y. bucket		9.06	9.06
1450	10' deep, 1 C.Y. bucket		24.65	24.65
1480	18' deep, 2-1/2 C.Y. bucket		39.50	39.50
1520	6' wide, 6' deep, 5/8 C.Y. bucket w/trench box		23.95	23.95
1540	10' deep, 1 C.Y. bucket		32.50	32.50
1570	20' deep, 3-1/2 C.Y. bucket		56	56
1640	8' wide, 12' deep, 1-1/2 C.Y. bucket w/trench box		45.80	45.80
1680	24' deep, 3-1/2 C.Y. bucket		88	88
1730	10' wide, 20' deep, 3-1/2 C.Y. bucket w/trench box		78.50	78.50
1740	24' deep, 3-1/2 C.Y. bucket		102.50	102.50
3500	1 to 1 slope, 2' wide, 2' deep, 3/8 C.Y. bucket		5.19	5.19
3540	4' deep, 3/8 C.Y. bucket		15.43	15.43
3600	10' deep, 1 C.Y. bucket		69	69
3800	4' wide, 2' deep, 3/8 C.Y. bucket		8	8
3840	4' deep, 1/2 C.Y. bucket		17.94	17.94
3900	10' deep, 1 C.Y. bucket		87.50	87.50
4030	6' wide, 6' deep, 5/8 C.Y. bucket w/trench box		47.50	47.50
4050	10' deep, 1 C.Y. bucket		79	79
4080	20' deep, 3-1/2 C.Y. bucket		237.50	237.50
4500	8' wide, 12' deep, 1-1/2 C.Y. bucket w/trench box		117	117
4650	24' deep, 3-1/2 C.Y. bucket		359	359
4800	10' wide, 20' deep, 3-1/2 C.Y. bucket w/trench box		242.50	242.50
4850	24' deep, 3-1/2 C.Y. bucket		380	380

G1030 806	Trenching Loam & Sandy Clay	COST PER L.F.		
		MAT.	INST.	TOTAL
1310	Trenching, loam & sandy clay, no slope, 2' wide, 2' deep, 3/8 C.Y. bucket		2.59	2.59

G1030 807	Trenching Sand & Gravel	COST PER L.F.		
		MAT.	INST.	TOTAL
1310	Trenching, sand & gravel, no slope, 2' wide, 2' deep, 3/8 C.Y. bucket		2.43	2.43
1320	3' deep, 3/8 C.Y. bucket		4.02	4.02
1330	4' deep, 3/8 C.Y. bucket		4.80	4.80
1340	6' deep, 3/8 C.Y. bucket		6	6
1350	8' deep, 1/2 C.Y. bucket		7.95	7.95
1360	10' deep, 1 C.Y. bucket		8.55	8.55
1400	4' wide, 2' deep, 3/8 C.Y. bucket		5	5
1410	3' deep, 3/8 C.Y. bucket		7.35	7.35
1420	4' deep, 1/2 C.Y. bucket		8.40	8.40
1430	6' deep, 1/2 C.Y. bucket		13.10	13.10
1440	8' deep, 1/2 C.Y. bucket		18.65	18.65
1450	10' deep, 1 C.Y. bucket		18.90	18.90
1460	12' deep, 1 C.Y. bucket		23.50	23.50
1470	15' deep, 1-1/2 C.Y. bucket		28	28
1480	18' deep, 2-1/2 C.Y. bucket		31.50	31.50
1520	6' wide, 6' deep, 5/8 C.Y. bucket w/trench box		22	22
1530	8' deep, 3/4 C.Y. bucket		29.50	29.50
1540	10' deep, 1 C.Y. bucket		29	29
1550	12' deep, 1-1/2 C.Y. bucket		33	33
1560	16' deep, 2 C.Y. bucket		43	43
1570	20' deep, 3-1/2 C.Y. bucket		51	51
1580	24' deep, 3-1/2 C.Y. bucket		63	63

G1030 Site Earthwork

G1030 807	Trenching Sand & Gravel	COST PER L.F.		
		MAT.	INST.	TOTAL
1640	8' wide, 12' deep, 1-1/2 C.Y. bucket w/trench box		43	43
1650	15' deep, 1-1/2 C.Y. bucket		54	54
1660	18' deep, 2-1/2 C.Y. bucket		62.50	62.50
1680	24' deep, 3-1/2 C.Y. bucket		82.50	82.50
1730	10' wide, 20' deep, 3-1/2 C.Y. bucket w/trench box		82.50	82.50
1740	24' deep, 3-1/2 C.Y. bucket		103	103
1780	12' wide, 20' deep, 3-1/2 C.Y. bucket w/trench box		99	99
1790	25' deep, 3-1/2 C.Y. bucket		128	128
1800	1/2:1 slope, 2' wide, 2' deep, 3/8 C.Y. bucket		3.63	3.63
1810	3' deep, 3/8 C.Y. bucket		6.30	6.30
1820	4' deep, 3/8 C.Y. bucket		9.55	9.55
1840	6' deep, 3/8 C.Y. bucket		14.90	14.90
1860	8' deep, 1/2 C.Y. bucket		24	24
1880	10' deep, 1 C.Y. bucket		30	30
2300	4' wide, 2' deep, 3/8 C.Y. bucket		6.20	6.20
2310	3' deep, 3/8 C.Y. bucket		10	10
2320	4' deep, 1/2 C.Y. bucket		12.50	12.50
2340	6' deep, 1/2 C.Y. bucket		23	23
2360	8' deep, 1/2 C.Y. bucket		37.50	37.50
2380	10' deep, 1 C.Y. bucket		43	43
2400	12' deep, 1 C.Y. bucket		71.50	71.50
2430	15' deep, 1-1/2 C.Y. bucket		80.50	80.50
2460	18' deep, 2-1/2 C.Y. bucket		126	126
2840	6' wide, 6' deep, 5/8 C.Y. bucket w/trench box		33.50	33.50
2860	8' deep, 3/4 C.Y. bucket		48.50	48.50
2880	10' deep, 1 C.Y. bucket		54	54
2900	12' deep, 1-1/2 C.Y. bucket		66.50	66.50
2940	16' deep, 2 C.Y. bucket		101	101
2980	20' deep, 3-1/2 C.Y. bucket		138	138
3020	24' deep, 3-1/2 C.Y. bucket		191	191
3100	8' wide, 12' deep, 1-1/4 C.Y. bucket w/trench box		77	77
3120	15' deep, 1-1/2 C.Y. bucket		106	106
3140	18' deep, 2-1/2 C.Y. bucket		135	135
3180	24' deep, 3-1/2 C.Y. bucket		210	210
3270	10' wide, 20' deep, 3-1/2 C.Y. bucket w/trench box		169	169
3280	24' deep, 3-1/2 C.Y. bucket		230	230
3370	12' wide, 20' deep, 3-1/2 C.Y. bucket w/trench box		189	189
3380	25' deep, 3-1/2 C.Y. bucket		267	267
3500	1:1 slope, 2' wide, 2' deep, 3/8 C.Y. bucket		6.50	6.50
3520	3' deep, 3/8 C.Y. bucket		9	9
3540	4' deep, 3/8 C.Y. bucket		14.30	14.30
3560	6' deep, 3/8 C.Y. bucket		14.90	14.90
3580	8' deep, 1/2 C.Y. bucket		40	40
3600	10' deep, 1 C.Y. bucket		51	51
3800	4' wide, 2' deep, 3/8 C.Y. bucket		7.35	7.35
3820	3' deep, 3/8 C.Y. bucket		12.75	12.75
3840	4' deep, 1/2 C.Y. bucket		16.60	16.60
3860	6' deep, 1/2 C.Y. bucket		33	33
3880	8' deep, 1/2 C.Y. bucket		56	56
3900	10' deep, 1 C.Y. bucket		66.50	66.50
3920	12' deep, 1 C.Y. bucket		95.50	95.50
3940	15' deep, 1-1/2 C.Y. bucket		133	133
3960	18' deep, 2-1/2 C.Y. bucket		173	173
4030	6' wide, 6' deep, 5/8 C.Y. bucket w/trench box		44.50	44.50
4040	8' deep, 3/4 C.Y. bucket		68.50	68.50
4050	10' deep, 1 C.Y. bucket		78.50	78.50
4060	12' deep, 1-1/2 C.Y. bucket		99.50	99.50
4070	16' deep, 2 C.Y. bucket		159	159

G10 Site Preparation

G1030 Site Earthwork

G1030 807	Trenching Sand & Gravel	COST PER L.F.		
		MAT.	INST.	TOTAL
4080	20' deep, 3-1/2 C.Y. bucket		224	224
4090	24' deep, 3-1/2 C.Y. bucket		320	320
4500	8' wide, 12' deep, 1-1/2 C.Y. bucket w/trench box		111	111
4550	15' deep, 1-1/2 C.Y. bucket		159	159
4600	18' deep, 2-1/2 C.Y. bucket		209	209
4650	24' deep, 3-1/2 C.Y. bucket		335	335
4800	10' wide, 20' deep, 3-1/2 C.Y. bucket w/trench box		255	255
4850	24' deep, 3-1/2 C.Y. bucket		355	355
4950	12' wide, 20' deep, 3-1/2 C.Y. bucket w/trench box		272	272
4980	25' deep, 3-1/2 C.Y. bucket		405	405

G1030 815	Pipe Bedding	COST PER L.F.		
		MAT.	INST.	TOTAL
1440	Pipe bedding, side slope 0 to 1, 1' wide, pipe size 6" diameter	1.68	.92	2.60
1460	2' wide, pipe size 8" diameter	3.63	1.97	5.60
1500	Pipe size 12" diameter	3.78	2.05	5.83
1600	4' wide, pipe size 20" diameter	9.05	4.92	13.97
1660	Pipe size 30" diameter	9.50	5.15	14.65
1680	6' wide, pipe size 32" diameter	16.25	8.85	25.10
1740	8' wide, pipe size 60" diameter	32	17.45	49.45
1780	12' wide, pipe size 84" diameter	59	32	91

The Bituminous Roadway Systems are listed for pavement thicknesses between 3-1/2″ and 7″ and crushed stone bases from 3″ to 22″ in depth. Systems costs are expressed per linear foot for varying widths of two and multi-lane roads. Earth moving is not included. Granite curbs and line painting are added as required system components.

G2010 232	Bituminous Roadways Crushed Stone	COST PER L.F.		
		MAT.	INST.	TOTAL
1050	Bitum. roadway, two lanes, 3-1/2″ th. pvmt., 3″ th. crushed stone, 24′ wide	73	53.50	126.50
1100	28′ wide	80.50	54	134.50
1150	32′ wide	88	62.50	150.50
1200	4″ thick crushed stone, 24′ wide	75	51.50	126.50
1210	28′ wide	83	55.50	138.50
1220	32′ wide	91	59	150
1222	36′ wide	99	62.50	161.50
1224	40′ wide	107	66.50	173.50
1230	6″ thick crushed stone, 24′ wide	79.50	53.50	133
1240	28′ wide	88.50	57	145.50
1250	32′ wide	97	61.50	158.50
1252	36′ wide	106	65.50	171.50
1254	40′ wide	115	69.50	184.50
1256	8″ thick crushed stone, 24′ wide	84	55	139
1258	28′ wide	93.50	59.50	153
1260	32′ wide	103	63.50	166.50
1262	36′ wide	113	68	181
1264	40′ wide	122	72.50	194.50
1300	9″ thick crushed stone, 24′ wide	87	56	143
1350	28′ wide	96.50	61	157.50
1400	32′ wide	107	65.50	172.50
1410	10″ thick crushed stone, 24′ wide	89	56.50	145.50
1412	28′ wide	99	61	160
1414	32′ wide	109	66	175
1416	36′ wide	120	70.50	190.50
1418	40′ wide	130	75.50	205.50
1420	12″ thick crushed stone, 24′ wide	93.50	58.50	152
1422	28′ wide	104	63.50	167.50
1424	32′ wide	115	68.50	183.50
1426	36′ wide	126	73.50	199.50
1428	40′ wide	137	78	215
1430	15″ thick crushed stone, 24′ wide	100	61	161
1432	28′ wide	112	66.50	178.50
1434	32′ wide	124	71.50	195.50
1436	36′ wide	137	77	214
1438	40′ wide	149	82.50	231.50
1440	18″ thick crushed stone, 24′ wide	107	63.50	170.50
1442	28′ wide	120	69.50	189.50
1444	32′ wide	134	75.50	209.50
1446	36′ wide	147	81	228
1448	40′ wide	160	87	247
1550	4″ thick pavement., 4″ thick crushed stone, 24′ wide	81	52	133

511

G2010 Roadways

G2010 232	Bituminous Roadways Crushed Stone	COST PER L.F.		
		MAT.	INST.	TOTAL
1600	28' wide	90	56	146
1650	32' wide	99	60	159
1652	36' wide	107	63.50	170.50
1654	40' wide	116	67.50	183.50
1700	6" thick crushed stone, 24' wide	85.50	54	139.50
1710	28' wide	95	58	153
1720	32' wide	105	62.50	167.50
1722	36' wide	114	66	180
1724	40' wide	124	70.50	194.50
1726	8" thick crushed stone, 24' wide	90	55.50	145.50
1728	28' wide	100	60	160
1730	32' wide	111	64	175
1732	36' wide	121	69	190
1734	40' wide	132	73.50	205.50
1800	10" thick crushed stone, 24' wide	94.50	57.50	152
1850	28' wide	106	61	167
1900	32' wide	117	67	184
1902	36' wide	128	71.50	199.50
1904	40' wide	133	74.50	207.50

G20 Site Improvements

G2020 Parking Lots

G2020 210	Parking Lots Gravel Base	COST PER CAR		
		MAT.	INST.	TOTAL
1500	Parking lot, 90° angle parking, 3" bituminous paving, 6" gravel base	830	440	1,270
1540	10" gravel base	945	510	1,455
1560	4" bituminous paving, 6" gravel base	1,050	465	1,515
1600	10" gravel base	1,175	535	1,710
1620	6" bituminous paving, 6" gravel base	1,425	495	1,920
1660	10" gravel base	1,550	570	2,120
1800	60° angle parking, 3" bituminous paving, 6" gravel base	830	440	1,270
1840	10" gravel base	945	510	1,455
1860	4" bituminous paving, 6" gravel base	1,050	465	1,515
1900	10" gravel base	1,175	535	1,710
1920	6" bituminous paving, 6" gravel base	1,425	495	1,920
1960	10" gravel base	1,550	570	2,120
2200	45° angle parking, 3" bituminous paving, 6" gravel base	850	450	1,300
2240	10" gravel base	970	525	1,495
2260	4" bituminous paving, 6" gravel base	1,075	480	1,555
2300	10" gravel base	1,200	550	1,750
2320	6" bituminous paving, 6" gravel base	1,450	510	1,960
2360	10" gravel base	1,575	585	2,160

For customer support on your Square Foot Cost Data, call 877.694.8329.

513

G20 Site Improvements

G2040 Site Development

G2040 810	Flagpoles	COST EACH		
		MAT.	INST.	TOTAL
0110	Flagpoles, on grade, aluminum, tapered, 20' high	1,150	690	1,840
0120	70' high	9,225	1,750	10,975
0130	Fiberglass, tapered, 23' high	635	690	1,325
0140	59' high	5,300	1,550	6,850
0150	Concrete, internal halyard, 20' high	1,500	555	2,055
0160	100' high	19,400	1,375	20,775

G2040 950	Other Site Development, EACH	COST EACH		
		MAT.	INST.	TOTAL
0110	Grandstands, permanent, closed deck, steel, economy (per seat)			22
0120	Deluxe (per seat)			24.50
0130	Composite design, economy (per seat)			42
0140	Deluxe (per seat)			75

514

G3030 Storm Sewer

G3030 210	Manholes & Catch Basins	COST PER EACH		
		MAT.	INST.	TOTAL
1920	Manhole/catch basin, brick, 4' I.D. riser, 4' deep	1,300	1,775	3,075
1980	10' deep	2,900	4,200	7,100
3200	Block, 4' I.D. riser, 4' deep	1,175	1,450	2,625
3260	10' deep	2,625	3,500	6,125
4620	Concrete, cast-in-place, 4' I.D. riser, 4' deep	1,275	2,475	3,750
4680	10' deep	3,075	5,950	9,025
5820	Concrete, precast, 4' I.D. riser, 4' deep	1,525	1,325	2,850
5880	10' deep	3,125	3,050	6,175
6200	6' I.D. riser, 4' deep	3,325	1,975	5,300
6260	10' deep	6,675	4,650	11,325

515

For customer support on your Square Foot Cost Data, call 877.694.8329.

G4020 Site Lighting

Light Pole

G4020 210	Light Pole (Installed)	COST EACH		
		MAT.	**INST.**	**TOTAL**
0200	Light pole, aluminum, 20' high, 1 arm bracket	1,250	1,175	2,425
0240	2 arm brackets	1,375	1,175	2,550
0280	3 arm brackets	1,500	1,225	2,725
0320	4 arm brackets	1,650	1,225	2,875
0360	30' high, 1 arm bracket	2,200	1,500	3,700
0400	2 arm brackets	2,325	1,500	3,825
0440	3 arm brackets	2,450	1,525	3,975
0480	4 arm brackets	2,600	1,525	4,125
0680	40' high, 1 arm bracket	2,675	2,000	4,675
0720	2 arm brackets	2,825	2,000	4,825
0760	3 arm brackets	2,950	2,050	5,000
0800	4 arm brackets	3,100	2,050	5,150
0840	Steel, 20' high, 1 arm bracket	1,475	1,250	2,725
0880	2 arm brackets	1,525	1,250	2,775
0920	3 arm brackets	1,575	1,300	2,875
0960	4 arm brackets	1,625	1,300	2,925
1000	30' high, 1 arm bracket	1,725	1,600	3,325
1040	2 arm brackets	1,775	1,600	3,375
1080	3 arm brackets	1,800	1,650	3,450
1120	4 arm brackets	1,875	1,650	3,525
1320	40' high, 1 arm bracket	2,275	2,175	4,450
1360	2 arm brackets	2,325	2,175	4,500
1400	3 arm brackets	2,350	2,225	4,575
1440	4 arm brackets	2,425	2,225	4,650

Reference Section

All the reference information is in one section making it easy to find what you need to know . . . and easy to use the book on a daily basis. This section is visually identified by a vertical black bar on the edge of pages.

In this Reference Section, we've included General Conditions; Location Factors for adjusting costs to the region you are in; Historical Cost Indexes for cost comparisons over time; a Glossary; an explanation of all Abbreviations used in the book; and sample Estimating Forms.

Table of Contents

General Conditions, Overhead & Profit

The total building costs in the Commercial/Industrial/Institutional section include a 10% allowance for general conditions and a 15% allowance for the general contractor's overhead and profit and contingencies.

General contractor overhead includes indirect costs such as permits, Workers' Compensation, insurances, supervision and bonding fees. Overhead will vary with the size of project, the contractor's operating procedures and location. Profits will vary with economic activity and local conditions.

Contingencies provide for unforeseen construction difficulties which include material shortages and weather. In all situations, the appraiser should give consideration to possible adjustment of the 25% factor used in developing the Commercial/Industrial/Institutional models.

Architectural Fees

Tabulated below are typical percentage fees by project size, for good professional architectural service. Fees may vary from those listed depending upon degree of design difficulty and economic conditions in any particular area.

Rates can be interpolated horizontally and vertically. Various portions of the same project requiring different rates should be adjusted proportionately. For alterations, add 50% to the fee for the first $500,000 of project cost and add 25% to the fee for project cost over $500,000.

Architectural fees tabulated below include Engineering fees.

Insurance Exclusions

Many insurance companies exclude from coverage such items as architect's fees, excavation, foundations below grade, underground piping and site preparation. Since exclusions vary among insurance companies, it is recommended that for greatest accuracy each exclusion be priced separately using the unit-in-place section.

As a rule of thumb, exclusions can be calculated at 9% of total building cost plus the appropriate allowance for architect's fees.

Building Types	Total Project Size in Thousands of Dollars						
	100	250	500	1,000	5,000	10,000	50,000
Factories, garages, warehouses, repetitive housing	9.0%	8.0%	7.0%	6.2%	5.3%	4.9%	4.5%
Apartments, banks, schools, libraries, offices, municipal buildings	12.2	12.3	9.2	8.0	7.0	6.6	6.2
Churches, hospitals, homes, laboratories, museums, research	15.0	13.6	12.7	11.9	9.5	8.8	8.0
Memorials, monumental work, decorative furnishings	–	16.0	14.5	13.1	10.0	9.0	8.3

Location Factors

Costs shown in *RSMeans Square Foot Costs* are based on national averages for materials and installation. To adjust these costs to a specific location, simply multiply the base cost by the factor for that city.

The data is arranged alphabetically by state and postal zip code numbers. For a city not listed, use the factor for a nearby city with similar economic characteristics.

STATE/ZIP	CITY	Residential	Commercial
ALABAMA			
350-352	Birmingham	.88	.90
354	Tuscaloosa	.78	.83
355	Jasper	.72	.82
356	Decatur	.76	.82
357-358	Huntsville	.86	.88
359	Gadsden	.75	.83
360-361	Montgomery	.75	.81
362	Anniston	.75	.83
363	Dothan	.74	.77
364	Evergreen	.71	.78
365-366	Mobile	.80	.85
367	Selma	.71	.77
368	Phenix City	.73	.80
369	Butler	.72	.77
ALASKA			
995-996	Anchorage	1.25	1.19
997	Fairbanks	1.28	1.19
998	Juneau	1.25	1.19
999	Ketchikan	1.29	1.25
ARIZONA			
850,853	Phoenix	.86	.88
851,852	Mesa/Tempe	.86	.87
855	Globe	.83	.86
856-857	Tucson	.85	.87
859	Show Low	.87	.87
860	Flagstaff	.86	.88
863	Prescott	.82	.86
864	Kingman	.83	.86
865	Chambers	.83	.86
ARKANSAS			
716	Pine Bluff	.78	.84
717	Camden	.69	.77
718	Texarkana	.73	.77
719	Hot Springs	.70	.77
720-722	Little Rock	.81	.83
723	West Memphis	.77	.81
724	Jonesboro	.75	.80
725	Batesville	.73	.77
726	Harrison	.73	.76
727	Fayetteville	.69	.75
728	Russellville	.75	.77
729	Fort Smith	.81	.82
CALIFORNIA			
900-902	Los Angeles	1.09	1.07
903-905	Inglewood	1.05	1.03
906-908	Long Beach	1.04	1.04
910-912	Pasadena	1.04	1.04
913-916	Van Nuys	1.07	1.06
917-918	Alhambra	1.08	1.06
919-921	San Diego	1.05	1.05
922	Palm Springs	1.04	1.04
923-924	San Bernardino	1.05	1.02
925	Riverside	1.08	1.06
926-927	Santa Ana	1.06	1.03
928	Anaheim	1.08	1.06
930	Oxnard	1.09	1.06
931	Santa Barbara	1.08	1.06
932-933	Bakersfield	1.07	1.06
934	San Luis Obispo	1.08	1.05
935	Mojave	1.05	1.03
936-938	Fresno	1.12	1.07
939	Salinas	1.15	1.10
940-941	San Francisco	1.25	1.23
942,956-958	Sacramento	1.12	1.09
943	Palo Alto	1.19	1.14
944	San Mateo	1.23	1.16
945	Vallejo	1.16	1.12
946	Oakland	1.22	1.17
947	Berkeley	1.24	1.17
948	Richmond	1.24	1.15
949	San Rafael	1.24	1.18
950	Santa Cruz	1.17	1.12
951	San Jose	1.23	1.17
952	Stockton	1.13	1.08
953	Modesto	1.12	1.07

STATE/ZIP	CITY	Residential	Commercial
CALIFORNIA (CONT'D)			
954	Santa Rosa	1.20	1.16
955	Eureka	1.14	1.08
959	Marysville	1.13	1.08
960	Redding	1.18	1.12
961	Susanville	1.17	1.12
COLORADO			
800-802	Denver	.90	.93
803	Boulder	.91	.91
804	Golden	.87	.91
805	Fort Collins	.88	.92
806	Greeley	.86	.89
807	Fort Morgan	.89	.90
808-809	Colorado Springs	.88	.93
810	Pueblo	.90	.92
811	Alamosa	.85	.90
812	Salida	.88	.90
813	Durango	.89	.91
814	Montrose	.86	.90
815	Grand Junction	.92	.92
816	Glenwood Springs	.88	.92
CONNECTICUT			
060	New Britain	1.11	1.09
061	Hartford	1.10	1.10
062	Willimantic	1.11	1.09
063	New London	1.11	1.07
064	Meriden	1.11	1.08
065	New Haven	1.11	1.10
066	Bridgeport	1.12	1.10
067	Waterbury	1.11	1.10
068	Norwalk	1.13	1.13
069	Stamford	1.13	1.13
D.C.			
200-205	Washington	.93	.97
DELAWARE			
197	Newark	1.02	1.03
198	Wilmington	1.02	1.04
199	Dover	1.01	1.04
FLORIDA			
320,322	Jacksonville	.81	.85
321	Daytona Beach	.86	.88
323	Tallahassee	.78	.82
324	Panama City	.76	.82
325	Pensacola	.81	.85
326,344	Gainesville	.80	.86
327-328,347	Orlando	.86	.87
329	Melbourne	.86	.91
330-332,340	Miami	.84	.88
333	Fort Lauderdale	.84	.87
334,349	West Palm Beach	.85	.86
335-336,346	Tampa	.88	.91
337	St. Petersburg	.79	.86
338	Lakeland	.85	.90
339,341	Fort Myers	.84	.87
342	Sarasota	.88	.89
GEORGIA			
300-303,399	Atlanta	.87	.88
304	Statesboro	.73	.79
305	Gainesville	.77	.81
306	Athens	.76	.82
307	Dalton	.75	.81
308-309	Augusta	.81	.82
310-312	Macon	.77	.83
313-314	Savannah	.78	.83
315	Waycross	.77	.81
316	Valdosta	.73	.82
317,398	Albany	.75	.82
318-319	Columbus	.78	.84
HAWAII			
967	Hilo	1.21	1.17
968	Honolulu	1.25	1.20

For customer support on your Square Foot Cost Data, call 877.694.8329.

519

Location Factors

STATE/ZIP	CITY	Residential	Commercial
STATES & POSS.			
969	Guam	.97	1.02
IDAHO			
832	Pocatello	.87	.91
833	Twin Falls	.81	.87
834	Idaho Falls	.86	.89
835	Lewiston	1.00	.98
836-837	Boise	.88	.91
838	Coeur d'Alene	.98	.97
ILLINOIS			
600-603	North Suburban	1.20	1.15
604	Joliet	1.22	1.16
605	South Suburban	1.20	1.15
606-608	Chicago	1.22	1.17
609	Kankakee	1.14	1.11
610-611	Rockford	1.12	1.11
612	Rock Island	.98	.99
613	La Salle	1.11	1.10
614	Galesburg	1.03	1.02
615-616	Peoria	1.06	1.04
617	Bloomington	1.03	1.03
618-619	Champaign	1.04	1.04
620-622	East St. Louis	1.02	1.02
623	Quincy	1.01	.99
624	Effingham	1.03	1.01
625	Decatur	1.02	1.02
626-627	Springfield	1.03	1.03
628	Centralia	1.02	1.01
629	Carbondale	1.01	1.01
INDIANA			
460	Anderson	.92	.91
461-462	Indianapolis	.93	.93
463-464	Gary	1.06	1.04
465-466	South Bend	.91	.92
467-468	Fort Wayne	.89	.90
469	Kokomo	.91	.89
470	Lawrenceburg	.87	.87
471	New Albany	.87	.87
472	Columbus	.92	.90
473	Muncie	.92	.91
474	Bloomington	.94	.92
475	Washington	.90	.90
476-477	Evansville	.91	.93
478	Terre Haute	.91	.93
479	Lafayette	.93	.91
IOWA			
500-503,509	Des Moines	.91	.94
504	Mason City	.76	.84
505	Fort Dodge	.74	.81
506-507	Waterloo	.83	.87
508	Creston	.83	.89
510-511	Sioux City	.86	.88
512	Sibley	.74	.80
513	Spencer	.75	.81
514	Carroll	.80	.86
515	Council Bluffs	.86	.91
516	Shenandoah	.81	.86
520	Dubuque	.87	.91
521	Decorah	.78	.83
522-524	Cedar Rapids	.93	.93
525	Ottumwa	.87	.87
526	Burlington	.88	.88
527-528	Davenport	.97	.96
KANSAS			
660-662	Kansas City	.97	.98
664-666	Topeka	.78	.85
667	Fort Scott	.88	.87
668	Emporia	.83	.85
669	Belleville	.80	.84
670-672	Wichita	.79	.86
673	Independence	.88	.87
674	Salina	.80	.87
675	Hutchinson	.79	.83
676	Hays	.82	.85
677	Colby	.83	.85
678	Dodge City	.81	.87
679	Liberal	.80	.85
KENTUCKY			
400-402	Louisville	.90	.91
403-405	Lexington	.88	.90

STATE/ZIP	CITY	Residential	Commercial
KENTUCKY (CONT'D)			
406	Frankfort	.87	.91
407-409	Corbin	.81	.84
410	Covington	.91	.94
411-412	Ashland	.94	.97
413-414	Campton	.87	.88
415-416	Pikeville	.91	.93
417-418	Hazard	.86	.88
420	Paducah	.88	.90
421-422	Bowling Green	.89	.91
423	Owensboro	.91	.93
424	Henderson	.91	.90
425-426	Somerset	.85	.87
427	Elizabethtown	.87	.88
LOUISIANA			
700-701	New Orleans	.85	.88
703	Thibodaux	.82	.84
704	Hammond	.75	.78
705	Lafayette	.82	.83
706	Lake Charles	.83	.84
707-708	Baton Rouge	.82	.84
710-711	Shreveport	.79	.83
712	Monroe	.75	.81
713-714	Alexandria	.77	.81
MAINE			
039	Kittery	.92	.91
040-041	Portland	.96	.97
042	Lewiston	.96	.96
043	Augusta	.92	.92
044	Bangor	.95	.95
045	Bath	.92	.91
046	Machias	.95	.92
047	Houlton	.96	.92
048	Rockland	.95	.92
049	Waterville	.90	.90
MARYLAND			
206	Waldorf	.89	.90
207-208	College Park	.87	.91
209	Silver Spring	.88	.91
210-212	Baltimore	.90	.93
214	Annapolis	.89	.93
215	Cumberland	.88	.90
216	Easton	.83	.85
217	Hagerstown	.87	.91
218	Salisbury	.82	.83
219	Elkton	.88	.90
MASSACHUSETTS			
010-011	Springfield	1.08	1.06
012	Pittsfield	1.07	1.05
013	Greenfield	1.06	1.04
014	Fitchburg	1.14	1.10
015-016	Worcester	1.16	1.12
017	Framingham	1.19	1.13
018	Lowell	1.19	1.15
019	Lawrence	1.20	1.15
020-022, 024	Boston	1.22	1.18
023	Brockton	1.18	1.12
025	Buzzards Bay	1.16	1.09
026	Hyannis	1.15	1.11
027	New Bedford	1.17	1.12
MICHIGAN			
480,483	Royal Oak	1.03	1.00
481	Ann Arbor	1.03	1.02
482	Detroit	1.06	1.03
484-485	Flint	.96	.96
486	Saginaw	.93	.94
487	Bay City	.94	.95
488-489	Lansing	.95	.97
490	Battle Creek	.93	.94
491	Kalamazoo	.92	.93
492	Jackson	.93	.96
493,495	Grand Rapids	.91	.93
494	Muskegon	.90	.92
496	Traverse City	.85	.89
497	Gaylord	.89	.91
498-499	Iron mountain	.90	.94
MINNESOTA			
550-551	Saint Paul	1.09	1.09
553-555	Minneapolis	1.12	1.09
556-558	Duluth	1.04	1.04

For customer support on your Square Foot Cost Data, call 877.694.8329.

Location Factors

STATE/ZIP	CITY	Residential	Commercial
MINNESOTA (CONT'd)			
559	Rochester	1.03	1.02
560	Mankato	.99	.99
561	Windom	.93	.94
562	Willmar	.96	.98
563	St. Cloud	1.04	1.05
564	Brainerd	.95	.99
565	Detroit Lakes	.94	.97
566	Bemidji	.94	.98
567	Thief River Falls	.93	.95
MISSISSIPPI			
386	Clarksdale	.74	.78
387	Greenville	.80	.84
388	Tupelo	.75	.80
389	Greenwood	.77	.78
390-392	Jackson	.81	.84
393	Meridian	.80	.83
394	Laurel	.77	.80
395	Biloxi	.79	.81
396	Mccomb	.74	.78
397	Columbus	.75	.79
MISSOURI			
630-631	St. Louis	1.02	1.03
633	Bowling Green	.95	.96
634	Hannibal	.92	.95
635	Kirksville	.89	.95
636	Flat River	.95	.97
637	Cape Girardeau	.91	.96
638	Sikeston	.90	.95
639	Poplar Bluff	.89	.94
640-641	Kansas City	1.04	1.03
644-645	St. Joseph	.97	.98
646	Chillicothe	.96	.96
647	Harrisonville	.98	.99
648	Joplin	.91	.92
650-651	Jefferson City	.94	.97
652	Columbia	.92	.98
653	Sedalia	.94	.99
654-655	Rolla	.96	.97
656-658	Springfield	.91	.94
MONTANA			
590-591	Billings	.90	.92
592	Wolf Point	.87	.90
593	Miles City	.88	.89
594	Great Falls	.91	.92
595	Havre	.84	.89
596	Helena	.87	.90
597	Butte	.87	.92
598	Missoula	.88	.89
599	Kalispell	.86	.89
NEBRASKA			
680-681	Omaha	.90	.91
683-685	Lincoln	.87	.89
686	Columbus	.87	.89
687	Norfolk	.89	.90
688	Grand Island	.88	.90
689	Hastings	.90	.91
690	Mccook	.83	.87
691	North Platte	.88	.90
692	Valentine	.84	.89
693	Alliance	.83	.88
NEVADA			
889-891	Las Vegas	1.03	1.04
893	Ely	1.02	1.01
894-895	Reno	.95	.97
897	Carson City	.96	.97
898	Elko	.93	.92
NEW HAMPSHIRE			
030	Nashua	.98	.98
031	Manchester	.97	.98
032-033	Concord	.98	.97
034	Keene	.88	.87
035	Littleton	.92	.88
036	Charleston	.86	.85
037	Claremont	.86	.84
038	Portsmouth	.97	.96

STATE/ZIP	CITY	Residential	Commercial
NEW JERSEY			
070-071	Newark	1.14	1.13
072	Elizabeth	1.16	1.11
073	Jersey City	1.12	1.11
074-075	Paterson	1.13	1.12
076	Hackensack	1.13	1.11
077	Long Branch	1.13	1.10
078	Dover	1.13	1.11
079	Summit	1.14	1.11
080,083	Vineland	1.11	1.09
081	Camden	1.13	1.11
082,084	Atlantic City	1.15	1.10
085-086	Trenton	1.13	1.12
087	Point Pleasant	1.12	1.11
088-089	New Brunswick	1.15	1.12
NEW MEXICO			
870-872	Albuquerque	.84	.88
873	Gallup	.83	.88
874	Farmington	.84	.88
875	Santa Fe	.85	.89
877	Las Vegas	.84	.87
878	Socorro	.83	.87
879	Truth/Consequences	.82	.85
880	Las Cruces	.81	.84
881	Clovis	.83	.87
882	Roswell	.83	.88
883	Carrizozo	.83	.88
884	Tucumcari	.84	.87
NEW YORK			
100-102	New York	1.35	1.32
103	Staten Island	1.31	1.29
104	Bronx	1.30	1.28
105	Mount Vernon	1.16	1.17
106	White Plains	1.19	1.17
107	Yonkers	1.21	1.20
108	New Rochelle	1.20	1.17
109	Suffern	1.15	1.11
110	Queens	1.31	1.30
111	Long Island City	1.34	1.31
112	Brooklyn	1.37	1.31
113	Flushing	1.33	1.31
114	Jamaica	1.32	1.30
115,117,118	Hicksville	1.23	1.24
116	Far Rockaway	1.32	1.31
119	Riverhead	1.23	1.22
120-122	Albany	1.00	1.02
123	Schenectady	1.02	1.02
124	Kingston	1.05	1.10
125-126	Poughkeepsie	1.22	1.17
127	Monticello	1.04	1.07
128	Glens Falls	.93	.96
129	Plattsburgh	1.00	.98
130-132	Syracuse	.98	.99
133-135	Utica	.96	.97
136	Watertown	.94	.98
137-139	Binghamton	.99	1.00
140-142	Buffalo	1.06	1.03
143	Niagara Falls	1.03	1.02
144-146	Rochester	1.00	1.00
147	Jamestown	.93	.95
148-149	Elmira	.95	.97
NORTH CAROLINA			
270,272-274	Greensboro	.84	.81
271	Winston-Salem	.85	.82
275-276	Raleigh	.84	.80
277	Durham	.84	.82
278	Rocky Mount	.80	.78
279	Elizabeth City	.79	.79
280	Gastonia	.85	.81
281-282	Charlotte	.87	.82
283	Fayetteville	.90	.84
284	Wilmington	.86	.81
285	Kinston	.81	.79
286	Hickory	.80	.79
287-288	Asheville	.84	.81
289	Murphy	.83	.77
NORTH DAKOTA			
580-581	Fargo	.80	.87
582	Grand Forks	.74	.80
583	Devils Lake	.78	.83
584	Jamestown	.72	.77
585	Bismarck	.79	.87

Location Factors

STATE/ZIP	CITY	Residential	Commercial
NORTH DAKOTA (CONT'D)			
586	Dickinson	.76	.85
587	Minot	.84	.89
588	Williston	.76	.84
OHIO			
430-432	Columbus	.94	.95
433	Marion	.91	.91
434-436	Toledo	.99	.98
437-438	Zanesville	.91	.92
439	Steubenville	.95	.96
440	Lorain	.96	.96
441	Cleveland	1.00	1.00
442-443	Akron	.98	.97
444-445	Youngstown	.95	.95
446-447	Canton	.93	.93
448-449	Mansfield	.91	.93
450	Hamilton	.92	.92
451-452	Cincinnati	.92	.92
453-454	Dayton	.93	.92
455	Springfield	.94	.92
456	Chillicothe	.97	.95
457	Athens	.90	.91
458	Lima	.92	.94
OKLAHOMA			
730-731	Oklahoma City	.83	.85
734	Ardmore	.78	.82
735	Lawton	.80	.84
736	Clinton	.79	.84
737	Enid	.76	.83
738	Woodward	.79	.83
739	Guymon	.78	.82
740-741	Tulsa	.78	.82
743	Miami	.84	.83
744	Muskogee	.74	.79
745	Mcalester	.75	.80
746	Ponca City	.78	.81
747	Durant	.77	.81
748	Shawnee	.76	.81
749	Poteau	.76	.80
OREGON			
970-972	Portland	.99	1.00
973	Salem	.99	1.00
974	Eugene	.99	.99
975	Medford	.97	.99
976	Klamath Falls	.98	.99
977	Bend	1.00	.99
978	Pendleton	.98	.98
979	Vale	.98	.95
PENNSYLVANIA			
150-152	Pittsburgh	1.00	1.02
153	Washington	.96	1.00
154	Uniontown	.94	.99
155	Bedford	.90	.95
156	Greensburg	.96	.99
157	Indiana	.93	.98
158	Dubois	.92	.98
159	Johnstown	.92	.98
160	Butler	.93	.97
161	New Castle	.92	.95
162	Kittanning	.94	.97
163	Oil City	.90	.94
164-165	Erie	.94	.95
166	Altoona	.88	.93
167	Bradford	.91	.96
168	State College	.90	.94
169	Wellsboro	.91	.95
170-171	Harrisburg	.95	.98
172	Chambersburg	.88	.93
173-174	York	.92	.96
175-176	Lancaster	.91	.92
177	Williamsport	.84	.89
178	Sunbury	.93	.96
179	Pottsville	.93	.97
180	Lehigh Valley	1.01	1.04
181	Allentown	1.03	1.03
182	Hazleton	.92	.97
183	Stroudsburg	.92	.98
184-185	Scranton	.97	.99
186-187	Wilkes-Barre	.94	.97
188	Montrose	.91	.96
189	Doylestown	1.06	1.06

STATE/ZIP	CITY	Residential	Commercial
PENNSYLVANIA (CONT'D)			
190-191	Philadelphia	1.16	1.15
193	Westchester	1.09	1.07
194	Norristown	1.09	1.10
195-196	Reading	.98	.99
PUERTO RICO			
009	San Juan	.76	.80
RHODE ISLAND			
028	Newport	1.09	1.07
029	Providence	1.10	1.08
SOUTH CAROLINA			
290-292	Columbia	.85	.81
293	Spartanburg	.85	.80
294	Charleston	.93	.85
295	Florence	.81	.80
296	Greenville	.84	.80
297	Rock Hill	.80	.77
298	Aiken	.96	.86
299	Beaufort	.79	.76
SOUTH DAKOTA			
570-571	Sioux Falls	.76	.82
572	Watertown	.73	.77
573	Mitchell	.74	.76
574	Aberdeen	.75	.78
575	Pierre	.78	.83
576	Mobridge	.73	.77
577	Rapid City	.78	.83
TENNESSEE			
370-372	Nashville	.85	.88
373-374	Chattanooga	.83	.85
375,380-381	Memphis	.84	.87
376	Johnson City	.71	.81
377-379	Knoxville	.81	.84
382	Mckenzie	.73	.81
383	Jackson	.76	.83
384	Columbia	.77	.84
385	Cookeville	.70	.81
TEXAS			
750	Mckinney	.82	.83
751	Waxahackie	.83	.84
752-753	Dallas	.85	.86
754	Greenville	.83	.84
755	Texarkana	.80	.81
756	Longview	.80	.81
757	Tyler	.84	.84
758	Palestine	.80	.82
759	Lufkin	.82	.84
760-761	Fort Worth	.83	.85
762	Denton	.84	.83
763	Wichita Falls	.85	.83
764	Eastland	.83	.82
765	Temple	.82	.80
766-767	Waco	.84	.82
768	Brownwood	.80	.82
769	San Angelo	.79	.82
770-772	Houston	.84	.87
773	Huntsville	.81	.84
774	Wharton	.81	.85
775	Galveston	.83	.85
776-777	Beaumont	.85	.85
778	Bryan	.80	.83
779	Victoria	.82	.84
780	Laredo	.82	.84
781-782	San Antonio	.82	.84
783-784	Corpus Christi	.86	.85
785	Mc Allen	.85	.83
786-787	Austin	.81	.84
788	Del Rio	.82	.84
789	Giddings	.81	.83
790-791	Amarillo	.81	.84
792	Childress	.83	.84
793-794	Lubbock	.83	.85
795-796	Abilene	.83	.83
797	Midland	.86	.85
798-799,885	El Paso	.80	.82
UTAH			
840-841	Salt Lake City	.81	.87
842,844	Ogden	.80	.86
843	Logan	.80	.87

Location Factors

<table>
<tr><th>STATE/ZIP</th><th>CITY</th><th>Residential</th><th>Commercial</th></tr>
<tr><td colspan="4">UTAH (CONT'D)</td></tr>
<tr><td>845</td><td>Price</td><td>.78</td><td>.85</td></tr>
<tr><td>846-847</td><td>Provo</td><td>.81</td><td>.86</td></tr>
<tr><td colspan="4">VERMONT</td></tr>
<tr><td>050</td><td>White River Jct.</td><td>.90</td><td>.90</td></tr>
<tr><td>051</td><td>Bellows Falls</td><td>.96</td><td>.95</td></tr>
<tr><td>052</td><td>Bennington</td><td>1.00</td><td>.96</td></tr>
<tr><td>053</td><td>Brattleboro</td><td>.96</td><td>.96</td></tr>
<tr><td>054</td><td>Burlington</td><td>.94</td><td>.94</td></tr>
<tr><td>056</td><td>Montpelier</td><td>.93</td><td>.92</td></tr>
<tr><td>057</td><td>Rutland</td><td>.94</td><td>.93</td></tr>
<tr><td>058</td><td>St. Johnsbury</td><td>.91</td><td>.90</td></tr>
<tr><td>059</td><td>Guildhall</td><td>.91</td><td>.89</td></tr>
<tr><td colspan="4">VIRGINIA</td></tr>
<tr><td>220-221</td><td>Fairfax</td><td>1.02</td><td>.93</td></tr>
<tr><td>222</td><td>Arlington</td><td>1.04</td><td>.94</td></tr>
<tr><td>223</td><td>Alexandria</td><td>1.05</td><td>.94</td></tr>
<tr><td>224-225</td><td>Fredericksburg</td><td>1.01</td><td>.92</td></tr>
<tr><td>226</td><td>Winchester</td><td>1.00</td><td>.92</td></tr>
<tr><td>227</td><td>Culpeper</td><td>1.01</td><td>.91</td></tr>
<tr><td>228</td><td>Harrisonburg</td><td>.87</td><td>.86</td></tr>
<tr><td>229</td><td>Charlottesville</td><td>.90</td><td>.86</td></tr>
<tr><td>230-232</td><td>Richmond</td><td>.95</td><td>.87</td></tr>
<tr><td>233-235</td><td>Norfolk</td><td>.96</td><td>.86</td></tr>
<tr><td>236</td><td>Newport News</td><td>.96</td><td>.86</td></tr>
<tr><td>237</td><td>Portsmouth</td><td>.90</td><td>.85</td></tr>
<tr><td>238</td><td>Petersburg</td><td>.93</td><td>.86</td></tr>
<tr><td>239</td><td>Farmville</td><td>.84</td><td>.79</td></tr>
<tr><td>240-241</td><td>Roanoke</td><td>.95</td><td>.86</td></tr>
<tr><td>242</td><td>Bristol</td><td>.83</td><td>.80</td></tr>
<tr><td>243</td><td>Pulaski</td><td>.83</td><td>.81</td></tr>
<tr><td>244</td><td>Staunton</td><td>.86</td><td>.83</td></tr>
<tr><td>245</td><td>Lynchburg.</td><td>.93</td><td>.85</td></tr>
<tr><td>246</td><td>Grundy</td><td>.80</td><td>.79</td></tr>
<tr><td colspan="4">WASHINGTON</td></tr>
<tr><td>980-981,987</td><td>Seattle</td><td>1.03</td><td>1.03</td></tr>
<tr><td>982</td><td>Everett</td><td>1.04</td><td>1.01</td></tr>
<tr><td>983-984</td><td>Tacoma</td><td>1.01</td><td>1.01</td></tr>
<tr><td>985</td><td>Olympia</td><td>.98</td><td>1.00</td></tr>
<tr><td>986</td><td>Vancouver</td><td>.96</td><td>.98</td></tr>
<tr><td>988</td><td>Wenatchee</td><td>.93</td><td>.95</td></tr>
<tr><td>989</td><td>Yakima</td><td>.98</td><td>.99</td></tr>
<tr><td>990-992</td><td>Spokane</td><td>1.00</td><td>.94</td></tr>
<tr><td>993</td><td>Richland</td><td>1.00</td><td>.97</td></tr>
<tr><td>994</td><td>Clarkston</td><td>.94</td><td>.92</td></tr>
<tr><td colspan="4">WEST VIRGINIA</td></tr>
<tr><td>247-248</td><td>Bluefield</td><td>.94</td><td>.96</td></tr>
<tr><td>249</td><td>Lewisburg</td><td>.93</td><td>.97</td></tr>
<tr><td>250-253</td><td>Charleston</td><td>.97</td><td>.98</td></tr>
<tr><td>254</td><td>Martinsburg</td><td>.89</td><td>.94</td></tr>
<tr><td>255-257</td><td>Huntington</td><td>.98</td><td>.99</td></tr>
<tr><td>258-259</td><td>Beckley</td><td>.94</td><td>.96</td></tr>
<tr><td>260</td><td>Wheeling</td><td>.95</td><td>.98</td></tr>
<tr><td>261</td><td>Parkersburg</td><td>.93</td><td>.97</td></tr>
<tr><td>262</td><td>Buckhannon</td><td>.95</td><td>.98</td></tr>
<tr><td>263-264</td><td>Clarksburg</td><td>.95</td><td>.98</td></tr>
<tr><td>265</td><td>Morgantown</td><td>.95</td><td>.98</td></tr>
<tr><td>266</td><td>Gassaway</td><td>.94</td><td>.97</td></tr>
<tr><td>267</td><td>Romney</td><td>.93</td><td>.96</td></tr>
<tr><td>268</td><td>Petersburg</td><td>.94</td><td>.96</td></tr>
<tr><td colspan="4">WISCONSIN</td></tr>
<tr><td>530,532</td><td>Milwaukee</td><td>1.07</td><td>1.03</td></tr>
<tr><td>531</td><td>Kenosha</td><td>1.03</td><td>1.01</td></tr>
<tr><td>534</td><td>Racine</td><td>1.02</td><td>1.01</td></tr>
<tr><td>535</td><td>Beloit</td><td>1.00</td><td>1.00</td></tr>
<tr><td>537</td><td>Madison</td><td>1.00</td><td>.99</td></tr>
<tr><td>538</td><td>Lancaster</td><td>.97</td><td>.95</td></tr>
<tr><td>539</td><td>Portage</td><td>.96</td><td>.96</td></tr>
<tr><td>540</td><td>New Richmond</td><td>.98</td><td>.96</td></tr>
<tr><td>541-543</td><td>Green Bay</td><td>1.01</td><td>.98</td></tr>
<tr><td>544</td><td>Wausau</td><td>.96</td><td>.95</td></tr>
<tr><td>545</td><td>Rhinelander</td><td>.95</td><td>.96</td></tr>
<tr><td>546</td><td>La Crosse</td><td>.96</td><td>.96</td></tr>
<tr><td>547</td><td>Eau Claire</td><td>.98</td><td>.98</td></tr>
<tr><td>548</td><td>Superior</td><td>.96</td><td>.97</td></tr>
<tr><td>549</td><td>Oshkosh</td><td>.96</td><td>.95</td></tr>
<tr><td colspan="4">WYOMING</td></tr>
<tr><td>820</td><td>Cheyenne</td><td>.81</td><td>.86</td></tr>
<tr><td>821</td><td>Yellowstone Nat. Pk.</td><td>.81</td><td>.85</td></tr>
<tr><td>822</td><td>Wheatland</td><td>.79</td><td>.85</td></tr>
</table>

<table>
<tr><th>STATE/ZIP</th><th>CITY</th><th>Residential</th><th>Commercial</th></tr>
<tr><td colspan="4">WYOMING (CONT'D)</td></tr>
<tr><td>823</td><td>Rawlins</td><td>.82</td><td>.87</td></tr>
<tr><td>824</td><td>Worland</td><td>.81</td><td>.85</td></tr>
<tr><td>825</td><td>Riverton</td><td>.81</td><td>.86</td></tr>
<tr><td>826</td><td>Casper</td><td>.80</td><td>.85</td></tr>
<tr><td>827</td><td>Newcastle</td><td>.81</td><td>.86</td></tr>
<tr><td>828</td><td>Sheridan</td><td>.81</td><td>.86</td></tr>
<tr><td>829-831</td><td>Rock Springs</td><td>.86</td><td>.87</td></tr>
<tr><td colspan="4">CANADIAN FACTORS (reflect Canadian currency)</td></tr>
<tr><td colspan="4">ALBERTA</td></tr>
<tr><td></td><td>Calgary</td><td>1.11</td><td>1.11</td></tr>
<tr><td></td><td>Edmonton</td><td>1.10</td><td>1.11</td></tr>
<tr><td></td><td>Fort McMurray</td><td>1.12</td><td>1.10</td></tr>
<tr><td></td><td>Lethbridge</td><td>1.10</td><td>1.07</td></tr>
<tr><td></td><td>Lloydminster</td><td>1.05</td><td>1.03</td></tr>
<tr><td></td><td>Medicine Hat</td><td>1.06</td><td>1.03</td></tr>
<tr><td></td><td>Red Deer</td><td>1.06</td><td>1.03</td></tr>
<tr><td colspan="4">BRITISH COLUMBIA</td></tr>
<tr><td></td><td>Kamloops</td><td>1.04</td><td>1.04</td></tr>
<tr><td></td><td>Prince George</td><td>1.03</td><td>1.04</td></tr>
<tr><td></td><td>Vancouver</td><td>1.05</td><td>1.06</td></tr>
<tr><td></td><td>Victoria</td><td>1.04</td><td>1.03</td></tr>
<tr><td colspan="4">MANITOBA</td></tr>
<tr><td></td><td>Brandon</td><td>1.10</td><td>1.04</td></tr>
<tr><td></td><td>Portage la Prairie</td><td>1.00</td><td>.96</td></tr>
<tr><td></td><td>Winnipeg</td><td>.98</td><td>1.00</td></tr>
<tr><td colspan="4">NEW BRUNSWICK</td></tr>
<tr><td></td><td>Bathurst</td><td>.93</td><td>.92</td></tr>
<tr><td></td><td>Dalhousie</td><td>.93</td><td>.93</td></tr>
<tr><td></td><td>Fredericton</td><td>.98</td><td>.98</td></tr>
<tr><td></td><td>Moncton</td><td>.94</td><td>.94</td></tr>
<tr><td></td><td>Newcastle</td><td>.93</td><td>.93</td></tr>
<tr><td></td><td>St. John</td><td>1.02</td><td>.98</td></tr>
<tr><td colspan="4">NEWFOUNDLAND</td></tr>
<tr><td></td><td>Corner Brook</td><td>1.03</td><td>1.03</td></tr>
<tr><td></td><td>St Johns</td><td>1.06</td><td>1.06</td></tr>
<tr><td colspan="4">NORTHWEST TERRITORIES</td></tr>
<tr><td></td><td>Yellowknife</td><td>1.13</td><td>1.13</td></tr>
<tr><td colspan="4">NOVA SCOTIA</td></tr>
<tr><td></td><td>Bridgewater</td><td>.96</td><td>.96</td></tr>
<tr><td></td><td>Dartmouth</td><td>1.04</td><td>1.04</td></tr>
<tr><td></td><td>Halifax</td><td>1.00</td><td>1.03</td></tr>
<tr><td></td><td>New Glasgow</td><td>1.03</td><td>1.03</td></tr>
<tr><td></td><td>Sydney</td><td>1.02</td><td>1.01</td></tr>
<tr><td></td><td>Truro</td><td>.96</td><td>.96</td></tr>
<tr><td></td><td>Yarmouth</td><td>1.03</td><td>1.03</td></tr>
<tr><td colspan="4">ONTARIO</td></tr>
<tr><td></td><td>Barrie</td><td>1.12</td><td>1.07</td></tr>
<tr><td></td><td>Brantford</td><td>1.13</td><td>1.07</td></tr>
<tr><td></td><td>Cornwall</td><td>1.12</td><td>1.06</td></tr>
<tr><td></td><td>Hamilton</td><td>1.09</td><td>1.08</td></tr>
<tr><td></td><td>Kingston</td><td>1.12</td><td>1.06</td></tr>
<tr><td></td><td>Kitchener</td><td>1.05</td><td>1.02</td></tr>
<tr><td></td><td>London</td><td>1.08</td><td>1.06</td></tr>
<tr><td></td><td>North Bay</td><td>1.19</td><td>1.12</td></tr>
<tr><td></td><td>Oshawa</td><td>1.09</td><td>1.04</td></tr>
<tr><td></td><td>Ottawa</td><td>1.09</td><td>1.07</td></tr>
<tr><td></td><td>Owen Sound</td><td>1.11</td><td>1.06</td></tr>
<tr><td></td><td>Peterborough</td><td>1.10</td><td>1.05</td></tr>
<tr><td></td><td>Sarnia</td><td>1.13</td><td>1.07</td></tr>
<tr><td></td><td>Sault Ste Marie</td><td>1.06</td><td>1.02</td></tr>
<tr><td></td><td>St. Catharines</td><td>1.07</td><td>1.02</td></tr>
<tr><td></td><td>Sudbury</td><td>1.04</td><td>1.01</td></tr>
<tr><td></td><td>Thunder Bay</td><td>1.09</td><td>1.02</td></tr>
<tr><td></td><td>Timmins</td><td>1.09</td><td>1.05</td></tr>
<tr><td></td><td>Toronto</td><td>1.12</td><td>1.11</td></tr>
<tr><td></td><td>Windsor</td><td>1.08</td><td>1.02</td></tr>
<tr><td colspan="4">PRINCE EDWARD ISLAND</td></tr>
<tr><td></td><td>Charlottetown</td><td>.91</td><td>.95</td></tr>
<tr><td></td><td>Summerside</td><td>.98</td><td>.99</td></tr>
<tr><td colspan="4">QUEBEC</td></tr>
<tr><td></td><td>Cap-de-la-Madeleine</td><td>1.11</td><td>1.02</td></tr>
<tr><td></td><td>Charlesbourg</td><td>1.11</td><td>1.02</td></tr>
<tr><td></td><td>Chicoutimi</td><td>1.14</td><td>1.03</td></tr>
<tr><td></td><td>Gatineau</td><td>1.10</td><td>1.01</td></tr>
</table>

For customer support on your Square Foot Cost Data, call 877.694.8329.

523

Location Factors

STATE/ZIP	CITY	Residential	Commercial
QUEBEC (CONT'D)			
	Granby	1.11	
	Hull	1.10	1.01
	Joliette	1.11	1.02
	Laval	1.11	1.01
	Montreal	1.11	1.07
	Quebec	1.10	1.07
	Rimouski	1.14	1.04
	Rouyn-Noranda	1.10	1.01
	Saint Hyacinthe	1.11	1.01
	Sherbrooke	1.11	1.02
	Sorel	1.11	1.02
	St Jerome	1.10	1.01
	Trois Rivieres	1.22	1.09
SASKATCHEWAN			
	Moose Jaw	.93	.93
	Prince Albert	.92	.92
	Regina	1.06	1.07
	Saskatoon	1.05	1.02
YUKON			
	Whitehorse	1.03	1.06

Historical Cost Indexes

The following tables are the estimated Historical Cost Indexes based on a 30-city national average with a base of 100 on January 1, 1993.

The indexes may be used to:

1. Estimate and compare construction costs for different years in the same city.

2. Estimate and compare construction costs in different cities for the same year.

3. Estimate and compare construction costs in different cities for different years.

4. Compare construction trends in any city with the national average.

EXAMPLES

1. Estimate and compare construction costs for different years in the same city.

   A. To estimate the construction cost of a building in Lexington, KY in 1970, knowing that it cost $915,000 in 2015.

   Index Lexington, KY in 1970 = 26.9
   Index Lexington, KY in 2015 = 186.2

   $$\frac{\text{Index } 1970}{\text{Index } 2015} \quad x \quad \text{Cost } 2015 \quad = \quad \text{Cost } 1970$$

   $$\frac{26.9}{186.2} \quad x \quad \$915,000 \quad = \quad \$132,000$$

   Construction Cost in Lexington, KY in 1970 = $132,000

   B. To estimate the current construction cost of a building in Boston, MA that was built in 1980 for $900,000.

   Index Boston, MA in 1980 = 64.0
   Index Boston, MA in 2015 = 244.3

   $$\frac{\text{Index } 2015}{\text{Index } 1980} \quad x \quad \text{Cost } 1980 \quad = \quad \text{Cost } 2015$$

   $$\frac{244.3}{64.0} \quad x \quad \$900,000 \quad = \quad \$3,435,500$$

   Construction Cost in Boston in 2015 = $3,435,500

2. Estimate and compare construction costs in different cities for the same year.

   To compare the construction cost of a building in Topeka, KS in 2015 with the known cost of $800,000 in Baltimore, MD in 2015

   Index Topeka, KS in 2015   = 175.7
   Index Baltimore, MD in 2015 = 191.5

   $$\frac{\text{Index Topeka}}{\text{Index Baltimore}} \quad x \quad \text{Cost Baltimore} \quad = \quad \text{Cost Topeka}$$

   $$\frac{175.7}{191.5} \quad x \quad \$800,000 \quad = \quad \$734,000$$

   Construction Cost in Topeka in 2015 = $734,000

3. Estimate and compare construction costs in different cities for different years.

   To compare the construction cost of a building in Detroit, MI in 2015 with the known construction cost of $5,000,000 for the same building in San Francisco, CA in 1980.

   Index Detroit, MI in 2015   = 213.0
   Index San Francisco, CA in 1980 = 75.2

   $$\frac{\text{Index Detroit } 2015}{\text{Index San Francisco } 1980} \quad x \text{ Cost San Francisco } 1980 = \text{Cost Detroit } 2015$$

   $$\frac{213.0}{75.2} \quad x \quad \$5,000,000 \quad = \quad \$14,162,000$$

   Construction Cost in Detroit in 2015   = $14,162,000

4. Compare construction trends in any city with the national average.

   To compare the construction cost in Las Vegas, NV from 1975 to 2015 with the increase in the National Average during the same time period.

Index Las Vegas, NV for 1975 =	42.8	For 2015 =	215.7
Index 30 City Average for 1975 =	43.7	For 2015 =	206.7

   A. National Average escalation   = $\dfrac{\text{Index} - 30 \text{ City } 2015}{\text{Index} - 30 \text{ City } 1975}$
   From 1975 to 2015

   $$= \quad \frac{206.7}{43.7}$$

   National Average escalation
   From 1975 to 2015   =   4.73 or increased by 473%

   B. Escalation for Las Vegas, NV = $\dfrac{\text{Index Las Vegas, NV } 2015}{\text{Index Las Vegas, NV } 1975}$
   From 1975 to 2015

   $$= \quad \frac{215.7}{42.8}$$

   Las Vegas escalation
   From 1975 to 2015   =   5.04 or increased by 504%

   Conclusion: Construction costs in Las Vegas are higher than National average costs and increased at a greater rate from 1975 to 2015 than the National Average.

Historical Cost Indexes

Year	National 30 City Average	Birmingham	Huntsville	Mobile	Montgomery	Tuscaloosa	Anchorage	Phoenix	Tuscon	Fort Smith	Little Rock	Anaheim	Bakersfield	Fresno	Los Angeles	Oxnard
		Alabama					Alaska	Arizona		Arkansas		California				
Jan 2015	206.7E	186.4E	181.5E	175.5E	168.0E	171.6E	245.5E	182.2E	178.9E	168.9E	172.5E	219.6E	219.9E	220.7E	221.6E	219.3E
2014	203.0	180.3	175.8	170.6	163.1	165.9	241.0	180.0	177.0	165.1	168.4	214.2	215.2	217.0	217.3	214.8
2013	196.9	173.3	168.7	165.7	158.7	161.6	235.3	174.1	169.2	161.2	163.2	207.0	205.6	210.3	210.7	207.8
2012	194.0	169.1	162.7	163.2	153.8	154.6	232.5	172.2	166.4	158.7	161.0	204.1	202.9	207.3	207.2	204.7
2011	185.7	163.0	156.3	156.9	147.6	148.3	225.1	163.8	159.5	152.0	154.1	196.1	194.6	199.8	199.2	197.0
2010	181.6	159.6	152.9	153.1	144.4	144.9	218.3	160.7	157.3	150.3	152.5	192.8	190.8	195.0	194.9	192.8
2009	182.5	162.7	157.2	155.6	148.8	149.4	222.9	161.3	156.8	149.2	156.3	194.5	192.3	195.0	196.6	194.5
2008	171.0	150.3	146.9	143.7	138.4	138.8	210.8	152.2	148.4	138.6	145.4	182.7	179.8	183.2	184.7	182.6
2007	165.0	146.9	143.3	140.3	134.8	135.4	206.8	147.7	143.1	134.8	141.7	175.1	173.7	176.7	177.5	176.2
2006	156.2	135.7	133.6	126.7	124.0	122.2	196.4	137.9	134.8	123.2	127.2	166.8	164.9	169.8	167.3	167.4
2005	146.7	127.9	125.4	119.3	116.6	114.6	185.6	128.5	124.1	115.4	119.4	156.5	153.0	157.9	157.1	156.4
2004	132.8	115.9	112.9	107.0	104.9	102.9	167.0	116.5	113.6	103.8	108.6	142.1	139.5	143.8	142.0	140.9
2003	129.7	113.1	110.7	104.8	102.6	100.8	163.5	113.9	110.6	101.8	105.8	139.4	137.2	142.1	139.6	138.7
2002	126.7	110.0	103.4	103.6	101.7	99.7	159.5	113.3	110.2	100.4	102.1	136.5	133.4	136.0	136.4	136.3
2001	122.2	106.0	100.5	100.6	98.3	95.9	152.6	109.0	106.4	97.3	98.7	132.5	129.3	132.8	132.4	132.4
2000	118.9	104.1	98.9	99.1	94.2	94.6	148.3	106.9	104.9	94.4	95.7	129.4	125.2	129.4	129.9	129.7
1999	116.6	101.2	97.4	97.7	92.5	92.8	145.9	105.5	103.3	93.1	94.4	127.9	123.6	126.9	128.7	128.2
1998	113.6	96.2	94.0	94.8	90.3	89.8	143.8	102.1	101.0	90.6	91.4	125.2	120.3	123.9	125.8	124.7
1997	111.5	94.6	92.4	93.3	88.8	88.3	142.0	101.8	100.7	89.3	90.1	124.0	119.1	122.3	124.6	123.5
1996	108.9	90.9	91.4	92.3	87.5	86.5	140.4	98.3	97.4	86.5	86.7	121.7	117.2	120.2	122.4	121.5
1995	105.6	87.8	88.0	88.8	84.5	83.5	138.0	96.1	95.5	85.0	85.6	120.1	115.7	117.5	120.9	120.0
1990	93.2	79.4	77.6	82.7	78.4	75.2	125.8	86.4	87.0	77.1	77.9	107.7	102.7	103.3	107.5	107.4
1985	81.8	71.1	70.5	72.7	70.9	67.9	116.0	78.1	77.7	69.6	71.0	95.4	92.2	92.6	94.6	96.6
1980	60.7	55.2	54.1	56.8	56.8	54.0	91.4	63.7	62.4	53.1	55.8	68.7	69.4	68.7	67.4	69.9
1975	43.7	40.0	40.9	41.8	40.1	37.8	57.3	44.5	45.2	39.0	38.7	47.6	46.6	47.7	48.3	47.0
1970	27.8	24.1	25.1	25.8	25.4	24.2	43.0	27.2	28.5	24.3	22.3	31.0	30.8	31.1	29.0	31.0
1965	21.5	19.6	19.3	19.6	19.5	18.7	34.9	21.8	22.0	18.7	18.5	23.9	23.7	24.0	22.7	23.9
1960	19.5	17.9	17.5	17.8	17.7	16.9	31.7	19.9	20.0	17.0	16.8	21.7	21.5	21.8	20.6	21.7
1955	16.3	14.8	14.7	14.9	14.9	14.2	26.6	16.7	16.7	14.3	14.5	18.2	18.1	18.3	17.3	18.2
1950	13.5	12.2	12.1	12.3	12.3	11.7	21.9	13.8	13.8	11.8	11.6	15.1	15.0	15.1	14.3	15.1
1945	8.6	7.8	7.7	7.8	7.8	7.5	14.0	8.8	8.8	7.5	7.4	9.6	9.5	9.6	9.1	9.6
1940	6.6	6.0	6.0	6.1	6.0	5.8	10.8	6.8	6.8	5.8	5.7	7.4	7.4	7.4	7.0	7.4

Year	National 30 City Average	Riverside	Sacramento	San Diego	San Francisco	Santa Barbara	Stockton	Vallejo	Colorado Springs	Denver	Pueblo	Bridge-Port	Bristol	Hartford	New Britain	New Haven
		California							Colorado			Connecticut				
Jan 2015	206.7E	219.3E	224.9E	216.1E	254.0E	219.2E	223.3E	231.4E	192.9E	191.3E	190.4E	227.2E	226.1E	227.2E	225.6E	227.4E
2014	203.0	214.1	221.5	211.4	248.0	214.7	218.4	227.9	187.8	189.1	184.7	223.8	222.8	223.9	222.4	224.1
2013	196.9	207.0	215.0	203.2	241.0	207.9	211.8	222.2	180.7	183.4	180.0	217.8	217.2	218.4	216.8	218.5
2012	194.0	204.1	211.9	198.9	237.6	204.8	208.8	218.9	179.4	182.2	177.9	213.9	213.3	214.4	212.9	214.7
2011	185.7	195.8	203.1	192.3	227.9	196.2	201.3	210.3	171.7	174.1	170.7	205.2	204.2	205.1	203.9	205.8
2010	181.6	192.7	196.7	188.2	223.0	192.9	195.9	204.6	170.5	172.6	168.1	198.8	199.1	199.8	198.8	200.7
2009	182.5	193.1	198.1	191.6	224.9	193.6	194.5	202.7	168.4	172.0	166.7	199.0	197.7	198.7	197.4	199.3
2008	171.0	181.3	186.0	179.9	210.6	181.5	183.2	191.6	158.8	161.9	157.9	185.8	184.8	185.7	184.5	186.1
2007	165.0	174.7	179.1	173.6	201.1	175.2	178.2	185.9	152.6	155.9	152.3	179.6	178.3	179.6	178.0	179.5
2006	156.2	166.0	172.2	164.0	191.2	166.4	171.0	177.9	146.1	149.2	145.0	169.5	168.0	169.5	167.7	169.8
2005	146.7	155.4	161.1	153.8	179.7	155.7	160.0	167.2	138.3	141.1	136.4	160.8	159.4	159.6	159.1	160.8
2004	132.8	141.0	147.2	139.0	163.6	141.4	144.2	148.6	125.6	127.2	123.5	143.6	142.9	142.4	142.7	144.3
2003	129.7	138.7	144.9	136.6	162.1	139.2	141.8	146.4	123.1	123.7	120.7	141.3	140.2	140.3	140.0	141.6
2002	126.7	135.3	138.4	134.2	157.9	135.9	136.9	143.8	118.7	121.3	116.7	133.8	133.5	133.1	133.3	134.8
2001	122.2	131.4	135.5	129.7	151.8	131.5	134.1	140.2	113.3	117.2	113.1	128.6	128.5	128.5	128.3	128.6
2000	118.9	128.0	131.5	127.1	146.9	128.7	130.7	137.1	109.8	111.8	108.8	122.7	122.6	122.9	122.4	122.7
1999	116.6	126.5	129.4	124.9	145.1	127.2	127.9	135.3	107.0	109.1	107.1	121.1	121.3	121.2	121.1	121.3
1998	113.6	123.7	125.9	121.3	141.9	123.7	124.7	132.5	103.3	106.5	103.7	119.1	119.4	120.0	119.7	120.0
1997	111.5	122.6	124.7	120.3	139.2	122.4	123.3	130.6	101.1	104.4	102.0	119.2	119.5	119.9	119.7	120.0
1996	108.9	120.6	122.4	118.4	136.8	120.5	121.4	128.1	98.5	101.4	99.7	117.4	117.6	117.9	117.8	118.1
1995	105.6	119.2	119.5	115.4	133.8	119.0	119.0	122.5	96.1	98.9	96.8	116.0	116.5	116.9	116.3	116.5
1990	93.2	107.0	104.9	105.6	121.8	106.4	105.4	111.4	86.9	88.8	88.8	96.3	95.9	96.6	95.9	96.5
1985	81.8	94.8	92.2	94.2	106.2	93.2	93.7	97.0	79.6	81.0	80.4	86.1	86.2	87.2	86.1	86.3
1980	60.7	68.7	71.3	68.1	75.2	71.1	71.2	71.9	60.7	60.9	59.5	61.5	60.7	61.9	60.7	61.4
1975	43.7	47.3	49.1	47.7	49.8	46.1	47.6	46.5	43.1	42.7	42.5	45.2	45.5	46.0	45.2	46.0
1970	27.8	31.0	32.1	30.7	31.6	31.3	31.8	31.8	27.6	26.1	27.3	29.2	28.2	29.6	28.2	29.3
1965	21.5	23.9	24.8	24.0	23.7	24.1	24.5	24.5	21.3	20.9	21.0	22.4	21.7	22.6	21.7	23.2
1960	19.5	21.7	22.5	21.7	21.5	21.9	22.3	22.2	19.3	19.0	19.1	20.0	19.8	20.0	19.8	20.0
1955	16.3	18.2	18.9	18.2	18.0	18.4	18.7	18.6	16.2	15.9	16.0	16.8	16.6	16.8	16.6	16.8
1950	13.5	15.0	15.6	15.0	14.9	15.2	15.4	15.4	13.4	13.2	13.2	13.9	13.7	13.8	13.7	13.9
1945	8.6	9.6	10.0	9.6	9.5	9.7	9.9	9.8	8.5	8.4	8.4	8.8	8.7	8.8	8.7	8.9
1940	6.6	7.4	7.7	7.4	7.3	7.5	7.6	7.6	6.6	6.5	6.5	6.8	6.8	6.8	6.8	6.8

For customer support on your Square Foot Cost Data, call 877.694.8329.

Historical Cost Indexes

Year	National 30 City Average	Connecticut			Delaware	D.C.	Florida						Georgia			
		Norwalk	Stamford	Water-bury	Wilming-ton	Washing-ton	Fort Lau-derdale	Jackson-ville	Miami	Orlando	Talla-hassee	Tampa	Albany	Atlanta	Colum-bus	Macon
Jan 2015	206.7E	233.6E	233.8E	226.5E	214.4E	200.7E	180.5E	175.1E	182.4E	180.4E	169.0E	187.5E	170.2E	180.9E	173.8E	171.6E
2014	203.0	230.3	230.5	223.2	210.2	197.0	176.5	171.0	178.4	176.5	164.4	183.2	166.6	177.5	170.2	168.2
2013	196.9	223.1	224.2	217.3	203.6	191.7	173.2	168.4	175.8	174.8	160.8	180.2	163.2	173.3	166.7	164.9
2012	194.0	219.8	220.0	213.4	201.0	189.6	170.9	165.8	173.7	172.8	158.5	177.8	158.3	170.7	160.6	159.3
2011	185.7	211.0	211.2	204.6	193.3	182.3	164.4	159.4	167.0	166.1	151.7	171.4	151.8	164.0	154.3	152.7
2010	181.6	198.3	203.8	199.4	187.7	179.0	160.3	155.4	163.3	162.5	148.1	167.5	148.3	160.1	150.9	149.4
2009	182.5	198.8	204.3	198.4	188.9	181.4	160.7	152.0	165.2	163.9	145.1	165.1	152.0	164.5	155.6	153.5
2008	171.0	185.2	189.2	185.3	177.5	170.4	149.6	142.5	152.9	153.0	135.3	155.6	140.4	153.2	143.9	142.5
2007	165.0	178.6	183.3	179.1	173.4	163.1	145.6	139.0	148.9	148.1	131.7	152.0	135.8	148.0	140.1	138.1
2006	156.2	169.7	173.7	169.1	158.3	153.0	136.0	126.5	136.7	134.3	118.4	136.6	123.6	139.9	126.3	124.3
2005	146.7	160.9	164.1	160.5	149.6	142.7	126.0	119.2	127.1	126.0	110.7	127.7	115.7	131.7	118.9	116.1
2004	132.8	143.6	146.5	143.2	136.1	126.8	114.8	107.8	115.6	113.2	100.6	116.6	102.8	119.6	102.0	105.0
2003	129.7	140.6	145.1	140.8	133.0	124.6	109.0	105.6	110.5	107.8	98.0	103.5	100.5	115.8	99.1	102.7
2002	126.7	133.5	136.2	133.9	129.4	120.3	107.1	104.6	107.4	106.7	97.3	102.8	99.7	113.0	98.3	101.9
2001	122.2	128.2	131.5	128.8	124.8	115.9	104.3	100.8	104.6	103.8	94.8	100.3	96.4	109.1	95.5	99.0
2000	118.9	121.5	126.4	123.2	117.4	113.8	102.4	99.0	101.8	101.2	93.6	98.9	94.9	106.1	94.1	97.0
1999	116.6	120.0	122.0	121.2	116.6	111.5	101.4	98.0	100.8	100.1	92.5	97.9	93.5	102.9	92.8	95.8
1998	113.6	118.8	120.8	120.1	112.3	109.6	99.4	96.2	99.0	98.4	90.9	96.3	91.4	100.8	90.4	93.3
1997	111.5	118.9	120.9	120.2	110.7	106.4	98.6	95.4	98.2	97.2	89.9	95.5	89.9	98.5	88.7	91.6
1996	108.9	117.0	119.0	118.4	108.6	105.4	95.9	92.6	95.9	95.1	88.0	93.6	86.6	94.3	83.9	88.3
1995	105.6	115.5	117.9	117.3	106.1	102.3	94.0	90.9	93.7	93.5	86.5	92.2	85.0	92.0	82.6	86.8
1990	93.2	96.3	98.9	95.1	92.5	90.4	83.9	81.1	84.0	82.9	78.4	85.0	75.0	80.4	74.0	76.9
1985	81.8	85.3	86.8	85.6	81.1	78.8	76.7	73.4	78.3	73.9	70.6	77.3	66.9	70.3	66.1	68.1
1980	60.7	60.7	60.9	62.3	58.5	59.6	55.3	55.8	56.5	56.7	53.5	57.2	51.7	54.0	51.1	51.3
1975	43.7	44.7	45.0	46.3	42.9	43.7	42.1	40.3	43.2	41.5	38.1	41.3	37.5	38.4	36.2	36.5
1970	27.8	28.1	28.2	28.9	27.0	26.3	25.7	22.8	27.0	26.2	24.5	24.2	23.8	25.2	22.8	23.4
1965	21.5	21.6	21.7	22.2	20.9	21.8	19.8	17.4	19.3	20.2	18.9	18.6	18.3	19.8	17.6	18.0
1960	19.5	19.7	19.7	20.2	18.9	19.4	18.0	15.8	17.6	18.3	17.2	16.9	16.7	17.1	16.0	16.4
1955	16.3	16.5	16.5	17.0	15.9	16.3	15.1	13.2	14.7	15.4	14.4	14.1	14.0	14.4	13.4	13.7
1950	13.5	13.6	13.7	14.0	13.1	13.4	12.5	11.0	12.2	12.7	11.9	11.7	11.5	11.9	11.0	11.3
1945	8.6	8.7	8.7	8.9	8.4	8.6	7.9	7.0	7.8	8.1	7.6	7.5	7.4	7.6	7.0	7.2
1940	6.6	6.7	6.7	6.9	6.4	6.6	6.1	5.4	6.0	6.2	5.8	5.7	5.7	5.8	5.5	5.6

Year	National 30 City Average	Georgia	Hawaii	Idaho		Illinois						Indiana				
		Savan-nah	Hono-lulu	Boise	Poca-tello	Chicago	Decatur	Joliet	Peoria	Rock-ford	Spring-field	Ander-son	Evans-ville	Fort Wayne	Gary	Indian-apolis
Jan 2015	206.7E	170.7E	247.9E	187.6E	187.7E	242.4E	210.3E	240.6E	215.9E	228.7E	212.8E	188.7E	191.7E	185.2E	214.3E	192.1E
2014	203.0	167.2	239.7	184.1	184.4	238.1	206.0	236.8	212.5	224.4	208.2	184.5	188.1	180.8	210.1	188.4
2013	196.9	163.5	231.5	179.6	179.9	231.1	200.8	228.7	207.2	217.2	203.2	177.8	182.5	175.9	203.6	182.7
2012	194.0	158.9	227.9	170.4	171.5	226.2	195.7	222.7	199.7	211.8	197.2	175.2	177.8	171.8	197.9	180.6
2011	185.7	151.7	219.0	162.8	163.6	217.6	187.9	216.8	193.4	205.8	189.4	168.4	169.7	165.0	191.0	173.0
2010	181.6	147.9	214.6	161.7	162.4	210.6	181.2	208.3	186.2	197.6	182.2	162.8	165.9	160.0	183.7	168.1
2009	182.5	152.8	218.6	162.6	163.4	209.0	182.2	205.9	186.2	196.5	184.0	164.9	166.9	163.0	185.2	170.3
2008	171.0	141.1	205.5	153.2	152.8	195.7	168.7	185.5	172.2	180.2	168.5	152.4	155.8	151.0	169.0	159.2
2007	165.0	136.9	200.4	148.0	148.1	186.5	160.5	177.7	166.0	175.0	159.2	149.0	152.6	147.7	165.4	154.1
2006	156.2	124.0	191.9	141.6	141.2	177.8	153.8	168.7	155.2	163.3	152.8	139.8	144.4	139.1	154.5	144.6
2005	146.7	116.9	181.2	133.8	132.5	164.4	145.0	158.3	146.5	151.2	145.4	131.4	136.2	131.0	146.4	137.4
2004	132.8	105.6	161.2	122.0	120.5	149.3	129.3	145.1	133.9	138.1	130.3	120.9	123.5	120.1	132.2	125.0
2003	129.7	103.0	159.4	120.2	118.7	146.2	127.4	143.5	132.4	137.3	128.4	119.7	121.4	118.2	131.4	122.5
2002	126.7	102.0	157.2	118.3	117.1	141.2	123.8	138.5	129.6	132.9	125.3	117.5	119.0	116.4	129.1	120.5
2001	122.2	99.0	150.0	114.3	113.4	135.8	120.1	133.7	124.3	127.8	119.8	113.4	115.6	112.1	123.4	116.4
2000	118.9	97.5	144.8	112.9	112.1	131.2	115.1	124.6	119.0	122.2	116.2	109.8	111.5	108.4	117.8	113.2
1999	116.6	96.0	143.0	110.2	109.6	129.6	113.0	122.6	116.4	120.7	113.8	107.1	109.3	106.8	112.8	110.6
1998	113.6	93.7	140.4	107.4	107.0	125.2	110.1	119.8	113.8	115.5	111.1	105.0	107.2	104.5	111.1	108.0
1997	111.5	92.0	139.8	104.6	104.7	121.3	107.8	117.5	111.5	113.1	108.9	101.9	104.4	101.8	110.3	105.2
1996	108.9	88.6	134.5	102.2	102.1	118.8	106.6	116.2	109.3	111.5	106.5	100.0	102.1	99.9	107.5	102.7
1995	105.6	87.4	130.3	99.5	98.2	114.2	98.5	110.5	102.3	103.6	98.1	96.4	97.2	95.0	100.7	100.1
1990	93.2	77.9	104.7	88.2	88.1	98.4	90.9	98.4	93.7	94.0	90.1	84.6	89.3	83.4	88.4	87.1
1985	81.8	68.9	94.7	78.0	78.0	82.4	81.9	83.4	83.7	83.0	81.5	75.2	79.8	75.0	77.8	77.1
1980	60.7	52.2	68.9	60.3	59.5	62.8	62.3	63.4	64.5	61.6	61.1	56.5	59.0	56.7	59.8	57.9
1975	43.7	36.9	44.6	40.8	40.5	45.7	43.1	44.5	44.7	42.7	42.4	39.5	41.7	39.9	41.9	40.6
1970	27.8	21.0	30.4	26.7	26.6	29.1	28.0	28.6	29.0	27.5	27.6	25.4	26.4	25.5	27.1	26.2
1965	21.5	16.4	21.8	20.6	20.5	22.7	21.5	22.1	22.4	21.2	21.3	19.5	20.4	19.7	20.8	20.7
1960	19.5	14.9	19.8	18.7	18.6	20.2	19.6	20.0	20.3	19.2	19.3	17.7	18.7	17.9	18.9	18.4
1955	16.3	12.5	16.6	15.7	15.6	16.9	16.4	16.8	17.0	16.1	16.2	14.9	15.7	15.0	15.9	15.5
1950	13.5	10.3	13.7	12.9	12.9	14.0	13.6	13.9	14.0	13.3	13.4	12.3	12.9	12.4	13.1	12.8
1945	8.6	6.6	8.8	8.3	8.2	8.9	8.6	8.9	9.0	8.5	8.6	7.8	8.3	7.9	8.4	8.1
1940	6.6	5.1	6.8	6.4	6.3	6.9	6.7	6.8	6.9	6.5	6.6	6.0	6.4	6.1	6.5	6.3

Historical Cost Indexes

Year	National 30 City Average	Indiana Muncie	Indiana South Bend	Indiana Terre Haute	Iowa Cedar Rapids	Iowa Davenport	Iowa Des Moines	Iowa Sioux City	Iowa Waterloo	Kansas Topeka	Kansas Wichita	Kentucky Lexington	Kentucky Louisville	Louisiana Baton Rouge	Louisiana Lake Charles	Louisiana New Orleans
Jan 2015	206.7E	188.3E	190.5E	192.2E	191.7E	199.4E	194.1E	181.8E	180.9E	175.7E	177.1E	186.2E	187.8E	173.2E	173.1E	181.0E
2014	203.0	185.0	185.5	189.2	188.3	195.9	188.5	178.9	177.4	173.6	173.7	183.7	186.1	171.3	170.9	173.1
2013	196.9	179.3	179.3	182.6	183.4	190.8	181.8	173.2	171.3	168.8	167.3	179.1	182.4	166.3	166.2	171.7
2012	194.0	175.8	174.3	178.9	180.3	184.6	179.2	168.8	169.2	162.3	161.4	174.8	177.5	164.4	163.8	162.1
2011	185.7	169.2	168.1	171.3	173.8	178.1	172.7	162.3	162.8	156.3	155.3	167.8	170.8	156.9	156.5	160.3
2010	181.6	163.1	163.5	167.8	170.4	175.6	170.2	159.0	160.2	152.8	151.3	158.7	166.9	153.7	153.4	
2009	182.5	164.5	167.1	168.0	165.9	172.0	162.1	156.3	147.6	154.5	152.4	160.7	167.5	156.9	154.4	161.9
2008	171.0	153.0	153.1	155.9	157.3	163.4	152.4	147.8	139.2	144.9	143.1	152.0	156.8	144.0	141.5	149.3
2007	165.0	149.7	149.5	152.7	152.4	157.6	148.9	144.0	135.0	141.0	138.7	147.6	150.9	139.4	138.0	145.2
2006	156.2	141.2	141.3	144.0	145.8	151.1	142.9	137.3	128.4	134.7	132.9	129.0	143.0	129.4	129.8	135.8
2005	146.7	131.0	133.8	135.2	135.7	142.1	133.5	129.4	120.0	125.9	125.6	121.5	135.1	120.8	117.7	126.2
2004	132.8	120.0	120.6	123.0	122.7	128.6	121.6	116.2	108.3	112.6	113.4	110.1	120.3	105.6	109.0	115.3
2003	129.7	118.7	119.2	121.7	120.9	126.4	120.2	114.6	105.9	109.4	111.3	107.8	118.7	103.3	106.1	112.4
2002	126.7	116.7	117.1	119.9	116.0	122.1	116.7	111.4	103.7	107.8	109.6	106.2	116.4	102.5	105.0	110.6
2001	122.2	112.8	111.6	115.5	112.5	117.5	113.2	107.7	100.6	104.3	105.2	103.3	112.7	99.4	101.3	104.6
2000	118.9	109.1	106.9	110.8	108.8	112.6	108.9	99.0	98.1	101.0	101.1	101.4	109.3	97.8	99.7	102.1
1999	116.6	105.7	103.7	107.9	104.1	109.2	107.6	96.7	96.4	98.7	99.3	99.7	106.6	96.1	97.6	99.5
1998	113.6	103.6	102.3	106.0	102.5	106.8	103.8	95.1	94.8	97.4	97.4	97.4	101.5	93.3	96.1	97.7
1997	111.5	101.3	101.2	104.6	101.2	105.6	102.5	93.8	93.5	96.3	96.2	96.1	99.9	91.5	96.8	96.2
1996	108.9	99.4	99.1	102.1	99.1	102.0	99.1	90.9	90.7	93.6	93.8	94.4	98.3	89.6	94.9	94.3
1995	105.6	95.6	94.6	97.2	95.4	96.6	94.4	88.2	88.9	91.1	91.0	91.6	94.6	92.7	91.6	
1990	93.2	83.9	85.1	89.1	87.1	86.5	86.9	80.4	80.3	83.2	82.6	82.8	82.6	82.0	84.1	84.5
1985	81.8	75.2	76.0	79.2	75.9	77.6	77.1	72.1	72.3	75.0	74.7	75.5	74.5	59.1	60.0	57.2
1980	60.7	56.1	58.3	59.7	62.7	59.6	61.8	59.3	57.7	58.9	58.0	59.3	59.8	40.4	40.6	41.5
1975	43.7	39.2	40.3	41.9	43.1	41.0	43.1	41.0	40.0	42.0	43.1	42.7	42.5	25.0	26.7	27.2
1970	27.8	25.3	26.2	27.0	28.2	26.9	27.6	26.6	25.9	27.0	25.5	26.9	25.9			
1965	21.5	19.5	20.2	20.9	21.7	20.8	21.7	20.5	20.0	20.8	19.6	20.7	20.3	19.4	20.6	20.4
1960	19.5	17.8	18.3	18.9	19.7	18.8	19.5	18.6	18.2	18.9	17.8	18.8	18.4	17.6	18.7	18.5
1955	16.3	14.9	15.4	15.9	16.6	15.8	16.3	15.6	15.2	15.8	15.0	15.8	15.4	14.8	15.7	15.6
1950	13.5	12.3	12.7	13.1	13.7	13.1	13.5	12.9	12.6	13.1	12.4	13.0	12.8	12.2	13.0	12.8
1945	8.6	7.8	8.1	8.4	8.7	8.3	8.6	8.2	8.0	8.3	7.9	8.3	8.1	7.8	8.3	8.2
1940	6.6	6.0	6.3	6.5	6.7	6.4	6.6	6.3	6.2	6.4	6.1	6.4	6.3	6.0	6.4	6.3

Year	National 30 City Average	Louisiana Shreveport	Maine Lewiston	Maine Portland	Maryland Baltimore	Massachusetts Boston	Massachusetts Brockton	Massachusetts Fall River	Massachusetts Lawrence	Massachusetts Lowell	Massachusetts New Bedford	Massachusetts Pittsfield	Massachusetts Springfield	Massachusetts Worcester	Michigan Ann Arbor	Michigan Dearborn
Jan 2015	206.7E	171.6E	197.5E	200.3E	191.5E	244.3E	232.0E	232.0E	238.4E	237.4E	231.2E	216.7E	218.4E	230.8E	210.2E	213.2E
2014	203.0	167.7	192.4	195.4	187.6	239.1	229.1	229.3	234.8	233.9	228.5	213.3	214.8	227.3	206.1	208.4
2013	196.9	159.0	187.9	189.5	183.2	232.9	221.6	221.8	225.3	223.4	221.1	205.9	206.5	219.3	201.2	202.4
2012	194.0	153.2	180.7	182.4	180.8	229.3	217.2	217.8	221.6	219.3	217.0	200.2	201.9	215.2	196.1	199.2
2011	185.7	146.7	167.4	169.1	173.6	219.3	208.2	206.0	211.6	210.1	205.2	191.2	194.5	206.6	186.7	189.4
2010	181.6	145.1	162.1	163.7	168.3	214.3	202.7	202.7	204.9	204.7	200.0	185.9	189.4	201.1	183.3	186.8
2009	182.5	147.6	159.4	161.8	169.1	211.8	199.2	197.6	201.9	201.9	196.9	185.1	187.1	198.2	179.2	186.9
2008	171.0	135.9	149.1	150.9	157.7	198.6	186.4	185.1	188.7	189.1	184.5	171.3	173.7	184.5	169.6	176.4
2007	165.0	132.4	146.4	148.2	152.5	191.8	180.4	179.7	182.0	181.7	179.1	166.8	169.1	180.4	166.8	172.5
2006	156.2	125.2	140.2	139.8	144.9	180.4	171.6	170.4	172.3	173.5	170.4	157.3	159.7	171.6	161.1	165.5
2005	146.7	117.4	131.7	131.4	135.8	169.6	161.3	159.6	162.2	162.3	159.6	147.6	150.7	158.8	147.5	155.8
2004	132.8	105.7	119.8	119.4	121.4	154.1	144.4	143.6	146.4	146.7	143.6	131.9	136.5	143.2	135.9	142.1
2003	129.7	103.9	117.7	117.3	118.1	150.2	139.6	140.6	143.3	143.6	140.5	128.9	133.8	139.0	134.1	139.0
2002	126.7	102.1	117.5	117.1	115.6	145.6	136.0	135.0	136.9	137.3	134.9	124.7	128.3	134.9	131.4	134.3
2001	122.2	98.3	114.6	114.3	111.7	140.9	132.1	131.4	133.1	132.9	131.3	120.3	124.5	130.1	126.8	129.8
2000	118.9	96.2	105.7	105.4	107.7	138.9	129.4	128.4	129.7	130.3	128.3	116.7	121.0	127.3	124.3	125.8
1999	116.6	94.8	105.0	104.7	106.4	136.2	126.7	126.3	127.3	127.4	126.2	115.0	118.8	123.8	117.5	122.7
1998	113.6	92.0	102.8	102.5	104.1	132.8	125.0	124.7	125.0	125.3	124.6	114.2	117.1	122.6	116.0	119.7
1997	111.5	90.1	101.7	101.4	102.2	132.1	124.4	124.7	125.1	125.6	124.7	114.6	117.4	122.9	113.5	117.9
1996	108.9	88.5	99.5	99.2	99.6	128.7	121.6	122.2	121.7	122.1	122.2	112.7	115.0	120.2	112.9	116.2
1995	105.6	86.7	96.8	96.5	96.1	128.6	119.6	117.7	120.5	119.6	117.2	110.7	112.7	114.8	106.1	110.5
1990	93.2	80.3	88.5	88.6	85.6	110.9	103.7	102.9	105.2	102.8	102.6	98.7	101.4	103.2	93.2	96.5
1985	81.8	73.6	76.7	77.0	72.7	92.8	88.8	88.7	89.4	88.2	88.6	85.0	85.6	86.7	80.1	82.7
1980	60.7	58.7	57.3	58.5	53.6	64.0	63.7	64.1	63.4	62.7	63.1	61.8	62.0	62.3	62.9	64.0
1975	43.7	40.5	41.9	42.1	39.8	46.6	45.8	45.7	46.2	45.7	46.1	45.5	45.8	46.0	44.1	44.9
1970	27.8	26.4	26.5	25.8	25.1	29.2	29.1	29.2	29.1	28.9	29.0	28.6	28.5	28.7	28.5	28.9
1965	21.5	20.3	20.4	19.4	20.2	23.0	22.5	22.5	22.5	22.2	22.4	22.0	22.4	22.1	22.0	22.3
1960	19.5	18.5	18.6	17.6	17.5	20.5	20.4	20.4	20.4	20.2	20.3	20.0	20.1	20.1	20.0	20.2
1955	16.3	15.5	15.6	14.7	14.7	17.2	17.1	17.1	17.1	16.9	17.1	16.8	16.9	16.8	16.7	16.9
1950	13.5	12.8	12.9	12.2	12.1	14.2	14.1	14.1	14.1	14.0	14.1	13.8	14.0	13.9	13.8	14.0
1945	8.6	8.1	8.2	7.8	7.7	9.1	9.0	9.0	9.0	9.0	9.0	8.9	9.0	8.9	8.8	8.9
1940	6.6	6.3	6.3	6.0	6.0	7.0	7.0	7.0	7.0	6.9	6.9	6.8	6.9	6.8	6.8	6.9

For customer support on your Square Foot Cost Data, call 877.694.8329.

Historical Cost Indexes

Year	National 30 City Average	Michigan Detroit	Flint	Grand Rapids	Kala-mazoo	Lansing	Sagi-naw	Minnesota Duluth	Minne-apolis	Roches-ter	Mississippi Biloxi	Jackson	Missouri Kansas City	St. Joseph	St. Louis	Spring-field
Jan 2015	206.7E	213.0E	199.5E	193.0E	192.3E	201.2E	195.3E	215.0E	225.4E	210.9E	168.2E	173.6E	213.7E	201.7E	212.4E	194.0E
2014	203.0	208.7	196.0	187.9	188.4	195.7	191.7	211.1	220.7	207.5	164.6	169.8	210.8	199.3	208.6	189.5
2013	196.9	203.1	190.6	182.4	182.4	190.2	186.5	205.1	216.3	201.4	160.9	164.3	204.7	193.1	202.4	181.6
2012	194.0	200.2	186.3	176.4	178.9	185.9	182.8	203.1	214.7	199.7	157.6	160.5	200.8	189.0	198.0	177.7
2011	185.7	190.3	178.5	169.9	171.9	178.0	174.6	195.7	208.1	193.8	151.4	154.3	191.1	178.8	190.6	168.3
2010	181.6	187.5	176.2	160.6	167.3	175.8	171.9	193.1	203.8	188.9	148.0	151.0	186.1	174.2	185.9	164.3
2009	182.5	187.8	176.0	156.7	166.3	173.7	168.6	191.8	203.1	188.0	152.8	156.6	185.6	172.3	187.2	163.7
2008	171.0	177.0	164.4	140.2	155.7	161.6	158.7	177.4	190.6	175.0	141.8	147.1	175.5	164.2	176.2	152.9
2007	165.0	172.9	161.7	137.2	152.6	158.7	156.1	174.2	184.5	171.4	137.7	131.2	169.0	157.8	170.6	145.3
2006	156.2	165.8	153.8	131.1	146.0	152.5	150.6	166.2	173.9	161.9	123.3	117.1	162.0	152.4	159.3	145.3
2005	146.7	156.2	145.1	123.1	134.9	142.0	140.6	157.3	164.6	152.8	116.0	109.9	151.3	143.1	149.4	131.7
2004	132.8	141.9	129.6	112.8	122.5	129.7	127.6	139.1	150.1	137.0	105.3	99.3	135.1	126.9	135.5	116.3
2003	129.7	138.7	127.9	110.9	120.8	127.7	126.0	136.7	146.5	134.5	101.4	96.7	131.9	124.8	133.3	113.9
2002	126.7	134.2	126.9	107.4	119.9	125.3	124.4	131.9	139.5	130.0	101.0	96.3	128.4	122.7	129.6	112.5
2001	122.2	129.4	122.4	104.3	115.2	120.1	119.7	131.4	136.1	124.9	98.7	93.9	121.8	114.7	125.5	106.7
2000	118.9	125.3	119.8	102.7	111.6	117.6	115.9	124.1	131.1	120.1	97.2	92.9	118.2	111.9	122.4	104.3
1999	116.6	122.6	113.7	100.9	106.1	111.5	110.1	120.3	126.5	117.1	95.7	91.8	114.9	106.1	119.8	101.1
1998	113.6	119.5	112.3	99.7	104.9	110.1	108.7	117.7	124.6	115.5	92.1	89.5	108.2	104.2	115.9	98.9
1997	111.5	117.6	110.8	98.9	104.2	108.7	107.5	115.9	121.9	114.1	90.9	88.3	106.4	102.4	113.2	97.3
1996	108.9	116.0	109.7	93.9	103.5	107.8	106.9	115.0	120.4	113.5	87.1	85.4	103.2	99.9	110.1	94.6
1995	105.6	110.1	104.1	91.1	96.4	98.3	102.3	100.3	111.9	102.5	84.2	83.5	99.7	96.1	106.3	89.5
1990	93.2	96.0	91.7	84.0	87.4	90.2	88.9	94.3	100.5	95.6	76.4	75.6	89.5	86.8	94.0	82.6
1985	81.8	81.6	80.7	75.8	79.4	80.0	80.5	85.2	87.9	86.5	69.4	68.5	78.2	77.3	80.8	73.2
1980	60.7	62.5	63.1	59.8	61.7	60.3	63.3	63.4	64.1	64.8	54.3	54.3	59.1	60.4	59.7	56.7
1975	43.7	45.8	44.9	41.8	43.7	44.1	44.8	45.2	46.0	45.3	37.9	37.7	42.7	45.4	44.8	41.1
1970	27.8	29.7	28.5	26.7	28.1	27.8	28.7	28.9	29.5	28.9	24.4	21.3	25.3	28.3	28.4	26.0
1965	21.5	22.1	21.9	20.6	21.7	21.5	22.2	22.3	23.4	22.3	18.8	16.4	20.6	21.8	21.8	20.0
1960	19.5	20.1	20.0	18.7	19.7	19.5	20.1	20.3	20.5	20.2	17.1	14.9	19.0	19.8	19.5	18.2
1955	16.3	16.8	16.7	15.7	16.5	16.3	16.9	17.0	17.2	17.0	14.3	12.5	16.0	16.6	16.3	15.2
1950	13.5	13.9	13.8	13.0	13.6	13.5	13.9	14.0	14.2	14.0	11.8	10.3	13.2	13.7	13.5	12.6
1945	8.6	8.9	8.8	8.3	8.7	8.6	8.9	8.9	9.0	8.9	7.6	6.6	8.4	8.8	8.6	8.0
1940	6.6	6.8	6.8	6.4	6.7	6.6	6.9	6.9	7.0	6.9	5.8	5.1	6.5	6.7	6.7	6.2

Year	National 30 City Average	Montana Billings	Great Falls	Nebraska Lincoln	Omaha	Nevada Las Vegas	Reno	New Hampshire Man-chester	Nashua	New Jersey Camden	Jersey City	Newark	Pater-son	Trenton	NM Albu-querque	NY Albany
Jan 2015	206.7E	189.3E	190.2E	184.8E	187.7E	215.7E	200.3E	202.9E	201.9E	228.6E	229.5E	233.8E	231.4E	230.6E	181.7E	209.4E
2014	203.0	185.4	186.3	181.1	183.7	210.9	195.7	198.8	197.3	224.6	226.6	230.6	227.8	226.6	178.0	205.1
2013	196.9	180.5	181.1	175.5	180.6	206.8	190.4	193.8	192.5	217.7	219.3	223.6	221.2	218.7	173.6	195.3
2012	194.0	178.6	179.5	169.3	177.1	202.6	186.8	189.2	188.3	213.6	215.2	219.2	217.5	213.7	171.0	191.2
2011	185.7	168.8	170.8	162.8	169.6	195.7	179.8	176.8	176.1	205.5	207.0	210.3	209.2	206.4	163.3	180.9
2010	181.6	166.5	168.6	159.3	165.8	193.7	175.6	172.5	172.0	201.2	203.3	206.3	205.3	200.8	162.5	177.8
2009	182.5	165.3	166.0	161.7	165.0	191.5	176.5	174.0	173.5	196.8	200.5	203.6	202.0	198.6	163.0	178.1
2008	171.0	153.1	154.6	152.0	154.8	176.2	167.0	162.5	161.9	184.0	187.3	189.6	188.7	185.6	152.6	166.0
2007	165.0	147.7	147.8	147.7	150.3	166.7	161.4	159.3	158.7	179.3	183.3	185.2	184.6	181.7	146.7	159.4
2006	156.2	140.5	140.8	132.4	140.7	160.0	154.5	146.7	146.4	167.7	171.0	173.6	172.2	169.9	140.1	151.6
2005	146.7	131.9	132.4	124.1	132.8	149.5	145.4	136.8	136.5	159.0	162.4	163.9	163.0	161.5	130.4	142.2
2004	132.8	118.2	117.9	112.1	119.4	137.1	130.5	124.2	123.9	143.4	145.4	147.2	146.6	146.0	118.3	128.6
2003	129.7	115.8	115.8	110.2	117.3	133.8	128.3	122.4	122.2	142.0	144.3	145.7	145.0	143.0	116.4	126.8
2002	126.7	114.6	114.5	108.2	115.0	131.9	126.4	119.9	119.6	135.9	138.5	140.2	140.1	137.6	114.6	122.6
2001	122.2	117.5	117.7	101.3	111.6	127.8	122.5	116.2	116.2	133.4	136.7	136.9	136.8	136.0	111.4	119.2
2000	118.9	113.7	113.9	98.8	107.0	125.8	118.2	111.9	111.9	128.4	130.5	132.6	132.4	130.6	109.0	116.5
1999	116.6	112.1	112.7	96.6	104.8	121.9	114.4	109.6	109.6	125.3	128.8	131.6	129.5	129.6	106.7	114.6
1998	113.6	109.7	109.1	94.9	101.2	118.1	111.4	110.1	110.0	124.3	127.7	128.9	128.5	127.9	103.8	113.0
1997	111.5	107.4	107.6	93.3	99.5	114.6	109.8	108.6	108.6	121.9	125.1	126.4	126.4	125.2	100.8	110.0
1996	108.9	108.0	107.5	91.4	97.4	111.8	108.9	106.5	106.4	119.1	122.5	122.5	123.8	121.8	98.6	108.3
1995	105.6	104.7	104.9	85.6	93.4	108.5	105.0	100.9	100.8	107.0	112.2	111.9	112.1	111.2	96.3	103.6
1990	93.2	92.9	93.8	78.8	85.0	96.3	94.5	86.3	86.3	93.0	93.5	93.8	97.2	94.5	84.9	93.2
1985	81.8	83.9	84.3	71.5	77.5	87.6	85.0	78.1	78.1	81.3	83.3	83.5	84.5	82.8	76.7	79.5
1980	60.7	63.9	64.6	58.5	63.5	64.6	63.0	56.6	56.0	58.6	60.6	60.1	60.0	58.9	59.0	59.5
1975	43.7	43.1	43.8	40.9	43.2	42.8	41.9	41.3	40.8	42.3	43.4	44.2	43.9	43.8	40.3	43.9
1970	27.8	28.5	28.9	26.4	26.8	29.4	28.0	26.2	25.6	27.2	27.8	29.0	27.8	27.4	26.4	28.3
1965	21.5	22.0	22.3	20.3	20.6	22.4	21.6	20.6	19.7	20.9	21.4	23.8	21.4	21.3	20.6	22.3
1960	19.5	20.0	20.3	18.5	18.7	20.2	19.6	18.0	17.9	19.0	19.4	19.4	19.4	19.2	18.5	19.3
1955	16.3	16.7	17.0	15.5	15.7	16.9	16.4	15.1	15.0	16.0	16.3	16.3	16.3	16.1	15.6	16.2
1950	13.5	13.9	14.0	12.8	13.0	14.0	13.6	12.5	12.4	13.2	13.5	13.5	13.5	13.3	12.9	13.4
1945	8.6	8.8	9.0	8.1	8.3	8.9	8.7	8.0	7.9	8.4	8.6	8.6	8.6	8.5	8.2	8.5
1940	6.6	6.8	6.9	6.3	6.4	6.9	6.7	6.2	6.1	6.5	6.6	6.6	6.6	6.6	6.3	6.6

For customer support on your Square Foot Cost Data, call 877.694.8329.

529

Historical Cost Indexes

Year	National 30 City Average	New York Binghamton	New York Buffalo	New York New York	New York Rochester	New York Schenectady	New York Syracuse	New York Utica	New York Yonkers	North Carolina Charlotte	North Carolina Durham	North Carolina Greensboro	North Carolina Raleigh	North Carolina Winston-Salem	N. Dakota Fargo	Ohio Akron
Jan 2015	206.7E	206.4E	212.9E	272.7E	207.3E	210.3E	203.8E	200.8E	248.7E	170.1E	170.0E	168.5E	166.0E	168.9E	180.8E	201.5E
2014	203.0	202.4	207.8	267.7	202.2	204.2	198.8	196.4	245.6	165.5	165.2	163.9	161.6	164.3	177.1	199.2
2013	196.9	195.5	200.7	259.4	194.3	195.5	193.6	188.5	230.0	159.6	158.4	158.7	157.5	159.1	169.6	191.8
2012	194.0	192.9	198.1	256.3	191.2	191.4	189.6	185.2	227.1	156.4	155.6	156.3	155.4	157.2	163.3	187.7
2011	185.7	176.1	188.1	245.6	181.9	181.5	180.8	173.5	218.9	142.6	143.8	142.2	142.8	140.9	160.3	180.8
2010	181.6	173.8	184.0	241.4	179.6	179.0	176.7	170.5	217.1	140.5	141.6	140.0	140.6	138.8	156.2	174.5
2009	182.5	172.9	184.4	239.9	179.6	178.8	176.9	171.3	217.5	145.1	145.8	144.0	145.1	142.7	154.3	175.7
2008	171.0	160.7	173.9	226.8	168.0	167.2	165.9	161.6	202.1	135.8	136.6	134.6	135.3	133.8	145.0	165.4
2007	165.0	155.7	168.6	215.2	163.5	159.8	159.8	155.0	196.5	132.8	133.7	131.7	132.1	131.0	139.8	159.0
2006	156.2	147.5	159.2	204.5	155.2	151.1	150.9	146.4	186.0	125.1	124.6	123.8	124.2	123.0	133.2	152.8
2005	146.7	137.5	149.4	194.0	147.3	142.2	141.9	136.9	176.1	110.5	112.1	112.1	112.1	111.1	125.3	144.4
2004	132.8	123.5	136.1	177.7	132.0	128.4	127.3	124.4	161.8	98.9	100.0	100.1	100.3	99.4	113.0	131.9
2003	129.7	121.8	132.8	173.4	130.3	126.7	124.7	121.7	160.0	96.2	97.5	97.5	97.8	96.8	110.6	129.6
2002	126.7	119.0	128.5	170.1	127.1	123.0	121.8	118.4	154.8	94.8	96.1	96.1	96.3	95.5	106.3	127.2
2001	122.2	116.0	125.2	164.4	123.1	120.1	118.6	115.3	151.4	91.5	92.9	92.9	93.3	92.3	103.1	123.5
2000	118.9	112.4	122.3	159.2	120.0	117.5	115.1	112.4	144.8	90.3	91.6	91.6	91.9	91.0	97.9	117.8
1999	116.6	108.6	120.2	155.9	116.8	114.7	113.7	108.5	140.6	89.3	90.2	90.3	90.5	90.1	96.8	116.0
1998	113.6	109.0	119.1	154.4	117.2	114.2	113.6	108.6	141.4	88.2	89.1	89.2	89.4	89.0	95.2	113.1
1997	111.5	107.2	115.7	150.3	115.2	111.2	110.6	107.0	138.8	86.8	87.7	87.8	87.9	87.6	93.5	110.6
1996	108.9	105.5	114.0	148.0	113.3	109.5	108.2	105.2	137.2	85.0	85.9	86.0	86.1	85.8	91.7	107.9
1995	105.6	99.3	110.1	140.7	106.6	104.6	104.0	104.0	129.4	81.8	82.6	82.6	82.7	82.6	88.4	103.4
1990	93.2	87.0	94.1	118.1	94.6	93.9	91.0	85.6	111.4	74.8	75.4	75.5	75.5	75.3	81.3	94.6
1985	81.8	77.5	83.2	94.9	81.6	80.1	81.0	77.0	92.8	66.9	67.6	67.7	67.6	67.5	73.4	86.8
1980	60.7	58.0	60.6	66.0	60.5	60.3	61.6	58.5	65.9	51.1	52.2	52.5	51.7	50.8	57.4	62.3
1975	43.7	42.5	45.1	49.5	44.8	43.7	44.8	42.7	47.1	36.1	37.0	37.0	37.3	36.1	39.3	44.7
1970	27.8	27.0	28.9	32.8	29.2	27.8	28.5	26.9	30.0	20.9	23.7	23.6	23.4	23.0	25.8	28.3
1965	21.5	20.8	22.2	25.5	22.8	21.4	21.9	20.7	23.1	16.0	18.2	18.2	18.0	17.7	19.9	21.8
1960	19.5	18.9	19.9	21.5	19.7	19.5	19.9	18.8	21.0	14.4	16.6	16.6	16.4	16.1	18.1	19.9
1955	16.3	15.8	16.7	18.1	16.5	16.3	16.7	15.8	17.6	12.1	13.9	13.9	13.7	13.5	15.2	16.6
1950	13.5	13.1	13.8	14.9	13.6	13.5	13.8	13.0	14.5	10.0	11.5	11.5	11.4	11.2	12.5	13.7
1945	8.6	8.3	8.8	9.5	8.7	8.6	8.8	8.3	9.3	6.4	7.3	7.3	7.2	7.1	8.0	8.8
1940	6.6	6.4	6.8	7.4	6.7	6.7	6.8	6.4	7.2	4.9	5.6	5.6	5.6	5.5	6.2	6.8

Year	National 30 City Average	Ohio Canton	Ohio Cincinnati	Ohio Cleveland	Ohio Columbus	Ohio Dayton	Ohio Lorain	Ohio Springfield	Ohio Toledo	Ohio Youngstown	Oklahoma Lawton	Oklahoma Oklahoma City	Oklahoma Tulsa	Oregon Eugene	Oregon Portland	PA Allentown
Jan 2015	206.7E	193.0E	190.9E	206.0E	195.4E	190.5E	199.2E	190.7E	203.6E	196.0E	174.1E	176.3E	170.0E	204.6E	205.9E	212.1E
2014	203.0	190.8	188.1	203.1	192.4	185.8	196.7	187.2	199.7	193.8	169.3	170.9	166.7	200.4	201.4	209.0
2013	196.9	183.9	181.9	195.5	187.1	180.7	187.6	181.5	193.5	186.5	162.7	165.5	161.7	194.7	195.6	203.2
2012	194.0	180.2	178.6	191.4	183.5	177.6	184.1	178.0	190.4	183.3	158.1	158.0	151.3	190.1	191.8	199.3
2011	185.7	172.1	171.3	184.0	175.7	168.4	176.8	168.4	183.1	176.5	151.7	151.6	144.8	184.9	186.6	192.2
2010	181.6	167.5	166.5	180.2	170.3	163.4	173.0	164.6	176.5	171.7	149.5	149.4	143.1	181.5	182.9	187.9
2009	182.5	168.4	168.1	181.6	171.2	165.3	174.0	166.0	178.5	172.8	152.5	153.3	146.8	181.3	183.6	188.3
2008	171.0	158.4	159.0	170.6	160.2	157.5	163.9	157.5	168.2	161.5	141.3	140.6	136.2	172.3	174.7	175.7
2007	165.0	153.2	152.4	164.9	155.1	149.4	158.1	150.3	161.6	156.7	136.2	135.6	132.5	168.5	169.5	169.4
2006	156.2	147.0	144.9	156.9	146.9	142.9	151.8	143.4	154.5	150.3	129.5	129.5	125.7	160.5	161.7	160.1
2005	146.7	137.7	136.8	147.9	138.0	134.5	143.3	134.8	145.5	141.1	121.6	121.6	117.8	150.9	152.2	150.2
2004	132.8	125.6	124.3	135.6	126.2	121.1	131.5	122.0	132.8	129.3	109.3	109.4	107.3	136.8	137.8	133.9
2003	129.7	123.7	121.3	132.7	123.7	119.4	126.1	120.0	130.4	126.2	107.5	107.8	105.0	134.2	135.9	130.3
2002	126.7	120.9	119.2	130.0	120.9	117.5	124.3	118.0	128.4	123.6	106.3	106.0	104.1	132.2	133.9	128.1
2001	122.2	117.7	115.9	125.9	117.1	114.0	120.4	114.7	123.8	119.9	102.0	102.1	99.7	129.8	131.2	123.5
2000	118.9	112.7	110.1	121.3	112.5	109.0	115.0	109.2	115.7	114.1	98.7	98.9	97.5	126.3	127.4	119.9
1999	116.6	110.6	107.9	118.7	109.5	107.1	112.0	106.4	113.6	111.9	97.5	97.4	96.2	120.9	124.3	117.8
1998	113.6	108.5	105.0	114.8	106.8	104.5	109.4	104.1	111.2	109.0	94.3	94.5	94.5	120.0	122.2	115.7
1997	111.5	106.4	102.9	112.9	104.7	102.6	107.4	102.2	108.5	107.2	92.9	93.2	91.2	114.4	116.1	112.0
1996	108.9	103.2	100.2	110.1	101.1	98.8	103.8	98.4	105.1	104.1	90.9	91.3	89.0	112.2	114.3	108.3
1995	105.6	98.8	97.1	106.4	99.1	94.6	97.1	92.0	100.6	100.1	85.3	88.0		98.0	99.4	95.0
1990	93.2	92.1	86.7	95.1	88.1	85.9	89.4	83.4	92.9	91.8	77.1	79.9	80.0	88.7	90.1	81.9
1985	81.8	83.7	78.2	86.2	77.7	76.3	80.7	74.3	84.7	82.7	71.2	73.9	73.8	68.9	68.4	58.7
1980	60.7	60.6	59.8	61.0	58.6	56.6	60.8	56.2	64.0	61.5	52.2	55.5	57.2	44.6	44.9	42.3
1975	43.7	44.0	43.7	44.6	42.2	40.5	42.7	40.0	44.9	44.5	38.6	38.6	39.2	30.4	30.0	27.3
1970	27.8	27.8	28.2	29.6	26.7	27.0	27.5	25.4	29.1	29.6	24.1	23.4	24.1	23.4	22.6	21.0
1965	21.5	21.5	20.7	21.1	20.2	20.0	21.1	19.6	21.3	21.0	18.5	17.8	20.6	23.4	22.6	21.0
1960	19.5	19.5	19.3	19.6	18.7	18.1	19.2	17.8	19.4	19.1	16.9	16.1	18.1	21.2	21.1	19.1
1955	16.3	16.3	16.2	16.5	15.7	15.2	16.1	14.9	16.2	16.0	14.1	13.5	15.1	17.8	17.7	16.0
1950	13.5	13.5	13.3	13.6	13.0	12.5	13.3	12.3	13.4	13.2	11.7	11.2	12.5	14.7	14.6	13.2
1945	8.6	8.6	8.5	8.7	8.3	8.0	8.5	7.8	8.6	8.4	7.4	7.1	8.0	9.4	9.3	8.4
1940	6.6	6.7	6.5	6.7	6.4	6.2	6.5	6.1	6.6	6.5	6.5	5.5	6.1	7.2	7.2	6.5

For customer support on your Square Foot Cost Data, call 877.694.8329.

Historical Cost Indexes

Year	National 30 City Average	Pennsylvania Erie	Harrisburg	Philadelphia	Pittsburgh	Reading	Scranton	RI Providence	South Carolina Charleston	Columbia	South Dakota Rapid City	Sioux Falls	Tennessee Chattanooga	Knoxville	Memphis	Nashville
Jan 2015	206.7E	196.6E	202.7E	236.8E	210.9E	205.4E	204.5E	223.3E	175.4E	166.8E	170.8E	168.7E	176.3E	173.1E	179.8E	181.7E
2014	203.0	191.4	200.3	232.1	207.1	201.4	201.4	220.3	171.0	162.5	168.6	166.8	174.1	170.9	176.5	178.2
2013	196.9	187.2	192.5	223.6	201.4	196.1	196.0	213.4	166.3	157.9	159.6	161.2	167.4	159.3	169.3	172.8
2012	194.0	182.0	187.9	221.8	196.8	190.8	191.5	209.3	155.3	148.1	156.7	158.8	163.8	155.8	166.3	167.5
2011	185.7	176.0	180.9	212.8	187.7	184.3	185.3	196.8	149.4	142.3	149.5	151.3	157.3	149.3	158.4	160.7
2010	181.6	171.0	175.2	209.3	182.3	179.8	180.8	192.8	147.4	140.2	147.3	149.0	152.4	144.7	154.2	156.7
2009	182.5	171.9	176.5	209.5	181.7	180.7	181.3	192.9	150.8	144.3	148.2	151.1	155.7	148.0	157.2	159.8
2008	171.0	161.0	165.7	196.2	169.2	169.3	167.9	178.2	142.1	134.8	138.4	141.3	136.7	133.4	146.0	147.5
2007	165.0	156.0	159.4	188.4	163.2	164.9	163.1	174.4	139.3	131.7	129.5	133.0	133.7	130.6	143.1	144.4
2006	156.2	148.7	150.7	177.8	155.5	155.0	153.9	163.6	122.4	118.7	122.8	126.6	127.0	123.9	136.5	135.5
2005	146.7	140.6	141.5	166.4	146.6	145.5	142.9	155.6	110.2	109.6	115.2	118.5	116.0	115.1	128.6	127.9
2004	132.8	127.0	126.2	148.4	133.1	128.7	129.0	138.8	98.9	98.4	103.9	107.3	105.0	104.5	115.6	115.8
2003	129.7	124.5	123.8	145.5	130.5	126.3	126.0	135.9	96.1	95.7	101.3	104.0	102.8	102.3	111.1	110.9
2002	126.7	122.1	121.1	142.0	127.8	123.4	124.2	131.1	94.8	93.9	100.2	103.2	102.2	101.2	107.6	109.2
2001	122.2	118.9	118.4	136.9	124.1	120.4	121.1	127.8	92.1	91.3	96.8	99.7	98.2	96.5	103.3	104.3
2000	118.9	114.6	114.0	132.1	120.9	115.9	117.7	122.8	89.9	89.1	94.2	98.0	96.9	95.0	100.8	100.8
1999	116.6	113.8	112.8	129.8	119.6	114.8	116.3	121.6	89.0	88.1	92.4	95.8	95.9	93.4	99.7	98.6
1998	113.6	109.8	110.5	126.6	117.1	112.5	113.7	120.3	88.0	87.2	90.7	93.6	94.9	92.4	97.2	96.8
1997	111.5	108.5	108.8	123.3	113.9	110.8	112.3	118.9	86.5	85.6	89.3	92.1	93.6	90.8	96.1	94.9
1996	108.9	106.8	105.7	120.3	110.8	108.7	109.9	117.2	84.9	84.1	87.2	90.0	91.6	88.8	94.0	91.9
1995	105.6	99.8	100.6	117.1	106.3	103.6	103.8	111.1	82.7	82.2	84.0	84.7	89.2	86.0	91.2	87.6
1990	93.2	88.5	89.5	98.5	91.2	90.8	91.3	94.1	73.7	74.5	77.2	78.4	79.9	75.1	81.3	77.1
1985	81.8	79.6	77.2	82.2	81.5	77.7	80.0	83.0	65.9	66.9	69.7	71.2	72.5	67.7	74.3	66.7
1980	60.7	59.4	56.9	58.7	61.3	57.7	58.4	59.2	50.6	51.1	54.8	57.1	55.0	51.6	55.9	53.1
1975	43.7	43.4	42.2	44.5	44.5	43.6	41.9	42.7	35.0	36.4	37.7	39.6	39.1	37.3	40.7	37.0
1970	27.8	27.5	25.8	27.5	28.7	27.1	27.0	27.3	22.8	22.9	24.8	25.8	24.9	22.0	23.0	22.8
1965	21.5	21.1	20.1	21.7	22.4	20.9	20.8	21.9	17.6	17.6	19.1	19.9	19.2	17.1	18.3	17.4
1960	19.5	19.1	18.6	19.4	19.7	19.0	18.9	19.1	16.0	16.0	17.3	18.1	17.4	15.5	16.6	15.8
1955	16.3	16.1	15.6	16.3	16.5	15.9	15.9	16.0	13.4	13.4	14.5	15.1	14.6	13.0	13.9	13.3
1950	13.5	13.3	12.9	13.5	13.6	13.2	13.1	13.2	11.1	11.1	12.0	12.5	12.1	10.7	11.5	10.9
1945	8.6	8.5	8.2	8.6	8.7	8.4	8.3	8.4	7.1	7.1	7.7	8.0	7.7	6.8	7.3	7.0
1940	6.6	6.5	6.3	6.6	6.7	6.4	6.5	6.5	5.4	5.5	5.9	6.2	6.0	5.3	5.6	5.4

Year	National 30 City Average	Texas Abilene	Amarillo	Austin	Beaumont	Corpus Christi	Dallas	El Paso	Fort Worth	Houston	Lubbock	Odessa	San Antonio	Waco	Wichita Falls	Utah Ogden
Jan 2015	206.7E	172.3E	173.8E	173.8E	175.5E	174.8E	176.7E	169.9E	174.7E	179.5E	175.7E	173.0E	172.6E	170.3E	171.3E	177.5E
2014	203.0	160.9	167.2	167.5	165.2	167.3	172.4	156.2	169.5	176.0	166.9	159.2	169.5	161.7	160.6	174.2
2013	196.9	155.2	161.7	158.7	160.2	157.7	167.4	151.2	161.7	169.4	159.3	151.5	164.3	156.1	155.9	167.7
2012	194.0	152.5	158.6	154.3	158.2	151.8	165.2	149.1	159.4	167.8	156.6	149.1	160.7	153.8	152.9	165.6
2011	185.7	145.9	151.9	147.6	152.4	145.0	157.9	142.5	152.8	160.8	150.0	142.5	152.6	147.3	146.6	158.7
2010	181.6	144.2	150.1	144.7	150.5	142.1	155.1	140.8	149.6	157.3	148.1	140.7	147.9	145.8	145.0	157.5
2009	182.5	143.5	147.3	146.4	149.8	141.7	155.5	141.1	150.1	160.9	144.6	144.6	150.8	146.7	146.1	155.1
2008	171.0	132.4	137.0	137.5	140.3	133.2	144.1	130.7	138.3	149.1	134.7	139.1	141.0	136.1	136.1	144.8
2007	165.0	128.5	132.2	131.6	137.1	128.8	138.6	126.2	134.8	146.0	130.2	125.0	136.4	132.1	132.1	140.0
2006	156.2	121.6	125.7	125.5	130.1	122.0	131.1	120.3	127.6	138.2	123.2	118.3	129.8	125.2	125.3	140.0
2005	146.7	113.8	117.3	117.9	121.3	114.4	123.7	112.5	119.4	129.0	115.5	110.5	121.3	116.1	117.3	133.4 126.0
2004	132.8	102.9	106.3	105.7	108.5	102.9	112.0	101.5	108.4	115.9	104.7	99.9	108.4	104.8	105.1	115.2
2003	129.7	100.5	104.4	103.9	106.1	100.4	109.3	99.5	105.5	113.4	102.3	97.6	104.8	102.7	103.0	112.7
2002	126.7	99.9	101.9	102.4	104.5	98.9	107.9	98.7	104.6	111.5	100.5	95.9	103.7	100.5	101.4	110.8
2001	122.2	93.4	98.4	99.8	102.3	96.6	103.8	95.5	100.9	107.8	97.6	93.4	100.5	97.8	97.3	107.7
2000	118.9	93.4	98.1	99.1	101.6	96.9	102.7	92.4	99.9	106.0	97.6	93.4	99.4	97.3	96.9	104.6
1999	116.6	91.8	94.5	96.0	99.7	94.0	101.0	90.7	97.6	104.6	96.0	92.1	98.0	94.8	95.5	103.3
1998	113.6	89.8	92.7	94.2	97.9	91.8	97.9	88.4	94.5	101.3	93.3	90.1	94.8	92.7	92.9	98.5
1997	111.5	88.4	91.3	92.8	96.8	90.3	96.1	87.0	93.3	100.1	91.9	88.8	93.4	91.4	91.5	96.1
1996	108.9	86.8	89.6	90.9	95.3	88.5	94.1	86.7	91.5	97.9	90.2	87.1	92.3	89.7	89.9	94.1
1995	105.6	85.2	87.4	89.3	93.7	87.4	91.4	85.2	89.5	95.9	88.4	85.6	88.9	86.4	86.8	92.2
1990	93.2	78.0	80.1	81.3	86.5	79.3	84.5	76.7	82.1	85.4	81.5	78.6	80.7	79.6	80.3	83.4
1985	81.8	71.1	72.5	74.5	79.3	72.3	77.6	69.4	75.1	79.6	74.0	71.2	73.9	71.7	73.3	75.2
1980	60.7	53.4	55.2	54.5	57.6	54.5	57.9	53.1	57.0	59.4	55.6	57.2	55.0	54.9	55.4	62.2
1975	43.7	37.6	39.0	39.0	39.6	38.1	40.7	38.0	40.4	41.2	38.9	37.9	39.0	38.6	38.0	40.0
1970	27.8	24.5	24.9	24.9	25.7	24.5	25.5	23.7	25.9	25.4	25.1	24.6	23.3	24.8	24.5	26.8
1965	21.5	18.9	19.2	19.2	19.9	18.9	19.9	19.0	19.9	20.0	19.4	19.0	18.5	19.2	18.9	20.6
1960	19.5	17.1	17.4	17.4	18.1	17.1	18.2	17.0	18.1	18.2	17.6	17.3	16.8	17.4	17.2	18.8
1955	16.3	14.4	14.6	14.6	15.1	14.4	15.3	14.3	15.2	15.2	14.8	14.5	14.1	14.6	14.4	15.7
1950	13.5	11.9	12.1	12.1	12.5	11.9	12.6	11.8	12.5	12.6	12.2	12.0	11.6	12.1	11.9	13.0
1945	8.6	7.6	7.7	7.7	8.0	7.6	8.0	7.5	8.0	8.0	7.8	7.6	7.4	7.7	7.6	8.3
1940	6.6	5.9	5.9	5.9	6.1	5.8	6.2	5.8	6.0	6.2	6.0	5.9	5.7	5.9	5.8	6.4

Historical Cost Indexes

Year	National 30 City Average	Utah — Salt Lake City	Vermont — Burlington	Vermont — Rutland	Virginia — Alexandria	Virginia — Newport News	Virginia — Norfolk	Virginia — Richmond	Virginia — Roanoke	Washington — Seattle	Washington — Spokane	Washington — Tacoma	West Virginia — Charleston	West Virginia — Huntington	Wisconsin — Green Bay	Wisconsin — Kenosha
Jan 2015	206.7E	180.9E	194.9E	191.8E	195.0E	178.2E	178.8E	180.1E	177.1E	212.7E	194.8E	208.7E	202.0E	203.8E	202.3E	208.2E
2014	203.0	177.1	191.3	188.6	190.5	174.4	175.7	176.4	173.5	209.8	190.8	205.2	197.0	199.5	198.9	205.2
2013	196.9	171.1	178.7	176.6	185.0	170.0	171.5	170.7	168.3	203.4	184.3	199.1	187.2	193.2	193.9	201.2
2012	194.0	168.8	172.4	170.5	182.4	168.3	169.7	169.4	166.4	201.9	181.6	194.3	183.4	186.4	190.0	195.5
2011	185.7	161.8	158.8	157.1	174.5	159.8	161.1	159.0	154.2	194.1	175.5	187.6	177.5	180.3	181.4	187.6
2010	181.6	160.9	156.3	154.6	170.9	156.7	158.0	156.6	151.8	191.8	171.5	186.0	171.6	176.1	179.0	185.1
2009	182.5	160.4	158.3	156.7	172.5	160.0	161.9	160.7	155.6	188.5	173.0	186.3	174.4	177.4	175.5	182.5
2008	171.0	149.7	147.2	145.8	161.8	150.8	150.8	150.9	145.9	176.9	162.4	174.9	162.8	165.7	165.6	172.2
2007	165.0	144.6	144.4	143.2	155.0	146.2	146.1	147.3	142.9	171.4	156.7	168.7	158.7	160.1	159.4	164.8
2006	156.2	137.7	131.7	130.7	146.2	133.7	134.6	134.8	129.7	162.9	150.1	160.7	149.4	151.4	152.7	157.1
2005	146.7	129.4	124.6	123.8	136.5	123.5	124.4	125.4	112.2	153.9	141.9	151.5	140.8	141.0	144.4	148.3
2004	132.8	117.8	113.0	112.3	121.5	108.9	110.2	110.9	99.7	138.0	127.6	134.5	124.8	125.6	128.9	132.4
2003	129.7	116.0	110.7	110.1	119.5	104.8	106.2	108.6	97.0	134.9	125.9	133.0	123.3	123.6	127.4	130.9
2002	126.7	113.7	109.0	108.4	115.1	102.9	104.1	106.6	95.2	132.7	123.9	131.4	121.2	120.9	123.0	127.3
2001	122.2	109.1	105.7	105.2	110.8	99.8	100.3	102.9	92.1	127.9	120.3	125.7	114.6	117.5	119.1	123.6
2000	118.9	106.5	98.9	98.3	108.1	96.5	97.6	100.2	90.7	124.6	118.3	122.9	111.5	114.4	114.6	119.1
1999	116.6	104.5	98.2	97.7	106.1	95.6	96.5	98.8	89.8	123.3	116.7	121.6	110.6	113.4	112.1	115.8
1998	113.6	99.5	97.8	97.3	104.1	93.7	93.9	97.0	88.3	119.4	114.3	118.3	106.7	109.0	109.5	112.9
1997	111.5	97.2	96.6	96.3	101.2	91.6	91.7	92.9	86.9	118.1	111.7	117.2	105.3	107.7	105.6	109.1
1996	108.9	94.9	95.1	94.8	99.7	90.2	90.4	91.6	85.5	115.2	109.2	114.3	103.1	104.8	103.8	106.4
1995	105.6	93.1	91.1	90.8	96.3	86.0	86.4	87.8	82.8	113.7	107.4	112.8	95.8	97.2	97.6	97.9
1990	93.2	84.3	83.0	82.9	86.1	76.3	76.7	77.6	76.1	100.1	98.5	100.5	86.1	86.8	86.7	87.8
1985	81.8	75.9	74.8	74.9	75.1	68.7	68.8	69.5	67.2	88.3	89.0	91.2	77.7	77.7	76.7	77.4
1980	60.7	57.0	55.3	58.3	57.3	52.5	52.4	54.3	51.3	67.9	66.3	66.7	57.7	58.3	58.6	58.3
1975	43.7	40.1	41.8	43.9	41.7	37.2	36.9	37.1	37.1	44.9	44.4	44.5	41.0	40.0	40.9	40.5
1970	27.8	26.1	25.4	26.8	26.2	23.9	21.5	22.0	23.7	28.8	29.3	29.6	26.1	25.8	26.4	26.5
1965	21.5	20.0	19.8	20.6	20.2	18.4	17.1	17.2	18.3	22.4	22.5	22.8	20.1	19.9	20.3	20.4
1960	19.5	18.4	18.0	18.8	18.4	16.7	15.4	15.6	16.6	20.4	20.8	20.8	18.3	18.1	18.4	18.6
1955	16.3	15.4	15.1	15.7	15.4	14.0	12.9	13.1	13.9	17.1	17.4	17.4	15.4	15.2	15.5	15.6
1950	13.5	12.7	12.4	13.0	12.7	11.6	10.7	10.8	11.5	14.1	14.4	14.4	12.7	12.5	12.8	12.9
1945	8.6	8.1	7.9	8.3	8.1	7.4	6.8	6.9	7.3	9.0	9.2	9.2	8.1	8.0	8.1	8.2
1940	6.6	6.3	6.1	6.4	6.2	5.7	5.3	5.3	5.7	7.0	7.1	7.1	6.2	6.2	6.3	6.3

Year	National 30 City Average	Wisconsin — Madison	Wisconsin — Milwaukee	Wisconsin — Racine	Wyoming — Cheyenne	Canada — Calgary	Canada — Edmonton	Canada — Hamilton	Canada — London	Canada — Montreal	Canada — Ottawa	Canada — Quebec	Canada — Toronto	Canada — Vancouver	Canada — Winnipeg
Jan 2015	206.7E	205.0E	213.3E	207.9E	177.2E	230.3E	230.3E	223.6E	219.9E	221.4E	221.5E	220.8E	229.4E	219.7E	205.8E
2014	203.0	202.4	210.2	205.3	173.6	228.2	229.1	216.0	213.7	219.8	214.8	218.3	227.4	218.8	202.9
2013	196.9	197.8	204.6	200.5	167.6	222.7	223.3	211.1	208.2	214.5	214.1	212.6	220.9	216.2	200.2
2012	194.0	191.2	202.2	194.9	165.0	219.1	219.6	211.1	200.1	209.6	209.7	207.5	216.6	214.0	200.7
2011	185.7	183.4	191.9	187.0	155.3	212.6	212.4	204.2	193.7	202.1	202.0	200.6	209.6	207.1	192.6
2010	181.6	181.3	187.1	184.4	154.4	200.6	200.7	197.7	193.7	193.9	195.6	192.7	200.7	192.7	182.9
2009	182.5	180.0	187.3	182.8	155.2	205.4	206.6	203.0	198.9	198.8	200.1	197.4	205.6	199.6	188.5
2008	171.0	168.4	176.3	173.0	146.5	190.2	191.0	194.3	188.6	186.7	188.0	186.8	194.6	184.9	174.4
2007	165.0	160.7	168.9	164.7	141.3	183.1	184.4	186.8	182.4	181.7	182.4	182.0	187.7	180.6	170.1
2006	156.2	152.8	158.8	157.2	128.0	163.6	164.8	169.0	164.9	158.6	163.7	159.0	170.6	169.3	155.9
2005	146.7	145.2	148.4	147.9	118.9	154.4	155.7	160.0	156.2	149.2	155.1	150.1	162.6	159.4	146.9
2004	132.8	129.7	134.2	132.7	104.7	138.8	139.4	142.6	140.4	134.5	141.0	135.7	146.0	141.7	129.7
2003	129.7	128.4	131.1	131.5	102.6	133.6	134.2	139.1	137.0	130.2	137.4	131.4	142.3	137.0	124.4
2002	126.7	124.5	128.2	126.5	101.7	122.5	122.2	136.3	134.1	127.2	134.9	128.4	139.8	134.6	121.4
2001	122.2	120.6	123.9	123.3	99.0	117.5	117.4	131.5	129.1	124.4	130.2	125.5	134.7	130.2	117.2
2000	118.9	116.6	120.5	118.7	98.1	115.9	115.8	130.0	127.5	122.8	128.8	124.1	133.1	128.4	115.6
1999	116.6	115.9	117.4	115.5	96.9	115.3	115.2	128.1	125.6	120.8	126.8	121.9	131.2	127.1	115.2
1998	113.6	110.8	113.4	112.7	95.4	112.5	112.4	126.3	123.9	119.0	124.7	119.6	128.5	123.8	113.7
1997	111.5	106.1	110.1	109.3	93.2	110.7	110.6	124.4	121.9	114.6	122.8	115.4	125.7	121.9	111.4
1996	108.9	104.4	107.1	106.5	91.1	109.1	109.0	122.6	120.2	112.9	121.0	113.6	123.9	119.0	109.7
1995	105.6	96.5	103.9	97.8	87.6	107.4	107.4	119.9	117.5	110.8	118.2	111.5	121.6	116.2	107.6
1990	93.2	84.3	88.9	87.3	79.1	98.0	97.1	103.8	101.4	99.0	102.7	96.8	104.6	103.2	95.1
1985	81.8	74.3	77.4	77.0	72.3	90.2	89.1	85.8	84.9	82.4	83.9	79.9	86.5	89.8	83.4
1980	60.7	56.8	58.8	58.1	56.9	64.9	63.3	63.7	61.9	59.2	60.9	59.3	60.9	65.0	61.7
1975	43.7	40.7	43.3	40.7	40.6	42.2	41.6	42.9	41.6	39.7	41.5	39.0	42.2	42.4	39.2
1970	27.8	26.5	29.4	26.5	26.0	28.9	28.6	28.5	27.8	25.6	27.6	26.0	25.6	26.0	23.1
1965	21.5	20.6	21.8	20.4	20.0	22.3	22.0	22.0	21.4	18.7	21.2	20.1	19.4	20.5	17.5
1960	19.5	18.1	19.0	18.6	18.2	20.2	20.0	20.0	19.5	17.0	19.3	18.2	17.6	18.6	15.8
1955	16.3	15.2	15.9	15.6	14.0	15.2	17.0	16.8	16.3	14.3	16.2	15.3	14.8	15.5	13.3
1950	13.5	12.5	13.2	12.9	12.6	14.0	13.9	13.8	13.5	11.8	13.4	12.6	12.2	12.8	10.9
1945	8.6	8.0	8.4	8.2	8.0	9.0	8.8	8.8	8.6	7.5	8.5	8.0	7.8	8.2	7.0
1940	6.6	6.2	6.5	6.3	6.2	6.9	6.9	6.8	6.8	5.8	6.6	6.2	6.0	6.3	5.4
1940	6.6	6.2	6.5	6.3	6.2	6.9	6.9	6.8	6.8	5.8	6.6	6.2	6.0	6.3	5.4

Estimating Tips

- The cost figures in this section were derived from approximately 11,000 projects contained in the RSMeans database of completed construction projects. They include the contractor's overhead and profit, but do not generally include architectural fees or land costs. The figures have been adjusted to January of the current year. New projects are added to our files each year, and outdated projects are discarded. For this reason, certain costs may not show a uniform annual progression. In no case are all subdivisions of a project listed.

- These projects were located throughout the U.S. and reflect a tremendous variation in square foot (S.F.) and cubic foot (C.F.) costs. This is due to differences, not only in labor and material costs, but also in individual owners' requirements. For instance, a bank in a large city would have different features than one in a rural area. This is true of all the different types of buildings analyzed. Therefore, caution should be exercised when using these square foot costs. For example, for courthouses, costs in the database are local courthouse costs and will not apply to the larger, more elaborate federal courthouses. As a general rule, the projects in the 1/4 column do not include any site work or equipment, while the projects in the 3/4 column may include both equipment and site work.

The median figures do not generally include site work.

- None of the figures "go with" any others. All individual cost items were computed and tabulated separately. Thus, the sum of the median figures for plumbing, HVAC, and electrical will not normally total up to the total mechanical and electrical costs arrived at by separate analysis and tabulation of the projects.

- Each building was analyzed as to total and component costs and percentages. The figures were arranged in ascending order with the results tabulated as shown. The 1/4 column shows that 25% of the projects had lower costs and 75% had higher. The 3/4 column shows that 75% of the projects had lower costs and 25% had higher. The median column shows that 50% of the projects had lower costs and 50% had higher.

- There are two times when square foot costs are useful. The first is in the conceptual stage when no details are available. Then, square foot costs make a useful starting point. The second is after the bids are in and the costs can be worked back into their appropriate categories for information purposes. As soon as details become available in the project design, the square foot approach should be discontinued and the project priced as to its particular components. When more precision is required, or

for estimating the replacement cost of specific buildings, the current edition of *RSMeans Square Foot Costs* should be used.

- In using the figures in this section, it is recommended that the median column be used for preliminary figures if no additional information is available. The median figures, when multiplied by the total city construction cost index figures (see City Cost Indexes) and then multiplied by the project size modifier at the end of this section, should present a fairly accurate base figure, which would then have to be adjusted in view of the estimator's experience, local economic conditions, code requirements, and the owner's particular requirements. There is no need to factor the percentage figures, as these should remain constant from city to city. All tabulations mentioning air conditioning had at least partial air conditioning.

- The editors of this book would greatly appreciate receiving cost figures on one or more of your recent projects, which would then be included in the averages for next year. All cost figures received will be kept confidential, except that they will be averaged with other similar projects to arrive at square foot cost figures for next year's book. See the last page of the book for details and the discount available for submitting one or more of your projects. ∎

50 17 00 | S.F. Costs		UNIT	UNIT COSTS			% OF TOTAL		
			1/4	MEDIAN	3/4	1/4	MEDIAN	3/4
01 0010	**APARTMENTS Low Rise (1 to 3 story)**	S.F.	74.50	94	125			
0020	Total project cost	C.F.	6.70	8.85	10.95			
0100	Site work	S.F.	5.45	8.70	15.30	6.05%	10.55%	13.95%
0500	Masonry		1.47	3.62	5.95	1.54%	3.92%	6.50%
1500	Finishes		7.90	10.85	13.45	9.05%	10.75%	12.85%
1800	Equipment		2.44	3.70	5.50	2.71%	3.99%	5.95%
2720	Plumbing		5.80	7.45	9.50	6.65%	8.95%	10.05%
2770	Heating, ventilating, air conditioning		3.70	4.56	6.70	4.20%	5.60%	7.60%
2900	Electrical		4.33	5.75	7.80	5.20%	6.65%	8.35%
3100	Total: Mechanical & Electrical		15.40	19.95	24.50	16.05%	18.20%	23%
9000	Per apartment unit, total cost	Apt.	69,500	106,000	156,500			
9500	Total: Mechanical & Electrical	"	13,100	20,700	27,000			
02 0010	**APARTMENTS Mid Rise (4 to 7 story)**	S.F.	99	119	147			
0020	Total project costs	C.F.	7.70	10.65	14.55			
0100	Site work	S.F.	3.95	8	15.45	5.25%	6.70%	9.20%
0500	Masonry		4.13	9.05	12.40	5.10%	7.25%	10.50%
1500	Finishes		12.95	17.35	20.50	10.70%	13.50%	17.70%
1800	Equipment		2.77	4.30	5.65	2.54%	3.47%	4.31%
2500	Conveying equipment		2.24	2.76	3.33	2.05%	2.27%	2.69%
2720	Plumbing		5.80	9.30	9.85	5.70%	7.20%	8.95%
2900	Electrical		6.50	8.85	10.75	6.35%	7.20%	8.95%
3100	Total: Mechanical & Electrical		21	26	31.50	18.25%	21%	23%
9000	Per apartment unit, total cost	Apt.	112,000	132,000	218,500			
9500	Total: Mechanical & Electrical	"	21,100	24,400	25,600			
03 0010	**APARTMENTS High Rise (8 to 24 story)**	S.F.	112	129	155			
0020	Total project costs	C.F.	10.90	12.65	14.95			
0100	Site work	S.F.	4.07	6.60	9.20	2.58%	4.84%	6.15%
0500	Masonry		6.50	11.80	14.65	4.74%	9.65%	11.05%
1500	Finishes		12.45	15.55	18.35	9.75%	11.80%	13.70%
1800	Equipment		3.61	4.44	5.85	2.78%	3.49%	4.35%
2500	Conveying equipment		2.55	3.87	5.25	2.23%	2.78%	3.37%
2720	Plumbing		7.15	9.75	11.95	6.80%	7.20%	10.45%
2900	Electrical		7.70	9.75	13.15	6.45%	7.65%	8.80%
3100	Total: Mechanical & Electrical		23	29.50	35.50	17.95%	22.50%	24.50%
9000	Per apartment unit, total cost	Apt.	116,500	128,500	178,000			
9500	Total: Mechanical & Electrical	"	23,200	28,800	30,500			
04 0010	**AUDITORIUMS**	S.F.	117	161	234			
0020	Total project costs	C.F.	7.30	10.15	14.55			
2720	Plumbing	S.F.	6.90	10.20	12.20	5.85%	7.20%	8.70%
2900	Electrical		9.40	13.55	22.50	6.85%	9.50%	11.40%
3100	Total: Mechanical & Electrical		62.50	83	101	24.50%	27.50%	31%
05 0010	**AUTOMOTIVE SALES**	S.F.	86.50	119	146			
0020	Total project costs	C.F.	5.70	6.80	8.85			
2720	Plumbing	S.F.	3.93	6.85	7.45	2.89%	6.05%	6.50%
2770	Heating, ventilating, air conditioning		6.05	9.25	10	4.61%	10%	10.35%
2900	Electrical		6.95	10.90	16	7.25%	8.80%	12.15%
3100	Total: Mechanical & Electrical		19.35	31	37.50	17.30%	20.50%	22%
06 0010	**BANKS**	S.F.	170	211	268			
0020	Total project costs	C.F.	12.10	16.45	21.50			
0100	Site work	S.F.	19.40	30.50	43	7.90%	12.95%	17%
0500	Masonry		7.70	16.50	29.50	3.36%	6.90%	10.05%
1500	Finishes		15.15	23	28.50	5.85%	8.65%	11.70%
1800	Equipment		6.35	14	28.50	1.34%	5.55%	10.50%
2720	Plumbing		5.30	7.55	11.05	2.82%	3.90%	4.93%
2770	Heating, ventilating, air conditioning		10.05	13.45	17.90	4.86%	7.15%	8.50%
2900	Electrical		16.05	21.50	28	8.25%	10.20%	12.20%
3100	Total: Mechanical & Electrical		38.50	51.50	62	16.55%	19.45%	24%
3500	See also division 11 22 00							

50 17 00 | S.F. Costs

		UNIT	UNIT COSTS			% OF TOTAL		
			1/4	MEDIAN	3/4	1/4	MEDIAN	3/4
13 0010	**CHURCHES**	S.F.	114	145	190			
0020	Total project costs	C.F.	7.05	8.90	11.75			
1800	Equipment	S.F.	1.22	3.19	6.80	.83%	2.04%	4.30%
2720	Plumbing		4.43	6.20	9.15	3.51%	4.96%	6.25%
2770	Heating, ventilating, air conditioning		10.35	13.50	19.15	7.50%	10%	12%
2900	Electrical		9.60	13.20	18	7.30%	8.80%	10.95%
3100	Total: Mechanical & Electrical		29.50	39.50	53	18.30%	22%	25%
3500	See also division 11 91 00							
15 0010	**CLUBS, COUNTRY**	S.F.	122	147	185			
0020	Total project costs	C.F.	9.85	12	16.55			
2720	Plumbing	S.F.	7.40	10.95	25	5.60%	7.90%	10%
2900	Electrical		9.60	13.15	17.15	7%	8.95%	11%
3100	Total: Mechanical & Electrical		51	64	67.50	19%	26.50%	29.50%
17 0010	**CLUBS, SOCIAL Fraternal**	S.F.	103	141	188			
0020	Total project costs	C.F.	6.10	9.25	11			
2720	Plumbing	S.F.	6.15	7.65	11.55	5.60%	6.90%	8.55%
2770	Heating, ventilating, air conditioning		7.15	10.70	13.75	8.20%	9.25%	14.40%
2900	Electrical		7.15	12.05	13.75	5.95%	9.30%	10.55%
3100	Total: Mechanical & Electrical		38	41	52	21%	23%	23.50%
18 0010	**CLUBS, Y.M.C.A.**	S.F.	134	167	252			
0020	Total project costs	C.F.	5.65	9.45	14.05			
2720	Plumbing	S.F.	7.75	15.40	17.25	5.65%	7.60%	10.85%
2900	Electrical		10.05	13	22.50	6.45%	7.95%	8.90%
3100	Total: Mechanical & Electrical		41	46.50	76	20.50%	21.50%	28.50%
19 0010	**COLLEGES Classrooms & Administration**	S.F.	118	167	216			
0020	Total project costs	C.F.	7.10	12.50	20			
0500	Masonry	S.F.	9.30	17.15	21	5.10%	8.05%	10.50%
2720	Plumbing		5.30	12.90	24	5.10%	6.60%	8.95%
2900	Electrical		10.05	15.85	22	7.70%	9.85%	12%
3100	Total: Mechanical & Electrical		42.50	58.50	77.50	23%	28%	31.50%
21 0010	**COLLEGES Science, Engineering, Laboratories**	S.F.	230	269	310			
0020	Total project costs	C.F.	13.20	19.25	22			
1800	Equipment	S.F.	6.25	29	31.50	2%	6.45%	12.65%
2900	Electrical		18.95	27	41.50	7.10%	9.40%	12.10%
3100	Total: Mechanical & Electrical		70.50	83.50	129	28.50%	31.50%	41%
3500	See also division 11 53 00							
23 0010	**COLLEGES Student Unions**	S.F.	147	199	241			
0020	Total project costs	C.F.	8.20	10.70	13.25			
3100	Total: Mechanical & Electrical	S.F.	55	59.50	70.50	23.50%	26%	29%
25 0010	**COMMUNITY CENTERS**	S.F.	122	150	203			
0020	Total project costs	C.F.	7.90	11.30	14.65			
1800	Equipment	S.F.	2.45	4.80	7.90	1.47%	3.01%	5.30%
2720	Plumbing		5.75	10.05	13.70	4.85%	7%	8.95%
2770	Heating, ventilating, air conditioning		8.65	13.35	19.15	6.80%	10.35%	12.90%
2900	Electrical		10.25	13.55	19.90	7.15%	8.90%	10.40%
3100	Total: Mechanical & Electrical		34	43	61.50	18.80%	23%	30%
28 0010	**COURT HOUSES**	S.F.	175	206	284			
0020	Total project costs	C.F.	13.40	16	20			
2720	Plumbing	S.F.	8.30	11.60	13.15	5.95%	7.45%	8.20%
2900	Electrical		18.55	21	30.50	9.05%	10.65%	12.15%
3100	Total: Mechanical & Electrical		52	68	74.50	22.50%	26.50%	30%
30 0010	**DEPARTMENT STORES**	S.F.	64.50	87.50	110			
0020	Total project costs	C.F.	3.46	4.48	6.10			
2720	Plumbing	S.F.	2.01	2.54	3.85	1.82%	4.21%	5.90%
2770	Heating, ventilating, air conditioning		4.88	9.05	13.65	8.20%	9.10%	14.80%

For customer support on your Square Foot Cost Data, call 877.694.8329.

535

		50 17 00 | S.F. Costs	UNIT	UNIT COSTS			% OF TOTAL		
				1/4	MEDIAN	3/4	1/4	MEDIAN	3/4
30	2900	Electrical	S.F.	7.40	10.15	12	9.05%	12.15%	14.95%
	3100	Total: Mechanical & Electrical	↓	13.05	16.65	29	13.20%	21.50%	50%
31	0010	**DORMITORIES Low Rise (1 to 3 story)**	S.F.	122	170	212			
	0020	Total project costs	C.F.	6.90	11.20	16.75			
	2720	Plumbing	S.F.	7.35	9.85	12.45	8.05%	9%	9.65%
	2770	Heating, ventilating, air conditioning		7.80	9.35	12.45	4.61%	8.05%	10%
	2900	Electrical		8.10	12.35	16.90	6.40%	8.65%	9.50%
	3100	Total: Mechanical & Electrical	↓	42	45	70.50	22%	25%	27%
	9000	Per bed, total cost	Bed	52,000	57,500	123,500			
32	0010	**DORMITORIES Mid Rise (4 to 8 story)**	S.F.	150	196	242			
	0020	Total project costs	C.F.	16.55	18.20	22			
	2900	Electrical	S.F.	15.95	18.15	24.50	8.20%	10.20%	11.95%
	3100	Total: Mechanical & Electrical	"	44.50	89	90.50	25%	30.50%	35.50%
	9000	Per bed, total cost	Bed	21,400	48,800	282,500			
34	0010	**FACTORIES**	S.F.	57	85	131			
	0020	Total project costs	C.F.	3.65	5.45	9.05			
	0100	Site work	S.F.	6.50	11.85	18.75	6.95%	11.45%	17.95%
	2720	Plumbing		3.07	5.70	9.45	3.73%	6.05%	8.10%
	2770	Heating, ventilating, air conditioning		5.95	8.55	11.55	5.25%	8.45%	11.35%
	2900	Electrical		7.05	11.20	17.05	8.10%	10.50%	14.20%
	3100	Total: Mechanical & Electrical	↓	20	32.50	41	21%	28.50%	35.50%
36	0010	**FIRE STATIONS**	S.F.	113	156	213			
	0020	Total project costs	C.F.	6.60	9.05	12.05			
	0500	Masonry	S.F.	15.40	30	38	7.50%	10.95%	15.55%
	1140	Roofing		3.68	9.95	11.35	1.90%	4.94%	5.05%
	1580	Painting		2.86	4.27	4.37	1.37%	1.57%	2.07%
	1800	Equipment		1.40	2.69	4.97	.62%	1.63%	3.42%
	2720	Plumbing		6.30	10.05	14.25	5.85%	7.35%	9.45%
	2770	Heating, ventilating, air conditioning		6.25	10.15	15.65	5.15%	7.40%	9.40%
	2900	Electrical		8.15	14.35	19.35	6.90%	8.60%	10.60%
	3100	Total: Mechanical & Electrical	↓	41.50	53	60	18.40%	23%	27%
37	0010	**FRATERNITY HOUSES & Sorority Houses**	S.F.	113	145	199			
	0020	Total project costs	C.F.	11.25	11.70	14.10			
	2720	Plumbing	S.F.	8.55	9.75	17.90	6.80%	8%	10.85%
	2900	Electrical	↓	7.45	16.10	19.70	6.60%	9.90%	10.65%
38	0010	**FUNERAL HOMES**	S.F.	119	162	295			
	0020	Total project costs	C.F.	12.15	13.55	26			
	2900	Electrical	S.F.	5.25	9.65	10.55	3.58%	4.44%	5.95%
39	0010	**GARAGES, COMMERCIAL (Service)**	S.F.	67.50	104	144			
	0020	Total project costs	C.F.	4.43	6.55	9.50			
	1800	Equipment	S.F.	3.80	8.55	13.30	2.21%	4.62%	6.80%
	2720	Plumbing		4.67	7.20	13.10	5.45%	7.85%	10.65%
	2730	Heating & ventilating		4.60	7.95	10.15	5.25%	6.85%	8.20%
	2900	Electrical		6.40	9.75	14.10	7.15%	9.25%	10.85%
	3100	Total: Mechanical & Electrical	↓	12.10	27	40	12.35%	17.40%	26%
40	0010	**GARAGES, MUNICIPAL (Repair)**	S.F.	105	135	192			
	0020	Total project costs	C.F.	6.20	7.85	13.50			
	0500	Masonry	S.F.	5.20	18.20	28	4.03%	9.15%	12.50%
	2720	Plumbing		4.44	8.55	16.05	3.59%	6.70%	7.95%
	2730	Heating & ventilating		7.60	11	21	6.15%	7.45%	13.50%
	2900	Electrical		7.65	12.35	22.50	6.90%	9.40%	13%
	3100	Total: Mechanical & Electrical	↓	34.50	56.50	68	21.50%	25.50%	35.50%
41	0010	**GARAGES, PARKING**	S.F.	39.50	56	96.50			
	0020	Total project costs	C.F.	3.62	4.92	7.15			

50 17 00 | S.F. Costs

			UNIT	UNIT COSTS			% OF TOTAL			
				1/4	MEDIAN	3/4	1/4	MEDIAN	3/4	
41	2720	Plumbing	S.F.	.66	1.69	2.61	1.72%	2.70%	3.85%	**41**
	2900	Electrical		2.12	2.80	4.08	4.52%	5.40%	6.35%	
	3100	Total: Mechanical & Electrical	↓	4.11	6.05	7.50	7%	8.90%	11.05%	
	3200									
	9000	Per car, total cost	Car	16,300	20,400	26,000				
43	0010	**GYMNASIUMS**	S.F.	108	143	196				**43**
	0020	Total project costs	C.F.	5.35	7.25	8.90				
	1800	Equipment	S.F.	2.55	4.78	8.60	1.76%	3.26%	6.70%	
	2720	Plumbing		5.90	7.90	10.40	4.65%	6.40%	7.75%	
	2770	Heating, ventilating, air conditioning		6.40	11.10	22.50	5.15%	9.05%	11.10%	
	2900	Electrical		8.50	11.25	14.95	6.75%	8.50%	10.30%	
	3100	Total: Mechanical & Electrical	↓	29	40.50	49.50	19.75%	23.50%	29%	
	3500	See also division 11 66 00								
46	0010	**HOSPITALS**	S.F.	206	258	355				**46**
	0020	Total project costs	C.F.	15.60	19.40	28				
	1800	Equipment	S.F.	5.20	10.05	17.30	.80%	2.53%	4.80%	
	2720	Plumbing		17.70	25	32	7.60%	9.10%	10.85%	
	2770	Heating, ventilating, air conditioning		26	33.50	46.50	7.80%	12.95%	16.65%	
	2900	Electrical		22.50	30.50	45.50	10%	11.75%	14.10%	
	3100	Total: Mechanical & Electrical	↓	66.50	92.50	137	28%	33.50%	37%	
	9000	Per bed or person, total cost	Bed	238,000	328,500	378,500				
	9900	See also division 11 71 00								
48	0010	**HOUSING For the Elderly**	S.F.	101	128	157				**48**
	0020	Total project costs	C.F.	7.20	10	12.80				
	0100	Site work	S.F.	7.05	10.95	16.05	5.05%	7.90%	12.10%	
	0500	Masonry		1.91	11.55	16.85	1.30%	6.05%	11%	
	1800	Equipment		2.45	3.37	5.35	1.88%	3.23%	4.43%	
	2510	Conveying systems		2.29	3.31	4.49	1.78%	2.20%	2.81%	
	2720	Plumbing		7.50	9.60	12.10	8.15%	9.55%	10.50%	
	2730	Heating, ventilating, air conditioning		3.86	5.45	8.15	3.30%	5.60%	7.25%	
	2900	Electrical		7.55	10.25	13.10	7.30%	8.50%	10.25%	
	3100	Total: Mechanical & Electrical	↓	26	32	41	18.10%	22.50%	29%	
	9000	Per rental unit, total cost	Unit	94,000	110,000	122,500				
	9500	Total: Mechanical & Electrical	"	21,000	24,100	28,100				
50	0010	**HOUSING Public (Low Rise)**	S.F.	85	118	154				**50**
	0020	Total project costs	C.F.	6.75	9.45	11.75				
	0100	Site work	S.F.	10.85	15.60	25.50	8.35%	11.75%	16.50%	
	1800	Equipment		2.31	3.77	5.75	2.26%	3.03%	4.24%	
	2720	Plumbing		6.15	8.10	10.25	7.15%	9.05%	11.60%	
	2730	Heating, ventilating, air conditioning		3.08	6	6.55	4.26%	6.05%	6.45%	
	2900	Electrical		5.15	7.65	10.65	5.10%	6.55%	8.25%	
	3100	Total: Mechanical & Electrical	↓	24.50	31.50	35	14.50%	17.55%	26.50%	
	9000	Per apartment, total cost	Apt.	93,500	106,500	133,500				
	9500	Total: Mechanical & Electrical	"	19,900	24,600	27,200				
51	0010	**ICE SKATING RINKS**	S.F.	72.50	170	187				**51**
	0020	Total project costs	C.F.	5.35	5.45	6.30				
	2720	Plumbing	S.F.	2.72	5.10	5.20	3.12%	3.23%	5.65%	
	2900	Electrical	↓	7.80	11.95	12.65	6.30%	10.15%	15.05%	
52	0010	**JAILS**	S.F.	189	286	370				**52**
	0020	Total project costs	C.F.	19.95	28	32.50				
	1800	Equipment	S.F.	8.65	25.50	43.50	2.80%	6.95%	9.80%	
	2720	Plumbing		21.50	28.50	37.50	7%	8.90%	13.35%	
	2770	Heating, ventilating, air conditioning		19.95	26.50	51.50	7.50%	9.45%	17.75%	
	2900	Electrical		23.50	31.50	40	9.40%	11.70%	15.25%	
	3100	Total: Mechanical & Electrical	↓	62	110	131	28%	31%	36%	
53	0010	**LIBRARIES**	S.F.	143	190	248				**53**
	0020	Total project costs	C.F.	9.55	12	15.30				

50 17 00 | S.F. Costs

	Code	Description	UNIT	UNIT COSTS 1/4	MEDIAN	3/4	% OF TOTAL 1/4	MEDIAN	3/4	
53	0500	Masonry	S.F.	9.50	19.35	32.50	5.60%	7.15%	10.95%	53
	1800	Equipment		1.91	5.15	7.75	.28%	1.39%	4.07%	
	2720	Plumbing		5.15	7.30	10.20	3.38%	4.60%	5.70%	
	2770	Heating, ventilating, air conditioning		11.45	19.35	23.50	7.80%	10.95%	12.80%	
	2900	Electrical		14.40	19	25.50	8.40%	10.55%	12%	
	3100	Total: Mechanical & Electrical		47	57	67	21%	24%	28%	
54	0010	LIVING, ASSISTED	S.F.	129	154	180				54
	0020	Total project costs	C.F.	10.90	12.70	14.45				
	0500	Masonry	S.F.	3.75	4.53	5.55	2.36%	3.16%	3.86%	
	1800	Equipment		2.87	3.42	4.38	2.12%	2.45%	2.87%	
	2720	Plumbing		10.85	14.50	15.05	6.05%	8.15%	10.60%	
	2770	Heating, ventilating, air conditioning		12.85	13.45	14.75	7.95%	9.35%	9.70%	
	2900	Electrical		12.70	14.25	16.50	9%	9.95%	10.65%	
	3100	Total: Mechanical & Electrical		35	41	48	24%	28.50%	31.50%	
55	0010	MEDICAL CLINICS	S.F.	132	162	207				55
	0020	Total project costs	C.F.	9.65	12.50	16.60				
	1800	Equipment	S.F.	1.68	7.50	11.65	1.05%	2.94%	6.35%	
	2720	Plumbing		8.75	12.30	16.50	6.15%	8.40%	10.10%	
	2770	Heating, ventilating, air conditioning		10.45	13.70	20	6.65%	8.85%	11.35%	
	2900	Electrical		11.30	16.10	21	8.10%	10%	12.20%	
	3100	Total: Mechanical & Electrical		36	49	67.50	22%	27%	33.50%	
	3500	See also division 11 71 00								
57	0010	MEDICAL OFFICES	S.F.	124	154	190				57
	0020	Total project costs	C.F.	9.25	12.50	16.95				
	1800	Equipment	S.F.	4.09	8.10	11.35	.66%	4.87%	6.50%	
	2720	Plumbing		6.85	10.55	14.25	5.60%	6.80%	8.50%	
	2770	Heating, ventilating, air conditioning		8.25	11.95	15.75	6.10%	8%	9.70%	
	2900	Electrical		9.90	14.50	20.50	7.50%	9.80%	11.70%	
	3100	Total: Mechanical & Electrical		27	39.50	56	19.30%	22.50%	27.50%	
59	0010	MOTELS	S.F.	78	113	148				59
	0020	Total project costs	C.F.	6.95	9.30	15.25				
	2720	Plumbing	S.F.	7.90	10.10	12.05	9.45%	10.60%	12.55%	
	2770	Heating, ventilating, air conditioning		4.83	7.20	12.90	5.60%	5.60%	10%	
	2900	Electrical		7.40	9.35	11.65	7.45%	9.05%	10.45%	
	3100	Total: Mechanical & Electrical		22.50	31.50	54	18.50%	24%	25.50%	
	5000									
	9000	Per rental unit, total cost	Unit	39,800	75,500	82,000				
	9500	Total: Mechanical & Electrical	"	7,750	11,700	13,600				
60	0010	NURSING HOMES	S.F.	123	158	197				60
	0020	Total project costs	C.F.	9.65	12.05	16.45				
	1800	Equipment	S.F.	3.34	5.10	8.55	2%	3.62%	4.99%	
	2720	Plumbing		10.50	15.90	19.20	8.75%	10.10%	12.70%	
	2770	Heating, ventilating, air conditioning		11.05	16.80	22.50	9.70%	11.45%	11.80%	
	2900	Electrical		12.15	15.20	21	9.40%	10.60%	12.50%	
	3100	Total: Mechanical & Electrical		29	40.50	68	26%	28%	30%	
	9000	Per bed or person, total cost	Bed	54,500	68,000	88,000				
61	0010	OFFICES Low Rise (1 to 4 story)	S.F.	103	135	175				61
	0020	Total project costs	C.F.	7.35	10.15	13.40				
	0100	Site work	S.F.	8.30	14.45	21.50	5.95%	9.70%	13.55%	
	0500	Masonry		4	8.05	14.70	2.61%	5.40%	8.45%	
	1800	Equipment		.93	2.15	5.85	.57%	1.50%	3.42%	
	2720	Plumbing		3.69	5.70	8.35	3.66%	4.50%	6.10%	
	2770	Heating, ventilating, air conditioning		8.15	11.40	16.65	7.20%	10.30%	11.70%	
	2900	Electrical		8.45	12.10	17.20	7.45%	9.65%	11.40%	
	3100	Total: Mechanical & Electrical		23.50	32.50	48.50	18.20%	22%	27%	
62	0010	OFFICES Mid Rise (5 to 10 story)	S.F.	109	132	180				62
	0020	Total project costs	C.F.	7.75	9.90	14				

50 17 00 | S.F. Costs

			UNIT	UNIT COSTS			% OF TOTAL			
				1/4	MEDIAN	3/4	1/4	MEDIAN	3/4	
62	2720	Plumbing	S.F.	3.30	5.10	7.35	2.83%	3.74%	4.50%	62
	2770	Heating, ventilating, air conditioning		8.30	11.85	18.95	7.65%	9.40%	11%	
	2900	Electrical		8.10	10.40	15.70	6.35%	7.80%	10%	
	3100	Total: Mechanical & Electrical		21	27	51.50	18.95%	21%	27.50%	
63	0010	**OFFICES High Rise (11 to 20 story)**	S.F.	134	169	208				63
	0020	Total project costs	C.F.	9.40	11.75	16.85				
	2900	Electrical	S.F.	8.15	9.95	14.80	5.80%	7.85%	10.50%	
	3100	Total: Mechanical & Electrical		26.50	35.50	59.50	16.90%	23.50%	34%	
64	0010	**POLICE STATIONS**	S.F.	161	212	270				64
	0020	Total project costs	C.F.	12.85	15.70	21.50				
	0500	Masonry	S.F.	16.10	26.50	33.50	7.80%	9.10%	11.35%	
	1800	Equipment		2.31	11.25	17.80	.98%	3.35%	6.70%	
	2720	Plumbing		9	17.95	22.50	5.65%	6.90%	10.75%	
	2770	Heating, ventilating, air conditioning		13.20	18.70	26.50	5.85%	10.55%	11.70%	
	2900	Electrical		17.60	25.50	33	9.80%	11.70%	14.50%	
	3100	Total: Mechanical & Electrical		67	70.50	94	28.50%	31.50%	32.50%	
65	0010	**POST OFFICES**	S.F.	127	157	200				65
	0020	Total project costs	C.F.	7.65	9.70	11				
	2720	Plumbing	S.F.	5.15	7.10	8.95	4.24%	5.30%	5.60%	
	2770	Heating, ventilating, air conditioning		8	11.05	12.30	6.65%	7.15%	9.35%	
	2900	Electrical		10.50	14.75	17.50	7.25%	9%	11%	
	3100	Total: Mechanical & Electrical		30.50	39.50	45	16.25%	18.80%	22%	
66	0010	**POWER PLANTS**	S.F.	880	1,175	2,150				66
	0020	Total project costs	C.F.	24.50	53	113				
	2900	Electrical	S.F.	62.50	132	197	9.30%	12.75%	21.50%	
	8100	Total: Mechanical & Electrical		155	505	1,125	32.50%	32.50%	52.50%	
67	0010	**RELIGIOUS EDUCATION**	S.F.	103	137	170				67
	0020	Total project costs	C.F.	5.70	8.15	10.20				
	2720	Plumbing	S.F.	4.28	6.05	8.60	4.40%	5.30%	7.10%	
	2770	Heating, ventilating, air conditioning		10.80	12.25	17.30	10.05%	11.45%	12.35%	
	2900	Electrical		8.15	11.65	17.10	7.70%	9.05%	10.35%	
	3100	Total: Mechanical & Electrical		35.50	45	53	22%	23.50%	26%	
69	0010	**RESEARCH Laboratories & Facilities**	S.F.	157	222	325				69
	0020	Total project costs	C.F.	11.40	22.50	27				
	1800	Equipment	S.F.	6.80	13.40	32.50	1.22%	4.80%	9.35%	
	2720	Plumbing		11.20	19.70	31.50	6.15%	8.30%	10.80%	
	2770	Heating, ventilating, air conditioning		13.80	46.50	55	7.25%	16.50%	17.50%	
	2900	Electrical		18.45	29.50	48.50	9.45%	11.15%	14.60%	
	3100	Total: Mechanical & Electrical		59	104	147	29.50%	36%	41%	
70	0010	**RESTAURANTS**	S.F.	149	192	249				70
	0020	Total project costs	C.F.	12.50	16.40	21.50				
	1800	Equipment	S.F.	8.80	23.50	35.50	6.10%	13%	15.65%	
	2720	Plumbing		11.80	14.30	18.75	6.10%	8.15%	9%	
	2770	Heating, ventilating, air conditioning		14.95	21	25	9.20%	12%	12.40%	
	2900	Electrical		15.70	19.40	25	8.35%	10.55%	11.55%	
	3100	Total: Mechanical & Electrical		48.50	52	67.50	21%	25%	29.50%	
	9000	Per seat unit, total cost	Seat	5,450	7,275	8,600				
	9500	Total: Mechanical & Electrical	"	1,375	1,825	2,150				
72	0010	**RETAIL STORES**	S.F.	69.50	93.50	124				72
	0020	Total project costs	C.F.	4.71	6.70	9.35				
	2720	Plumbing	S.F.	2.52	4.20	7.15	3.26%	4.60%	6.80%	
	2770	Heating, ventilating, air conditioning		5.45	7.45	11.20	6.75%	8.75%	10.15%	
	2900	Electrical		6.25	8.55	12.35	7.25%	9.90%	11.60%	
	3100	Total: Mechanical & Electrical		16.65	21.50	28.50	17.05%	21%	23.50%	
74	0010	**SCHOOLS Elementary**	S.F.	114	141	172				74
	0020	Total project costs	C.F.	7.40	9.50	12.25				
	0500	Masonry	S.F.	8	17.30	25.50	4.89%	10.50%	14%	
	1800	Equipment		2.63	4.96	9.40	1.83%	3.13%	4.61%	

50 17 00 | S.F. Costs

			UNIT	UNIT COSTS			% OF TOTAL		
				1/4	MEDIAN	3/4	1/4	MEDIAN	3/4
74	2720	Plumbing	S.F.	6.50	9.20	12.30	5.70%	7.15%	9.35%	74
	2730	Heating, ventilating, air conditioning		9.80	15.55	22.50	8.15%	10.80%	14.90%	
	2900	Electrical		10.70	14.25	18.20	8.45%	10.05%	11.85%	
	3100	Total: Mechanical & Electrical	↓	39	47.50	60	25%	27.50%	30%	
	9000	Per pupil, total cost	Ea.	13,000	19,300	43,300				
	9500	Total: Mechanical & Electrical	"	3,675	4,650	11,700				
76	0010	**SCHOOLS Junior High & Middle**	S.F.	117	145	177				76
	0020	Total project costs	C.F.	7.40	9.60	10.75				
	0500	Masonry	S.F.	12.55	18.80	23	7.55%	11.10%	14.30%	
	1800	Equipment		3.20	6	8.80	1.80%	3.03%	4.80%	
	2720	Plumbing		6.80	8.40	10.40	5.30%	6.80%	7.25%	
	2770	Heating, ventilating, air conditioning		13.60	16.55	29	8.90%	11.55%	14.20%	
	2900	Electrical		11.90	14.70	18.60	8.05%	9.55%	10.60%	
	3100	Total: Mechanical & Electrical	↓	37.50	48	59.50	23%	25.50%	29.50%	
	9000	Per pupil, total cost	Ea.	14,800	19,400	26,100				
78	0010	**SCHOOLS Senior High**	S.F.	123	150	188				78
	0020	Total project costs	C.F.	7.35	10.30	17.40				
	1800	Equipment	S.F.	3.24	7.55	10.45	1.88%	2.67%	4.30%	
	2720	Plumbing		6.80	10.25	18.70	5.60%	6.90%	8.30%	
	2770	Heating, ventilating, air conditioning		11.75	15.95	24.50	8.95%	11.60%	15%	
	2900	Electrical		12.30	16.40	24	8.70%	10.35%	12.50%	
	3100	Total: Mechanical & Electrical	↓	40.50	48	77.50	24%	26.50%	28.50%	
	9000	Per pupil, total cost	Ea.	11,500	23,300	29,200				
80	0010	**SCHOOLS Vocational**	S.F.	98.50	143	177				80
	0020	Total project costs	C.F.	6.10	8.80	12.15				
	0500	Masonry	S.F.	4.33	14.25	22	3.20%	4.61%	10.95%	
	1800	Equipment		3.08	7.65	10.60	1.24%	3.10%	4.26%	
	2720	Plumbing		6.30	9.35	13.80	5.40%	6.90%	8.55%	
	2770	Heating, ventilating, air conditioning		8.80	16.40	27.50	8.60%	11.90%	14.65%	
	2900	Electrical		10.25	14	19.15	8.45%	11%	13.20%	
	3100	Total: Mechanical & Electrical	↓	38.50	67.50	86.50	27.50%	29.50%	31%	
	9000	Per pupil, total cost	Ea.	13,700	36,700	54,500				
83	0010	**SPORTS ARENAS**	S.F.	86	115	177				83
	0020	Total project costs	C.F.	4.67	8.35	10.80				
	2720	Plumbing	S.F.	4.31	7.55	15.95	4.35%	6.35%	9.40%	
	2770	Heating, ventilating, air conditioning		10.75	12.70	17.60	8.80%	10.20%	13.55%	
	2900	Electrical	↓	7.95	12.15	15.70	7.65%	9.75%	11.90%	
85	0010	**SUPERMARKETS**	S.F.	79.50	92	108				85
	0020	Total project costs	C.F.	4.42	5.35	8.10				
	2720	Plumbing	S.F.	4.44	5.60	6.50	5.40%	6%	7.45%	
	2770	Heating, ventilating, air conditioning		6.55	8.65	10.55	8.60%	8.65%	9.60%	
	2900	Electrical		9.30	11.35	13.50	10.40%	12.45%	13.60%	
	3100	Total: Mechanical & Electrical	↓	25.50	27	36.50	23.50%	27.50%	28.50%	
86	0010	**SWIMMING POOLS**	S.F.	129	300	460				86
	0020	Total project costs	C.F.	10.30	12.85	14				
	2720	Plumbing	S.F.	11.90	13.60	18.25	4.80%	9.70%	20.50%	
	2900	Electrical		9.75	15.65	34.50	6.05%	6.95%	7.60%	
	3100	Total: Mechanical & Electrical	↓	61	80.50	106	11.15%	14.10%	23.50%	
87	0010	**TELEPHONE EXCHANGES**	S.F.	160	243	315				87
	0020	Total project costs	C.F.	10.65	17.05	23.50				
	2720	Plumbing	S.F.	7.20	11.15	16.30	4.52%	5.80%	6.90%	
	2770	Heating, ventilating, air conditioning		16.75	33.50	42	11.80%	16.05%	18.40%	
	2900	Electrical		17.40	27.50	49	10.90%	14%	17.85%	
	3100	Total: Mechanical & Electrical	↓	51.50	97.50	138	29.50%	33.50%	44.50%	
91	0010	**THEATERS**	S.F.	102	138	203				91
	0020	Total project costs	C.F.	4.96	7.35	10.80				

50 17 00 | S.F. Costs		UNIT	UNIT COSTS			% OF TOTAL		
			1/4	MEDIAN	3/4	1/4	MEDIAN	3/4
91 2720	Plumbing	S.F.	3.34	3.88	15.85	2.92%	4.70%	6.80%
2770	Heating, ventilating, air conditioning		10.45	12.65	15.65	8%	12.25%	13.40%
2900	Electrical		9.40	12.70	26	8.05%	9.35%	12.25%
3100	Total: Mechanical & Electrical		24	32.50	38.50	21.50%	25.50%	27.50%
94 0010	**TOWN HALLS City Halls & Municipal Buildings**	S.F.	112	152	199			
0020	Total project costs	C.F.	8.10	12.60	18.45			
2720	Plumbing	S.F.	5	9.35	16.45	4.31%	5.95%	7.95%
2770	Heating, ventilating, air conditioning		9.05	17.95	26.50	7.05%	9.05%	13.45%
2900	Electrical		11.40	15.70	21.50	8.05%	9.45%	11.65%
3100	Total: Mechanical & Electrical		39.50	50	76.50	22%	26.50%	31%
97 0010	**WAREHOUSES & Storage Buildings**	S.F.	44.50	63.50	95			
0020	Total project costs	C.F.	2.34	3.67	6.05			
0100	Site work	S.F.	4.62	9.15	13.80	6.05%	12.95%	19.55%
0500	Masonry		2.54	6.35	13.70	3.60%	7.35%	12%
1800	Equipment		.68	1.55	8.70	.72%	1.69%	5.55%
2720	Plumbing		1.49	2.68	5	2.90%	4.80%	6.55%
2730	Heating, ventilating, air conditioning		1.70	4.80	6.45	2.41%	5%	8.90%
2900	Electrical		2.65	4.98	8.20	5.15%	7.20%	10.05%
3100	Total: Mechanical & Electrical		7.40	11.35	22.50	13.30%	18.90%	26%
99 0010	**WAREHOUSE & OFFICES Combination**	S.F.	55	73.50	101			
0020	Total project costs	C.F.	2.82	4.09	6.05			
1800	Equipment	S.F.	.95	1.84	2.74	.42%	1.19%	2.40%
2720	Plumbing		2.13	3.70	5.50	3.74%	4.76%	6.30%
2770	Heating, ventilating, air conditioning		3.36	5.25	7.35	5%	5.65%	10.05%
2900	Electrical		3.75	5.55	8.65	5.85%	8%	10%
3100	Total: Mechanical & Electrical		10.45	15.85	25	14.55%	19.95%	24.50%

For customer support on your Square Foot Cost Data, call 877.694.8329.

547

Square Foot Project Size Modifier

One factor that affects the S.F. cost of a particular building is the size. In general, for buildings built to the same specifications in the same locality, the larger building will have the lower S.F. cost. This is due mainly to the decreasing contribution of the exterior walls, plus the economy of scale usually achievable in larger buildings. The area conversion scale shown below will give a factor to convert costs for the typical size building to an adjusted cost for the particular project.

The square foot base size table lists the median costs, most typical project size in our accumulated data, and the range in size of the projects.

The size factor for your project is determined by dividing your project area in S.F. by the typical project size for the particular building type. With this factor, enter the area conversion scale at the appropriate size factor and determine the appropriate cost multiplier for your building size.

Example: Determine the cost per S.F. for a 100,000 S.F. Mid-rise apartment building.

$$\frac{\text{Proposed building area} = 100,000 \text{ S.F.}}{\text{Typical size from below} = 50,000 \text{ S.F.}} = 2.00$$

Enter area conversion scale at 2.0, intersect curve, read horizontally the appropriate cost multiplier of .94. Size adjusted cost becomes .94 × $119.00 = $112.00 based on national average costs.

Note: For size factors less than .50, the cost multiplier is 1.1
For size factors greater than 3.5, the cost multiplier is .90

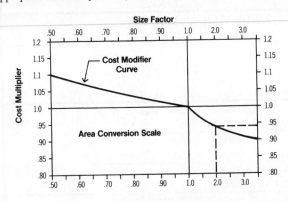

Square Foot Base Size

Building Type	Median Cost per S.F.	Typical Size Gross S.F.	Typical Range Gross S.F.	Building Type	Median Cost per S.F.	Typical Size Gross S.F.	Typical Range Gross S.F.
Apartments, Low Rise	$ 94.00	21,000	9,700 - 37,200	Jails	$ 286.00	40,000	5,500 - 145,000
Apartments, Mid Rise	119.00	50,000	32,000 - 100,000	Libraries	190.00	12,000	7,000 - 31,000
Apartments, High Rise	129.00	145,000	95,000 - 600,000	Living, Assisted	154.00	32,300	23,500 - 50,300
Auditoriums	161.00	25,000	7,600 - 39,000	Medical Clinics	162.00	7,200	4,200 - 15,700
Auto Sales	119.00	20,000	10,800 - 28,600	Medical Offices	154.00	6,000	4,000 - 15,000
Banks	211.00	4,200	2,500 - 7,500	Motels	113.00	40,000	15,800 - 120,000
Churches	145.00	17,000	2,000 - 42,000	Nursing Homes	158.00	23,000	15,000 - 37,000
Clubs, Country	147.00	6,500	4,500 - 15,000	Offices, Low Rise	135.00	20,000	5,000 - 80,000
Clubs, Social	141.00	10,000	6,000 - 13,500	Offices, Mid Rise	132.00	120,000	20,000 - 300,000
Clubs, YMCA	167.00	28,300	12,800 - 39,400	Offices, High Rise	169.00	260,000	120,000 - 800,000
Colleges (Class)	167.00	50,000	15,000 - 150,000	Police Stations	212.00	10,500	4,000 - 19,000
Colleges (Science Lab)	269.00	45,600	16,600 - 80,000	Post Offices	157.00	12,400	6,800 - 30,000
College (Student Union)	199.00	33,400	16,000 - 85,000	Power Plants	1175.00	7,500	1,000 - 20,000
Community Center	150.00	9,400	5,300 - 16,700	Religious Education	137.00	9,000	6,000 - 12,000
Court Houses	206.00	32,400	17,800 - 106,000	Research	222.00	19,000	6,300 - 45,000
Dept. Stores	87.50	90,000	44,000 - 122,000	Restaurants	192.00	4,400	2,800 - 6,000
Dormitories, Low Rise	170.00	25,000	10,000 - 95,000	Retail Stores	93.50	7,200	4,000 - 17,600
Dormitories, Mid Rise	196.00	85,000	20,000 - 200,000	Schools, Elementary	141.00	41,000	24,500 - 55,000
Factories	85.00	26,400	12,900 - 50,000	Schools, Jr. High	145.00	92,000	52,000 - 119,000
Fire Stations	156.00	5,800	4,000 - 8,700	Schools, Sr. High	150.00	101,000	50,500 - 175,000
Fraternity Houses	145.00	12,500	8,200 - 14,800	Schools, Vocational	143.00	37,000	20,500 - 82,000
Funeral Homes	162.00	10,000	4,000 - 20,000	Sports Arenas	115.00	15,000	5,000 - 40,000
Garages, Commercial	104.00	9,300	5,000 - 13,600	Supermarkets	92.00	44,000	12,000 - 60,000
Garages, Municipal	135.00	8,300	4,500 - 12,600	Swimming Pools	300.00	20,000	10,000 - 32,000
Garages, Parking	56.00	163,000	76,400 - 225,300	Telephone Exchange	243.00	4,500	1,200 - 10,600
Gymnasiums	143.00	19,200	11,600 - 41,000	Theaters	138.00	10,500	8,800 - 17,500
Hospitals	258.00	55,000	27,200 - 125,000	Town Halls	152.00	10,800	4,800 - 23,400
House (Elderly)	128.00	37,000	21,000 - 66,000	Warehouses	63.50	25,000	8,000 - 72,000
Housing (Public)	118.00	36,000	14,400 - 74,400	Warehouse & Office	73.50	25,000	8,000 - 72,000
Ice Rinks	170.00	29,000	27,200 - 33,600				

Glossary

Accent lighting–Fixtures or directional beams of light arranged to bring attention to an object or area.

Acoustical material–A material fabricated for the sole purpose of absorbing sound.

Addition–An expansion to an existing structure generally in the form of a room, floor(s) or wing(s). An increase in the floor area or volume of a structure.

Aggregate–Materials such as sand, gravel, stone, vermiculite, perlite and fly ash (slag) which are essential components in the production of concrete, mortar or plaster.

Air conditioning system–An air treatment to control the temperature, humidity and cleanliness of air and to provide for its distribution throughout the structure.

Air curtain–A stream of air directed downward to prevent the loss of hot or cool air from the structure and inhibit the entrance of dust and insects. Generally installed at loading platforms.

Anodized aluminum–Aluminum treated by an electrolytic process to produce an oxide film that is corrosion resistant.

Arcade–A covered passageway between buildings often with shops and offices on one or both sides.

Ashlar–A square-cut building stone or a wall constructed of cut stone.

Asphalt paper–A sheet paper material either coated or saturated with asphalt to increase its strength and resistance to water.

Asphalt shingles–Shingles manufactured from saturated roofing felts, coated with asphalt and topped with mineral granules to prevent weathering.

Assessment ratio–The ratio between the market value and assessed valuation of a property.

Back-up–That portion of a masonry wall behind the exterior facing-usually load bearing.

Balcony–A platform projecting from a building either supported from below or cantilevered. Usually protected by railing.

Bay–Structural component consisting of beams and columns occurring consistently throughout the structure.

Beam–A horizontal structural framing member that transfers loads to vertical members (columns) or bearing walls.

Bid–An offer to perform work described in a contract at a specified price.

Board foot–A unit or measure equal in volume to a board one foot long, one foot wide and one inch thick.

Booster pump–A supplemental pump installed to increase or maintain adequate pressure in the system.

Bowstring roof–A roof supported by trusses fabricated in the shape of a bow and tied together by a straight member.

Brick veneer–A facing of brick laid against, but not bonded to, a wall.

Bridging–A method of bracing joists for stiffness, stability and load distribution.

Broom finish–A method of finishing concrete by lightly dragging a broom over freshly placed concrete.

Built-up roofing–Installed on flat or extremely low-pitched roofs, a roof covering composed of plies or laminations of saturated roofing felts alternated with layers of coal tar pitch or asphalt and surfaced with a layer of gravel or slag in a thick coat of asphalt and finished with a capping sheet.

Caisson–A watertight chamber to expedite work on foundations or structure below water level.

Cantilever–A structural member supported only at one end.

Carport–An automobile shelter having one or more sides open to the weather.

Casement window–A vertical opening window having one side fixed.

Caulking–A resilient compound generally having a silicone or rubber base used to prevent infiltration of water or outside air.

Cavity wall–An exterior wall usually of masonry having an inner and outer wall separated by a continuous air space for thermal insulation.

Cellular concrete–A lightweight concrete consisting of cement mixed with gas-producing materials which, in turn, forms bubbles resulting in good insulating qualities.

Chattel mortgage–A secured interest in a property as collateral for payment of a note.

Clapboard–A wood siding used as exterior covering in frame construction. It is applied horizontally with grain running lengthwise. The thickest section of the board is on the bottom.

Clean room–An environmentally controlled room usually found in medical facilities or precision manufacturing spaces where it is essential to eliminate dust, lint or pathogens.

Close studding–A method of construction whereby studs are spaced close together and the intervening spaces are plastered.

Cluster housing–A closely grouped series of houses resulting in a high density land use.

Cofferdam–A watertight enclosure used for foundation construction in waterfront areas. Water is pumped out of the cofferdam allowing free access to the work area.

Collar joint–Joint between the collar beam and roof rafters.

Column–A vertical structural member supporting horizontal members (beams) along the direction of its longitudinal axis.

Combination door (windows)–Door (windows) having interchangeable screens and glass for seasonal use.

Common area–Spaces either inside or outside the building designated for use by the occupant of the building but not the general public.

Common wall–A wall used jointly by two dwelling units.

Compound wall–A wall constructed of more than one material.

Concrete–A composite material consisting of sand, coarse aggregate (gravel, stone or slag), cement and water that when mixed and allowed to harden forms a hard stone-like material.

Concrete floor hardener–A mixture of chemicals applied to the surface of concrete to produce a dense, wear-resistant bearing surface.

Conduit–A tube or pipe used to protect electric wiring.

Coping–The top cover or capping on a wall.

Craneway–Steel or concrete column and beam supports and rails on which a crane travels.

Curtain wall–A non-bearing exterior wall not supported by beams or girders of the steel frame.

Dampproofing–Coating of a surface to prevent the passage of water or moisture.

Dead load–Total weight of all structural components plus permanently attached fixtures and equipment.

Decibel–Unit of acoustical measurement.

Decorative block–A concrete masonry unit having a special treatment on its face.

Deed restriction–A restriction on the use of a property as set forth in the deed.

Depreciation–A loss in property value caused by physical aging, functional or economic obsolescence.

Direct heating–Heating of spaces by means of exposed heated surfaces (stove, fire, radiators, etc.).

Distributed load–A load that acts evenly over a structural member.

Dock bumper–A resilient material attached to a loading dock to absorb the impact of trucks backing in.

Dome–A curved roof shape spanning an area.

Dormer–A structure projecting from a sloping roof.

Double-hung window–A window having two vertical sliding sashes-one covering the upper section and one covering the lower.

Downspout–A vertical pipe for carrying rain water from the roof to the ground.

Drain tile–A hollow tile used to drain water-soaked soil.

Dressed size–In lumber about 3/8″ - 1/2″ less than nominal size after planing and sawing.

Drop panel–The depressed surface on the bottom side of a flat concrete slab which surrounds a column.

Dry pipe sprinkler system–A sprinkler system that is activated with water only in case of fire. Used in areas susceptible to freezing or to avoid the hazards of leaking pipes.

Drywall–An interior wall constructed of gypsum board, plywood or wood paneling. No water is required for application.

Duct–Pipe for transmitting warm or cold air. A pipe containing electrical cable or wires.

Dwelling–A structure designed as living quarters for one or more families.

Easement–A right of way or free access over land owned by another.

Eave–The lower edge of a sloping roof that overhangs the side wall of the structure.

Economic rent–That rent on a property sufficient to pay all operating costs exclusive of services and utilities.

Economic life–The term during which a structure is expected to be profitable. Generally shorter than the physical life of the structure.

Economic obsolescence–Loss in value due to unfavorable economic influences occurring from outside the structure itself.

Effective age–The age a structure appears to be based on observed physical condition determined by degree of maintenance and repair.

Elevator–A mechanism for vertical transport of personnel or freight equipped with car or platform.

Ell–A secondary wing or addition to a structure at right angles to the main structure.

Eminent domain–The right of the state to take private property for public use.

Envelope–The shape of a building indicating volume.

Equity–Value of an owner's interest calculated by subtracting outstanding mortgages and expenses from the value of the property.

Escheat–The assumption of ownership by the state, of property whose owner cannot be determined.

Facade–The exterior face of a building sometimes decorated with elaborate detail.

Face brick–Brick manufactured to present an attractive appearance.

Facing block–A concrete masonry unit having a decorative exterior finish.

Feasibility study–A detailed evaluation of a proposed project to determine its financial potential.

Felt paper–Paper sheathing used on walls as an insulator and to prevent infiltration of moisture.

Fenestration–Pertaining to the density and arrangement of windows in a structure.

Fiberboard–A building material composed of wood fiber compressed with a binding agent, produced in sheet form.

Fiberglass–Fine spun filaments of glass processed into various densities to produce thermal and acoustical insulation.

Field house–A long structure used for indoor athletic events.

Finish floor–The top or wearing surface of the floor system.

Fireproofing–The use of fire-resistant materials for the protection of structural members to ensure structural integrity in the event of fire.

Fire stop–A material or member used to seal an opening to prevent the spread of fire.

Flashing–A thin, impervious material such as copper or sheet metal used to prevent air or water penetration.

Float finish–A concrete finish accomplished by using a flat tool with a handle on the back.

Floor drain–An opening installed in a floor for removing excess water into a plumbing system.

Floor load–The live load the floor system has been designed for and which may be applied safely.

Flue–A heat-resistant enclosed passage in a chimney to remove gaseous products of combustion from a fireplace or boiler.

Footing–The lowest portion of the foundation wall that transmits the load directly to the soil.

Foundation–Below grade wall system that supports the structure.

Foyer–A portion of the structure that serves as a transitional space between the interior and exterior.

Frontage–That portion of a lot that is placed facing a street, public way or body of water.

Frost action–The effects of freezing and thawing on materials and the resultant structural damage.

Functional obsolescence–An inadequacy caused by outmoded design, dated construction materials or over or undersized areas, all of which cause excessive operating costs.

Furring strips–Wood or metal channels used for attaching gypsum or metal lath to masonry walls as a finish or for leveling purposes.

Gambrel roof (mansard)–A roof system having two pitches on each side.

Garden apartment–Two or three story apartment building with common outside areas.

General contractor–The prime contractor responsible for work on the construction site.

Girder–A principal horizontal supporting member.

Girt–A horizontal framing member for adding rigid support to columns which also acts as a support for sheathing or siding.

Grade beam–That portion of the foundation system that directly supports the exterior walls of the building.

Gross area–The total enclosed floor area of a building.

Ground area–The area computed by the exterior dimensions of the structure.

Ground floor–The floor of a building in closest proximity to the ground.

Grout–A mortar containing a high water content used to fill joints and cavities in masonry work.

Gutter–A wood or sheet metal conduit set along the building eaves used to channel rainwater to leaders or downspouts.

Gunite–A concrete mix placed pneumatically.

Gypsum–Hydrated calcium sulphate used as both a retarder in Portland cement and as a major ingredient in plaster of Paris. Used in sheets as a substitute for plaster.

Hall–A passageway providing access to various parts of a building–a large room for entertainment or assembly.

Hangar–A structure for the storage or repair of aircraft.

Head room–The distance between the top of the finished floor to the bottom of the finished ceiling.

Hearth–The floor of a fireplace and the adjacent area of fireproof material.

Heat pump–A mechanical device for providing either heating or air conditioning.

Hip roof–A roof whose four sides meet at a common point with no gabled ends.

I-beam–A structural member having a cross-section resembling the letter "I".

Insulation–Material used to reduce the effects of heat, cold or sound.

Jack rafter–An unusually short rafter generally found in hip roofs.

Jalousie–Adjustable glass louvers which pivot simultaneously in a common frame.

Jamb–The vertical member on either side of a door frame or window frame.

Joist–Parallel beams of timber, concrete or steel used to support floor and ceiling.

Junction box–A box that protects splices or joints in electrical wiring.

Kalamein door–Solid core wood doors clad with galvanized sheet metal.

Kip–A unit of weight equal to 1,000 pounds.

Lally column–A concrete-filled steel pipe used as a vertical support.

Laminated beam–A beam built up by gluing together several pieces of timber.

Landing–A platform between flights of stairs.

Lath–Strips of wood or metal used as a base for plaster.

Lead-lined door, sheetrock–Doors or sheetrock internally lined with sheet lead to provide protection from X-ray radiation.

Lean-to–A small shed or building addition with a single pitched roof attached to the exterior wall of the main building.

Lintel–A horizontal framing member used to carry a load over a wall opening.

Live load–The moving or movable load on a structure composed of furnishings, equipment or personnel weight.

Load bearing partition–A partition that supports a load in addition to its own weight.

Loading dock leveler–An adjustable platform for handling off-on loading of trucks.

Loft building–A commercial/industrial type building containing large, open unpartitioned floor areas.

Louver window–A window composed of a series of sloping, overlapping blades or slats that may be adjusted to admit varying degrees of air or light.

Main beam–The principal load bearing beam used to transmit loads directly to columns.

Mansard roof–A roof having a double pitch on all four sides, the lower level having the steeper pitch.

Mansion–An extremely large and imposing residence.

Masonry–The utilization of brick, stone, concrete or block for walls and other building components.

Mastic–A sealant or adhesive compound generally used as either a binding agent or floor finish.

Membrane fireproofing–A coating of metal lath and plaster to provide resistance to fire and heat.

Mesh reinforcement–An arrangement of wire tied or welded at their intersection to provide strength and resistance to cracking.

Metal lath–A diamond-shaped metallic base for plaster.

Mezzanine–A low story situated between two main floors.

Mill construction–A heavy timber construction that achieves fire resistance by using large wood structural members, noncombustible bearing and non-bearing walls and omitting concealed spaces under floors and roof.

Mixed occupancy–Two or more classes of occupancy in a single structure.

Molded brick–A specially shaped brick used for decorative purposes.

Monolithic concrete–Concrete poured in a continuous process so there are no joints.

Movable partition–A non-load bearing demountable partition that can be relocated and can be either ceiling height or partial height.

Moving walk–A continually moving horizontal passenger carrying device.

Multi-zone system–An air conditioning system that is capable of handling several individual zones simultaneously.

Net floor area–The usable, occupied area of a structure excluding stairwells, elevator shafts and wall thicknesses.

Non-combustible construction–Construction in which the walls, partitions and the structural members are of non-combustible materials.

Nurses call system–An electrically operated system for use by patients or personnel for summoning a nurse.

Occupancy rate–The number of persons per room, per dwelling unit, etc.

One-way joist connection–A framing system for floors and roofs in a concrete building consisting of a series of parallel joists supported by girders between columns.

Open web joist–A lightweight prefabricated metal chord truss.

Parapet–A low guarding wall at the point of a drop. That portion of the exterior wall that extends above the roof line.

Parging–A thin coat of mortar applied to a masonry wall used primarily for waterproofing purposes.

Parquet–Inlaid wood flooring set in a simple design.

Peaked roof–A roof of two or more slopes that rises to a peak.

Penthouse–A structure occupying usually half the roof area and used to house elevator, HVAC equipment, or other mechanical or electrical systems.

Pier–A column designed to support a concentrated load.

Pilaster–A column usually formed of the same material and integral with, but projecting from, a wall.

Pile–A concrete, wood or steel column usually less than 2' in diameter which is driven or otherwise introduced into the soil to carry a vertical load or provide vertical support.

Pitched roof–A roof having one or more surfaces with a pitch greater than 10°.

Plenum–The space between the suspended ceiling and the floor above.

Plywood–Structural wood made of three or more layers of veneer and bonded together with glue.

Porch–A structure attached to a building to provide shelter for an entranceway or to serve as a semi-enclosed space.

Post and beam framing–A structural framing system where beams rest on posts rather than bearing walls.

Post-tensioned concrete–Concrete that has the reinforcing tendons tensioned after the concrete has set.

Pre-cast concrete–Concrete structural components fabricated at a location other than in-place.

Pre-stressed concrete–Concrete that has the reinforcing tendons tensioned prior to the concrete setting.

Purlins–A horizontal structural member supporting the roof deck and resting on the trusses, girders, beams or rafters.

Rafters–Structural members supporting the roof deck and covering.

Resilient flooring–A manufactured interior floor covering material in either sheet or tile form, that is resilient.

Ridge–The horizontal line at the junction of two sloping roof edges.

Rigid frame–A structural framing system in which all columns and beams are rigidly connected; there are no hinged joints.

Roof drain–A drain designed to accept rainwater and discharge it into a leader or downspout.

Rough floor–A layer of boards or plywood nailed to the floor joists that serves as a base for the finished floor.

Sanitary sewer–A sewer line designed to carry only liquid or waterborne waste from the structure to a central treatment plant.

Sawtooth roof–Roof shape found primarily in industrial roofs that creates the appearance of the teeth of a saw.

Semi-rigid frame–A structural system wherein the columns and beams are so attached that there is some flexibility at the joints.

Septic tank–A covered tank in which waste matter is decomposed by natural bacterial action.

Service elevator–A combination passenger and freight elevator.

Sheathing–The first covering of exterior studs by boards, plywood or particle board.

For customer support on your Square Foot Cost Data, call 877.694.8329.

545

Shoring–Temporary bracing for structural components during construction.

Span–The horizontal distance between supports.

Specification–A detailed list of materials and requirements for construction of a building.

Storm sewer–A system of pipes used to carry rainwater or surface waters.

Stucco–A cement plaster used to cover exterior wall surfaces usually applied over a wood or metal lath base.

Stud–A vertical wooden structural component.

Subfloor–A system of boards, plywood or particle board laid over the floor joists to form a base.

Superstructure–That portion of the structure above the foundation or ground level.

Terrazzo–A durable floor finish made of small chips of colored stone or marble, embedded in concrete, then polished to a high sheen.

Tile–A thin piece of fired clay, stone or concrete used for floor, roof or wall finishes.

Unit cost–The cost per unit of measurement.

Vault–A room especially designed for storage.

Veneer–A thin surface layer covering a base of common material.

Vent–An opening serving as an outlet for air.

Wainscot–The lower portion of an interior wall whose surface differs from that of the upper wall.

Wall bearing construction–A structural system where the weight of the floors and roof are carried directly by the masonry walls rather than the structural framing system.

Waterproofing–Any of a number of materials applied to various surfaces to prevent the infiltration of water.

Weatherstrip–A thin strip of metal, wood or felt used to cover the joint between the door or window sash and the jamb, casing or sill to keep out air, dust, rain, etc.

Wing–A building section or addition projecting from the main structure.

X-ray protection–Lead encased in sheetrock or plaster to prevent the escape of radiation.

A	Area Square Feet; Ampere	Brk., brk	Brick	Csc	Cosecant
AAFES	Army and Air Force Exchange Service	brkt	Bracket	C.S.F.	Hundred Square Feet
		Brng.	Bearing	CSI	Construction Specifications Institute
ABS	Acrylonitrile Butadiene Stryrene; Asbestos Bonded Steel	Brs.	Brass	CT	Current Transformer
		Brz.	Bronze	CTS	Copper Tube Size
A.C., AC	Alternating Current; Air-Conditioning; Asbestos Cement; Plywood Grade A & C	Bsn.	Basin	Cu	Copper, Cubic
		Btr.	Better	Cu. Ft.	Cubic Foot
		Btu	British Thermal Unit	cw	Continuous Wave
		BTUH	BTU per Hour	C.W.	Cool White; Cold Water
ACI	American Concrete Institute	Bu.	bushels	Cwt.	100 Pounds
ACR	Air Conditioning Refrigeration	BUR	Built-up Roofing	C.W.X.	Cool White Deluxe
ADA	Americans with Disabilities Act	BX	Interlocked Armored Cable	C.Y.	Cubic Yard (27 cubic feet)
AD	Plywood, Grade A & D	°C	degree centegrade	C.Y./Hr.	Cubic Yard per Hour
Addit.	Additional	c	Conductivity, Copper Sweat	Cyl.	Cylinder
Adj.	Adjustable	C	Hundred; Centigrade	d	Penny (nail size)
af	Audio-frequency	C/C	Center to Center, Cedar on Cedar	D	Deep; Depth; Discharge
AFUE	Annual Fuel Utilization Efficiency	C-C	Center to Center	Dis., Disch.	Discharge
AGA	American Gas Association	Cab	Cabinet	Db	Decibel
Agg.	Aggregate	Cair.	Air Tool Laborer	Dbl.	Double
A.H., Ah	Ampere Hours	Cal.	caliper	DC	Direct Current
A hr.	Ampere-hour	Calc	Calculated	DDC	Direct Digital Control
A.H.U., AHU	Air Handling Unit	Cap.	Capacity	Demob.	Demobilization
A.I.A.	American Institute of Architects	Carp.	Carpenter	d.f.t.	Dry Film Thickness
AIC	Ampere Interrupting Capacity	C.B.	Circuit Breaker	d.f.u.	Drainage Fixture Units
Allow.	Allowance	C.C.A.	Chromate Copper Arsenate	D.H.	Double Hung
alt., alt	Alternate	C.C.F.	Hundred Cubic Feet	DHW	Domestic Hot Water
Alum.	Aluminum	cd	Candela	DI	Ductile Iron
a.m.	Ante Meridiem	cd/sf	Candela per Square Foot	Diag.	Diagonal
Amp.	Ampere	CD	Grade of Plywood Face & Back	Diam., Dia	Diameter
Anod.	Anodized	CDX	Plywood, Grade C & D, exterior glue	Distrib.	Distribution
ANSI	American National Standards Institute	Cefi.	Cement Finisher	Div.	Division
APA	American Plywood Association	Cem.	Cement	Dk.	Deck
Approx.	Approximate	CF	Hundred Feet	D.L.	Dead Load; Diesel
Apt.	Apartment	C.F.	Cubic Feet	DLH	Deep Long Span Bar Joist
Asb.	Asbestos	CFM	Cubic Feet per Minute	dlx	Deluxe
A.S.B.C.	American Standard Building Code	CFRP	Carbon Fiber Reinforced Plastic	Do.	Ditto
Asbe.	Asbestos Worker	c.g.	Center of Gravity	DOP	Dioctyl Phthalate Penetration Test (Air Filters)
ASCE.	American Society of Civil Engineers	CHW	Chilled Water; Commercial Hot Water	Dp., dp	Depth
A.S.H.R.A.E.	American Society of Heating, Refrig. & AC Engineers	C.I., CI	Cast Iron	D.P.S.T.	Double Pole, Single Throw
		C.I.P., CIP	Cast in Place	Dr.	Drive
ASME	American Society of Mechanical Engineers	Circ.	Circuit	DR	Dimension Ratio
ASTM	American Society for Testing and Materials	C.L.	Carload Lot	Drink.	Drinking
		CL	Chain Link	D.S.	Double Strength
Attchmt.	Attachment	Clab.	Common Laborer	D.S.A.	Double Strength A Grade
Avg., Ave.	Average	Clam	Common maintenance laborer	D.S.B.	Double Strength B Grade
AWG	American Wire Gauge	C.L.F.	Hundred Linear Feet	Dty.	Duty
AWWA	American Water Works Assoc.	CLF	Current Limiting Fuse	DWV	Drain Waste Vent
Bbl.	Barrel	CLP	Cross Linked Polyethylene	DX	Deluxe White, Direct Expansion
B&B, BB	Grade B and Better; Balled & Burlapped	cm	Centimeter	dyn	Dyne
B&S	Bell and Spigot	CMP	Corr. Metal Pipe	e	Eccentricity
B.&W.	Black and White	CMU	Concrete Masonry Unit	E	Equipment Only; East; emissivity
b.c.c.	Body-centered Cubic	CN	Change Notice	Ea.	Each
B.C.Y.	Bank Cubic Yards	Col.	Column	EB	Encased Burial
BE	Bevel End	CO$_2$	Carbon Dioxide	Econ.	Economy
B.F.	Board Feet	Comb.	Combination	E.C.Y	Embankment Cubic Yards
Bg. cem.	Bag of Cement	comm.	Commercial, Communication	EDP	Electronic Data Processing
BHP	Boiler Horsepower; Brake Horsepower	Compr.	Compressor	EIFS	Exterior Insulation Finish System
		Conc.	Concrete	E.D.R.	Equiv. Direct Radiation
B.I.	Black Iron	Cont., cont	Continuous; Continued, Container	Eq.	Equation
bidir.	bidirectional	Corr.	Corrugated	EL	elevation
Bit., Bitum.	Bituminous	Cos	Cosine	Elec.	Electrician; Electrical
Bit., Conc.	Bituminous Concrete	Cot	Cotangent	Elev.	Elevator; Elevating
Bk.	Backed	Cov.	Cover	EMT	Electrical Metallic Conduit; Thin Wall Conduit
Bkrs.	Breakers	C/P	Cedar on Paneling		
Bldg., bldg	Building	CPA	Control Point Adjustment	Eng.	Engine, Engineered
Blk.	Block	Cplg.	Coupling	EPDM	Ethylene Propylene Diene Monomer
Bm.	Beam	CPM	Critical Path Method		
Boil.	Boilermaker	CPVC	Chlorinated Polyvinyl Chloride	EPS	Expanded Polystyrene
bpm	Blows per Minute	C.Pr.	Hundred Pair	Eqhv.	Equip. Oper., Heavy
BR	Bedroom	CRC	Cold Rolled Channel	Eqlt.	Equip. Oper., Light
Brg.	Bearing	Creos.	Creosote	Eqmd.	Equip. Oper., Medium
Brhe.	Bricklayer Helper	Crpt.	Carpet & Linoleum Layer	Eqmm.	Equip. Oper., Master Mechanic
Bric.	Bricklayer	CRT	Cathode-ray Tube	Eqol.	Equip. Oper., Oilers
		CS	Carbon Steel, Constant Shear Bar Joist	Equip.	Equipment
				ERW	Electric Resistance Welded

For customer support on your Square Foot Cost Data, call 877.694.8329.

547

E.S.	Energy Saver	H	High Henry	Lath.	Lather
Est.	Estimated	HC	High Capacity	Lav.	Lavatory
esu	Electrostatic Units	H.D., HD	Heavy Duty; High Density	lb.; #	Pound
E.W.	Each Way	H.D.O.	High Density Overlaid	L.B., LB	Load Bearing; L Conduit Body
EWT	Entering Water Temperature	HDPE	High density polyethelene plastic	L. & E.	Labor & Equipment
Excav.	Excavation	Hdr.	Header	lb./hr.	Pounds per Hour
excl	Excluding	Hdwe.	Hardware	lb./L.F.	Pounds per Linear Foot
Exp., exp	Expansion, Exposure	H.I.D., HID	High Intensity Discharge	lbf/sq.in.	Pound-force per Square Inch
Ext., ext	Exterior; Extension	Help.	Helper Average	L.C.L.	Less than Carload Lot
Extru.	Extrusion	HEPA	High Efficiency Particulate Air	L.C.Y.	Loose Cubic Yard
f.	Fiber stress		Filter	Ld.	Load
F	Fahrenheit; Female; Fill	Hg	Mercury	LE	Lead Equivalent
Fab., fab	Fabricated; fabric	HIC	High Interrupting Capacity	LED	Light Emitting Diode
FBGS	Fiberglass	HM	Hollow Metal	L.F.	Linear Foot
F.C.	Footcandles	HMWPE	high molecular weight	L.F. Nose	Linear Foot of Stair Nosing
f.c.c.	Face-centered Cubic		polyethylene	L.F. Rsr	Linear Foot of Stair Riser
f'c.	Compressive Stress in Concrete;	HO	High Output	Lg.	Long; Length; Large
	Extreme Compressive Stress	Horiz.	Horizontal	L & H	Light and Heat
F.E.	Front End	H.P., HP	Horsepower; High Pressure	LH	Long Span Bar Joist
FEP	Fluorinated Ethylene Propylene	H.P.F.	High Power Factor	L.H.	Labor Hours
	(Teflon)	Hr.	Hour	L.L., LL	Live Load
F.G.	Flat Grain	Hrs./Day	Hours per Day	L.L.D.	Lamp Lumen Depreciation
F.H.A.	Federal Housing Administration	HSC	High Short Circuit	lm	Lumen
Fig.	Figure	Ht.	Height	lm/sf	Lumen per Square Foot
Fin.	Finished	Htg.	Heating	lm/W	Lumen per Watt
FIPS	Female Iron Pipe Size	Htrs.	Heaters	LOA	Length Over All
Fixt.	Fixture	HVAC	Heating, Ventilation & Air-	log	Logarithm
FJP	Finger jointed and primed		Conditioning	L-O-L	Lateralolet
Fl. Oz.	Fluid Ounces	Hvy.	Heavy	long.	longitude
Flr.	Floor	HW	Hot Water	L.P., LP	Liquefied Petroleum; Low Pressure
FM	Frequency Modulation;	Hyd.; Hydr.	Hydraulic	L.P.F.	Low Power Factor
	Factory Mutual	Hz	Hertz (cycles)	LR	Long Radius
Fmg.	Framing	I.	Moment of Inertia	L.S.	Lump Sum
FM/UL	Factory Mutual/Underwriters Labs	IBC	International Building Code	Lt.	Light
Fdn.	Foundation	I.C.	Interrupting Capacity	Lt. Ga.	Light Gauge
FNPT	Female National Pipe Thread	ID	Inside Diameter	L.T.L.	Less than Truckload Lot
Fori.	Foreman, Inside	I.D.	Inside Dimension; Identification	Lt. Wt.	Lightweight
Foro.	Foreman, Outside	I.F.	Inside Frosted	L.V.	Low Voltage
Fount.	Fountain	I.M.C.	Intermediate Metal Conduit	M	Thousand; Material; Male;
fpm	Feet per Minute	In.	Inch		Light Wall Copper Tubing
FPT	Female Pipe Thread	Incan.	Incandescent	M²CA	Meters Squared Contact Area
Fr	Frame	Incl.	Included; Including	m/hr.; M.H.	Man-hour
F.R.	Fire Rating	Int.	Interior	mA	Milliampere
FRK	Foil Reinforced Kraft	Inst.	Installation	Mach.	Machine
FSK	Foil/scrim/kraft	Insul., insul	Insulation/Insulated	Mag. Str.	Magnetic Starter
FRP	Fiberglass Reinforced Plastic	I.P.	Iron Pipe	Maint.	Maintenance
FS	Forged Steel	I.P.S., IPS	Iron Pipe Size	Marb.	Marble Setter
FSC	Cast Body; Cast Switch Box	IPT	Iron Pipe Threaded	Mat; Mat'l.	Material
Ft., ft	Foot; Feet	I.W.	Indirect Waste	Max.	Maximum
Ftng.	Fitting	J	Joule	MBF	Thousand Board Feet
Ftg.	Footing	J.I.C.	Joint Industrial Council	MBH	Thousand BTU's per hr.
Ft lb.	Foot Pound	K	Thousand; Thousand Pounds;	MC	Metal Clad Cable
Furn.	Furniture		Heavy Wall Copper Tubing, Kelvin	MCC	Motor Control Center
FVNR	Full Voltage Non-Reversing	K.A.H.	Thousand Amp. Hours	M.C.F.	Thousand Cubic Feet
FVR	Full Voltage Reversing	kcmil	Thousand Circular Mils	MCFM	Thousand Cubic Feet per Minute
FXM	Female by Male	KD	Knock Down	M.C.M.	Thousand Circular Mils
Fy.	Minimum Yield Stress of Steel	K.D.A.T.	Kiln Dried After Treatment	MCP	Motor Circuit Protector
g	Gram	kg	Kilogram	MD	Medium Duty
G	Gauss	kG	Kilogauss	MDF	Medium-density fibreboard
Ga.	Gauge	kgf	Kilogram Force	M.D.O.	Medium Density Overlaid
Gal., gal.	Gallon	kHz	Kilohertz	Med.	Medium
gpm, GPM	Gallon per Minute	Kip	1000 Pounds	MF	Thousand Feet
Galv., galv	Galvanized	KJ	Kiljoule	M.F.B.M.	Thousand Feet Board Measure
GC/MS	Gas Chromatograph/Mass	K.L.	Effective Length Factor	Mfg.	Manufacturing
	Spectrometer	K.L.F.	Kips per Linear Foot	Mfrs.	Manufacturers
Gen.	General	Km	Kilometer	mg	Milligram
GFI	Ground Fault Interrupter	KO	Knock Out	MGD	Million Gallons per Day
GFRC	Glass Fiber Reinforced Concrete	K.S.F.	Kips per Square Foot	MGPH	Thousand Gallons per Hour
Glaz.	Glazier	K.S.I.	Kips per Square Inch	MH, M.H.	Manhole; Metal Halide; Man-Hour
GPD	Gallons per Day	kV	Kilovolt	MHz	Megahertz
gpf	Gallon per flush	kVA	Kilovolt Ampere	Mi.	Mile
GPH	Gallons per Hour	kVAR	Kilovar (Reactance)	MI	Malleable Iron; Mineral Insulated
GPM	Gallons per Minute	KW	Kilowatt	MIPS	Male Iron Pipe Size
GR	Grade	KWh	Kilowatt-hour	mj	Mechanical Joint
Gran.	Granular	L	Labor Only; Length; Long;	m	Meter
Grnd.	Ground		Medium Wall Copper Tubing	mm	Millimeter
GVW	Gross Vehicle Weight	Lab.	Labor	Mill.	Millwright
GWB	Gypsum wall board	lat	Latitude	Min., min.	Minimum, minute

548

Abbreviations

Misc.	Miscellaneous	PDCA	Painting and Decorating Contractors of America	SC	Screw Cover	
ml	Milliliter, Mainline	P.E., PE	Professional Engineer; Porcelain Enamel; Polyethylene; Plain End	SCFM	Standard Cubic Feet per Minute	
M.L.F.	Thousand Linear Feet			Scaf.	Scaffold	
Mo.	Month			Sch., Sched.	Schedule	
Mobil.	Mobilization	P.E.C.I.	Porcelain Enamel on Cast Iron	S.C.R.	Modular Brick	
Mog.	Mogul Base	Perf.	Perforated	S.D.	Sound Deadening	
MPH	Miles per Hour	PEX	Cross linked polyethylene	SDR	Standard Dimension Ratio	
MPT	Male Pipe Thread	Ph.	Phase	S.E.	Surfaced Edge	
MRGWB	Moisture Resistant Gypsum Wallboard	P.I.	Pressure Injected	Sel.	Select	
		Pile.	Pile Driver	SER, SEU	Service Entrance Cable	
MRT	Mile Round Trip	Pkg.	Package	S.F.	Square Foot	
ms	Millisecond	Pl.	Plate	S.F.C.A.	Square Foot Contact Area	
M.S.F.	Thousand Square Feet	Plah.	Plasterer Helper	S.F. Flr.	Square Foot of Floor	
Mstz.	Mosaic & Terrazzo Worker	Plas.	Plasterer	S.F.G.	Square Foot of Ground	
M.S.Y.	Thousand Square Yards	plf	Pounds Per Linear Foot	S.F. Hor.	Square Foot Horizontal	
Mtd., mtd., mtd	Mounted	Pluh.	Plumbers Helper	SFR	Square Feet of Radiation	
Mthe.	Mosaic & Terrazzo Helper	Plum.	Plumber	S.F. Shlf.	Square Foot of Shelf	
Mtng.	Mounting	Ply.	Plywood	S4S	Surface 4 Sides	
Mult.	Multi; Multiply	p.m.	Post Meridiem	Shee.	Sheet Metal Worker	
M.V.A.	Million Volt Amperes	Pntd.	Painted	Sin.	Sine	
M.V.A.R.	Million Volt Amperes Reactance	Pord.	Painter, Ordinary	Skwk.	Skilled Worker	
MV	Megavolt	pp	Pages	SL	Saran Lined	
MW	Megawatt	PP, PPL	Polypropylene	S.L.	Slimline	
MXM	Male by Male	P.P.M.	Parts per Million	Sldr.	Solder	
MYD	Thousand Yards	Pr.	Pair	SLH	Super Long Span Bar Joist	
N	Natural; North	P.E.S.B.	Pre-engineered Steel Building	S.N.	Solid Neutral	
nA	Nanoampere	Prefab.	Prefabricated	SO	Stranded with oil resistant inside insulation	
NA	Not Available; Not Applicable	Prefin.	Prefinished			
N.B.C.	National Building Code	Prop.	Propelled	S-O-L	Socketolet	
NC	Normally Closed	PSF, psf	Pounds per Square Foot	sp	Standpipe	
NEMA	National Electrical Manufacturers Assoc.	PSI, psi	Pounds per Square Inch	S.P.	Static Pressure; Single Pole; Self-Propelled	
		PSIG	Pounds per Square Inch Gauge			
NEHB	Bolted Circuit Breaker to 600V.	PSP	Plastic Sewer Pipe	Spri.	Sprinkler Installer	
NFPA	National Fire Protection Association	Pspr.	Painter, Spray	spwg	Static Pressure Water Gauge	
NLB	Non-Load-Bearing	Psst.	Painter, Structural Steel	S.P.D.T.	Single Pole, Double Throw	
NM	Non-Metallic Cable	P.T.	Potential Transformer	SPF	Spruce Pine Fir; Sprayed Polyurethane Foam	
nm	Nanometer	P. & T.	Pressure & Temperature			
No.	Number	Ptd.	Painted	S.P.S.T.	Single Pole, Single Throw	
NO	Normally Open	Ptns.	Partitions	SPT	Standard Pipe Thread	
N.O.C.	Not Otherwise Classified	Pu	Ultimate Load	Sq.	Square; 100 Square Feet	
Nose.	Nosing	PVC	Polyvinyl Chloride	Sq. Hd.	Square Head	
NPT	National Pipe Thread	Pvmt.	Pavement	Sq. In.	Square Inch	
NQOD	Combination Plug-on/Bolt on Circuit Breaker to 240V.	PRV	Pressure Relief Valve	S.S.	Single Strength; Stainless Steel	
		Pwr.	Power	S.S.B.	Single Strength B Grade	
N.R.C., NRC	Noise Reduction Coefficient/ Nuclear Regulator Commission	Q	Quantity Heat Flow	sst, ss	Stainless Steel	
		Qt.	Quart	Sswk.	Structural Steel Worker	
N.R.S.	Non Rising Stem	Quan., Qty.	Quantity	Sswl.	Structural Steel Welder	
ns	Nanosecond	Q.C.	Quick Coupling	St.; Stl.	Steel	
nW	Nanowatt	r	Radius of Gyration	STC	Sound Transmission Coefficient	
OB	Opposing Blade	R	Resistance	Std.	Standard	
OC	On Center	R.C.P.	Reinforced Concrete Pipe	Stg.	Staging	
OD	Outside Diameter	Rect.	Rectangle	STK	Select Tight Knot	
O.D.	Outside Dimension	recpt.	receptacle	STP	Standard Temperature & Pressure	
ODS	Overhead Distribution System	Reg.	Regular	Stpi.	Steamfitter, Pipefitter	
O.G.	Ogee	Reinf.	Reinforced	Str.	Strength; Starter; Straight	
O.H.	Overhead	Req'd.	Required	Strd.	Stranded	
O&P	Overhead and Profit	Res.	Resistant	Struct.	Structural	
Oper.	Operator	Resi.	Residential	Sty.	Story	
Opng.	Opening	RF	Radio Frequency	Subj.	Subject	
Orna.	Ornamental	RFID	Radio-frequency identification	Subs.	Subcontractors	
OSB	Oriented Strand Board	Rgh.	Rough	Surf.	Surface	
OS&Y	Outside Screw and Yoke	RGS	Rigid Galvanized Steel	Sw.	Switch	
OSHA	Occupational Safety and Health Act	RHW	Rubber, Heat & Water Resistant; Residential Hot Water	Swbd.	Switchboard	
				S.Y.	Square Yard	
Ovhd.	Overhead	rms	Root Mean Square	Syn.	Synthetic	
OWG	Oil, Water or Gas	Rnd.	Round	S.Y.P.	Southern Yellow Pine	
Oz.	Ounce	Rodm.	Rodman	Sys.	System	
P.	Pole; Applied Load; Projection	Rofc.	Roofer, Composition	t.	Thickness	
p.	Page	Rofp.	Roofer, Precast	T	Temperature; Ton	
Pape.	Paperhanger	Rohe.	Roofer Helpers (Composition)	Tan	Tangent	
P.A.P.R.	Powered Air Purifying Respirator	Rots.	Roofer, Tile & Slate	T.C.	Terra Cotta	
PAR	Parabolic Reflector	R.O.W.	Right of Way	T & C	Threaded and Coupled	
P.B., PB	Push Button	RPM	Revolutions per Minute	T.D.	Temperature Difference	
Pc., Pcs.	Piece, Pieces	R.S.	Rapid Start	Tdd	Telecommunications Device for the Deaf	
P.C.	Portland Cement; Power Connector	Rsr	Riser			
P.C.F.	Pounds per Cubic Foot	RT	Round Trip	T.E.M.	Transmission Electron Microscopy	
PCM	Phase Contrast Microscopy	S.	Suction; Single Entrance; South	temp	Temperature, Tempered, Temporary	
		SBS	Styrene Butadiere Styrene	TFFN	Nylon Jacketed Wire	

Abbreviations

TFE	Tetrafluoroethylene (Teflon)
T. & G.	Tongue & Groove;
	Tar & Gravel
Th., Thk.	Thick
Thn.	Thin
Thrded	Threaded
Tilf.	Tile Layer, Floor
Tilh.	Tile Layer, Helper
THHN	Nylon Jacketed Wire
THW.	Insulated Strand Wire
THWN	Nylon Jacketed Wire
T.L., TL	Truckload
T.M.	Track Mounted
Tot.	Total
T-O-L	Threadolet
tmpd	Tempered
TPO	Thermoplastic Polyolefin
T.S.	Trigger Start
Tr.	Trade
Transf.	Transformer
Trhv.	Truck Driver, Heavy
Trlr	Trailer
Trlt.	Truck Driver, Light
TTY	Teletypewriter
TV	Television
T.W.	Thermoplastic Water Resistant
	Wire
UCI	Uniform Construction Index
UF	Underground Feeder
UGND	Underground Feeder
UHF	Ultra High Frequency
U.I.	United Inch

U.L., UL	Underwriters Laboratory
Uld.	unloading
Unfin.	Unfinished
UPS	Uninterruptible Power Supply
URD	Underground Residential
	Distribution
US	United States
USGBC	U.S. Green Building Council
USP	United States Primed
UTMCD	Uniform Traffic Manual For Control
	Devices
UTP	Unshielded Twisted Pair
V	Volt
VA	Volt Amperes
VAT	Vinyl Asbestos Tile
V.C.T.	Vinyl Composition Tile
VAV	Variable Air Volume
VC	Veneer Core
VDC	Volts Direct Current
Vent.	Ventilation
Vert.	Vertical
V.F.	Vinyl Faced
V.G.	Vertical Grain
VHF	Very High Frequency
VHO	Very High Output
Vib.	Vibrating
VLF	Vertical Linear Foot
VOC	Volitile Organic Compound
Vol.	Volume
VRP	Vinyl Reinforced Polyester
W	Wire; Watt; Wide; West

w/	With
W.C., WC	Water Column; Water Closet
W.F.	Wide Flange
W.G.	Water Gauge
Wldg.	Welding
W. Mile	Wire Mile
W-O-L	Weldolet
W.R.	Water Resistant
Wrck.	Wrecker
W.S.P.	Water, Steam, Petroleum
WT., Wt.	Weight
WWF	Welded Wire Fabric
XFER	Transfer
XFMR	Transformer
XHD	Extra Heavy Duty
XHHW,	Cross-Linked Polyethylene Wire
XLPE	Insulation
XLP	Cross-linked Polyethylene
Xport	Transport
Y	Wye
yd	Yard
yr	Year
Δ	Delta
%	Percent
~	Approximately
∅	Phase; diameter
@	At
#	Pound; Number
<	Less Than
>	Greater Than
Z	zone

RESIDENTIAL
COST ESTIMATE

OWNER'S NAME: _____

APPRAISER: _____

RESIDENCE ADDRESS: _____

PROJECT: _____

CITY, STATE, ZIP CODE: _____

DATE: _____

CLASS OF CONSTRUCTION
- ☐ ECONOMY
- ☐ AVERAGE
- ☐ CUSTOM
- ☐ LUXURY

RESIDENCE TYPE
- ☐ 1 STORY
- ☐ 1-1/2 STORY
- ☐ 2 STORY
- ☐ 2-1/2 STORY
- ☐ 3 STORY
- ☐ BI-LEVEL
- ☐ TRI-LEVEL

CONFIGURATION
- ☐ DETACHED
- ☐ TOWN/ROW HOUSE
- ☐ SEMI-DETACHED

OCCUPANCY
- ☐ ONE STORY
- ☐ TWO FAMILY
- ☐ THREE FAMILY·
- ☐ OTHER _____

EXTERIOR WALL SYSTEM
- ☐ WOOD SIDING—WOOD FRAME
- ☐ BRICK VENEER—WOOD FRAME
- ☐ STUCCO ON WOOD FRAME
- ☐ PAINTED CONCRETE BLOCK
- ☐ SOLID MASONRY (AVERAGE & CUSTOM)
- ☐ STONE VENEER—WOOD FRAME
- ☐ SOLID BRICK (LUXURY)
- ☐ SOLID STONE (LUXURY)

*LIVING AREA (Main Building)

First Level _____ S.F.
Second Level _____ S.F.
Third Level _____ S.F.

Total _____ S.F.

*Basement Area is not part of living area.

*LIVING AREA (Wing or Ell) (   )

First Level _____ S.F.
Second Level _____ S.F.
Third Level _____ S.F.

Total _____ S.F.

*LIVING AREA (Wing or Ell) (   )

First Level _____ S.F.
Second Level _____ S.F.
Third Level _____ S.F.

Total _____ S.F.

MAIN BUILDING	COSTS PER S.F. LIVING AREA
Cost per Square Foot of Living Area, from Page _____	$
Basement Addition: _____ % Finished, _____ % Unfinished	+
Roof Cover Adjustment: _____ Type, Page _____ (Add or Deduct)	( )
Central Air Conditioning: ☐ Separate Ducts ☐ Heating Ducts, Page _____	+
Heating System Adjustment: _____ Type, Page _____ (Add or Deduct)	( )
Main Building: Adjusted Cost per S.F. of Living Area	$

MAIN BUILDING
TOTAL COST

$ _____ /S.F. x _____ S.F. x _____ = $ _____

Cost per S.F. Living Area — Living Area — Town/Row House Multiplier (Use 1 for Detached) — TOTAL COST

WING OR ELL ( ) _____ STORY	COSTS PER S.F. LIVING AREA
Cost per Square Foot of Living Area, from Page _____	$
Basement Addition: _____ % Finished, _____ % Unfinished	+
Roof Cover Adjustment: _____ Type, Page _____ (Add or Deduct)	( )
Central Air Conditioning: ☐ Separate Ducts ☐ Heating Ducts, Page _____	+
Heating System Adjustment: _____ Type, Page _____ (Add or Deduct)	( )
Wing or Ell ( ): Adjusted Cost per S.F. of Living Area	$

WING OR ELL (   )
TOTAL COST

$ _____ /S.F. x _____ S.F. = $ _____

Cost per S.F. Living Area — Living Area — TOTAL COST

WING OR ELL ( ) _____ STORY	COSTS PER S.F. LIVING AREA
Cost per Square Foot of Living Area, from Page _____	$
Basement Addition: _____ % Finished, _____ % Unfinished	+
Roof Cover Adjustment: _____ Type, Page _____ (Add or Deduct)	( )
Central Air Conditioning: ☐ Separate Ducts ☐ Heating Ducts, Page _____	+
Heating System Adjustment: _____ Type, Page _____ (Add or Deduct)	( )
Wing or Ell ( ): Adjusted Cost per S.F. of Living Area	$

WING OR ELL (   )
TOTAL COST

$ _____ /S.F. x _____ S.F. = $ _____

Cost per S.F. Living Area — Living Area — TOTAL COST

TOTAL THIS PAGE _____

Page 1 of 2

RESIDENTIAL COST ESTIMATE

		QUANTITY	UNIT COST	$
Total Page 1				
Additional Bathrooms: _____ Full, _____ Half				
Finished Attic: _____ Ft. x _____ Ft.		S.F.		+
Breezeway: ☐ Open ☐ Enclosed _____ Ft. x _____ Ft.		S.F.		+
Covered Porch: ☐ Open ☐ Enclosed _____ Ft. x _____ Ft.		S.F.		+
Fireplace: ☐ Interior Chimney ☐ Exterior Chimney ☐ No. of Flues ☐ Additional Fireplaces				+
Appliances:				+
Kitchen Cabinets Adjustment:	(+/−)			
☐ Garage ☐ Carport: _____ Car(s) Description _____ (+/−)				+
Miscellaneous:				

ADJUSTED TOTAL BUILDING COST $ _____

REPLACEMENT COST

ADJUSTED TOTAL BUILDING COST	$ _____
Site Improvements	
(A) Paving & Sidewalks	$ _____
(B) Landscaping	$ _____
(C) Fences	$ _____
(D) Swimming Pool	$ _____
(E) Miscellaneous	$ _____
TOTAL	$ _____
Location Factor	x _____
Location Replacement Cost	$ _____
Depreciation	−$ _____
LOCAL DEPRECIATED COST	$ _____

INSURANCE COST

ADJUSTED TOTAL BUILDING COST	$ _____
Insurance Exclusions	
(A) Footings, Site Work, Underground Piping	−$ _____
(B) Architects' Fees	−$ _____
Total Building Cost Less Exclusion	$ _____
Location Factor	x _____
LOCAL INSURABLE REPLACEMENT COST	$ _____

SKETCH AND ADDITIONAL CALCULATIONS

CII APPRAISAL

1. SUBJECT PROPERTY: _____
2. BUILDING: _____
3. ADDRESS: _____
4. BUILDING USE: _____

5. DATE: _____
6. APPRAISER: _____

7. YEAR BUILT: _____

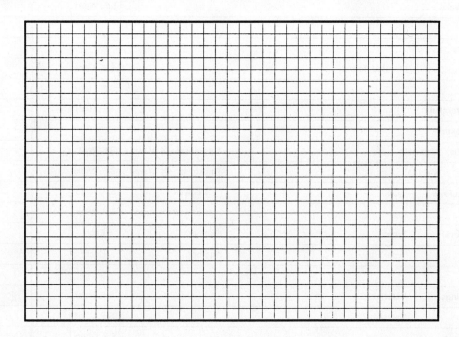

8. EXTERIOR WALL CONSTRUCTION: _____
9. FRAME: _____
10. GROUND FLOOR AREA: _____ S.F.   11. GROSS FLOOR AREA (EXCL. BASEMENT): _____ S.F.
12. NUMBER OF STORIES: _____   13. STORY HEIGHT: _____
14. PERIMETER: _____ L.F.   15. BASEMENT AREA: _____ S.F.
16. GENERAL COMMENTS: _____
_____
_____
_____
_____
_____
_____

	NO.	DESCRIPTION				UNIT	UNIT COST	NEW SF COST	MODEL SF COST	+/- CHANGE
A	**SUBSTRUCTURE**									
A	1010	Standard Foundations		Bay size:		S.F. Gnd.				
A	1030	Slab on Grade	Material:	Thickness:		S.F. Slab				
A	2010	Basement Excavation	Depth:	Area:		S.F. Gnd.				
A	2020	Basement Walls				L.F. Walls				
B	**SHELL**									
	B10 Superstructure									
B	1010	Floor Construction	Elevated floors:			S.F. Floor				
						S.F. Floor				
B	1020	Roof Construction				S.F. Roof				
	B20 Exterior Enclosure									
B	2010	Exterior walls	Material:	Thickness: % of wall		S.F. Walls				
			Material:	Thickness: % of wall		S.F. Walls				
B	2020	Exterior Windows	Type:	% of wall		S.F. Wind.				
			Type:	% of wall	Each	S.F. Wind.				
B	2030	Exterior Doors	Type:	Number:		Each				
			Type:	Number:		Each				
	B30 Roofing									
B	3010	Roof Coverings	Material:			S.F. Roof				
			Material:			S.F. Roof				
B	3020	Roof Openings				S.F. Opng.				
C	**INTERIORS**									
C	1010	Partitions:	Material:	Density:		S.F. Part.				
			Material:	Density:						
C	1020	Interior Doors	Type:	Number:		Each				
C	1030	Fittings				Each				
C	2010	Stair Construction				Flight				
C	3010	Wall Finishes	Material:	% of Wall		S.F. Walls				
			Material:			S.F. Walls				
C	3020	Floor Finishes	Material:			S.F. Floor				
			Material:			S.F. Floor				
			Material:			S.F. Floor				
C	3030	Ceiling Finishes	Material:			S.F. Ceil.				

NO.	SYSTEM/COMPONENT	DESCRIPTION	UNIT	UNIT COST	NEW S.F. COST	MODEL S.F. COST	+/- CHANGE
D SERVICES							
D10 Conveying							
D 1010	Elevators & Lifts	Type: Capacity: Stops:	Each				
D 1020	Escalators & Moving Walks	Type:	Each				
D20 Plumbing							
D 2010	Plumbing		Each				
D 2020	Domestic Water Distribution		S.F. Floor				
D 2040	Rain Water Drainage		S.F. Roof				
D30 HVAC							
D 3010	Energy Supply		Each				
D 3020	Heat Generating Systems	Type:	S.F. Floor				
D 3030	Cooling Generating Systems	Type:	S.F. Floor				
D 3090	Other HVAC Sys. & Equipment		Each				
D40 Fire Protection							
D 4010	Sprinklers		S.F. Floor				
D 4020	Standpipes		S.F. Floor				
D50 Electrical							
D 5010	Electrical Service/Distribution		S.F. Floor				
D 5020	Lighting & Branch Wiring		S.F. Floor				
D 5030	Communications & Security		S.F. Floor				
D 5090	Other Electrical Systems		S.F. Floor				
E EQUIPMENT & FURNISHINGS							
E 1010	Commercial Equipment		Each				
E 1020	Institutional Equipment		Each				
E 1030	Vehicular Equipment		Each				
E 1090	Other Equipment		Each				
F SPECIAL CONSTRUCTION							
F 1020	Integrated Construction		S.F.				
F 1040	Special Facilities		S.F.				
G BUILDING SITEWORK							

ITEM		Total Change
17	Model 020 total ____ $ _____	
18	Adjusted S.F. cost ____ $ _____ item 17 +/- changes	
19	Building area - from item 11 _____ S.F. x adjusted S.F. cost.............................	$...
20	Basement area - from item 15 _____ S.F. x S.F. cost $	$..
21	Base building sub-total - item 20 + item 19	$.....
22	Miscellaneous addition (quality, etc.)	$......
23	Sub-total - item 22 + 21	$.......
24	General conditions -25 % of item 23	$.....
25	Sub-total - item 24 + item 23...	$......
26	Architects fees _____ % of item 25	$
27	Sub-total - item 26 + item 27	$.....
28	Location modifier	x....
29	Local replacement cost - item 28 x item 27	$....
30	Depreciation _____ % of item 29	$
31	Depreciated local replacement cost - item 29 less item 30	$....
32	Exclusions	$.......
33	Net depreciated replacement cost - item 31 less item 32	$....

Base Cost per square foot floor area: *(from square foot table)*

Specify source: Page: _____ , Model # _____ , Area _____ S.F.,
Exterior wall _____ , Frame _____

Adjustments for exterior wall variation:

Size Adjustment:
  (Interpolate)

Height Adjustment: _____ + _____ = _____

Adjusted Base Cost per square foot:

Building Cost $ _____ x _____ = _____
               Adjusted Base Cost   *Floor Area*
               per square foot

Basement Cost $ _____ x _____ = _____
               Basement Cost   *Basement Area*

Lump Sum Additions

TOTAL BUILDING COST *(Sum of above costs)* _____

Modifications: *(complexity, workmanship, size)* +/- _____ % _____

Location Modifier: City _____ Date _____ x _____

  Local cost of replacement _____

Less depreciation: _____ - _____

  Local cost of replacement less depreciation $ _____

Other RSMeans Products & Services

Table of Contents
Annual Cost Guides
RSMeans Online and *CostWorks* CD Comparison Matrix
Seminars
Reference Books
Order Form

RSMeans—
a tradition of excellence in construction cost information and services since 1942

For more information visit the RSMeans website at **www.rsmeans.com**

Unit prices according to the latest MasterFormat!

Book Selection Guide
The following table provides definitive information on the content of each cost data publication. The number of lines of data provided in each unit price or assemblies division, as well as the number of crews, is listed for each book. The presence of other elements such as reference tables, square foot models, equipment rental costs, historical cost indexes, and city cost indexes, is also indicated. You can use the table to help select the RSMeans book that has the quantity and type of information you most need in your work.

Unit Cost Divisions	Building Construction	Mechanical	Electrical	Commercial Renovation	Square Foot	Site Work Landsc.	Green Building	Interior	Concrete Masonry	Open Shop	Heavy Construction	Light Commercial	Facilities Construction	Plumbing	Residential
1	573	393	410	506		509	191	312	456	572	501	246	1042	403	178
2	777	279	85	733		993	178	399	213	776	734	481	1221	286	257
3	1690	340	230	1085		1471	987	355	2032	1689	1685	481	1788	316	388
4	957	21	0	922		720	179	612	1155	925	612	529	1172	0	441
5	1893	158	155	1090		841	1799	1095	720	1893	1037	979	1909	204	721
6	2452	18	18	2110		110		589	1528	281	2448	123	2140	2124	22	2660
7	1584	215	128	1623		581	757	532	524	1581	26	1317	1685	227	1037
8	2082	81	45	2609		265	1123	1753	105	2077	0	2193	2879	0	1526
9	1971	72	26	1767		313	449	2056	390	1911	15	1659	2220	54	1435
10	1026	17	10	642		215	27	842	158	1026	29	511	1118	235	224
11	1087	208	165	546		124	54	932	28	1071	0	230	1107	169	110
12	544	0	2	319		217	137	1654	14	532	0	352	1704	23	310
13	719	140	137	242		343	111	248	66	695	244	83	746	94	79
14	273	36	0	221		0	0	257	0	273	0	12	293	16	6
21	90	0	16	37		0	0	250	0	87	0	68	426	431	220
22	1157	7544	150	1190		1568	1067	838	20	1121	1677	857	7440	9347	712
23	1177	6971	580	928		158	902	776	38	1101	111	878	5187	1918	469
26	1354	454	10081	1021		793	611	1136	55	1349	562	1298	10030	399	617
27	72	0	297	34		13	0	71	0	72	39	52	279	0	22
28	96	58	136	71		0	21	78	0	99	0	40	151	44	25
31	1498	735	610	803		3217	289	7	1208	1443	3256	600	1560	660	611
32	822	53	8	886		4426	356	405	291	793	1841	420	1701	165	468
33	530	1079	538	252		2178	41	0	237	522	2159	128	1699	1286	154
34	107	0	47	4		190	0	0	31	62	193	0	128	0	0
35	18	0	0	0		327	0	0	0	18	442		84	0	0
41	60	0	0	32		7	0	22	0	60	30	0	67	14	0
44	75	79	0	0		0	0	0	0	0	0	0	75	75	0
46	23	16	0	0		274	261	0	0	23	264	0	33	33	0
48	12	0	25	0		0	25	0	0	12	17	12	12	0	12
Totals	24719	18967	13899	19673		19853	10154	16158	8022	24231	15597	15566	49880	16421	12682

Assem Div	Building Construction	Mechanical	Electrical	Commercial Renovation	Square Foot	Site Work Landscape	Assemblies	Green Building	Interior	Concrete Masonry	Heavy Construction	Light Commercial	Facilities Construction	Plumbing	Asm Div	Residential
A		15	0	188	150	577	598	0	0	536	571	154	24	0	1	376
B		0	0	848	2498	0	5658	56	329	1975	368	2089	174	0	2	211
C		0	0	647	926	0	1304	0	1629	146	0	816	250	0	3	588
D		1067	941	712	1859	72	2538	330	825	0	0	1345	1105	1088	4	851
E		0	0	86	260	0	300	0	5	0	0	257	5	0	5	392
F		0	0	0	114	0	114	0	0	0	0	114	3	0	6	357
G		527	447	318	249	3365	729	0	0	535	1350	205	293	677	7	307
															8	760
															9	80
															10	0
															11	0
															12	0
Totals		1609	1388	2799	6056	4014	11241	386	2788	3192	2289	4980	1854	1765		3922

Reference Section	Building Construction Costs	Mechanical	Electrical	Commercial Renovation	Square Foot	Site Work Landscape	Assem.	Green Building	Interior	Concrete Masonry	Open Shop	Heavy Construction	Light Commercial	Facilities Construction	Plumbing	Resi.
Reference Tables	yes	yes	yes	yes	no	yes	yes	yes	yes	yes	yes	yes	yes	yes	yes	yes
Models					111			25					50			28
Crews	571	571	571	550		571		571	571	571	548	571	548	550	571	548
Equipment Rental Costs	yes	yes	yes	yes		yes		yes	yes	yes	yes	yes	yes	yes	yes	yes
Historical Cost Indexes	yes	yes	yes	yes	yes	yes	yes	yes	yes	yes	yes	yes	yes	yes	yes	no
City Cost Indexes	yes	yes	yes	yes	yes	yes	yes	yes	yes	yes	yes	yes	yes	yes	yes	yes

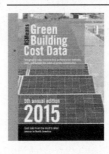

RSMeans Building Construction Cost Data 2015

Offers you unchallenged unit price reliability in an easy-to-use format. Whether used for verifying complete, finished estimates or for periodic checks, it supplies more cost facts better and faster than any comparable source. More than 24,700 unit prices have been updated for 2015. The City Cost Indexes and Location Factors cover more than 930 areas, for indexing to any project location in North America. Order and get *RSMeans Quarterly Update Service* FREE.

$206.95 | Available Sept. 2014 | Catalog no. 60015

RSMeans Mechanical Cost Data 2015

Total unit and systems price guidance for mechanical construction . . . materials, parts, fittings, and complete labor cost information. Includes prices for piping, heating, air conditioning, ventilation, and all related construction.

Plus new 2015 unit costs for:

- Thousands of installed HVAC/controls, sub-assemblies and assemblies
- "On-site" Location Factors for more than 930 cities and towns in the U.S. and Canada
- Crews, labor, and equipment

$203.95 | Available Oct. 2014 | Catalog no. 60025

RSMeans Square Foot Costs 2015
Accurate and Easy-to-Use

- **Updated price information** based on nationwide figures from suppliers, estimators, labor experts, and contractors.
- Green building models
- Realistic graphics, offering true-to-life illustrations of building projects
- Extensive information on using square foot cost data, including sample estimates and alternate pricing methods

$218.95 | Available Oct. 2014 | Catalog no. 60055

RSMeans Electrical Cost Data 2015

Pricing information for every part of electrical cost planning. More than 13,800 unit and systems costs with design tables; clear specifications and drawings; engineering guides; illustrated estimating procedures; complete labor-hour and materials costs for better scheduling and procurement; and the latest electrical products and construction methods.

- A variety of special electrical systems, including cathodic protection
- Costs for maintenance, demolition, HVAC/mechanical, specialties, equipment, and more

$208.95 | Available Oct. 2014 | Catalog no. 60035

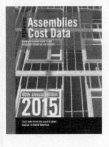

RSMeans Assemblies Cost Data 2015

RSMeans Assemblies Cost Data takes the guesswork out of preliminary or conceptual estimates. Now you don't have to try to calculate the assembled cost by working up individual component costs. We've done all the work for you.

Presents detailed illustrations, descriptions, specifications, and costs for every conceivable building assembly—over 350 types in all—arranged in the easy-to-use UNIFORMAT II system. Each illustrated "assembled" cost includes a complete grouping of materials and associated installation costs, including the installing contractor's overhead and profit.

$334.95 | Available Sept. 2014 | Catalog no. 60065

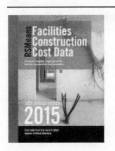

RSMeans Green Building Cost Data 2015

Estimate, plan, and budget the costs of green building for both new commercial construction and renovation work with this fifth edition of *RSMeans Green Building Cost Data*. More than 10,000 unit costs for a wide array of green building products plus assemblies costs. Easily identified cross references to LEED and Green Globes building rating systems criteria.

$170.95 | Available Nov. 2014 | Catalog no. 60555

RSMeans Facilities Construction Cost Data 2015

For the maintenance and construction of commercial, industrial, municipal, and institutional properties. Costs are shown for new and remodeling construction and are broken down into materials, labor, equipment, and overhead and profit. Special emphasis is given to sections on mechanical, electrical, furnishings, site work, building maintenance, finish work, and demolition.

More than 49,800 unit costs, plus assemblies costs and a comprehensive Reference Section are included.

$524.95 | Available Nov. 2014 | Catalog no. 60205

RSMeans Commercial Renovation Cost Data 2015
Commercial/Multi-family Residential

Use this valuable tool to estimate commercial and multi-family residential renovation and remodeling.

Includes: Updated costs for hundreds of unique methods, materials, and conditions that only come up in repair and remodeling, PLUS:

- Unit costs for more than 19,600 construction components
- Installed costs for more than 2,700 assemblies
- More than 930 "on-site" localization factors for the U.S. and Canada

$166.95 | Available Oct. 2014 | Catalog no. 60045

RSMeans Electrical Change Order Cost Data 2015

RSMeans Electrical Change Order Cost Data provides you with electrical unit prices exclusively for pricing change orders—based on the recent, direct experience of contractors and suppliers. Analyze and check your own change order estimates against the experience others have had doing the same work. It also covers productivity analysis, and change order cost justifications. With useful information for calculating the effects of change orders and dealing with their administration.

$199.95 | Available Dec. 2014 | Catalog no. 60235

RSMeans Open Shop Building Construction Cost Data 2015

The latest costs for accurate budgeting and estimating of new commercial and residential construction . . . renovation work . . . change orders . . . cost engineering.

RSMeans Open Shop "BCCD" will assist you to:

- Develop benchmark prices for change orders.
- Plug gaps in preliminary estimates and budgets.
- Estimate complex projects.
- Substantiate invoices on contracts.
- Price ADA-related renovations.

$175.95 | Available Dec. 2014 | Catalog no. 60155

Annual Cost Guides

Unit prices according to the latest MasterFormat!

Annual Cost Guides

Unit prices according to the latest MasterFormat!

RSMeans Residential
Cost Data 2015

Contains square foot costs for 28 basic home models with the look of today, plus hundreds of custom additions and modifications you can quote right off the page. Includes more than 3,900 costs for 89 residential systems. Complete with blank estimating forms, sample estimates, and step-by-step instructions.

Contains line items for cultured stone and brick, PVC trim, lumber, and TPO roofing.

$148.95 | Available Oct. 2014 | Catalog no. 60175

RSMeans Facilities Maintenance & Repair
Cost Data 2015

RSMeans Facilities Maintenance & Repair Cost Data gives you a complete system to manage and plan your facility repair and maintenance costs and budget efficiently. Guidelines for auditing a facility and developing an annual maintenance plan. Budgeting is included, along with reference tables on cost and management, and information on frequency and productivity of maintenance operations.

The only nationally recognized source of maintenance and repair costs. Developed in cooperation with the Civil Engineering Research Laboratory (CERL) of the Army Corps of Engineers.

$461.95 | Available Nov. 2014 | Catalog no. 60305

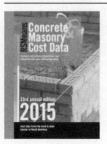

RSMeans Concrete & Masonry
Cost Data 2015

Provides you with cost facts for virtually all concrete/masonry estimating needs, from complicated formwork to various sizes and face finishes of brick and block—all in great detail. The comprehensive Unit Price Section contains more than 8,000 selected entries. Also contains an Assemblies [Cost] Section, and a detailed Reference Section that supplements the cost data.

$188.95 | Available Dec. 2014 | Catalog no. 60115

RSMeans Construction
Cost Indexes 2015

What materials and labor costs will change unexpectedly this year? By how much?

• Breakdowns for 318 major cities
• National averages for 30 key cities
• Expanded five major city indexes
• Historical construction cost indexes

$359.95 per year (subscription) | Catalog no. 50145
$94.95 individual quarters | Catalog no. 60145 A,B,C,D

RSMeans Heavy Construction
Cost Data 2015

A comprehensive guide to heavy construction costs. Includes costs for highly specialized projects such as tunnels, dams, highways, airports, and waterways. Information on labor rates, equipment, and materials costs is included. Features unit price costs, systems costs, and numerous reference tables for costs and design.

$204.95 | Available Dec. 2014 | Catalog no. 60165

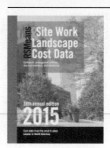

RSMeans Site Work & Landscape
Cost Data 2015

Includes unit and assemblies costs for earthwork, sewerage, piped utilities, site improvements, drainage, paving, trees and shrubs, street openings/repairs, underground tanks, and more. Contains more than 60 types of assemblies costs for accurate conceptual estimates.

Includes:

• Estimating for infrastructure improvements
• Environmentally-oriented construction
• ADA-mandated handicapped access
• Hazardous waste line items

$199.95 | Available Dec. 2014 | Catalog no. 60285

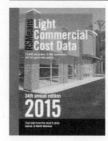

RSMeans Light Commercial
Cost Data 2015

Specifically addresses the light commercial market, which is a specialized niche in the construction industry. Aids you, the owner/designer/contractor, in preparing all types of estimates—from budgets to detailed bids. Includes new advances in methods and materials.

Assemblies Section allows you to evaluate alternatives in the early stages of design/planning.

More than 15,500 unit costs ensure that you have the prices you need, when you need them.

$153.95 | Available Nov. 2014 | Catalog no. 60185

RSMeans Labor Rates for the
Construction Industry 2015

Complete information for estimating labor costs, making comparisons, and negotiating wage rates by trade for more than 300 U.S. and Canadian cities. With 46 construction trades in each city, and historical wage rates included for comparison. Each city chart lists the county and is alphabetically arranged with handy visual flip tabs for quick reference.

$450.95 | Available Dec. 2014 | Catalog no. 60125

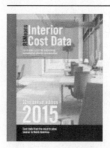

RSMeans Interior
Cost Data 2015

Provides you with prices and guidance needed to make accurate interior work estimates. Contains costs on materials, equipment, hardware, custom installations, furnishings, and labor for new and remodel commercial and industrial interior construction, including updated information on office furnishings, and reference information.

$208.95 | Available Nov. 2014 | Catalog no. 60095

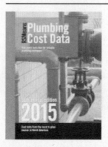

RSMeans Plumbing
Cost Data 2015

Comprehensive unit prices and assemblies for plumbing, irrigation systems, commercial and residential fire protection, point-of-use water heaters, and the latest approved materials. This publication and its companion, *RSMeans Mechanical Cost Data*, provide full-range cost estimating coverage for all the mechanical trades.

Contains updated costs for potable water, radiant heat systems, and high efficiency fixtures.

$206.95 | Available Oct. 2014 | Catalog no. 60215

RSMeans Online and RSMeans CostWorks CD Comparison Matrix

Both the RSMeans CostWorks CD package and the RSMeans Online estimating tool provide the same comprehensive, up-to-date, and localized cost information to improve planning, estimating, and budgeting with RSMeans—the industry's most-trusted source of construction cost data.

In addition to key differences noted in the comparison matrix below, both products let you do the following:

• Quickly locate costs in the searchable RSMeans database.

• Instantly adjust the data to your specific location for accurate costs anywhere in the U.S. and Canada.

• Create reliable, updated cost estimates in minutes.

• Manage, share, and export your estimates to MS Excel® with ease.

• Access a wide range of reference materials, including crew rates and national average labor rates.

Both the CD and Online options are available in a convenient one-year subscription!

Use the handy comparison matrix below to help you decide the best product for your cost estimating needs!

Why us?

For more than 70 years, RSMeans has provided the quality cost data construction professionals everywhere depend on to estimate with confidence, save time, improve decision-making, and increase profits.

Improve planning and decision-making

Back your estimates with complete, accurate, and up-to-date cost data for informed business decisions.
• Verify construction costs from third parties.
• Check validity of subcontractor proposals.
• Evaluate material and assembly alternatives.

Increase profits

Use RSMeans to estimate projects quickly and accurately, so you can gain an edge over your competition.
• Create accurate and competitive bids.
• Minimize the risk of cost overruns.
• Reduce variability.
• Gain control over costs.

The following matrix summarizes the key comparison points between RSMeans' two electronic cost data products.

	Question	CostWorks CD	RSMeans Online.com
1	How is the data presented?	• The data is organized by book title and sold at comparable book cost. • Several packaged bundles of five titles are available at a cost savings. • Available in a convenient one-year subscription	• The data is organized by book title and sold at comparable book cost. • Several packaged bundles of five titles are available at a cost savings. • RSMeans Online also offers a Complete Library title with all online data at a cost savings. • Available in a convenient one-year subscription
2	Where is the data stored?	With a typical install, the data is read off of your hard drive or off of the CD.	All data is stored on our secure server.
3	How are updates delivered?	Updates are available on a quarterly basis for download from the Costworks HomePort website.	Data is updated automatically on a quarterly basis, and users are notified when updates occur.
4	Where are my projects and estimates stored?	Upon installation, a "My CostWorks Projects" folder is created within the "My Documents" directory. All projects are defaulted to save here unless otherwise specified.	Estimates are stored on our secure server.
5	Does my IT team need to assist me in installing and running the program?	Yes, depending on the level of your permission. See #11 for system requirements.	While the product is entirely Web-based, you may need IT support with permissions for certain tasks. See #11 for system requirements.
6	Can users collaborate on projects/estimates?	Yes, you can copy project folders between computers.	You can create shared folders that can be viewed, copied, and edited by anyone within your account workgroup.

Seminar Schedule and Professional Development

Comparison matrix continued

7	Can I create and use my own data with the RSMeans database?	Custom lines cannot be added to the CD database. With the Estimator tool, however, you can create and add custom cost lines to any estimate. You can also save custom lines as an Estimator template for use in creating new estimates.	You can create • a custom line within an estimate • a custom cost book consisting of custom unit (not assembly) cost lines, no more than once per quarter **Note** that custom data is searchable using the Include My Custom Data feature.
8	Can I enter my own trade/labor rates and markups?	Although you cannot adjust the trade labor rates, you can add custom markups for each report template using the Estimator tool.	For each quarterly update, you can create one custom unit cost book. You can edit the labor rates by individual trade and the crew equipment costs, and also edit the associated material and equipment markups.
9	What are my reporting options?	The Estimator tool lets you create editable custom reports and report templates. Reports can also be exported to MS Excel®.	You can generate pre-formatted estimate summaries and cost lists in both PDF and Excel formats.
10	How do I find specific cost data?	You can search by CSI line number and by keyword within a division.	You can search by CSI line number and by keyword within a division. There are several additional search options, including keyword and advanced search using special characters.
11	What are the system requirements?	• Operating system – Windows XP or later • PC requirements - 256 MB RAM (recommended), 1024 x 768 High Color Monitor, 800MHz Computer (recommended), 24X CD-ROM, 250 MB Hard Disk Space	• Resolution – Recommended resolution is 1024 x 768 or higher • Compatibility - Internet Explorer version 8.x, 9.x and Mozilla Firefox version 10 and higher • iPad - iOS 5 and higher

2015 RSMeans Seminar Schedule

Note: call for exact dates and details.

Location	Dates	Location	Dates
Seattle, WA	January and August	Jacksonville FL	September
Dallas/Ft. Worth, TX	January	Dallas, TX	September
Austin, TX	February	Charleston, SC	October
Anchorage, AK	March and September	Houston, TX	October
Las Vegas, NV	March and October	Atlanta, GA	November
Washington, DC	April and September	Baltimore, MD	November
Phoenix, AZ	April	Orlando, FL	November
Kansas City, MO	April	San Diego, CA	December
Toronto	May	San Antonio, TX	December
Denver, CO	May	Raleigh, NC	December
San Francisco, CA	June		
Bethesda, MD	June		
Columbus, GA	June		
El Segundo, CA	August		

1-800-334-3509, Press 1

Professional Development

eLearning Training Sessions

Learn how to use *RSMeans Online®* or *RSMeans CostWorks®* CD from the convenience of your home or office. Our eLearning training sessions let you join a training conference call and share the instructors' desktops, so you can view the presentation and step-by-step instructions on your own computer screen. The live webinars are held from 9 a.m. to 4 p.m. eastern standard time, with a one-hour break for lunch.

For these sessions, you must have a computer with high speed Internet access and a compatible Web browser. Learn more at www.rsmeansonline.com or call for a schedule: 1-800-334-3509 and press 1. Webinars are generally held on selected Wednesdays each month.

RSMeans Online®	**RSMeans CostWorks® CD**
$299 per person	**$299 per person**

RSMeans Online™ Training

Construction estimating is vital to the decision-making process at each state of every project. RSMeansOnline works the way you do. It's systematic, flexible and intuitive. In this one day class you will see how you can estimate any phase of any project faster and better.

Some of what you'll learn:
• Customizing RSMeansOnline
• Making the most of RSMeans "Circle Reference" numbers
• How to integrate your cost data
• Generate reports, exporting estimates to MS Excel, sharing, collaborating and more

Also available: RSMeans Online® training webinar

Maintenance & Repair Estimating for Facilities

This *two-day* course teaches attendees how to plan, budget, and estimate the cost of ongoing and preventive maintenance and repair for existing buildings and grounds.

Some of what you'll learn:
• The most financially favorable maintenance, repair, and replacement scheduling and estimating
• Auditing and value engineering facilities
• Preventive planning and facilities upgrading
• Determining both in-house and contract-out service costs
• Annual, asset-protecting M&R plan

Who should attend: facility managers, maintenance supervisors, buildings and grounds superintendents, plant managers, planners, estimators, and others involved in facilities planning and budgeting.

Facilities Construction Estimating

In this *two-day* course, professionals working in facilities management can get help with their daily challenges to establish budgets for all phases of a project.

Some of what you'll learn:
• Determining the full scope of a project
• Identifying the scope of risks and opportunities
• Creative solutions to estimating issues
• Organizing estimates for presentation and discussion
• Special techniques for repair/remodel and maintenance projects
• Negotiating project change orders

Who should attend: facility managers, engineers, contractors, facility tradespeople, planners, and project managers.

Practical Project Management for Construction Professionals

In this *two-day* course, acquire the essential knowledge and develop the skills to effectively and efficiently execute the day-to-day responsibilities of the construction project manager.

Covers:
• General conditions of the construction contract
• Contract modifications: change orders and construction change directives
• Negotiations with subcontractors and vendors
• Effective writing: notification and communications
• Dispute resolution: claims and liens

Who should attend: architects, engineers, owner's representatives, project managers.

Construction Cost Estimating: Concepts and Practice

This *one-day* introductory course to improve estimating skills and effectiveness starts with the details of interpreting bid documents, ending with the summary of the estimate and bid submission.

Some of what you'll learn:
• Using the plans and specifications for creating estimates
• The takeoff process—deriving all tasks with correct quantities
• Developing pricing, using various sources; how subcontractor pricing fits in
• Summarizing the estimate to arrive at the final number
• Formulas for area and cubic measure, adding waste and adjusting productivity to specific projects
• Evaluating subcontractors' proposals and prices
• Adding Insurance and bonds
• Understanding how labor costs are calculated
• Submitting bids and proposals

Who should attend: project managers, architects, engineers, owner's representatives, contractors, and anyone who's responsible for budgeting or estimating construction projects.

Mechanical & Electrical Estimating

This *two-day* course teaches attendees how to prepare more accurate and complete mechanical/electrical estimates, avoiding the pitfalls of omission and double-counting, while understanding the composition and rationale within the RSMeans mechanical/electrical database.

Some of what you'll learn:
• The unique way mechanical and electrical systems are interrelated
• M&E estimates—conceptual, planning, budgeting, and bidding stages
• Order of magnitude, square foot, assemblies, and unit price estimating
• Comparative cost analysis of equipment and design alternatives

Who should attend: architects, engineers, facilities managers, mechanical and electrical contractors, and others who need a highly reliable method for developing, understanding, and evaluating mechanical and electrical contracts.

Professional Development

RSMeans data titles are also available in online format! Go to RSMeansOnline.com for more details.

For more information visit the RSMeans website at **www.rsmeans.com**

Unit Price Estimating

This interactive *two-day* seminar teaches attendees how to interpret project information and process it into final, detailed estimates with the greatest accuracy level.

The most important credential an estimator can take to the job is the ability to visualize construction and estimate accurately.

Some of what you'll learn:
- Interpreting the design in terms of cost
- The most detailed, time-tested methodology for accurate pricing
- Key cost drivers—material, labor, equipment, staging, and subcontracts
- Understanding direct and indirect costs for accurate job cost accounting and change order management

Who should attend: corporate and government estimators and purchasers, architects, engineers, and others who need to produce accurate project estimates.

RSMeans CostWorks® CD Training

This one-day course helps users become more familiar with the functionality of *RSMeans CostWorks* program. Each menu, icon, screen, and function found in the program is explained in depth. Time is devoted to hands-on estimating exercises.

Some of what you'll learn:
- Searching the database using all navigation methods
- Exporting RSMeans data to your preferred spreadsheet format
- Viewing crews, assembly components, and much more
- Automatically regionalizing the database

This training session requires you to bring a laptop computer to class.

When you register for this course you will receive an outline for your laptop requirements.

Also offering web training for RSMeans CostWorks CD!

Facilities Estimating Using RSMeans CostWorks® CD

Combines hands-on skill building with best estimating practices and real-life problems. Brings you up-to-date with key concepts, and provides tips, pointers, and guidelines to save time and avoid cost oversights and errors.

Some of what you'll learn:
- Estimating process concepts
- Customizing and adapting RSMeans cost data
- Establishing scope of work to account for all known variables
- Budget estimating: when, why, and how
- Site visits: what to look for—what you can't afford to overlook
- How to estimate repair and remodeling variables

This training session requires you to bring a laptop computer to class.

Who should attend: facility managers, architects, engineers, contractors, facility tradespeople, planners, project managers and anyone involved with JOC, SABRE, or IDIQ.

Conceptual Estimating Using RSMeans CostWorks® CD

This *two-day* class uses the leading industry data and a powerful software package to develop highly accurate conceptual estimates for your construction projects. All attendees must bring a laptop computer loaded with the current year *Square Foot Costs* and the *Assemblies Cost Data* CostWorks titles.

Some of what you'll learn:
- Introduction to conceptual estimating
- Types of conceptual estimates
- Helpful hints
- Order of magnitude estimating
- Square foot estimating
- Assemblies estimating

Who should attend: architects, engineers, contractors, construction estimators, owner's representatives, and anyone looking for an electronic method for performing square foot estimating.

Assessing Scope of Work for Facility Construction Estimating

This *two-day* practical training program addresses the vital importance of understanding the SCOPE of projects in order to produce accurate cost estimates for facility repair and remodeling.

Some of what you'll learn:
- Discussions of site visits, plans/specs, record drawings of facilities, and site-specific lists
- Review of CSI divisions, including means, methods, materials, and the challenges of scoping each topic
- Exercises in SCOPE identification and SCOPE writing for accurate estimating of projects
- Hands-on exercises that require SCOPE, take-off, and pricing

Who should attend: corporate and government estimators, planners, facility managers, and others who need to produce accurate project estimates.

Unit Price Estimating Using RSMeans CostWorks® CD

Step-by-step instruction and practice problems to identify and track key cost drivers—material, labor, equipment, staging, and subcontractors—for each specific task. Learn the most detailed, time-tested methodology for accurately "pricing" these variables, their impact on each other and on total cost.

Some of what you'll learn:
- Unit price cost estimating
- Order of magnitude, square foot, and assemblies estimating
- Quantity takeoff
- Direct and indirect construction costs
- Development of contractor's bill rates
- How to use *RSMeans Building Construction Cost Data*

This training session requires you to bring a laptop computer to class.

Who should attend: architects, engineers, corporate and government estimators, facility managers, and government procurement staff.

Professional Development

564

For more information visit the RSMeans website at **www.rsmeans.com**

Registration Information

Register early . . . Save up to $100! Register 30 days before the start date of a seminar and save $100 off your total fee. *Note: This discount can be applied only once per order. It cannot be applied to team discount registrations or any other special offer.*

How to register Register by phone today! The RSMeans toll-free number for making reservations is **1-800-334-3509, Press 1.**

Two-day seminar registration fee - $935. One-day *RSMeans CostWorks*® training registration fee - $375. To register by mail, complete the registration form and return, with your full fee, to: RSMeans Seminars, 700 Longwater Drive, Norwell, MA 02061.

Government pricing All federal government employees save off the regular seminar price. Other promotional discounts cannot be combined with the government discount.

Team discount program for two to four seminar registrations. Call for pricing: 1-800-334-3509, Press 1

Multiple course discounts When signing up for two or more courses, call for pricing.

Refund policy Cancellations will be accepted up to ten business days prior to the seminar start. There are no refunds for cancellations received later than ten working days prior to the first day of the seminar. A $150 processing fee will be applied for all cancellations. Written notice of cancellation is required. Substitutions can be made at any time before the session starts. **No-shows are subject to the full seminar fee.**

AACE approved courses Many seminars described and offered here have been approved for 14 hours (1.4 recertification credits) of credit by the AACE International Certification Board toward meeting the continuing education requirements for recertification as a Certified Cost Engineer/ Certified Cost Consultant.

AIA Continuing Education We are registered with the AIA Continuing Education System (AIA/CES) and are committed to developing quality learning activities in accordance with the CES criteria. Many seminars meet the AIA/CES criteria for Quality Level 2. AIA members may receive (14) learning units (LUs) for each two-day RSMeans course.

Daily course schedule The first day of each seminar session begins at 8:30 a.m. and ends at 4:30 p.m. The second day begins at 8:00 a.m. and ends at 4:00 p.m. Participants are urged to bring a hand-held calculator, since many actual problems will be worked out in each session.

Continental breakfast Your registration includes the cost of a continental breakfast, and a morning and afternoon refreshment break. These informal segments allow you to discuss topics of mutual interest with other seminar attendees. (You are free to make your own lunch and dinner arrangements.)

Hotel/transportation arrangements RSMeans arranges to hold a block of rooms at most host hotels. To take advantage of special group rates when making your reservation, be sure to mention that you are attending the RSMeans seminar. You are, of course, free to stay at the lodging place of your choice. **(Hotel reservations and transportation arrangements should be made directly by seminar attendees.)**

Important Class sizes are limited, so please register as soon as possible.

Note: Pricing subject to change.

Registration Form

ADDS-1000

Call 1-800-334-3509, Press 1 to register or FAX this form 1-800-632-6732. Visit our website: www.rsmeans.com

Please register the following people for the RSMeans construction seminars as shown here. We understand that we must make our own hotel reservations if overnight stays are necessary.

☐ Full payment of $_____ enclosed.

☐ Bill me.

Please print name of registrant(s).

(To appear on certificate of completion)

_____

_____

_____

_____

P.O. #: _____
GOVERNMENT AGENCIES MUST SUPPLY PURCHASE ORDER NUMBER OR TRAINING FORM.

Firm name_____

Address_____

City/State/Zip_____

Telephone no._____ Fax no._____

E-mail address_____

Charge registration(s) to: ☐ MasterCard ☐ VISA ☐ American Express

Account no._____ Exp. date_____

Cardholder's signature_____

Seminar name_____

Seminar City _____

Please mail check to: RSMeans Seminars, 700 Longwater Drive, Norwell, MA 02061 USA

Professional Development

565

RSMeans data titles are also available in online format! Go to RSMeansOnline.com for more details.

For more information visit the RSMeans website at www.rsmeans.com

Managing the Profitable Construction Business: The Contractor's Guide to Success and Survival Strategies, 2nd Ed.

by Thomas C. Scheifler, Ph.D.

Learn from a team of construction business veterans led by Thomas C. Schleifer, who is commonly referred to as a construction business "turnaround" expert due to the number of construction companies he has rescued from financial distress. His financial acumen, combined with his practical, hands-on experience, has made him a sought-after private consultant.

$60.00 | 288 pages, hardover | Catalog no. 67370

Construction Project Safety

by John Schaufelberger, Ken-Yu Lin

This essential introduction to construction safety for construction management personnel takes a project-based approach to present potential hazards in construction and their mitigation or prevention.

Beginning with an introduction to Accident Prevention Programs and OSHA compliance requirements, the book integrates safety instruction into the building process by following a building project from site construction through interior finish. Reinforcing this applied approach are photographs, drawings, contract documentation, and an online 3D BIM model to help visualize the on-site scenarios.

$85.00 | 320 pages, hardover | Catalog no. 67368

Project Control: Integrating Cost & Schedule in Construction

by Wayne J. Del Pico

Written by a seasoned professional in the field, *Project Control: Integrating Cost and Schedule in Construction* fills a void in the area of project control as applied in the construction industry today. It demonstrates how productivity models for an individual project are created, monitored, and controlled, and how corrective actions are implemented as deviations from the baseline occur.

$65.00 | 240 pages, softcover | Catalog no. 67366

Project Scheduling and Management for Construction, 4th Edition

by David R. Pierce

First published in 1988 by RSMeans, the new edition of *Project Scheduling and Management for Construction* has been substantially revised for both professionals and students enrolled in construction management and civil engineering programs. While retaining its emphasis on developing practical, professional-level scheduling skills, the new edition is a relatable, real world case study.

$95.00 | 272 pages, softcover | Catalog no. 67367

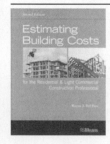

Estimating Building Costs, 2nd Edition
For the Residential & Light Commercial Construction Professional

by Wayne J. Del Pico

Explains, in detail, how to put together a reliable estimate that can be used not only for budgeting, but also for developing a schedule, managing a project, dealing with contingencies, and, ultimately, making a profit.

Completely revised and updated to reflect the CSI MasterFormat 2010 system.

$75.00 | 528 pages, softcover | Catalog no. 67343A

The Gypsum Construction Handbook, 7th Edition

by USG

The tried-and-true Gypsum Construction Handbook is a systematic guide to selecting and using gypsum drywall, veneer plaster, tile backers, ceilings, and conventional plaster building materials. A widely respected training text for aspiring architects and engineers, the book provides detailed product information and efficient installation methodology.

The 7th edition features updates in gypsum products, including ultralight panels, glass-mat panels, paper-faced plastic bead, and ultralightweight joint compound, and modern specialty acoustical and ceiling product guidelines. This comprehensive reference also incorporates the latest in sustainable products.

$40.00 | 576 pages, softcover | Catalog no. 67357B

Solar Energy: Technologies & Project Delivery for Buildings

by Andy Walker

An authoritative reference on the design of solar energy systems in building projects, with applications, operating principles, and simple tools for the construction, engineering, and design professionals, the book simplifies the solar design and engineering process and provides sample documentation for the complete design of a solar energy system for buildings.

$85.00 | 320 pages, hardover | Catalog no. 67365

RSMeans Cost Data, Student Edition

Provides a thorough introduction to cost estimating in a print and online package. Clear explanations and example-driven. The ideal reference for students and new professionals. Features include:

- Commercial and residential construction cost data in print and online formats
- Complete how-to guidance on the essentials of cost estimating
- A supplemental website with plans, problem sets, and a full sample estimate

$110.00 | 512 pages, softcover | Catalog no. 67363

How to Estimate with Means Data & CostWorks, 4th Edition

by RSMeans and Saleh A. Mubarak, Ph.D.

This step-by-step guide takes you through all the major construction items with extensive coverage of site work, concrete and masonry, wood and metal framing, doors and windows, and other divisions. The only construction cost estimating handbook that uses the most popular source of construction data, RSMeans, this indispensible guide features access to the instructional version of *CostWorks* in electronic form, enabling you to practice techniques to solve real-world estimating problems.

$75.00 | 320 pages, softcover | Includes CostWorks CD | Catalog no. 67324C

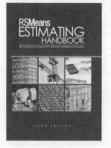

RSMeans Estimating Handbook, 3rd Edition

Widely used in the industry for tasks ranging from routine estimates to special cost analysis projects. This handbook will help construction professionals:

- Evaluate architectural plans and specifications
- Prepare accurate quantity takeoffs
- Compare design alternatives and costs
- Perform value engineering
- Double-check estimates and quotes
- Estimate change orders

$110.00 | Over 976 pages, hardcover | Catalog No. 67276B

Reference Books

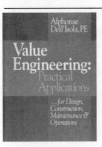

Value Engineering: Practical Applications

For Design, Construction, Maintenance & Operations

by Alphonse Dell'Isola, PE

A tool for immediate application—for engineers, architects, facility managers, owners, and contractors. Includes making the case for VE—the management briefing; integrating VE into planning, budgeting, and design; conducting life cycle costing; using VE methodology in design review and consultant selection; and case studies.

$90.00 | Over 427 pages, illustrated, softcover | Catalog no. 67319A

Risk Management for Design and Construction

Introduces risk as a central pillar of project management and shows how a project manager can prepare to deal with uncertainty. Includes:

• Integrated cost and schedule risk analysis

• An introduction to a ready-to-use system of analyzing a project's risks and tools to proactively manage risks

• A methodology that was developed and used by the Washington State Department of Transportation

• Case studies and examples on the proper application of principles

• Combining value analysis with risk analysis

$125.00 | Over 288 pages, softcover | Catalog no. 67359

How Your House Works, 2nd Edition

by Charlie Wing

Knowledge of your home's systems helps you control repair and construction costs and makes sure the correct elements are being installed or replaced. This book uncovers the mysteries behind just about every major appliance and building element in your house. See-through, cross-section drawings in full color show you exactly how these things should be put together and how they function, including what to check if they don't work. It just might save you having to call in a professional.

$22.95 | 208 pages, softcover | Catalog no. 67351A

Construction Business Management

by Nick Ganaway

Only 43% of construction firms stay in business after four years. Make sure your company thrives with valuable guidance from a pro with 25 years of success as a commercial contractor. Find out what it takes to build all aspects of a business that is profitable, enjoyable, and enduring. With a bonus chapter on retail construction.

$60.00 | 201 pages, softcover | Catalog no. 67352

The Practice of Cost Segregation Analysis

by Bruce A. Desrosiers and Wayne J. Del Pico

This expert guide walks you through the practice of cost segregation analysis, which enables property owners to defer taxes and benefit from "accelerated cost recovery" through depreciation deductions on assets that are properly identified and classified.

With a glossary of terms, sample cost segregation estimates for various building types, key information resources, and updates via a dedicated website, this book is a critical resource for anyone involved in cost segregation analysis.

$105.00 | Over 240 pages | Catalog no. 67345

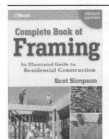

Complete Book of Framing, 2nd Edition

by Scot Simpson

This updated, easy-to-learn guide to rough carpentry and framing is written by an expert with more than thirty years of framing experience. Starting with the basics, this book begins with types of lumber, nails, and what tools are needed, followed by detailed, fully illustrated steps for framing each building element. Framer-Friendly Tips throughout the book show how to get a task done right— and more easily.

$29.95 | 368 pages, softcover | Catalog no. 67353A

Green Home Improvement

by Daniel D. Chiras, PhD

With energy costs rising and environmental awareness increasing, people are looking to make their homes greener. This book, with 65 projects and actual costs and projected savings, helps homeowners prioritize their green improvements.

Projects range from simple water savers that cost only a few dollars, to bigger-ticket items such as HVAC systems. With color photos and cost estimates, each project compares options and describes the work involved, the benefits, and the savings.

$34.95 | 320 pages, illustrated, softcover | Catalog no. 67355

Life Cycle Costing for Facilities

by Alphonse Dell'Isola and Dr. Stephen Kirk

Guidance for achieving higher quality design and construction projects at lower costs! Cost-cutting efforts often sacrifice quality to yield the cheapest product. Life cycle costing enables building designers and owners to achieve both. The authors of this book show how LCC can work for a variety of projects—from roads to HVAC upgrades to different types of buildings.

$110.00 | 396 pages, hardcover | Catalog no. 67341

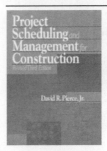

Project Scheduling and Management for Construction, 3rd Edition

by David R. Pierce, Jr.

A comprehensive, yet easy-to-follow guide to construction project scheduling and control—from vital project management principles through the latest scheduling, tracking, and controlling techniques. The author is a leading authority on scheduling, with years of field and teaching experience at leading academic institutions. Spend a few hours with this book and come away with a solid understanding of this essential management topic.

$64.95 | Over 286 pages, illustrated, hardcover | Catalog no. 67247B

Preventive Maintenance for Multi-Family Housing

by John C. Maciha

Prepared by one of the nation's leading experts on multi-family housing.

This complete PM system for apartment and condominium communities features expert guidance, checklists for buildings and grounds maintenance tasks and their frequencies, and a dedicated website featuring customizable electronic forms. A must-have for anyone involved with multi-family housing maintenance and upkeep.

$95.00 | 290 pages | Catalog no. 67346

Reference Books

For more information visit the RSMeans website at **www.rsmeans.com**

Green Building: Project Planning & Cost Estimating, 3rd Edition

Since the widely read first edition of this book, green building has gone from a growing trend to a major force in design and construction.

This new edition has been updated with the latest in green building technologies, design concepts, standards, and costs. Full-color with all new case studies—plus a new chapter on commercial real estate.

$110.00 | 480 pages, softcover | Catalog no. 67338B

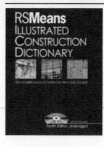

RSMeans Illustrated Construction Dictionary Unabridged 4th Edition, with CD-ROM

Long regarded as the industry's finest, *Means Illustrated Construction Dictionary* is now even better. With nearly 20,000 terms and more than 1,400 illustrations and photos, it is the clear choice for the most comprehensive and current information. The companion CD-ROM that comes with this new edition adds extra features, such as larger graphics and expanded definitions.

$110.00 | Over 880 pages, illust., hardcover | Catalog no. 67292B

Construction Supervision

Inspires excellence with proven tactics and techniques applied by thousands of construction supervisors. Leadership guidelines carve out a practical blueprint for motivating work performance and increasing productivity through effective communication. Features:

- A unique focus on field supervision and crew management
- Coverage of supervision from the foreman to the superintendent level
- An overview of technical skills whose mastery will build confidence and success for the supervisor
- A detailed view of "soft" management and communication skills

$95.00 | 464 pages, softcover | Catalog no. 67358

Building & Renovating Schools

This all-inclusive guide covers every step of the school construction process—from initial planning, needs assessment, and design, right through moving into the new facility. A must-have resource for anyone concerned with new school construction or renovation. With square foot cost models for elementary, middle, and high school facilities, and real-life case studies of recently completed school projects.

The contributors to this book—architects, construction project managers, contractors, and estimators who specialize in school construction—provide start-to-finish, expert guidance on the process.

$110.00 | Over 412 pages, hardcover | Catalog no. 67342

The Homeowner's Guide to Mold

By Michael Pugliese

Mold, whether caused by leaks, humidity or flooding, is a real health and financial issue—for homeowners and contractors. This full-color book explains:

- Construction and maintenance practices to prevent mold
- How to inspect for and remove mold
- Mold remediation procedures and costs
- What to do after a flood
- How to deal with insurance companies

$21.95 | 144 pages, softcover | Catalog no. 67344

Universal Design Ideas for Style, Comfort & Safety

by RSMeans and Lexicon Consulting, Inc.

Incorporating universal design when building or remodeling helps people of any age and physical ability more fully and safely enjoy their living spaces. This book shows how universal design can be artfully blended into the most attractive homes. It discusses specialized products like adjustable countertops and chair lifts, as well as simple ways to enhance a home's safety and comfort. With color photos and expert guidance, every area of the home is covered. Includes budget estimates that give an idea how much projects will cost.

$21.95 | 160 pages, illustrated, softcover | Catalog no. 67354

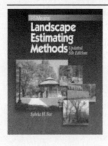

Landscape Estimating Methods, 5th Edition

Answers questions about preparing competitive landscape construction estimates, with up-to-date cost estimates and the new MasterFormat classification system. Expanded and revised to address the latest materials and methods, including new coverage on approaches to green building. Includes:

- Step-by-step explanation of the estimating process
- Sample forms and worksheets that save time and prevent errors

$75.00 | 336 pages, softcover | Catalog no. 67295C

Plumbing Estimating Methods, 3rd Edition

by Joseph J. Galeno and Sheldon T. Greene

Updated and revised! This practical guide walks you through a plumbing estimate, from basic materials and installation methods through change order analysis. *Plumbing Estimating Methods* covers residential, commercial, industrial, and medical systems, and features sample takeoff and estimate forms and detailed illustrations of systems and components.

$65.00 | 381+ pages, softcover | Catalog no. 67283B

Understanding & Negotiating Construction Contracts

by Kit Werremeyer

Take advantage of the author's 30 years' experience in small-to-large (including international) construction projects. Learn how to identify, understand, and evaluate high risk terms and conditions typically found in all construction contracts—then, negotiate to lower or eliminate the risk, improve terms of payment, and reduce exposure to claims and disputes. The author avoids "legalese" and gives real-life examples from actual projects.

$80.00 | 320 pages, softcover | Catalog no. 67350

Reference Books

For more information visit the RSMeans website at www.rsmeans.com

Qty.	Book no.	COST ESTIMATING BOOKS	Unit Price	Total
	60065	Assemblies Cost Data 2015	$334.95	
	60015	Building Construction Cost Data 2015	206.95	
	60045	Commercial Renovaion Cost Data 2015	166.95	
	60115	Concrete & Masonry Cost Data 2015	188.95	
	50145	Construction Cost Indexes 2015 (subscription)	359.95	
	60145A	Construction Cost Index–January 2015	94.95	
	60145B	Construction Cost Index–April 2015	94.95	
	60145C	Construction Cost Index–July 2015	94.95	
	60145D	Construction Cost Index–October 2015	94.95	
	60345	Contr. Pricing Guide: Resid. R & R Costs 2015	39.95	
	60235	Electrical Change Order Cost Data 2015	199.95	
	60035	Electrical Cost Data 2015	208.95	
	60205	Facilities Construction Cost Data 2015	524.95	
	60305	Facilities Maintenance & Repair Cost Data 2015	461.95	
	60555	Green Building Cost Data 2015	170.95	
	60165	Heavy Construction Cost Data 2015	204.95	
	60095	Interior Cost Data 2015	208.95	
	60125	Labor Rates for the Const. Industry 2015	450.95	
	60185	Light Commercial Cost Data 2015	153.95	
	60025	Mechanical Cost Data 2015	203.95	
	60155	Open Shop Building Const. Cost Data 2015	175.95	
	60215	Plumbing Cost Data 2015	206.95	
	60175	Residential Cost Data 2015	148.95	
	60285	Site Work & Landscape Cost Data 2015	199.95	
	60055	Square Foot Costs 2015	218.95	
	62014	Yardsticks for Costing (2014)	189.95	
	62015	Yardsticks for Costing (2015)	201.95	
		REFERENCE BOOKS		
	67329	Bldrs Essentials: Best Bus. Practices for Bldrs	45.00	
	67307	Bldrs Essentials: Plan Reading & Takeoff	50.00	
	67342	Building & Renovating Schools	110.00	
	67261A	The Building Prof. Guide to Contract Documents	75.00	
	67353A	Complete Book of Framing, 2nd Ed.	29.95	
	67146	Concrete Repair & Maintenance Illustrated	75.00	
	67352	Construction Business Management	60.00	
	67364	Construction Law	115.00	
	67358	Construction Supervision	95.00	
	67363	Cost Data Student Edition	110.00	
	67314	Cost Planning & Est. for Facil. Maint.	95.00	
	67230B	Electrical Estimating Methods, 3rd Ed.	75.00	
	67343A	Est. Bldg. Costs for Resi. & Lt. Comm., 2nd Ed.	75.00	
	67276B	Estimating Handbook, 3rd Ed.	110.00	
	67318	Facilities Operations & Engineering Reference	115.00	
	67338B	Green Building: Proj. Planning & Cost Est., 3rd Ed.	110.00	
	67355	Green Home Improvement	34.95	
	67357B	The Gypsum Construction Handbook	40.00	
	67148	Heavy Construction Handbook	110.00	
	67308E	Home Improvement Costs–Int. Projects, 9th Ed.	24.95	
	67309E	Home Improvement Costs–Ext. Projects, 9th Ed.	24.95	
	67344	Homeowner's Guide to Mold	21.95	
	67324C	How to Est.w/Means Data & CostWorks, 4th Ed.	75.00	

Qty.	Book no.	REFERENCE BOOKS (Cont.)	Unit Price	Total
	67351A	How Your House Works, 2nd Ed.	22.95	
	67282A	Illustrated Const. Dictionary, Condensed, 2nd Ed.	65.00	
	67292B	Illustrated Const. Dictionary, w/CD-ROM, 4th Ed.	110.00	
	67362	Illustrated Const. Dictionary, Student Ed.	60.00	
	67348	Job Order Contracting	130.00	
	67295C	Landscape Estimating Methods, 5th Ed.	75.00	
	67341	Life Cycle Costing for Facilities	110.00	
	67370	Managing the Profitable Construction Business	60.00	
	67294B	Mechanical Estimating Methods, 4th Ed.	75.00	
	67345	Practice of Cost Segregation Analysis	105.00	
	67366	Project Control: Integrating Cost & Sched.	95.00	
	67346	Preventive Maint. for Multi-Family Housing	95.00	
	67367	Project Sched. and Manag. for Constr., 4th Ed.	95.00	
	67322B	Resi. & Light Commercial Const. Stds., 3rd Ed.	65.00	
	67359	Risk Management for Design and Construction	125.00	
	67365	Solar Energy: Technology & Project Delivery	85.00	
	67145B	Sq. Ft. & Assem. Estimating Methods, 3rd Ed.	75.00	
	67321	Total Productive Facilities Management	85.00	
	67350	Understanding and Negotiating Const. Contracts	80.00	
	67303B	Unit Price Estimating Methods, 4th Ed.	75.00	
	67354	Universal Design	21.95	
	67319A	Value Engineering: Practical Applications	90.00	

MA residents add 6.25% state sales tax

Shipping & Handling**

Total (U.S. Funds)*

Prices are subject to change and are for U.S. delivery only. *Canadian customers may call for current prices. **Shipping & handling charges: Add 8% of total order for check and credit card payments. Add 8% of total order for invoiced orders.

Send order to: ADDV-1000

Name (please print)

Company

☐ Company

☐ Home Address

City/State/Zip

Phone # _____ P.O. # _____

(Must accompany all orders being billed)

Mail to: RSMeans, 700 Longwater Drive, Norwell, MA 02061

By filling out this report, your project data will contribute to the database that supports the RSMeans® Project Cost Square Foot Data. When you fill out this form, RSMeans will provide a $50 discount off one of the RSMeans products advertised in the preceding pages. Please complete the form including all items where you have cost data, and all the items marked (☑).

$50.00 Discount per product for each report you submit

Project Description (NEW construction only, please)

☑ Building Use (Office School . . .) _____

☑ Address (City, State)_____

_____

☑ Total Building Area (SF) _____

☑ Ground Floor (SF) _____

☑ Frame (Wood, Steel . . .) _____

☑ Exterior Wall (Brick, Tilt-up . . .) _____

☑ Basement: (check one) ☐ Full ☐ Partial ☐ None

☑ Number of Stories _____

☑ Floor-to-Floor Height _____

☑ Volume (C.F.) _____

% Air Conditioned _____ Tons_____

Total Project Cost $ _____

Owner _____

Architect_____

General Contractor _____

☑ Bid Date _____

☑ Typical Bay Size _____

☑ Occupant Capacity_____

☑ Labor Force: _____ % Union _____ % Non-Union

☑ Project Description (Circle one number in each line.)

   1. Economy 2. Average 3. Custom 4. Luxury

   1. Square 2. Rectangular 3. Irregular 4. Very Irregular

Comments _____

_____

			$
A	☑	General Conditions	$
B	☑	Site Work	$
C	☑	Concrete	$
D	☑	Masonry	$
E	☑	Metals	$
F	☑	Wood & Plastics	$
G	☑	Thermal & Moisture Protection	$
GR		Roofing & Flashing	$
H	☑	Doors and Windows	$
J	☑	Finishes	$
JP		Painting & Wall Covering	$

			$
K	☑	Specialties	$
L	☑	Equipment	$
M	☑	Furnishings	$
N	☑	Special Construction	$
P	☑	Conveying Systems	$
Q	☑	Mechanical	$
QP		Plumbing	$
QB		HVAC	$
R	☑	Electrical	$
S	☑	Mech./Elec. Combined	$

Please specify the RSMeans product you wish to receive.
Complete the address information.

Product Name _____

Product Number _____

Your Name _____

Title _____

Company_____
   ☐ Company
   ☐ Home   Street Address_____

City, State, Zip _____

Email Address _____
Return by mail or fax 888-492-6770.

Method of Payment:

Credit Card # _____

Expiration Date _____

Check_____

Purchase Order _____

Phone Number _____
(if you would prefer a sales call)

RSMeans
Square Foot Costs Department
700 Longwater Drive
Norwell, MA 02061
www.RSMeans.com